Metallurgical Technologies, Energy Conversion, and Magnetohydrodynamic Flows

Edited by
Herman Branover
Yeshajahu Unger
Ben-Gurion University of the Negev
Beer-Sheva, Israel

Volume 148
PROGRESS IN
ASTRONAUTICS AND AERONAUTICS

A. Richard Seebass, Editor-in-Chief
University of Colorado at Boulder
Boulder, Colorado

Technical papers from the Proceedings of the Sixth Beer-Sheva International Seminar on Magnetohydrodynamic Flows and Turbulence, Ben-Gurion University of the Negev, Beer-Sheva, Israel, February 25–March 2, 1990, and subsequently revised for this volume.

Published by the American Institute of Aeronautics and Astronautics, Inc., 370 L'Enfant Promenade, SW, Washington, DC, 20024-2518

Copyright © 1993 by the American Institute of Aeronautics and Astronautics, Inc. Printed in the United States of America. All rights reserved. Reproduction or translation of any part of this work beyond that permitted by Sections 107 and 108 of the U.S. Copyright Law without the permission of the copyright owner is unlawful. The code following this statement indicates the copyright owner's consent that copies of articles in this volume may be made for personal or internal use, on condition that the copier pay the per-copy fee ($2.00) plus the per-page fee ($0.50) through the Copyright Clearance Center, Inc., 21 Congress Street, Salem, Massachusetts 01970. This consent does not extend to other kinds of copying, for which permission requests should be addressed to the publisher. Users should employ the following code when reporting copying from this volume to the Copyright Clearance Center:

1-56347-019-5/93 $2.00 + .50

Data and information appearing in this book are for informational purposes only. AIAA is not responsible for any injury or damage resulting from use or reliance, nor does AIAA warrant that use or reliance will be free from privately owned rights.

ISSN 0079-6050

Progress in Astronautics and Aeronautics

Editor-in-Chief
A. Richard Seebass
University of Colorado at Boulder

Editorial Board

Richard G. Bradley
General Dynamics

Allen E. Fuhs
Carmel, California

George J. Gleghorn
*TRW Space
and Technology Group*

Dale B. Henderson
Los Alamos National Laboratory

Carolyn L. Huntoon
NASA Johnson Space Center

Reid R. June
Boeing Military Airplane Company

John L. Junkins
Texas A&M University

John E. Keigler
*General Electric Company
Astro-Space Division*

Daniel P. Raymer
*Conceptual Research
Corporation*

Martin Summerfield
*Princeton Combustion Research
Laboratories, Inc.*

Charles E. Treanor
*Arvin/Calspan
Advanced Technology Center*

Jeanne Godette
Director
Book Publications
AIAA

Preface

This volume contains a collection of papers devoted to metallurgical and energy conversion technologies that utilize magnetohydrodynamic (MHD) phenomena. These papers were presented at the Sixth Beer-Sheva International Seminar on Magnetohydrodynamic (MHD) Flows and Turbulence.

The Beer-Sheva Seminars have been held every three years since 1975, when a group of researchers from different countries, interested in both MHD flows and different aspects of turbulence in electoconductive as well as nonconductive fluids, came together for a week-long seminar. All participants felt that the meeting was very fruitful in detecting promising areas of research and developing new ideas. Therefore, the seminar was repeated in 1978 and ultimately became an ongoing event, taking place once every three years. It also became a tradition to publish the papers presented at the Beer-Sheva Seminar after a thorough review and editing. The first two volumes were published by John Wiley and Sons, and the subsequent volumes were published by AIAA (Volumes 84, 100, 111, and 112 in the Progress in Astronautics and Aeronautics series).

The Beer-Sheva Seminars became well known in the international scientific community, and the number of researchers willing to participate kept growing steadily. This forced the organizers, who wanted to preserve the intimate atmosphere of the seminars and to keep substantial time for informal discussions, to introduce more and more severe criteria in the paper selection process. The intention was to keep the number of participants in each event at approximately 100. The Sixth Seminar, held from February 25 to March 2, 1990, hosted 128 participants from 18 countries. For the first time, East European countries participated, including massive delegations from the CIS. The number of papers that passed all of the reviews reached 72, which is many more than at previous seminars. In addition, a number of extended invited review papers were presented. Therefore, the papers on MHD and turbulence are published in two separate volumes.

The papers of the present volume, *Metallurgical Technologies, Energy Conversion, and Magnetohydrodynamic Flows,* reflect most of the main trends in contemporary applied MHD research worldwide.

In this volume the papers are organized in four parts. The reader will find a number of articles that present in detail the results related to metallurgical technologies, MHD energy conversion and MHD ship propulsion, liquid-metal systems as well as plasma MHD systems, and MHD flow studies of liquid metals and two-phase flow studies related to MHD technologies.

To give the reader an overview of the content of this volume, we will present a brief survey of a number of papers.

The names of authors who actually presented their papers are printed in italics.

Some review papers open Chapter 1, Metallurgical Technologies. In the first paper, *Fautrelle* reviews metallurgical applications of MHD in France. These applications fall under one of the following categories: 1) heating and melting of conducting materials, 2) stirring control, and 3) free surface control. Among the specific applications reviewed are 1) induction furnaces; 2) skull melting furnaces (they feature no contact between the melt and the walls and, hence, low chemical pollution); 3) large size channel furnace; 4) free surface control, including dome formation and instability suppression in Al reduction cells and in cold crucibles; 5) surface stirring; 6) shaping and control of liquid-metal jets; and 7) conduction melting of glasses and of ceramic materials.

Ievlev and *Baranov* present an overview of the MHD-related devices and techniques that have been commercialized in the CIS for industrial applications. Nearly four dozen devices and techniques were developed. These include a large variety of MHD pumps (including the world's largest MHD pump, which is capable of pumping 3500 m^3/h of 300°C sodium; a high-temperature pump capable of pumping 600°C alkali metals; and submersible MHD pumps); a large variety of metallurgical applications (including stirring, preheating, refining, and cleaning of liquid metals; confinement of the melt; delivery of the melt to the desirable spot; and exact batching and control over the process of crystallization); MHD granulators; techniques for the production of composite materials; and MHD separators (used for cleaning liquid metals of impurities as well as for the separation of multiphase systems into their components).

Gelfgat presents a complementary review on practical applications of MHD devices and techniques developed at the Institute of Physics of the Latvian Academy of Sciences, emphasizing recent developments. He groups these applications under four categories: 1) MHD techniques and devices for controlling the flow and treating the metals (including electromagnetic pumps, valves, etc. for transporting, batching, and stirring melts; effecting heat and mass transfer in crystallization processes, etc.); 2) power engineering (including use of MHD effects in electrical furnaces, electrolytic reduction, electroslag remelting, electrowelding, etc.); 3) protection devices (such as startup control devices, switching system, and instrument protection); and 4) new MHD techniques and devices for the treatment and control of metal flows (such as crucible free melting, separation and purification of multicomponent alloys, production of composite materials, and MHD techniques for space). Special novel applications elaborated upon include 1) an MHD facility for producing thin wall casting having a complex shape; 2) MHD stirrers featuring 30–50% increase in productivity for producing aluminum and its alloys, a hundred-fold increase in the purity of mercury, 3–5 times increase in the rate of zone melting of superconducting materials, and 50–100% increase in the rate of homogenization of optical glasses; 3) metal counterflow refining by slag using an inclined MHD trough; 4) hot anticorrosive metallic coatings on long-size steel products; 5) MHD throttles for controlling the flow rate in liquid-metal nuclear reactors; 6) MHD wave generators for soldering printed circuits; 7) induction furnaces in which electrovortex flows are created by a monodirectional loop of induced current; 8) hermetically sealed power supplies, current regulators, current stabilizers, etc.; 9) creation of space conditions in terrestrial processes, using EM forces to compensate the effects of gravity (such as for semiconductor single crystal growth); and 10) production of composite materials consisting of immiscible components with a tendency of gravitational sedimentation.

Slepian et al. describe a modified method for electromagnetic modulation of molten metal flow. It consists of a high-frequency coil surrounding a column of liquid metal flowing through a nonconducting pipe, and an obstruction in the flow channel section surrounded by the coil.

Del Vecchio et al. report upon a study aimed at optimizing an electromagnetic pump for liquid metals featuring a superconducting magnet.

Poinsot et al. discuss the technical and potential advantages and disadvantages of induction electromagnetic pumps for fast breeder reactors relative to mechanical pumps.

Vivès reports upon grain refinement in 1085 and 2214 aluminum alloys by an electromagnetic vibrational method.

Takeuchi et al. describe how MHD effects can be used to improve the quality of metals by controlling the rate and direction of their solidification. Their approach involves the simultaneous application of cooling to one face of the casting mold and of a magnetic field in a direction normal to the cooled surface. The field suppresses turbulence and convective heat transfer along the field lines, so that the solidification temperature propagates relatively slowly in the direction normal to the cooled face. This causes elongated grain formation in the direction of the temperature gradient. In the absence of the field, the heat transfer rate is very high, and the solidification occurs almost simultaneously and is characterized by small grains of random orientation. The experiments were conducted using an Sn-10% Pb alloy with a magnetic field strength of up to 4 T.

Gelfgat and *Gorbunov* discuss at length MHD means for affecting the hydrodynamics and heat and mass transfer for the production of single crystals. This approach is illustrated for a Czochralski single crystal puller. Thus, for example, a 0.2-T magnetic field can lead to a 25-fold decrease in the convective heat and mass transfer due to suppression of thermoconvective effects. The absence of convective stirring results in the increase of both the radial and axial temperature difference in the melt. Through its effect on mass transfer, the magnetic field can be used to control impurity (such as oxygen) concentration in the single crystal and the uniformity of the dopant distribution in the crystal.

In Chapter 2, MHD energy conversion is addressed. The new area of MHD ship propulsion is described by *Meng* of the Naval Underwater Systems Center in Newport, Rhode Island, including a detailed engineering-physics optimization of an MHD seawater thruster using superconducting magnets. The superconducting electromagnetic thruster is composed of four major components: a hydropropulsor, a seawater electrode, a superconducting magnet, and a cryostat. The design optimization of each of these components is reviewed, while considering three figures of merit: 1) the maximum overall efficiency of the hydropropulsor; 2) the maximum Lorentz force given the overall system outer diameter; and 3) the minimum system total weight given the required system thrust. Major parameters are presented for three magnet configurations, including the toroid racetrack, the cluster dipole, and the double helix solenoid. The use of superconducting magnets with composite materials for the magnet structural support is found essential for the viability of the seawater MHD thruster. It is estimated that low efficiency (~20%) MHD thrusters based on state-of-the-art technology can now be realized.

Power conversion systems are presented further in this chapter. *Branover* reviews the liquid-metal (LM) MHD program in Israel. An intensive research and development program established the database for a reliable design of the natural circulation OMACON systems. The present effort is aimed at the development of OMACON plants for the cogeneration market in the 1–20 MW_e power range. A couple of complete OMACON-type experimental power conversion systems—ER-4 and Etgar-3—have been built and successfully operated at the Ben-Gurion University. The commercialization of the OMACON technology is planned in three phases: 1) Etgar-5, to provide about 1 MW_e to the grid along with 480°C steam to a phosphates industry; its construction could start within a year or so; 2) Etgar-6, to deliver 3.2 MW_e AC along with 13.4 MW_{th} of 5-bar steam; and 3) Etgar-7, representing a mature technology. Compared with conventional steam turbine plants designed to provide the same process heat, Etgar-7 is found to have a higher efficiency for electricity generation and lower specific capital cost, and, thus, to be more economically attractive.

Barak and Greenspan assess the promise of the OMACON technology for seawater desalination along with electricity generation. A couple of novel desalination methods are being proposed. One proposal is an open cycle featuring the evaporation of seawater fed by direct contact with the liquid metal of the first (i.e., high-pressure) OMACON loop. The second proposal is a closed cycle in which superheated steam exiting from the expansion in the OMACON system is brought into direct contract with seawater, heating it to evaporation while being desuperheated. Both approaches are found to feature a relatively low specific energy investment and are free of special heat exchangers (i.e., metallic heat transfer surfaces) for the desalination process. A preliminary economic analysis indicates that both methods have the potential of desalinating sea water with relatively small (few MW_e) OMACON plants at least as economically as large-size dual purpose plants using the best commercially available desalination technologies. Of the two novel approaches, the direct injection of sea water into the OMACON system offers almost an order of magnitude more desalinated water per MW of electricity than the evaporation by desuperheating of the exhaust steam. The compatibility of the liquid-metal and structural material with seawater must, however, be established before the practicality of the first method can be reliably ascertained.

Kirko discusses LMMHD systems for waste heat utilization in Northern climates, proposing an interesting modification of the OMACON concept. Rather than designing the riser to feature as fine a suspension of bubbles as possible (so as to reduce the slip ratio), establishing large bubbles that occupy the entire cross section of the riser and act as pistons on the liquid-metal slugs in between the bubbles is proposed. To create the bubbly pistons, Kirko proposes an addition to the liquid-metal ferromagnetic particles in the form of long needles of iron or strips of amorphous steel and an application of a longitudinal magnetic field to the riser. Utilizing his modified OMACON systems for converting waste heat into electricity in the

Perm region is also suggested, where the heat sink temperature can be in the range from $-15°$ to $-25°C$. Eutectic NaK is proposed for the liquid metal, whereas methylamine (CH_3NH_2) is proposed for the volatile fluid.

Marty et al. demonstrate experimentally the possibility of self-excitation of an LMMHD induction generator working with a sodium-potassium eutectic.

End effects are presented by Thiyagarajan et al. as well as by *Blumenau*, who proposes a novel approach for eliminating end losses in linear MHD channels featuring a high magnetic Reynolds number. A theoretical analysis shows that the end losses can be highly reduced by applying a proper step change in the magnetic field at the edges of the electrodes, commensurate with the load factor. It is predicted that, using the proposed approach, MHD channels offering 94% efficiency could be realized. Moreover, the high efficiency is expected to persist for a wide range of flow rates in a given channel. No expermental confirmation of the theoretical analysis has so far been attempted.

Material compatibility for LMMHD channels with lead and mercury as working fluids are addressed by Venkataraman et al. and by *Kirshenbaum*.

A number of papers deal with plasma MHD, including a short review of activities in the field in Poland by *Zaporowski* and Sroka.

MHD flows are presented in Chapter 3. *Pigny* and Moreau discuss interfacial instabilities in the presence of electric currents and magnetic fields. They extend previous stability models and develop a linear stability analysis for problems featuring two liquid layers, separated by an interface, that are crossed by an electric current and are subjected to a magnetic field. Their model takes into account motion in the fluids and nonuniformities in the electric current density and in the magnetic field. The model is applicable to situations where the thickness of the two layers is small and the magnetic Reynolds number is $\ll 1$.

Lielausis et al. describe the first series of MHD experiments performed on a loop with a low-temperature, nonaggressive, nontoxic eutectic melt containing 20% In, 67% Ga, and 13% Sn (melting point 10.5°C). The loop contained 50 liters of the molten metal and operated with a 5-T superconducting magnet. Pressure drops, velocities, and electrical potentials were measured.

Evtushenko et al. report upon a combined theoretical, numerical, and experimental investigation of hydrodynamic and heat transfer phenomena in thin-layer liquid-metal flows in the presence of a strong perpendicular magnetic field. Whereas in the absence of side walls there is no interaction between the field and the flow, there exists interaction in the presence of side walls, unless the liquid-metal layer is very thin. The flow stability is found to increase with the increase in the magnetic field strength and the liquid-metal layer thickness. Sufficiently strong coplanar magnetic fields suppress surface disturbances and level the free surface along the field lines. The application of a thin liquid-metal film flow for the protection of divertor plates of fusion reactors is analyzed. It is found that effective protection of divertor plates is possible with relatively thick (e.g. 8-mm) fast flowing (e.g. 0.9 m/s) liquid-metal layers.

A set of papers relating to different cases of MHD flows around a cylinder are presented by Josserand et al., Thess et al. (both papers presented by *Marty*), and *Gerbeth*.

In Chapter 4, two-phase flow studies are presented. *Lykoudis* and Tokuhiro describe a natural convection heat transfer experiment in mercury using a vertical heated flat plate subject to gas injection. In the laminar regime, the heat transfer coefficient was enhanced by a factor of two to three in the low heat flux case when gas injection was used. Heat transfer is promoted by the presence of bubble-generated turbulence inside the laminar boundary layer. This enhancement was less pronounced in the turbulent regime, occurring only at the highest gas injection rates. Here, the boundary layer is too thin to contain most of the bubbles. The bubbles that lie outside the boundary layer suppress the stratification of the fluid. In the presence of a transverse magnetic field with gas injection, the heat transfer coefficient decreased significantly for the laminar regime. The thermally-induced motion as well as the bubble-generated turbulence within the boundary layer is suppressed by the electromagnetic body force. In the turbulent regime, the decrease was less dramatic. Here, the effect of the electromagnetic body force is felt primarily by the thermally-induced turbulence. Lykoudis and Tokuhiro also analyze the influence of stratification on natural convection by introducing a stratification parameter into an order of magnitude analysis. The data reduced to a single line when both the Nusselt and Boussinesq numbers were modified using correction factors based upon the stratification parameter.

Using a double conductivity probe, *Lykoudis* and Takahashi measure the frequency, geometry, and growth velocity of bubbles generated during pool boiling of mercury.

A two-phase MHD energy converter generates mechanical energy through the energy transfer between the liquid-metal flow and the gas flow. *Mathes* and Alemany analyze the performance of an LMMHD converter using a one-dimensional model for annular dispersed flow which accounts for mass and heat transfer between the two phases. The flow consists of a thin film along the tube wall; a core flow, comprised of gas, treated as a perfect gas; and liquid droplets, assumed to be solid spheres whose diameters vary. The model also includes the interfacial forces between the gas flow and liquid film, the heat transfer from the core flow to the liquid film, and the entrainment of droplets from the liquid film by the core flow. Experiments were performed using air and water. The theoretical calculations of the temperature and entrainment rate were in reasonable agreement with the experimental results; however, the pressure losses, and therefore the velocity distributions, were not. Agreement between the theoretical pressure losses and the experimental pressure losses should improve as the flow rate is increased and, as a consequence, the void fraction decreased.

Thibault et al. describe a newly installed experimental facility intended for experimental investigation of pressures, temperatures, void fraction, and electrical conductivity in two-phase mercury-nitrogen flow with and without a magnetic field. The data from experiments on this facility are intended for establishing semiempirical laws that are necessary for closure of the two-phase flow equations system.

Kushelevsky presents a detailed analysis of problems related to void fraction measurement in two-phase organic-liquid metal MHD generators using gamma ray transmission. He discusses both the theoretical aspects and the practical arrangements. Problems of scattered radiation that introduces a systematic error into the measurement in large diameter ducts and the effect of intense magnetic field on the behavior of the detector are also discussed.

Farchi et al. present results of an extensive study of the performance of six different mixers installed in a vertical two-phase (water-air) flow system. This system simulated a liquid-metal MHD power system of the OMACON type. It was found that the average void fraction and the cross-sectional distribution of bubbles in a two-phase vertical upflow are rather indifferent to the type of mixer. However, pressure losses related to the mixer are lowest for a jet mixer. A parametric study including jet mixer losses is also presented.

In their totality, the papers referred to above and the papers of this volume not mentioned here present quite a complete picture of the contemporary trends in liquid-metal MHD research and its practical applications. The extensive references given in most papers of this volume should be an additional aid to the reader who wishes to make himself familiar with the present status of the field.

The authors and editors are very pleased that AIAA is continuing to publish the proceedings of the Beer-Sheva International Seminars on MHD Flows and Turbulence in their Progress in Astronautics and Aeronautics series.

The editors would like to express their sincerest gratitude to Dr. Martin Summerfield for his continuing enthusiastic interest in the Beer-Sheva Seminars and for his permanent guidance, and to Ms. Jeanne Godette, Director of the Progress in Astronautics and Aeronautics series, and her staff members for their continuing creative assistance, as well as to the entire team of the Progress in Astronautics and Aeronautics series.

We now look forward to the next volumes of this series, and we invite the reader to participate in the 7th Beer-Sheva International Seminar on MHD Flows and Turbulence, which is planned for February 1993.

Herman Branover
Yeshajahu Unger
November 1992

Table of Contents

Preface

Chapter 1. Metallurgical Technologies

Metallurgical Applications of Magnetohydrodynamics 3
    Yves Fautrelle, *Madylam-Institut National Polytechnique de Grenoble, St. Martin d'Héres, France*

Research and Development in the Field of MHD Devices Utilizing Liquid
    Working Medium for Process Applications 24
    V. M. Ievlev and N. N. Baranov, *Russian Academy of Sciences, Moscow, Russia*

Application of MHD Facilities to Technology 32
    Yu. M. Gelfgat, *Latvian Academy of Sciences, Riga-Salaspils, Latvia*

Electromagnetic Modulation of Molten Metal Flow 50
    R. M. Slepian, *Westinghouse Science & Technology Center, Pittsburgh, Pennsylvania*, P. A. Davidson, *Imperial College, London, England*, and A. R. Keeton, *Westinghouse Science & Technology Center, Pittsburgh, Pennsylvania*

Superconducting MHD Devices: Parametric Study of an Electromagnetic
    Pump Performance 66
    P. Del Vecchio, A. Geri, and G. M. Veca, *University "La Sapienza," Rome, Italy*

Induction Electromagnetic Pumps for Alkali Metals: Status and
    Perspectives. 76
    S. Poinsot and R. Rossignol, *Centre d'Etudes de Cadarache, Saint Paul Lez Durance, France*, F. Werkoff and A. Marechal, *Centre d'Etudes de Grenoble, Grenoble, France*, and J. Rapin, *FRAMATOME–Direction NOVATOME, Lyon, France*

Physical Model for Electromagnetically Driven Flow in Channel
    Induction Furnaces 92
    René Ricou and Charles Vivès, *Université d'Avignon, Avignon, France*

Grain Refinement in Aluminum Alloys by an Electromagnetic
    Vibrational Method 107
    Charles Vivès, *Université d'Avignon, Avignon, France*

Dendrite Growth of Solidifying Metal in DC Magnetic Field 125
    Eiichi Takeuchi, Ikuto Miyoshino, Kouichi Takeda, and Yutaka Kishida, *Nippon Steel Corporation, Chiba, Japan*

MHD Means for Affecting Hydrodynamics, Heat Transfer, and Mass
  Transfer at Single Crystal Melt Growth 138
    Yu. M. Gelfgat and L. A. Gorbunov, *Latvian Academy of Sciences,
    Riga-Salaspils, Latvia*

Inverse Electromagnetic Shaping Problem 158
    T. P. Felici and J. P. Brancher, *Institut National Polytechnique de Lorraine,
    Nancy, France*

Chapter 2. Energy Conversion

Major Engineering Physics for Optimization of the Seawater
  Superconducting Electromagnetic Thruster 183
    J. C. S. Meng, *Naval Underwater Systems Center, Newport, Rhode Island*

Liquid-Metal MHD Research and Development in Israel 209
    H. Branover, *Ben-Gurion University of the Negev, Beer-Sheva, Israel*

OMACON Technology for Seawater Desalination 222
    A. Barak and E. Greenspan, *Israel Atomic Energy Commission, Tel-Aviv, Israel*

MHD Generator for Waste Heat Utilization in Northern Conditions.... 239
    I. M. Kirko, *Perm Institute of Technical Physics, Perm, Russia*

Recent Results on LMMHD Induction Generators 244
    Ph. Marty, L. Leboucher, A. Alemany, and A. Pilaud, *Institut de Mécanique de
    Grenoble, Grenoble, France*, Ph. Massé, *Madylam-Institut National
    Polytechnique de Grenoble, St. Martin d'Héres, France*, and H. Branover,
    A. El Boher, and Y. Kaplan, *Ben-Gurion University of the Negev,
    Beer-Sheva, Israel*

Analytical and Experimental Studies of End Effects in an LMMHD
  Generator 261
    T. K. Thiyagarajan, P. Satyamurthy, N. S. Dixit, and N. Venkatramani, *Bhabha
    Atomic Research Center, Bombay, India*, and V. K. Rohatgi, *University of
    Poona, Pune, India*

Theoretical Magnetic Field Distributions Eliminating End Losses in
  Linear High Magnetic Reynolds Number MHD Channels 284
    L. Blumenau, *Ben-Gurion University of the Negev, Beer-Sheva, Israel*

Embrittlement of Steels by Lead 310
    Rangarajan Venkataraman, Michael D. Baldwin, and Glen R. Edwards,
    Colorado School of Mines, Golden, Colorado

Materials Compatibility of Mercury for Practical Applications at
  Elevated Temperatures 335
    Noel W. Kirshenbaum, *Placer Dome U.S. Inc., San Francisco, California*

Status of MHD Energy Conversion Research in Poland 343
B. Zaporowski, *Technical University of Posnan, Posnan, Poland*

Open Cycle Disk Generator Operating Conditions 348
H. K. Messerle, *University of Sydney, Sydney, Australia*

Conceptual Design of an MHD Retrofit of the Corette Plant in Billings, Montana 361
R. Labrie and N. Egan, *MHD Development Corporation, Butte, Montana,* and F. Walter, *Montana Power Company, Butte, Montana*

Constricted Discharges in Ar-Cs MHD Generators 373
W. F. H. Merck, A. P. C. Holten, A. Veefkind, and E. M. van Veldhuizen, *Eindhoven University of Technology, Eindhoven, The Netherlands,* V. A. Bityurin and A. P. Likhachev, *USSR Academy of Sciences, Moscow, Russia,* and B. Stefanov and L. Zarkova, *Institute of Electronics, Sofia, Bulgaria*

Pseudo Two-Phase Flow in an Open-Cycle MHD Generator 398
V. A. Bityurin and A. P. Likhachev, *USSR Academy of Sciences, Moscow, Russia*

Analysis of Flow Parameters in MHD Channel at Various Load Conditions 410
B. Zaporowski and K. Sroka, *Technical University of Posnan, Posnan, Poland*

Simulation and Comparison with the Experiment: The Dynamic Processes in an MHD Facility Flow Train 422
A. M. Levints, V. R. Satanovsky, and V. N. Zatelepin, *Russian Academy of Sciences, Moscow, Russia*

Acceleration of Gas-Liquid Piston Flows for Molten-Metal MHD Generators 431
A. Kolesnichencko and V. Malakhov, *Ukranian Academy of Sciences, Kiev, Ukraine*

Recent Developments in Liquid-Metal MHD Thermoacoustic Engines .. 441
D. Hamann, *Dresden University of Technology, Dresden, Germany,* and G. Gerbeth, *Central Institute for Nuclear Research, Rossendorf, Germany*

Chapter 3. Magnetohydrodynamic Flows

Interfacial Instabilities in the Presence of Electric Current and Magnetic Field 457
Sylvain Pigny and René Moreau, *Madylam-Institut National Polytechnique de Grenoble, St. Martin d'Héres, France*

Survey of Liquid-Metal MHD Activities in Dresden 470
G. Gerbeth and G. Uhlmann, *Central Institute for Nuclear Research, Rossendorf, Germany,* and D. Hamann, *Dresden University of Technology, Dresden, Germany*

Experiments with a Superconducting Magnet on an InGaSn Loop 476
O. Lielausis, E. Platacis, I. Platnieks, M. Pukis, and A. Shishko, *Latvian Academy of Sciences, Riga-Salaspils, Latvia*

Comparison of the Core Flow Solution and the Full Solution for MHD Flow 482
Lutz Lenhart, *Kernforschungszentrum Karlsruhe GmbH, Karlsruhe, Germany,* and Kathy McCarthy, *Idaho National Engineering Laboratory, Idaho Falls, Idaho*

Hydrodynamics and Heat Transfer of Thin Liquid-Metal Films in a Magnetic Field 500
I. A. Evtushenko, E. M. Kirillina, S. Y. Smolentzev, and A. V. Tananaev, *Leningrad Polytechnic Institute, St. Petersburg, Russia*

Effects of a Vertical Magnetic Field on Rayleigh-Bénard Convection in Mercury 509
Joel Stavans, *Weizmann Institute of Science, Rehovot, Israel*

MHD Flow Around a Cylinder in an Aligned Magnetic Field 519
J. Josserand, Ph. Marty, and A. Alemany, *Institut de Mécanique de Grenoble, Grenoble, France,* and G. Gerbeth, *Central Institute for Nuclear Research, Rossendorf, Germany*

Electromagnetically Driven Flow Around a Cylinder 535
A. Thess and G. Gerbeth, *Central Institute for Nuclear Research, Rossendorf, Germany,* and Ph. Marty, *Institut de Mécanique de Grenoble, Grenoble, France*

New Results for MHD Drag Coefficients 551
G. Gerbeth, *Central Institute for Nuclear Research, Rossendorf, Germany*

Heat Transfer in an MHD Flow Inside a Channel with Walls of Finite Thickness 566
Sergio Cuevas, *Instituto de Investigaciones Eléctricas, Cuernavaca, Mexico,* and Eduardo Ramos, *Laboratorio de Energia Solar, Temixco, Mexico*

Instability of a Liquid-Metal Surface in a Low-Frequency Alternating Magnetic Field 580
Jean-Marie Galpin, *Madylam-Institut National Polytechnique de Grenoble, St. Martin d'Héres, France,* Alfred Sneyd, *University of Waikato, Hamilton, New Zealand,* and Yves Fautrelle, *Madylam-Institut National Polytechnique de Grenoble, St. Martin d'Héres, France*

Chapter 4. Two-Phase Flows

Natural Convection over a Vertical Heated Flat Plate with Gas Injection and in the Presence of a Magnetic Field 601
Paul S. Lykoudis and Akira T. Tokuhiro, *Purdue University, West Lafeyette, Indiana*

Nucleate Boiling of Mercury in the Presence of a Horizontal Magnetic Field 626
  Paul S. Lykoudis and M. Takahashi, *Purdue University, West Lafeyette, Indiana*

Direct Contact Heat Transfer in Two-Phase Gas Liquid Flow 635
  C. Pisoni, C. Schenone, and L. Tagliafico, *University of Genoa, Genoa, Italy*

Heat and Kinetic Energy Transfer in Two-Phase Flow: Theoretical Aspects 649
  R. Mathes and A. Alemany, *Institut de Mécanique de Grenoble, Grenoble, France*

Nuclear Void Fraction Gaging in Large Two-Phase Organic Liquid-Metal MHD Generators 662
  A. P. Kushelevsky, *Ben-Gurion University of the Negev, Beer-Sheva, Israel*

Liquid-Metal Magnetohydrodynamic Two-Phase Flow Experiment 667
  J.-P. Thibault, B. Seck, and A. Cartellier, *Institut de Méchanique de Grenoble, Grenoble, France*

Investigation of Two-Phase Liquid Gas Mixers for MHD Energy Conversion Systems 678
  D. Farchi, A. El Boher, S. Lesin, Y. Unger, and H. Branover, *Ben-Gurion University of the Negev, Beer-Sheva, Israel*

Two-Phase Flow Measurements in Reaction Systems. 705
  Yaakov M. Timnat, *Technion—Israel Institute of Technology, Haifa, Israel*

Author Index for Volume 148 725

List of Series Volumes 727

Table of Contents for Companion Volume 149

Preface

Vortex Reconnection, Cascade, and Mixing in Turbulent Flows 1
F. Hussain, D. Virk, and M. V. Melander, *University of Houston, Houston, Texas*

Intermittent Turbulence from Closures 17
Robert H. Kraichnan, *369 Montezuma 108, Santa Fe, New Mexico*

Plane Mixing Layer Between Parallel Streams of Different Velocities and Different Densities 40
H. E. Fiedler, M. Lummer, and K. Nottmeyer, *Technische Universität Berlin, Berlin, Germany*

Instabilities in the Axisymmetric Jet: Subharmonic Resonance 53
C. O. Paschereit and I. J. Wygnanski, *University of Arizona, Tucson, Arizona*

Three-Dimensional Vortex MHD Flows at High Reynolds Numbers in Thin Layers of Conducting Incompressible Fluid 65
V. D. Zimin and N. Ju. Kolpakov, *Institute of Continuous Media Mechanics, Perm, Russia*

Tearing Instabilities in Two-Dimensional MHD Turbulence 81
H. Politano, A. Pouquet, and P. L. Sulem, *Observatoire de la Côte d'Azur, Nice, France*

Axisymmetric Hydromagnetic Dynamo 87
M. A. Goldshtik and V. N. Shtern, *Siberian Branch of the Academy of Sciences, Novosibirsk, Siberia*

Bifurcations of Self-Oscillating and Almost Periodical Regimes in an Azimuthal MHD Jet 103
K. Sergeev and V. N. Shtern, *Siberian Branch of the Academy of Sciences, Novosibirsk, Siberia*

Bifurcations in MHD Flow Generated by Electric Current Discharge 116
A. A. Petrunin and V. N. Shtern, *Siberian Branch of the Academy of Sciences, Novosibirsk, Siberia*

Homogeneous MHD Turbulence at Weak Magnetic Reynolds Numbers: Approach to Angular-Dependent Spectra 131
C. Cambon, *Ecole Centrale de Lyon, Ecully, France*

Inverse Cascades Generated by Alpha Dynamo and Anisotropic Kinetic Alpha Effect 146
P. L. Sulem and B. Galanti, *Observatoire de la Côte d'Azur, Nice, France*, and A. D. Gilbert, *Cambridge University, Cambridge, England*

Renormalization Group Analysis of MHD Turbulence with Low Magnetic Reynolds Number 151
S. Sukoriansky, I. Staroselsky, B. Galperin, S. Roy, and S. A. Orszag, *Princeton University, Princeton, New Jersey*

Renormalization Group Approach to Two-Dimensional Turbulence and the ε-Expansion for the Vorticity Equation 159
I. Staroselsky and S. Sukoriansky, *Princeton University, Princeton, New Jersey*

Liquid-Metal Flows in Sliding Electric Contacts: Solution for Turbulent Primary Azimuthal Velocity 165
G. Talmage and J. S. Walker, *University of Illinois at Champain-Urbana, Urbana, Illinois*, S. H. Brown and N. A. Sondergaard, *David Taylor Research Center, Annapolis, Maryland*, and H. Branover and S. Sukoriansky, *Ben-Gurion University of the Negev, Beer-Sheva, Israel*

Anisotropic Turbulence: Analogies Between Geophysical and Hydromagnetic Flows 190
C. Henoch and M. Hoffert, *New York University, New York, New York*, and H. Branover and S. Sukoriansky, *Ben-Gurion University of the Negev, Beer-Sheva, Israel*

Turbulent Electrically-Induced Vortical Flows 210
    V. N. Vlasyuk, *Vinica Institute of Teaching, Vinica, Ukraine,* and E. V. Shcherbinin, *Latvian Academy of Sciences,* Riga-Salaspils, Latvia

Dissipation Length Scale Dynamics 221
    D. Naot, *Center for Technological Education Holon, Holon, Israel,* and N. Yacoub and D. Maron Moalem, *Tel Aviv University, Ramat Aviv, Israel*

Towards Quasi-Isotropic Algebraic Stress Model for Magnetohydrodynamic Channel Flow 235
    D. Naot and J. Tanny, *Center for Technological Education Holon, Holon, Israel*

Two-Phase Grid Turbulence 254
    Th. Panidis and D. D. Papailiou, *University of Patras, Patras-Rion, Greece*

Abridged Octave Wavenumber Ring Models for Two-Dimensional Turbulence 268
    J. Lee, *Flight Dynamics Laboratory, Wright-Patterson Air Force Base, Ohio*

Solving Partial Differential Equations via Boolean Automata: Statistical and Deterministic Approaches 295
    K. Dang Tran, A. Cosnuau, J. Ryan, and Y. Morchoisne, *Office National d'Etudes et de Recherches Aerospatiales, Châtillon, France*

Rag Theory of Magnetic Fluctuations in Turbulent Flow 309
    A. A. Ruzmaikin, *Institute of Terrestrial Magnetism, Ionosphere, and Radio Wave Propagation, Troitsk, Russia*

Ballooning Instability in Fluid Dynamics 317
    M. Mond, *Ben-Gurion University of the Negev, Beer-Sheva, Israel,* and E. Hameiri, *New York University, New York, New York*

Instabilities of the Nonuniform Flows of a Low-Temperature Plasma in MHD Channels 328
    I. M. Rutkevich, *Russian Academy of Sciences, Moscow, Russia*

Author Index for Volume 149 342

List of Series Volumes 343

Chapter 1. Metallurgical Technologies

Metallurgical Applications of Magnetohydrodynamics

Yves Fautrelle*
*Madylam-Institut National Polytechnique de Grenoble,
St. Martin d' Héres, France*

Abstract

Liquid metal magnetohydrodynamics (MHD) principles are widely used in various metallurgical processing partly to melt solid materials, and partly to control both the liquid phase motion and shape. The explosion of interest in MHD, has generated new ideas very recently, and a new field called "the electromagnetic metallurgy" is born. After a period of euphoria, it has been realized that MHD is not a miraculous solution to all metallurgical problems and will not revolutionize the traditional processes. Its role is rather to enlarge the possibility of casting and elaborating materials of better quality thanks to a better control of the process. Nevertheless, MHD has been a way of generating in some cases new technologies in new materials manufacturing that would have been impossible without the help of magnetic fields. The purpose of this paper is to review some magnetohydrodynamic phenomena which have lead to industrial applications or to developing researches in metallurgy. These may classified as follows: melting, levitation and surface control, and stirring.

1.0 Introduction

Electromagnetic processing of liquid materials is a new field of research and application that originates

---

Copyright © 1991 by the American Institute of Aeronautics and Astronautics, Inc. All rights reserved.
*Professor.

from old induction heating techniques. Electromagnetic processing of liquid materials is the result of a collective effort that was made during the 15 past years not only by researchers but also by industrial companies whose aim was to improve conventional metallurgical processes and to develop new ones. The Symposia of the International Union of Theoretical and Applied Mechanics, held at Cambridge (United Kingdom) in 1982 and Riga (Soviet Union)[1,2] in 1988, provided an interesting forum for the interaction of industrial metallurgists and theoretical and experimental researchers in magneto-hydrodynamics. The scientists have demonstrated the vitality of this new field called "the electromagnetic metallurgy." The development of that field is also linked to the emergence of numerical tools, which has facilitated the solution of coupled problems involving electromagnetism, fluid mechanics, and metallurgy.

Some 10 years later, it is of interest to evaluate the situation. Many ideas have germed, but few have really emerged as industrial processes. It has been realized that MHD is not a miraculous solution to all metallurgical problems and will not yet revolutionize the traditional processes. The scientist's role is rather to enlarge the possibility of casting and elaborating better quality materials thanks to a better control of the process. Moreover, in some cases MHD has been a way of generating new technologies in new materials manufacturing that would have been impossible without the help of magnetic fields. It is noticeable that MHD applications have also generated progresses in some connected fields. The most striking example is the electromagnetic stirring in continuous casting. By creating an asymptotic situation, namely, the well-homogenized fluid, stirring has made easier the understanding of the mechanisms of the columnar-equiaxed transition in solidification. The aim of this paper is to illustrate some typical examples of the various challenges where MHD has brought elegant solutions. We have distinguished three main types of processes corresponding to the main functions exerted by electromagnetic fields: 1) heating and melting processes, 2) surface control and levitation, and 3) stirring and mixing processes.

The preceding classification may be in some cases quite arbitrary. In many applications, the aforementioned functions are actually coupled. Nevertheless, the system will be used for classification in this paper.

2.0 Heating and Melting Processes

2.1 Induction Furnaces

One of the most efficient ways to heat a liquid metal is to create eddy currents generated by alternating magnetic fields. This phenomenon has lead to the invention of high electrical efficiency devices such as the well known induction coreless or channel furnaces (Figs. 1 and 2). Many studies have been performed on such devices.[3-9] If this technique is now widely industrially used, the industrial units are still restricted to small or medium sized induction furnaces. Large power units are not widespread, although the industrial need for large capacity furnaces is great. The limitation mainly comes from a MHD phenomenon, namely, the electromagnetic stirring, which combines with physicochemical phenomena to shorten by erosion the lifetime of the refractory wall. Indeed, the characteristic velocity U in the liquid metal is a growing function of the Joule power P_J dissipated in the bath in following the following form:

$$U = P_J^{1/2} g(R_\omega), \quad R_\omega = \mu\sigma\omega a^2, \quad \omega = 2\pi f \qquad (1)$$

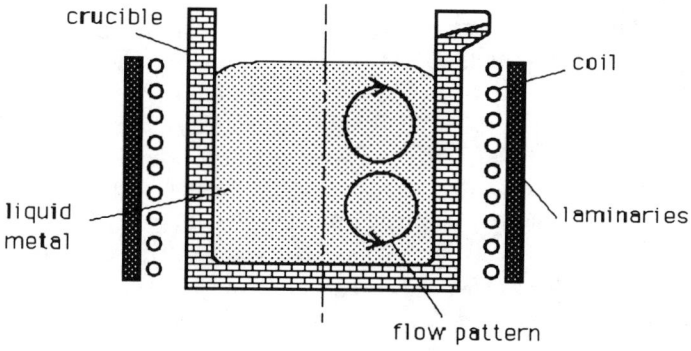

Fig. 1 Scheme of a coreless induction furnace.

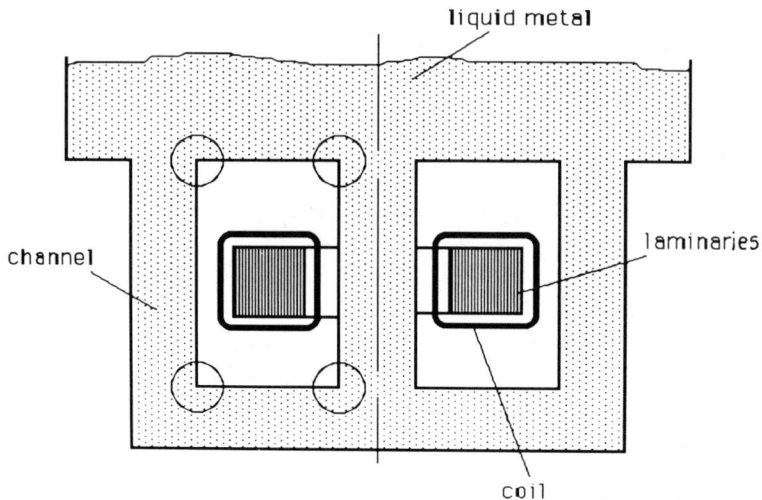

Fig. 2 Scheme of a channel furnace.

where R_ω, μ, σ, f, and a denote the shield parameter, the magnetic permeability, the electrical conductivity of the melt, the frequency of the coil current, and the radius of the melt, respectively. In a simple coreless furnace geometry Fig. 3 shows that, for large R_ω values, variations of g are not significant, and U may be quite large for high-power induction furnaces. The stirring phenomena will be discussed in more detail in Section 4.0.

2.2 Skull Melting Furnaces

To avoid any pollution or wall erosion problems, the induction furnace technique has been coupled with a skull melting process by replacing the refractory walls either by a cold crucible or by the inductor itself (cf. Figs. 4a and 4b). Two related devices have been developed : 1)   the direct skull melting device (Fig. 4a) and 2)   the cold crucible melting process (Fig. 4b). The overall advantages of both techniques are well known.[10] The most interesting ones are the following: high heating rates that allow increased productivity, no pollution by the crucible and no limitations for the melting of high-melting-point materials.

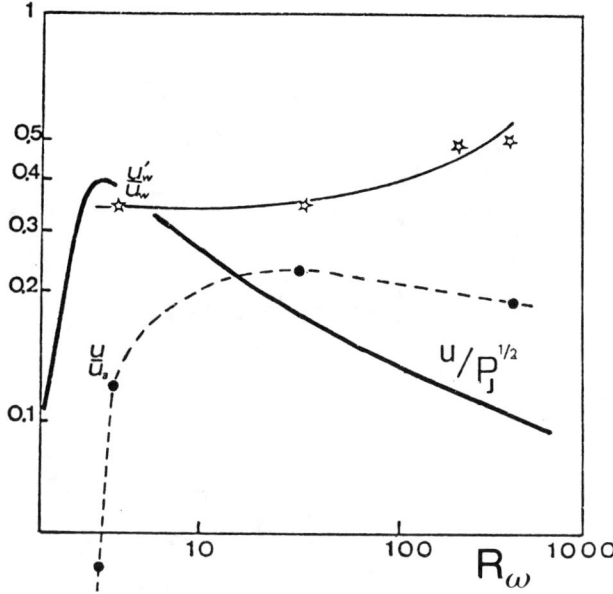

Fig. 3 Evolution of the computed characteristic velocity U with respect to the shield parameter R_ω proportional to the frequency; the velocity is divided by either $(P_J)^{1/2}$ or u_A, with P_J and u_A denoting the Joule power in the bath and the Alfven speed, respectively, and u'_w/u_w is the turbulent intensity near the wall.

Direct Skull Melting Furnaces

The direct skull melting device is well fitted for melting or heating special materials such as oxides or glasses. Indeed, the feasability of the process is based on the fact that the solidified crust must insulate electrically the liquid materials from the inductor. Accordingly, the only possible applications concern materials whose electrical conductivity decreases rapidly with the temperature, e.g., oxides and glasses. In such material the electrically insulating skull acts as a thermal barrier as well, since the thermal and electrical conductivities behave in the same way. This leads to two important consequences. First, high-melting-point materials may be melted without limitations. Second, the bath overheat is generally large, which may be a

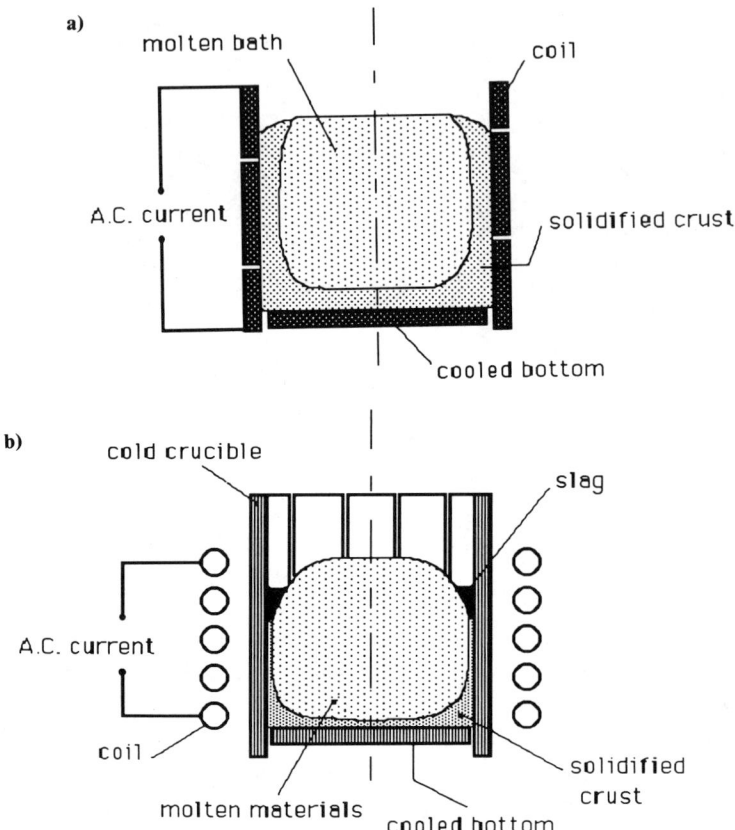

Fig. 4 Scheme of a skull melting device: a) direct skull melting, b) cold crucible melting.

drawback in some cases, e.g., single crystal pulling, where it is needed to adjust the bath temperature near the melting point.[2] The electromagnetic forces are usually negligible with respect to buoyancy, and the convective motion may only be controled by a suitable choice of the distribution of the Joule losses. In the simple geometry of Fig. 4a the mean convective motion consists of two vortices in a half meridian plane leading to a temperature distribution shown in Fig. 5.

As for the electrical coupling, the electrical efficiency of the process may exceed 60-70%.[2] However, such efficiencies are reached only if the shield parameter R_ω is at least of the order of one. Because of the weak values of the electrical conductivity of oxides

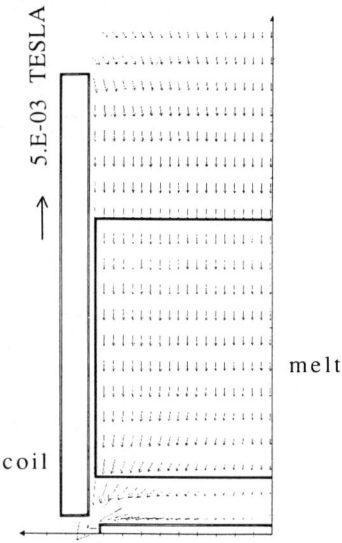
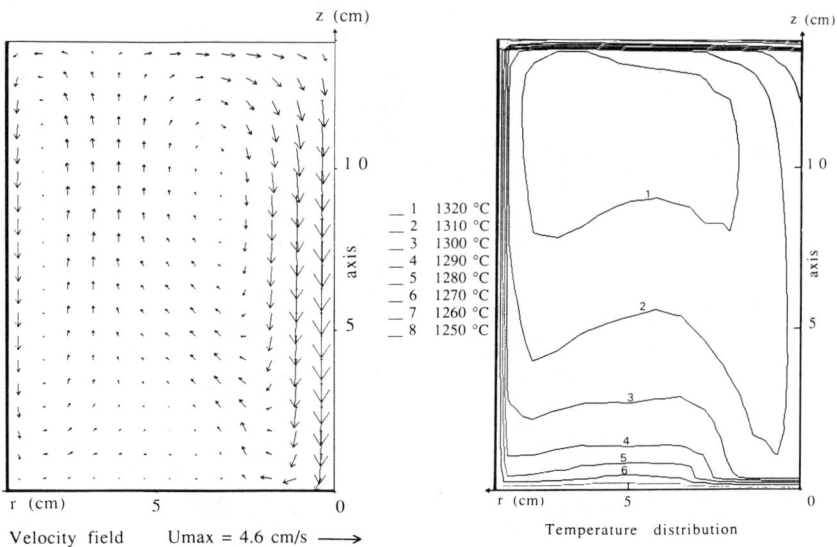

Fig. 5   Computed velocity and temperature in a lithium-niobate bath molten by a direct skull melting furnace from Ref. 2 ; the inductor consists in a single-turn coil.

and glasses, the device requires the use of high-frequency power sources.

The extension to large furnaces remains questionable. The main limitations come from electrical short-circuit risks due to the thinness of the gap between the coil and the conducting bath. Those risks increase with the size of the furnaces, since larger furnaces require higher voltage supply. Another electrical drawback is the existence of a quite high bath floating electrical potential.[2]

Cold Crucible Induction Furnaces

The cold crucible technique is very similar to the previous one. The main difference is the existence of an additional component between the melt and the coil, the so-called cold crucible. The cold crucible is obviously a protecting device of the coil, but its other main function is to replace the coil by playing the role of a "transfer coil." The electrical currents induced in the internal wall of the cold crucible must be a replica of those circulating in the coil (cf. Fig. 6). Accordingly, the design of the cold crucible (e.g., its shape and the number of sectors) is a key factor for the correct operating of the process.[11,15] An industrial device, the so-called 4C process (cold crucible continuous casting) has been developed for the remelting of titanium alloys.[11]

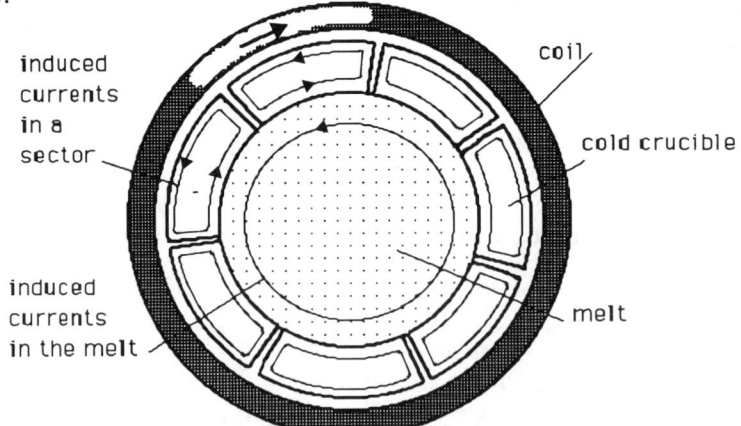

Fig. 6 Distribution of the induced electric currents in the cross section of a cold crucible furnace.

From an electrical point of view, the efficiency of such a device is lower, since a part of the input power is lost in the cold crucible. However, the device can be made safer by fixing the floating electrical potential of the cold crucible to zero ; thus short-circuit risks will no longer exist.

The main applications are relevant to metals, oxide, and glass processing. As for the metals, strong heat losses from the melt to the cold crucible limit the bath overheat to approximately 50°C at most. Accordingly, such a process is well fitted for continuous casting where the overheat of the bath must not exceed 20-40° C.

3.0 - Surface Control

3.1 Levitation

In liquid metal processing it is often necessary to use a liquid free surface either to avoid contact with a wall or to shape and stabilize it.

Minimizing any contact with a wall may be desirable for various reasons. In aluminium continuous casting, efforts have been made to reduce totally (EMC process[12]) or partially (CREM process[13]) the contact between the liquid metal and the mold. The expected result is a reduction of the depth of the chill zones, with important savings on scalping and edge trimming in the rolling mill. The principles of such processes are illustrated in Figs. 7a and 7b. The melts are located inside a single-phase ac current coil consisting of a few turns. The distribution of the electromagnetic forces must be adjusted to equilibrate the metallostatic pressure. The free surface has a meniscus shape whose height and position may be precisely controled. The inclination of the electromagnetic forces with respect to the normal of the meniscus requires a strong tangential variation of the magnetic field along the free surface. In the 4C process (cf. Section 2), minimization of the contact between the liquid metal and the cold crucible leads to a reduction of the heat losses of the bath, and higher overheat may be reached (Fig. 7c).

Control of the free surface may be quite difficult in the 4C process. The free surface is submitted to azimuthal electromagnetic force perturbations due to the existence of higher magnetic field in the vicinity of

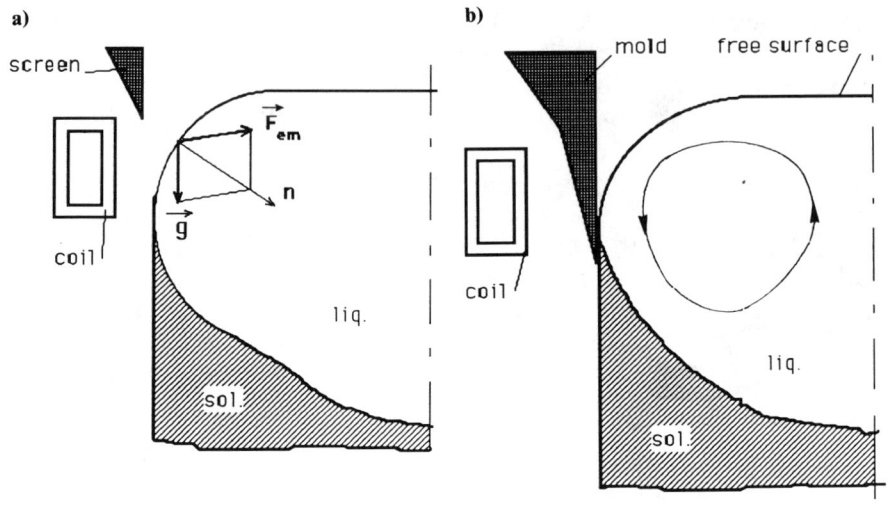
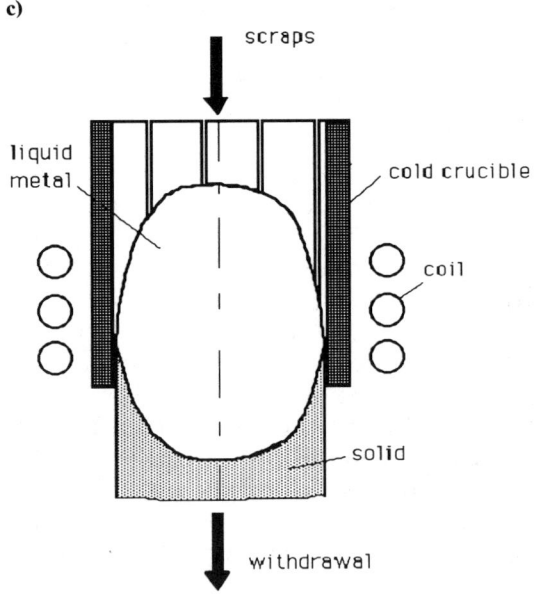

Fig. 7 Free surface deformations: a) in a moldless electromagnetic casting of aluminium (EMC) ; b) in a CREM-electromagnetic casting device ; c) in a cold crucible continuous casting device.

Fig. 8   Top view of the free surface of a semi levitated liquid nickel bath in a levitation cold crucible furnace (from Ref. 15).

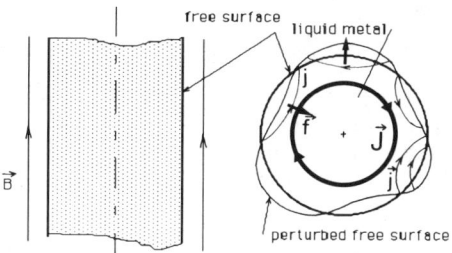

Fig. 9   Scheme of the free electromagnetic deformation of the surface of an infinitely long liquid metal rod in a uniform axial magnetic field and the corresponding electric and forces perturbations.

the gap between two sectors. Experimental observations have shown the emergence of very strong azimuthal free surface deformations,[14,15] as shown in Fig. 8. To understand that phenomenon, let us consider, for example, an infinite liquid metal cylindrical domain in a uniform axial magnetic field (see Fig. 9). In the absence of gravity the equilibrium shape of the free surface is an infinite cylinder of circular cross section. It is argued that the free surface exhibits free damped oscillation modes in analogy with classical surface waves. The free surface azimuthal deformations may be a result of a resonant forcing due to the forced azimuthal magnetic field perturbations.

When high superheat is needed together with high purity, the only solution is levitation. Levitation can be achieved by classical conical inductors or by a "basket-shaped" cold crucible. That process is limited to small-scale melting units. An interesting example of the industrial use of the levitation cold crucible is the elaboration of tungsten carbides in a graphite cold crucible by achieving semilevitation. In that case the liquid bath is almost entirely levitated, except a small-contact zone between the melt and the bottom part of the cold crucible. That contact zone allows the chemical reaction between tungsten and the carbon of the graphite.

Levitation may also be obtained by crossed dc magnetic fields and electrical currents creating a vertical electromagnetic force. This principle has been used to develop a new process of horizontal electromagnetic moldless casting (HEMC[2]). As in the CREM or CEM processes, it enables the elimination of surface defects of slab caused by the contact between the wall and the solidifying metal. The HEMC apparatus is illustrated in Fig. 10. The length of the levitated zone must be adjusted to avoid the growth of surface instabilities.[17]

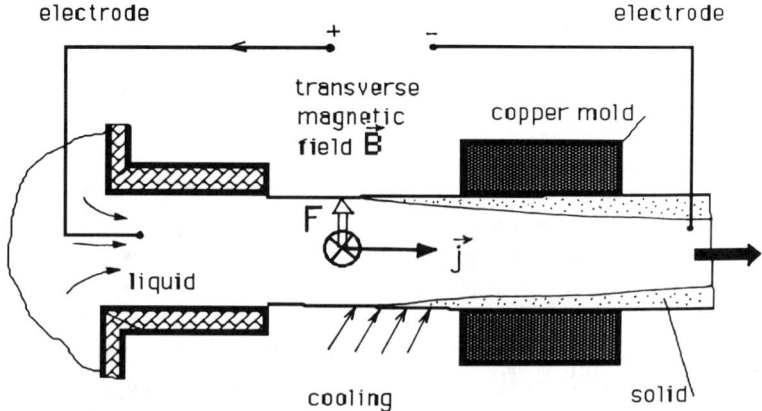

Fig. 10 Scheme a horizontal electromagnetic moldless casting device (HEMC from Ref. 2).

3.2 Instability control

The presence of free surfaces in many casting processes leads to problems linked to the emergence of a wide variety of free surface instabilities that worsen the quality of the solidified product. Use of dc or ac magnetic fields may be excellent solutions to those difficulties.

If dc and ac magnetic fields have stabilizing properties, the damping mechanisms are quite different. In dc magnetic field processes the damping action is obtained by the braking Lorentz force created by the liquid motion $\vec{u}$ across the magnetic lines of forces, namely,

$$\vec{F} = \sigma (\vec{u} \times \vec{B}) \times \vec{B} \qquad (2)$$

That principle is used for example to suppress wave motion at the meniscus of the liquid metal in conventional casting of steel (see, for example, Ref. 2). As for ac magnetic field devices, the electromagnetic pressure $p = B^2/2\mu$ is used to shape and to control the free surface deformations as shown in Ref. 1.

It may be desired in some metallurgical processes, e.g., steel processing, to create surface motions. Part of the refining is achieved by a surface slag. Enhancement of the slag-metal reaction kinetics may be obtained by creating strong surface wave motions. Surface waves may be generated by the use of low-frequency ac magnetic fields. The principle consists of adjusting the frequency of the coil currents, hence the Lorentz forces, to excite by resonant effects the free surface eigenmodes. Some examples of the expected wave patterns are illustrated by Galpin and others.[18]

3.3 Shaping of Liquid Metal Jets

The electromagnetic pressure created by ac magnetic fields is reponsible for a wide variety of very interesting phenomena on liquid metal jets.[19] Combinations of inductors of various shapes and positions generate various types of effects : 1) shaping of a round jet into a thin sheet, 2) coalescence of several round jets to form a wider sheet, and 3) guiding or deflection of a liquid metal jet. These phenomena are illustrated in Fig. 11. The potential metallurgical

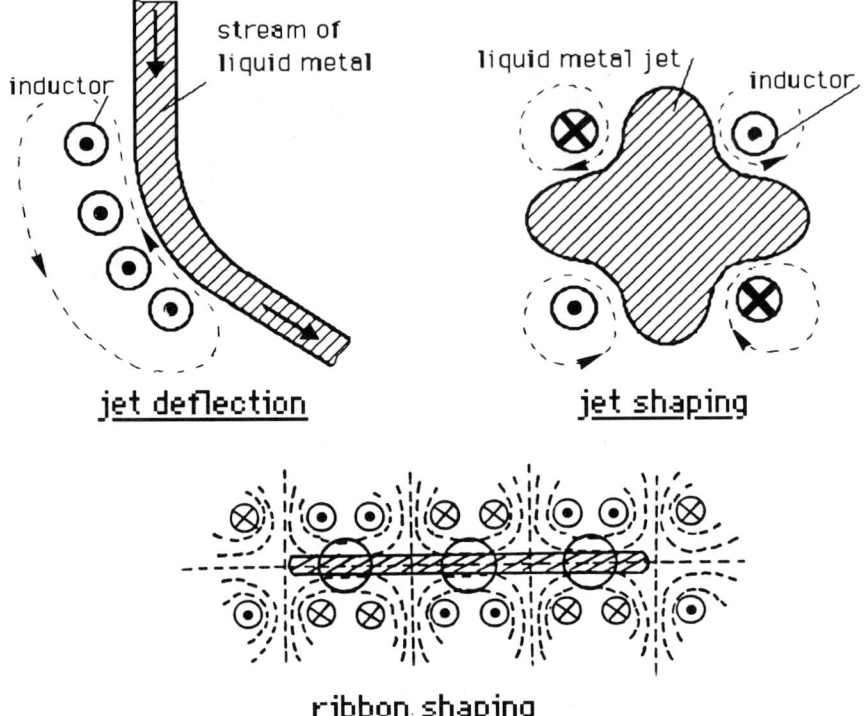

Fig. 11 Various devices aimed at shaping or guiding a liquid metal jet (from Ref. 18).

applications of such devices are numerous. The most important one is the direct (moldless) casting of metal to produce strips or thin rods. Development of such techniques is now limited by the difficulty to cool the liquid metal in the absence of mold.

4.0 - Stirring Control

The ability to control liquid metal motions by magnetic fields has been widely used in many metallurgical processes either to eliminate or to promote stirring.

4.1 Turbulence Control

Reduction of (turbulent) velocity fluctuations in liquid metal is achieved preferably by dc magnetic field.

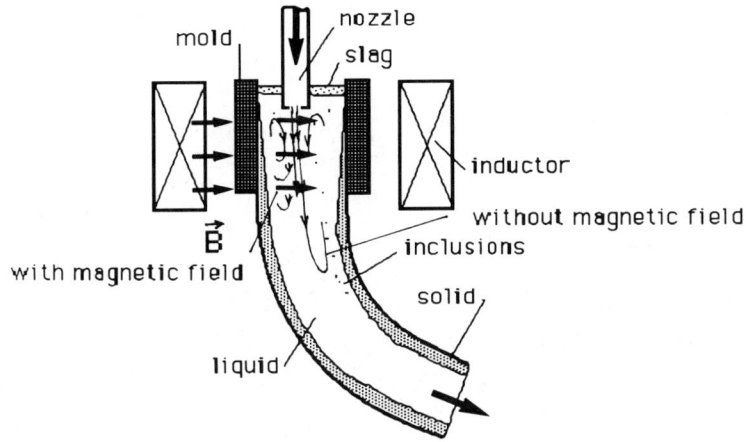

Fig. 12 Principle of the electromagnetic brake (EMBR) in curved-type continuous casters of steel.

As in free surface stabilization, the damping of the motion is caused by the braking Lorentz forces [cf. Eq. (2)]. This principle has been mainly applied to crystal growth and continuous casting of steel (see, for example, Ref. 2).

In single crystal growth by a Czokralski method, velocity and hence temperature fluctuations are responsible for structure defects in the pulled single crystal. The use of strong dc magnetic fields allows an efficient damping of the instabilities that arise both from the temperature gradients and the imposed rotation of the crystal and the melt.[2]

In the molds of continuous casting of steel, the turbulent velocities created by the nozzle may cause entrainments of inclusions in the liquid metal bulk. That problem arises when the casting speed in curved-type continuous casters is increased. The inclusions are deeply entrained by the nozzle velocity and are trapped in the upper half of the curved part of the strand. A solution of this problem is proposed by ASEA, which develops a dc magnetic field device, the so-called electromagnetic brake (EMBR). The electromagnetic braking effect consists in applying a magnetic field perpendicular to the nozzle jet. The $\sigma \vec{u} \times \vec{B}$ induced electrical current interacts with the applied magnetic field to create a Lorentz force opposed to the jet velocity (cf. Fig 12).

4.2 Electromagnetic Stirring

Electromagnetic stirring may be both an advantage and a drawback. Its main advantages are : 1) homogenisation of the bath, 2) increase of the heat and mass transfers in the bath, 3) inclusion removal due to the electromagnetic forces, and 4) influence on the solidification structures. However, it leads to some important drawbacks, such as rapid erosions of the refractory wall in induction furnaces and possible entrainment of surface particles in the bath.

Most of the electromagnetic stirring devices use an ac magnetic field. The absence of any contacts between the inductor and the bath is a major advantage of the technique, but it may be pointed out that such a device possesses a large number of degrees of freedom, which allows a precise control of the stirring. By acting on both the frequency, the intensity, the phase shifts of the applied electric currents, and the geometry of the inductor, it is possible to obtain a large variety of fluid flows.

Two main configurations may be distinguished : 1) single phase coil devices and 2) multi-phase coil devices.

Single-phase inductors are mainly used for melting or heating, but they may be efficient for stirring or levitation as well. The origin of the large success of that technique lies in its great simplicity and consequently its low investment cost. A striking example is the CREM process (cf. Section 3.0). Beside the levitation effect, the single-phase coil is responsible for a vigorous stirring of the liquid aluminium, which promotes the grain refinement of the solidified structure.

The efficiency of the electromagnetic stirring in such a device lies on the end effects and asymmetry that enhance the rotational part of the Lorentz forces.[5] Hence, in single-phase induction processes the geometry of both the inductor and the liquid metal pool are is of primary importance. In coreless induction furnaces vorticity is classically generated in the corner regions of the pool where locally strong rotational electromagnetic forces exist.[1,2,6] Another striking and

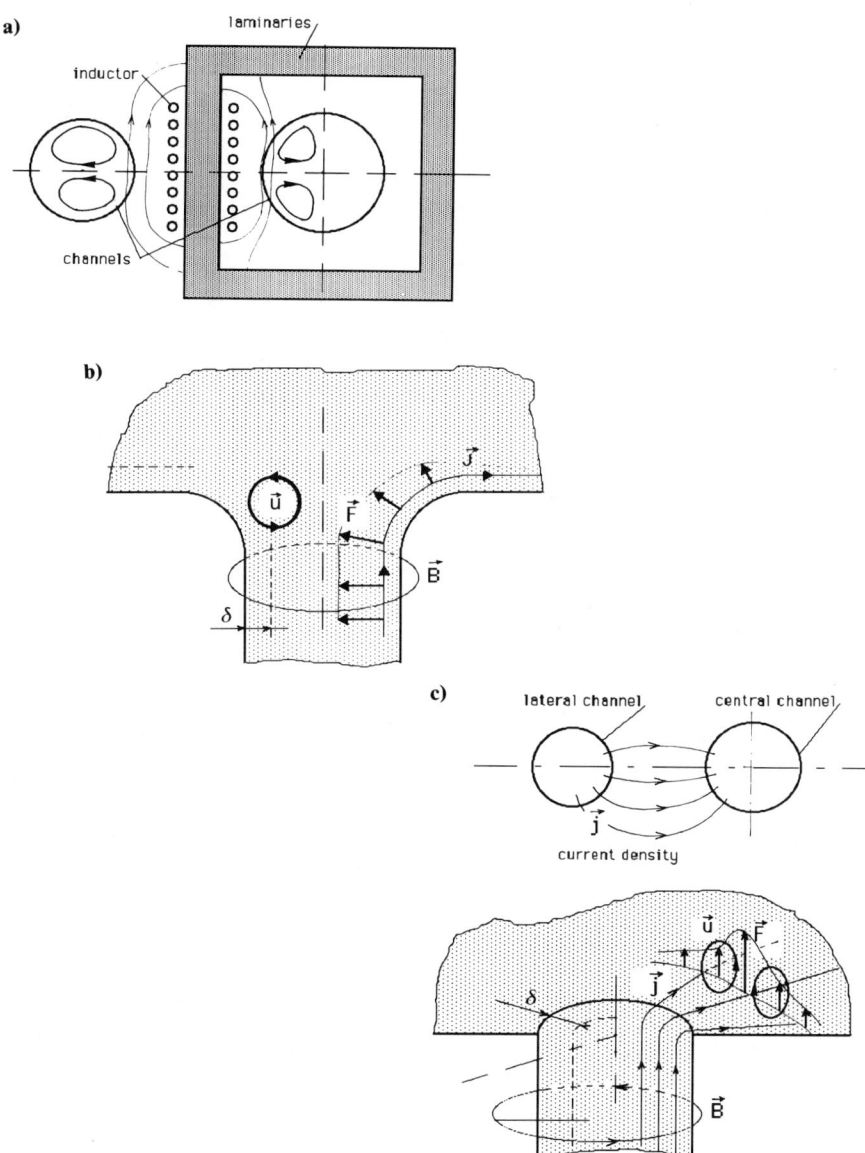

Fig. 13 Sketch of the electromagnetic force distribution in the channel expansion of a channel furnace and the corresponding vorticity: a) distribution in a cross section; b) case of an axisymmetric channel expansion; c) case of a nonsymmetric channel expansion.

complex example is the channel furnace, which possesses both asymmetry and various geometrical singularities such as abrupt expansions with corners, channel elbows, and dividing channels. Each singularity creates a particular rotational Lorentz force which generates local vorticity[7-9] in a similar way as divergent dc current situation.[2] An illustration is given in Fig. 13 which shows the vorticity created by : 1) an asymmetric magnetic field distribution in a channel cross section (cf. Fig. 13a), 2) an axisymmetric corner (cf. Fig. 13b), and 3) an asymmetric corner (cf. Fig. 13c). It is noticeable that in large channel furnaces the rotational part of the electromagnetic forces is of the order of $\rho u_A^2/\delta$ in the corner regions, with u_A, ρ, δ, denoting the Alfvén speed, the density and the electromagnetic skin depth, respectively. The local expected strong vorticity is partly responsible for the erosion problems encountered in industrial furnaces.

Multiphase inductors are preferred when only the function of stirring is needed. The efficiency of the inductor may be achieved by a suitable choice of the

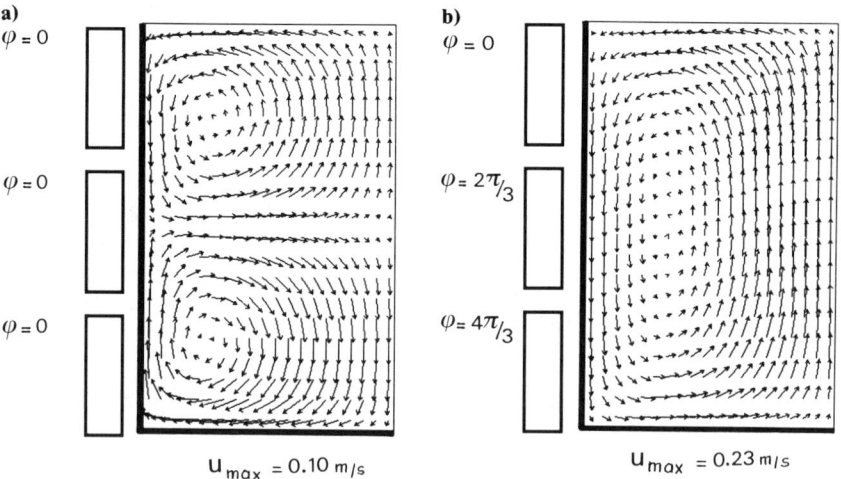

Fig. 14 Computed electromagnetic stirring in a 200-mm-diam mercury pool induced by a three turn coil (CEPHISE package) supplied with 20 Hz-ac currents and I=12,130 Amp.: a) single-phase supply; b) three phase-supply.

phase distribution and end effects are no longer required to create strong rotational forces. An example is given in Fig. 14, which illustrates the computed effects of the phase distribution on the liquid motion in a 200-mm mercury pool. It may be noticed that for the same coil currents the velocity magnitude is much greater in a three-phase configuration than in a single-phase one. Multiphase electromagnetic stirrers is now commonly used, and various types are available according to the necessities, namely,[2] mold or strand electromagnetic stirrers in continuous casting of steel, and ladle electromagnetic stirrers in steel processing.

That technology is now quite well mastered. Thanks to the large international research effort on the subject, precise rules now exist and are used by manufacturers to determine the type of stirrer that will gives the best result according to the nature of the product.

5.0 - Conclusion

Many industrial processes using alternating, traveling, rotating or steady magnetic fields are now available and are used in metallurgical industry. The rapid progress of the last few years is due to the industrial need of developing well-controled processes and reducing the energy consumptions. There still exists many potential applications in the processing of materials such as plasmas, oxides, glasses, or ceramics, where induction may revolutionize the traditional technologies.

References

1 Moffatt, H. K. and Proctor, P. R. E., "Metallurgical Applications of Magnetohydrodynamics," <u>Proceedings of the IUTAM Symposium</u>, Metals Society, 1984.

2 Lielpeteris, J. and Moreau, R., <u>Liquid Metal Magnetohydrodynamics</u>, Kluwer, London, 1989.

3 Tarapore, E. D., and Evans, J. W., "Fluid Velocities in Induction Melting Furnaces. Partl: Theory and Laboratory Experiments," Metallurgical Transactions B., Vol. 7, 1976, pp. 343-351.

4 Mikelson, Y. Y., Yakovitch, A. T. and Pavlov, S. I., "Numerical Investigation of Averaged MHD-flow in Cylindrical Regions with the Adoption of Working Hypotheses for Turbulent Stresses," Magnitnaya Gidrodinamika, Vol. 14, No. 1, 1978, pp. 51-58.

5 Fautrelle, Y. R., "Analytical and Numerical Aspects of the Electromagnetic Stirring Induced by Alternating Magnetic Fields," Journal of Fluid Mechanics, Vol. 102, 1981, pp. 405-430.

6 Moore, D. J., and Hunt, J. C. R., "Flow, Turbulence and Unsteadiness in Coreless Induction Furnaces," Proceedings of the 2nd Beer-Sheva Seminar on MHD Flows and Turbulence , edited by H. Branover and A. Yakhot, Israel Univ. Press, 1984, pp. 65-82.

7 Butsenieks, I. E., Levina, M. Y., Stolov, M. Y., and Shcherbinin, E. V., "Motion of Metal in a Channel Type Induction Oven," Magnitnaya Gidrodinamika, Vol. 16, No. 3, 1980, pp. 324-330.

8 Moros, A. and Hunt, J. C. R., "Modeling Recirculating Flows in Channel Induction Furnaces", Liquid Metal Flows : Magnetohydrodynamics and Applications, edited by H. Branover, M. Mond, and Y. Unger, Progress in Astronautics and Aeronautics, series AIAA, Washington DC, 1988, Vol. 111, pp. 421-441.

9 Mühlbauer, A. Fricke, R., and Walther, A., "Directional Melt-Flow in Channel Induction Furnaces," Liquid Metal Magnetohydrodynamics, edited by J. Lielpeteris and R. Moreau, Kluwer, London, 1989, pp. 247-254.

10 Garnier, M., "Metallurgical + MHD = Innovative Technologies," Liquid Metal Flows: Magnetohydro-dynamics and Applications, edited

by H.Branover, and M. Mond, and Y. Unger Eds., Progress in Astronautics and Aeronautics, series AIAA Inc., Washington DC, Vol. 111, 1988, pp. 377-399.

11 Garnier, M., Leclercq, I., Paillere, P., and Wadier, J. F., Proceedings of the 6th International Iron and Steel Congress, edited by The Iron and Steel Institute of Japan, Vol.4, 1990, pp. 260-266.

12 Getselev, Z. N. and Martynov, G. I., "Calculation of the Velocity Induced in the Liquid Phase of a Casting by Electromagnetic Forces," Magnitnaya Gidrodinamika, Vol. 11, No. 2, 1975, pp. 106-111.

13 Riquet, J. P. and Meyer, J. L., "CREM a New Casting Process, part II: Industrial Aspects," Light Metals, 1989, pp. 779-784.

14 Leclercq, I., "Conception d'une Installation Pilote de Fusion en Creuset Froid," PhD dissertation, Institut National Polytechnique de Grenoble, France, 1989.

15 Boussant-Roux, Y., "Les procédés de refusion en creuset froid : analyse des paramètres-clés", PhD Dissertation, Institut National Polytechnique de Grenoble, France, 1990.

16 Brunet, P., "Fusion en Creuset Froid et Pulvérisation de Carbure de Tungstene", PhD Dissertation, Institut National Polytechnique de Grenoble, France, 1987.

17 Kozuka, T., Asai, S. and Muchi, I., "Horizontal Electromagnetic Casting Process of Thin Plate and its Stability Analysis," Tetsu To Hagane, Vol. 74, 1988, pp.1793-1802.

18 Galpin, J. M., Gillon, P., Gelfert, Y., and Fautrelle, Y., "Effects of a Low Frequency Alternating Magnetic Field," Proceedings of the 6th International Iron and Steel Congress, edited by The Iron and Steel Institute of Japan, Vol.4, 1990, pp. 362-369.

19 Etay, J., "Le problème de frontières libres en magnétodynamique des liquides avec champs magnétiques alternatifs," PhD Dissertation, Institut National Polytechnique de Grenoble, France, 1988.

Research and Development in the Field of MHD Devices Utilizing Liquid Working Medium for Process Applications

V. M. Ievlev* and N. N. Baranov†
Russian Academy of Sciences, Moscow, Russia

Abstract

The use of electromagnetic forces to stimulate liquid media is important from the standpoint of both power generation (MHD generators of electrical power) and improvement of various production processes in many branches of industry. The MHD approach helps expand the knowledge of production process and find ways for its optimum control. The available experience shows that the MHD technology in a number of spatial applications, such as the metallurgy of ferrous, nonferrous, and rare metals and alloys, foundry, industrial power engineering, mechanical engineering, and chemical industry, considerably improves the quality of products and assists in solving environmental problems and for some processes it creates new possibilities. This paper briefly describes the main principles and uniqueness of realization of MHD flows, employed in the development of process devices. Based on these principles, the organizations coordinated by the Academy of Sciences' Scientific Council on Methods of Direct Energy Conversion developed several dozens of prototypes of MHD equipment intended for commercialization in various branches of industry.

Introduction

One of the promising activities related to applied MHD involves research and development associated with process MHD devices designed to provide for contactless force stimulation of various electrically conducting liquid working media such as liquid metals and electrolytes. The operation of such devices is based on the utilization of forces of interaction between currents flowing in working media and magnetic fields. In so doing, a magnetic field may be due either to the current flowing in the medium or to external sources. The current,

---

Copyright © 1992 by the American Institute of Aeronautics and Astronautics, Inc. All rights reserved.
*Institute of High Temperature, Scientific Council on Methods of Direct Energy Conversion.
†Senior Researcher, Institute of High Temperature, Scientific Council on Methods of Direct Energy Conversion.

in turn, can be delivered to the working medium with the aid of electrodes, or it can be generated by induction (without electrodes).

The forces emerging in the liquid as a result of MHD interaction make it possible to set the liquid in motion in various directions (including the upward vertical direction) or, conversely, to keep the melt from spreading, stir and heat the liquid, control its motion, pump over large volumes of liquid, finely proportion its flow rate, break up a jet of melt into droplets (when acted upon by a variable force), separate from each other the liquids of different electrical conductivity dissolved one in the other, or separate them from solid particles.

In a number of research organizations in the USSR, investigations have been underway for quite some time into the peculiarities of electromagnetic stimulation of electrically conducting liquid media and accompanying MHD effects. In particular, experiments are performed in magnetic hydrodynamics of liquid metals by use of experimental liquid-metal test rigs with different working media such as mercury, alkali metals, and alloys of indium, gallium, and tin. Physical and mathematical simulation consists of various metallurgical processes (heating, stirring, conveyance, etc.); equipment is tested under conditions of MHD stimulation of the melt. Investigations are made into the distribution of electric, magnetic, and hydrodynamic fields, as well as the conditions of heat and mass transfer in melts.

Based on the results of those studies, prototypes are developed of new MHD equipment that, in the form of pumps, flowmeters, valves, batchers, stirrers, separators, and other devices, can be used effectively for continuous stimulation of various working media including high-temperature and chemically aggressive media (e.g., mercury, liquid sodium, and molten zinc). The use of the thus developed devices to perform new production processes in major spheres of production (industrial power engineering, foundry, metallurgy of ferrous, nonferrous and rare metals and alloys, mechanical engineering, etc.) makes it possible to improve the quality of products and substantially raise the productivity of labor, helps solve the energy conservation and environmental problems, and, in the case of some processes, offers new possibilities.

At present, organizations coordinated by the USSR Academy of Sciences' Scientific Council on Methods of Direct Energy Conversion have developed several dozens of process MHD devices designed for diverse purposes.

Consider some spheres in which such devices can be used and their principal advantages.

First, both conduction (with electrodes) and induction (without electrodes), should be described. Large MHD pumps can have an efficiency of about 40% and higher.

The advantages of MHD pumps over various mechanical pumps include the following:

1) the possibility of manufacturing a fully hermetically sealed structure of MHD pump and entire liquid circulation loop (this is especially important in the cases when even minor leakage of liquid from the loop cannot be permitted or

when even the slightest contamination of liquid upon interaction with ambient medium must be avoided);
2) the absence of moving parts, noiseless operation; and
3) relative simplicity of control, whereby such pumps can be made unattended if necessary.

The Institute of Physics of the Latvian Academy of Sciences and its special design bureau (SKB MGD) have developed diverse modifications of induction MHD pumps (screws, plane, and cylindrical), including those for mercury rated for a pressure of up to 60 atm, flow rate of up to 250 ton/hr, and temperature of metal being pumped of up to 500°C. These machines are incorporated in mercury production complexes to provide for the highest purity of metal and the absence of harmful gas and waste emissions (Fig. 1).

MHD pumps for the sodium loops of nuclear power plants with fast reactors have been developed at the Efremov Institute of Electrophysical Equipment in St. Petersburg. Operating experience shows their high efficiency and reliability. The world's largest MHD pump is presently being tested; this pump has a flow rate of 3,500 m^3/h and a pressure of 0.3 MPa, and the temperature of sodium being pumped is 300°C. (Fig. 2).

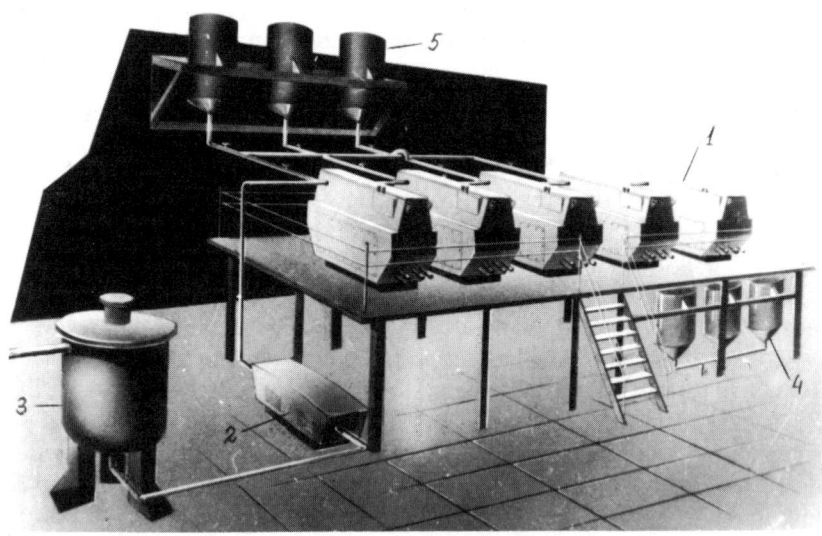

Fig. 1 Universal electromagnetic facility for chemical cleaning of mercury. 1, MHD device for mercury cleaning in reagents; 2, MHD pump/batcher for batched delivery of mercury; 3, starting container; 4, container for finished product; 5, container for reagents.

Fig. 2 TsLIN-3/3500 pump (cylindrical, induction) with a flow rate of 3,500 m^3/h and pressure of 3x10^5 Pa.

Heat-resistant MHD pumps with no other cooling except by the metal being pumped with a temperature of up to 600°C have been developed at the MHD Laboratory of the Krzhizhanovski Institute of Power Research. These pumps are intended for pumping alkali metals in various processes. Such pumps include a 10-channel MHD pump with plane channels and unified magnet system, with a flow rate per channel of 2.6 liters/s (Fig. 3). This pump is advantageous in that the total pressure/flow rate characteristic of the pump varies little when one of the inductors fails, and the inductor may be replaced without shutdown of the loop.

A submersible MHD pump for zinc melt has been developed in the same laboratory; this pump is intended for periodic rapid transfers of large masses of molten zinc (400-700 tons) (Fig. 4). The pump is submerged in molten metal and allows the bath with metal to be emptied in 1 hr, whereas other techniques require several days for this purpose.

Second, the use of MHD devices in conjunction with metallurgical units allows a whole complex of production operations to be performed: the stirring of liquid metal in the bath and in the flow (thereby improving its quality), preheating of the metal, confinement of the melt, refining and cleaning of metals and alloys from slags, delivery of the melt to the desired spot, exact batching of liquid metal in discrete batches, and control over the process of crystallization of ingots.

The use of MHD stirrers, which were developed at the special design bureau of the Institute of Physics of the Latvian Academy of Sciences, in a number of aluminum-making plants helped sharply reduce the time of metal preparation (in the case of Silumin, by a factor of 1.5), reduce the losses of aluminum and alloying metals, improve the metal quality (thanks to elimination of harmful impurities), and raise the productivity of labor.

MHD facilities for automatic pouring of pig iron in molds (Fig. 5), a process developed at the Institute of Foundry Studies of the Ukrainian Academy of Sciences and currently operating in the factories of the Kievtraktordetal production association, helped reduce the labor input of the puring process by 90-100% and permitted raising the yield of good castings by 3-5%, the output of automatic foundry lines by 10-25%, and the labor productivity by 25-50% because of the reduction of time required to perform auxiliary operations.

MHD facilities for processing lead, aluminum, and cooper alloys (cast shapes, liquid stamping, etc.) have been developed, ensuring a high quality of products, improved mechanical operating properties of parts prepared from

Fig. 3 RIN-10 10-module radial induction pump.

DEVICES FOR PROCESS APPLICATIONS 29

Fig. 4 AMNTs-7 MHD pump for zinc melt. 1, suction of melt; 2, electric lead; 3, hydraulic pipe.

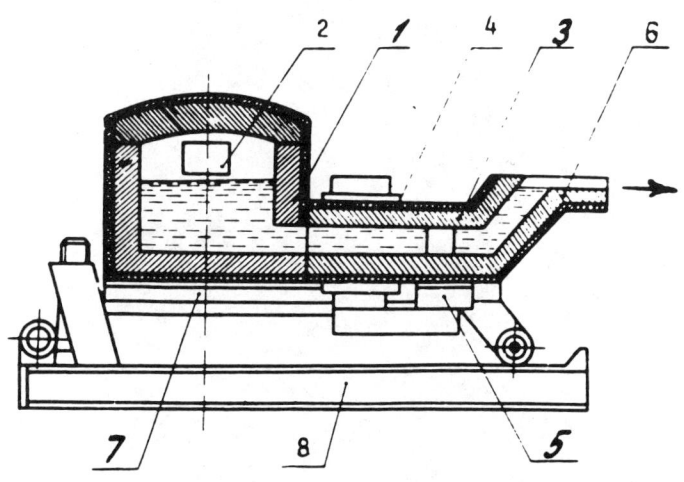

Fig. 5 MHD facility for puring pig iron. 1, crucible with liquid metal; 2, port; 3, induction magnetic unit; 4, inductor; 5, electromagnetic; 6, metal overflow lip; 7, frame; 8, bed plate.

castings and ingots, and smaller amount of castings rejected because of nonmetallic inclusions and gas porosity.

Third, in the Institute of Electrodynamics of the Ukrainian Academy of Sciences MHD granulators have been developed, i.e., devices designed to control the shape of free surface of a jet of liquid metal when it disintegrates into drops. These devices allow the preparation of spherical granules from tin, lead, zinc, and other metals having a diameter of 0.5-5 mm with monodispersity of 96% and output of 0.5-2 ton/hr. The size and shape of granules are defined by the law of variation in time of the external magnetic field; the pulse shape of this field is maintained by a special supply source (Figs. 6a and 6b). Such granules find application in metallurgical, chemical, aviation, and other branches of industry.

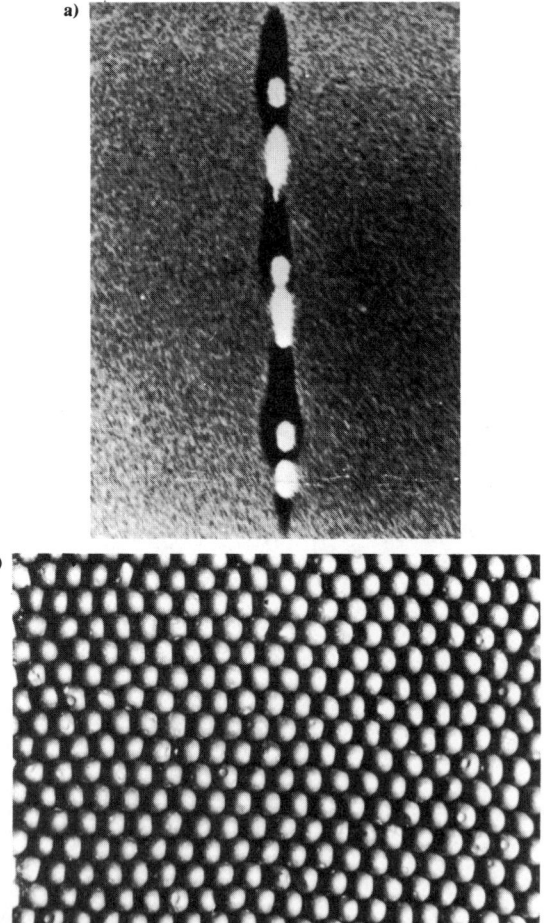

Fig. 6 Preparation of spherical granules using MHD technology: a) disintegration of free liquid-metal jet under the effect of magnetic field; b) finished product.

Fourth, MHD technology may be used to prepare principally novel composite materials. For instance, for a mixture of mutually insoluble molten components, using an appropriate selection and orientation of forces due to the interaction of the current and the magnetic field (directed vertically upward and balancing out the gravitational forces), one can develop conditions of "quasi-weightlessness," providing for uniform and finely divided distribution of phases in a solidifying ingot. Therefore, MHD technology helps develop new alloys possessing important practical properties consisting of components that cannot be "combined" under regular conditions (including antifriction alloys based on the aluminum-lead, zinc-lead, copper-lead systems; aluminum-graphite, aluminum-mica, aluminum-silicon carbide, and other composites).

Other process possibilities exist that are analogous to those investigated during materials studies in space stations under conditions of weightlessness.

Fifth, in solving the problems of a number of industries, involving the cleaning from impurities of electrically conducting media, as well as the separation of multiphase systems into components, effective use can be made of MHD separators, which include devices for recovery of metals from molten and crushed slags and cleaning of liquid metals from nonmetallic inclusions. In addition, there are some developments related to primary processing of geological samples with a view to extracting minerals and to concentration of ores and coals.

Along with the foregoing, other uses of MHD technology are presently being developed.

Therefore, one can make a general conclusion of the existence of ample possibilities for practical applications of process MHD devices based on the results of domestic research. Further research in this field (including international scientific cooperation), more extensive commercial utilization of the existing developments, as well as new developments, may considerably contribute toward acceleration of scientific and technical progress.

Application of MHD Facilities to Technology

Yu. M. Gelfgat*

Latvian Academy of Sciences, Riga-Salaspils, Latvia

Abstract

This paper gives a description of various aspects of applications of both MHD technologies and facilities. Moreover, the applications of MHD technology (liquid metal transport, mixing, batching, mold formation, etc.) have been reviewed briefly. The main attention has been focused on MHD processes and devices that have been developed recently and thus have not been studied thoroughly. In particular, results are given of a study on MHD techniques of a counterflow refining of metals and alloys, the effect on the structure of crystallizing melts, the production of thin-wall casting, the applications of MHD effects to stimulating and implementing the elements of space technology on the earth, the production of metallic composite materials under MHD weightlessness, the growing of semiconductor single crystals from melts, the soldering of printed circuits of radio electronic equipment, etc.

Introduction

The objective of magnetohydrodynamics of viscous incompressible conducting media is liquids that are, as a rule, rarely encountered in nature. In most cases these are melts of various metals and alloys as well as electrolytes and some organic liquid media employed in man-created technologies. Hence, there is a close link between the development of magnetohydrodynamics and its applications. Both MHD techniques and facilities have been successfully used to intensify and control the basic parameters of production relevant to various fields of modern technology: metallurgy and foundry work, power engineering and chemical industry, atomic power stations, and nuclear fusion devices. Moreover, the various effects of MHD on conducting media (by using dc, ac, travelling and other magnetic fields) result in the alteration of the characteristics of transfer of motion, heat, and mass in liquids as well as on the interface under the effect of magnetic fields. Several of the most widespread manifestations of MHD effects are listed in Fig. 1.

There is a wide range of applications of MHD effects to the technology; these applications can be conventionally divided into several large groups (Fig. 2):

1) MHD techniques and devices for controlling the flow and treating the

---

Copyright © 1992 by the American Institute of Aeronautics and Astronautics, Inc. All rights reserved.
*Professor, Institute of Physics.

APPLICATION TO TECHNOLOGY

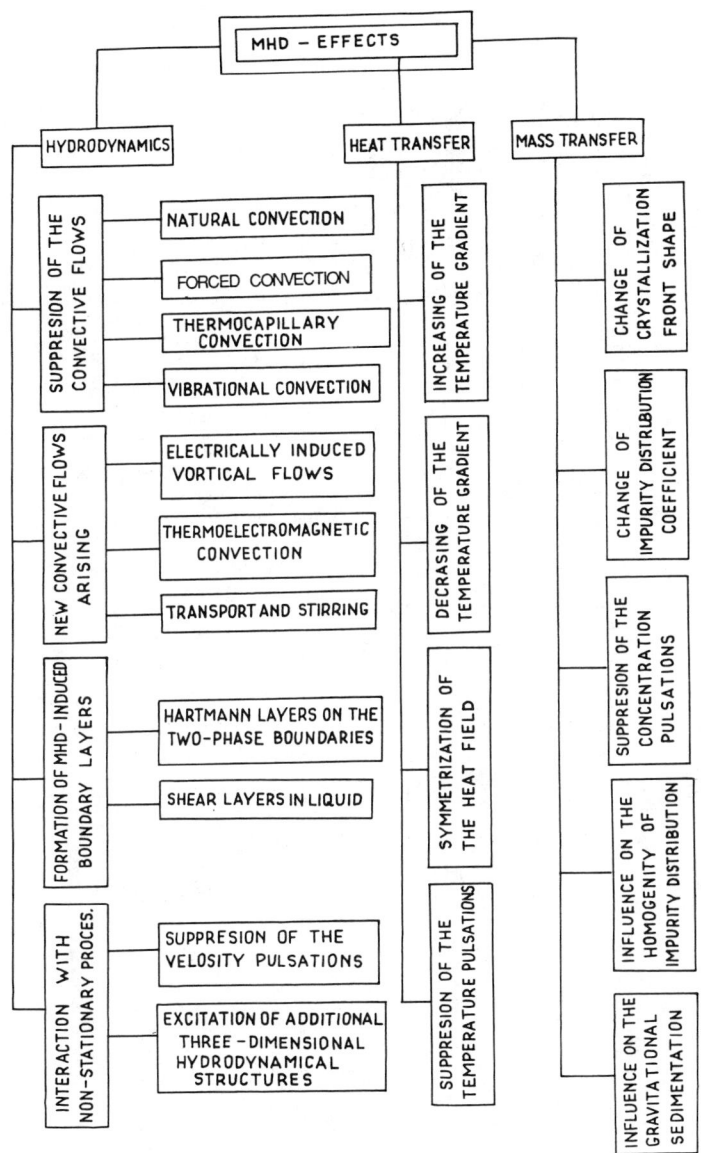

Fig. 1   MHD effects on the transfer of motion, heat, and mass.

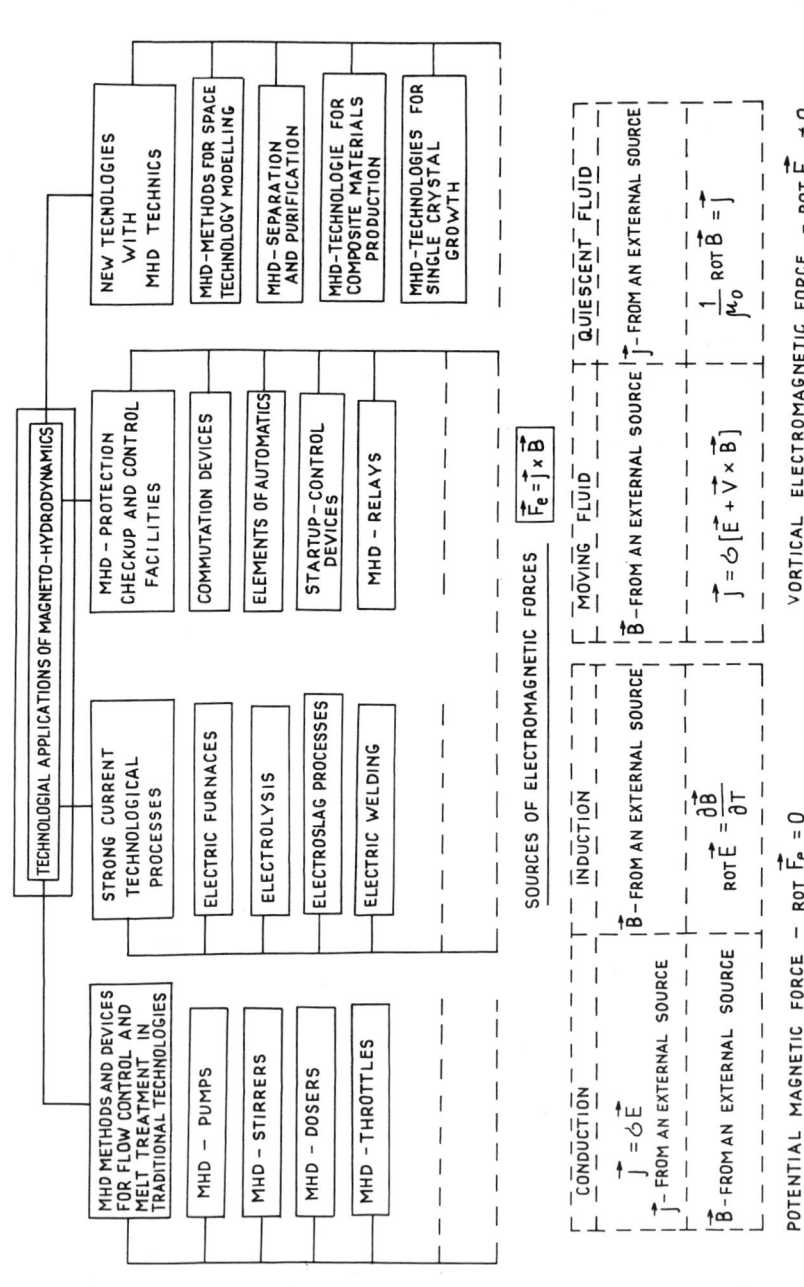

Fig. 2 Classification of magnetohydrodynamics applications and techniques of creating electromagnetic forces in conducting liquids.

melts by an electromagnetic force caused in a conducting medium by external electromagnetic fields. This would include various types of electromagnetic pumps, valves, etc., for transportation, batching, and stirring melts, which effect the convective heat and mass transfer in crystallization processes in metallurgy, in liquid metal cooling systems for nuclear technology.

2) Power engineering devices, metallurgical and electrical technologies in which MHD effects play an essential role and, in many ways, determine the efficiency of operation when strong electromagnetic fields are used. These devices include various types of electrical furnaces, electrolytic reduction devices, and in electroslag remelting, different types of electrowelding.

3) MHD protection, checking, and control devices for instrument engineering, switching systems, and startup control devices. This field of MHD technology comprises various types of MHD relays, communication switches, and units of automatics with a liquid working body.

4) New MHD techniques and devices for the treatment and control of metal flows that provide novel original technologies. Crucible-free melting, MHD separation and purifying multicomponent alloys, MHD production techniques of composite single crystal materials, MHD techniques for space technology simulation, etc., could serve as examples.

Both the general electrodynamic and hydrodynamic problems as well as specific problems relevant to production conditions must be solved in each of the aforementioned groups of applications of MHD effects. Thus, in the first case it is necessary to find the best parameters of MHD aggregates with an account of changes in fluid due to the effect of an additional field of the forces; in the second case it is necessary to specify the conditions under which a smooth technological process could occur with the least power losses and maximum yield for given parameters; in the third case the problems of stable operation of MHD devices, their reliability, lifetime and high speed rate are of a primary importance; in the fourth case design methods of the new devices must be developed and the best regimes and parameters of technological operations must be determined.

The aforementioned solutions and processes are influenced by the volume electromagnetic forces; the primary methods of excitation are presented in Fig. 2. It should be noted that the electromagnetic force can be either potential or vortical. In the former case its action is balanced by the pressure gradient, whereas in the later case it causes the origin of additional flows in liquid.

One distinguishing feature of technical applications of magnetohydrodynamics is that they must have a complex nature and combine not only MHD problems but also a number of physico-chemical and specific (metallurgical, materiological) questions relevant to actual conditions. The lack of cooperation among various specialists explains the large time gap between theoretical proposals and possible MHD applications to technology.

This work is concerned with various aspects of applications of MHD technologies and facilities exemplified by a number of projects and studies developed a the Institute of Physics of the Latvian Academy of Sciences. Moreover, we do not consider here conventional proposals of the MHD technology application, for example, to metallurgy for transportation, batching, stirring, shaping, and other processes that have been well described in the literature[1]. Attention is centered primarily on MHD processes and devices that have recently emerged and have thus far been used in practice rather inefficiently.

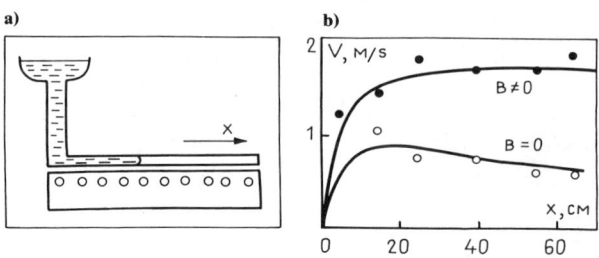

Fig. 3 Change of metal flow velocity in a mold under the effect of travelling magnetic field: a) diagram of the process; b) dependencies v = v(t).

Our aim is to provide a number of examples of the use of MHD effects within the framework of the four groups given earlier.

First we will touch upon some novel technological proposals concerning electromagnetic pumps and their modifications. Figure 3a presents a diagram of a MHD facility for producing thin-wall casting from nonferrous and ferrous metals with a complex and extended transversal cross-sectional profiles.[2] The production of such components using standard foundry techniques is a problem. The facility consists of a casting mold with a system of traveling magnetic field inductors placed underneath. These inductors excite electromagnetic forces, providing fast filling of the whole volume of the mold before the solidification in a liquid flowing into the mold begins.

A distinguishing feature of the previously mentioned MHD casting technology compared with the conventional ones is a different kind of physical conditions of filling the mold. If in a conventional situation metal is kept in motion in the mold due to the effect of hydrostatic pressure of liquid column, then in our case it is also affected by the electromagnetic volume force, which increases as the metal moves on. As a result, the rate of filling the mold has been drastically changed (Fig. 3b), and this provides production of high-quality castings with thin and extended walls (Fig. 4).

Stirring of the liquid metal alloys has proved to be one of the most popular applications of MHD technology. This MHD technique has been most extensively used in large steel-making furnaces[3] and continuous metal casting facilities[4]. Moreover, MHD stirring is likewise essential, for example, in ferrous metal alloys in mixers and furnaces to provide their homogenization, refining, alloying, etc. Figure 5 presents two modifications of MHD stirrers for aluminum and its alloys.[5] Forced convective flows excited by the traveling field inductors are employed that provide stirring in either a horizontal or vertical plane. (In the latter case the MHD device can perform two actions: 1) alloy stirring and 2) batched output from the mixer of the furnace). Also, it follows from investigations[6] that the rate of preparing melts is mainly determined by the regime of flows (laminar or turbulent) and depends considerably less on the modification of the MHD effect used. The results of physical modeling of stirring by means of a rotating electromagnetic field (Fig. 6) illustrates the fact that when reaching a turbulent regime alloy homogenization is increased

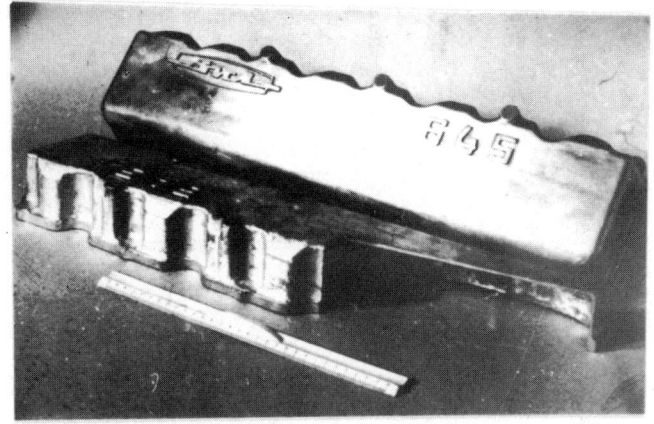

Fig. 4   Samples of produced thin-wall castings.

repeatedly; moreover, its quantitative characteristics are comparable for various modifications of MHD stirring devices.[6]

We were able to obtain the following results using MHD stirrers:
1) an increase of 30-50% in the productivity of mixers for producing aluminium and its alloys;
2) an increase of ~100 times in the rate of chemical purification of mercury;
3) an increase of 3-5 times in the rate of zone remelting of semiconductor materials; and
4) an increase of 1.5-2 times in the homogeneity and rate of homogenization of optical glasses.

The MHD technique of metal counterflow refining by slag using an inclined MHD trough[7] is of great interest. Figure 7d shows refining slag flowing

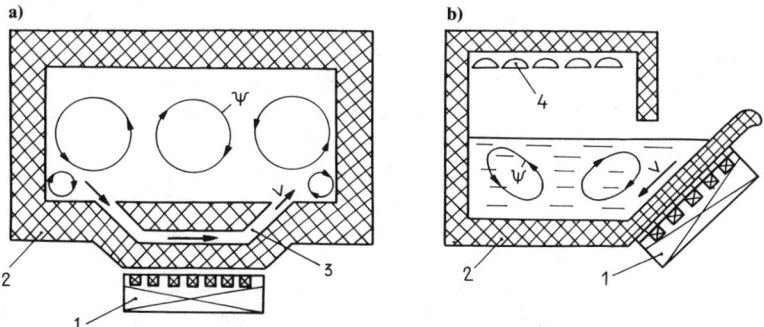

Fig. 5   Diagrams of MHD stirrers for aluminium and its alloys:
a) stirring in a horizontal plane; b) stirring in a vertical plane;
1, travelling magnetic field inductor; 2, mixer;
3, working channels; 4, heating elements.

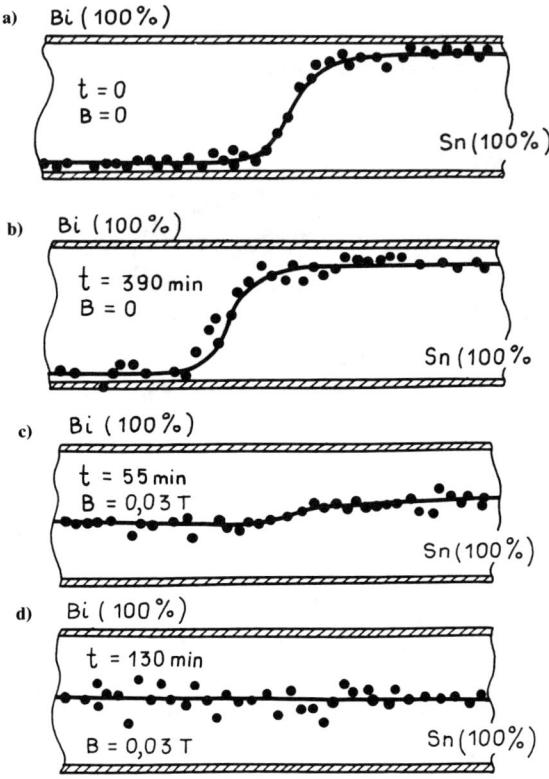

Fig. 6 Stirring of bismuth-lead system by a rotating magnetic field. The alloy conductivity at various moments is indicated by the curve and dots.

downward along the metal flow under the force of gravity and carried to the surface of metal moving upward along the trough. Because of the specific profiling of a traveling magnetic field inductor (Fig. 7), additional vortices are formed in the liquid metal flow, greatly intensifying the mass transfer between the slag and the metal. As a result, the rate of refining grows significantly (Fig. 8), even in comparison with the known diagram of a MHD trough in which a conventional traveling magnetic field inductor is used. At optimum electromagnetic and hydrodynamic parameters of the previously mentioned devices, the efficiency of metal treatment by slag can be increased up to 60%, which allows the length of the MHD trough to be decreased to 40% without worsening its operational characteristics.

A number of technologies use MHD devices for controlling the liquid metal flows: MHD stoppers, valves, throttles, etc.[1] In particular, by using MHD stoppers (gates), we can solve the problem of hot coating of anticorrosive metallic coatings (aluminium, zinc, and their alloys) on long steel products (angles, channel bars, pipes, etc.) via continuous baths (see Fig. 9)[8] Unlike the conventional metallization procedure of immersing in melt, the continuous bath

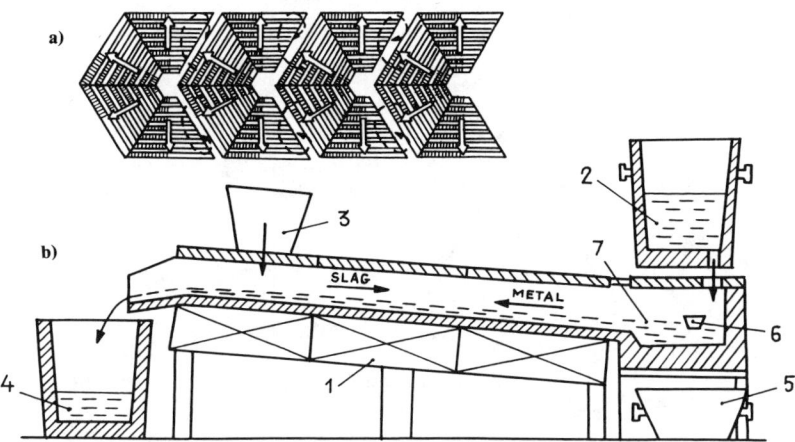

Fig. 7 Diagram of counterflow refining by a MHD trough (a) with the combined inductor of a travelling magnetic field; (b) 1, trough inductor; 2, vessel with starting metal; 3, vessel with slag; 4, vessel with refined metal; 5, collector of spent slag; 6, opening for slag discharge; 7, a trough with a metal container.

has openings positioned below the liquid metal level through which the units to be treated pass. The melt is held in the bath windows by means of MHD gates consisting of a set traveling magnetic field inductors that excite the force toward the metal volume. Such MHD devices have a significantly smaller size and metal volume (5-10 times), which provides the treatment of units having unlimited length, full automation of the process, and better quality of coating. We have designed MHD facilities for aluminization of angles of a size up to 160 x 160 mm, with the pipes with diameters of up to 180 mm and sheets of widths up to 700 mm.

Figure 10 shows a diagram and layout of a variant of MHD throttles for controlling the flowrate in liquid metal loops of nuclear power plants.[9] The figure shows that throttling of metal flow is reached due to its interaction with a steady radial magnetic field in an annular cross-sectional channel. This interaction brings about the excitation of electromagnetic forces in liquid, greatly increasing its resistance to motion and decreasing the initial flow rate by 3-10 times (Fig. 11). We were able to devise many modifications of MHD throttling devices for various technical purposes. The results of both research and applied investigations relevant to the aforementioned problem have been generalized in Ref. 9.

A wide application of MHD technology to industry can be illustrated by MHD wave generators for soldering the printed circuits in electronic industry.[10] These are various types of MHD devices providing the supply of melted solder to a contact with the surface of a printed circuit. Figure 12 presents a diagram of a type of MHD generator with a helical channel and a rotating magnetic field inductor. As seen, its operation is as follows: liquid metal rotating under the effect of electromagnetic forces is directed downward along the helical channel

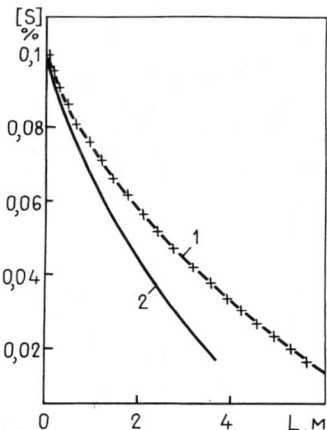

Fig. 8 Reduction of sulfur along the length of MHD trough during refining of cast iron by soda slag: 1, using a traveling field linear inductor; 2, using a combined inductor.

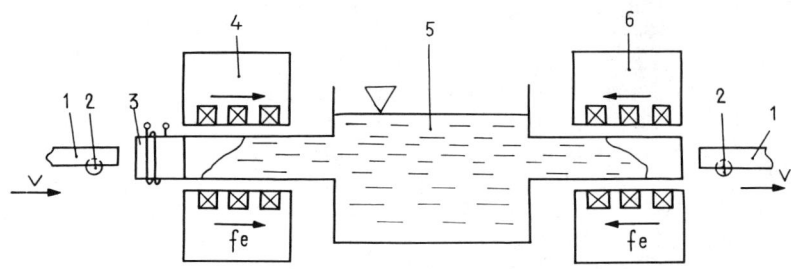

Fig. 9 Diagram of aluminization by the use of continuous baths provided with MHD gates: 1, a unit to be treated; 2, guiding roller table; 3, preliminary heating chamber; 4, entrance MHD lock; 5, bath with liquid aluminium; 6, exit MHD lock.

and is delivered to a central cylindrical cavity and forming at its outlet a profiled solder wave, washing the printed circuits passing above the wave. Compared with conventional devices with mechanical drive, MHD wave generators have numerous advantages (including the absence of contact with components moving in solder, decreased oxidation of solder, small power capacity, etc.), and also provide soldering of circuits with widths of up to 500 mm at optimum parameters of technological processes.

Next we will consider examples relevant to the second group of MHD applications: power processes and technologies in which MHD effects have a decisive role in the characteristics and regimes of operation. In accordance with investigations, the efficiency of various types of electric furnaces for metal and alloy melting (induction crucible, electric-arc, vacuum, induction channel), electrolyzers, electroslag, smelting, and electric welding devices greatly depends

Fig. 10 Diagram and general view of MHD throttles with a radial magnetic field: 1, pole terminal of a constant magnetic field inductor; 2, magnetic inductor elements; 3, excitation winding; 4, internal core; 5, external channel wall.

on the nature of specific hydrodynamic flows caused by the action of electromagnetic forces in liquid. Their physico-mathematical and applied aspects are discussed in greater detail in Ref. 11. To illustrate this branch of MHD technology, let us consider two typical ways of solving practical problems.

In induction channel furnaces (Fig. 13) molten metal can be considered as a secondary winding of the transformer in which currents of up to 10^4 A are induced. The heat released in channels by these currents must be transferred to the bath by convective flows. Obviously, the mass-transfer intensity between the masses of alloy in channels and the main volume of the furnace depends on the motion of the liquid, which would be difficult to excite in a conventional

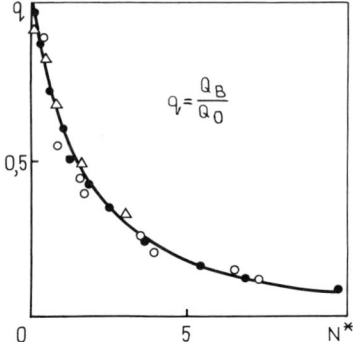

Fig. 11 Flow rate characteristics of MHD throttles with a radial steady magnetic field.

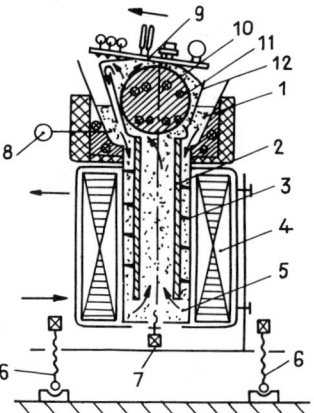

Fig. 12 Diagram of MHD wave generator with a helical flow duct:
1, Bath with solder; 2, inner container; 3, helical guide; 4, rotating magnetic field inductor; 5, outer container; 6, adjustable support; 7, plug; 8, thermocouple; 9, printed circuit to be soldered; 10, heaters; 11, 12, shape-forming waves of solder.

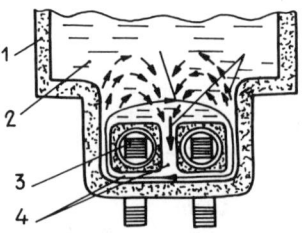

Fig. 13 Diagram of induction channel furnace:
1, casing of furnace; 2, metal melt; 3, inductor; 4, furnace channel.

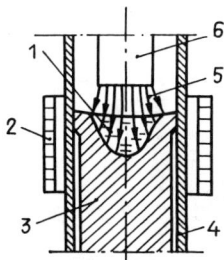

Fig. 14 Diagram of vacuum arc remelting with electromagnetic stirring of crystallizing melt: 1, metal melt; 2, magnetic field inductor; 3, solidifying metal; 4, crystallizer; 5, electric arc.

design. To solve this problem[12] we propose to employ electrovortex flows caused by vortical electromagnetic force due to the interaction of the spatially nonuniform electric current with its own magnetic field. It was sufficient to create a monodirectional loop of the induced current in the channels and then to increase its spatial nonuniformity by design measures to cause intense transit metal flows through furnace channels without the use of any additional power sources. This, in turn, allowed the efficiency, lifetime, and power of the aforementioned devices to be increased significantly.

When electrically induced vortical flows (which, as a rule, accompany power processes) do not satisfy the requirements of technology, it is useful to employ additional external magnetic fields to solve the problem. Thus, in a vacuum arc smelting process (Fig. 14), to improve the structure of an alloy solidifying in a crystallizer, one may use the external vertical magnetic field by means of a cylindrical solenoid surrounding the crystallizer.[13] The field interaction with the radial components of the current results in melt rotation in the crystallizer with a secondary flow in the meridional plane. The latter opposes the convective and electrovortex flows, which allows the heat and mass transfer on the solid-liquid interface to be controlled. A significant improvement of the crystalline structure of the ingot (Fig. 15) can be reached by the given technique[13] at an optimum regime of azimuthal rotation of the melt.

The use of the MHD devices in switching systems and automatic elements (the third group of MHD technology applications) has the following advantages in comparison with the conventional solutions:

1) wide functional possibilities that permit several problems of control to be solved in one device;
2) high reliability due to the use of only one mobile element (liquid), excluding wear of components, welding, and sealing of contacts, etc.;
3) hermeticity of the design, allowing operation under specific conditions and
4) good service without maintenance.

As a rule, these MHD devices are constructed as hermetically sealed steel devices; inside these devices there are channels and chambers filled with conducting fluid with a low melting temperature (mercury, gallium) together

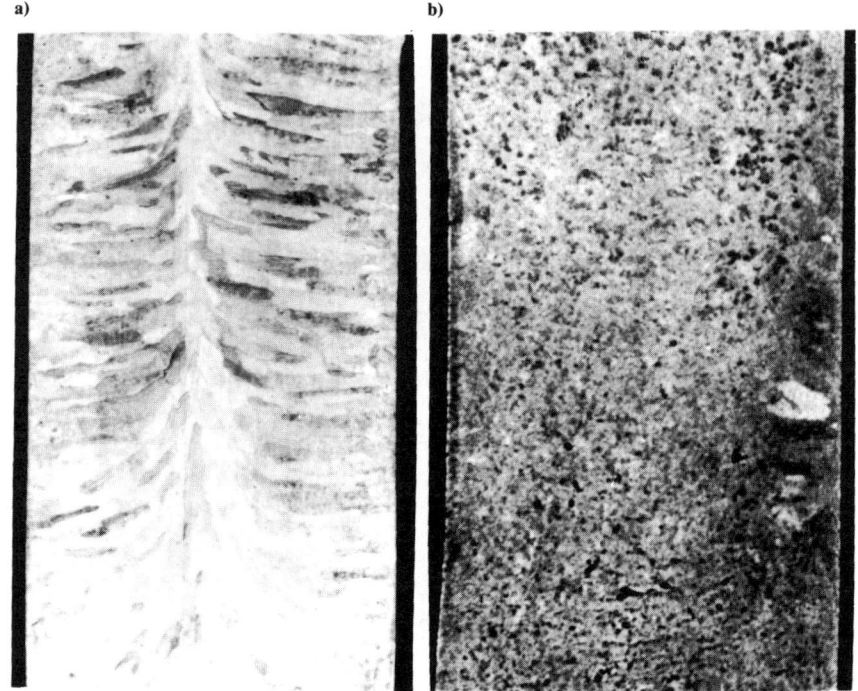

Fig. 15 Macrostructure of steel ingot produced by a vacuum arc melting: a) under conventional conditions; b) using electromagnetic stirring.

with magnetic systems whose design and parameters have been specifically determined. We have designed a number of modifications of similar MHD technology[14] for both dc and ac power supply. These include voltage, current, power, phase shift and break and time relays, current regulators and stabilizers, commutators for measurement circuits, and nonelectrical quantity sensors.

Now consider two modifications of similar devices. Figure 16 gives a diagram of a MHD device for adjusting and stabilizing large currents.[15] It consists of a tank (1), which is divided into two parts by a partition (2) and a nozzle (3); the lower region is filled with a conducting liquid (4), and the upper region is connected with the atmosphere. Electrodes (5) are mounted into opposite end walls of the lower region, which, in turn, is placed between the magnet poles (6). This electromagnet has two coils (7 and 8): one coil is connected to the electrodes (5) and the external load (9) and the other is parallel to the electrodes (5).

During an operation, the electromagnetic force f_e directed downward affects the liquid located in the lower chamber as a result of the field interaction with the current. Moreover, the level of the liquid column in the nozzle (3) increases and compensates for the electromagnetic pressure. If the current in the load (9) grows, the current of power supply and the field in the electromagnetic cap also

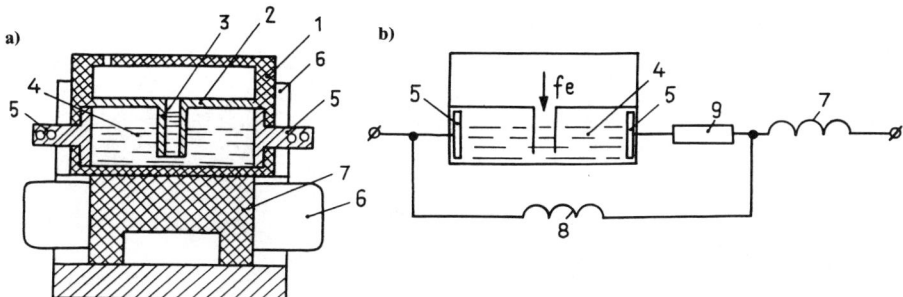

Fig. 16 MHD device for adjusting and stabilizing of large currents: a) design diagram; b) electric circuit.

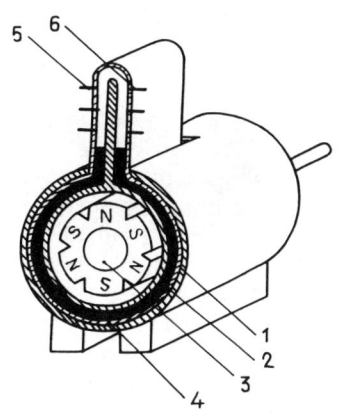

Fig. 17 MHD angular velocity relay.

increases. This causes the growth of f_e, which results in the decrease of fluid level in the lower region. The decrease of the liquid level makes electric resistance in the circuit between electrodes (5) greater, which results in a decrease (stabilization) of the current in the load (9). The current decrease in the coil (9) brings about a reverse situation, and in this way the necessary current value is preserved in the circuit.

Figure 17 presents a diagram of the MHD angular velocity relay,[14] where the two coaxial cylinders (1) and (2) form a channel (4) filled with a conducting liquid. Inside the cylinder (1) there is a star-shaped rotor (3) with permanent magnets that is mechanically connected to the device whose angular velocity is controlled. Rotation of the rotor (3) produces a rotating magnetic field, which induces electrical currents in the liquid, where the electromagnetic pressure arises. This current causes liquid to move into the channel (4) proportionally to the input signal (the angular velocity) and contact (5 or 6) switches the output circuit in accordance with rotation direction and velocity.

Furthermore, we will consider the fourth group of MHD technology application comprising original technologies based on the use of specific MHD

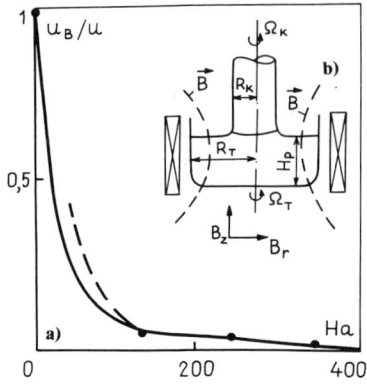 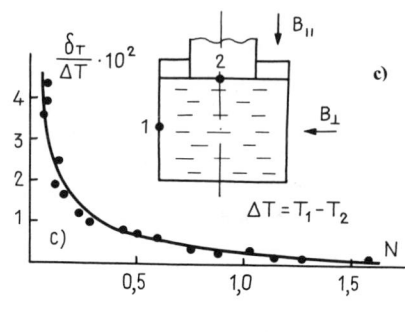

Fig. 18 The change of convective flow velocity (a) and temperature oscillations (b) during the Czochralski process.

effects. One of the latest developments is the MHD technique of modelling and implementing space technology elements on the earth.[16]

As known, one of the main problems of space technology is the production of new materials that exhibit very good operational characteristics. Moreover, an important condition is weightlessness, which permits the positive results to be reached when there must be no gravitational separation and component sedimentation in complex systems. Crystallization and heat transfer processes are determined solely by diffusion mechanisms due to the absence of buoyancy-drive flows, and it is possible to eliminate the harmful effect of the alloy contact with the container walls as well as to suppress various nonstationary perturbations in the fluid. However, numerous studies relevant to space research show that it is difficult to reach the necessary quality of materials under these conditions[17] because of residual thermoconvective (at $g = 10^{-2}g_o - 10^{-4}g_o$) and thermocapillary flows.

At the same time, MHD techniques affect conducting media by means of electromagnetic forces excited in them, which allows conditions similar to those observed under weightlessness to be reproduced in a number of cases on the earth. For instance, suppression of natural and forced convective flows in melts by means of a constant magnetic field, suppression of various stationary and nonstationary perturbations of flows, and determination of velocity gradients along the direction of the magnetic field vector permit to the heat- and mass-transfer processes relevant to space technology to be partially simulated. In multicomponent media, the mass and volume nature of gravitational and electromagnetic forces (they act on each elementary volume of matter) allows one to compensate for the effect of gravity and to reach neutral equilibrium (quasi-weightlessness) of components in the melt as well as to eliminate gravitational sedimentation and phase segregation on the earth.

Let us now consider some technologies. We will analyze the possible use of steady magnetic fields in the experiments on semiconductor single crystal growth by the Bridgman technique under microgravitation to reduce and convert residual convective flows.[18] For a qualitative analysis of the problem of determining dimensionless parameters, the degree of the field effect on the intensity of thermogravitational convection is characterized at the Grashoff numbers Gr > 1, by the Lykoudis number: $Ly = Ha^2/\sqrt{Gr}$ (Ha = $BL\sqrt{\sigma/\rho\nu}$). At B = 0, heat transfer is determined by the Peclet number Pe = $\sqrt{Gr \cdot Pr}$ (where Pr is the Prandlt number); and at B ≠ 0 by $Pe_B = GrPr/Ha^2$. In the same way mass transfer will be found from the diffusion Peclet numbers $Pe_D = \sqrt{Gr \cdot Sc}$ (where Sc is the Schmidt number) and $Pe_{D,B} = GrSc/Ha^2$ at B = 0 and B ≠ 0, respectively.

Similar parameters can be written for the estimation of the effect on the heat and mass transfer of thermocapillary convection characterized by the Marangoni number Ma. At B = 0, heat and diffusion Peclet numbers are $Pe^M = Pr(Ma/Pr)^{1/3}$ and $Pe_D^M = Sc(\frac{Ma}{Pr})^{1/3}$. If B ≠ 0, they can be written as

$$Pe_B^M = \frac{Pr}{Ha^2}\left(\frac{Ma}{Pr}\right)^{2/3} \quad \text{and} \quad Pe_{D,B}^M = \frac{Sc}{Ha^2}\left(\frac{Ma}{Pr}\right)^{2/3}$$

Since in space technology experiments g=$10^{-3} \div 10^{-5} g_0$ but the magnetic field values reached under these conditions result in Ha = 10÷50, it is probably possible to decrease the role of gravitational and thermocapillary (to a lesser extent) convections in the previously mentioned technological processes at relatively small values of a magnetic field. The values of actual space experiments relevant to growth of germanium, indium antimonide, and other single crystals reveal that, by applying the magnetic field B = 0.05÷0.1T, the heat-transfer process approaches that of diffusion, whereas the mass transfer still preserves its convective character but at much smaller transfer velocities.

It should be noted that some of the conditions of space experiments require forced convective homogenization of the melt to be carried out before crystallization. This can be done by means of an alternating (rotating, traveling) magnetic field. An evaluation of the stirring velocity Re ~ $Ha^2 \omega R^2$ reveals that the two functions (reduction of convective stirring and its intensification) can be combined into one device having comparatively small dimensions and required power, which is very essential in space technology.

It is clear that the aforementioned MHD effects of affecting the melt in production of single crystals can also be reproduced on the earth. The MHD technique of controlling hydrodynamic, heat, and mass transfer during the production of semiconductor single crystals[19] is based on a complex of similar technical solutions. The first investigations in this field have yielded positive results, even when the most widespread technologies of crystal production, such as the Czochralski technique,[20] have been employed. As has been anticipated, the flow pattern in the melt is directly connected to the corresponding changes of the heat and mass transfer in liquid and on the crystallization front, and the suppression of nonstationary velocity perturbation causes the suppression of temperature and concentration fluctuations in an alloy. As a result, when MHD effects are used on the processes of single crystal production, an optimum technological condition might be achieved and the production of high-quality materials could be realized.

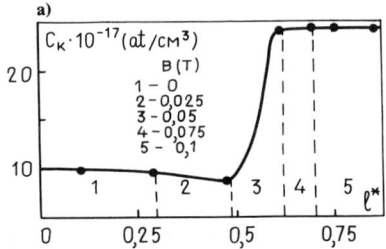 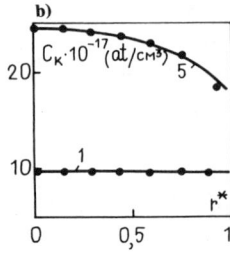

Fig. 19 Axial (a) and radial (b) oxygen distribution of silicon single crystal grown by the Czochralski technique for various values of an axial magnetic field.

On earth, MHD technologies of single crystal growth effect both: the steady (axial, transversal, and axial-radial) magnetic fields, and MHD methods of exciting various forced convective flows (rotating, traveling, alternating electromagnetic fields). Figure 18 presents the data of the magnetic field effect on both the velocities of convective flows and the pulsations of temperature in the melt characterizing the efficiency of MHD techniques in the single crystal growth techniques. These data show that even moderate values of the magnetic field (B = 0.1-0.2T) have a considerable influence on the hydrodynamics and heat transfer in the melt and allow single crystals with improved characteristics to be produced. (Fig. 19).

The preceding examples of various applications of MHD techniques to different industries give evidence of their future development. There should be a close collaboration between MHD specialists and those working in other fields who could contribute to the development of novel technologies.

References

[1] "Metallurgical Applications of Magnetohydrodynamics," *Proceedings of IUTAM Symposium*, Shercliff, J.A., Ed., 1982.

[2] Rabinovitch, B.V., and Wolkow, W.M., "Forfullung und Kristallisation von Schmelzen in Wandernden Magnetfeld," *Giessereitechnik*, Vol. 1 and 2, Nos. 3 and 4, 1972, pp. 242-250.

[3] Sundberg, Y., "Metallurgical Aspects of Induction Stirring," *Proceedings of IUTAM Symposium*, Shercliff, J.A., Ed., 1982, pp. 217-223.

[4] Birat, J.P. and Chone, J., "Electromagnetic Stirring on Billet, Bloom and Slab Continuous Casters: State of the Art in 1982," *Iron Making and Steelmaking*, No. 4, 1983, pp. 269-281.

[5] Gelfgat, Yu.M., "Metallurgical Applications of Magnetohydrodynamics," *Magnitnaya Gidrodinamika*, No. 3, 1987, pp. 120-137.

[6] Gelfgat, Yu.M. and Gorbunov, L.A., "On Homogenization of Multi-Component Melts under Weightlessness," *Izvestiya Akademii Nauk SSSR, Seriya Fizicheskaya*, Vol. 9, No. 4, 1985, pp. 667-672.

[7] "Electromagnetic Trough," USSR Authors' Certificate No. 1049690, Otkrytiya, Izobreteniya, Moscow, No. 39, 1983, pp. 140-141.

[8] Strekalov, G.N., Birger, B.L. and Kurlyk, N.P., *Metallic Coating: Techniques and Equipment*, Latviyskiy Nauchno Issledovatelskiy Institut Nauchno Technicheskoy Infirmacii, Riga, USSR, 1986.

[9]Gelfgat, Yu.M., Gorbunov, L.A. and Vitkovskij, I.V., *Magnetohydrodynamic Throttling and Control of Liquid Metal Flows*, Zinatne, Riga, USSR, 1989, 312p.

[10]Simsons, J. and Gelfgat, Yu.M., "MHD Wave Converters for Automated Soldering of Printed Circuits," *Automatic Welding Journal*, No. 14, (1986, p. 404.

[11]Bojarevics, V., Freiburg, J., Shilova, E.I. and Shcherbinin, E.V., *Electrovortex Flows*, Zinatne, Riga, USSR, 1985.

[12]Bucenieks, I., Levina, M.J., Stolov, M.J. and Shcherbinin, E.V., "Study of Metal Flow in Induction Channel Furnaces," *Magnitnaya Gidrodinamika*, No. 3, 1960, pp. 123-130.

[13]Abrika, M., Mikelson, A., Moshnyaga, V.N., Chernov, Y.V., and Scherbakov, A.I., "The Effect of Constant Magnetic Fields on the Motion of Liquid Metal During Vacuum Arc Remelting," *Magnitnaya Gidrodinamika*, No. 3, 1979, pp. 105-110.

[14]Barinberg, A.D., *Magnetohydrodynamic Devices of Protection Checkup and Control*, Energija, Moscow, 1978.

[15]Birger, B.L., Gelfgat, Yu.M., Gorovitz, V.S. and Sorkin, M.Z., "Electrodynamics Pump," USSR Authors' Certificate No. 550749, Otkrytiya, Izobreteniya, Moscow, No. 10, 1977, p. 143.

[16]Gelfgat, Yu.M. and Mikelson, A., "Electromagnetic Modelling Techniques and Effect on Space Technology Elements," *Nauchnye Chtenija po Aviatsii i Kosmonavtike*, Nauka, Moscow, 1981.

[17]*Hydrodynamics and Heat Transfer Under Weightlessness*, Ed. Avduyevskiy V.S. and Polezhaev, V.I., Nauka, Moscow, 1982, pp. 163-241.

[18]Gelfgat, Yu.M. and Sorkin, M.Z., "On Magnetohydrodynamical Effect on Convective Transfer Parameters in Space Technology Experiments," *Gagarinskije Chtenija po Kosmonavtike i Aviatsii*, Nauka, Moscow, 1986, pp. 289-290.

[19]Gelfgat, Yu.M., Zemskov, V.S., and Rauhman, M.R., "Semiconductor Single Crystal Growth with Electromagnetic Effect upon Melt," Sbornik Nauchnyh Trudov Protsessy Rosta Poluprovodnikovyh Kristallov i Pljenok, Nauka, Novosibirsk, USSR, Sib. Otdelenie, 1988, pp. 38-55.

[20]Gelfgat, Yu.M., Gorbunov, L.A. and Sorkin, M.Z., "Electromagnetic Means to Effect Hydrodynamics and Heat Transfer During Single Crystal Volume Growth," *Rost Kristallov*, Vol. XVI, 1988, pp. 234-247.

[21]Gelfgat, Yu.M., "MHD Techniques of Composite Material Production under Quasi-weightlessness," *Izvestiya Akademii*, Latvia, SSR, No. 10, 1980, pp. 75-92.

[22]Abramov, O.V., Bushe, N.A., Gelfgat, Yu.M., Markova, T.F. et al., "Structure and Properties of Aluminum-Lead Antifriction Alloys Produced Under Gravitational Liquation Compensation by Electromagnetic Forces," *Metalurgiya i Termoograbotka*, No. 4, 1982, pp. 54-60.

Electromagnetic Modulation of Molten Metal Flow

R. M. Slepian*
Westinghouse Science & Technology Center, Pittsburgh, Pennsylvania 15235
P. A. Davidson†
Imperial College, London, England
and
A. R. Keeton‡
Westinghouse Science & Technology Center, Pittsburgh, Pennsylvania 15235

Abstract

It is well known that high frequency electromagnetic fields may be used to constrict a liquid metal jet. This constriction provides a means of modulating the flow of metal through an orifice, and such a valve has many applications in the metallurgical industries. Conventionally, these valves have a "pinch" type configuration, in which the exclusion of the magnetic field from the core of the jet causes an overpressure in the liquid. However, such configurations are difficult to implement, because of: (1) the uncertainty in the position of the separation point within the outlet nozzle, (2) the excessive joule dissipation which results from the high frequencies and coil current densities required for effective modulation, and (3) the necessity to select different frequencies for different orifice sizes. In order to circumvent these problems, we have developed an alternative method of electromagnetic flow control. This method relies not on a radial pinch effect, but on the generation of a direct axial force, using the field arising from single phase excitation. Such an approach does not rely on separation of the jet from the nozzle

---

Copyright © 1991 by the American Institute of Aeronautics and Astronautics, Inc. All rights reserved.
* Senior Engineer, Engineering Science.
† Associate Professor, Mechanical Engineering.
‡ Retired.

wall and is less sensitive to the choice of field frequency. Computational and experimental studies of this configuration are presented which show the the degree of flow modulation as a function of magnetic field strength and frequency. Also, its characteristics will be compared with those of a conventional pinch valve.

Introduction

One of the most useful applications of magnetofluidynamics (MFD) is flow control of liquid metals. Examples range from pumping of alkali liquid metals in nuclear reactors to foundry pouring control of molten aluminum and flow control of copper and steel in continuous casting. Many different implementations of liquid metal flow control exist, but they can generally be grouped into three broad classes based on the type of magnetic fields used. These classes are d.c. devices, which use d.c. magnetic fields and may use d.c. currents; traveling-wave devices, which use a traveling magnetic field produced by balanced excitation of multiphase coils; and single-phase devices, which use stationary magnetic fields which vary in time. Each of these classes of devices has its own advantages and disadvantages that influence the choice of device for specific applications. This paper is primarily concerned with single-phase devices, which are particularly useful when relatively small hydraulic forces are required, and space, particularly in the flow direction, is at a premium. One application that has been proposed for single-phase systems is flow control of molten metal through an orifice exiting to atmosphere, an electromagnetic valve. Although single-phase systems are well suited for this application, some practical difficulties exist with the most commonly suggested embodiment. In this paper an alternative embodiment of a single-phase device for flow control is presented that circumvents most of these difficulties.

The rest of this paper consist of five short sections, the first of which presents a discussion on electromagnetic flow control with special emphasis on single-phase systems. Following that discussion, the new embodiment is described qualitatively. A more rigorous analysis of the device is then presented, followed by a description of the experimental verification of the device. Finally, a summary discussion of the new device is given.

Electromagnetic Flow Control

Consider the discharge of liquid metal from a container, driven by a hydrostatic head, H. The metal exits through a refractory nozzle in the form of a jet. We are concerned with regulating this flow using electromagnetic body forces applied within the outlet nozzle.

There are at least four different methods of controlling the flow using magnetic fields. Perhaps the simplest, in a conceptual sense, is to apply steady, mutually perpendicular electric and magnetic fields, arranged across the stream. Another obvious candidate is to use multiphase ac magnetic fields, arranged as a traveling wave and directed along the axis of the stream. Both of these configurations are, in essence, types of magnetic pump. They operate by generating an <u>axial</u> force on the liquid metal stream.[1,2]

The third and perhaps most familiar technique for this application operates by pinching the stream as it emerges from the nozzle. Here a solenoid-like induction coil is placed around the base of the nozzle and is fed with single phase, high-frequency current. An axial magnetic field is generated along the surface of the liquid metal stream, which, in turn, induces circumferential eddy-currents in the jet. The imposed axial field and induced currents interact to produce a <u>radial</u> magnetic body force on the stream.

These currents also tend to exclude the imposed field from the center of the stream, so that the centerline field strength B_c is less than the surface field strength B_o. Integrating the electromagnetic body force from the surface of the jet to the axis, gives the magnetic "overpressure" P_m. When curvature of the steamlines in the jet is ignored, this magnetic pressure is equal to the difference in the fluid pressure between the jet surface and the axis. Its magnitude is given by the following familiar expression:

$$P_m = \frac{B_o^2 - B_c^2}{4\mu} \qquad (1)$$

where μ is the permeability of free space.

As far as a fluid element on the axis is concerned, the jet appears to emerge at an ambient pressure of $P_a + P_m$, where P_a is the atmospheric pressure. Consequently, application of the magnetic field <u>reduces the</u> centerline velocity at the exit, from $\sqrt{2gH}$ to $\sqrt{2gH - 2P_m/\rho}$, where ρ is the fluid density. Note that the value of P_m in this expression corresponds to the field strength at the point at which the jet exits the nozzle, or more strictly, the point at which the jet leaves the nozzle wall. This concept has been studied in detail by, for example, Garnier.[3]

A fourth method of controlling the flow, the method that was adopted in this experimental program, has many similarities to the "pinch valve." It also uses a single-phase, high-frequency field, generated by a solenoid-like induction coil. However, like a magnetic pump, it relies on generating a direct axial force, rather than a radial, pinching force. We shall return to this later.

In principle, any of these four configurations could be used as the basis for a commercial electromagnetic valve. Such a valve would find application in, for example, regulating the flow of steel from a tundish. In practice, however, all of these concepts are limited to varying degrees by the high levels of Ohmic heating generated by such a valve in both the metal stream and the inductors. For example, if a pinch valve were required to produce a magnetic pressure equivalent to 1m of steel, the resulting average Ohmic dissipation throughout a jet of radius 1 cm would be of the order of 10^{10} Watts/m^3 [see Eq. (8)]. Consequently, an important measure of the utility of a particular design is the ratio of the magnetic pressure generated to the power dissipated in the liquid metal stream.

Other desirable features of a commercial electromagnetic valve are that it is compact, so that it may be readily installed in, say, a steel caster, and that relatively few components are in contact with the liquid metal stream, so that the metal remains uncontaminated and the valve is protected from the high metal temperature.

We shall now consider the relative merits of the D.C., traveling wave, and pinch valves. In particular, we shall comment on the power dissipation in each.

Since it is the most familiar for this application, we shall start by examining a pinch valve. For the purposes of estimating the dissipation, we shall consider the simplified case of a metal stream of constant radius R, subject to an axial magnetic field B_0. For simplicity, we shall assume that B_0 varies only slowly with axial position.

Let the frequency of the applied field be denoted ω, the conductivity of the metal be denoted σ, and the skin depth, $\sqrt{2/\mu\sigma\omega}$, be denoted σ. The magnetic field distribution within the metal stream is determined by the induction equation,

$$\nabla^2 B = \sigma\mu \frac{\partial B}{\partial t} \tag{2}$$

Solving this equation in a cylinder of radius R, subject to the boundary condition $B = B_0$ on the surface, yields the familiar solution:

$$B = B_0 \frac{ber_0(y) + j\, bei_0(y)}{ber_0(Y) + j\, bei_0(Y)} \tag{3}$$

where ber_0 and bei_0 are Kelvin functions, y is the dimensionless radial coordinate $\sqrt{2}r/\delta$, and Y is $\sqrt{2}R/\delta$.

It is convenient to introduce the function

$$f(Y) = ber_0^2(Y) + bei_0^2(Y)$$

Then the magnetic pressure is given by Eq. (1) as

$$P = \frac{B_0^2}{4\mu}\left\{1 - \frac{1}{f(Y)}\right\} \tag{4}$$

The current density, J, can be calculated from Eq. (3) using Amperes law and the total power dissipation can be obtained from integrating J^2/σ over the volume of the jet. The resulting average dissipation rate per unit volume is

$$\dot{W} = \frac{B_0^2}{4\mu} 2\omega \frac{f'(Y)}{Y\, f(Y)} \tag{5}$$

The ratio of the magnetic pressure to the dissipation is then given by

$$\eta = \frac{P_m}{\dot{W}} = \frac{1}{2\omega} \frac{Yf(Y) - Y}{f'(Y)} \qquad (6)$$

In the limit of low frequency ($\delta >> R$), the magnetic pressure per unit dissipation becomes

$$\eta_\ell = 1/8 \; \mu\sigma R^2 \qquad (7a)$$

and for high frequency ($\delta << R$), the ratio becomes

$$\eta_h = 1/8 \; \mu\sigma R^2 \; \{2\delta/R\} \qquad (7b)$$

Examination of Eq. (6) shows that there is, in fact, a rather rapid transition from low- to high-frequency behavior, at around $\delta \sim R/2$.

Equations (7a) and (7b) show that the pressure to dissipation ratio is largest at low frequencies and decreases at a rate proportional to $\omega^{-1/2}$ at high frequencies. From the point of view of minimizing losses in the steel, it is preferential to operate at a frequency where $\delta \geq R$.

However, we must also consider the resistive losses in the inductors. These are proportional to the square of the coil current, and hence to B_o^2. Equation (4) shows that, in the low-frequency regime,

$$P_m = \frac{B_o^2}{4\mu} \frac{R^4}{8\delta^4}$$

whereas at high frequency,

$$P_m = \frac{B_o^2}{4\mu}$$

The field required to produce a given magnetic pressure decreases with ω^{-2} at low frequency and is constant at high frequency. Thus, the losses in the inductors (for a given magnetic pressure) are a minimum for a value of δ somewhat greater than R.

Clearly, there is a conflicting requirement on frequency with respect to minimizing the total Joule dissipation. Typically, the optimum skin depth, which minimizes the overall losses, is of the order of R, say, $\delta = R/2k$, where k is close to unity. Thus, we might take a characteristic ratio of magnetic pressure to dissipation for a pinch valve to be

$$\eta_p = 1/8 \; \mu\sigma R^2 \qquad (8)$$

Note that, since we require $\delta = R/2k$ for minimum dissipation, jets of different diameters will exhibit different optimum operating frequencies.

This is undesirable from a practical standpoint, since it is inconvenient to have to match the coil frequency to the stream size.

We shall now consider a D.C. valve. Here the magnetic field and current levels may be specified independently, and we can achieve high values of η by keeping the current low and using a large magnetic field. Let ℓ be the axial length of the D.C. valve. Then $p_m \sim BJ\ell$ and, consequently,

$$\eta_{dc} \sim \frac{\sigma B_{dc}^2 \ell^2}{p_m} \qquad (9)$$

Comparing this with Eq. (8), and taking ℓ to be of the order of 2R, we see that a D.C. valve is more efficient than a pinch valve, provided that

$$\frac{B_{dc}^2}{\mu} > \frac{p_m}{32}$$

For a magnetic pressure equivalent to 1m of steel, this requirement is that the field is greater than 0.05 T, a relatively modest value. Consequently, a D.C. valve will generally be more efficient than a pinch valve. The principle difficulty with a D.C. valve lies in the construction of the electrodes. These must not melt or erode while in contact with the stream, nor must the liquid metal freeze on the electrode face.

A traveling-wave valve may offer no advantage over a pinch valve in terms of efficiency, since they both employ relatively high-frequency fields for this application. Indeed, it has the additional disadvantage of requiring several magnetic pole pairs to achieve high efficiencies, which implies a long active length leading to a less compact design.

In summary, then, a pinch valve can induce large, unwanted losses in the liquid metal, whereas a D.C. valve is more efficient, but introduces the complexity of electrodes in contact with the melt. A traveling-wave valve offers no improvement in efficiency and is necessarily much larger than a pinch valve.

There is one further disadvantage of a pinch valve that we have not yet touched on. This relates to the use of a radial force, rather than an axial force, to control the stream.

We have already noted that the modulation of the flow depends on the magnetic pressure at the point where the stream leaves the nozzle wall. It is important, therefore, to ensure that this separation occurs at the midplane of the induction coil, where the magnetic field is largest. To some extent, this can be achieved by placing the coil midplane at the base of the nozzle. However, this does not preclude the possibility of the stream separating from the nozzle wall inside the nozzle. Such a separation would then occur at a lower magnetic pressure, resulting in less modulation of the flow.

At high frequencies, where the magnetic pressure is independent of jet radius, this effect is less important, since separation tends to occur at the point of maximum magnetic pressure anyway. (If internal separation occurred before the point of maximum magnetic pressure, then below that point the average jet velocity would tend to decrease as the magnetic pressure increased, and so the jet radius would grow, causing re-attachment to the wall.) However, at low or intermediate frequencies we cannot guarantee that this is the case.

Ideally, an electromagnetic valve would retain the benefits of the pinch concept (simple, compact, only passive components in contact with the liquid metal) but employ a direct axial force, rather than a radial force, thus avoiding the need for careful control of the separation point.

New Single-Phase Alternative

In order to overcome some of the practical difficulties described in the previous section, an alternative single-phase device has been developed that can be used to modulate the flow of liquid metals. This device, which is described in the following paragraph, is similar to the conventional "pinch" device, but differs in several key areas. The differences are the following: (1) a direct axial force is produced; (2) the point of action for the force is removed from any connection to atmospheric pressure; (3) force is produced independent of motion; and (4) ohmic dissipation in the liquid metal is reduced. Difference (1) removes the need for separation of the liquid metal from the wall of the flow passage. Since difference (2) uncouples the force action from the exit orifice, it also eliminates the sensitivity of frequency to the exit orifice.

This device[4] consists of a nonconducting shell through which liquid metal flows and around that is a single-phase electrical coil. Inside the shell is an insert that is located axially in the same position as the coil. This insert divides the axial length of the flow passage into two equal regions. In one region circumferential eddy currents, which are generated in response to the a.c. excitation of the coil, are allowed to flow unimpeded. In the second region a series of radial fins breaks up the circumferential flow of the eddy currents. As a result, the magnetic field generated by the coil can diffuse further into the liquid metal in the finned region than in the region without fins. In the junction between the finned and unfinned regions there is then, by necessity, a radial component of the magnetic field. This radial component of the magnetic field interacts with the circumferential currents in the unfinned region to produce a direct axial force. If the finned region is in the downstream half of the coil's active length, a retarding force is produced in the fluid, whereas an accelerating force is produced is the fins occupy the upstream half of the active length. The operation of this device is shown conceptually in Figure 1.

Theoretical Analysis

The operating principle of the electromagnetic valve is illustrated in Fig. 1. A circular coil is generating high frequency magnetic flux in its interior which contains the valve. The valve consists of an annular

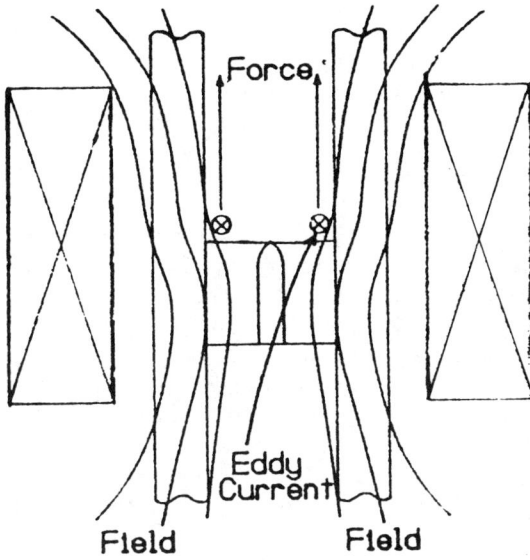

Fig. 1   Operating principle of new MFD valve.

region containing the liquid metal above the plane passing through the center of the coil, perpendicular to the z axis. This annular region is defined by a nonconducting central cylinder and by an outer cylindrical shell (not shown) between the coil and liquid metal. Eddy currents are free to flow about the z axis through this liquid annulus. Below the center plane the annular region or any region containing liquid metal has fins or vanes which block the eddy currents that would otherwise flow about the z axis.

The magnetic flux lines can penetrate the nonconducting region below the center plane, but above the center plane the eddy currents tend to exclude the flux. In the transition region near the center plane, the flux lines develop a radial component, as shown in Fig. 2. The force on the eddy currents flowing about the z axis due to this radial magnetic field is directed along the z axis and provides a braking or pumping action on the fluid flowing in this z direction. This force is given by

$$F_< = \frac{2\pi}{T} \int_0^T dt \int_{r_1}^{r_2} \int_0^Z J_\theta B_r r \, dr \, dz \qquad (10)$$

where J_θ is the azimuthal current density, which is a function of r, z, and time t. B_r is the radial flux density that is also a function of the same variables. Azimuthal symmetry is assumed, and the time dependence is assumed to be sinusoidal of period T. The terms r_1 and r_2 are the inner and outer annular radii of the liquid metal, and Z is some sufficiently large z value beyond which the force essentially vanishes. Because we are dealing with fluids, this force is assumed to give rise to an effective pressure given by

$$P_z = \frac{F_z}{A_z} \qquad (11)$$

where $A_z = \pi (r_2^2 - r_1^2)$ is the cross-sectional area of the annulus.

If the fins were not present, the force given by (1) would still exist, but B_r would be very small near the center plane where J_θ is large. B_r becomes larger as z increases, but then J_θ becomes small. Moreover, the force above the center plane would be canceled by an equal and opposite force below the center plane, resulting in zero net axial force. Calculations using our electromagnetic finite element code WEMAP[5,6] bear this out.

There is another force component produced by the axial field B_z. This is a radially directed force of density F_r (force per unit volume):

$$F_r = -J_\theta B^z \qquad (12)$$

As discussed in the previous section, this pinch pressure can lead to flow modulation on its own. Whether the magnetic pressure P_m is the result of axial or radial forces, it can be used to modulate the flow

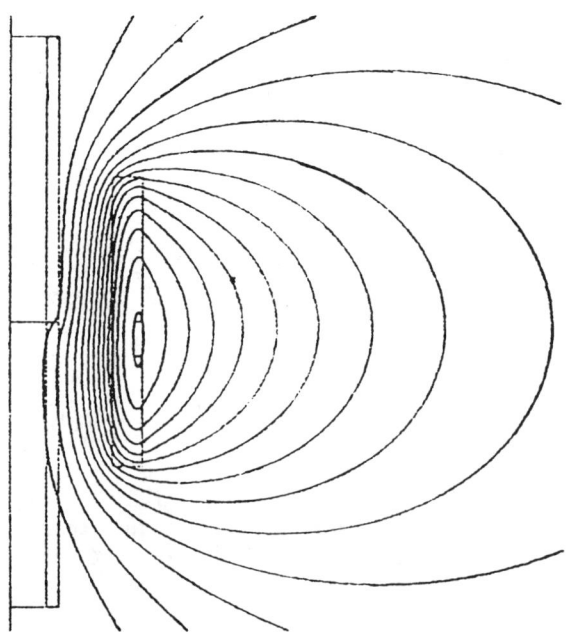

Fig. 2   Flux plot.

according to the well known

$$Q/Q_o = \sqrt{1 - \frac{P_m}{\rho g H}} \tag{13}$$

For our valve alternative P_m is generally taken to be P_z, which can be obtained from numerical analyses.

In order to gain some insight into the parameter dependences of the forces and pressures produced by this valve, it is useful to perform a simplified analysis that can be done analytically. Based on numerous finite element calculations, it was found that the eddy currents induced in the liquid steel are concentrated close to the top of the finned region. For instance Fig. 2 shows a flux plot for a particular valve-coil geometry, and Fig. 3 shows the eddy current contours in the liquid metal. Note the concentration near the center plane, just above the finned region. We therefore approximate the annular liquid metal region as a single-turn thin-wire coil of radius r_o = the average annulus radius and wire radius Δr = the skin depth in the liquid metal or 1/2 the annular thickness, whichever is smaller.

This thin-coil assumption can be used to produce an approximate analytic solution that can be used to write the same figure of merit used in the previous section. The analysis is based on coupled circuits and is not reproduced here for sake of brevity. The result, which uses

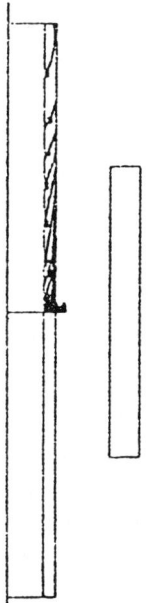

Fig. 3   Current contours.

Eq. (10) and (11), is the following:

$$\eta = \frac{P_z}{\dot{W}} \frac{\mu_0 (\Delta r)^2}{4\pi r_0^2 \rho_e \Delta r_a} \left[\ln\left(\frac{8 r_0}{\Delta r}\right) - \frac{7}{4} \right] \left(\frac{B_{r,o}}{B_{z,o}}\right) \qquad (14)$$

with

$$\Delta r = \min\left\{ \Delta r_a, \sqrt{\frac{2 \rho_e}{\mu \, \omega_0}} \right\} \qquad (15)$$

The expression for Δr should indicate a more gradual transition between the two values in a more realistic treatment. The preceding expressions can, at best, provide some qualitative insight into how parameter changes can affect valve performance. For instance, Eq. (14) implies that η, and in reality, P_z, is proportional to r_0. (The logarithm term is very slowly varying.) We found in a WEMAP study that doubling r_0 multiplied by P_z by 1.6. The frequency dependence of P_z is more complicated. It saturates, i.e. approaches a constant value, at high frequency. This behavior was also observed in a WEMAP frequency study. On the other hand, the losses increase monotonically with frequency, and this behavior was also exhibited in our WEMAP frequency study. Thus, the frequency should be carefully chosen.

This frequency dependence is concerned only with the geometry of the device directly above the fins, which is the point of application for the electromagnetic force. Since in this device the exit orifice can be well removed from the point of application, there is no need to select different excitation frequencies for different orifices. As long as the exit orifice and the flow path through the device are sized so the device is always full of the liquid metal, the valving action will be present.

Experimental Evidence

In order to prove the principles behind the new device, two series of tests were conducted using low-melting-point liquid metals. One series of tests used NaK, a eutectic sodium-potassium mixture. These tests were designed to demonstrate the device using a low-density, high-conductivity material in geometries where the conventional "pinch" valve clearly could not operate. Several different valve geometries were investigated with this system. The second series of tests were run with a eutectic form of Wood's Metal, a mixture of bismuth, lead, tin, and cadmium. Tests with this high-density, low-conductivity material were designed to operate the device in a valve configuration that would be similar to tests run using conventional "pinch" valves. Only the best configuration from the NaK tests was used with the Wood's Metal. Both of these tests series are briefly described in the following. In both cases there was good agreement between the data and performance predictions made using finite element modeling.

NaK Tests

The NaK test system is shown schematically in Fig. 4 and the various components shown are connected together with 1/2-in.-o.d. polyethylene tubing. The test device consisted of a housing of G-11 material measuring 12.0 in. long by 1.75 in. with flange connections at each end. A 1.0 in. dia.-hole running axially the full 12 in. was designed for inserts of the various valve configurations. A water-cooled, 20-turn copper coil around the outside of the housing provided the magnetic field for valve operation. Variations in valve configuration included the shape of the flow passage directly above the fins, cylindrical or annular, and the number of fins. With the annular flow passage the dimensions of the annulus were also varied.

The test system was designed for operation in two different modes: static and dynamic. In the static mode the loop was filled with NaK to a level just above the test valve, and that level was recorded at the static test scale (see Fig. 4). The coil was energized at the desired current and frequency, and the static head displacement between the two legs of the U-tube monometer was recorded. In the dynamic mode the loop was completely filled above the pump (see Fig. 4) to the center of

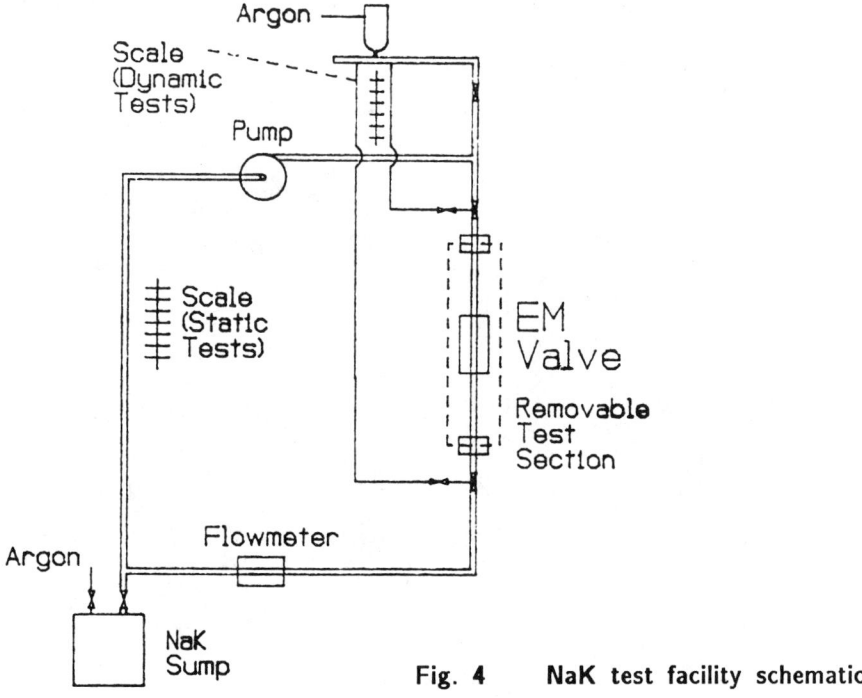

Fig. 4   NaK test facility schematic.

the flow test scale. The pump was started, and the pressure drop across the test valve was recorded at the test scale while the flow rate was recorded at the flowmeter. The coil was energized at the desired current and frequency while the difference in pressure drop across the test device was recorded.

NaK Results

There were many results from the tests using the NaK facility, and each will be discussed briefly in the following paragraphs. Before the actual result is presented, the consistency and repeatability of the results deserves mention. The results were very repeatable between successive tests and from day to day. The repeatability of the results contributes to the confidence in the following results. The primary result of the NaK tests was the demonstration of the electromagnetic force on the fluid. In both the static and dynamic (flowing liquid) tests, the forces, as measured by a change in fluid level or increased pressure drop, were present. These forces were found to be directly proportional to the square of the current flowing in the coil. The theory behind the valve that is outlined in the previous section predicts this result. By comparing the proportionality constant between the effect and the square of the current, it is possible to determine the relative effectiveness of the different geometries that were tried in the experimental program. It is also possible to calculate the equivalent proportionality constant from the finite element analysis of the valve. Figure 5 shows the

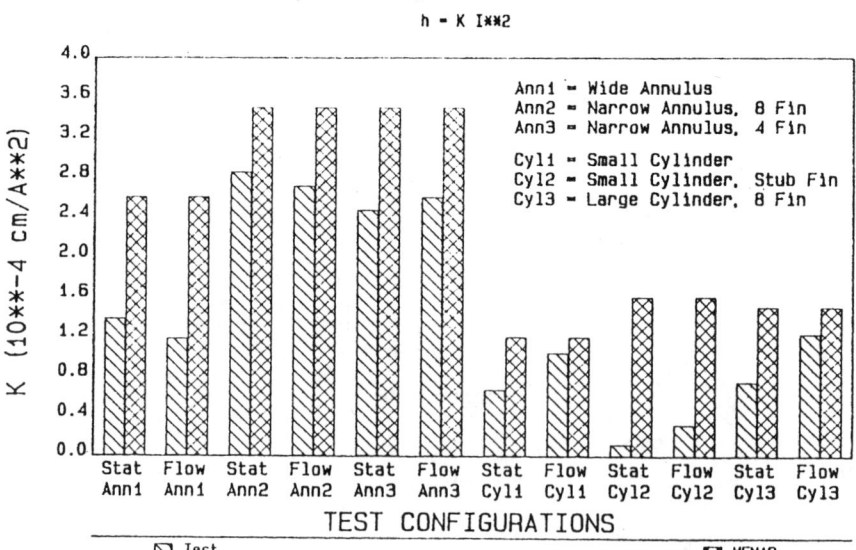

Fig. 5   Proportionality constants for NaK tests.

proportionality constants for the different geometries; both the experimental and theoretical constants are shown.

As can be seen from the figure, the theoretical predictions for the annular geometry are, in general, better than the cylindrical cases. The experimental data also support this conclusion, although the effect is more evident in the static tests. In general the experiments show nearly the same result that for the annular flow cases as that for the annular static cases. Such is not the case for the cylindrical geometry, where the flow cases are markedly closer to the WEMAP predictions than are the static ones. It is felt that an extra viscous fluid loss term is introduced in the cylindrical geometry when the electromagnetic forces are introduced and that these extra fluid losses make the electromagnetic forces look more effective than they really are.

Wood's Metal Tests

The Wood's Metal tests used a eutectic alloy of bismuth, lead, tin, and cadmium as the working fluid. This eutectic alloy is known as Lipowitz metal and is more like heavy molten metals in physical properties than was the NaK of the earlier test series. The Wood's metal test facility consists of a simulated tundish for supplying a constant head of liquid to the test valve, a removable test valve for quickly installing different configurations, a simulated mold for collecting the Wood's metal and acting as a reservoir, a mechanical pump for pumping the liquid metal back to the tundish, and a flowmeter for measuring the Wood's metal flow rate. The test system contains ~355 lbs. (4 1/2 gallons) of Wood's metal and is trace heated to maintain the Wood's metal in the liquid state (i.e., >70°C).

Wood's Metal Test Results

The results of the Wood's metal experiments also demonstrated the valves ability to produce electromagnetic forces. Furthermore, because the hydraulic system included accumulators, it was possible to show that the flow rate could be modulated using the valve. The variations between successive tests were larger for these tests than they were for the NaK tests. As a result, it was necessary to repeat tests several times in order to determine a meaningful data point.

The data for the annular valve are presented in Fig. 6, where the dimensionless electromagnetic head is plotted as a function of the square of the excitation current. This representation of the data is very similar to that used in the NaK tests. As was the case with the NaK test, the effect, as measured by the dimensionless head, is linearly proportional to the square of the current. Also shown on the curve is the WEMAP prediction for the test geometry. For these tests the range of experimental data spans the predictions. This situation is different from the NaK tests, where the predicted values always exceeded the experimental results. The difference is attributed to the presence of the pinch pressure effect of the flow. In this representation the

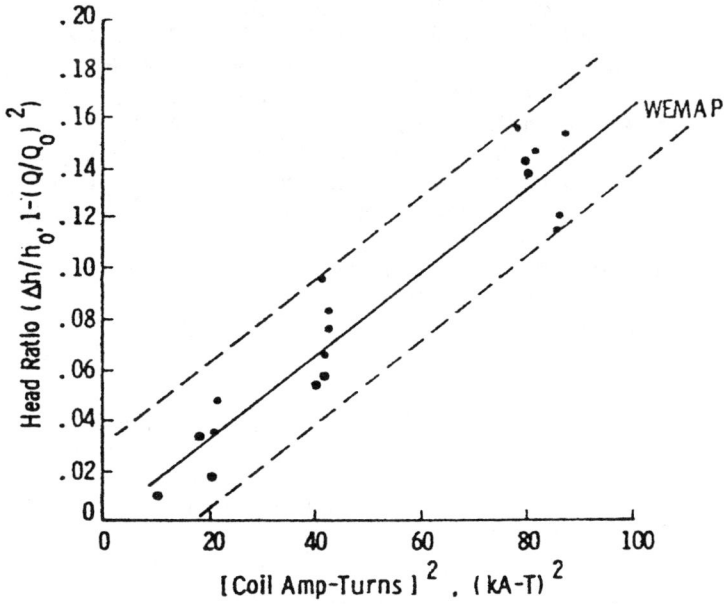

Fig. 6 Results for Wood's metal test.

dimensionless head is a calculated quantity based on the measured flow ratio. It is calculated according to the simplified fluid model commonly used in producing Eq. (13).

Summary

In simplest form this new embodiment of the single-phase electromagnetic valve consists of a high-frequency coil surrounding a column of liquid metal flowing through a nonconducting pipe. An obstruction is deliberately placed in the flow channel in a position corresponding to the active length of the coil bore. The obstruction allows circumferential eddy currents to flow in one-half of the the active length and substantially blocks the eddy currents in the other half. The difference in eddy currents ensures a radial component of magnetic field, which interacts with the circumferential eddy currents to produce a direct axial force. This force can be used to either retard or augment the flow of liquid metal.

This embodiment of the electromagnetic valve has a different set of strengths and weaknesses than the conventional single-phase valve. It also shares almost all of the same auxiliary equipment: high-frequency power supply, capacitor tuning station, cooled high-current buswork, and cooling system. The only significant weakness of this device is the requirement for an obstruction to be inserted into the flow channel. This obstruction is a mechanical complication and finding a suitable material may be difficult. The insert must be made of a nonconducting material, be rugged enough to remain in place, show high wear and corrosion resistance, and not be a source of clogging. The advantages of this device are, like the conventional device, a minimal space requirement, and, unlike the conventional device, an insensitivity of operating frequency to orifice size, no requirement for the fluid to separate from the flow channel wall, an insensitivity to flow direction, and a reduced ohmic dissipation in the fluid. It is felt that these advantages will make this single-phase device suitable for many different applications. Among the applications where this device is thought to be suitable are those where a wide range of exit orifices are required or where communication with the atmosphere is not desired. Both of these conditions are found in continuous steel casting conditions and where quality constraints sometimes require submerged pouring.

References

[1] Blake, L. R., "Conduction and Induction Pumps for Liquid Metals", Journal of the IEEE, Vol. 2, 1956, pp. 429-435.

[2] Baker, R. S., and Tessier, M. J., Handbook of Electromagnetic Pump Technology, Elsevier, New York, 1987.

[3] Garnier, M., "Electromagnetic Devices for Molten Metal Confinement", Liquid-Metal Flows and Magnetohydrodynamics, Progress in Astronautics and Aeronautics, Vol. 84, pp. 443-441.

[4] Slepian, R. M., and Delvecchio, R. M., "Liquid Metal Electromagnetic Flow Control Device Incorporating a Pumping Action", U.S. Patent No. 4842170, June 27, 1989.

[5] Barton, M., Garg, V., Ince, I., Sternheim, E., Weis, J., WEMAP: A General Purpose System for Electromagnetic Analysis and Design", IEEE Transaction of Magnetics, Vol. MAG-19, No. 6, Nov. 1983, pp. 2674-2677.

[6] Weiss, J., Barton, M., Garg, V., and Ince, I., Finite Element Analysis of Magnetic Fields," IEEE Transactions on Magnetics. Vol. MAG-20, No. 6, Nov. 1984, pp. 1933-1935.

Superconducting MHD Devices: Parametric Study of an Electromagnetic Pump Performance

P. Del Vecchio,* A. Geri,† and G. M. Veca‡
University "La Sapienza," Rome, Italy

Abstract

The authors have implemented a mathematical lumped-parameter model capable of pointing out the performance of a dc MHD device using rectangular or circular ducts. Magnetohydrodynamic flow is subjected to the action of a high magnetic field. The mathematical model presented is a bidimensional one, but it takes into account the edge effects and the magnetic field distortion caused by electric currents in the fluid. The computation program enables one to carry out a parametric study of the efficiency, head, and electric currents in fluid, since an already optimized shape of the magnetic field is obtained.

Once the model of the duct offering the highest efficiency was found, the authors have examined the feasibility of the superconducting magnet to that geometry. Then they have tested how the pump operated when it was subjected to a magnetic field produced by a magnet that closely approaches the ideal magnetic field.

Introduction

The Fast Nuclear Reactor Department of the Italian National Council for Nuclear and Alternative Energy (ENEA), because of the introduction of new technologies creating high-field magnets (II- type superconducting wires are the technologies of the last 10 years; meanwhile high-temperature superconducting materials will be the technologies of the future), has drawn a certain attention to the MHD device fitted with a superconducting magnet.

The authors obtained a research contract with ENEA in order to optimize an electromagnetic pump for liquid metals (see Fig. 1) both for the geometry of duct section to be chosen and for the feasibility of the superconducting magnet. The

---

Copyright © 1992 by the American Institute of Aeronautics and Astronautics, Inc. All rights reserved.
 * Associate Professor, Electrical Engineering Department.
 † Assistant Professor, Electrical Engineering Department.
 ‡ Professor, Electrical Engineering Department.

authors have taken into account two different geometries for the duct section: a rectangular geometry and a circular one. In this study they have pointed out that the best performances are obtained making use of rectangular geometries limited to the electrodynamic optimization of duct.

The authors, therefore, have carried out a study related to the height/width α ratio of duct. It follows that in a pumping device the best performances are obtained when the α ratio is the highest possible. They have also taken into account a magnet capable of carrying out a field distribution that is the closest possible to the optimal one for the device.

A thorough parametric study [1,2] has proven that the external magnetic field must extend beyond the length of the electrodes: to be more precise, it must be extended longitudinally by a length nearly equivalent to the distance between electrodes, both at the device inlet and the outlet. In other words, the ideal profile of the magnetic flux density along the median plane of the channel would be constituted by a flat top for the entire length of the electrodes, linearly declining to zero at the two extremes, as later shown in Fig. 9, curve C.

Turning to usual techniques of superconducting magnets construction, the authors could not exactly obtain the desired proceeding. The greatest shifts especially occur in the outer areas of electrodes. On the basis of a preliminary design of a superconducting magnet, the authors have compared the performance of a pump excited by the real field and one excited by the ideal magnetic field.

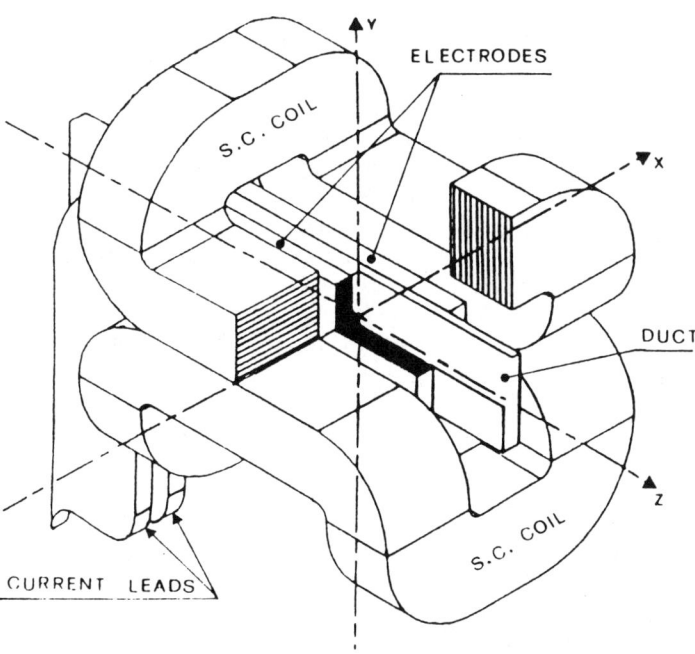

Fig. 1 Schematic drawing of a dc electromagnetic pump.

Outline of the Model

The model takes into account the armature reaction causing a distortion of the magnetic field generated in the pump duct, as well as Hartmann's velocity profile of the magnetohydrodynamic flow.

As fully explained in Refs. 3 and 4, the model splits fluid flow into various sheets independent of one another. An analytic iterative method is used to calculate the electric current and the generated magnetic field due to each sheet.[5]

The mathematical model also takes into account the fringing currents on the edges of electrodes both as to the additional effect of electromechanical conversion and as to Joule losses. The calculation code works out a lumped-parameter electrical network as shown in Figs. 2a and 2b. In this network the inductive parameters are ignored since the model works in steady-state conditions. The complexity of such a network depends on the accuracy to be obtained and it is limited only by the computing time. As shown in Fig. 2, the resistance network is divided into three parts:

1) A central one, referred to as the "active part," which simulates the fluid performance between electrodes.

2) A middle one, referred to as the "front part," which simulates the performance of those fluid parts being outside the electrodes but still subject to the magnetic field action. Fringing currents will be curbed here by the electromotive forces induced by the magnetic field, and they will increase the performance of the device.

3) An external part, referred to as "the passive part," where fringing currents are outside the magnetic field dissipating energy.

The network created by the computing code refers to only a sheet of the fluid (see Fig. 3). The overall fluid performance results from all of the networks among which the fluid has been shared. The purpose is to make the model versatile and, therefore, adaptable to the various shapes of magnetohydrodynamic flow ducts. For MHD devices using a circular duct, the velocity of each fluid sheet is related to the velocity of the other sheets by means of Hartmann's profile. To each electrical network corresponding to a fluid sheet it is given an input voltage that can be changed according to a prearranged program whose purpose is to produce a constant electric field inside the fluid and equal for all shets. All that is designed to reduce to a minimum both passive resistances to the fluid flow and transverse currents among the various fluid sheets. In fact, these transverse currents increase the forces opposite to the motion of the flow and deeply trouble the laminar flow. Figure 3b and Fig. 4 show how it has been possible to study the performance of an MHD device fitted with a circular duct.

Numerical Results

Numerical results have been obtained following the data shown in Table 1.

Figures 5−7 show in the form of diagrams the results achieved with MHD devices having the same cross-sectional area but different duct shapes (rectangular or circular), the length of the device being equal. As to a rectangular duct, the authors have studied the device performance related to three different α ratios, between height and width.

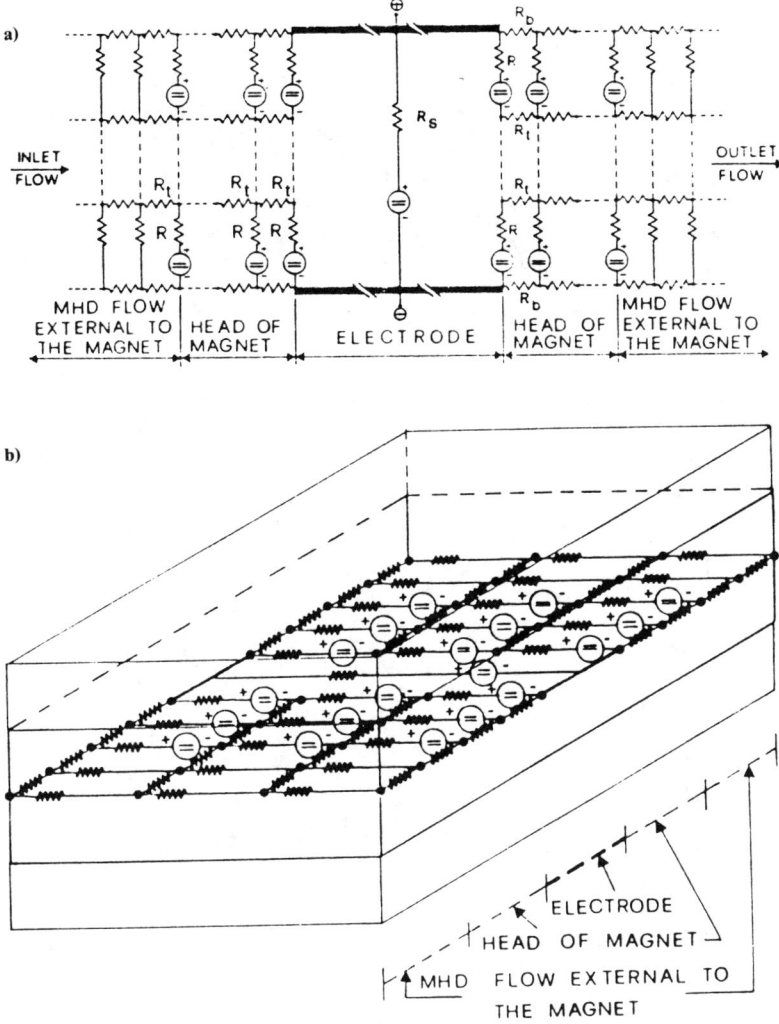

Fig. 2 Lumped-parameter network created by calculation code: a) Plan; and b) isometric drawing of a sheet.

Figure 5 indicates head and efficiency as a function of the pump flow rate. It presents six curves referring to three α ratios. It must be noted that the highest efficiency is obtained with the highest α ratio. The shorter L (distance between electrodes) is, the lower electrode voltage is, the electric field inside the fluid being the same. When the duct area is not changed, the electric field constancy implies the pump head constancy, as pointed out by virtually-coinciding curves 1',2' and 3' in Fig. 5.

Figure 6 presents the evolutions of head and efficiency as a function of the flow rate in a device fitted with a circular duct, which has the same cross-section area

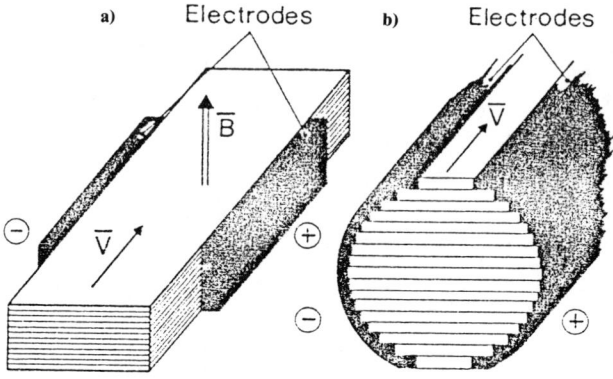

Fig. 3 Fluid sheets into which the flow is divided as to an MHD device having an a) rectangular, or b) circular duct. Each sheet corresponds to the electrical network shown in Fig. 2.

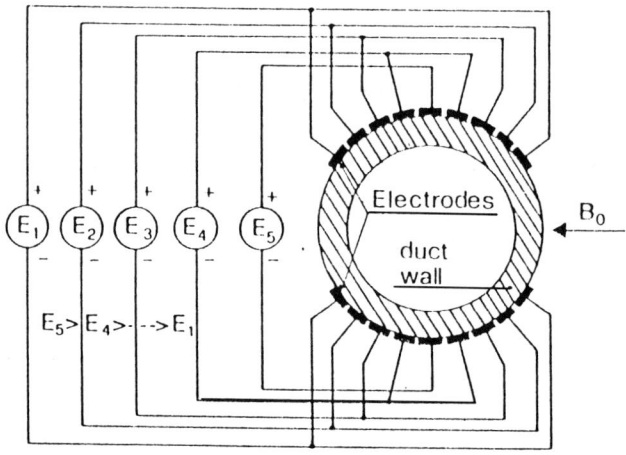

Fig. 4 Cross section of a circular duct; electrodes fed by different voltages.

Table 1 Data used for an MHD sodium pumping device

Type of fluid	sodium
Operating temperature	400 °C
Operating fluid velocity	4 m/s
Duct cross section	1 m^2 or 0.03 m^2
Duct shape	rectangular or circular
Active part length	0.4 m or 0.8 m
Magnetic field profile	ideal
Maximum value of ideal magnetic flux density on the flow	5 T

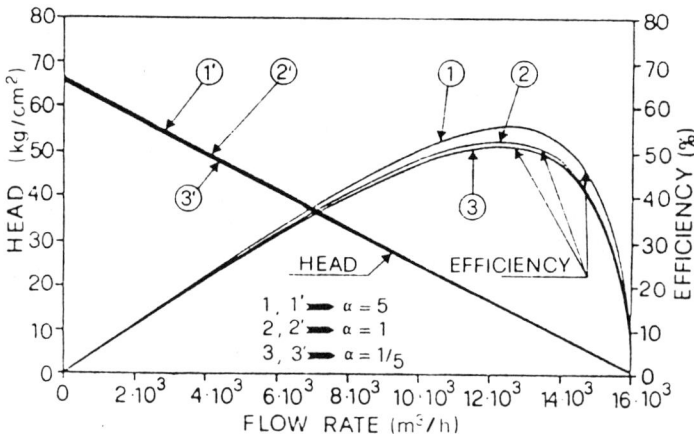

Fig. 5 The evolution of efficiency and head as a function of the flow rate in a device fitted with rectangular duct.

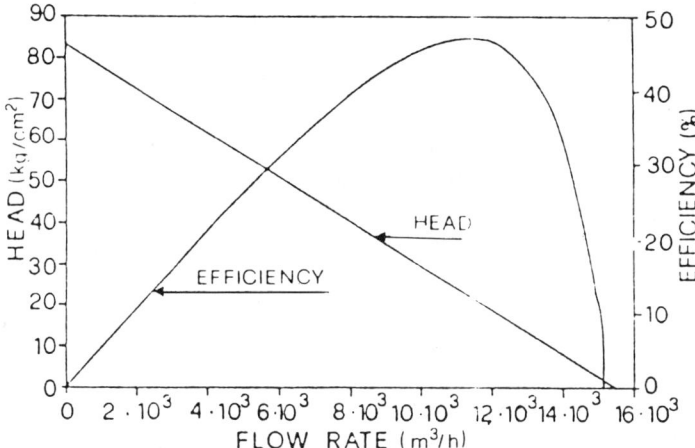

Fig. 6 The evolution of efficiency and head as a function of the flow rate in a device fitted with circular duct.

as the rectangular one. Comparing the data of Figs. 5 and 6 one should note, for the circular duct, an increase in the device head when flow rates are low but, on the other hand, a drop in the efficiency.

Figure 7 presents the curves referring to the efficiency and head as a function of the mechanical power supplied to the fluid of a device having a circular duct and another having a rectangular duct ($\alpha = 5$). As a result, the efficiency of the circular device is decidedly lower, whereas the device using a circular duct operates with higher heads, the mechanical power supplied to the fluid being equal.

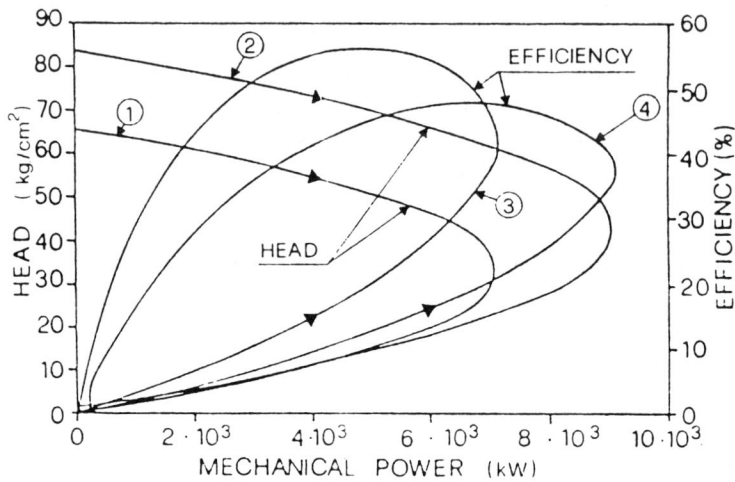

Fig. 7 Comparison with the results (efficiency and head vs mechanical power) computed for a device using rectangular ($\alpha = 5$) (curves 1 and 3) and circular ducts (curves 2 and 4), the area being equal.

Device Performance with the Real Magnet

Once the authors obtained the duct shape for optimum efficiency, they have computed with more accuracy the performance of the device with a cross section of 300 cm^2 and a length of 80 cm, taking into account the magnetic field produced by a real magnet that closely approaches the ideal profile and the contributions due to the magnetic field because of to the current in the leads.[6]

The saddle coil type magnet closely optimizes pump performance. The magnetic field is a dipole; it must be as uniform as possible in the area included between the electrodes that carry the electrical current to the fluid. Since the forces of attraction between the coils can be very strong, the authors have chosen the saddle coil shape as shown in Fig. 8.

The saddle coil configuration has an advantage over others such as the race track: it has a low relationship between the maximum magnetic field in the superconductor and the magnetic field in the useful region. This allows for sufficiently high current density in the superconductor wire.

The magnetic field that involves the device is all in the air. Therefore, to compute the contribution from the two superconducting coils and from the bars leading the current to the electrodes, the "VECTOR FIELDS" code GFUN ® was used.[7]

The current leads are chosen specifically to create a magnetic field with the same polarization as the magnet. In this way, the force that is generated on each bar for the interaction with the field of the magnet (F/l = 500 kN/m) is directed in such a way as to distance the bars, and it is easily containable with the proper external tying system. If the two magnetic systems (magnet and bars) produced opposing fields, the two bars would tend to crush the channel and it would be

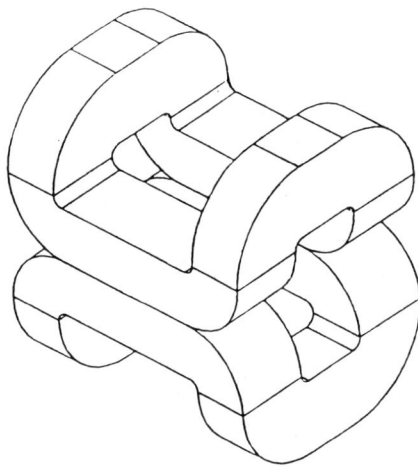

Fig. 8 A dipole magnet saddle coil type.

more difficult to contain the forces. Theoretically, it would have been easier to feed the electrodes with the bars positioned perpendicularly to the electrodes to avoid producing a magnetic field on the fluid. This would have complicated the construction of the dewar without producing any appreciable gains; the value of the magnetic field generated by the bars in fact is only 4 percent of that of the principal field.

In Table 2 the main parameters of the magnet are shown.

Figure 9 shows the magnetic field produced by the magnet. Curve A indicates the profile of the magnetic field on the axis of the duct as a function of the length of the device. Curve B indicates the profile of the magnetic field on the top of the duct near the lateral wall, still as a function of the length of the device. Curve C is the most convenient theoretical profile.

The profile of the magnetic field produced by the real magnet differs from the ideal one chiefly in the outer areas of electrodes, where a field proceeding declining quickly to zero is expected. This difference involves a decline in pump performance since in the areas close to the electrodes, even with slight fluid flows, a braking effect because of induced currents occurs.

Figure 10 shows this phenomenon and points out the contribution of mechanical power due to currents in the outer areas of electrodes related to the flow rate of the pump. Yet the decline in performance is absolutely slight since, for the chosen geometry, the contribution of fringing currents to the pump performance is slight on the whole.

Table 2 Principal parameters of the magnet

Internal width	$X = 0.40$ m
Length of the rettilinear tract	$Z = 0.80$ m
Section of a coil	(0.35×0.30) m^2
Max magnetic field on the axis	5.8 T
Max magnetic field on the coils	6.8 T
Average current density	$J = 3 \times 10^7$ A/m^2

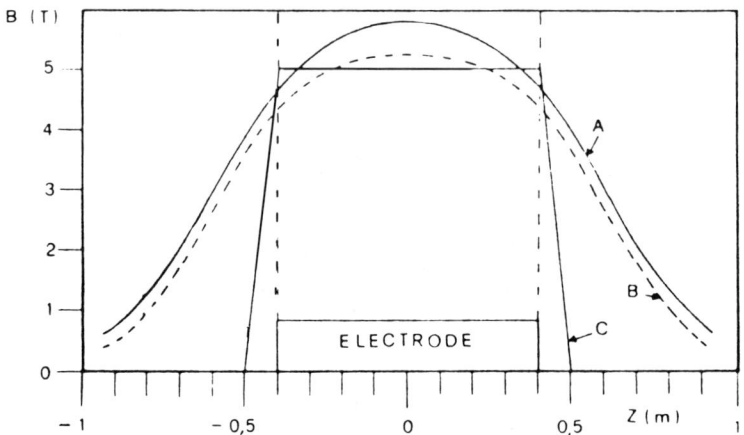

Fig. 9 Magnetic field produced by the magnet along the z axis: curve A = on the channel axis, curve B = on a corner of the duct, curve C = theoretical profile.

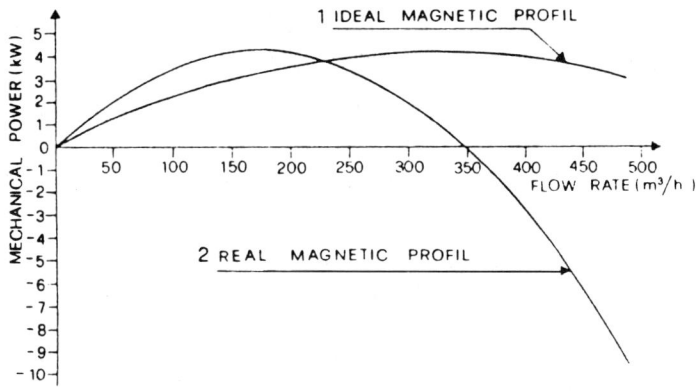

Fig. 10 Mechanical power of the fluid flow in those parts of the pump under the magnetic field and outside the electrodes: 1) pattern of the mechanical power for the profile C of the magnetic flux density of Fig. 9; 2) pattern of mechanical power for the profile A of the magnetic flux density of Fig. 9

Conclusions and Remarks

Numerical results show the advantage of rectangular ducts having close electrodes that can operate at reduced input voltage, the electric field inside the duct being equal.

Referring to the study carried out on a device provided with a circular duct, it can be pointed out how, the area of the duct being equal, performance considerably drops even if a different input voltage of electrodes is used and a constant electric field is generated inside the fluid. Such a drop in performance is, however, to be considered together with the advantage of building this type of device.

The performance illustrated refers to a pumping device in which it may be possible to insulate the electrodes from the duct and the walls of the duct from the fluid. In instances where this would not be possible, the performance of the pump deteriorates to a very low level.

In addition, a geometry of the magnet carrying out an ideal profile of the magnetic field would cause a small percentage improvement both in pump performance and in mechanical power. Therefore, a special effort to carry out a magnetic field with heads declining quickly seems to be unnecessary.

Acknowledgments

The present work is supported by the Fast Nuclear Reactor Department of ENEA and by the Italian Ministry of the Scientific Research (MURST) with a contract in the year 1988. We wish to thank Giorgio Pasotti of ENEA Frascati Laboratories, who took an active part in the preliminary design of the superconducting magnet.

References

[1] Del Vecchio P., Geri A., and Veca G.M., "Superconducting Magnets for Electromagnetic dc Pumps," IEEE Transaction on Magnetics, Vol. Mag-21, NO. 2, March 1985.

[2] Hughes W.F., and McNab R., "A Quasi-One-Dimensional Analysis of an Electromagnetic Pump Including End Effects," 3rd Beer-Sheva International Seminar on Magnetohydrodynamic Flows and Turbulence, Ben-Gurion University of the Negev, Beer-Sheva, Israel, March 23-27, 1981.

[3] Del Vecchio P., and Veca G.M., "Eddy Currents and Field Computation in a Superconducting Electromagnetic Pump," IEEE Transactions on Magnetics, Vol.Mag-19, NO. 5, September 1983.

[4] Del Vecchio P., and Veca G.M., "Eddy Current Computation in a Superconducting Electromagnetic Pump," IEEE Transaction on Magnetics, Vol. Mag-19, NO. 6, November 1983.

[5] Shercliff J. A., Magnetohydrodynamics, Pergamon, Oxford, England, U.K., 1965, pp. 143-149.

[6] Del Vecchio P., Pasotti G., and Veca G.M., "The Influence of the Magnet on the Performance of a Cold MHD device," 11th International Conference on Magnet Technology, Tsukuba, Japan, 28 August - 1 September 1989.

[7] Newmann, M.J. et al. "GFUN: an Interactive Program as an Aid to Magnet Design," 4th International Conference on Magnet Technology, Brookhaven National Lab., 1972.

Induction Electromagnetic Pumps for Alkali Metals: Status and Perspectives

S. Poinsot* and R. Rossignol*
Centre d'Etudes de Cadarache, Saint Paul Lez Durance, France
F. Werkoff† and A. Marechal†
Centre d'Etudes de Grenoble, Grenoble, France
and
J. Rapin‡
FRAMATOME-Direction NOVATOME, Lyon, France

Abstract

In the context of the research and development programs of fast breeder reactors, we discuss the technical and potential advantages of induction electromagnetic pumps and compared to mechanical pumps (e.g., simplicity, high reliability and reduced sizes) and of their disadvantages (e.g., relatively low energetic efficiencies and absence of experiences for the high deliveries needed in the mains loops of reactors). As an illustration we present the BPX-HT prototype, which is expected to be immerged in the hot sodium of the primary loop of the PHENIX reactor. We discuss existing high deliveries or EMPs and give some more details on a French prototype of 600 m³/h delivery, which allowed us to reach the highest energetic efficiency ≃ 48 %. We consider the possibility of increasing the performances of electromagnetic pumps by controlling stability and reducing Joule losses in the flow.

Introduction

The basic principles on induction Electromagnetic pumps have been known since before world war II.[1] As shown in the proceedings of the preceding Beer-Sheva Seminar,[2] these principles are quite similar to those of induction generators. With the same notations, the induced currents J_y in the direction perpendicular to both the flow velocity and the magnetic field are

$$J_y = \sigma \left[\left(\frac{\omega}{k} \right) - V \right] B_z \qquad (1)$$

For EMPs the phase velocity is greater than the flow velocity ($\omega/k > V$), whereas for the generator $\omega/k < V$.

Copyright © 1992 by the American Institute of Aeronautics and Astronautics, Inc. All rights reserved.
*Project Engineer, Département de Recherche Physique, Service de Technologie des Métaux Liquides.
†Project Engineer, Département de Transfert d'Energie/Service des Transferts Thermiques.
‡Project Engineer.

INDUCTION PUMPS

Among the various liquid metals, sodium has the best electrical conductivity σ. It is not a surprise that, during the last 30 years, the development of EMPs was parallel to that of sodium-cooled reactors. Today, for alkali metals, in fast reactors as well as in the laboratory loops, EMPs with annular channels have almost completely replaced mechanical pumps where the deliveries are lower than 200 m^3/h. The main technical advantages provided by EMPs consist of the following[3]:

1. There is not moving solid part ; therefore, there is no wear, no required lubrication, almost no maintenance, and a reduction of noise and vibration sources.
2. There is total and warranted tightness at the level of the pumping conduit (achieved by weld).
3. Because of the simplicity of construction and the reduced number of their constituent elements, these pumps are highly reliable (a failure rate of less than 10^{-6} failure per hour).
4. There is high control flexibility ; the pumping rate can be continuously adjusted by making the supply voltage fluctuate.
5. There is high versatility of installations ; these pumps can be mounted in line, in any possible position, and are provided with compact overall dimensions in the plans perpendicular to flow.
6. There is a better resistance to pressure and to thermal shocks.
7. The pumping forces appear as soon as the power is turned on. Moreover, this requires a progressive voltage increase.
8. The stator winding is easy to manufacture.
9. No neutral gas blanket is required.

On the other hand, if one considers the energetic efficiency, EMPs appear to be less favorable. The value of the sodium electrical conductivity is considerable lower than the copper one, which corresponds to most electromagnetic machines ; consequently, for obtaining the same induced current it is necessary to increase the inducing magnetic field, size of the coils and subsequent Joule losses in the primary are then much more important for EMPs than for other electric machines. In some cases those losses are not really restricting: It is necessary to maintain sodium in laboratory loops at temperatures greater than 150°C in order to avoid solidification, and fortunately most of the coils joule losses are transformed to heat into the sodium flow. Moreover, the situation is very different for the important power needed in the main pumps of reactors: The sodium heating is provided by nuclear reactions, and one cannot accept too high of an energy dissipation. The relation between inducing B_1 and induced B_i magnetic fields is such that

$$[B_i] \sim [B_i] R_m \qquad (2)$$

where

$$R_m = \frac{\mu \sigma}{k} \left(\frac{\omega}{k} - V \right) \qquad (3)$$

is the magnetic Reynolds number (μ is the magnetic permeability).

One possibility for reducing the joule losses in the winding coils of the magnet is to increase R_m. But one creates a situation propitious to the apparition of an MHD instability[4,5] which produces a flow inhomogeneity with pressure, current, and delivery oscillations and, finally, a reduction of the machine performances. Moreover, for high values of R_m the convection of the magnetic field in the flow tends to overcome its diffusion.[6]

This is sometimes[2] expressed by the fact that an adimensionnal parameter $\epsilon = V^2 \mu\sigma/4\omega \; \delta_h/\delta_m$ becomes greater than 1 where δ_h is the height of the channel and δ_m the magnetic gap. Because of the finite length of linear induction machines, electromagnetic edge waves are generated in the liquid metal and increase the ohmic losses all the more because ϵ increases.[7]

Most of the EMPs in operation today are characterized by low R_m and ϵ values. However, recent developments for high-delivery EMPs[3,5,8] have led to more precise studies of $R_m >> 1$ situations and consequently to more thorough studies of edge waves and MHD instability.

The next section of this paper is devoted to the various aspects of EMPs. First, as an example of the flexibility of low-delivery EMP, we shall present the BPX-HT prototype. Second, we shall exhibit results obtained on a 600-m^3/h-delivery EMP with $R_m > 3$, which allowed the highest energetic efficiency for an EMP to be reached. We shall compare these results with those for other high-delivery EMPs. Finally, we shall discuss the possibilities for increasing the performances of EMPs.

BPX-HT Low-Delivery Prototype

A closed sodium loop called BAUPHIX is intended to be immerged in the primary loop of the PHENIX reactor. Because of geometrical constraints, an EMP immerged in hot sodium ($\simeq 550°C$), with very small external radius, appears to be the only possible solution for ensuring the flow circulation. An annular prototype BPX, without specific cooling, was developed and tested on an autonomous loop in a laboratory context.

The characteristics and description of BPX-HT are given in Table 1, Fig.1, and Fig. 1a.

Although the magnetic Reynolds number is greater than unity ($R_m = 1.4$), the pump is stable. Because of its configuration, the crititical value, taking into account geometrical correction,[10]

$$R_{mc} = 1 + \frac{1}{k^2 R^2} \qquad (5)$$

indicates that in this case $R_{mc} = 2.09$.

Large R_m, 600 m^3/h, EMP

Design and Sizing

In F2 and T2 the main features of a prototype are given, which is presently the biggest realization of a French research and development program initiated in the last 30 years. Two

Table 1 BPX HT electromagnetic pump

External length, mm	2200
External diametre, mm	149
Weight, kg	150
Temperature, °C	560
Delivery Q, m^3/h	18
Frequency f, Hz	50
Flow velocity V, m/s	4.2
Pole pitch τ, mm	126
Wave number $k = \pi/\tau$	24.9
ϵ (with $\delta_m = \delta_h$)	0.057
R_m	1.4
R_{mc}	2.09
Duct height h, mm	5
Mean radius of channel R, mm	38.3
Length of inductor L, mm	1792

INDUCTION PUMPS

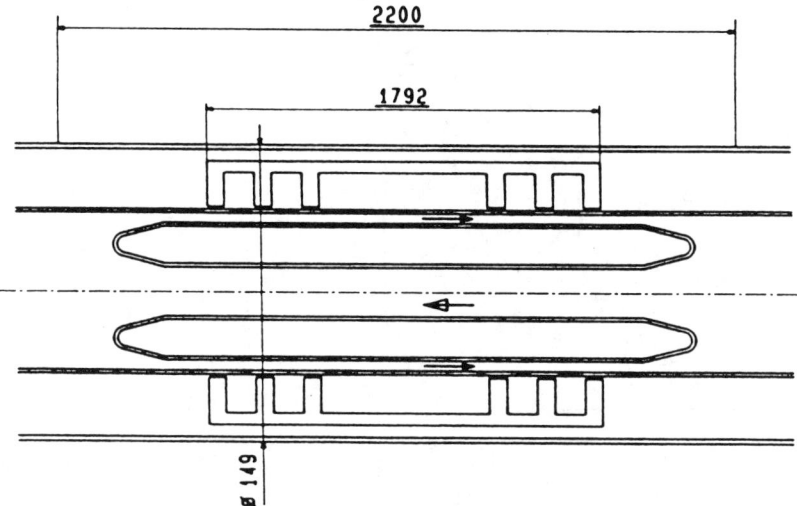

Fig. 1 Schematic of the BPX HT EMP.

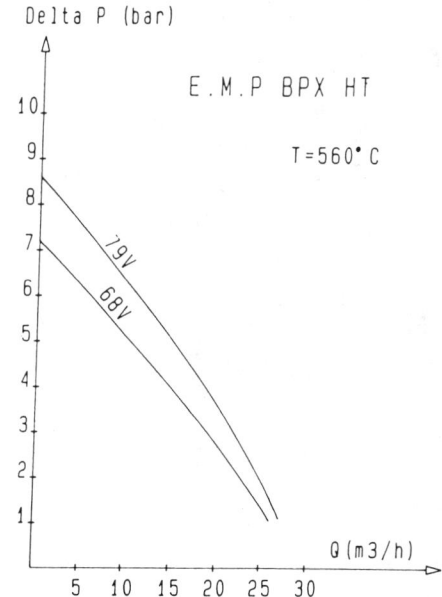

Fig. 1a Characteristic curves of the BPX HT EMP.

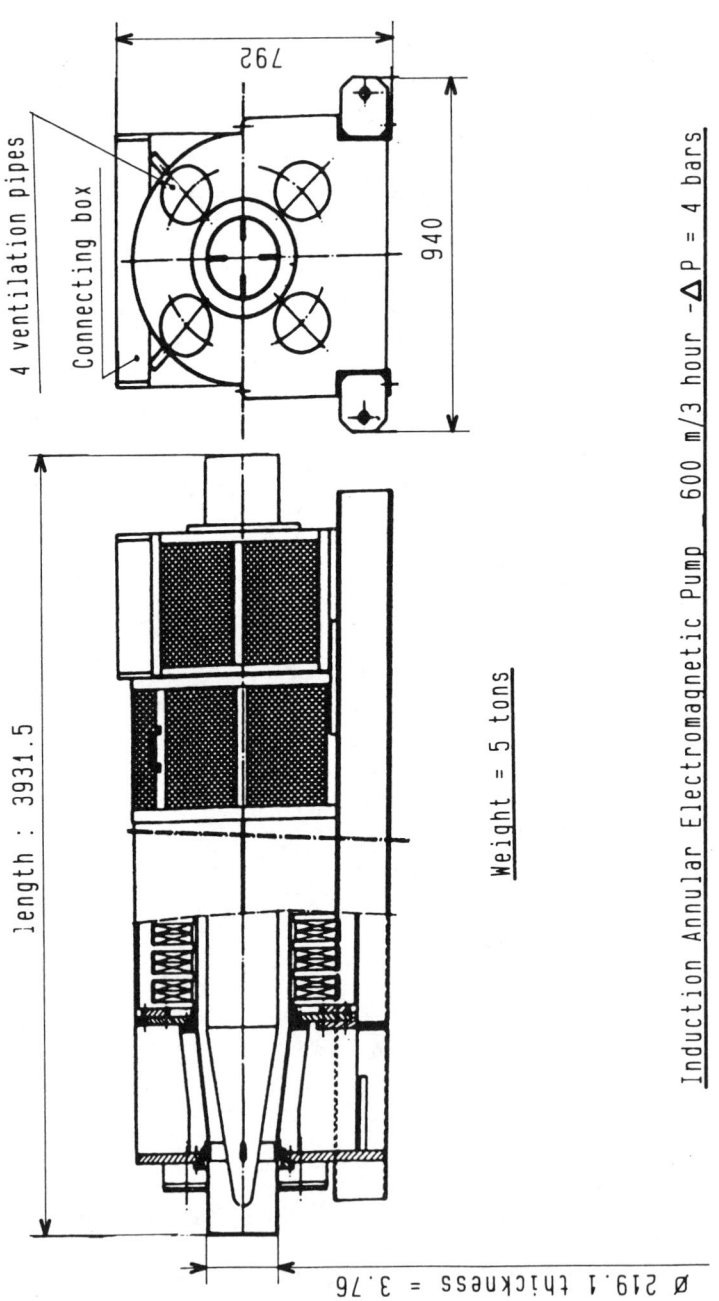

Fig. 2 Schematic of the French 600-m³/h prototype.

Table 2 French 600-m³/h prototype

Geometric features	
Overall length, mm	3930
Overall transverse dimensions, mm	740 x 790
Pump weight + frame, tons	4.5
Connecting duct diameter, mm	219.1 x 3.76
Cooling	
Mode	Axial air cooling
Air flowrate, m³/hour	4990
Head loss, mbar	34
Power supply	
Type	Cycloconverter 460 kVA
	Frequency range 0-15 Hz

computational codes validated during that program were used for the electrical sizing of the pump:

1. An analytical code, "PIANO", for the parametric study allowing one to define an optimized configuration with respect to the desired properties; and
2. a code with with finite elements, "LIGURE," using Galerkin's methode to achieve an accurate evaluation of the pump performances within the retained geometric configuration.

Three options have been chosen in order to improve performances:

1. the use of a three-phase coil type allowing a better induction distribution within the yoke teeth:
2. the setting up of end poles in order to reduce finite-length effects and to improve performances significantly: and
3. the adoption of a cycloconverter for the low-frequency pump.

Sodium Test Results

The sodium tests of the 600-m³/h pump were performed at the end of 1987 at the Arny test center on a 200 mm - diametre loop (see Figs. 3 and 4). A first series of tests sought to determine the maximum efficiency points at a full load for two sodium temperatures: 200 and 400°C. The experimental characteristic curves were obtained for those two temperatures and frequencies between 4 and 12 Hz. The pump efficiency η_e, which is the product of the pressure increase by the delivery over the supply electric power, has a maximum for T = 400°C and 625 m³/h of delivery. Hence, we include the power needed by the cycloconverter and find that η_e = 0.46, or we do not include it and then find that η_e = 0.485 (see Figs. 5 and 6).

Figure 7 gives a comparison between the experimental results and those obtained with the LIGURE code for T = 400°C and ν = 10 Hz. The magnetic Reynolds number linearly decreases from R_m = 7.1 at Q = 0 to Rm = 0 at Q = 820 m³/h. At the present stage of experimentation, it seems to us that instabilities[10] do not hinder normal running conditions, whereas the original analytical theory on azimuthal flow stability[4] indicates that instability must occur for Q < 600 m³/h. If in some cases we observed mechanical vibrations of the whole hydraulic loop, these vibrations can be due to the fact that the electromagnetic forces are no longer averaged over a period.[10] For instance, with ν = 4 Hz, T = 200°C, and a

Fig. 3 Test loop for the 600-m³/h prototype

Fig. 4 French 600-m³/h prototype

typical 0.15 value of the effective magnetic field B, the interaction parameter built with the pulsation ω is

$$N_\omega = \frac{\sigma B^2}{\rho \omega} \sim 7 \tag{6}$$

where ρ is the sodium specific mass.

Our results seem to contradict Soviet findings,[5,8] where frequently $R_m > 1$ conditions have led to important modifications of the characteristic curves and fluctuations of current, pressure, and delivery. This suggested to us that the stability problem should be reexamined by taking into account the influence of edge waves.

Model for Edge Waves and Stability

In the slit channel approximation[2,7] the induction equation that describes the transverse component of the magnetic field along the z axis can be obtained by solving a second-order ordinary differential equation, following the flow direction (Ox):

$$\frac{d^2 B_z}{dx^2} - \mu \sigma \left(j \omega B_z + V \frac{\partial B_z}{\partial x} \right) = \frac{\mu}{\delta_m} \cdot \frac{dJ}{dx} \tag{7}$$

with

$$J = \sqrt{2}\, J_0\, e^{j(\omega t - kx)} \quad \text{and} \quad B_z = B_z^* e^{j\omega t} \tag{8}$$

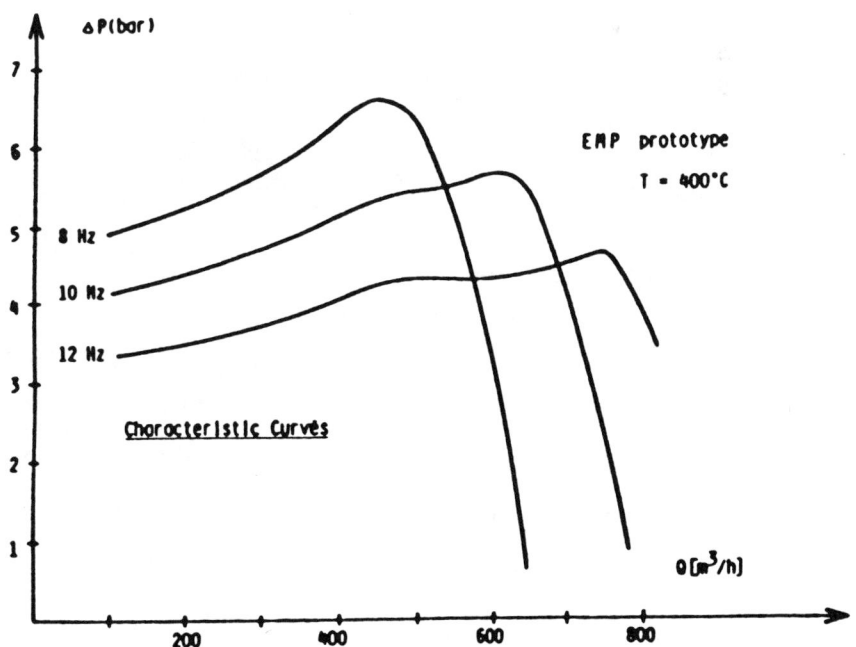

Fig. 5 Characteristic curves for the EMP prototype at T = 400 °C.

where, following the Ampere theorem J is the current sheet located at the magnet surface equivalent to the inducing currents in the slots of the stator and J_0 is constant and $\neq 0$ only on a finite-length:

$$J = 0 \text{ if } x < 0 \quad \text{or} \quad x > L \tag{9}$$

The solution of [Eq. (7)] is a combination of a particular solution of the form $B_0 \exp(j\omega t - kx)$ and of two solutions of Eq. (8) without the second member but which satisfy the two boundary conditions: $B^*_z = 0$ if $x = \pm \infty$.

The solution of [Eq. (7)] is then of the following form[9]:

$$B_z = B_0 e^{j(\omega t - kx)} + B_1 e^{j\omega t} e^{jkx} + B_2 e^{j\omega t} e^{jk'x} \tag{10}$$

where

$$\begin{Bmatrix} K \\ K' \end{Bmatrix} = \frac{\mu_0 V}{2} \left(-j \pm \sqrt{-1 - \frac{j}{\varepsilon}} \right) \tag{11}$$

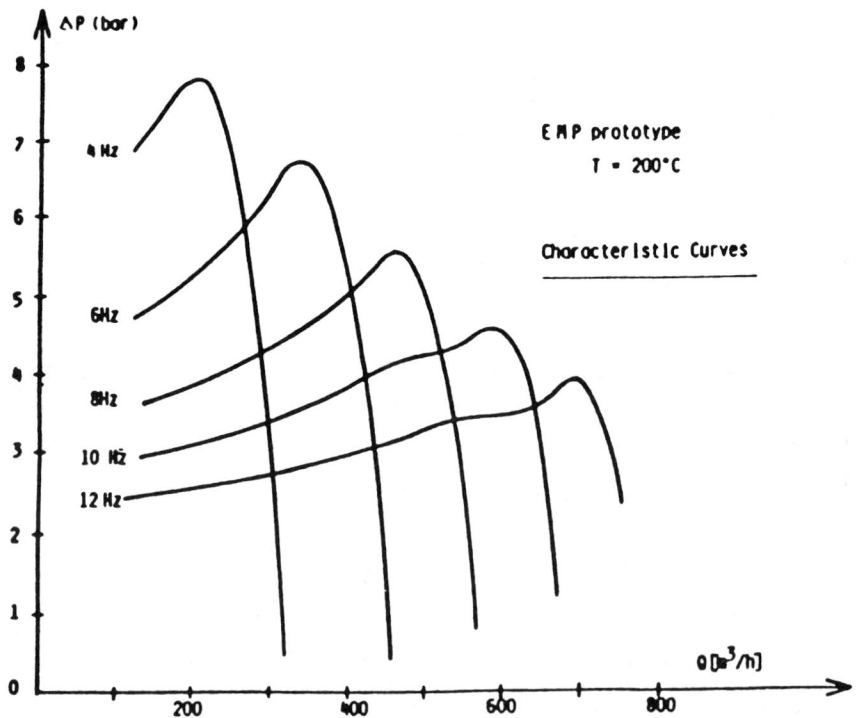

Fig. 6 Characteristic curves for the EMP prototype at T = 200°C.

with ϵ given by [Eq. (4)]. Depending on the value of ϵ, two different situations can occur[6,7]: If $\epsilon \ll 1$, the two edge waves are symmetric and damped on a short distance. On the contrary, if $\epsilon > 1$, the exit wave is very quickly damped, whereas the entrance wave (which corresponds to K) is weakly damped and greatly increases the ohmic losses in the flow, principally when the slip ratio $g = (\omega/k - V)/(\omega/k)$ tends toward zero.

The coupling between the motions of the liquid metal and the inducing magnetic field B_l generates an induced field B_i proportional to Rm [Eq. (2)]. Of course, if $R_m > 1$, B_i must be greater than B_l. Since R_m depends on the local value of the velocity [see Eq. (3)], a very small modification on the velocity can have an important effect on the total field $(B_i + B_l)$ and appreciably modify the Laplace force and thus the local velocity. This unstable situation[4] results in a flow inhomogenity, following the y direction for a flat machine and following the azimuth $\phi = y/R$ for an annular configuration.

The original linear analysis[4] of this stability problem considers an EMP of infinite-length. Then the projection of the Navier-Stokes equation in the x direction can be written as

$$\rho \left(\frac{\partial V}{\partial t} + V \frac{\partial V}{\partial x} + V_y \frac{\partial V}{\partial y} \right) + \frac{\partial P}{\partial x} = B_z \overline{B}_z \left(\frac{\omega}{k} - V \right) \quad (12)$$

where $\partial P/\partial x$ is the pressure gradient in the x direction and $\overline{B}_z$ the complex conjugate of B_z. The infinite-length assumption implies that B_z has a structure of progressive wave in the x direction. The corresponding induction equation including possible variations in the y direction is then

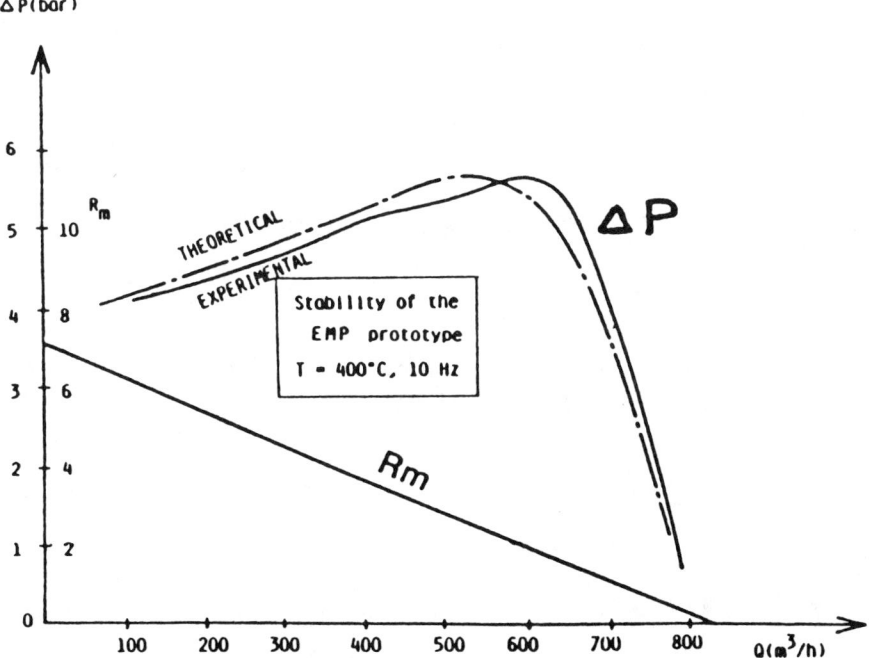

Fig. 7 EMP prototype, with end poles: comparison computed for characteristic test results (ν = 10 Hz, T = 400°C).

86   S. POINSOT ET AL.

$$\frac{\partial^2 B_o}{\partial y^2} - k^2 B_o - \mu\sigma j(\omega B_o - kVB_o) = 0 \tag{13}$$

with

$$B_z = B_o e^{j(\omega t - kx)} \tag{14}$$

If one removes the infinite-length assumtion, the simplified expression of B_z given by [Eq. (14)] must be replaced by the more complete equation [Eq. (10)], which includes the two edge waves. Moreover, if one restricts the study to high-delivery machines with $\epsilon > 1$, the exit wave has a quite small effect and B_2 is negligible compared to B_1. The induction equation must be modified to take B_1 into account. Because of the linear structure B_0 is still governed by [Eq. (13)], whereas B_1 obeys

$$\frac{\partial^2 B_1}{\partial y^2} + B_1[-K_2 - j\omega\sigma(\omega + KV)] = 0 \tag{15}$$

Of course, in the expression of the Laplace force in [Eq. (12)] both B_0 and B_1 must be included. The system of the three coupled equations [Eqs. (12), (13) and (15)] depends on the position of the x axis; consequently, the solutions will be only local solutions. However, let us point out that the present method could be easily extended to situations where $\epsilon << 1$ by simply adding a fourth equation for the exit wave of the same structure as that of [Eq. (15)].

The basic (unperturbed) solutions V^0, B_0, and B^0_1 of [Eqs. (12-14)] correspond to a homogeneous flow:

$$\frac{\partial B^o_{0,1}}{\partial y} = 0 \qquad \frac{\partial V^o}{\partial y} = 0 \tag{16}$$

The next step of the procedure is to consider small perturbations of the following form:

$$B^1_{0,1} = \sum_m B_{(0,1),m} e^{j(p_m t - m y/R)} << B^o_{0,1} \tag{17a}$$

$$V^1 = \sum_m V_m e^{j(p_m t - m y/R)} << V^o \tag{17b}$$

where m is integer for an annular configuration, and p_m is complex: $p_m = p_{rm} + j p_{im}$. After we linearize [Eqs. (12), (13), and (15)], we obtain a stability condition for each mode, corresponding to $p_{im} > 0$ with

$$p_{im} = p^*_{im} + \frac{\sigma}{\rho} \left[B_1 e^{kx}\right]^2 \left(\frac{1}{2} - A\right) \tag{18}$$

where p^*_{im} is, in the finite length assumption, the imaginary part of the solution to the problem. This implies neglecting the inlet wave :

$$p^*_{im} = [B_o]^2 \frac{\sigma}{\rho} \left(\frac{1}{2} - \frac{R^2_m k^4}{\left[R^2_m k^4 + \left(\frac{m^2}{R^2} + k^2\right)^2\right]} \right) \tag{19}$$

In [Eq. (18)] A is a constant function on all of the machine length. After separating the real and imaginary part in the expression of the complex wave number ($K = K_1 + j K_2$), we can write

$$A = \frac{\mu \sigma R^2}{m^2} \left(\frac{\omega K_1}{K_1^2 + K_2^2} - V \right) K^2 \qquad (20)$$

Depending on the value of A, it appears that the inlet wave can have two opposite effects:
1. if $A < 1/2$, it increases the stability, as the amplitude of that wave decreases along the flow, the stability continuously decreases in the flow direction.
2. if $A > 1/2$, the inlet wave decreases the stability, and the flow goes from unstable regions toward more stable ones.

Application to Existing EMP

We have developed a one-dimensional numerical model that allows a given induction machine to calculate at each point on the flow axis the amplitude of the edge wave, the p_{im} value (for m = 1), and the local stability condition. We have applied this model to experimental data from three recent large EMPs: the Soviet CLIP-3/3500,[8] the French prototype, and the German NWA.[9] We present in Table 3 some characteristics of the three pumps, as well as the values of the various adimensional parameters that correspond for each EMP to delivery of optimized running conditions. In Figs. 8-10, first we give the local value of the ratio between the amplitude of the entrance wave and of the wave of the infinite-length configuration, and second we give $p_{im} \rho/\sigma$ [see Eq. (18)], in order to demarcate precisely the stability regions. One can see that the damping of the entrance wave is much more important for NWA and CLIP-3/3500 than for the French prototype, to which a greater value of ϵ corresponds. The different values of A also significantly influence the stability.

For CLIP-3/3500, $R_m < 1$ and in regard to the previous studies,[4] this must be sufficient to guarantee the stability everywhere, but Fig. 8 shows an unstable region near the entrance in the pump.

For the French prototype R_m is appreciably greater than unity, but the entrance wave has a stabilizing effect, and there is a stable region near the entrance (Fig. 9).

Table 3 Comparison of three high-delivery EMPs

	French prototype	CLIP 3/3500	Interatom NWA
Temperature, °C	200	300	350
Delivery Q, m³/h	600	3600	609
Frequency f, Hz	10	50	40
Velocity V, m/s	6.5	12.9	9.52
Pole pitch τ, mm	450	156	210
Wave number $k = \pi/\tau$, m⁻¹	6.98	20.14	14.96
ϵ (with $\delta_m = \delta_h$)	1.56	0.95	0.573
A	0.03	5.59	1.7
R_m	3.33	0.96	3.09
Duct height h, mm	32.4	26	14.9
Mean radius R, mm	125.8	475	190
Length of the inductor L, mm	2270	5000	1000

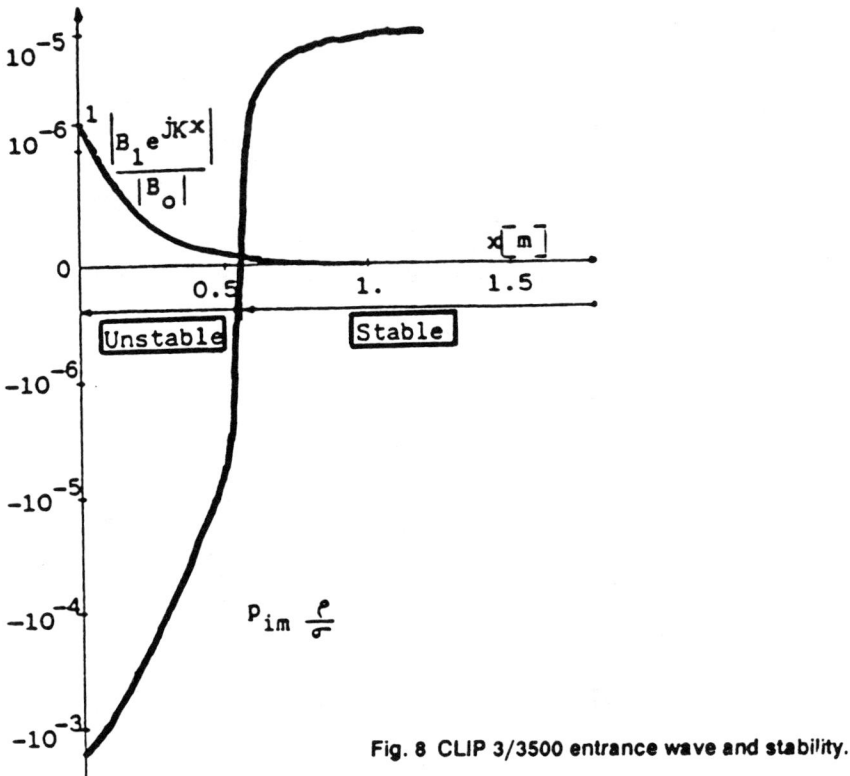

Fig. 8 CLIP 3/3500 entrance wave and stability.

For the German NWA, to which the smaller value of ϵ corresponds, the entrance wave is damped on a short distance and does not modify the instability condition appreciably. Figure 10 shows that our model predicts an inhomogeneous flow over all of the pump length.

Possibilities for Increasing Performances

We have found that the edge waves induced by the finite length of EMPs can either increase or decrease the stability. It appears that, for increasing EMP efficiency, it would be good to increase the R_m and ϵ values. It will be necessary to predetermine carefully the onset and the effects of the instability. Specifically local measurements inside the flow of electromagnetic fields and fluid velocities appear to be necessary. For this we have developed and built local velocity measurement probes, based on pitot tubes principles (see Fig. 11). The measurement of the static pressure is always fixed, whereas the dynamic pressure measurement can be moved along the duct height and rotated around the probe axis. By this manner one can obtain, for various radial positions, the values of two components of the velocity. That probe is to be calibrated in sodium and fitted in a next step on the French 600-m³/h prototype. Of course, the knowledge of local velocities on the various high-delivery EMPs[5,9] is very important to determine precisely the real effects of the instability.

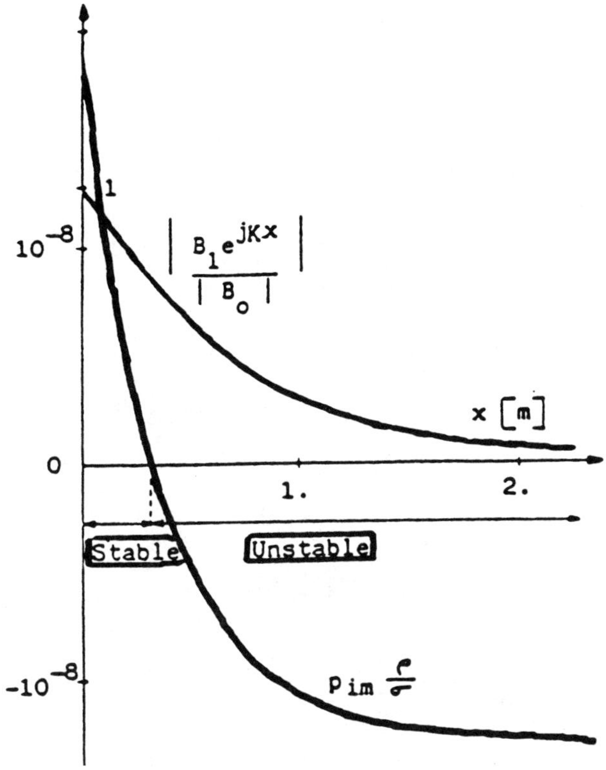

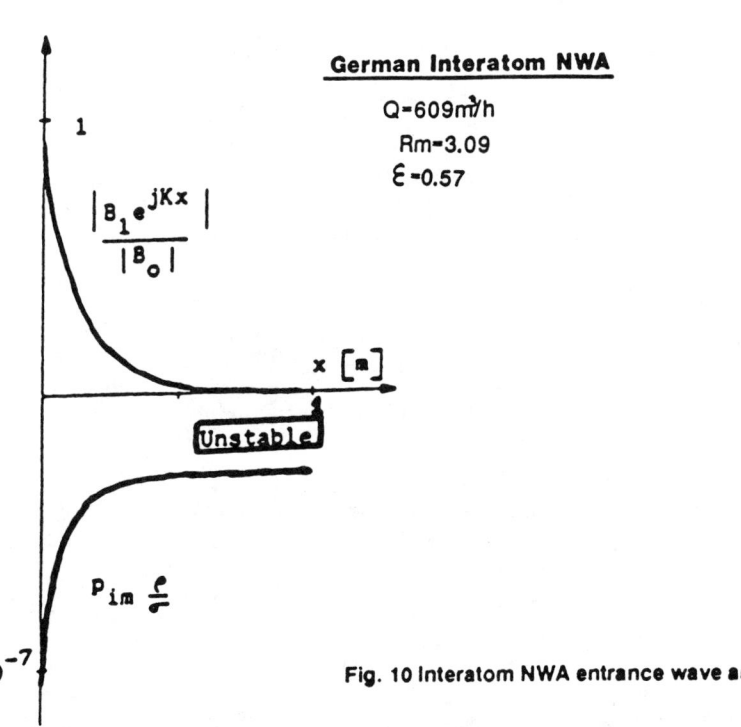

Fig. 9 French prototype entrance wave and stability.

Fig. 10 Interatom NWA entrance wave and stability.

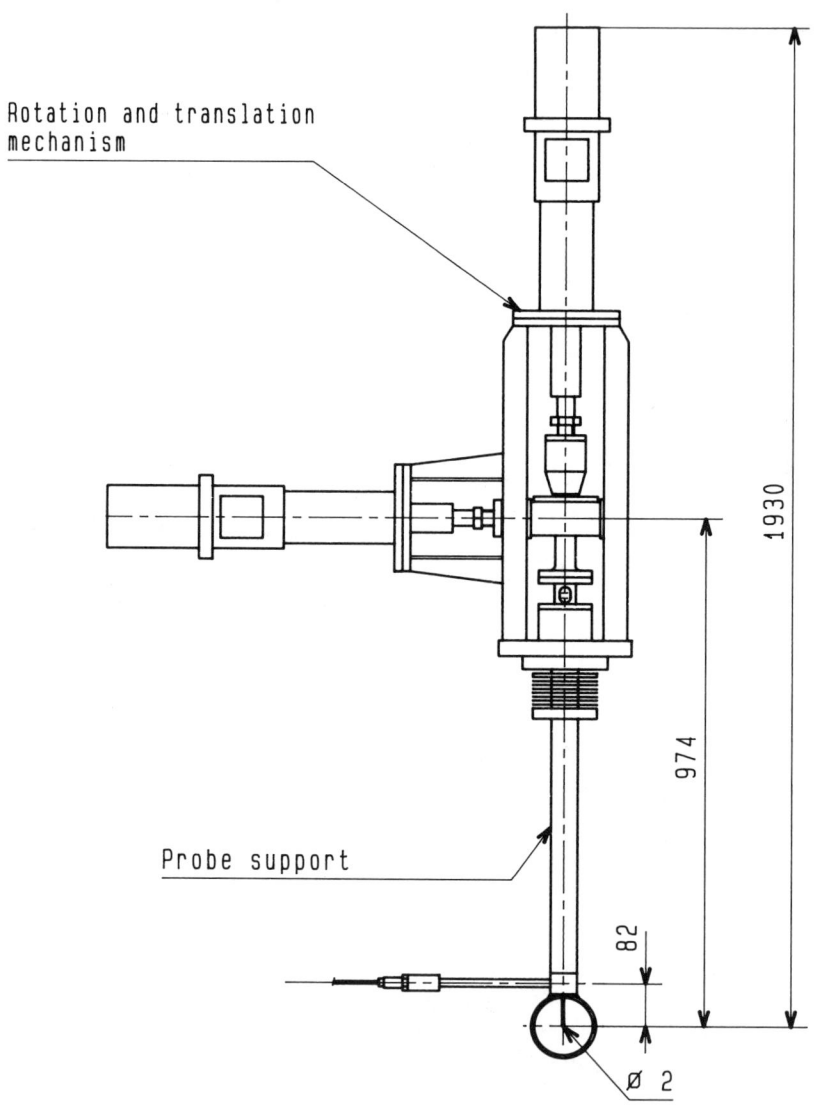

Fig. 11 Local velocity measurements probe.

The next stage for improving EMP efficiency should be, for fixed R_m and ϵ values, to reduce joule losses in the flow. For this, flexible autonomous electricity generators that allow frequency as well as phase equilibration to be controlled must be developed. Moreover, studies and tests on the repartition of the inductor currents in the stator coils[6] must be continued. Furthermore, due to scale effects, one important feature for future very-high-delivery EMPs ($Q \simeq 10{,}000$ m^3/h) will be the cooling systems.

References

1. Einstein, A., and Szilard, L., British Patent Application 303, 065, Dec. 24, 1928 ; British Patent Specification 344, 881, March 1931.
2. Jousselin, F., Marty, Ph., Alemany, A., and Werkoff, F., "Some Aspects of a Liquid-Metal Induction Generator," Liquid Metal Flows: Magnetohydrodynamics and Applications, Progress in Astronautics and Aeronautics series, Vol. 111, edited by H. Branover, M. Mond, and Y. Unger, AIAA, Washington DC, 1988, pp. 265-285.
3. Poinsot, S., and Rapin, J., "Construction and Test in Sodium of a 600 m^3/Hour Annular Induction Electromagnetic Pump," Proceedings of the 4th International Conference on Liquid-Metal Engineering and Technology, Vol. 1, Paper 104, 1988. Société Française d'Energie Atomique - 48, rue de la Procession - 75724 Paris Cedex 15, France.
4. Gailitis, A., and Leilausis, O., "Instability of Homogeneous Velocity Distribution in an Induction-type MHD Machine," Magnitnaya Gidrodinamika, Vol. 1, 1975, pp. 87-101.
5. Kirillov, I. R., and Ostapenko, V. P., "The Integral Characteristics of a Cylindrical Induction Pump at $R_m > 1$," Magnitnaya Gidrodinamika, Vol. 3, 1987, pp. 115-119.
6. Biedinger, J. M., and Kant, M., "On the Optimization of Magnetic Field Sources in Electromechanical Energy Conversion," Journal of Applied Physics, Vol. 53, N°10, Oct. 1982, pp. 7061-7070.
7. Poloujadoff, M., The Theory of Linear Induction Machinery, Clarendon, Oxford, England, 1980.
8. Baranov, N. N., "Research and Developement in the Field of Liquid MHD Devices for Industrial Applications." Israel - Academy of Sciences and Humanities.
9. Werkoff, F., "Finite-length Effects and Stability of Electromagnetic Pumps", Experimental Thermal and Fluid Science, Elsevier Science Publishing CO. Inc., 665 Avenue of Americas, New York, NY 10010. Vol. 4, 1991 pp. 166-170.
10. Rapin, J., Vaillant, P., and Werkoff, F., "Experimental and Theoretical Studies on the Stability of Induction Pumps at Large R_m Numbers," Liquid Metal Magnetohydrodynamics, Mechanics of Fluids and Transport Processes, Vol. 10, Kluwer Academic Publishers, P.O. Box 17, 3300 AA Dordrecht, The Netherlands, 1989, pp. 325-331.

Physical Model for Electromagnetically Driven Flow in Channel Induction Furnaces

René Ricou* and Charles Vivès†
Université d'Avignon, Avignon, France

Abstract

Experiments performed in a four-tenth-scale physical model of a 1300-kW inductor unit are described. The approach was, first, to measure the electromagnetic parameters, namely, the magnetic field and current density components, as well as the phase shift between these periodic vectors, and then, to find the electromagnetic force field from these parts. These results were connected with the experimentally determined liquid metal flowfields. This present work, which reveals the impact of the magnetic field leaks on the velocity pattern, may be considered as a preliminary investigation, in an attempt to control the magnetohydrodynamic flows in channel induction furnaces.

Introduction

From an electrical point of view, channel induction furnaces are originally 50-(or 60-) Hz transformers with a closed secondary, consisting of a single turn of molten metal. A magnetic iron core and primary coil assembly thread a refractory block containing the annular secondary cavity, which is called channel, for the melt. Later, an inductor was developed with two coils on a common core. Its molten secondary consists electrically of two parallel single turns that flow together into a common center channel located between coils. This design, known as a twin coil inductor, was developed to meet increasing melt rate requirements.

With regard to the other types of furnaces, this process presents manifold advantages: better temperature control, lower melt

---
Copyright © 1991 by the American Institute of Aeronautics and Astronautics, Inc. All rights reserved.
* Assistant Professor, Laboratoire de Magnetohydrodynamique.
† Professor, Laboratoire de Magnetohydrodynamique.

losses, lower energy costs, reduced gas absorption by the metal, etc.

Moreover, in a channel induction furnace with inductors of under 400-kW, the problem of the channel clogging made it necessary to rod out the channel very frequently. Such an operation results in loss of time and, consequently, of productive capacity; moreover, furnaces can even be lost from metal clogging in the channels. On the contrary, when the power rating of the inductor is of the order of several thousand kilowatts, the presence of high-speed horizontal vortices and, in certain cases the presence of a possible cavitation phenomenon, is mainly responsible for a fast erosion of the channel and dramatically shortens the inductor life.[1]

These inconveniences can probably be overcome by a better knowledge of the basic electromagnetic and hydrodynamic phenomena that appear in the molten secondary. Actually, these phenomena are still not totally understood for two main reasons:

1. The electromagnetic force distribution and the fluid flow pattern are extremely difficult to compute, because of the three-dimensionnal aspects of the problem and the rather complicated geometry of the molten secondary.

2. Because of the high temperature of the melt and the cost of such an operation, it is practically impossible to carry out a comprehensive direct experimentation on an actual channel furnace.

Several experiments using either horizontal or vertical laboratory models have already been performed [2-1]; on the other hand, some previous theoretical works must be mentioned [3,5,7,8]. These investigations were particularly useful for the improvement of the understanding of the electromagnetic force field and velocity field patterns in channel furnaces. Nevertheless, it seems that, up to now, a realistic solution to reduce efficiently the refractory erosion in high-powered channel inductors has not been devised.

This paper presents some findings obtained from electromagnetic and fluid flow measurements inside a four-tenth-scale physical model of a 1300-kW channel induction furnace. This work may be considered as an investigation of reference intended to be compared later with other new processes, which are currently being studied, with a view to control the recirculating flows of molten metal, particularly in the channel legs.

Experimental Apparatus

The principal features of the laboratory model are schematically presented in Fig. 1, which has been nearly drawn to scale, and both the main dimensions and the characteristic parameters are shown in

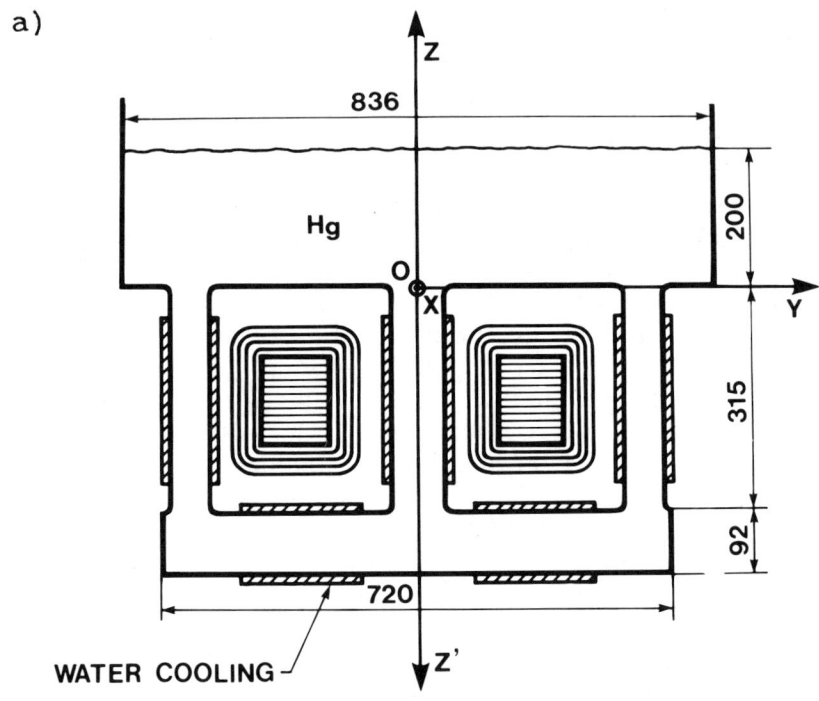

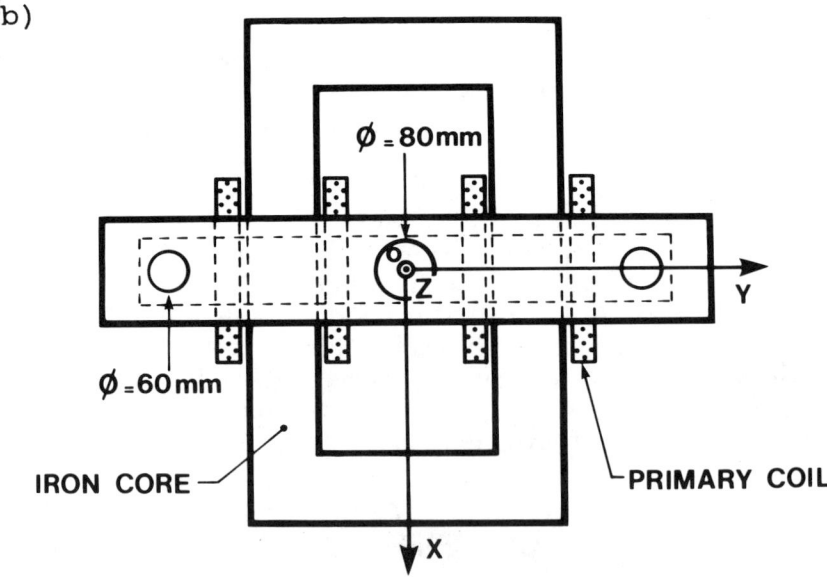

Fig. 1 Schematic diagram of the apparatus: a) front view; b) view from above.

Table 1 Parameters of the Model

Parameters	Values
Diameter of lateral channel	60 mm
Diameter of lateral channel	80 mm
Diameter of lower horizontal channel	92 mm
Depth of mercury	607 mm
Weight of mercury	420 kg
Shielding parameter: $R_\omega = \mu\omega\sigma R^2$	0.63
Applied frequency	50 Hz
Magnetizing force	3900 Ampere-turns
Voltage	320 V
Overall electric power consumption	2.55-kW
Power factor	0.92

Table 1. In essence, the experimental apparatus consisted of three vertical and one horizontal cylindrical legs made of stainless steel, having a wall thickness of 2 mm and topped by a parallelepipedic vessel, made of plexiglass and playing the role of the crucible. This secondary was filled with about 420 kg of mercury. In order to avoid the superheating of the bath and also to operate under isothermal conditions, the channel legs were cooled by water, which was flowing inside annular ducts formed by the legs and the concentric cylinders made of plexiglass (Fig. 1); the upper vessel was also cooled, by means of two narrow tanks traversed by water and placed vertically against the two larger walls, so that the temperature in the mercury was always lower than 25°C. Accordingly, in these experiments the fluid motion in the channel furnace may be considered as exclusively generated by the electromagnetic forces.

A laminated iron core of rectangular cross section (130 x 100 mm), constituted by the stacking of sheets made of soft magnetic material (iron-silicium-oriented crystals), was magnetized by two primary coils made up of 250 jointed turns of 4.55 mm diam., distributed in five layers, and traversed by an electric current of 50-Hz frequency. The horizontal plane of symmetry of the laminated core was located at the midheight of the vertical branches of the channel. The magnetizing force was modulated at will, from 0 to 3900 A-turns, by an autotransformer.

Three-dimensional electromagnetic and fluid flow measurements were taken virtually at any point within the melt, except for the lower horizontal channel, because the displacement of the various probes is obviously not easy inside this area. The

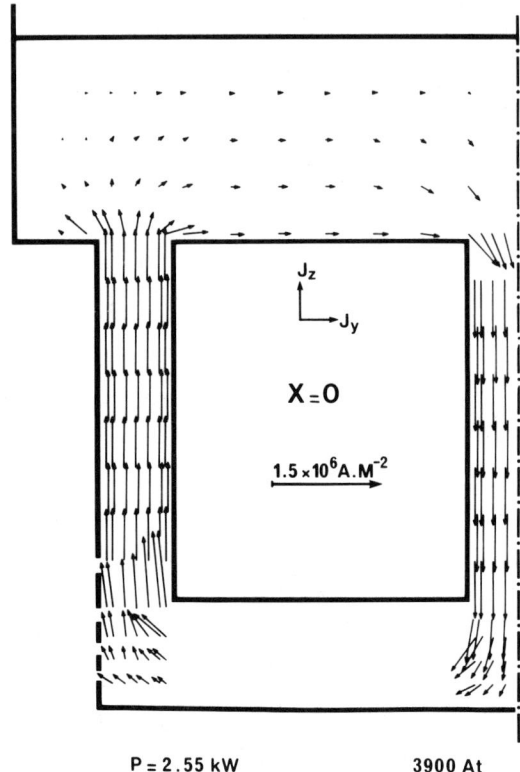

Fig. 2 **Electrical current density distributions.**

magnetic field and current density components were measured using a search coil and a potential sensor, respectively, whereas a phase meter allowed the measurement of the phase angle between these periodic vectors [9,10]. The local velocity was determined in magnitude and direction by means of an incorporated magnet probe [11,12]. Last, displacements of the probes along the X, Y-and Z coordinates (Fig. 1) were ensured by an appropriate traversing system.

Electromagnetic Parameter Measurements

Current Density Measurements

Measurements, using successively a potential sensor and a current transformer, allowed us to verify that the current intensity in the central leg is the sum of the electric current in the side branches; moreover, the secondary currents are practically equal to the ampere-turns that were flowing through the primary coils. Furthermore, it is seen in Fig. 2 that the current density distribution

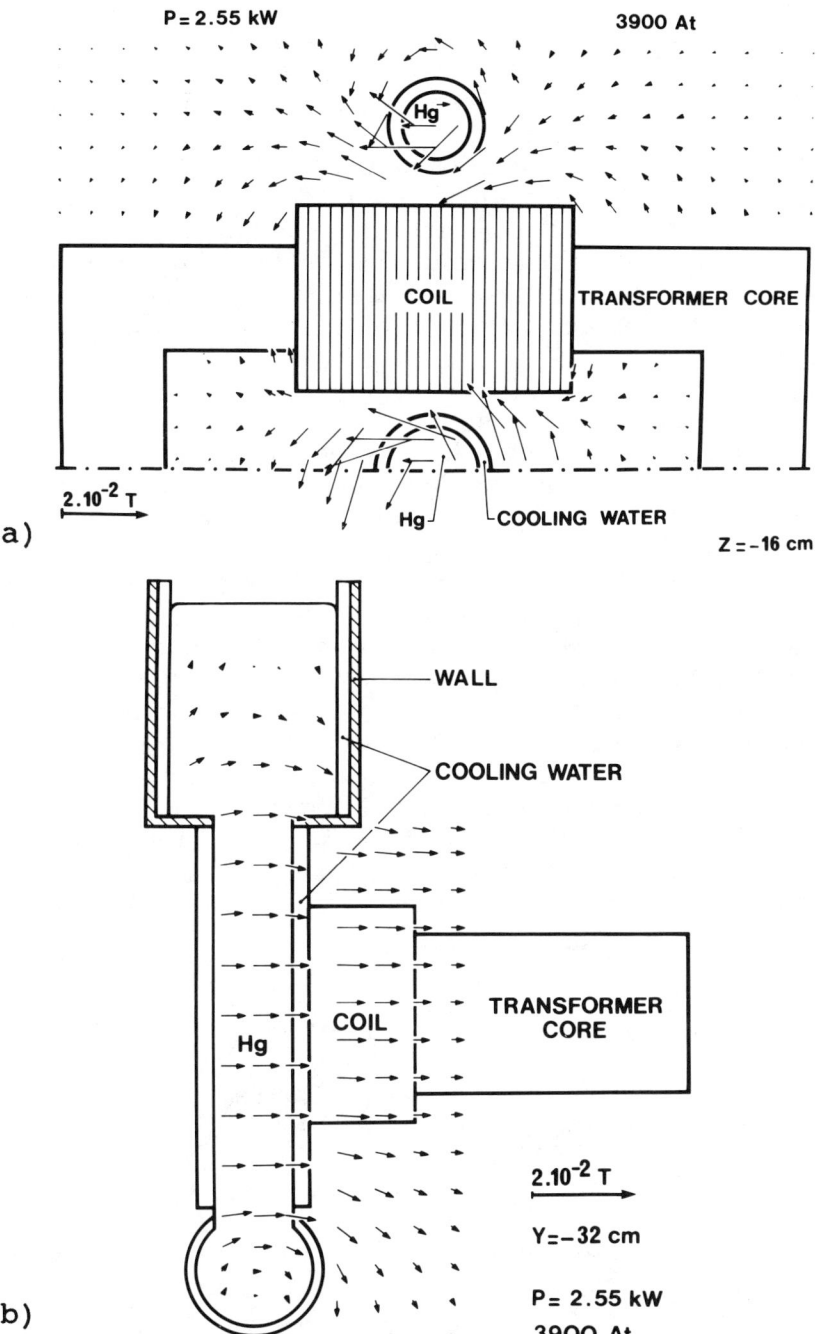

Fig. 3 Magnetic field pattern plotted in air and in mercury: a) inside the horizontal plane of symmetry ; b) inside the vertical plane of symmetry of a lateral channel.

is very nearly one-dimensional (J_Z component) inside the channels, except for the corner regions, where the vertical branches connect with the upper reservoir and the lower horizontal leg. It should be pointed out that the current density field, which exists virtually throughout the upper tank, is nonuniform and increases in magnitude, from top to bottom, within this container (Fig. 2). Accordingly, with respect to the horizontal plane of symmetry of the laminated core, the current density distributions inside the lower horizontal leg and the main bath are asymmetrical; such a situation may have a repercussion on the weak axial motions existing in the channel legs.

Magnetic Field Measurements

Figure 3 conveys magnetic field patterns, plotted both in air and mercury, inside the horizontal plane of symmetry of the transformer iron core (Fig. 3a) and inside the vertical plane of symmetry of a lateral channel (Fig. 3b). The magnetic field leaks, arising both from the ferromagnetic yoke and the primary coils, are of the order of 10^{-2} T in the vicinity of the channel legs. There, the impact of leakage from the inductor on the electromagnetic force field pattern and, in turn, on the flow structure, must be taken into consideration, because of its interaction with the strong electric currents that circulate through the furnace.

Examples of magnetic field distribution, plotted over a horizontal cross section of the lateral and central channels, are shown in Fig. 4. The arrows indicate the direction and magnitude of the r.m.s. values of the fluctuating magnetic field at the tail of the arrow. Because of the variation in space of the phase shift between

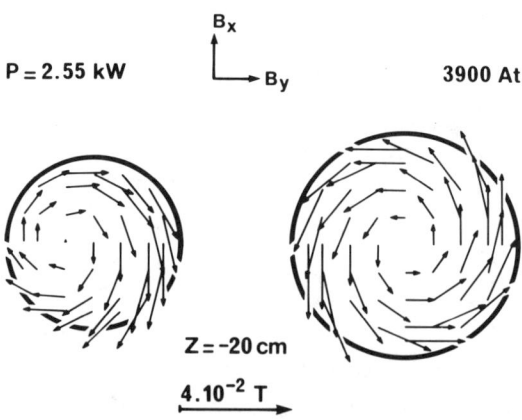

Fig. 4   Magnetic field distribution plotted over a horizontal cross section of the lateral and central channel (actual case).

the magnetic field vectors, it should be specified that these patterns do not represent a shapshot.

It is well known that the ideal case of an electric current passing through a cylindrical and rectilinear conductor of infinite length is characterized by concentric circular magnetic field lines (Fig. 5). On the contrary, inspection of Fig. 4 reveals that the magnetic field distributions in the X,Y plane are neither uniform along circular lines nor symmetrical with respect to the revolution axis of the lateral branches. This dissymetry is due to the superposition of the magnetic field produced by the electric currents that are flowing through the vertical channels, and of the magnetic leaks, which come mainly from the primary coils.

For the case of a lateral channel the leakage flux lines are primarily issued from the nearest coil and the strength of the corresponding magnetic field decreases from this winding (Fig. 3a). Under the effect of these losses the resultant magnetic field has a tendency to be enhanced inside the right-hand-side half cross section, located in the vicinity of the iron yoke and reduced in the left-hand-side one; however, the symmetry about the Y axis is maintained.

In the central channel, since the magnetic escapes that are flowing from the two coils are opposite in phase, it follows that their actions cancel each other along the diameter parallel to the X axis. Accordingly, their effects may be considered as insignificant inside a central region, which has a thickness of about 2 cm and is located on both sides of the diameter parallel to the X axis. In contrast, outside this area, it is seen that the impact of the leakage

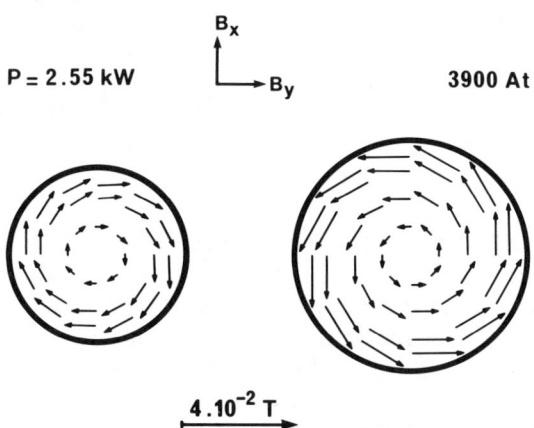

Fig. 5 Magnetic field distribution inside a horizontal cross section of the lateral and central channel (ideal case).

results in a quite marked enhancement of the magnetic field. Accordingly, and contrary to the case shown in Fig. 5, the magnetic field varies in amplitude along circular lines situated in horizontal cross sections, but remains symmetrical with respect to the revolution axis of the central leg. It should be pointed out that the findings are corroborated by the fact that the magnitude of the magnetic field vectors, plotted along the diameter parallel to the X axis and obtained either from measurements within the laboratory model (Fig. 4) or by calculations in the case corresponding to the absence of magnetic field leaks (Fig. 5), are in very good agreement.

Electromagnetic Force Field

The $J \times B$ force consists of a time-independent component and an oscillatory component of frequency 2N, where N is the frequency of the applied coil current. The possible impact of the vibrating force on the erosion of the refractory has already been examined ; hence, we will confine this study to the inspection of the time-mean electromagnetic body force patterns, exclusively [1].

Inside the channel legs, because of the one-dimensional character of the electric current density which was pointed out earlier, the time-smoothed components are only two dimensional and are given by

$$F_x = J_z B_y \cos(J_z, B_y)$$
$$F_y = J_z B_x \cos(J_z, B_x)$$

whereas in the gate areas, due to the curvature of the electric current lines in these zones, the forces are three dimensional and are practically given by

$$F_x = J_y B_z \cos(J_y, B_z) - J_z B_y \cos(J_z, B_y)$$
$$F_y = J_z B_x \cos(J_z, B_x)$$
$$F_z = J_y B_x \cos(J_y, B_x)$$

These relations reveals the impact of the values of the phase shift, measured between the components of B and J, on the magnitude and direction of the electromagnetic forces.

An example of electromagnetic force distribution, plotted over a horizontal cross section of the lateral and central legs, is depicted in Fig. 6. Once again, in contrast with the ideal case, already defined which gives rise to the classical "pinch effect" (Fig. 7), it appears in Fig. 6 that the electromagnetic forces are neither centripetal nor symmetrical with respect to the vertical axis of the

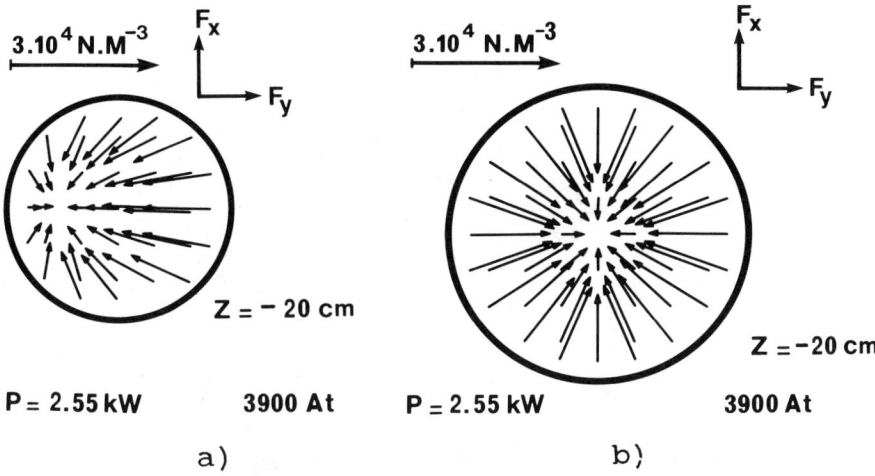

Fig. 6 Electromagnetic force distribution plotted over a horizontal cross section (actual case): a) lateral channel; b) central channel.

lateral channel. The predominant component F_y is parallel to the Y axis, which is also a symmetry axis, and the stronger forces are concentrated inside the right-hand-side half cross section, which is the nearest from the primary coil.

The impact of the magnetic flux losses is less pronounced in the central leg (Fig. 6), where the forces are symmetrical with regard to the Z axis. However, these forces are still not exactly directed toward the center of the circular section; this is particularly true for the forces where the X component prevails. Moreover, the

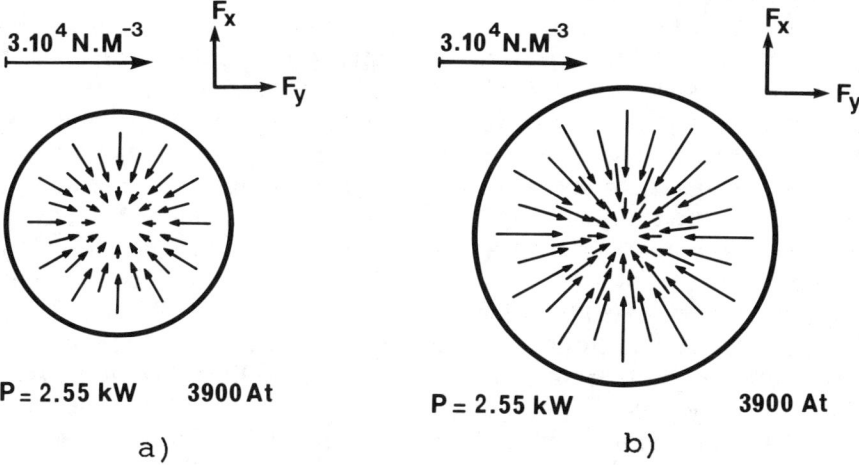

Fig. 7 Electromagnetic force distribution inside a horizontal cross section (ideal case): a) lateral channel; b) central channel.

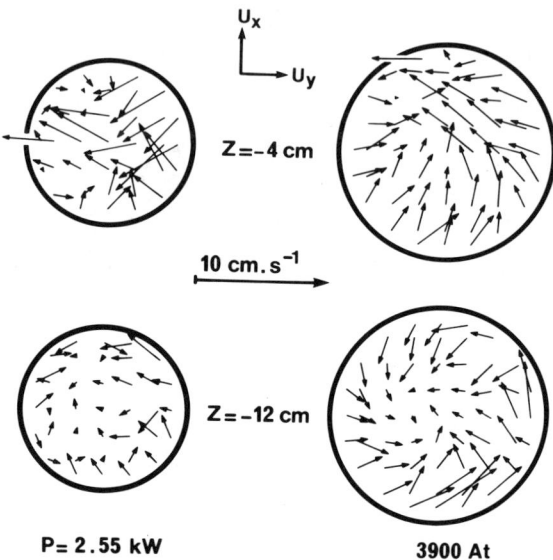

Fig. 8 Distribution of the horizontal component of the velocity inside channel cross sections.

magnitude of the Y components is generally greater than that of the X components.

Obviously, the asymmetrical trend of these patterns arises from the magnetic field distributions displayed in Fig. 4, which are influenced by the additional effect due to the magnetic escapes.

Velocity Field

In these experiments the velocity probe was connected to a micro-processor voltmeter that was able to take 200 readings per minute and also yielded their average, highest, and lowest values.

Figure 8 presents distributions of the horizontal components of the velocity inside channel cross sections, located at various vertical positions. In the central zone of the channels ($Z = -20$ cm), these transverse motions consist of the presence of two recirculating vortices in the outer branche and of a single cell in the inner one. In the channel bottom area ($Z = -32$ cm), the flow structure is made still more complicated because of its three-dimensional aspect (Fig. 9). It should be mentioned that, due to the random character of the fluid flow, the measurements inside the channel legs were not easy. Consequently, the degree of reliability of these data must be considered prudently, and their main interest was to yield an order of magnitude for the velocities. As an example, for an average velocity of 3 cm·s^{-1}, the corresponding highest value, measured at

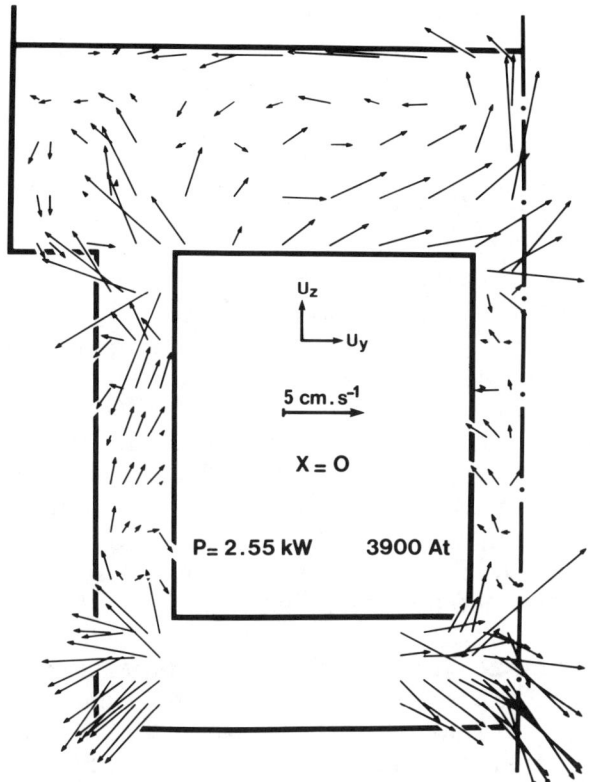

Fig. 9   Velocity field plotted inside the vertical plane of symmetry of the furnace.

the same point, can reach 10 cm·s^{-1}. Moreover, if the flow distribution in the lateral channel is rather consistent with the corresponding force field pattern (Fig. 6), it is difficult to explain coherently the tendency to an anticlockwise circulation in the central channel.

Figure 9 displays a velocity field plotted inside a half-vertical cross section. It appears from this figure that the primary flows start from the gate areas. This event is understood by the presence of strong driving forces in the corner regions, which are caused by the deflection of the electric current lines; the occurrence of the J_x and J_y components (Fig. 2), which interact with the magnetic field, gives rise to axial components F_z of the electromagnetic body forces in these zones. The main bath is occupied by two cells (i.e., four cells for the whole volume of the upper vessel), the smallest being located in the vicinity of the throat of the external leg.

In the vertical channels the net flow rate corresponding to the axial flow is nearly equal to zero. In fact, the axial velocity, measured at a given point, depends strongly on time and can change sign alternately. The average velocity was also measured inside the external and central legs by means of calibrated electromagnetic flowmeters. The results obtained from 600 readings and for an electric power consumption of 2.55-kW are presented in Table 2.

The weakness and the aleatory character of these axial motions can be understood by the following arguments: 1) F_Z is virtually non existent inside the vertical channel legs. 2) Since the furnace is very nearly symmetrical both from the geometrical and magnetic standpoints, it follows that, in the neighborhood of the upper and lower gates of each channel, the F_Z components are practically equal in magnitude and opposite in direction ; thus, their effects cancel approximately each other. Since the movements of the liquid metal in the legs result from the superposition of axial and transverse random motions, helical-type flows arise which, depending on the moment, rotate alternately clockwise or counterclockwise and are ascending or descending in an aleatory manner.

To summarize, this complex three-dimensional fluid flow structure, which exists everywhere inside the furnace, is due to the fact that the electromagnetic body force has a rotational part practically throughout in the melt. The peak velocities, included in the range of 5-10 $cm \cdot s^{-1}$, were found in the corner regions and sometimes inside the vertical channels (transverse motion). However, it should be emphasized that, in the case of an actual powerful furnace, the magnitude of the velocity vectors would be dramatically augmented and included in the range of 5-10 $cm \cdot s^{-1}$. In fact, the velocities are nearly proportional to the square root of the electric power consumption, which can reach 1300-kW, for an industrial scale unit for aluminum melting, and only 2.55-kW in the case of the physical model. Moreover, this tremendous discrepancy is still increased because of the large difference between the physical properties of the working fluids: Aluminum is five times lighter and four times more conducting than mercury.

Table 2   Average velocity ($cm \cdot s^{-1}$)

	Mean value	Higher value	Lower value
Outer channel	0.75	1.29	0.15
Inner channel	-0.25	-0.95	0.33

Conclusions

In this experimental study, on a cold physical model, the extent of the three-dimensional aspects of the electromagnetic and fluid flow phenomena, which occur in a channel type furnace was revealed. Furthermore, this investigation showed that the flow structure may be appreciably affected by the magnetic field leaks.

A new set of trials, using the main parts of this laboratory model, is currently in progress with a view to find new solutions allowing to overcome, at least partially, the two primary defects of this melting process:

1) In order to improve the heat transfer from the legs to the bulk of the melt, it is necessary to increase the net flow rate of the axial motion in the vertical branches.

2) The intensity of the transverse motions, which is one of the principal causes of the severe erosion of the channel refractory, must be lowered.

References

[1] Lillicrap, D. C., "Refractory Erosion in High Power Channel Inductors," Electrowärme International, Vol. 44, January 1986, pp. 116-122.

[2] Butsenieks, E., Levina, Y., Stolov, M., Sharamkin, V., and Shcherbinin, E., "Flow of Metal Through Induction Channel Furnaces due to Electromagnetic Forces," Magnitnaya Gidrodinamika, Vol. 4, Oct-Dec. 1977, pp. 103-106.

[3] Kolesnichenko, A., Gorislavets, Y., and Bundya, A., "Creation of a Undirectional Motion of a Liquid Metal in the Channels of Induction Melting Furnaces," Magnitnaya Gidrodinamika, Vol. 4, Oct-Dec. 1979, pp. 138-140.

[4] Butsenieks, I., Levina, M. Stolov, M., and Shcherbinin, E., "Motion of the Metal in Channel-Type Induction Oven," Magnitnaya Gidrodinamika, Vol. 3, July-Sept. 1980, pp. 123-130.

[5] Moros, A., Hunt, J. C. R., and Lillicrap, D. C., "Study of the Electromagnetic Features in Channel Induction Furnaces", Single and Multi-Phase Flows in an Electromagnetic Field, edited by H. Branover, P. Lykoudis and M. Mond, Progress in Astronautics and Aeronautics series, AIAA, New York, 1985, pp. 706-715.

[6] Stolov, M., "Research on MHD Processes in Induction-Type Channel Ovens as a Basis for Development," Magnitnaya Gidrodinamika, Vol. 2, April-June 1988, pp. 125-130.

[7] Moros, A., Hunt, J. C. R., and Lillicrap, D. C., "Modeling Recirculating Flows in Channel Induction Furnaces", Liquid Metal Flows : Magnetohydrodynamics and Applications, Vol. 119, edited by H. Branover, M. Mond, and Y. Unger, Progress in Astronautics and Aeronautics series, AIAA, Washington DC, 1988, pp. 421-441.

[8] Mestel, A., "On the Flow in a Channel Induction Furnace," Journal of Fluid Mechanics, Vol. 147, 1984, pp. 431-447.

[9] Vivès, C., and Ricou R., "Experimental Study of Continuous Electromagnetic Casting of Aluminum Alloys," Metallurgical Transactions, Vol. 16B, June 1985, pp. 377-384

[10] Vivès, C., and Ricou R., "Fluid Flow Phenomena in a Single Phase Coreless Induction Furnace," Metallurgical Transactions, Vol. 16B, June 1985, pp. 227-235.

[11] Ricou R., and Vivès, C., "Local Velocity and Mass Transfer Measurements in Molten Metals Using an Incorporated Magnet Probe," International Journal of Heat and Mass Transfer, Vol. 25, Oct. 1982, pp. 1579-1588.

[12] Lee, H.C., Evans, J.W., and Vivès, C., "Velocity Measurements in Wood's Metal Using an Incorporated Magnet Probe," Metallurgical Transactions, Vol. 7, Dec 1984, pp. 734-736.

Grain Refinement in Aluminum Alloys by an Electromagnetic Vibrational Method

Charles Vivès*
Université d'Avignon, Avignon, France

Abstract

The influence of electromagnetic vibrations imposed during solidification on grain refinement in the 1085 and 2214 aluminum alloys, respectively characterized by a narrow and a wide freezing range, has been examined. The vibrations were produced by the simultaneous application of a stationary magnetic field B'_0 and of a variable magnetic field B of 50 Hz frequency, in the sump of continuously cast ingots. Extensive grain refinement has been observed in both the alloys with increasing magnetizing force. This study shows that the mean grain size obtained by this vibrational technique is always smaller than that produced by the recently developed CREM process.

Introduction

The grain structure of metal is of great importance, since many of its mechanical properties are directly related to the size, shape, and distribution of the grains. It is well known that the production of a fine-grained equiaxed structure leads to a substantial improvement of the quality of the metal and allows an increase of the ingot drop rate and a reduction of craking. For instance, several dynamic methods, producing a vigorous forced convection in the melt during freezing, lead to substantial grain refinement.[1,2,3] In such processes heat and fluid flow are controlled by various externally applied forces. These methods primarily include the use of electromagnetic or mechanical stirring.

Copyright © 1991 by the American Institute of Aeronautics and Astronautics, Inc. All rights reserved.
* Professor, Laboratoire de Magnetohydrodynamique.

Moreover, many investigators have found that mechanical vibrations of both sonic and ultrasonic character, when applied during solidification of metals and alloys, modify conventionally obtained macrostructures and microstructures.[4,5] The most commonly observed effect is the suppression of undesirable dendritic and columnar zones and the development of a fine-grained equiaxed structure. In fact, the effects produced when high intensity sonic or ultrasonic waves are propagated through molten metals can be listed under three headings: grain refinement, dispersive effects, and degassing, with the result that porosity is reduced. There appear to be two distinct views regarding the mechanism.[1,6]

1. Ultrasonic irradiation at a suitable intensity and frequency gives rise to intense cavitation in the molten metal, and the forces associated with cavitation results in disruption of the growing crystals. This splitting up of crystals effectively produces many more nuclei around which new crystals can form. Thus, in this manner the crystals never grow beyond a certain size. Ultrasonic vibrations also give rise to considerable agitation of the melt and result in the newly formed nuclei being distributed throughout the molten pool so that the crystallization takes place uniformly inside the entire volume.

2. The other view is that vibration has much the same effects as turbulence, dispersing small crystals so that more of them grow, with the result that the grain size is reduced.

The sonic or ultrasonic irradiation of molten metals is mainly carried out with magnetostrictive transducers (Fig. 1). In some cases the frequency of the ultrasonic vibrations is likely to vary continuously in order to maintain the molten pool at resonance. Coupling rods made of quartz, graphite, and various ceramic materials have been used to communicate vibrations to a molten metal, and these materials are attached to the transducer by special cements.

However, such a technique presents several disavantages. The oscillating rods are very rapidly dissolved when they are immersed into molten aluminum alloys, and this circumstance provokes an undesirable pollution of the metal. Moreover, the intensity of cavitation is greatest near to the tranducer or the coupling rod face; thus, the use of such a system is principally justifies for the treatment of metal mixture on a small scale. Furthermore, in view both of the cost and of the bulkiness of the equipment, the production of large aluminum alloy ingots by continuous casting seems unrealistic.

The continuous casting process is shown schematically in Figs. 1-3, in which a mold, usually of rectangular or circular section, is used to produce the ingot, which is moved downward continuously by mechanical means. The molten metal is poured into the mold continuously in order for a constant level of liquid to be maintained in the mold. The mold is cooled, usually by water sprays or jets, and the metal that emerges from the mold is often also subjected to cooling. The most important characteristic of this process is that a steady state is maintained. This corresponds to a constant shape of the interface and a solidification rate that depends only on the position relative to the surface and not on time (Fig. 3).

In our work the vibrations are produced by an electromagnetic vibrational method and without any mechanical contact with the melt.[7] The principle consists of the simultaneous application of a stationary magnetic field B'_0 and of a variable magnetic field B, in the vicinity of the molten metal during the course of solidification (Figs. 2 and 4); these magnetic fields are nearly parallel to the vertical axis of the ingot. The stationary magnetic field is generated by at least one coil, which is supplied with direct current. The variable magnetic field is created by another inductor, which is of similar geometry, supplied with an alternating current of a frequency N. Under the effect of the periodic current, the inductor generates in the melt a variable field B, which, in turn, gives rise to an induced current of density J (Fig. 4).

The combined action of the colinear fields generates vibrations in the metal pool, which are of a double origin. The interaction of

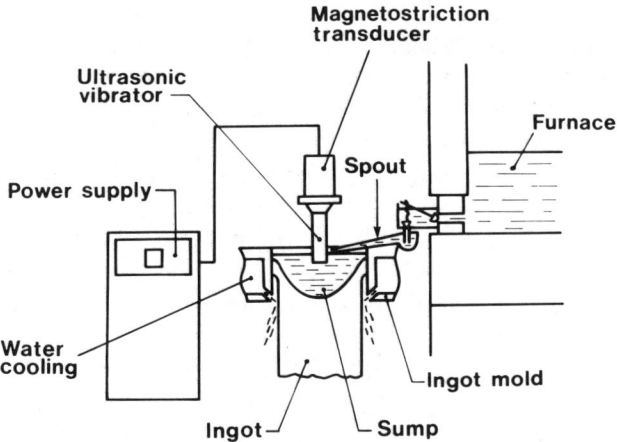

Fig. 1 Schematic diagram of continuous casting of aluminum alloy ingots with ultrasonic treatment of the molten pool.

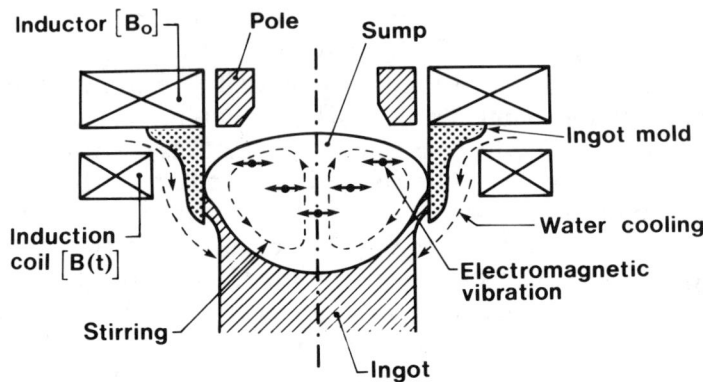

Fig. 2 Schematic diagram of continuous casting of aluminum alloy ingots using an electromagnetic vibrational method.

B'_0 and J engenders a vibrating force of frequency N, and the J x B force consists of a time-independent component and an oscillatory component of frequency 2N.

Experimental Apparatus

Experiments were made on ingots of circular and rectangular cross sections, and the main features of the apparatus are schematically presented in Fig. 2.

First, 320-mm-diam billets of 1085 and 2214 aluminum alloys were cast at a drop rate of 50 mm·mn^{-1} The induction electromagnetic field was generated by an inductor trasversed by an electric current of 50 Hz frequency and located below the top of the ingot mold; the magnetizing force was modulated at will, from O to 12,600 Ampere-turns (At), by an autotransformer. An externally imposed stationary magnetic field B'_0, mainly parallel to the gravitational force field vector, was created by a coil situated above the ingot mold and excited by a magnetizing force likely to vary between 0 and 50,000 At, whereas the corresponding magnetic field strength B'_0 was included between 0 and 0.2 T. Various inductors used for these trials are presented in Fig. 5.

Then slabs of aluminum alloys 2214 of 700 x 200 mm cross sections were cast with a drop rate of 50 mm·s^{-1} The device was similar to that adopted for the billets. The variable and stationary magnetic fields were being 11,000 and 60,000 At, respectively.

The set of probes for electric current density, magnetic field, and velocity measurements has already been described.[8-10]

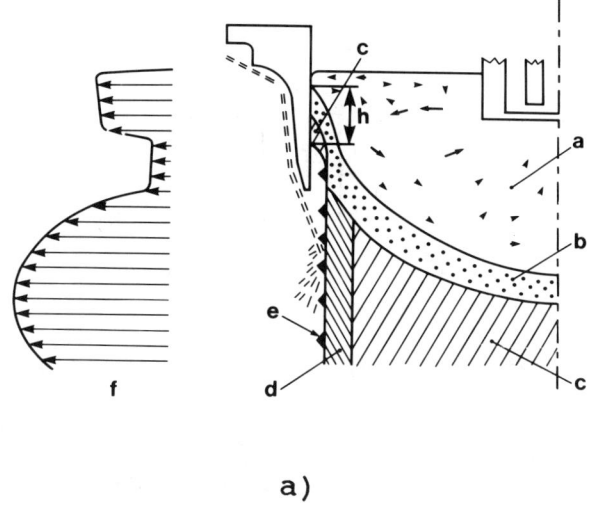

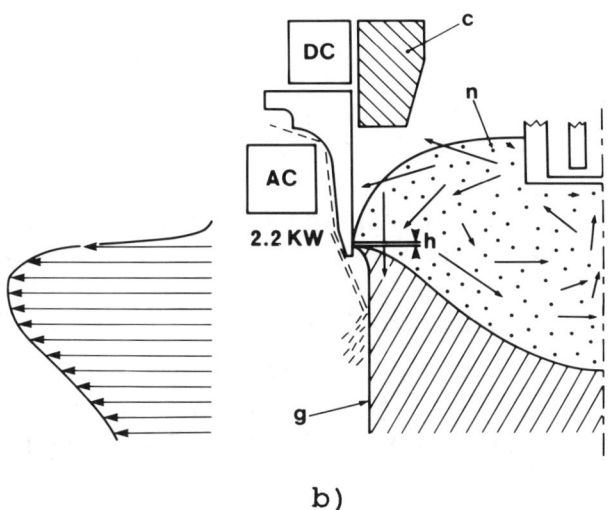

Fig. 3 Evolution of the casting with the total ac power input (2214 aluminum alloy billet of 320 mm diam.; a) conventional casting P = 0; b) electromagnetic process, P = 2.2 kW. a: liquid metal; b: pasty zone; c: solidified metal; d: segregation zone; e: exudations; f: estimated heat-flux profile; g: smooth surface; h: height of solidifying metal in contact with the ingot mold; i: inductors; n: nuclei.

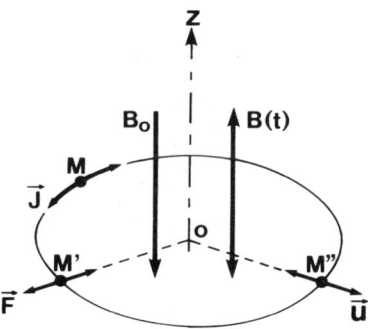

Fig. 4 Principle of production of electromagnetic vibrations.

Electromagnetic and Fluid Flow Phenomena

Electromagnetic Parameters Measurements

For sinusoidally varying fields one can assume[10] that the values at a given point and at a given time of the axial component of the current density are approximately given by

$$J_{(t)} = J_0 \, e^{-r/\delta} \cos|\omega t + r/\delta|$$

$$B_{(t)} = B_0 \, e^{-r/\delta} \cos|\omega t + r/\delta + \phi(J,B)|$$

were J_0 and B_0 are peak values on the wall, r is the distance along a radius with respect to the wall, and $\phi(J,B)$ is the phase difference between the periodic vectors **J** and **B**.

The results presented in the following corresponds to experimentation in the sump of an aluminum billet of radius R = 160 mm.

Figure 6 shows that both the parameters $B^*_z = B_z(r)/B_0$ and $J^* = J_z(r)/J_0$ follow the e^{-r/δ^*} exponential law. Hence the actual skin depth ($\delta^* = 45$ mm) is slightly different from the classical value corresponding to a plane electromagnetic wave penetrating into a plane conductor ($\delta = (2/\omega\sigma\mu)^{1/2} = 35.6$ mm) this slight discrepancy is explained by the importance of the fringe effect due to the predominance of the dimension of the diameter with respect to the bath and coil heights and also because the ratio $R/\delta = 3.5$ is not sufficiently high.

Figure 7 presents the variations of the axial component of the magnetic field B_z as a function of the vertical position z for a given value of the radius. It appears that B_z decreases as the depth z increases; this finding is readily understood by the fact that the

GRAIN REFINEMENT

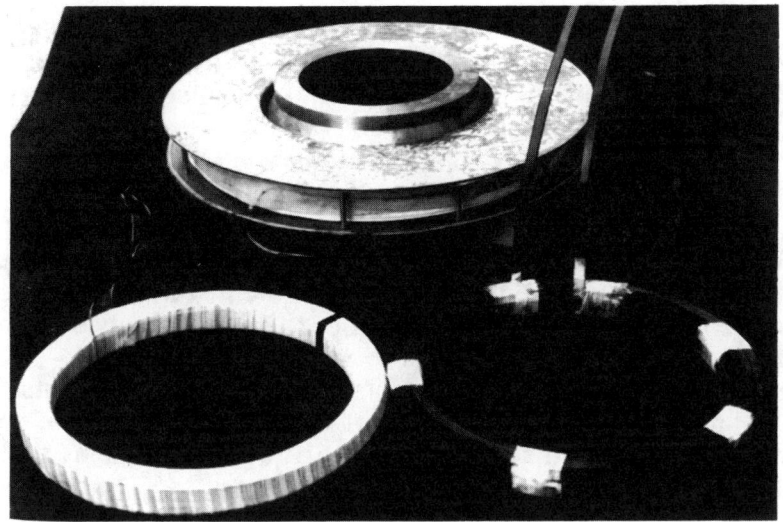

Fig. 5 Photograph of inductors for production of time-varying magnetic field (foreground) and of stationary magnetic field (background).

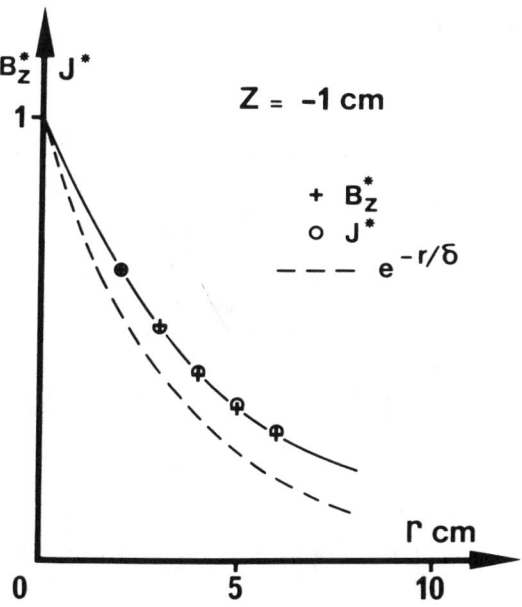

Fig. 6 Skin effect for the axial component of the magnetic field and the current density.

penetration of the electromagnetic wave in the melt is all the more subdued as the solidifying metal thickens. Furthermore, the skin depth is reduced in the solidified metal because the electrical conductivity σ is higher in this phase.

The distribution of B_Z along a radius are plotted in Fig. 8 for different values of the vertical position. The decreasing of the magnetic field inside the sump due to the exponential law of penetration and also to the shield effect caused by the solidified part of the metal is, once again, revealed. Complementary experiments show a similar evolution, with r and z, of the electric current density.

From methodical measurements of the current density inside a vertical and radial cross section of the sump, we reach, after graphical integration, the electric current flowing through the molten metal:

$$I = \iint J(r,z) \, dr \cdot dz$$

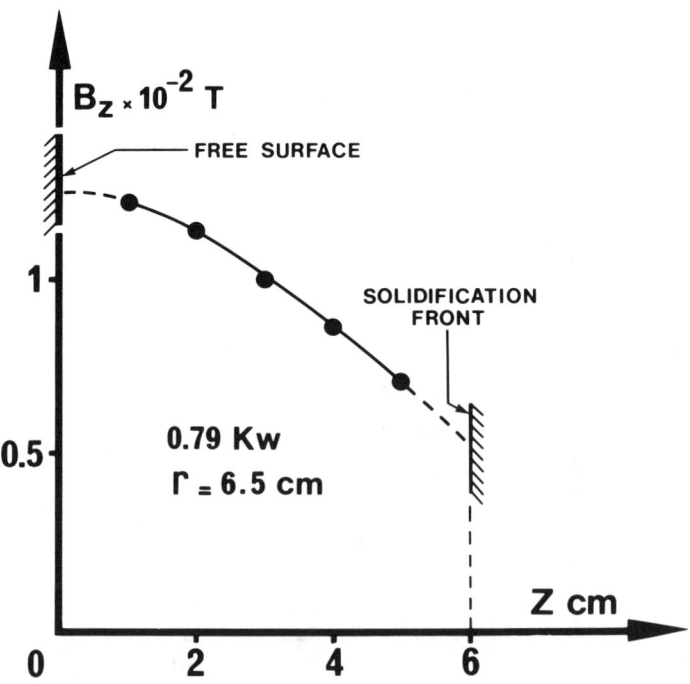

Fig. 7   Distribution of the axial component of the magnetic field as a function of the vertical position.

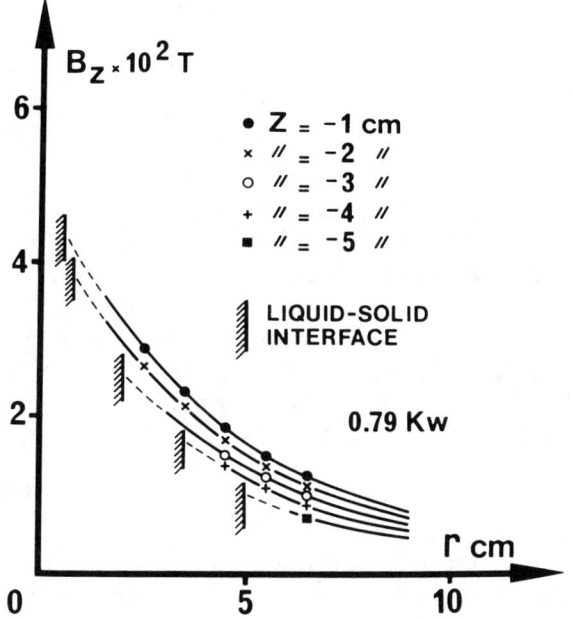

Fig. 8 Magnetic field profile for different depths.

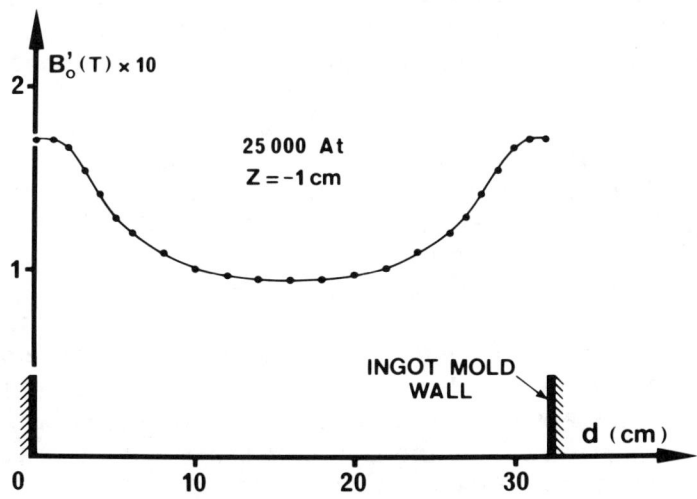

Fig. 9 Distribution of the vertical component of the stationary magnetic field along a billet diameter.

and the caloric power dissipated by the Joule effect within the aluminum pool :

$$W = \iiint J^2/\sigma \, dv$$

As an illustration, we find that a total power input $P = 0.79$ kW corresponds to $I = 1,400$ A and $W = 200$ W. The order of magnitude of I and W inside the solid part of the metal is analogous.

Furthermore, experiments shows that J, B, and I are proportional both to the magnetizing force and to the square root of the overall electric power dissipated P.

Figure 9 displays the evolution of the vertical component of the stationary magnetic field B'_0 along a diameter; inspection of this figure reveals that B'_0 is enhanced in the electromagnetic skin region. This effect is caused by the presence of an annular iron core, located above the melt, and fastened to the dc inductor (Figs. 2 and 5).

Electromagnetic Stirring

The effect of the induction electromagnetic field here is identical to that observed in the CREM process (Casting, Refining, ElectroMagnetic) [8-11]. The time mean electromagnetic body force (J x B) may be resolved into a radial component (principally irrotational) and a vertical component (primarily rotational). The potential forces, balanced by a pressure gradient, result in the formation of a convex surface meniscus, and the rotational forces are responsible for an electromagnetic stirring, similar to that encountered in a coreless induction furnace. As in the CREM process, the action of such a vigorous forced convection results in refining of the grain and a higher degree of homogeneity in crystallization.

Figure 3 depicts the behavior of a 2024 aluminum alloy billet of 320 mm diam. when the total ac power input (or the ac magnetizing force) increases. It appears that both the height of the dome-shaped free surface and the stirring intensity increase with the excitation while, at the same time, the height of contact h between the mold and the metal is gradually reduced.

In this new process the gradual enhancement of the magnetizing force allows a decrease in the level of the contact line between the liquid metal and the mold, until the metallic region in contact with the wall is practically reduced to a circular line (Fig. 3b). Figure 3 shows that the action of progressively lowering the contact line has a marked impact on the heat-flow distribution. Under these circumstances the ingot surface may become very smooth and the thickness of the segregated cortical layer tends to **zero**.

The mean velocity field has been experimentally explored in a half-vertical cross section of the sump, and the flow patterns, obtained for different values of the electric power input P, are schematically sketched in Fig. 3. Figure 3a corresponds to a conventional casting and shows the presence of two main vortices. The existence of these loops is mainly explained by a viscous friction phenomenon caused by the radial and horizontal pouring jet running out of the float. It should be mentioned that the velocities are low in the proximity of the mushy zone (of the order of 1 cm·s^{-1}) and the maximum velocities U (10 cm·s^{-1}) have been detected in the immediate vicinity of the dispenser outlet. When the power P increases (Fig. 3b) the relative importance of the viscous effect induced by the jet is progressively attenuated. The electromagnetically driven flow is now characterized by the occurrence of a main cell that fills the entire half cross section. Experiments show that the velocity, measured at a given point, is nearly proportional to the magnetizing force applied to the coil and to the square root of the total power P. At full power, the peak velocity can exceed 50 cm·s^{-1}; under this cricumstance the flow is obviously strongly turbulent. Experiments were carried out in the absence and presence of the stationary magnetic field B'$_0$, and it should be mentioned that the damping effect, due to B'$_0$, was practically insignificant for the high values of the ac magnetizing force.

Electromagnetic Vibrations

An example of penetration of the electromagnetic vibrating forces of 50 and 100 Hz frequency is showed in Fig. 10; it appears that the magnitude of the 50-Hz vibrations is largely predominant, particularly outside the electromagnetic skin depth, i.e., in the bulk liquid. Figure 11 depicts the evolution with time of the resultant of the fluctuating forces, measured at a given point (r = 2 cm, z = -1 cm); inspection of this figure once again reveals that the effects of the 50-Hz oscillating force prevails. This tendency is obviously increased inside the melt core.

An order-of-magnitude analysis of the vibration of 2N frequency has been developed by Hunt and Maxey, particularly inside the skin depth δ', corresponding to the electromagnetic forces of angular frequency 2ω (δ' = δ/2 ≈ 20 mm)[12]. According to these authors, the magnitude of the electromagnetically driven fluctuating velocity of 100-Hz frequency (Fig. 4) is approximately given at full power (B$_0$ = 0.16 T) by

$$u = B_0^2/\mu\sigma\omega L = 0.24 \text{ cm·s}^{-1}$$

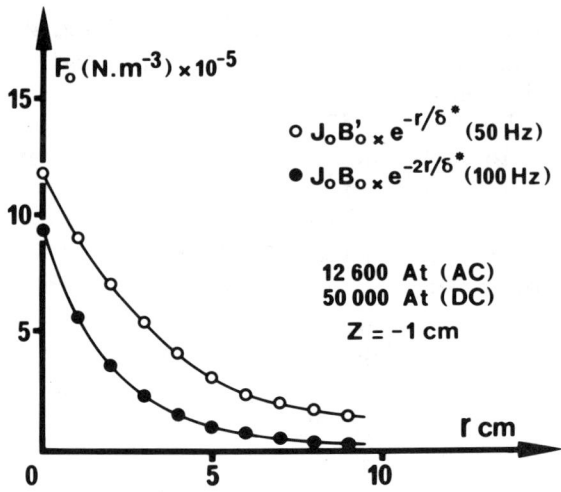

Fig. 10 Penetration of the electromagnetic vibrating forces of N and 2N frequency.

were L denotes the average sump depth, and the magnitude of the oscillating motion is given by

$$a = u/2\omega = 0.4 \text{ mm}$$

and the acceleration is given by

$$\gamma = 4a\omega^2 = 150 \text{ g}$$

Moreover, methodical measurements allow one to obtain a rough estimate of the peak of the oscillating electromagnetic pressure, which is of the order of 0.03 bar.

On the other hand, subsidiary measurements allowed the determination of the maximum of the fluctuating pressure of 50-Hz frequency, which was of the order of 0.16 bar, and the corresponding acceleration was more than 100 g.

It should be mentioned that some test using a cold laboratory model, have been performed. In these experiments the frequency of the electric current flowing through the induction coil was adjusted in order to reach the establishment of a resonance mode inside the liquid metal pool (mercury). This physical model was constituted by a 150-mm-dia billet made from stainless steel, in which a cavity of an identical shape to the liquid-solid interface was made, and the fluctuating pressure was measured by means of a piezoelectric sensor (Fig. 12). This study showed that, because of the conical shape of the cavity containing mercury, the resonance, reached for a frequency of the order of 30,000 Hz, was not sharp. The magnitude

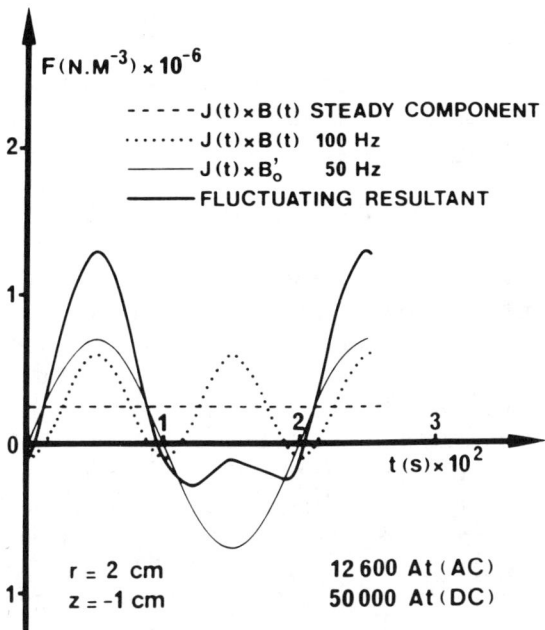

Fig. 11 Time-varying electromagnetic forces determined at a given point.

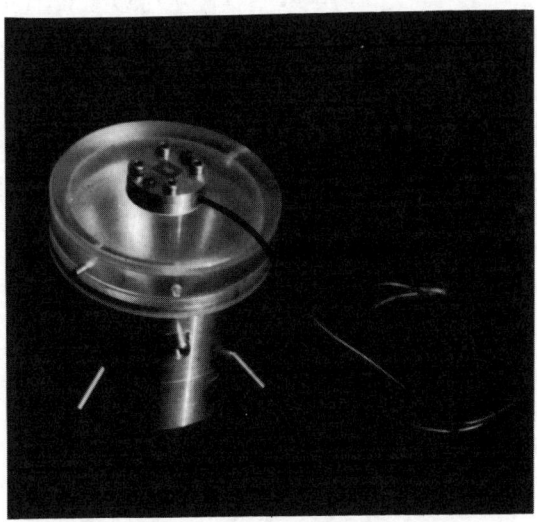

Fig. 12 A physical model, equipped with a piezoelectric sensor, for the study of the resonance phenomenon, produced by magnetoacoustic waves during casting.

of the pressure was only multiplied by about three with respect to the forced vibration mode. Thus, the difficulty in bringing this process into operation, the high cost of the corresponding equipment and, overall, its poor performance have led us to renounce to this way.

Metallurgical Results and Discussion

The macrostructures of the materials investigated have been revealed in order to provide information on variation in structure, such as grain size and columnar and equiaxed crystals. To this end, the ingot were sectioned, mechanically polished and then immersed into specific etching solutions.

Figure 13 displays macrographies yielded from 1085 aluminum alloy billets of 320 mm diam. Because of its relatively high degree of purity (99.85 %), this alloy, which is characterized by a very narrow freezing range (i.e., interval between liquidus and solidus temperature), is among the most difficult to grain refine. In agreement with previous works, it appears that the macrostructure is coarsened by the application of a steady magnetic field B'_0 (Fig. 13b), and inspection of Figs. 13c and 13d reveals that the refinement is substantially improved both by the CREM process and the electromagnetic vibration.[13,14] However, the best efficiency of the vibrational method is clearly apparent here, and this superiority is confirmed by methodical experiments carried out on billets and slabs of 2214 aluminum alloy.

Among the possible mechanisms related to the presence of sonic or ultrasonic vibrations, the cavitation phenomenon is well known for the erosion provoked on the solid walls of hydraulic facilities, and an interesting discusion, dealing with the conditions for cavitation in aluminum alloys, has been reported by Lillicrap.[6] According to Lillicrap, the wall pressure fluctuates at the same frequency as the electromagnetic force, and when a critical threshold is reached, the wall pressure can temporarily drop below the equilibrium vapor pressure of the most volatile constituent in the molten alloy. At this point, gas cavities can nucleate on the melt-solid interface and then shrink when the fluctuating forces decrease. When the cavities collapse, the impact of the liquid metal with the interface is likely to produce very high pressure (in the order of hundreds of atmospheres) over a small local region.

During the course of solidification, under the action of the strong stresses produced by such pressures, the dendrites arms are broken and uniformily distributed throughout the melt. The broken particles function as nuclei in the sump, and this increasing

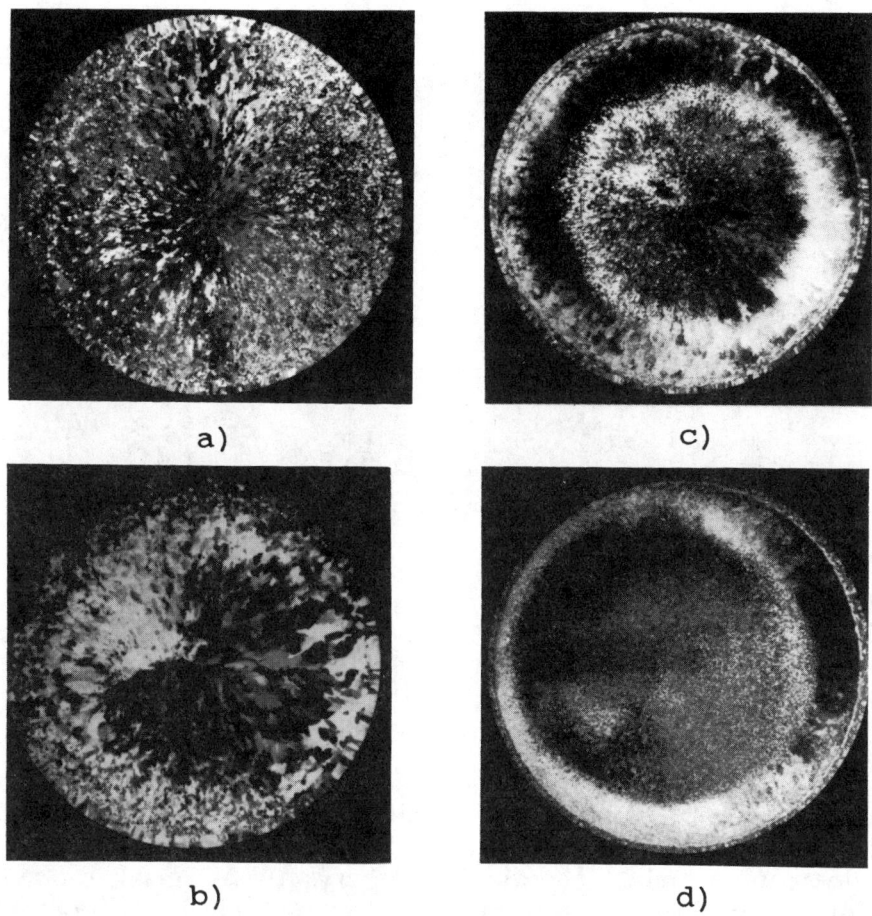

Fig. 13 Macrostructures of 1085 aluminum alloy billets of 320 mm diam: a) conventional casting; b) B'o, dc magnetizing force of 30,000 at; c) CREM process, B(t), ac magnetizing force of 5600 At; d) vibrational method: B'o, dc magnetizing force of 10,000 At; B(t), ac magnetizing force of 5600 At.

nucleation results in the production of a fine-grained equiaxed structure.

After Lillicrap, for a liquid metal temperature of 700°C the equilibrium hydrogen pressure corresponding to a level of 0.4 ppm is 0.3 bar, i.e., more than the peak pressure of 0.16 bar reached in the present investigation; thus, it seems that grain refinement does not occur here by cavitation. However, it would be expected that extra nuclei, submitted to the high accelerations already pointed out, could be produced by the detachment of the dendrite arms from the solidification front. These suspended nuclei are carried away by a

Fig. 14 Surface aspect of a 2214 aluminum alloy billets of 320 mm dia; conventional casting (left-hand side) and vibrational method (right-hand side), ac magnetizing force of 5600 At, dc magnetizing force of 10,000 At.

vigorous forced convection, which is mainly generated by the induction electromagnetic field, and dispersed in a slightly undercooled melt. Accordingly, the crystallization takes place simultaneously in most of the sump around a number of floating nuclei, and such a situation leads to a decided grain refinement.

Conclusions

Sonic vibrations were produced during molten metal castings by the simultaneous application of stationary and variable magnetic fields. It should be emphasized that, by this means, the intensity of vibration is modulated at will, without any mechanical contact with the melt and hence without risk of pollution.

Local measurement techniques for velocity, induced current, and magnetic field have been applied to the study of the intensity of vibrations and of electromagnetic and fluid flow phenomena inside the sump of continuously cast ingots of aluminum alloys. This investigation showed that the influence of the vibrations of N frequency was predominant.

To summarize, this technique has the main properties as the CREM process, i.e., deformation of the free surface (Fig. 3b), which results in an improvement of the surface aspect of the ingot (Fig. 14), and the presence of a strong forced convection, which promotes the growing of a fine equiaxed structure. However, due to the additional effect of sonic waves, the grain refinement obtained by this new vibrational method has proven to be, in any case, better than that achieved by the CREM process acting alone.

Acknowledgment

The author would like to thank the personnel management of the Aluminum Research Center of the Pechiney Society for their technical assistance with the work at the casthouse of Voreppe.

References

[1] Chalmers, B., "Principle of Solidification," Krieger, Malabar, India, 1982, pp. 253-281.

[2] Flemings, M. C., Solidification Processing, McGraw-Hill, New York, 1974, pp. 252-258.

[3] Vivès, C., Forest, B., and Riquet, J-P., "Method for Regulating the level of the Contact Line of the Free Surface with the Ingot Mold in a Continuous Vertical Casting." US Patent n° 511,398, 1986.

[4] Campbell, J., "Effects of Vibration During Solidification," International Metals Reviews, Vol. 2, Jan. 1981, pp. 71-108.

[5] Goel, D. B., Shunkla, D.P., Pandey, P. C., "Effect of Vibration During Solidification on Grain Refinement in Aluminium Alloys," Transaction of the Indian Institute of Metals, Vol. 33, March 1980, pp. 196-199.

[6] Lillicrap, D. C., "Refractory Erosion in High Powered Channel Inductors," Electrowärme International, Vol. 44, Jan. 1986, pp. 116-122.

[7] Vivès, C., "Process for Casting Metals in Which Magnetic Fields are Employed," U.S. Patent n° 4,530,404, 1984.

[8] Ricou, R., and Vivès, C., "Local Velocity and Mass Transfer Measurements in Molten Metals Using an Incorporated Magnet Probe," International Journal of Heat and Mass Transfer, Vol. 25, Oct. 1982, pp. 1579-1588.

[9] Lee, H. C., Evans, J. W., and Vivès, C., "Velocity Measurements in Wood's Metal Using an Incorporated Magnet Probe," Metallurgical Transactions, Vol. 7, Dec. 1984, pp. 734-736.

[10] Vivès, C., and Ricou R., "Experimental Study of Continuous Electromagnetic Casting of Aluminum Alloys," Metallurgical Transactions, Vol. 16, June 1985, pp.377-384.

[11] Vivès, C., and Forest, B.,."CREM : A New Casting Process, Part I: Fundamental Aspect," Light Metals, 1987, pp. 769-778.

[12] Hunt, J. C. R., "Estimating Velocities in Electromagnetically Driven Flows," Proceedings of 2nd Beer-Sheva Seminar on MHD Flows and Turbulence,", edited by H. Branover and A. Yakhot, Israel University Press, Jerusalem, 1980, pp. 249-269.

[13] Uhlmann, D. R., Seward, T. P., and Chalmers, B., "The Effects of Magnetic Fields on the Structure of Metal Alloy Castings," Transaction of the Metallurgical Society,", Vol. 236, June 1986, pp. 527-531.

[14] Vivès, C., and Perry, C., "Effects of Magnetically Damped Convection During the Controlled Solidification of Metals and Alloys," International Journal of Heat and Mass Transfer", Vol. 30, March 1987, pp. 479-496.

Dendrite Growth of Solidifying Metal in DC Magnetic Field

Eiichi Takeuchi,* Ikuto Miyoshino,† Kouichi Takeda,‡
and Yutaka Kishida§
Nippon Steel Corporation, Chiba 299-12, Japan

Abstract

Directional effect of magnetic flux on solidification of Pb-10%Sn alloy has been studied in a high magnetic field. A superconducting magnet was used to generate a magnetic field of 4.0 Tesla at maximum with the uniformity of 1% in which Sn-10%Pb alloy was poured into the mold designed to solidify the alloy unidirectionally (in a horizontal direction). Typical magnetic conditions for the solidification of alloy were as follows: 1) without a magnetic field for reference, 2) with a magnetic field perpendicular to the solidifying direction, and 3) with a magnetic field parallel to the solidifying direction.

The solidified structure of the alloy represents an equiaxed-dendrite structure without a magnetic field. This equiaxed structure changes to a columnar-dendrite structure in the parallel field of several tenths Tesla, while the equiaxed structure remains even at 4 Tesla in a case of with a perpendicular field.

Theoretical analysis predicts that the thermal convection during metal solidification can be suppressed when a DC magnetic field is imposed parallel to a plane of circulating convection, while no motion can be suppressed in a magnetic field perpendicular to the plane. A strong thermal convection during solidification of the alloy homogenizes temperature distributions of the metal pool to form nuclei of crystal. The convection also disperses them in the pool, which leads to the formation of equiaxed structure. Suppression of convection enhances the development of columnar structure. Thus, the dependency of the cast morphology on the direction of the magnetic flux has been elucidated as the anisotropic effect of magnetohydrodynamics on liquid alloy.

Copyright © 1992 by the American Institute of Aeronautics and Astronautics, Inc. All rights reserved.
 * Senior Researcher, Process Technology Research Laboratories.
 † Senior Researcher, Oita Works' Research & Development Laboratories.
 ‡ Chief Researcher, Advanced Materials & Technology Research Laboratories.
 § Senior Researcher, Advanced Materials & Technology Research Laboratories.

Introduction

A moving, electrically conducting fluid in a DC magnetic field induces electric currents, resulting in Lorentz force on the element of the liquid to suppress its own motion; several metallurgists have been enchanted with this effect in terms of controlling the solidification of metal.

Uhlmann et al.[1] first experimentally examined the influence of a DC magnetic field on the solidification structure of Cu-2%Al alloy. According to their results, the cast structure with a magnetic field of 0.2T (Tesla) was completely columnar, while the structure without a magnetic field was equiaxed. This change of morphology was understood as the result of suppression of the convection; less convection leads to the decrease of the remelting of dendrite branches and also to prevention of the nuclei from being carried to bulk liquid.

Utech and Flemings[2] experimented with crystal growth of tellurium-doped In-Sb in a DC magnetic field. They reported that the solute band, which usually appears in a crystal without a magnetic field, was eliminated in the presence of the magnetic field. This phenomenon was elucidated by the suppression of temperature fluctuations accompanied by a turbulent thermal convection.

Vivès and Perry[3] reported, according to their experimental results of unidirectional solidification of Sn-Al alloy in an annular mold, that the effect of a magnetic field was characterized by both the reduction of release of metal superheat and the acceleration of the solidification rate.

All these metallurgical studies have simply demonstrated the effect of electromagnetic dumping of thermal convection on the morphorogy of solidified structure in a relatively low magnetic field.

This study has been undertaken to clarify MHD effect on the solidification phenomena of liquid metal in a high magnetic field generated with a superconducting magnet.

Apparatus and Procedure for the Experimental Solidification

The experimental arrangement is shown in Figure 1. A solenoidal superconducting magnet was used to generate a magnetic field of 4.0T at maximum, with the uniformity of 1% in the central region of the room temperature bore. Figure 2 illustrates the configuration of the mold, which was designed to solidify the metal unidirectionally. One of the vertical walls was made of copper and cooled by water at 20°C. The other three walls and a bottom plate were made of a thermally insulating material of asbestos covered by fluorocarbon polymer film. The composition of the alloy used in the experiments is Sn-10%Pb, the properties of which are listed in Table 1. The alloy was superheated to a temperature of 260°C and then poured into the mold outside the magnet. Within 10 s, the mold was inserted to the center of the bore of the magnet. Typical magnetic conditions for the solidification of metal are as follows:

1) without a magnetic field for reference,

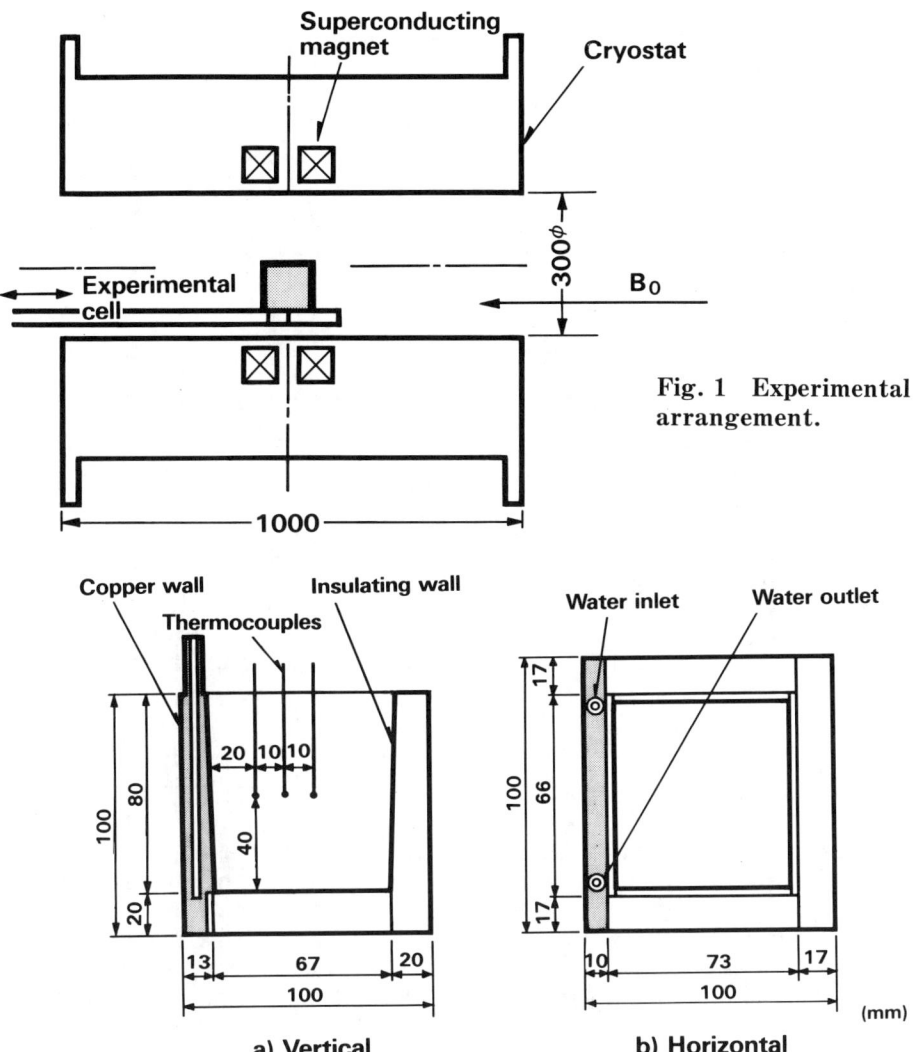

Fig. 1 Experimental arrangement.

a) Vertical   b) Horizontal

Fig. 2 Schematic view of the mold.

2) with a magnetic field perpendicular to the solidifying direction, and
3) with a magnetic field parallel to the solidifying direction.

The direction of the field was changed by turning the mold around the vertical axis. The variations of temperature were measured by three chromel-alumel thermocouples which were located at 40mm in depth, and at 20, 30, and 40mm horizontally apart from the copper cooling wall, respectively. The cast samples were sectioned in the vertical plane parallel to the solidifying direction to reveal the morphology of cast structure. The sections so obtained were machined flat, polished, and etched with Nital and Oberhoffer's reagents.

Table 1 Properties of Sn-10%Pb alloy

Liquidus	215°C
Solidus	183°C
Specific heat (liquid)	2.3×10^2 J·kg/K
Latent heat	3.8×10^4 J/kg
Coefficient of thermal expansion	1.02×10^4 1/K
Thermal conductivity (liquid)	21.0 W/m·K
Thermal conductivity (solid)	50.2 W/m·K
Density	6.97×10^3 kg/m^3
Kinetic viscosity	3.16×10^{-7} m^2/s
Electric conductivity	2.08×10^6 S/m

Experimental Results

Solidification Structure
Castings without a Magnetic Field

Macrostructures of solidified ingots without a magnetic field are shown in Figures 3a and 4a. With a loss of temperature during the pouring of the melt into the mold, the melt's initial superheat in the mold did not exceed 10°C. As expected, in the solidification under the condition of low superheat, fine equiaxed structure was observed almost in all volume of the ingot except for a narrow chill zone just in front of the cooling plate. Microstructure of the equiaxed grains is shown in Figure 5a.

Castings with a Perpendicular Magnetic Field

Figure 3 shows macrostructures of the ingots in the presence of various intensities of the magnetic field perpendicular to the solidifying direction. The macrostructure remained equiaxed with the magnetic field at less than 2T. Columnar structure partially appeared among equiaxed grains above 2T, in which the size of equiaxed grains was larger than without a magnetic field. Even at 4T, the equiaxed structure was not entirely replaced by the columnar structure.

Castings with a Parallel Magnetic Field

Figure 4 shows macrostructures in the various magnetic fields parallel to the solidifying direction. The morphology of the solidified structure is columnar and stable compared with that of the cast structure in a perpendicular magnetic field. Figure 6 indicates the variation of the volume ratio of columnar structure in a low magnetic field. Columnar structure began appearing at intensities above 0.2T and completely supplanted equiaxed crystals at 0.7T or more. Figure 5b shows the microstructure of columnar dendrites, which grow in the direction of heat flux (the direction of magnetic flux).

Figure 5c represents the transition of solidification structure from equiaxed dendrite to columnar dendrite, in which the 4T parallel magnetic field was imposed 30 s after starting solidification. The

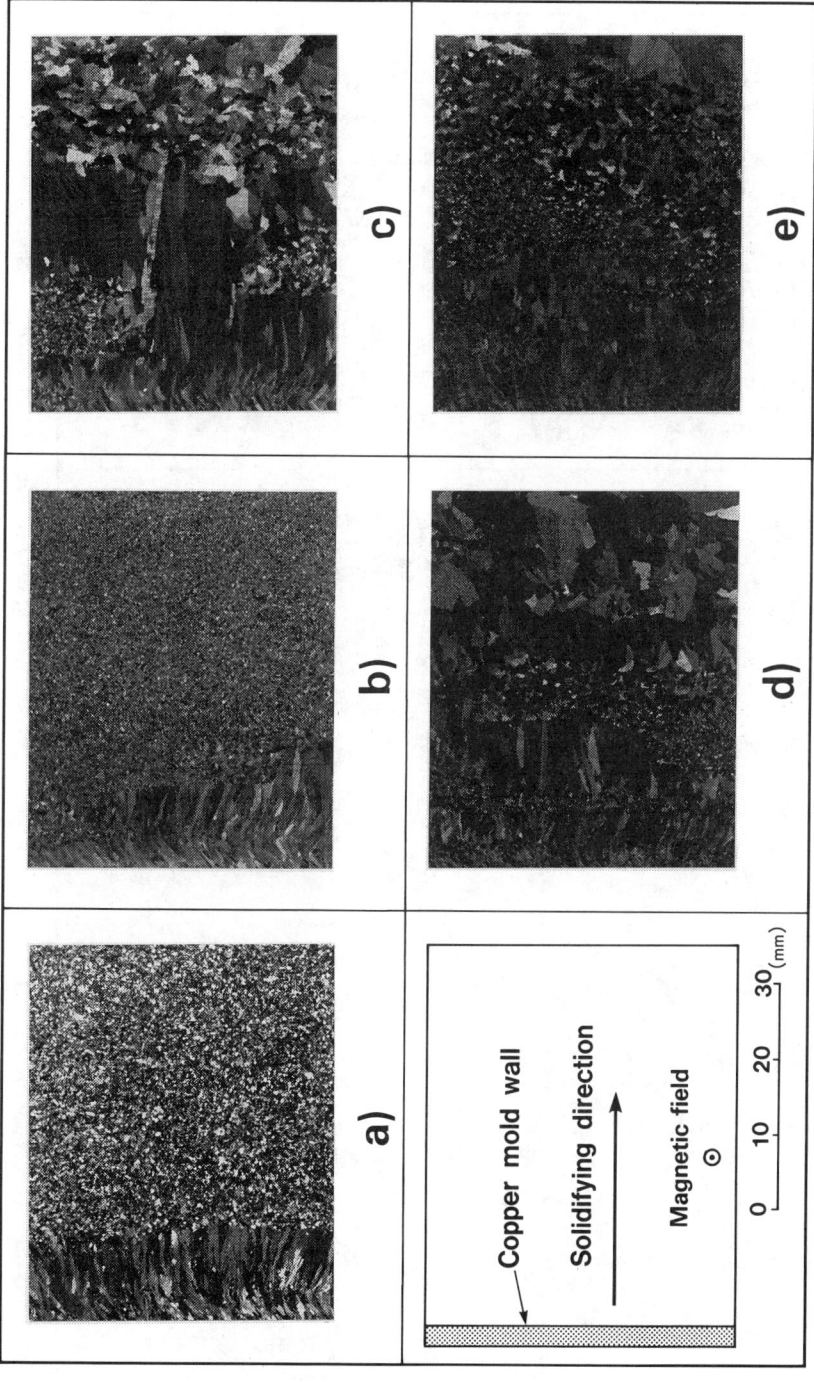

Fig. 3 Macrostructure of Sn–Pb alloy solidified under the DC magnetic flux perpendicular to the solidifying direction: (a) 0T, (b) 1T, (c) 2T, (d) 3T, (e) 4T.

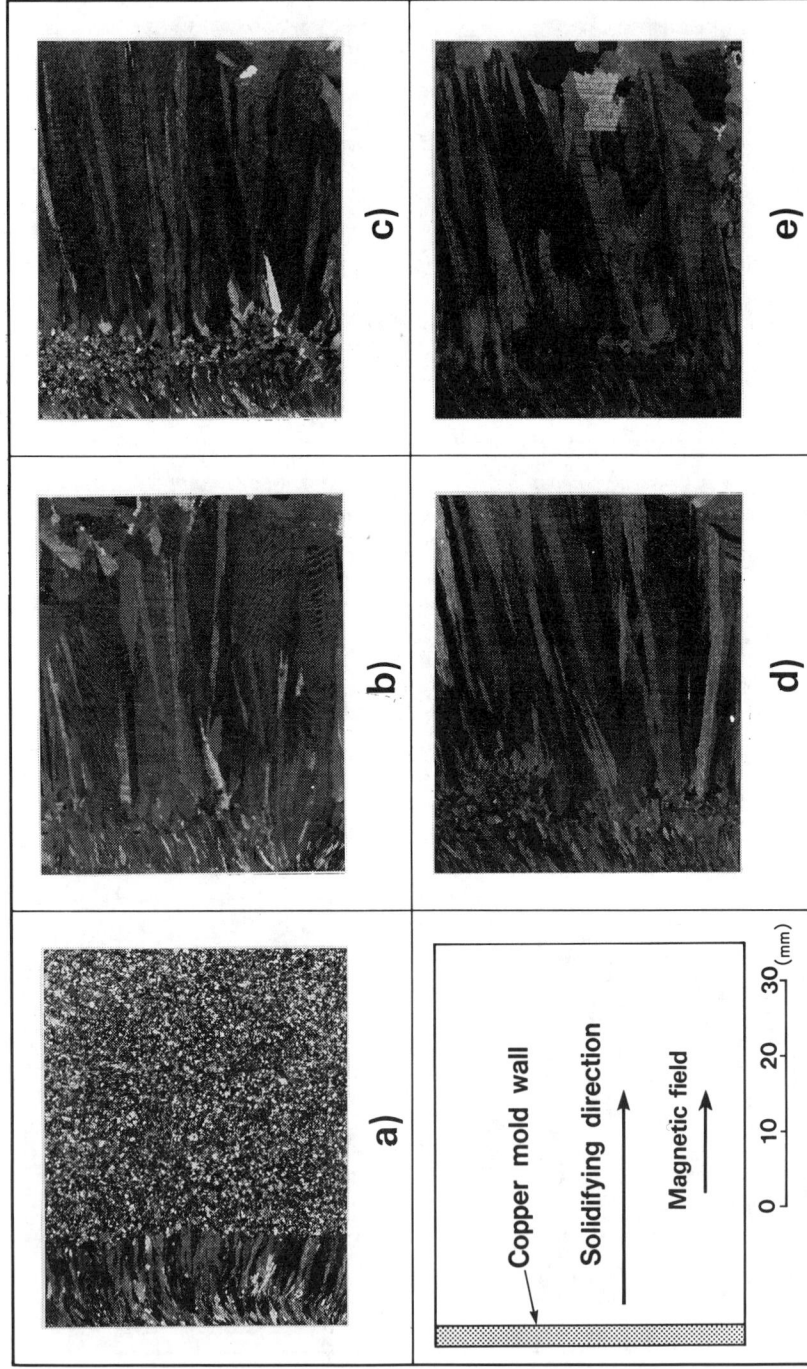

Fig. 4 Macrostructure of Sn–Pb alloy solidified under the DC magnetic flux parallel to the solidifying direction: (a) 0T, (b) 1T, (c) 2T, (d) 3T, (e) 4T.

DENDRITE GROWTH

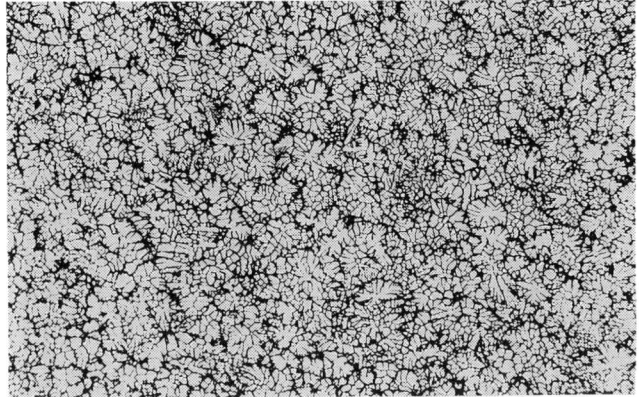

a) Equiaxed structure without a magnetic field.

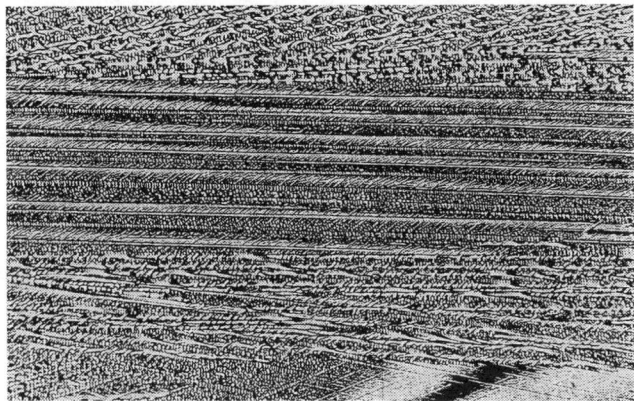

b) Columnar structure at a parallel magnetic field of 4T.

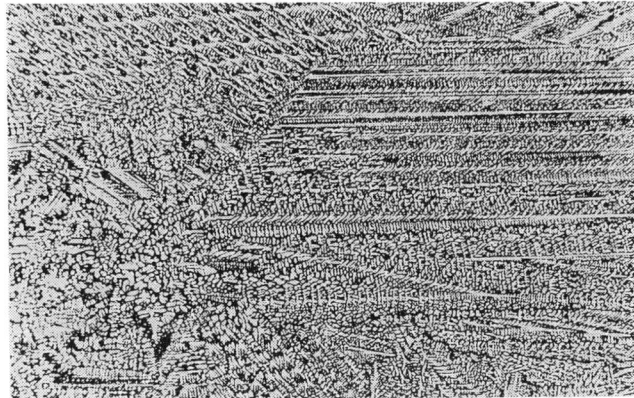

c) Transition of structure from equiaxed to columnar. A parallel magnetic field of 4T was imposed 30s after starting of solification.

Fig. 5 Microstructure of Sn-Pb alloy solidified under the DC magnetic field.

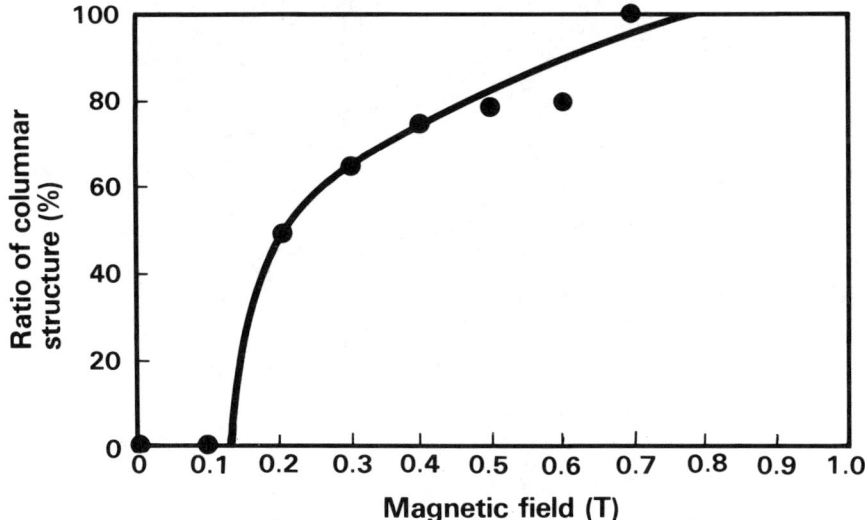

Fig. 6 Change of the columnar structure ratio with the magnetic intensity.

primary dendrites as well as the secondary dendrites of the nucleated equiaxed crystal, which oriented to the direction of imposed magnetic flux, preferably develop to form the columnar structure with a parallel magnetic field.

Heat Transfer

Figures 7a, 7b and 7c show the temperature variation of the alloy without a magnetic field, with a perpendicular magnetic field of 4T, and with a parallel magnetic field of 4T, respectively. Three curves in each figure were measured by thermocouples located at different points in the metal pool as shown in Figure 2. On a time axis of these figures, zero corresponds to the time when the temperature of the alloy becomes 225 °C at the point 20mm from the cooling wall. Figure 7a reveals that, in the case of no magnetic field, the melt at three different positions reached the liquidus temperature almost simultaneously. Temperature fluctuation of liquid metal was observed, which means turbulent flow developed in the metal pool due to thermal convection.

In the presence of a magnetic field, the time to reach the liquidus temperature was affected not only by the intensity but also by the direction of a magnetic field, as shown in Figure 8. Note from this figure that the magnetic field reduces the release of superheat and the magnitude of temperature fluctuation. The parallel magnetic field has a stronger effect on these phenomena than the perpendicular magnetic field.

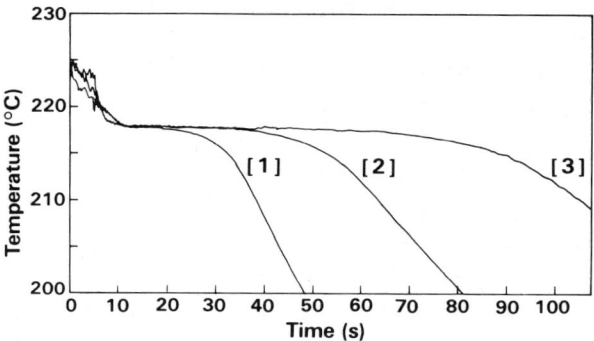

a) Without a magnetic field.

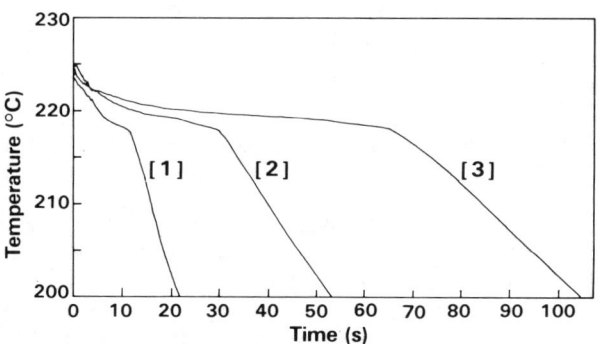

b) With a magnetic field perpendicular to the solidifying direction (4T).

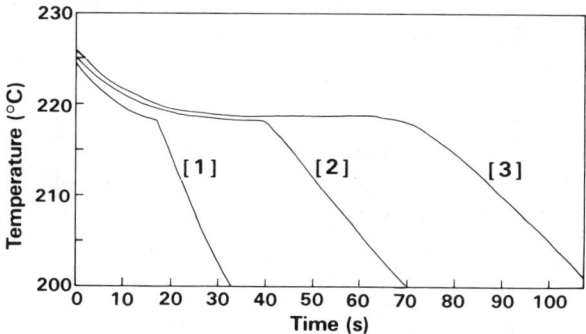

c) With a magnetic field parallel to the solidifying direction (4T).

Fig. 7 Changes of local temperature in the solidifying alloy at 20mm [1], 30mm [2], and 40mm [3] from the cooling plate.

Discussion

Solidification without a Magnetic Field

The magnitude of velocity of thermal convection W can be estimated by using Rayleigh number Ra and Prandtl number Pr[4]:

$$W = \alpha/L \, (RaPr)^{1/2} \quad (1)$$

where α = thermal diffusivity
L = the characteristic length of the mold.
Ra is defined as

$$Ra = \beta g \, \Delta T L^3 / \alpha \nu \quad (2)$$

where g = acceleration of gravity
ΔT = the characteristic temperature difference
β = cubic expansion coefficient
ν = kinematic viscosity of the melt.

Ra and W are estimated as follows in the case of $L = 70$mm and $\Delta T = 30°C$:

$$Ra = 2 \times 10^5$$
$$W = 1 \times 10^{-2} \text{m/s}$$

According to the diagram of Krishnamurit,[5] the convection with this value of Ra is turbulence. The Reynolds number Re is estimated to be 3×10^3, which also implies that the induced thermal convection is turbulence. That is confirmed by the temperature fluctuation shown in Figure 7a. The efficient heat transfer due to turbulent convection makes the temperature distribution in the melt uniform.

Several mechanisms have been proposed to explain the formation of equiaxed structure.[6,7] Although our experiment could not distinguish the governing mechanism, it was reconfirmed that the strong and **turbulent** convection without a magnetic field is favorable for the formation of equiaxed structure:

1) Strong convection may lead to sufficient transport of nuclei from the solidification front to the center region of the mold.
2) Homogenization of temperature distributions may result in heterogeneous growth of equiaxed crystals.
3) Temperature fluctuations due to turbulence might enhance the opportunities generate nuclei.

Solidification with a Magnetic Field

In order to elucidate the effect of a magnetic field including anisotropy on the solidification phenomenon, MHD analysis is introduced. According to the theory of Chandrasekhar,[8] vortical movement of the fluid can be suppressed when a DC magnetic field is applied parallel to a plane of circulation; on the other hand, no motion is suppressed in a magnetic field perpendicular to the plane. Thompson[9] studied the effect of a horizontal magnetic field on the Bénerd convection, which occurs in the presence of the temperature difference between top and bottom plates. His analysis gives the critical Rayleigh number Ra^* to suppress the disturbance in terms of Hartmann number M and wave numbers of the fluctuations as

$$Ra^* = M^2\, l^2 L^2\, [1+s^2/(l^2+m^2)] \qquad (3)$$

where
- l = the horizontal wave numbers of the fluid motion parallel to the applied magnetic field
- m = the wave numbers perpendicular to the field
- s = vertical wave number of disturbances.

The Hartmann number is defined as

$$M = \sqrt{\frac{\sigma}{\rho \nu}}\, L\, B \qquad (4)$$

where σ is the electrical conductivity of the fluid. When the Rayleigh number is smaller than Ra^*, convective motion is suppressed. The critical Rayleigh number varies not only with the intensity of a magnetic field but also with the wave number, as is known from Eq. (3). If the wave number l is reduced to zero, Ra^* becomes zero. This means that a magnetic field has no effect on the movement perpendicular to the field. On the other hand, Ra has a large value for the short range fluctuation, so that the turbulent motion which contains many disturbances of large wave number can be easily suppressed by a magnetic field. Although the system in Thompson's analysis is rather different from ours, the physical concept of the magnetic effect on the fluid motion is almost the same. Therefore, his criteria conditions to inhibit the fluctuations may be applicable to our system for the rough estimation.

In the experiment of the casting with a parallel magnetic field to the solidifying direction, the convective motion occurs in a plane parallel to the field. According to the analysis mentioned above, the suppression of the convective motion of liquid can be elucidative. By increasing the intensity of a magnetic field, a dominant mechanism of heat transfer changes from turbulent convection to a laminar one and ultimately to conduction. Deceleration of the temperature decay of the superheat shown in Figure 8 is inferred to be due to the suppression of turbulent convection.

Transition of the solidified structure can be also explained as follows: suppression of the convection leads to decrease of nucleation at the solidification front and to insufficient transport of nuclei to the center region of the molten pool, and the large temperature gradient prevents the equiaxed crystals from growing. As a result, the equiaxed structure turns to a columnar one with an increasing magnetic field. In equiaxed crystal growth, solidification occurs almost at the same time over the melt, while in columnar crystal growth solidification front proceeds successively.

The critical magnetic field B^* for the transition of morphology of cast can be estimated from Eq. (3). In the present work, B^* is predicted to be approximately 0.1T assuming that the wave number is

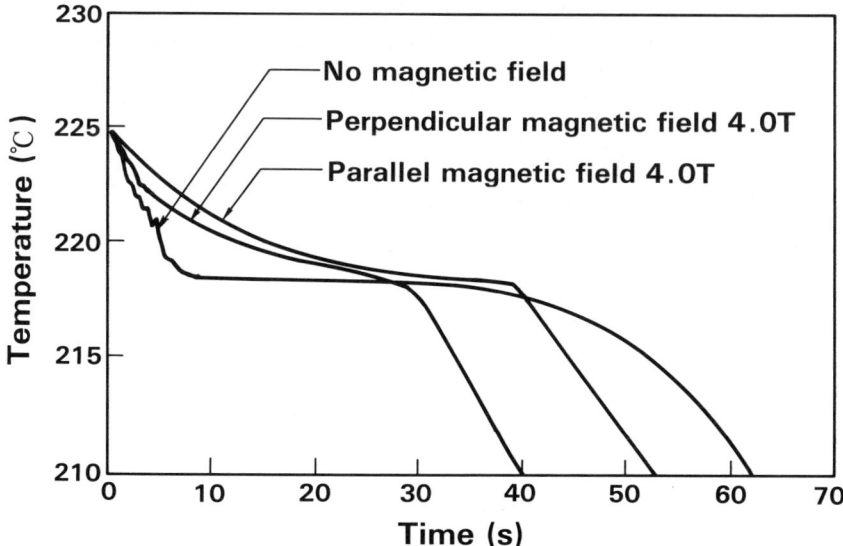

Fig. 8 Effect of magnetic field on the temperature variation at 30mm from the cooling plate.

$l/L = 1$. The critical magnetic field obtained experimentally is almost consistent with the predicted field.

On the other hand, as mentioned above, the magnetic field perpendicular to the solidifying direction hardly suppresses the convective motion. Hence, the convection, which is enough to transport heat and generates nuclei to the center of the molten pool, still exists. However, the convective motion hardly becomes turbulent at a high Hartmann number, where the dominant mechanism of transport becomes laminar. Thus, the deceleration of the temperature decrease of the melt in the case of perpendicular field lied between in no magnetic field and in parallel field. Temperature gradient along the solidifying direction in the perpendicular field is not steeper than in the parallel field. In this situation, both columnar and equiaxed crystals develop in the mold to form the mixed structure shown in Figure 3.

Conclusions

Experimental results show that the solidification is strongly affected not only by the intensity of the magnetic field but also by its direction. In the parellel field of several tenths Tesla, the equiaxed structural changes to a columnar one, whereas in the case of the perpendicular field, the equiaxed structure remains even at 4T.

Theoretical analysis predicts that the thermal convection can be suppressed when a DC magnetic field is imposed parallel to a plane of circulating convection (in the direction of solidification), while no motion can be suppressed in a magnetic field perpendicular to the plane. The critical magnetic flux density for the suppression of thermal convection predicted by the critical Rayleigh number is consistent with the metallurgical results; transition of alloy morphology from equiaxed to columnar structure.

From these findings, metallographical change was explained by these anisotropic magnetohydrodynamics effects on a transport mechanism during solidification, summarized as follows:

1) turbulent convection in no magnetic field,
2) laminar convection in a perpendicular field, and
3) suppressed convection in a parallel field.

Acknowledgment

The authors are grateful to Dr. M. C. Flemings, Professor at Massachusetts Institute of Technology, for his instructive suggestions.

References

[1] Uhlmann, D.R. et al., "The Effect of Magnetic Field on the Structure of Metal Alloy Castings."*Transactions of The Metallurgical Society of AIME*, Vol. 236, 1966, pp. 527–531.

[2] Utech, U.P. and Flemings, M.C., "Elimination of Solute Banding in Indium Antimonide Crystals by Growth in a Magnetic Field." *Journal of Applied Physics*, Vol. 5, 1966, pp. 2021–2024.

[3] Vivès, C., and Perry, C. "Effect of Magnetically Damped Convection during the Controlled Solidification of Metals and Alloys. "*International Journal of Heat and Mass Transfer*, Vol. 30, 1987, pp. 479–496.

[4] Bejon, A., "Convection Heat Transfer." Wily, New York, 1984.

[5] Krishnamurti, R., "On the Transition to Turbulent Convection, Part 2. The Transition to Time-dependent Flow."*Journal of Fluid Mechanics*, Vol. 42, 1970, pp. 309–320.

[6] Kurz, W., and Fisher, D.J., "Fundamentals of Solidification.", Trans. Tech Pub., New York, 1984.

[7] Flemings, M.C., "Solidification Processing", McGraw-Hill, New York, 1974.

[8] Chandrasekhar, S., "Hydrodynamics and Hydromagnetics Stability.", Clarendon Press, Oxford, England, 1961, p. 146.

[9] Thompson, W.B., "Thermal Convection in a Magnetic Field.", *Philosophical Magazine,* Vol. 42, 1951, pp. 1417–1432.

MHD Means for Affecting Hydrodynamics, Heat Transfer, and Mass Transfer at Single Crystal Melt Growth

Yu. M. Gelfgat* and L. A. Gorbunov†
Latvian Academy of Sciences, Riga-Salaspils, Latvia

Abstract

This paper presents the results of the effect of a steady magnetic field on hydrodynamics, heat transfer, and mass transfer in the process of growing silicon, germanium, indium antimonide single crystals from melt semiconductors. The investigation has been conducted by employing mathematical and physical simulation techniques of the relevant processes and by using natural experiments on growth facilities provided with the devices of electromagnetic effect on the alloy and crystallization front. The In-Ga-Sn eutectic served as a model alloy. The velocity structures have been measured by a conduction anemometer and optical fiber sensor, and high thermocouples have been used to measure the temperature characteristics. It has been found that in a situation in which the alloy had been affected by a steady magnetic field along with the classical MHD effects associated with the suppression of the averaged flow, various types of fluctuations of the velocity, temperature, and concentration there arises a number of additional MHD effects caused by rearrangement of flow structure, the formation of specific convective cells, and the origin of a MHD boundary and shift layer.

Introduction

Magnetohydrodynamic (MHD) means of creating forces by noncontact methods in liquid conducting media have been widely used in the production of volume semiconducting single crystals. The purpose of this method is to control convective transfer processes in the melt volume and on the crystallization front.[1-3] This is due to the sufficiently high conductivity of the semiconductor material melts ($\sigma = 10^4 \div 10^6 \Omega^{-1} m^{-1}$) and an obvious relationship between the hydrodynamics in melt and on interfaces and the conditions of the formation of single crystals and their quality.[1]

Copyright © 1992 by the American Institute of Aeronautics and Astronautics, Inc. All rights reserved.
*Professor and Head of Laboratory, Institute of Physics.
†Chief Scientist, Institute of Physics.

All types of MHD effects are based on the creation of volume electromagnetic forces in melt by various means as well as on the possibility of orienting them arbitrarily in space and easily adjusting their absolute values. Figure 1 presents different means of realizing the MHD effect in the single crystal melt growth. It should be emphasized that in all of the modifications given in Fig. 1 an additional energy introduced into melt cannot directly affect the process of crystallization since it is not comparable to the energy of the molecular forces. Both the influence and efficiency of the MHD means are determined solely by the change of motion, heat, and mass-transfer characteristics.

We will illustrate the preceding statements by using an example of a Czochralski single crystal puller when a growing crystal is rotated and pulled from the rotating crucible with melt (Fig. 2b), whereas the effect of MHD on the process takes place via a steady magnetic field that has one (axial) or two (axial and radial) components. As shown earlier[4] in a given situation various types of convective flows arise in the melt that are associated with the presence of the temperature gradient, differential rotation of crystal and crucible, vibrations, and other factors that determine the parameters of heat and mass transfer in liquid and on the phase interface. The application of a steady magnetic field to melt and crystallization front leads to a significant change of stationary and nonstationary

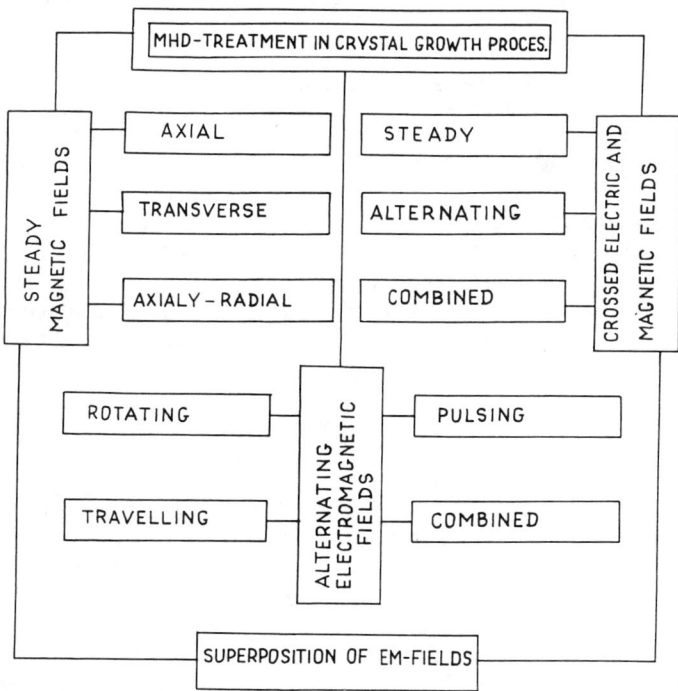

Fig. 1 Various types of MHD means of effecting the single crystal melt growth.

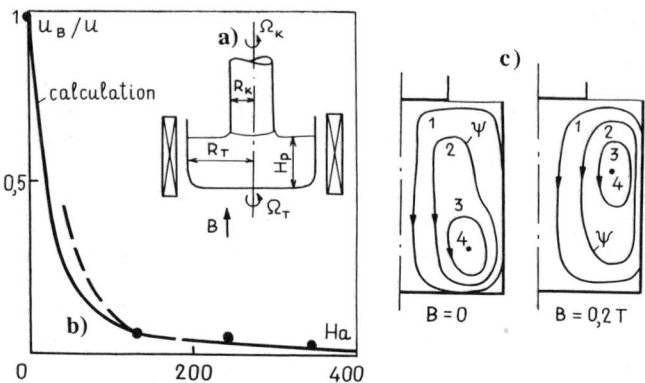

Fig. 2 Magnetic field effect on thermogravitational convection:
a) the Czochralski process diagram; b) change of flow velocity;
c) current functions at B = 0; 1, ψ = 2.3x10^{-8}; 2, 46.7x10^{-8};
3, 93.4x10^{-8}; 4, 140x10^{-8}; and at B = 0.2 T: 1, 23.4x10^{-10};
2, 140x10^{-10}; 3, 234x10^{-10}; 4, 310x10^{-10}.

hydrodynamical as well as heat- and mass-transfer characteristics of the process. This also allows MHD effects to be used for optimization of growth regimes. We discuss the results obtained in this field in the following.

Effect of a Steady Magnetic Field on Stationary Flows in Melt

The hydrodynamics of melt in the Czochralski process is characterized by four types of connective flows[4]: 1) thermogravitational (natural), 2) thermocapillary, 3) forced, and 4) thermoelectromagnetic.[5] The effect of the magnetic field on thermogravitational convection has been studied most extensively, both theoretically and experimentally. The pattern of convective flows has been found to change little (Fig. 2b) due to an axial magnetic field, whereas the intensity of liquid motion decreases (Fig. 2a) and is determined by the MHD interaction parameter:

$$\frac{u}{u_B} = \frac{Ha^2}{Re} = \frac{Ha^2}{\sqrt{Gr}} = N \qquad (1)$$

where u and u_B are the flow velocity at B = 0 and B ≠ 0; and Ha, Re and Gr are the Hartmann, Reynolds, and Grashoff numbers, respectively. A decrease of thermoconvection intensity obviously leads to the decrease of convective heat and mass transfer, which at B = 0.2T might decrease by ~25 times under actual conditions of silicon single crystal growth.

Thermocapillary convection in the Czocharalski process is caused by the gradient of surface tension on the free melt surface due to the radial temperature gradient. It has not been studied sufficiently even without an account of the MHD effect and thus remains a topic for further investigation. According to the results presented in Ref. 1, in a number of cases the intensity of thermocapillary

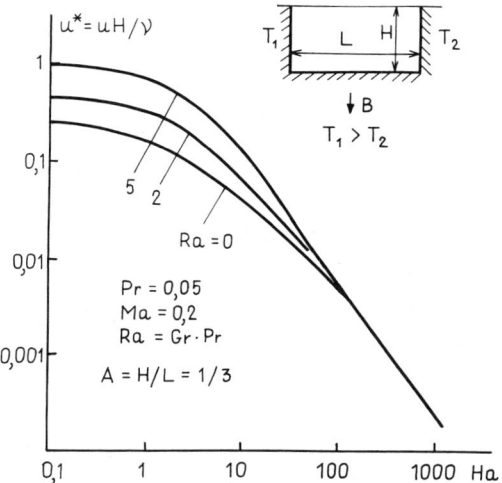

Fig. 3 Magnetic field effect on the velocity of thermocapillary convection flows on the melt surface.

convection can be comparable with the velocities of thermogravitational flows and might lead to both the change of hydrodynamic patterns in melt and the origin of oscillation components. An estimate of the axial steady magnetic field effect on thermocapillary convection presented in Ref. 2 and 4 has revealed that in this situation the ratio of flow velocities in the absence of a field and in its presence can be found from the following expression:

$$\frac{u}{u_B} = \frac{Ha}{Re^{1/3}} \qquad (2)$$

where $Re = |\partial\alpha/\partial T|\Delta TL|\rho v^2$, α is the melt surface tension coefficient, and ΔT is the temperature gradient on the characteristic length scale.

Expression (2) agrees well with the experimental data presented in Fig. 3. Suppression of thermocapillary flows is manifested quite clearly. In actual processes the degree of its manifestation probably depends greatly on such factors as the purity of melt surface, the presence of oxide films, and their properties.

The crucial factor in a conventional Czochralski process is forced convection in melt caused by the differential rotation of the crystal and crucible.

As seen from numerical simulations[7,8] and experiments[8] under the axial magnetic field effect at large Hartmann numbers $Ha = BL(\sigma/\rho v)^{1/2}$ and parameter $N = \frac{Ha^2}{Re}$, where $Re = (\Omega_k - \Omega_t) R_k^2/v$ in the case of a nonconducting crystal ($\sigma^* = \sigma_k/\sigma_p = 0$) and at a stationary crucible ($\Omega_t = 0$) in a region under the crystal, there arises a convective cell with a practically uniform angular velocity of melt rotation Ω, which is equal to half of the angular velocity of crystal rotation $\Omega/\Omega_k = 0.5$ (Fig. 4).

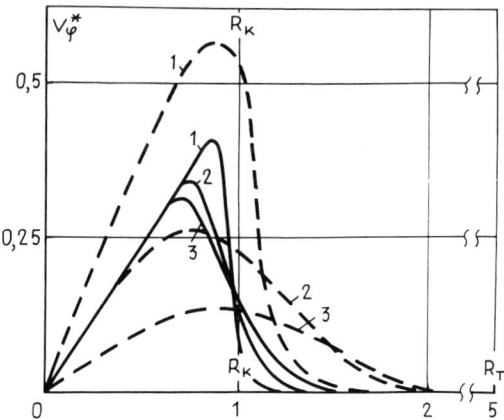

Fig. 4 Experimental dependencies of azimuthal velocity of melt in a magnetic field at various size ratios of a crystal and crucible. Solid lines: $N = Ha^2/Re = 24$; $R_T/R_k = 2.5$; $H_p/R_k = 1.67$. 1, $Z/H_p = 0.1$; 2, 0.5; 3, 0.9. Dashed lines: $Ha^2/Re = 0.65$; $R_T/R_k = 5$; $H_p/R_k = 3.33$; 1, $z/H_p = 0.1$; 2, 0.2; 3, 0.3.

Boundary layers with a large velocity gradient $\partial V_\varphi/\partial z$ and a width of the order of $\sim Ha^{-1}$ are formed on the surface of a crystal and the bottom of a crucible, whereas a shear layer with a width of $\sim Ha^{-1/2}$ forms on the side surface of the cell. In the shear layer the angular velocity depends on the two coordinates r and z; moreover, its thickness increases parabolically toward the bottom of the crucible volume, whereas in the rest of the crucible volume the azimuthal velocity is zero. As the number N decreases, the convective cell under the crystal changes, the width of the shear layer grows significantly, and at $N < 1$ the motion spreads within the whole volume of the crucible as it is at $B = 0$.

When both the crystal and crucible are rotating in the same direction or in the opposite direction, the pattern of forced convective flows does not change. Also at $Ha \gg 1$ there are the convective cells under the crystal, the shear layers, and the external region, in which melt rotates with a velocity of the crucible Ω_T.

The characteristics of flow patterns in the Czochralski process are different in the case in which the conductivities of the crystal and crucible are not equal to zero and the electric currents arising in the melt are also flowing through the crystal and crucible walls. Thus, for a well-conducting crystal ($\sigma^* = 17$), the convective cell under the crystal (Fig. 5) rotates with an angular velocity of the crystal $\Omega = \Omega_k$. The boundary layer on the crystal vanishes and is formed only at the bottom of the crucible; its thickness at $\sigma^* = 0$ is proportional to $\sim Ha^{-1}$. The shear layer on the side surface of the convective cells is formed in a similar way[4,8].

We will now consider the pattern of the secondary motion of the melt. The results of a numerical calculation[8] for the stream function give evidence that, when the convective cell is formed, the secondary motion is determined by the distribution of velocity (Fig. 5a). It is strongly suppressed by a magnetic field ($N \sim$ times) and localized near the shear-layer region.

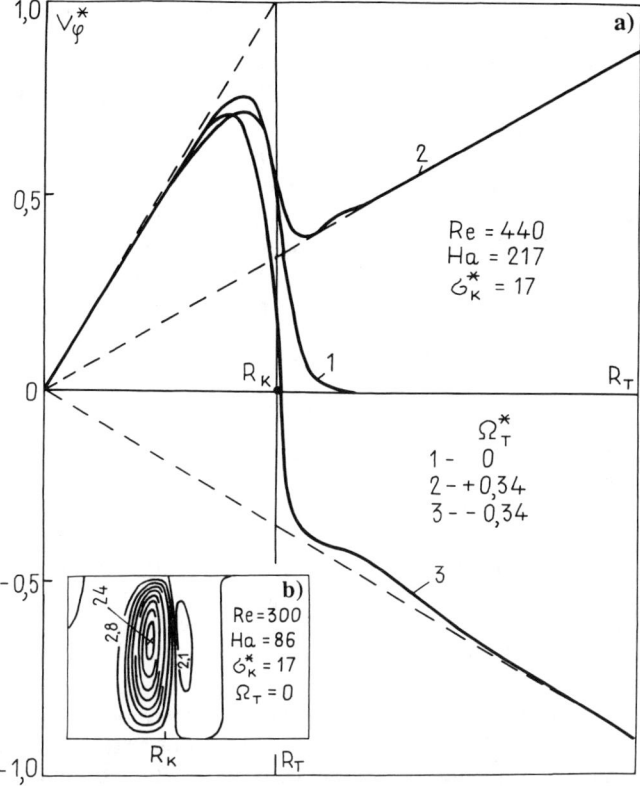

Fig. 5 Experimental distributions of azimuthal velocity of melt (a) and calculated values of the stream functions of meridional flows (b) at $\sigma_T^* \neq 0$ and $\Omega_T \neq 0$.

It should be noted that such a flow structure is a characteristic of the whole range of the parameter $H_p^* = H_p/R_k$ at the conductivity of the crucible walls $\sigma_t^* = 0$.

The presence of the crucible wall conductivity $\sigma_T \neq 0$ at their finite thickness significantly changes both the distribution of currents induced in the melt and the structure of the melt flow. Thus, at $Ha \gg 1$, $N \gg 1$, $\Omega_t = 0$, $\sigma_t^* \cong \sigma_k^*$, the velocity of melt rotation in a region under the crystal changes greatly along with the melt depth as the parameter Z/H_p grows. If there is a counter-rotation of the crucible, the convective cell can level the bottom of the crucible and can be positioned directly under the crystal at $Z/H_p \ll 1$. In this case the melt flow near the bottom of the crucible is determined by the crucible rotation, whereas the regions controlled by the crystal and crucible are divided by the shear layer.

The presence of a conducting crystal and crucible affects the development of thermoelectromagnetic convection in the melt.[9] This type of convective flow is caused by an interaction of an external field with thermoelectric currents induced in melt due to the difference in the thermoelectric power on the interface and proves to be an additional convection in the Czochralski process in a magnetic field.[10] The intensity of such flows is determined by the temperature gradient along the interface ΔT reaching 20-40°C under actual conditions and having a sufficiently large thermoelectric power for most of the semiconductor material melts ($\alpha = 30 \div 50 \mu V/K°grad$). If the paths of thermoelectric currents are closed via a conducting melt, crystal, or crucible, their interaction with an external magnetic field causes an electromagnetic force, which brings the melt into motion:

$$f_T = j_t B = \sigma \alpha \beta \Delta T / L \qquad (3)$$

To characterize thermoelectromagnetic convection (TEMC) using an analogy with the Grashoff number, the following criterion can be introduced:

$$Te = \alpha \sigma B \Delta T L^2 / \rho v^2 \qquad (4)$$

Then the ratio $Te/Gr = \alpha \sigma B / \rho \beta g L$ characterizes the relative contribution of the TEMC in comparison with thermogravitational convection. Table 1 shows the data for a number of semiconductor material melts at the magnetic field induction of $B = 0.2T$ and $L = 2 \times 10^{-2}$m.

The data in Table 1 shows that the TEMC can repeatedly exceed the intensity of free and forced convections in certain circumstances. This is also confirmed by the experimental dependencies given in Fig. 6 which shows the velocities of melt rotation in a cylindrical conducting crucible vs an interaction of thermoelectric currents with an axial external field. The nature of the curves in this figure can be explained by a breaking force in a rotating liquid having a square dependence on B; thus, the resulting value of an electromagnetic force has the following form:

$$f = j_t B - \sigma B^2 V_\varphi \qquad (5)$$

and as the melt rotation velocity grows, it must decrease as V_φ.

Table 1   Some semiconductor materials melt information

	Ge	InSb	GaAs
mkV/grad	-3.5	-20	-61
Te/Gr	19	16	66

It should be noted that the TEMC might also take place in the case of a conducting crucible directly in the melt-crystal boundary region. Then the origin of thermoelectric currents can be due to overcooling of the melt on the crystallization front depending on such factors as the growth direction, boundary shape, and temperature gradients in melts. In particular, there is a very strong overcooling in the presence of the facet effect. (For example, for indium antimonide it is $\Delta T \cong 13\text{-}20°C$.) In this situation the flows caused by the TEMC might have a significant influence on the parameters of heat and mass transfer in growing a crystal and lead to quite unexpected results (Fig. 7).

Figure 8 presents the effect of the TEMC on the stationary velocity structures in the Czochralski process. In this figure the isolines of nondimensional stream functions of melt secondary motion in the presence of both thermoelectromagnetic and thermogravitational convections are presented, with the crystal radius R_k serving as the characteristic length scale and the velocity on the edge of a crystal $n_k = 2$ rev/min being a characteristic velocity. The application of a magnetic field, which changes the structure of the melt motion leads at the same time to the suppression of its intensity. However, in the case of thermoelectromagnetic convection, suppression of the averaged melt motion takes place at larger induction values than in the case of thermogravitational convection.[11]

In such a way the TEMC significantly changes both the intensity and structure of the melt flow, especially in the region of small $B \cong 0.05T$; Fig. 8b).

Thus far we have been considering the hydrodynamical characteristics of the Czochralski in an axial magnetic field. If the magnetic field has not only an axial but also a radial component B_r, general trends of hydrodynamical structure transformation are the same. There is an analogical way of suppression of the averaged melt flow in the case of thermogravitational and thermocapillary convections. In formulas (1) and (2) the characteristic length scale must be the dimensions along which electromagnetic forces are induced in a liquid medium. For example, when an estimate is made using Eq. (1) for thermogravitational convection, the radius of a crucible R_t must be chosen as a characteristic length

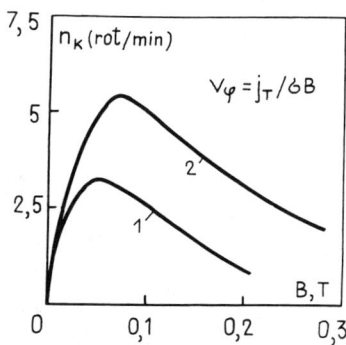

Fig. 6 Dependencies of melt rotation velocity in a conducting crucible on the interaction of thermal currents with an axial magnetic field: 1, experiment with melt In-Ga-Sn; 2, experiment with melt InSb.

Fig. 7 External view of a crystal InSb grown in the presence of thermoelectromagnetic convection.

in the case of a uniform field B_z, whereas the height of melt H_p must be taken as a characteristic length in a nonuniform field at $B_r > B_z$.

The forced convection of melt is characterized by a more expressible transformation of the hydrodynamic structure. As shown earlier, in the case of a uniform magnetic field B_z in a region under the crystal a specific convective cell is formed there (Fig. 4). In a nonuniform field at $B_z \geq B_r$ the flow structure of melt is similar to that when $B_r = 0$; however, here MHD interaction parameters are found from the value B_z on the crystallization front. At B_z on the crystallization front $B_z \to 0$, the convective cell vanishes in an area under the crystal (Fig. 9), the vortex is localized within a narrow region in the vicinity of the crystal, and the motion in the main volume of melt is determined by the crucible rotation.

In such a way, when one considers the hydrodynamics of melt during single crystal growth in axisymmetric magnetic fields using the Czochralski technique,

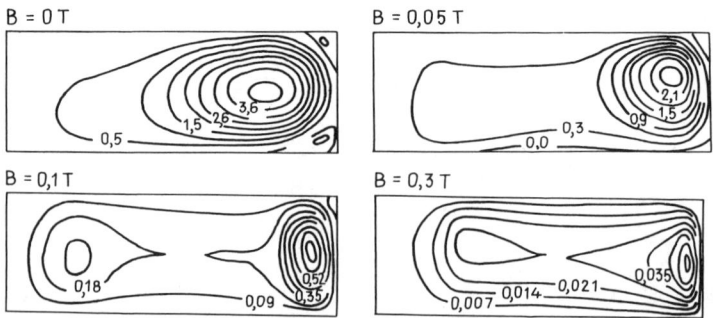

Fig. 8 Isolines of the stream functions of meridional flows in germanium melt in the presence of thermoelectromagnetic and thermogravitational convections.

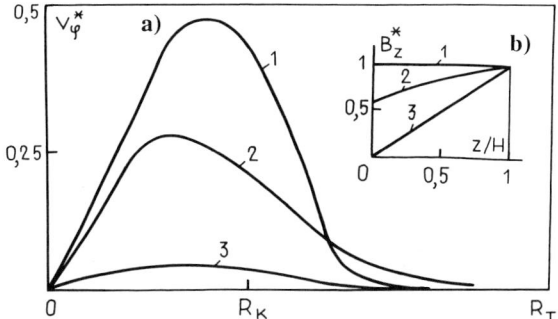

Fig. 9 Effect of a nonuniform magnetic field on azimuthal velocity of melt (a) at $n_k = 30$ rev/min and $Z/H_p = 0.1$; 1,2,3 for V_φ (a) corresponds to the cases 1,2,3 of the magnetic field induction distribution with respect to the height of melt (b).

one must take into account the significant effect on the velocity structures of finite conductivities of both the crystal and crucible as well as of the magnetic field pattern in the melt and on the crystallization front.

Effect of a Steady Magnetic Field on Heat Transfer in Melt

The convective mechanism is responsible for heat transfer in melt during single crystal growth by the Czochralski technique, and the corresponding Peclet numbers for free and forced convections range from $Pe = 10^2$ to $Pe = 10^3$. When an axial magnetic field with an induction of $B = 0.2 \div 0.3T$ in melt is applied, the convective flows, as indicated earlier, are suppressed, and the values of the Peclet numbers decrease by N times and are equal to $Pe/N < 1$ (Ref. 2). This means that in a sufficiently strong field the heat transfer in melt is determined only by conductivity laws, and the temperature distribution might be changed greatly.

To illustrate the preceding statements, Fig. 10 presents calculations of isotherms for thermogravitational melt convection for a nonrotated crystal and crucible. As seen, the isotherms correspond to diffusion-governed heat transfer in the field $B = 0.2T$.

The absence of convective stirring results in the increase of both the radial and axial temperature difference in melt. This effect is most vividly displayed when the crucible and crystal do not rotate ($n_t = n_k = 0$). In particular, in experiments relevant to the growth of single silicon crystals with a diameter of 60 mm from a crucible with a diameter of 152 mm and the depth of melt 40 mm, the vertical temperature difference changed from 13°C to 23°C at the field changes from $B = 0$ to $B = 0.2T$, and the radial temperature difference was that from $\Delta T = 11°C$ to $\Delta T = 15°C$. Rotation of a crystal and crucible decreases slightly the aforementioned temperature gradients, but at $B \neq 0$ they are larger than in the case of $B = 0$. At constant heating power this leads to the increase of the diameter of a growing crystal as well as to the formation of convex crystallization front.

The suppression of convective flows in melt by an axial magnetic field and the prevailing molecular heat transfer in it leads to pronounced symmetrization

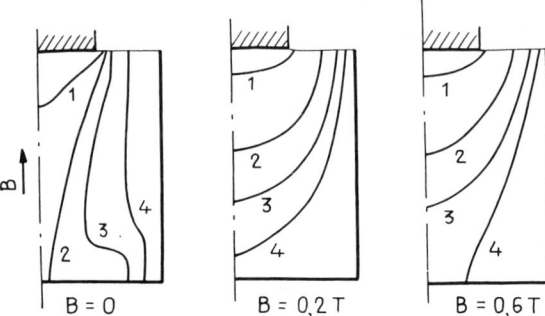

Fig. 10 Temperature field in a Czochralski melt in an axial magnetic field: 1, T = 28.5; 2, 29.5; 3, 29.8; 4, 29.9.

of heat field in melt even when $n_k = n_t = 0$ (Ref. 12). This effect starts manifesting itself already at 0.1T and allows to conduct growth at small rotation velocities of a crystal and crucible. The degree of its manifestation is reported in Fig. 11, which presents both the photograph and diagram of an axial cross section of a single crystal produced under specific conditions with pronounced asymmetry of a heat field. The first region of a single crystal has been grown at $n_t = 0$ and at $n_k = 10$ rev/min in the field of B = 0.15T, whereas in the diagram the second region depicts the position of heat center in the crucible when there is no crucible and crystal rotation in the field of B = 0.15T. The following regions reflect the heat field asymmetry in the crucible as the magnetic field induction decreases and there is a distortion of the shape of single crystal growing at

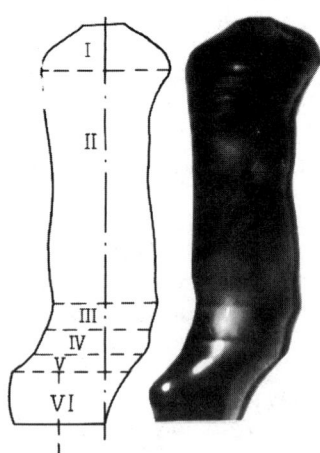

Fig. 11 Photograph and diagram of an axial cross section of germanium single crystal produced in a thermal field with pronounced asymmetry in melt: I, $n_t=0$, $n_k=10$rev/min, B=0.15T; II $n_t=n_k=0$, B=0.15T; III $n_t=n_k=0$, B=0.1T; IV $n_t=n_k=0$, B=0.05T; V $n_t=n_k=0$, B=0.025T; VI $n_t=n_k=0$, B=0.

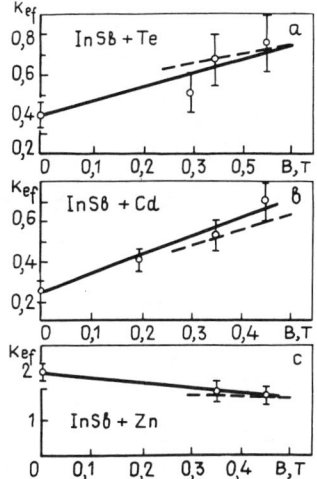

Fig. 12 Dependences of k_{ef} in a magnetic field for impurities Te, Cd, Zn in crystals InSb. Solid lines indicate experiment, and dashed lines indicate calculations.

$B \to 0$. It should be noted that a symmetrization of a heat field by a longitudinal magnetic magnetic field can take place only if the heat symmetry of heaters is satisfactory. Otherwise, even sufficiently high thermal conductivity of melt cannot balance the temperature throughout the whole volume of melt.[12]

When nonuniform magnetic fields are applied to the melt, the heat-transfer characteristics are similar to those described earlier. Regarding the quantitative characteristics, it is obvious that the integral characteristics of heat transfer are determined by the parameter Pe/N, and at Pe/N < 1 a diffusional mechanism of heat transfer prevails in the melt. Along with this one can assume that, in nonuniform magnetic fields, due to localizing of secondary convective flows in the regions with small volumes, situations might arise in which heat transfer has a diffusional character in one region of the melt, whereas in the other it is convective. This is likely to happen, for example, in an axisymmetric nonuniform magnetic field at $B_z = 0$ on the crystallization front. Here the secondary flows are localized under the crystal in a narrow region $Z/H_p < 1$ but are suppressed (due to $B_z = 0$ and $B_r \to 0$) much more weakly than they are in the rest of the melt region. This leads to the existence of regions with various heat-transfer mechanisms in melt, in particular, convective transfer in the region under the crystal and molecular transfer in the vicinity of the crucible.

Effect of a Magnetic Field on Mass Transfer in the Melt

In accordance with the estimates of the effect of an axial magnetic field of $B = 0.3$ T on mass transfer in the melt, the diffusion Pe_D Peclet number decreases N times, but unlike the thermal Peclet number, it does not reach the values that are characteristic of a situation in which a diffusion mass transfer $Pe_D/N \cong 10^3 \div 10^4$ prevails.[2] This means that the type of distribution of dopants and background impurities in a growing crystal depends, in many ways, on the structure and intensity of convective flows affected by a field in melt and near the crystallization front.

The change of mass transfer is displayed by the behavior of the effective distribution coefficient of impurities. As known, k_{ef} at B = 0 is found from the Barton-Prima-Slichter[13] relation and depends on the thickness of the diffusion boundary layer δ_D on the crystallization front. In a steady magnetic field the value δ_{DB} grows according to[13]

$$\delta_{DB} = \frac{\delta_D Ha}{\sqrt{Re}} \qquad (6)$$

where Re is found from the convective flow velocity at B = 0. In particular, at B = 0.25 T the calculated thickness of the diffusion boundary layer equals $\delta_{DB} < 6$ mm in comparison with $\delta_D \leq 1$ mm at B = 0.

Using expression (6) in the Barton-Prima-Slichter formula for δ_{DB}, we can obtain the theoretical dependence of k_{ef} on a magnetic field in the form of[13]

$$k_{ef} = \frac{k_o}{k_o + (1-k_o)\exp(-V\delta_D Ha/Re^{1/2}D)} \qquad (7)$$

where k_o is the equilibrium coefficient of distribution at B = 0. V is the pull velocity of a crystal, and D is the diffusion coefficient in melt. Expression (7) agrees well with the experimental data (Fig. 12) and shows that with the growth of B (and correspondingly δ_{DB}) the values k_{ef} tend to unity. Moreover, if $k_o < 1$, the growth of B results in the enrichment of impurity on the crystallization front, whereas it is depleted at $k_o > 1$. In any case, the tendency of k_{ef} to unity due to the magnetic field effect must lead to the increase of uniformity of impurity distribution in a growth crystal.

However, it should not be ignored that the mass transfer on the crystallization front determining the radial distribution of impurity depends, along with the above factors, on both the curvature of the interface and the nonuniformity of a diffusional boundary layer. These factors are, to a great extent, determined by the character of a convective flow on the crystallization front and the temperature field configuration in this region since there is, as a rule, sufficiently high thermal conductivity in semiconductor material melts. Experiments have revealed that the shape of the crystallization front is actually one of the factors determining the uniformity of the radial impurity distribution in a grown crystal. For example, Fig. 13 presents photographs of the shape of the crystallization front of germanium single crystals grown in a vertical magnetic field (the crystallization front shape has been obtained by an abrupt detachment of a crystal from melt). Depending on the growth regimes, the shape of the crystallization front can be convex to the melt as well as to the plane, or it can be concave (Fig. 13). The measurements of dopant radial distributions in crystals have revealed that the lowest nonuniformity of dopant concentration (~1÷2%) can be obtained with a plane crystallization front, when the diffusion boundary layer has the same thickness along the crystal radius. Thus, mass transfer on the crystallization front is determined not only by the structure of convective flows but also by the shape of the crystallization front and the structure of the temperature field.[12]

The effect of a steady magnetic field on mass transfer can be illustrated by the possibility of controlling the oxygen concentration in silicon single crystals[1,12]. In this case control is realized by the suppression of mean intensity of the melt flow and alteration of its pattern. For example, the decrease in oxygen concentration in single crystals down to $(2 \div 3) \times 10^{17}$ atoms/cm^3 might be reached by arranging the convective flows in such a way that melt flowing to the crystallization front has passed along the free surface, where it is depleted in oxygen.

It is clear that the uniformity of dopant distribution in crystals will be provided if the thickness of δ_{DB} grows so much that characteristic time for radial diffusion $\tau_1 = R_k^2/2D$ becomes less than the time of diffusion through the diffusion boundary layer $\tau_2 = \delta_{DB}^2/2D$, i.e., at $\delta_{DB}/R_k > 1$. Unfortunately, for

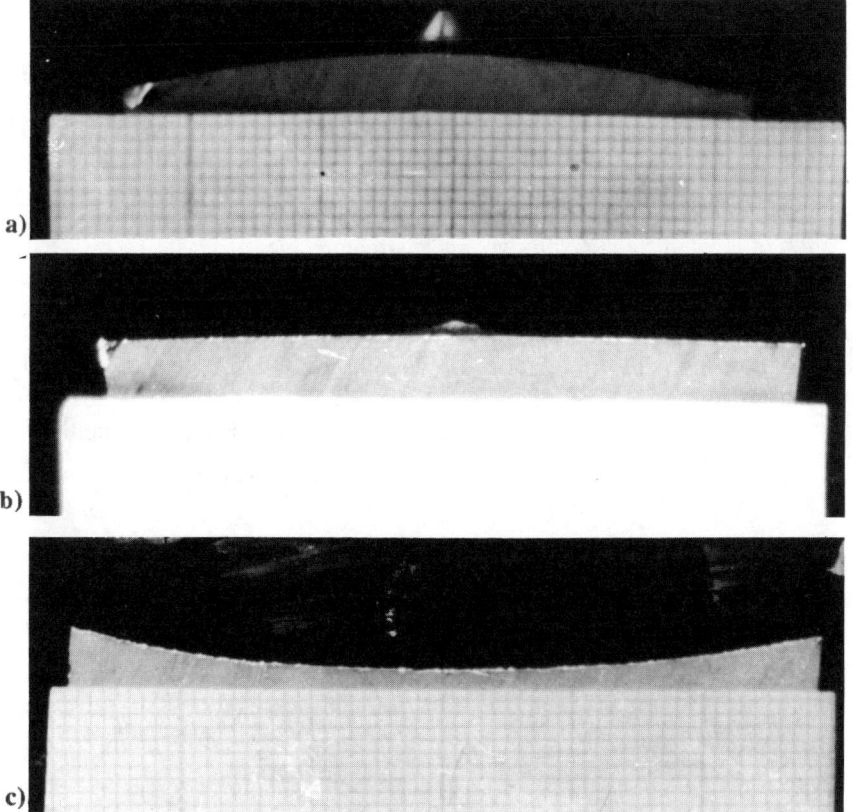

Fig. 13 Shape of the crystallization front during the production of germanium single crystals in various regimes of MHD interaction: a) convex; b) plane; c) concave.

the large-diameter crystals the necessary magnetic field induction cannot be reached.

The aforementioned features of mass transfer in an axial magnetic field are also observed in nonuniform magnetic fields. But there are some essential peculiarities. At real values of the magnetic field, induction mass transfer in melt is of a convective type (except the boundary layers on crucible walls and on the crystallization front) and is determined by the structure of melt flow, which, in turn, significantly depends (as shown earlier) on the field geometry. In particular, the effect of the magnetic field on the boundary layers on the crystallization front is mainly found from the value B_z. At $B_z = 0$ the boundary layers are formed on the crystallization front analogously to the case in which there is no MHD influence on melt. Thus, when $B_z = 0$ on the crystallization front, a magnetic field does not have a significant influence on the coefficients of impurity distribution.[14]

However, when silicon single crystals are grown in a nonuniform magnetic field, the presence of radial component induction B_r on the crucible walls can greatly change the structure of the hydrodynamical boundary layer and lead to the decrease of wall solubility of a quartz crucible.

Effect of a Magnetic Field on Nonstationary Processes in Melt

Convective flows in melt under real conditions of single crystal growth exhibit various types of instabilities not only at turbulent but also at laminar flow regimes[3]. In experiments the previously given phenomena manifest themselves as pulsations of velocity, temperature, and concentration, the formation of local unstable hydrodynamical structure, etc.[4].

The intensity of their manifestation and origin due to a magnetic field is determined by aspect ratios in the melt-crystal system, rotation velocities of crucible and seed, magnetic field configuration, flow pattern, etc.

In general, steady magnetic fields suppress all types of velocity perturbations in the melt and the associated heat and mass transfer processes. This can be well seen in Fig. 14a presenting the dependencies of amplitude

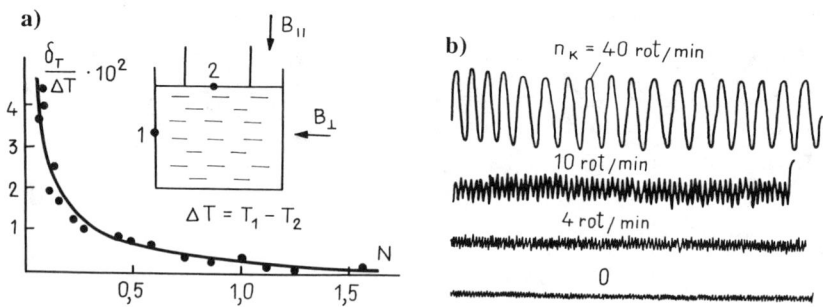

Fig. 14 Effect of a magnetic field with different orientation on the temperature pulsations in melt (a) and fluctuations of azimuthal velocity (b) during the Czochralski process.

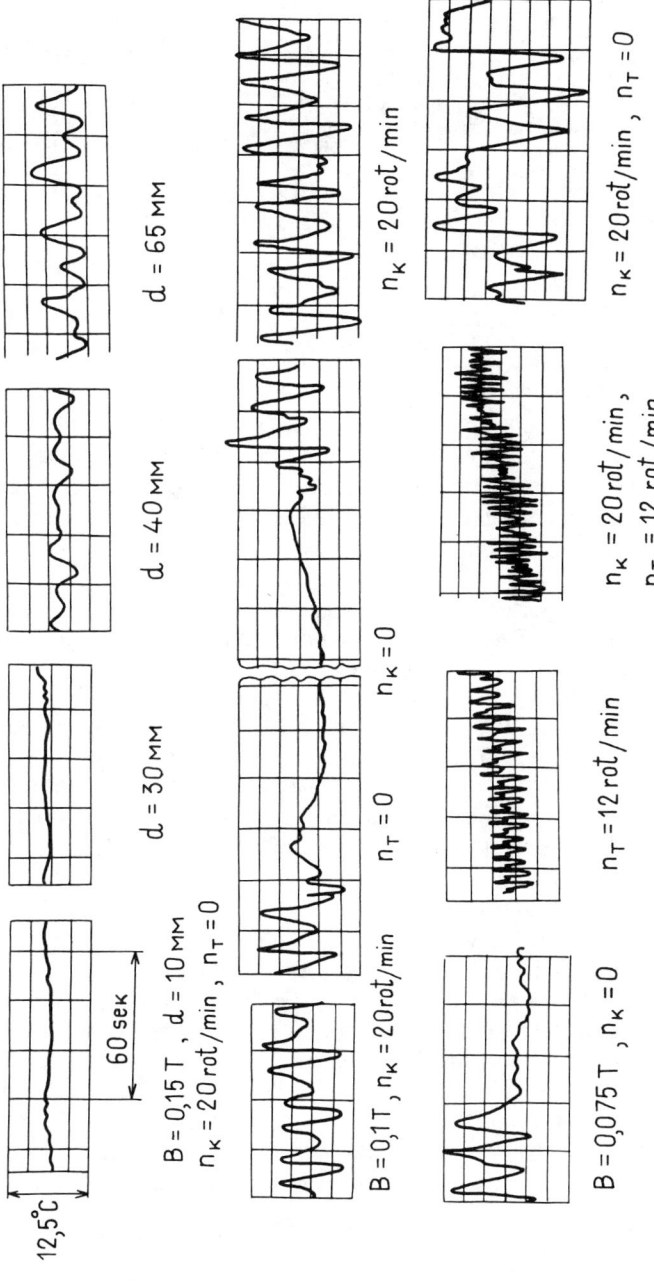

Fig. 15 Magnetic field effect on the temperature pulsations in silicon melt during the Czochralski growth.

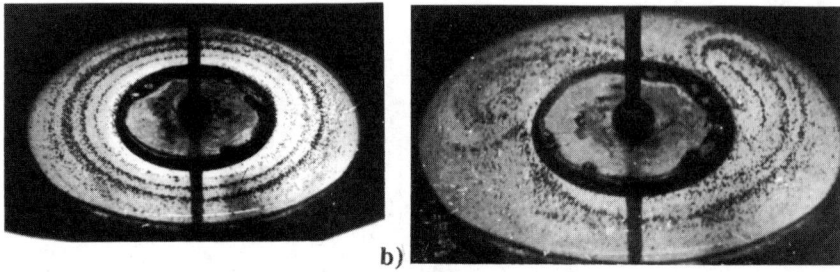

Fig. 16 Photograph of melt surface during the origin of secondary azimuthally periodic vortices caused by the field:
a) Re < Re_{cr} ; b) Re > Re_{cr}.

changes of temperature pulsations $\delta T^* = \delta T/\Delta T$ (where ΔT is the temperature difference in points 1 and 2) on the MHD interaction parameter $N = Ha^2/\sqrt{Gr}$ in the case of the non-rotated crystal and crucible.[15] At small values of 0.08-0.1T the intensity of temperature pulsations decreases sharply and vanishes with the growth of N. These data agree well with the results of pulsation measurement of the azimuthal component of melt velocity (Fig. 14b). At B = 0.08T and n_T = 5.5 rev/min, velocity pulsations decrease monotonously, reaching the zero value as the number of revolutions of a crystal decrease from n_k = 40 rev/min to n_k = 0 rev/min.

The behavior of perturbations in the presence of forced convection is rather complicated (Fig. 15). The data on temperature pulsations under real production of silicon single crystals with the diameter of 65 mm from the crucible with the diameter of 152 mm at ΔT = 15°C and the numbers $Gr = 2 \times 10^8$, $Ha = 1.7 \times 10^2$, and $N \cong 2$ show that as a crystal diameter grows in an axial magnetic field the amplitude of temperature pulsations grows and reaches the value δ = 5-6°C at $2R_k$ = 65mm. When the rotation of a crystal is stopped, temperature pulsations vanish but increase significantly with the growth of n_k. Analogous effect of increase is observed at n_k = 0 after rotation of a crucible or joint rotations of a crystal and crucible are started.[12]

The temperature fluctuations at $n_k = n_T \neq 0$ are connected with the instability of forced convective flows in an axial magnetic field.[4] According to pulsation measurements V_φ of the azimuthal component of melt velocity and visualization of liquid flows on its free surface at $Ha \gg 1$ and $Ha^2/Re > 1$ stationary axisymmetric flow with a pronounced convective cell and the shear layer of the angular velocity presented in Figs. 4 and 5, loses stability and changes into nonaxisymmetric three-dimensional flow. The origin of the aforementioned situation in an isothermic case (Gr = 0) occurs at Re > Re_{cr} due to the development of the Kelvin-Helmholtz instability in the shear layer[4] and is characterized by the emergence of an azimuthally periodic system of vortices whose axes are parallel to a magnetic field (Fig. 16).

Curve 1 in Fig. 17 characterizes the dependence of Re_{cr} on the parameter $(HaR_k/H_p)^{1/2}$, where $(HaR_k/H_p)^{-1/2}$ is proportional to the thickness of the free shear layer in the case when there is no thermogravitational convection. It is also preserved under nonisothermal conditions up to $Gr < 10^5$, so that the melt

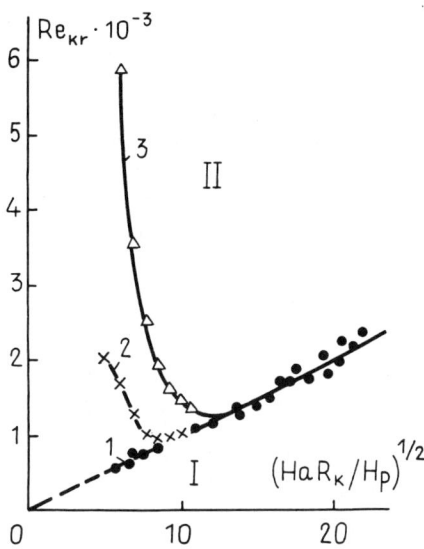

Fig. 17 Dependencies of Re_{cr} in an axial steady magnetic field during the Czochralski process.

flow will always be axisymmetric without pulsations of local velocity and temperature at the parameters fitting the region 1 under the curve 1.

At $Gr > 10^5$ the dependence Re_{cr} changes: curves 2 and 3 in Fig. 17 corresponds to $Gr = 3.7 \times 10^7$ and $Gr = 1.6 \times 10^8$, respectively. The regions of stable axisymmetric flow are growing; this can probably be explained by significant changes of the azimuthal velocity distribution in melt at $Ha^2/\sqrt{Gr} < 1$. At the regimes corresponding to the region II, nonaxisymmetric spatial flows arise with characteristic pulsations of velocity and temperature greatly increasing convective heat and mass transfer[4].

The axially symmetric nonuniform magnetic field with the components B_z and B_r strongly affects the structure of forced convective flows. Moreover, if on the crystallization on the axial component of the magnetic field induction is not zero, then a convective cell that is analogous to the case of a uniform magnetic field is formed in a region under the crystal. At the Reynolds number $Re > Re_{cr}$ the Kelvin-Helmholtz instability occurs again. In the presence of thermogravitational convection the critical Reynolds number, like in the previous case, starts depending on the Grashoff number. Here the number Ha found from the value of the magnetic field axial component on the crystallization front B_z should be employed as a typical parameter determining transfer to instability.

At $B_z \to 0$, the convective cell is not formed in the region under the crystal; hence, there is no mechanism of instability formation. As a result, the flow is axisymmetric in the whole range of the magnetic field. Thus, the stirring of melt on the crystallization front is increased and radial distribution of dopants becomes more uniform.

At the same time, the pulsations of velocity, temperature, and concentration associated with the instability of thermogravitational or capillary convection are strongly suppressed due to the effect of the B_r component, whereas the melt flow in the crucible volume becomes laminar.

The previously mentioned data on the effect of steady magnetic fields upon the hydrodynamics, heat, and mass transfer illustrate the fact that application of even the simplest MHD means can be highly beneficial for crystal melt growth. A thorough analysis of all of its parameters must be performed to reach positive results.

References

[1] Fiegl, G., "Recent Advances and Future Directions in CZ-Silicon Crystal Growth Technology," *Solid State Technology Journal*, Vol. 26, No. 8, 1983, pp. 121-131.

[2] Gelfgat, Yu.M., Gorbunov, L.A. and Sorkin, M.Z., "Electromagnetic Means to Affect Hydrodynamics and Heat and Mass Transfer During Production of Volume Single Crystals (Review)", *Rost Kirstallov*, Vol. XVI, Nauka, Moscow, 1988, pp. 234-247.

[3] Faifer, S.I., Zakharov, B.G., and Semenova, V.B., "The Present Situation and Prospective Developments of Production of High Quality Gallium Arsenide Single Crystals," *Elektronnaya Tekhnika*, Vol. 6, No. 2, 1988, pp. 3-9.

[4] Bojarevics, A., Gelfgat, Y.M. and Gorbunov, L.A., "Melt Magnetohydrodynamics of Single Crystal Growth," *Liquid Metal Magnetohydrodynamics*, edited by J. Lielpeteris and R. Moreau, Kluwer, Dordrectht, The Netherlands, 1988, pp. 127-134.

[5] Gelfgat, Yu.M., Gorbunov, L.A., Starshinov, I.V., Smirnov, V.A., and Friazinov, I.v., "The Effect of an Axial Magnetic Field upon Natural Convection During the Czochralski Process," *Prikladnie Zadachi Matematicheskoi Fiziki*, Latvian Univ., Riga, USSR, 1985, pp. 127-143.

[6] Ochiar, J., et al., "Experimental Study and Marangoni Convection," Committee on Space Research COSPAR, Vol. 25, 1984, pp. 5-14.

[7] Hjelming, L.N., and Walker, J.S., "Melt Motion in a Czochralski Crystal Puller with an Axial Magnetic Field: Isothermal Motion," *Journal of Fluid Mechanics*, Vol. 164, 1986, pp. 237-273.

[8] Bojarevics, A., Gorbunov, L.A. and Ljumkis, E.D., "Physical and Numerical Modelling of the Effect of a Vertical Magnetic Field on Forced Convection in the Single Crystal Growth by the Czochralski Technique," Magnitnaya Gidrodinamika, No. 2, 1989, pp. 81-87.

[9] Gelfgat, Yu.M., and Gorbunov, L.A., "On the Additional Source of Forced Convection in Semiconductor Material Melts During Single Crystal Growth in a Magnetic Field," *Doklady Akademii Nauk SSSR*, Vol. 306, No. 3, 1989, pp. 604-608.

[10] Gorbunov, L.A., "The Effect of a Thermoelectromagnetic Convection upon the Growth of Semiconductor Single Crystals from Melts in a Magnetic Field," *Magnitnaya Gidrodinamika*, No. 4, 1987, pp. 65-69.

[11] Gorbunov, L.A. and Ljumkis, E.D., "Mathematical Modelling of Melt Hydrodynamics in a Magnetic Field with an Account of Conductivity of a Crystal and Crucible," *Chislennie Metodi Modelirovaniya Tehnologicheskih Processov*, Latvian Univ., Riga, Latvia, 1989, pp. 94-95.

[12] Bochkarev, E.P., Foliforov, V.M., Gelfgat, Yu.M., et al., "The Effect of an Electromagnetic Field on Single Crystal Growth and Characteristics," *Liquid Metal*

Magnetohydrodynamics, 1st ed., edited by J. Lielpeteris and R. Moreau, Dordrecth, The Netherlands, 1989, pp. 135-144.

[13]Zemskov, V.S., Rauhman, M.P., Mgalobloshvili, D.P., Gelfgat, Yu.M., and Sorkin, M.Z., "The Coefficient of Impurity Distribution During the Growth of Indium Antinomide Single Crystals Under the Magnetic Field Effect," *Fiziki i Khimiya Obrabotki Materialov,* No. 2, 1986, pp. 64-67.

[14]Hicks, T.W., Organ, A.E. and Riley, N., "Oxygen Transport in Magnetic Czochralski Growth of Silicon with a Non-Uniform Magnetic Field," *Journal of Crystal Growth,* Vol. 94, 1989, pp. 213-228.

[15]Bojarevics, A. and Gorbunov, L.A., "The Study of the Effect of an Axial Magnetic Field upon the Melt Hydrodynamics in the Czochralski Process, Part II, "MHD Devices, *Proceedings of the 12th MHD Conference,* Institute of Physics, Riga-Salaspils, 1987, pp. 151-154.

Inverse Electromagnetic Shaping Problem

T. P. Felici* and J. P. Brancher†

Institut National Polytechnique de Lorraine, Nancy, France

Abstract

In this paper the authors present a study on the inverse problem in electromagnetic shaping, which consists of determining the necessary current distribution one needs to apply in order to shape a liquid metal into the required form. First, an overview is presented of previous work done for the two dimensional case. The autors then proceed to study the general three dimensional case. To begin with, a study of the field structure on the liquid surface is done, which gives an idea of the possible shapes that one can, or cannot obtain. Particular attention is given to the axisymmetric case. The complete problem is then presented, and an integral equation is derived for the required current distribution. The authors conclude this presentation by giving some numerical solutions of this integral equation for a variety of axisymmetric surfaces.

Introduction

In the various processes of elaboration of liquid metals, there arises the problem of controlling the equilibrium of the mass of metal, or of its interface (in the case of a liquid metal) by a high frequency oscillating magnetic field.

The direct problem has already been studied by using a magnetostatic approach. It consists of determining the shape of a mass of liquid metal or the cross section of a liquid jet falling in the vicinity of a given inductor, the main assumptions being that the magnetic Reynold's number $R_{mag} = \mu_0 \sigma U_0 L$ becomes small (μ_0 is the magnetic permeability of vacuum, σ is the electrical conductivity, U_0 is the typical velocity of stirring flow, and L

Copyright © 1992 by the American Institute of Aeronautics and Astronautics, Inc. All rights reserved.

*Post-Doctoral Researcher, Laboratoire d'Energetique et de Mecanique Theorique et Appliquee.

†Professor of Mechanics.

is the characteristic size of the liquid domain), and that the electromagnetic skin depth is negligible compared with the characteristic length scale of the metal shape ; that is the shield parameter $\mu_0 \sigma \omega L^2$ is large (ω being the field frequency). Under these hypotheses the magnetic field does not penetrate into the metal, and so the electromagnetic forces reduce to a magnetic pressure acting on the surface. Moreover, a considerable simplification is possible since we can treat this as a magnetostatic problem by considering the time average of the magnetic field, the current, and the surface pressure.

The free boundary equilibrium problem can be solved either by directly studying the equilibrium equations at the interface or by minimizing an appropriate energy functional. Several problems have been studied by one or the other of these methods. We mention the work done by Shercliff[1] and Brancher, Sero Guillaume and Etay [2,3] for the shaping of jets; that of Sneyd and Moffat [4], Brancher, Sero Guillaume, Gagnoud, Etay and Garnier [5,6,7], and Mestel [8] for the electromagnetic levitation.

The inverse problem consists in determining the inductor, given the shape of the liquid metal. This problem has already been studied for a horizontal two-dimensional configuration by Brancher, Henrot and Pierre [9]. Our aim in this paper is to complete this study, notably by the three-dimensional problem.

Definition of the problem

Let Ω be the regular domain occupied by the liquid metal and Ω_e the exterior region in which the inducting currents are found (figure 1). The equilibrium condition on the frontier $\partial \Omega$ is

$$\mu_0 \frac{H^2}{2} + \rho g z + \gamma C = const \qquad (1)$$

where
H = the root mean square of the magnetic field
μ_0 = the permeability of vacuum
ρ = the mass density

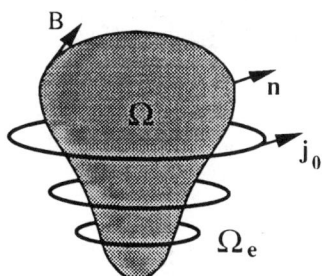

Fig. 1 Diagram for a typical levitation device.

γ = the surface tension coefficient
C = the curvature.
The oz axis is taken in the vertical upward direction.
The magnetic field satisfies:

$$\text{rot } \boldsymbol{H} = \mu_0 j_0 \quad \begin{array}{ll} \boldsymbol{H} = 0 & \text{in } \Omega \\ \text{div } \boldsymbol{H} = 0 & \text{in } \Omega_e \\ \boldsymbol{H}.\boldsymbol{n} = 0 & \text{on } \partial\Omega \end{array} \quad (2)$$

where j_0 is the current density in the inductors and $\boldsymbol{n}$ the normal on $\partial\Omega$ exterior to Ω.

Equation (1) can be written in an adimensional form:

$$H^2 + \tau_1 C + \tau_2 z = \text{const} \quad (3)$$

with
$$\tau_1 = \frac{2\gamma}{\mu_0 H_0^2 L_0} , \quad \tau_2 = \frac{2\rho g L_0}{\mu_0 H_0^2}$$

where L_0 is a characteristic length and H_0 a characteristic field.

For the inverse problem, Ω is given so the magnitude of the field $\boldsymbol{H}$ is known on $\partial\Omega$. It is also known that $\boldsymbol{H}$ is tangent to $\partial\Omega$, which enables us to evaluate the field on $\partial\Omega$ itself and to prolong it into Ω_e in such a way that div$\boldsymbol{H} = 0$. We can then look for a current distribution j_0 needed to produce such a field.

Two dimensional case

Properties and existence conditions

For the cross section of a vertical jet, gravity may be neglected, so:
$$\tau_2 = 0 \Rightarrow H^2 + \tau_1 C = \text{const} \equiv P$$

In this case $\boldsymbol{H} = \varepsilon\sqrt{P - \tau_1 C}\,\boldsymbol{t}$ with $\varepsilon = \pm 1$
where $\boldsymbol{t}$ is the oriented tangent to $\partial\Omega$.

The constant P is such that: $P \geq \tau_1 \underset{\partial\Omega}{\max} C$
if $\boldsymbol{H}$ has some singular points on $\partial\Omega$ then $P = \tau_1 C_m$
where C_m is the maximum curvature.
If $\boldsymbol{H}$ has no singular points then

$$P > \tau_1 C_m$$

It is clear that if $\boldsymbol{H}$ is continuous and remains nonzero, then $\varepsilon\sqrt{P-\tau_1 C}$ cannot change sign, so ε must be constant. This would imply that the liquid metal jet is crossed by a current with intensity

$$\oint \sqrt{P - \tau_1 C} \, dl$$

where dl is the length element of $\partial\Omega$.
Conversely, if the total current is zero, the magnetic field will have some singular points on $\partial\Omega$ and H on the frontier will be given by

$$H = \varepsilon\sqrt{\tau_1} \sqrt{C_m - C}) \, t \qquad (\varepsilon = \pm 1)$$

In ref. 9 it is proved that the field H can be prolonged into the exterior region Ω_e if and only if the curve $\partial\Omega$ is analytic. This property implies that there is at most only a finite number of singular points. If the multiplicity order of the singular point is even, ε remains constant in its neighborhood. If on the other hand ε is odd, it will change sign. Hence, by periodicity the number of singular points of odd order must be even. For example, any curve which has its curvature attaining its maximum value at an even number of points, and at which H has nondegenerate zeros, is in fact impossible to form.

Solving the problem

In the two-dimensional case, it is possible to reduce the problem to the shaping of a disc with unit radius, making it thus easy to prolong the field into a neighborhood of the disc. We can then consider a regular closed curve Γ_0 contained in the reciprocal image of this neighborhood (figure 2) and solve the exterior problem:

$$\Delta \Psi_e = 0 \text{ in the region exterior to } \Gamma_0$$
$$\Psi_e = \Psi_p \text{ on } \Gamma_0$$

where Ψ_p is the value of the prolonged function on Γ_0. As Ψ_e is required to be bounded, this problem has a unique solution, and the current density j_s to be placed on Γ_0 is then given by the jump in the normal derivative:

$$J_s = \frac{\partial \Psi_e}{\partial n} - \frac{\partial \Psi_p}{\partial n}$$

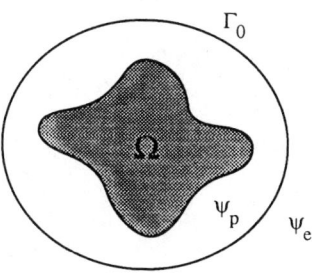

Fig. 2 Diagram for two-dimensional shaping.

In fact, here Ψz is the vector potential for the magnetic field (z is the unit vector in the oz direction), and so the jump in its normal derivative gives the jump in the tangential component of the surface current density on Γ_0.

In two dimensions, we do not have the problem of constructing the field H on the frontier since it is known up to a sign. The field in Ω_e is then obtained by a holomorphic extension of the complex potential in the image plain.

Inverse Problem in the Three-Dimensional Case

Consider a quantity of levitated liquid metal with a known shape. Our problem is to calculate the high frequency electromagnetic field or indeed the external inducing currents required to produce this shape.

The magnetic field must satisfy Eqs.(2) and (3), which we can rewrite as:

$$|H|^2 = K \qquad \text{where } K = P - \tau_1 C - \tau_2 Z \tag{4}$$

We will restrict our attention to surfaces $\Gamma = \partial\Omega$ homeomorphic to the sphere S^2, and we define Ω_e as a bounded exterior domain where $j_0 = 0$, that is, an open region in which a solution of

$$\text{rot } H = \text{div } H = 0$$

is assumed to exist. In this case, Ω_e is simply connected, so rot $H = 0$ implies that there exists an <u>univalued</u> function Φ unique up to an additive constant such that

$$H = \nabla\Phi$$

Hence, Φ satisfies

$$\Delta\Phi = 0 \qquad \text{in } \Omega_e, \quad \frac{\partial\Phi}{\partial n} \qquad \text{on } \Gamma$$

We assume that Γ is of class $C^{2,\alpha}$, i.e, that there is a parametrization of Γ whose second derivatives are Holder continuous. Then from the theory of elliptic partial differential equations we know that Γ has continuous second derivatives up to and including Γ. Hence, we can extend the field equations to Γ and write

$$H = \nabla_s \varphi \qquad \text{on } \Gamma$$

where $\varphi = \Phi|_\Gamma$ is univalued and ∇_s denotes the surface gradient. Then the surface equation becomes:

$$|\nabla_s \varphi|^2 = K \quad \text{on } \Gamma \tag{5}$$

Once we have found φ satisfying Eq.(5) we can proceed to find Φ in Ω_e by solving the Cauchy problem for Laplace's equation:

$$\Delta \Phi = 0 \quad in\ \Omega_e$$
$$\frac{\partial \Phi}{\partial n} = 0, \quad \Phi = \varphi \quad on\ \Gamma \tag{6}$$

Study of the magnetic field on Γ

<u>Existence conditions</u>

K must satisfy $k \geq 0$, with $K = 0$ at at least two isolated points on G. The continuous univalued function φ must attain a maximum and a minimum on the compact surface Γ. At these two distinct singular points we have $H = \nabla_s \varphi = 0$. So $K = 0$ and therefore the pressure P is fixed:

$$P = \max_{\Gamma} (\tau_1 C + \tau_2 z)$$

If the matrix of second derivatives of K is nondegenerate in any coordinate system on Γ defined in a neighborhood of these singular points, then the number of these points must be even, since on the sphere we have:

$$M + m - s = 2$$

where M, m and s are the number of maximums, minimums and saddle points respectively.

<u>Construction and uniqueness of the solution φ on Γ</u>

In any local coordinate system (u^1, u^2) on Γ with metric coefficients g_{ij}, the operator ∇_s becomes $g^{ij} \frac{\partial}{\partial u^j}$ and so the surface Eq. (5) becomes

$$g^{ij} \frac{\partial \varphi}{\partial u^i} \frac{\partial \varphi}{\partial u^j} = K$$

where now K is a function of u^1, u^2 and $(g^{ij}) = (g_{ij})^{-1}$.

This is a first-order nonlinear partial differential equation. Normally we would expect a solution to be dependent on some initial line data given on Γ. However, we will now show how φ can be constructed uniquely starting from a singular point ($K = 0$), providing Γ is analytic (i.e. such that Γ admits an analytical parametrization). We first state the following property:

<u>Property 1</u>

If F is a nonconstant analytic function defined on Γ (analytic), then
   a) all regions $\{p$ in $\Gamma : F(p) = $ const.$\}$ must either be isolated points or isolated lines which close on themselves.
   b) F will attain an extremum (a maximum or a minimum) at at least one <u>isolated</u> point

Proof:
   a) This is a direct consequence of the properties of analytic functions, and the fact that Γ is homeomorphic to the sphere.
   b) We know that F must attain an extremum on Γ. Assume this occurs on a <u>line</u> $F = $ const. Since all lines are closed, F must attain a maximum somewhere in the interior of this "loop". Again, this can occur either on a line or a point. We can repeat this argument in case it is a line only finitely many times because of (a). Hence eventually we must have an isolated point at which F attains an extremum.

Q.E.D.

We can immediately apply the above to the function K (which is analytic if Γ is analytic).

Construction of j around a singular point.

Now consider such an isolated singular point p_0. For simplicity, it is possible to choose an isothermal coordinate system (x,y) in which

a) $g = \lambda^2 Id$; Id is the identity matrix, g is the matrix (g_{ij}), λ is an analytic function.

b) The matrix of second derivatives $D^2K |_{p_0}$ is diagonal.

Then Eq. (5) reduces to

$$\left(\frac{\partial \varphi}{\partial x}\right)^2 + \left(\frac{\partial \varphi}{\partial y}\right)^2 = \lambda^2 K \equiv \tilde{K} \qquad (7)$$

We know from the theory of elliptic partial differential equations that if Γ is analytic, then Eq.(6) will admit a solution only if φ is analytic. With this in mind, express both K and φ in terms of a power series:

$$\varphi = \sum_{n,m=0}^{\infty} \varphi_{nm} X^n Y^m, \quad \tilde{K} = \sum_{n,m=2}^{\infty} \tilde{K}_{nm} X^n Y^m$$

(we set p_0 to be the origin).

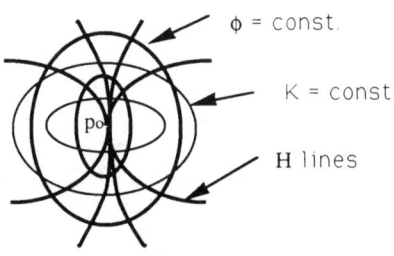

Fig. 3 Behavior or K, φ near a nondegenerate singular point.

Now substitute these into Eq.(7) and equate powers in x,y. The first nontrivial order is two:

<u>Order 2:</u>

$$4 \varphi_{20}^2 + \varphi_{11}^2 = \tilde{K}_{20}$$

$$4 \varphi_{11} (\varphi_{20} + \varphi_{02}) = \tilde{K}_{11} = 0$$

$$4 \varphi_{02}^2 + \varphi_{11}^2 = \tilde{K}_{02}$$

Let us first consider the nondegenerate case (K_{20}, K_{02} nonzero and positive). Then we obtain

$$\varphi_{11} = 0 \;,\; \varphi_{20} = \pm \frac{1}{2} \sqrt{\tilde{K}_{20}} \;,\; \varphi_{02} = \pm \frac{1}{2} \sqrt{\tilde{K}_{02}}$$

Hence, up to a sign, we have two solutions: an extremum or a saddle point. By hypothesis, choose the former one. We can continue to find the higher order coefficients:

<u>Order 3:</u>

$$12 \varphi_{20} \varphi_{30} = \tilde{K}_{30}$$

$$8 \varphi_{20} \varphi_{21} + 4 \varphi_{02} \varphi_{21} = \tilde{K}_{21}$$

$$8 \varphi_{02} \varphi_{12} + 4 \varphi_{20} \varphi_{12} = \tilde{K}_{12}$$

$$12 \varphi_{02} \varphi_{03} = \tilde{K}_{03}$$

Note that this is a set of complete linear equations for the terms φ_{30}, φ_{21}, φ_{12}, φ_{03} so we can solve this uniquely in terms of φ_{20}, φ_{02}, previously found. The same thing will happen for all successive orders. So, providing each resulting linear system remains nondegenerate, we can fix uniquely all terms of φ (see figure 3).

We now briefly discuss the case when $D^2 K \vert_{p_0} = 0$. Hence, $D^2 \varphi \vert_{p_0} = 0$ as well. Since φ is assumed to reach an extremum at p_0 then $D^3 \varphi \vert_{p_0}$ must also vanish. Hence, the coefficient equations can only be satisfied if $D^3 K \vert_{p_0}, D^4 K \vert_{p_0}, D^5 K \vert_{p_0}$ all vanish! So the next nontrivial equations are

$$\varphi_{13}^2 + 16 \varphi_{04}^2 = \tilde{K}_{06}$$

$$4\varphi_{13}\varphi_{22} + 24\varphi_{04}\varphi_{13} = \tilde{K}_{15}$$

$$6\varphi_{13}\varphi_{31} + 4\varphi_{22}^2 + 16\varphi_{04}\varphi_{22} + 9\varphi_{13}^2 = \tilde{K}_{24}$$

$$8\varphi_{13}\varphi_{40} + 12\varphi_{22}\varphi_{31} + 8\varphi_{04}\varphi_{31} + 12\varphi_{13}\varphi_{22} = \tilde{K}_{33}$$

$$6\varphi_{31}\varphi_{13} + 4\varphi_{22}^2 + 16\varphi_{40}\varphi_{22} + 9\varphi_{32}^2 = \tilde{K}_{42}$$

$$4\varphi_{31}\varphi_{22} + 24\varphi_{40}\varphi_{31} = \tilde{K}_{31}$$

$$\varphi_{31}^2 + 16\varphi_{40}^2 = \tilde{K}_{60}$$

Also note that the above system is <u>overdetermined</u>; hence, an additional restriction is imposed on the terms in the expansion for K. For example, a possible choice for K to this order is

$$K = K_{60} x^6 + K_{06} y^6 + \dots \qquad K_{60}, K_{06} > 0$$

which gives

$$\varphi = \varphi_{40} x^4 + \varphi_{04} y^4 + \dots$$

(figure 4). The same problem will occur for higher degeneracies.

So we conclude that the degenerate choices for K admitting a solution to Eq.(5) around a singular point is heavily restricted. Note that there is always a solution if Γ is axisymmetric (see below).

<u>Propagation of φ to the rest of Γ.</u>

Once we have found a solution for φ round P_o, we can then proceed to propagate it throughout Γ by the use of the chacteristic lines for Eq.(5), given by setting:

$$\frac{du^i}{dt} = H^i = g^{ij}\frac{\partial \varphi}{\partial u^j} \tag{8}$$

whence

$$\left|\frac{du}{dt}\right|^2 \equiv g_{ij}\frac{du^i}{dt}\frac{du^j}{dt} = K \tag{9}$$

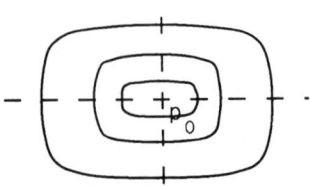

Fig. 4 Example of behavior of K, φ lines near a degenerate point.

By differentiating the latter we obtain, after various manipulations:

$$\frac{d^2 u^i}{dt^2} + \Gamma^i_{jk} \frac{du^j}{dt} \frac{du^k}{dt} = \frac{1}{2} g^{ij} \frac{\partial k}{\partial u^j} \qquad (10)$$

where Γ^k_{ij} are the Christoffel symbols for the metric coefficients g_{ij} defined by:

$$\Gamma^k_{ij} = \frac{1}{2} g^{kd} \left(\frac{\partial g_{ik}}{\partial u^j} + \frac{\partial g_{jk}}{\partial u^i} - \frac{\partial g_{ij}}{\partial u^k} \right)$$

This is the second order o.d.e which governs the characteristic lines. In the isothermal coordinate system defined above, Eqs.(9) and (10) reduce to

$$(9) \Rightarrow \left(\frac{dx}{dt}\right)^2 + \left(\frac{dy}{dt}\right)^2 = \tilde{K}(x,y)$$

$$(10) \Rightarrow \frac{d^2 x}{dt^2} = \frac{\partial \tilde{K}}{\partial x} , \frac{d^2 y}{dt^2} = \frac{\partial \tilde{K}}{\partial y}$$

Note that Eq.(8) states that the characteristic lines are none other than the H field lines themselves.

Then a curve $\{\varphi = \text{const.}\}$ selected round p_0, found by the method described above, together with Eq. (9) applied on this curve, provides Eq.(10) with the necessary conditons to generate a unique congruence of integral lines (figure 5). The function φ is then determined by integrating K along these lines since:

$$\frac{d\varphi}{dt} = \frac{\partial \varphi}{\partial u^i} \frac{du^i}{dt} = K$$

This congruence will in principle cover the whole surface Γ, provided it does not meet a singular line $\{K = 0\}$ on the way, which for instance will happen if p_0 lies inside a singular loop. In fact, if we choose a coordinate system (x,y) locally to such a line L, with y running along this line and x perpendicular to it, then K will be of the form

$$K = x^2 F_2(y) + x^3 F_3(y) + \ldots$$

so Eq. (10) implies

$$\frac{d^2 x}{dt^2} = \frac{\partial k}{\partial x} \sim -x$$

$$\frac{d^2 y}{dt^2} = \frac{\partial k}{\partial y} \sim 0$$

Hence $x \sim e^{-t}$, so that as t tends to infinity the field lines will tend to L but without going beyond it. However, we will now show that the congruence, and therefore φ, can be continued across such a line.

Property 2

Consider a line L on which $K = 0$. Then if φ is known on one side, it is uniquely determined on the other.

Proof:

a) First we show that if φ_1, φ_2 are two solutions to Eq. (5), then either $\varphi_1+\varphi_2$ or $\varphi_1-\varphi_2$ is constant in an open neighborhood of L. In fact:

$$|\nabla_s \varphi_1| = |\nabla_s \varphi_2| = K$$

$$\Rightarrow \quad |\nabla_s \varphi_1| - |\nabla_s \varphi_2| = 0$$

$$\Rightarrow \quad (\nabla_s \varphi_1 - \nabla_s \varphi_2) \cdot (\nabla_s \varphi_1 + \nabla_s \varphi_2) = 0$$

(11) $\quad \Rightarrow \quad \nabla_s p_+ \cdot \nabla_s p_- = 0$

where $p_{\pm} \equiv \varphi_1 \pm \varphi_2$. This means that the lines $\{p_- = \text{const.}\}$ must be orthogonal to $\{p_+ = \text{const.}\}$.

Now, since $\nabla\varphi_1, \nabla\varphi_2$ are zero only on L, then both $|\varphi_1|$, $|\varphi_2|$ must strictly increase only on L. Hence, on each side, either $|p_+|$ or $|p_-|$ must strictly increase in the same way. Assume that on one side $|p_+|$ increases: since both p_+ and p_- are constant on L, then Eq.(11) implies that p_- must be constant everywhere in this neighborhood of L. Hence

$$\varphi_1 - \varphi_2 = const.$$

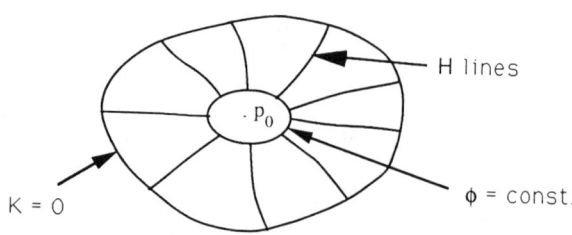

Fig. 5 Field lines near a singular point.

Conversely, if $|p_-|$ increases then

$$\varphi_1 + \varphi_2 = const.$$

b) Now select a coordinate system (u^1, u^2) local to L with $L = (u^1, 0)$. Let $(u_o^1, 0)$ be a point on L at which

$$K(u_o^1, u^2) = K_{2m} u^{2\,2m} + K_{2m+1} u^{2\,2m+1} + \ldots$$

where $m > 0$ and $K_{2m} > 0$ (since K must be a minimum on L). Then Eq. (5) gives

$$\varphi(u_o^1, u^2) = \varphi_m u^{2\,m} + \varphi_{m+1} u^{2\,m+1} + \ldots$$

with

$$\varphi_m = \pm\sqrt{K_{2m}/g^{22}} \neq 0$$

But by analiticity all the derivatives are continuous, so the sign of φ_m must be preserved everywhere. Hence, the "degree of degeneracy" m determines whether φ reaches an extremum at L (m even) or whether it inflects (m odd). This toghether with (a) ensures the unique continuation of φ across L.

Q.E.D

Hence, we can extend φ to completely cover Γ.

However, this will be possible in a well-defined manner only if the surface Γ, and therefore the function K, are such that

a) The H lines don't overlap (except at singular points).

b) The function φ thus constructed remains univalued and well defined all over Γ.

We can easily see that these conditions greatly restrict the class of surfaces admitting a well-defined (analytic) solution φ to Eq. (5).

As an illustration to this point, consider the case when our given liquid surface has $K = 0$ only at two isolated singular points A, B, where therefore the function φ must attain its maximum and minimum. The H field lines then run from A to B, so we can rewrite Eq. (5) in integral form:

$$\varphi(s) = \int^s \sqrt{K(\gamma(s))}\, ds'$$

where γ is a H line and s the distance along it. As φ is required to be a well-defined univalued function, it is clear that the surface must satisfy the condition that

$$\varphi(B) - \varphi(A) = \int_B^A \sqrt{K(\gamma(s'))} \qquad (12)$$

be independent of the chosen H field line γ.

<u>Uniqueness theorem</u>

Given two solutions φ_1, φ_2 to Eq.(5) then

<u>either</u> $\varphi_1 - \varphi_2 = \text{const.}$
<u>or</u> $\varphi_1 + \varphi_2 = \text{const.}$

globally on Γ, so that φ is determined up to a sign and additive constant.

<u>Proof:</u>

Again we have

(11) : $\nabla_s p_+ \cdot \nabla_s p_- = 0$

where $p_\pm = \varphi_1 \pm \varphi_2$. If φ_1 and φ_2 exist, then p_+, p_- also exist and are well defined all over Γ. Hence, from property 1 we know that p_+ reaches a maximum at an isolated point p_0. So the lines p_- must be a dense congruence of lines joining up at p_0. By consequence p_- must be constant in a neighborhood of p_0. The same reasoning applies if we interchange p_+ and p_-. Hence, $\varphi_1 + \varphi_2$ or $\varphi_1 - \varphi_2$ will be constant in this region. Moreover, φ_1 and φ_2 will also attain an extremum at p_0. Hence, the solution φ to Eq. (5) is propagated uniquely to the whole of Γ as described above by starting from this singular point.

Q.E.D

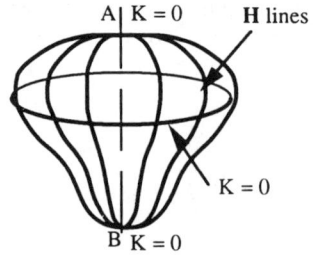

Fig. 6 Axisymmetric liquid shape showing field lines.

Axisymmetric surfaces

In this case, an axisymmetric solution to Eq.(5) can be immediately written down by integrating along the symmetry lines:

$$\varphi(s) = \pm \int^{s} \sqrt{K(s')}\, ds' \tag{13}$$

where s is now the distance along the lines of symmetry, which will in fact be the H field lines (figure 6). By the above theorem, we know that this is the only possible solution. Hence, the surface will admit solutions only if $K = 0$ at the end points A, B. The sign in Eq. (13) depends on the degeneracy of any circle of revolution L on which $K = 0$.

Conclusions on the General Properties of the Inverse Problem

The inverse problem for nonaxisymmetric shaping leads to a relatively complex study of the field on the surface Γ. In practice, the main difficulty lies in determining the surfaces that admit a well-defined solution to Eq. (5).

Note that the condition that φ be univalued shows that physically no current flux can go across the mass of liquid. If the shape of Γ were homeomorphic to a torus, currents would be allowed to circulate and so this condition would not be necessary.

To close this part we discuss the possibility of shaping some particular configurations.

1. Case of the sphere

In this case the function K reduces to

$$K = \tau_2 (z_0 - z) \tag{14}$$

since the curvature C is constant. Hence, K will only have one singular point at the top ($z = z_0$). Hence spheres cannot be formed unless $\tau_2 = 0$ (absence of gravity), which corresponds to the trivial situation where no field is applied. Physically this occurs because there can be no magnetic pressure to balance the hydrostatic pressure at the bottom.

2. Axisymmetric case

If $\tau_1 = 0$ (no surface tension), then K reduces to Eq. (14). Hence, by the the same reasoning no axisymmetric surface can be formed in the absence of surface tension.

3. Two isolated singular points and no surface tension

Again, K is given by Eq. (14). Note that the singular points must be positioned at the same height.

The minimum of $\int_{C_1} \sqrt{z_0 - z} \, ds$

is attainned on the curve C_1 (figure 7). Consider another H line C_2 and the parameter $t \in [0,1]$.

$$I = \int_0^1 \sqrt{z_0 - z_2(t)} \, \frac{ds_2}{dt} \, dt - \int_0^1 \sqrt{z_0 - z_1(t)} \, \frac{ds_1}{dt} \, dt$$

$$\Rightarrow I = \int_0^1 \left[\sqrt{z_0 - z_2(t)} - \sqrt{z_0 - z_1(t)} \right] \frac{ds_2}{dt} +$$

$$\int_0^1 \sqrt{z_0 - z_1(t)} \left[\frac{ds_2}{dt} - \frac{ds_1}{dt} \right] dt$$

$$\Rightarrow I > \sqrt{z_0 - z_1(t_0)} \, (l_2 - l_1) \qquad \text{for some } t_0 \in [0,1]$$

$$\Rightarrow I > 0$$

where l_1 and l_2 are the length of C_1 and C_2. Hence, the necessary condition Eq. (12) cannot be satisfied, and so it is impossible to obtain this configuration.

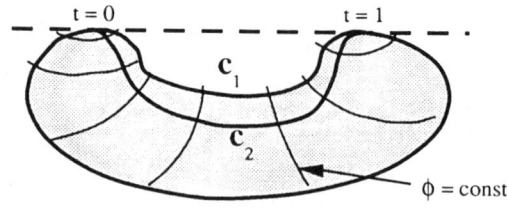

Fig. 7 Example of a liquid shape in levitation with two isolated singular points, in the absence of surface tension.

At first sight the shape in figure 8 may be obtained. Consider the curve C from A to B. Then

$$|\varphi_B - \varphi_A| = \left|\int_C \nabla\varphi \cdot t \, ds\right| \leq \sqrt{\tau_2} \int_C \sqrt{z_0 - z} \, ds$$

t is the unit direction vector of C. Denote C_0 the curves on which

$$\int_{C_0} \sqrt{z_0 - z} \, ds = \inf_{c \in \Gamma} \int_C \sqrt{z_0 - z} \, ds$$

If Γ is possible, a C_0 curve must pass by every point on it. Note that then C_0 will be a magnetic H line.

Integral Formulation of the Exterior Problem and Numerical Results

Our aim is to find a current distribution j_s on a given external surface S enveloping Γ which will produce the required magnetic field H needed to form Γ. Using the same potential formulation, we deduce from Maxwell's equations (applied in a neighborhood of S) that the normal derivative of Φ across S is continuous. Hence, once we have found $\Phi = \Phi^-$ on the <u>inside</u> of s by solving the Cauchy problem Eq. (6), we can find $\Phi = \Phi^+$ on the external side of S by solving the external Neuman problem (figure 9):

$$\Delta\Phi = 0 \text{ in } \Omega_2$$

$$\frac{\partial \Phi^+}{\partial n} = -\frac{\partial \Phi^-}{\partial n} \qquad (n \text{ is the outward normal on } S)$$

$$\Phi \to 0 \quad \text{at infinity}$$

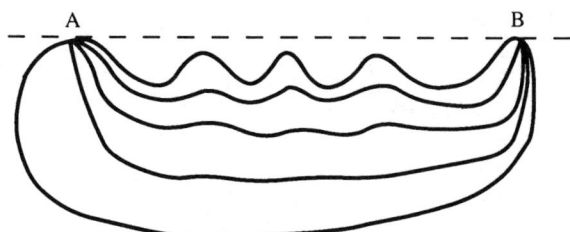

Fig. 8 Example of a possible liquid shape in levitation in the absence of surface tension.

The current j_s is then given by

$$j_s = \Gamma \wedge \nabla \varphi_s$$

where $\varphi_s = \Phi^+ - \Phi^-$: the jump in Φ across S. We call this the <u>stream function</u> of j_s.

The Cauchy-Kovalensky theorem guarantees the local existence of a solution to Eq. (6), providing Γ and $\varphi_\Gamma = \Phi/_\Gamma$ are analytic. Hence, a current j_s can always be found providing S is close enough to Γ. In practice it is much more convenient to find φ_s by an integral formulation of this problem. In general, the potential field produced by a current on a surface Σ with stream function φ_Σ is given by the surface integral

$$\int_{\xi \in \Sigma} \varphi_\Sigma(\xi) \frac{\partial H}{\partial n}(r - \xi) \quad ; H(\omega) = \frac{1}{|\omega|}$$

Hence, the field in our case is given by the sum of two such integrals over Γ and S:

$$\Phi(\Gamma) = \int_{\xi \in S} \varphi_s(\xi) \frac{\partial H}{\partial n}(r - \xi) \, d\varepsilon + \int_{\xi \in \Gamma} \varphi_\Gamma(\xi) \frac{\partial H}{\partial n}(r - \xi) \, d\varepsilon$$

The condition that Φ = const. on the inside of Γ gives us the desired integral equation:

$$\int_{\xi \in S} \varphi_s(\xi) \frac{\partial H}{\partial n}(r - \xi) d\varepsilon = F(\Gamma)$$

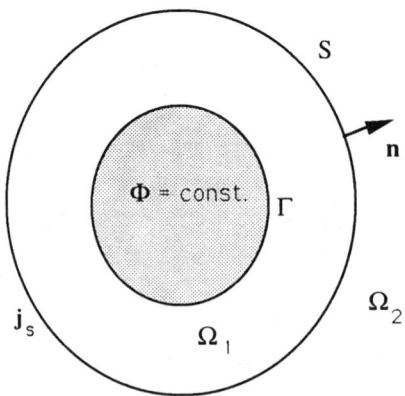

Fig. 9 Diagram showing the domain exterior to the liquid shape.

where

$$F(\Gamma) = \lim_{r \to \Gamma^-} \int_{\xi \in \Gamma} \varphi_\Gamma(\xi) \frac{\partial H}{\partial n}(r - \xi) \, d\varepsilon$$

$$= \frac{1}{2} \varphi_\Gamma(r) + \int_{\xi \in \Gamma} \varphi_\Gamma(\xi) \frac{\partial H}{\partial n}(r - \xi) \, d\varepsilon$$

Here " $\lim_{r \to \Gamma^-}$ " denotes the limit as r approaches Γ from the interior region.

In the axisymmetric case we can use spherical polar coordinates to rewrite Eq. (15) in the form

$$\int_{\theta'=0}^{\pi} \varphi_0(\theta') K(\theta, \theta') \, d\theta = G(\theta) \quad ; \theta \in [0, \pi]$$

where θ is the polar angle, and P, Q are known functions. We conlude this paper by giving some numerical results for a choice of axisymmetric surfaces Γ satisfying the existence conditions stated above (figures 10,11,12). The graph for each Γ shows the current distribution $j_s(\theta)$ for a range of external surfaces S, each being a magnified image of Γ. Note how the current peaks get stronger as the surface becomes larger. Also of interest is the case when Γ is an ellipse (figure 12): the current $j_s(\theta)$ seems to remain invariant! It could well prove motivating to make an analytical investigation of this case.

Appendix in Differential Geometry

Let (u^1, u^2) be a (nondegenerate) coordinate system describing an open subset of Γ by the three-component Cartesian vector function $r(u^1, u^2)$. The surface coordinate vectors $\frac{\partial r}{\partial u^1}, \frac{\partial r}{\partial u^2}$ form a complete basis (called the coordinate basis) for the space $T\Gamma$ of all vectors v tangent to Γ. Hence

$$v = v^i \frac{\partial r}{\partial u^i} \equiv v^1 \frac{\partial r}{\partial u^1} + v^2 \frac{\partial r}{\partial u^2} \quad \forall v \in T\Gamma$$

The vector v is then said to have components v^i (i = 1,2) in this basis.

Define the **induced surface metric** g as the symmetric tensor having components

$$g_{ij} = \frac{\partial r}{\partial u^i} \cdot \frac{\partial r}{\partial u^j}$$

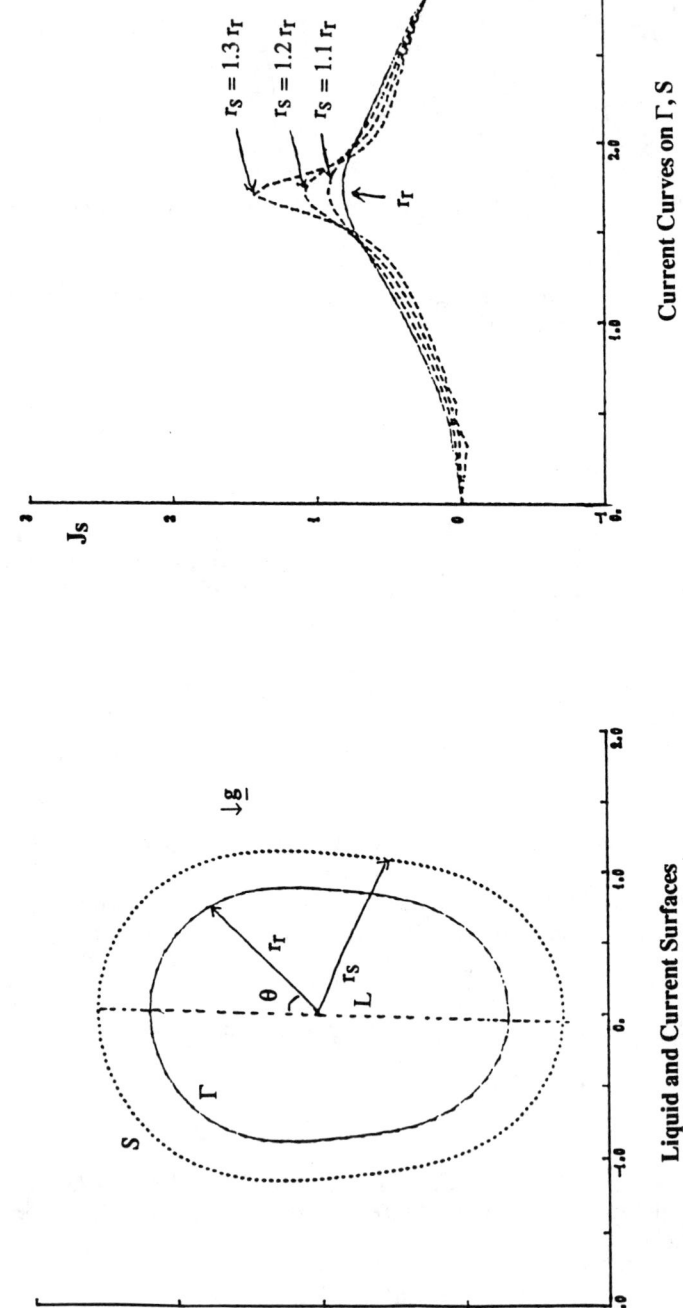

Fig. 10 Numerical results for a given liquid shape.

INVERSE SHAPING

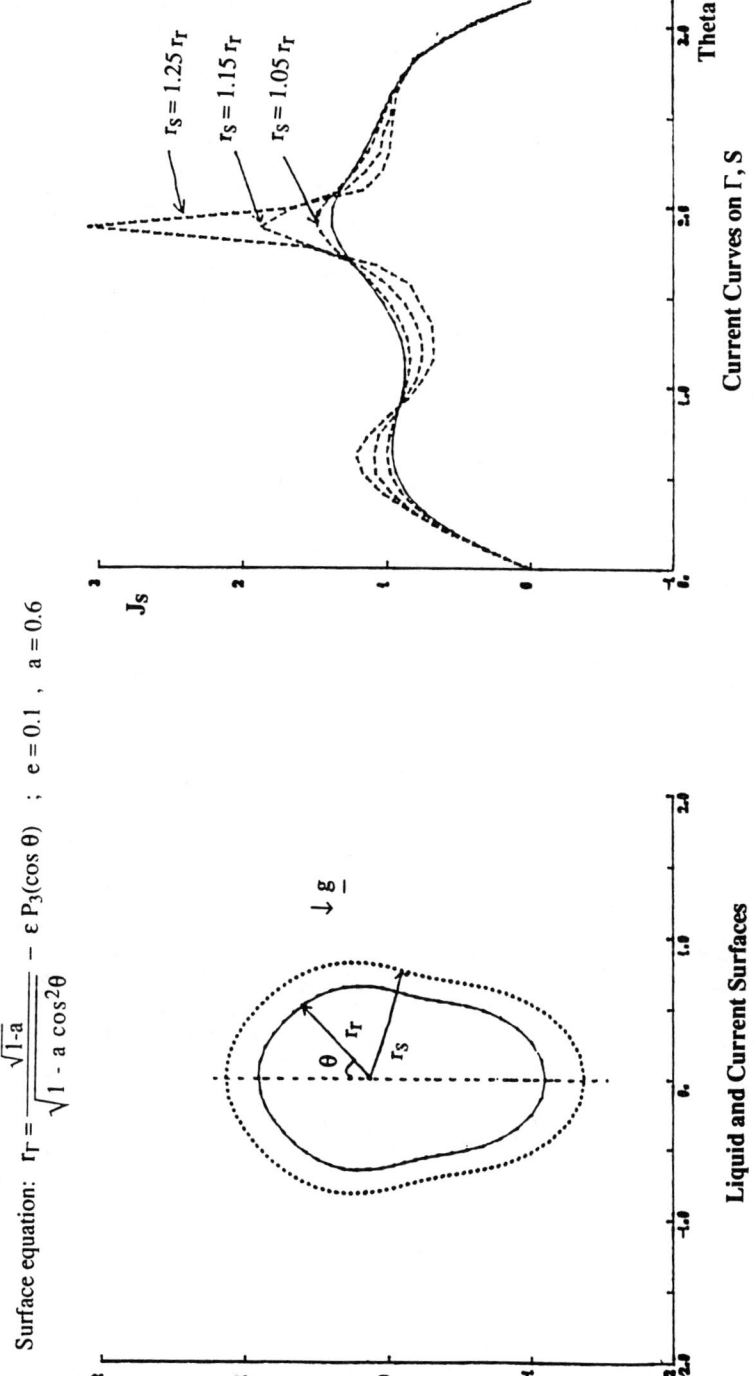

Surface equation: $r_\Gamma = \dfrac{\sqrt{1-a}}{\sqrt{1-a\cos^2\theta}} - \epsilon\, P_3(\cos\theta)$; $e = 0.1$, $a = 0.6$

Fig. 11 Numerical results for a given liquid shape.

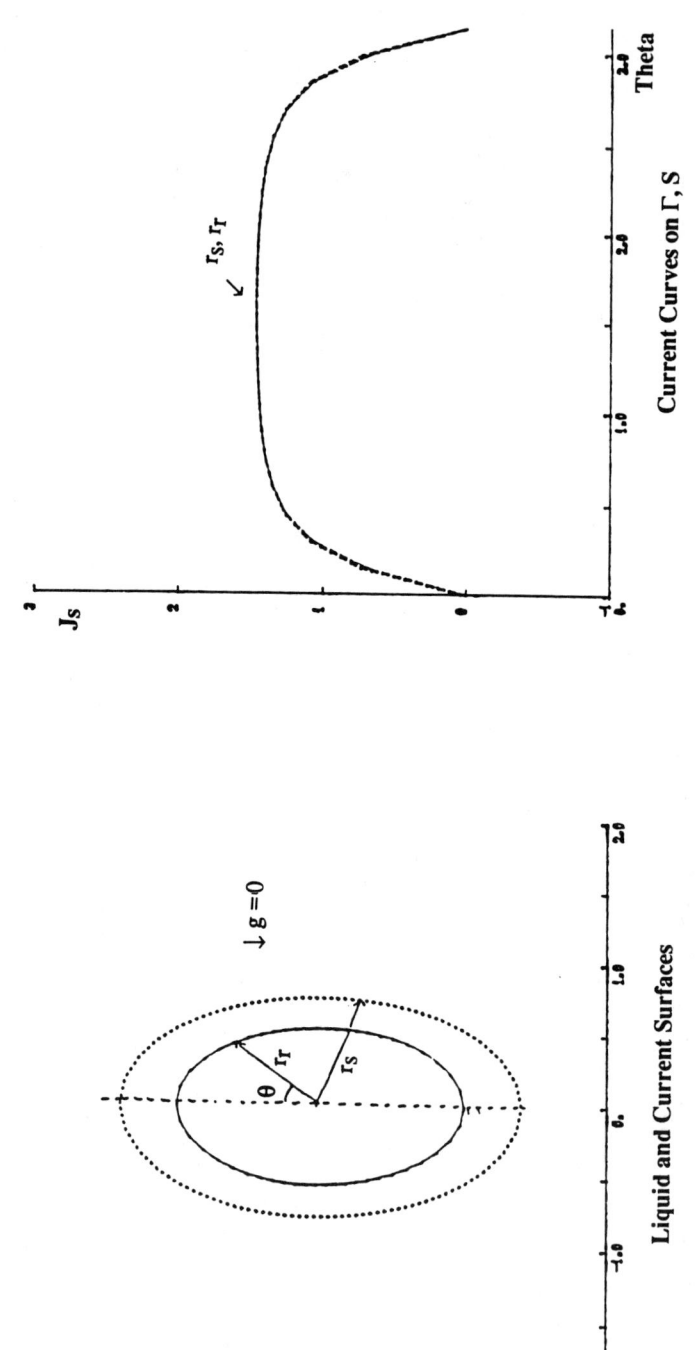

Fig. 12 Numerical results for a given liquid shape.

in this basis. The metric is positive definite if and only if the coordinate basis is nondegenerate. Then the norm of any vector v in $T\Gamma$ is given by

$$|v| = v^i \frac{\partial r}{\partial u^i} \cdot v^j \frac{\partial r}{\partial u^j} = g_{ij} v^i v^j$$

We define the surface gradient $\nabla_s \varphi$ of a function φ defined on Γ as the projection onto Γ of the usual gradient on the Euclidean three-dimensional space:

$$\nabla_s \varphi = \nabla \varphi - (n . \nabla \varphi) n$$

Hence $\nabla_s \varphi$ belongs to $T\Gamma$, so we can write

$$\nabla_s \varphi = \alpha^i \frac{\partial r}{\partial u^i}$$

Then

$$\alpha^i g_{ij} = [\nabla \varphi - (n . \nabla \varphi) n] \cdot \frac{\partial r}{\partial u^i} = \nabla \varphi \cdot \frac{\partial r}{\partial u^j} = \frac{\partial \varphi}{\partial r^i} \frac{\partial r^i}{\partial u^j} = \frac{\partial \varphi}{\partial u^i}$$

since $\varphi = \varphi(u^1, u^2)$. Hence in this coordinate basis $\nabla_s \varphi$ has components

$$\alpha^i = g^{ij} \frac{\partial \varphi}{\partial u^j}$$

where $(g^{ij}) = (g_{ij})^{-1}$.

References

[1] Shercliff,J.A, "Magnetic Shaping of Molten Metal Columns."*Proceedings of the Royal Society of London* A475 (1981)

[2] J.P.Brancher - O.Sero Guillaume "Sur l'equilibre des liquides magnetiques: Applications a la magnetostatique." *Journal de Mechanique théorique et appliqué* vol. 2 n° 2 (1983)

[3] J.P.Brancher - J.Etay - O.Sero Guillaume "Formage d'une lame metallique liquide:calcul et experiences."*Journal de Mechanique théorique et appliqué* vol. 2 (1983)

[4] A.D.Sneyd - H.K.Moffat "The fluid dynamics of the process of levitation melting."*Journal of Fluid Mechanics* Vol.117 (1982)

[5] Gagnoud,A,Sero Guillaume,O,."Le Creuset Froid de Levitation." *Bulletin de la Direction des Etudes et Recherches de l'Electricité de France*, 1985

[6] Brancher,JP,.Gagnoud,A,.*Conference on the computation of electromagnetic fields*, Fort Collins,CO, 1985.

[7] Etay, J,.Gagnoud,A and Garnier,M,. "Le Problème de Frontière Libre en Levitation Electromagnetique."*Journal de Mechanique théorique et appliqué* vol. 5 n° 6 (1986) pp. 911 - 934.

[8] Mestel,J, "Magnetic Levitation of Liquid Metals", *Journal of Fluid Mechanics,*.Vol.117, 1982.

[9] Henrot, A. - Pierre, M. & Brancher, J.P. (1989) "Un probleme inverse en formage des metaux liquides", *Mathematical modelling and numerical analysis* Vol. 23 No.1, p.155.

Chapter 2. Energy Conversion

Major Engineering Physics for Optimization of the Seawater Superconducting Electromagnetic Thruster

J. C. S. Meng*
Naval Underwater Systems Center, Newport, Rhode Island 02841

Abstract

The superconducting electromagnetic thruster (SCEMT) is composed of four major components: the hydropropulsor, the seawater electrode, the superconducting magnet, and the cryostat. This paper summarizes and discusses the issues of the basic engineering physics related to the optimization principles. The basic physics issues of the SCEMT are summarized, including those of the hydropropulsor, the seawater electrode, the superconducting magnet, and the size and weight of the system. Dynamic models of the individual components have been constructed. This paper summarizes the major physics, including meaningful results related to design optimization, which are based on 1) maximum overall efficiency of the hydro-propulsor, 2) maximum Lorentz force given overall system outer diameter, and 3) minimum system total weight given required system thrust.

Introduction

Numerous publications on the subject of magnetohydrodynamic (MHD) propulsors appear in the literature.[1-5]

This paper is declared the work of the U.S. Government and is not subject to copyright protection in the United States.
* Chief Scientist, Submarine Warfare Systems Directorate.

In this study the focus is on engineering physics issues not addressed before and on optimization requirements. The optimization approach is based on maximizing the total system efficiency, rather than only the propulsion efficiency, and on minimizing the system weight. The optimization parameters are the duct diameter or, equivalently, the ratio of inlet and exit velocities, the MHD interaction parameter, and the ratio of magnetic pressure to allowable stress.

The following sections describe the basic mathematics and sample results of the individual components. Analysis of these results leads to an optimization principle included in the overall system design code. Specific outputs from the model are major parameters for three configurations (i.e., the toroid racetrack, the cluster dipole, and double-helix solenoid) and the selected optimization parameter.

Hydropropulsor

An analysis of the existing empirical data base on convergent duct flow and underwater jet thrusters has been conducted. Specifically, this analysis focuses on the hydrodynamic design of the water inlet, duct, nozzle, and length of the SCEMT. The potential for using the lateral inlet technique is also indicated. Ultimately, the thruster hydrodynamics design can be optimized by use of parametric relationships derived from the overall system efficiency to maximize efficiency and minimize the jet noise, the peak electric current, the electrostatic potential, and the magnetic induction.

To prepare for the later discussion of frontal and lateral suction analyses and the relative contributions of momentum and pressure thrusts, it is useful to derive the drag and thrust relations from the basic integral equations. The control volume surrounding an axisymmetric body with either frontal or lateral suction inlet forms the surfaces s_1, s_2, and s_3; the large cylindrical surface is denoted by S, and the exit jet area is given as A_o. The in-flow plane A_{in} and the outflow plane are the planes upstream and downstream of the axisymmetric body under consideration. The thrust and drag balance and the pressure equations can be shown as follows:

$$\int_{A_o} \rho(u^2 - uU_o)dS + (p_{exit} - p_\infty)A_o$$
$$= -\int_{s_1+s_2+s_3} \vec{n} \cdot \vec{e}_x p\, dS + \iint_{s_1+s_2+s_3} \vec{e}_x \cdot (\tau\vec{n})dS \quad (1)$$

where ρ, u, U_o, and τ are the fluid density, velocity, freestream velocity, and the shear stress tensor, respectively; and P_{exit} and P_∞ are the pressures at exit plane and at infinity, respectively. The $\vec{e}_x$ and $\vec{n}$ are unit vectors in the axial directions of the axisymmetric body and normal to the body surfaces s_1, s_2, and s_3.

Equation (1) is the fundamental relationship between the thrust and drag terms and offers a clear definition of the thrust-drag balance. On the left-hand side of the equation are the thrust terms, the first of which is the commonly known momentum thrust, which will be used later in one-dimensional form as $\rho(V_o^2 - V_o U_o)A_o$. The second term, the pressure thrust, is unique to the thruster exit configuration for hydrodynamic applications on neutrally buoyant bodies.

On the right-hand side of the equation the pressure integrals always contribute a form drag. However, if the exit plane is replaced by a diffuser vs the nozzle, there would be a pressure thrust from the integrals over the s_2 and s_3 surfaces. For subsonic flow conditions this pressure term is small, but a tradeoff between the momentum and pressure thrust can be a significantly useful option. To maximize momentum thrust, the jet velocity must be maximized. While the pressure thrust is maximized, the jet velocity must be minimized, so that minimum turbulent flow noise will be generated. This issue has not been raised before and may have a significant impact on thruster design methodology. Also on the right-hand side are the "drag" terms. The drag includes the form drag, which in one-dimensional form will be $(p_\infty - p_{base}) \pi D^2/4$, where p_{base} is the pressure on the stern side of the thruster. Naturally, for the lateral inlet case, an additional suction penalty will be the suction power at the space between s_2 and s_3.

It is also clear that it makes no difference whether a frontal or lateral inlet is applied, as far as the total thrust is concerned. By using the lateral inlet, additional frictional drags will be incurred over surface s_2. However, the lateral inlet does offer two advantages. One advantage is that the inlet flow velocity can be tailored to a low value by use of a long suction surface; the other advantage is that there is additional freedom to place the inlet anywhere along the slender axisymmetric body surface.

One further step can be taken to derive the energy balance integral equation. Starting from the energy equation, apply the steady-state condition and the incompressibility condition. Then convert the volume integrals into surface integrals, applying over surfaces S, s, and A_{in}, combining the terms, and rearranging, it can be shown that

$$p_{exit} - p_\infty = \frac{1}{A_o V_o} \iiint \vec{u} \cdot (\vec{J} \times \vec{B}) dV$$

$$- \frac{1}{A_o A_o} \iiint_{A_o} \frac{\rho}{2} u (U_o^2 - u^2) dS - \Delta p_{loss} \quad (2)$$

where

$$\Delta p_{loss} \equiv - \frac{1}{A_o V_o} \iint_{s_1 + s_2} (\tau \vec{n}) \cdot \vec{u} dS$$

is friction-induced pressure loss. Equation (2) is the basic energy balance equation.

Equations (1) and (2) are the basic equations. Using the physical laws of hydrodynamics, magnetics, and electrics, they can be simplified into one-dimensional forms. From this analysis the propulsion efficiency, which is the ratio of the useful vehicle propulsion power to the total electric power input, can be shown as

$$\eta = 2\frac{U_o}{V_o} \frac{A_o}{d^2} \frac{(1 - U_o/V_o)}{[1 + K - (U_o/V_o)^2]\{1 + \xi(V_o/U_o)[1 + K - (U_o/V_o)^2]\}} \quad (3)$$

where V_o is jet exit speed, d is the jet diameter, and K is the inlet loss coefficient.

Equation (3) expresses the efficiency in terms of the geometry, flow parameters, and the inverse magnetohydrodynamic interaction parameter ξ. To start with, some insight can be obtained by setting K = 0, i.e., no Δp_{loss}. It is worth noting that when $U_o = V_o$, i.e., in the limiting condition of vanishing thrust, such as at the instant of starting up, the efficiency becomes

$$\eta = \pi/4$$

a pure geometric result. In another limit, i.e., $U_o/V_o \to 0$,

$$\eta \sim \frac{A_o}{2d^2} \frac{1}{1+\xi} \sim \frac{\pi}{2\xi}\left[\frac{U_o}{V_o}\right]^2$$

where the fact that

$$\xi \equiv \frac{\rho U_o}{2\sigma B^2 L_s} > 1$$

has been applied. The efficiency η, in this limit, can then be expressed as

$$\eta \to \pi \frac{\sigma B^2 L_s}{\rho U_o}\left[\frac{U_o}{V_o}\right]^2$$

This equation states that η increases quadratically with B and linearly with σ and L_s, while decreasing with V_o, a result that clearly indicates the importance of achieving higher B values.

To optimize the thruster design, the thruster diameter d is one of the most importance parameters. One way to choose the d is to maximize the efficiency η. Expressing

$$\frac{U_o}{V_o} = \left[\frac{d}{D_c}\right]^2 = \tilde{d}$$

where D_c is the capture area diameter, the following analysis can be done.

The efficiency has at least one optimum value of d that can be established examining the asymptotic behaviors as $\tilde{d} \to 0$ and $\tilde{d} \to 1$. Taking the limit of $\tilde{d} \to 0$, it is found that

$$\frac{\partial \eta}{\partial \tilde{d}} = \frac{2\pi \tilde{d}^3}{\xi(1+K)^3} \to 0$$

but > 0, and as $\tilde{d} \to 1$ results in

$$\frac{\partial \eta}{\partial \tilde{d}} = -\frac{\pi(1+K)}{K(1+K+\xi K)} < 0$$

Therefore, it is seen that $\partial \eta/\partial \tilde{d}$ as a function of $\tilde{d}$ changes sign as $\tilde{d}$ is increased from 0 to 1. This is an important part of the overall optimization procedure.

A dimensional analysis can easily be done to find the key dimensionless parameters controlling the system performance. Assuming that the system is isothermal and incompressible, the following set of variables is considered to be the most relevant:

Dynamic: E, B, U_o, V_o
Fluid property: σ, ρ, μ (viscosity)
Geometric: L_s, d

Choosing E, B, U_o, and ρ as the basic variables, the nondimensional parameters can be determined as the interaction parameter $\sigma B^2 L_s/\rho V_o$, the jet ratio V_o/U_o, the load parameter E/BV_o, the Reynolds number $\rho(U_o L/\mu)$, and the aspect ratio L_s/d. In principle, the efficiency should be expressed as a function of all the five aforementioned dimensionless parameters. However, the most significant dependence

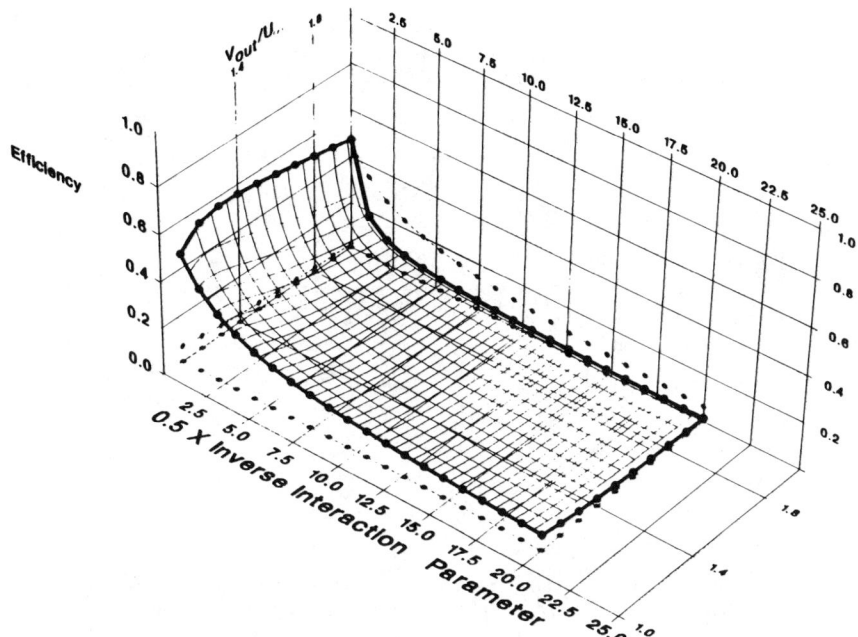

Fig. 1 Efficiency vs inverse MHD interaction parameter and velocity ratio.

has been found on the interaction parameter and the velocity ratio. Therefore, for illustration, the efficiency vs the velocity ratio and the inverse interaction parameter are shown in Fig. 1. This figure shows that there is a maximum relative to the velocity ratio. It also shows the quadratic increase due to the increase in B in the interaction parameter. Figure 2 illustrates the general phenomenology and design factors for the MHD channel dynamics.

Seawater Electrode

The main source of energy for the SCEMT system is derived from the direct current (dc) field applied to the electrodes. The electrolysis of the seawater takes place while an amount of direct current is transferred at a voltage. The mathematical representation of the phenomena that take place

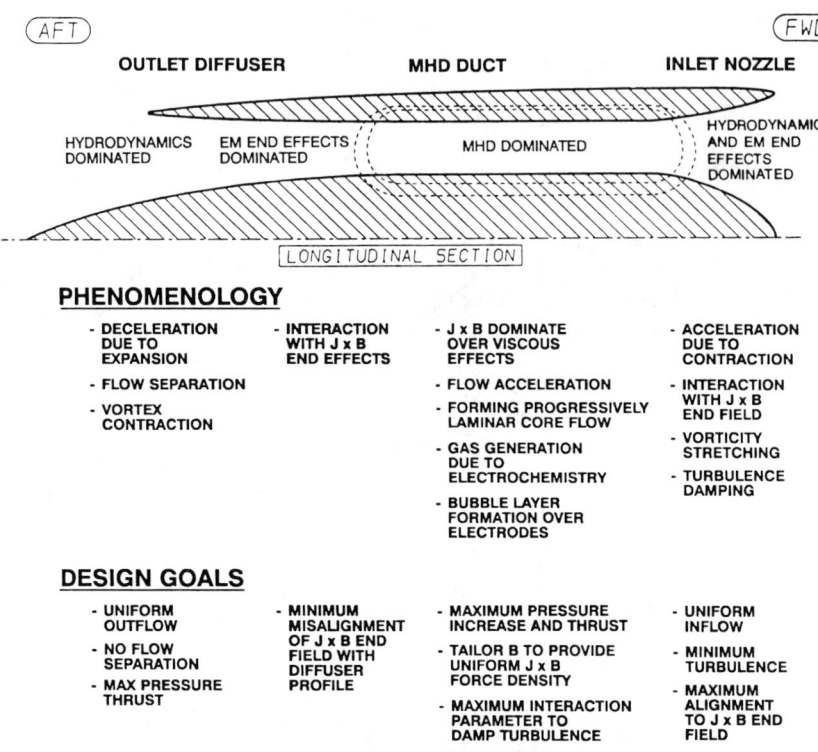

Fig. 2 Phenomenology and design factors for MHD channel dynamics.

during the transfer of dc current is the focus of this section.

The chemical reactions that take place at the electrodes during the dc electrolysis of a low flow speed aqueous alkaline solution of sodium chloride are as follows:

At the anode (oxidations)

$$2Cl^- \longrightarrow Cl_2 + 2e^- \tag{4}$$

$$2H_2O \longrightarrow O_2 + 4H^+ + 4e^- \tag{5}$$

At the cathode (reductions)

$$2H^+ + 2e^- \longrightarrow H_2 \tag{6}$$

$$2H_2O + 2e^- \longrightarrow 2OH^- + H_2 \qquad (7)$$

In the bulk of the liquid, then

$$Cl_2 + 2NaOH \longrightarrow NaClO + NaCl + H_2O \qquad (8)$$

In the absence of an alkaline medium, the generated chlorine undergoes immediate hydrolysis according to the following reactions:

$$Cl_2 + H_2O \longrightarrow HClO + Cl^- + H^+$$

$$HClO \longrightarrow ClO^- + H^+$$

The fundamental chemical and electrochemical principles of the electrolysis of sea water dictate the voltage requirements for the discharge of a given amount of current. There are different overvoltages that contribute to the overall cell voltage. They are: the reversible voltage, chemical polarization, concentration polarization, and the ohmic loss. The reversible voltage results from the net chemical effect from reactions (4) and (5) and is described by the Nernst equation. The chemical polarization voltage results from the charge transfer inhibitions at the electrodes. This is determined by the catalytic activity of the electrodes and the surface roughness and is given by the Tafel equation. The concentration polarization voltage results from the concentration gradients that develop on the boundary layer around the electrode during the electrolysis process.

In general, the concentration polarization contribution is small under turbulent flow conditions and relatively fast reactions (i.e., good catalysts as electrodes). For the turbulent flow conditions for marine applications, no concentration polarization effects are expected. The ohmic loss voltage is found to be by far the largest contribution to the overall cell voltage.

One of the key issues is the limiting current and voltage in seawater. This can be estimated by means of several approaches. Two of them are described briefly in the following paragraphs.

First, consider mass transfer based on the Reynolds analogy. Under turbulent conditions, the heat-transfer coefficient inside a pipe can be given in terms of the Reynolds and Prandtl numbers. Then the limiting current density value can be expressed in terms of the Faraday constant, the flow speed, and the number of electrons transferred due to chemical reactions. Results indicate that, for a seawater speed of 10 m/s, the expected limited current density value should be approximately 10^{14} A/m^2. This value is exceedingly higher than the expected operating values for the SCEMT system. Thus, no limiting current density values are expected from this consideration.

Second, consider mass transfer based on concentration polarization effects. When the ionic transport is not adequate to fulfill the reaction requirements at the surface of the electrode, then a variation of concentration with distance from the interface toward the bulk of the solution will be present for the seawater electrolysis at steady-state conditions. This concentration gradient at the interface then drives the diffusion flux of reacting species at the surface of the electrode. Thus, using Fick's law of diffusion, it can be found that this is not a limiting current value for chlorine generation at the anode. Similarly, experimental data for hydrogen generation at the cathode indicate again that this is not a current limiting factor at the cathode.

However, the issue of bubbly effects in a turbulent flowfield on the uniformity of electrical conductivity has not been addressed. Intuitively, one can conceive that the current density along a path between bubbles can vary drastically and cause localized vaporization, which may further increase the local current density. Such arcing instability has not yet been addressed.

Gas generation rate calculations have been conducted (at 25°C and atmospheric conditions), under the assumption that the current efficiency is 100% for both electrodes and assuming that either chlorine or oxygen is generated at the anode and hydrogen at the cathode. The results indicate that the volume of gas generated is small. Furthermore, the volumetric

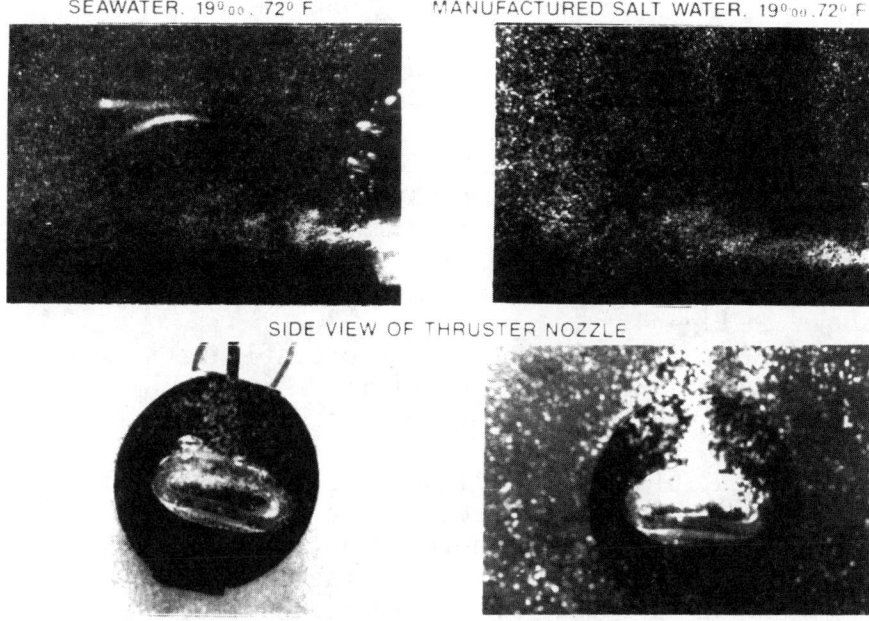

Fig. 3 Subscale EMT model comparision of seawater and manufactured saltwater.

percentage of this gas, as a fraction of the total volume of seawater flowing through the SCEMT duct, assuming uniform mixing, is a small number and corresponds to less than 10 ppm.

Some of the total energy provided by the SCEMT system is dissipated as heat. This heat is then transferred into the flowing seawater in front of the electrodes. The resulting temperature increase is less than 1°C, again assuming uniform mixing.

One interesting phenomenon is the apparent contrast of the gas and bubble generation rates in seawater and fresh water mixed with sea salt of equal salinity. Figure 3 shows the gas bubbles generated from the thruster in a magnetic field of 0.3 T in seawater and saltwater with the same salinity of 19 parts per thousand. The difference may be due to the different chemical composition and the fact that bubbles in seawater are coated with an organic film that is absent in artificial saltwater.

Superconducting Magnet

The use of three configurations of a superconducting magnet in MHD thrusters for direct propulsion of vessels in seawater has been examined. The object of this study has been the comparison of the characteristics of the three magnet configurations, which are (1) the racetrack toroid, (2) the clustered dipole, and (3) the shielded solenoid. The general arrangement of these configurations is shown in Fig. 4.

The characteristics of these configurations that affect their suitability as elements in SCEMT include field strength, uniformity, orientation, fraction of total field energy in the duct, specific thrust, confinement of magnetic field, and complexity of construction. Some of these are clearly quantifiable, whereas other characteristics are difficult to quantify but, nevertheless, have a significant influence on system performance, reliability, and cost.

The principle on which any SCEMT system depends on the passage of a current through seawater perpendicular to a magnetic field. The Lorentz force density is the vector product of local field and current density. In general, the thrust density will vary in magnitude and direction from point to point within the hydraulic duct. This variation arises from the nonuniformity of the field and current density. In all practical configurations either the current density in the water or the flux density or both will vary spatially. The significance of spatial variation of thrust density is that a shear will arise in the thrust density from which turbulence and reduction of duct efficiency may result. Therefore, the minimizing of this nonuniformity is critical in selecting the configurations of the magnet and the thruster duct.

It is convenient to consider the quantifiable characteristics in terms of a figure of merit (FOM). This permits simple comparison between configurations. Several suitable FOMs are, for example, the ratios of Lorentz force to total stored energy or quantity of conductor and the fraction of stored magnetic energy in the thruster duct. Each of

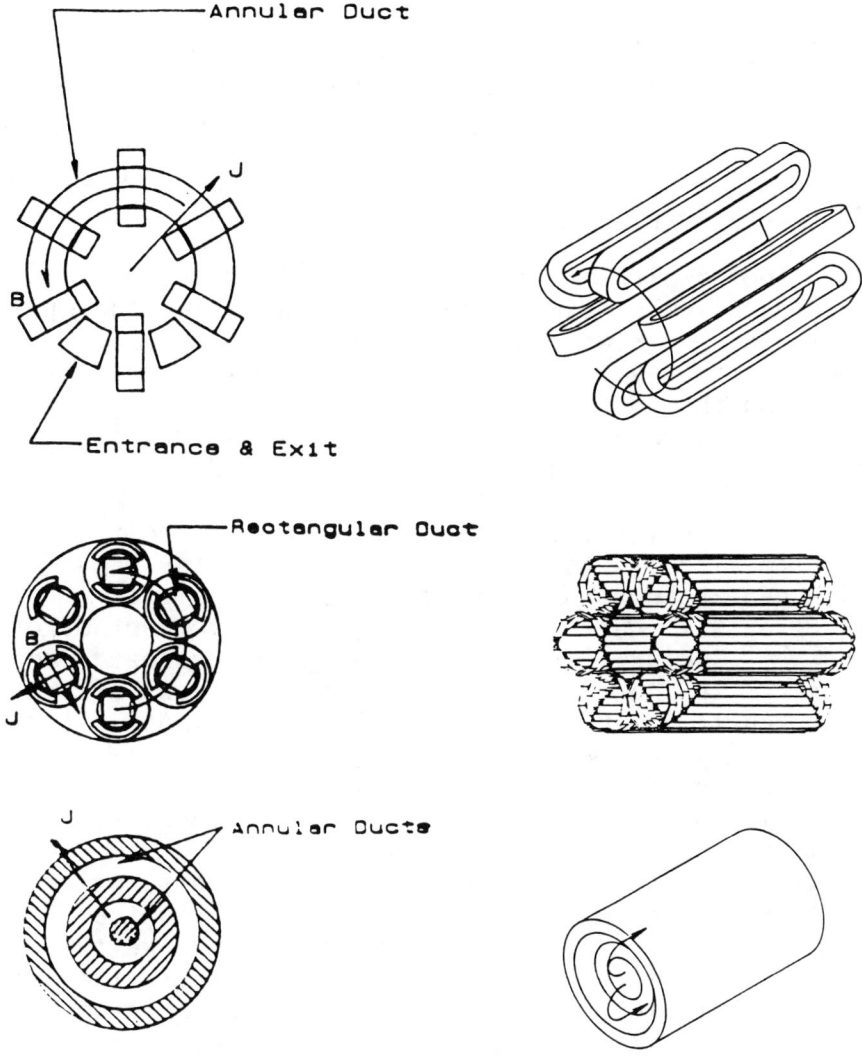

Fig. 4 Schematic configurations for the racetrack toroid, clustered dipole, and shielded solenoid.

the first two broadly describes the cost of the magnet and the last is a representation of the magnet, efficiency.

The nonquantifiable characteristics include the stability (against premature quenching) of the superconducting winding or the topological complexity of the winding. These, of course, are difficult to factor into a FOM but, nevertheless, play an important, if not pivotal, role in the selection of a preferred configuration.

In the following paragraphs each configuration is briefly described, and some sample results are given.

Racetrack Toroid

The simplest approximation to the racetrack toroid, allowing rapid computation of field, energy, and forces, is two concentric hollow conductors. However, this approximation underestimates the peak field on the inner straight sections. Thus, a detailed calculation allowing the current density in the superconductor as a function of field is necessary.

In principle, the great advantage of the racetrack toroid is that a large part of the magnetic flux generated within the coils is available for reaction with current to produce thrust (excepted, of course, is flux within the cryostat vacuum spaces and internal to the windings). In contrast, much of the flux generated by the clustered dipole is not in the ducts. However, the racetrack toroid has the disadvantage that the water flow paths are constricted and tortuous at the entry to and exit from the annular duct region where they are restricted by the coil subassemblies. The turbulence and pressure drop in these areas increase as the number of coils decrease. This occurs because the available flow cross section at the entry or exit decreases rapidly as the number of coils increases unless the width of each coil is decreased. However, that would lead to a greater peak field on the coil and a lower average field in the duct. Also, the average azimuthal component of duct field, which produces thrust, would be lower. The optimum number and width of coils can only be determined by an integrated coil-duct trade study, which is beyond the scope of this study.

The forces and stresses developed in the inner set of straight bars of the racetrack coils must be balanced. The cross section of structure in these straight bars is derived from a combination of the hoop force per meter, the radial force per meter, the hoop compression, the winding thickness, the tensile stress in the bars, and the allowed stress in the winding.

The tensile structure is wrapped around the outside of the winding that is, in turn, located in the annular space defined by the concentric cylinder of the cryostat. Therefore, there will be regions around and between the forms with neither structure nor windings. The fraction of the annular space occupied by winding and structure is the "utilization." In general, as previously described, the larger the number of coils the greater the utilization, but the more restricted the entrance and exit ducts.

Clustered Dipole

The dipole with a winding current gradient (amperes per meter of circumference), which is distributed in proportion to the cosine of the angle from the midplane, generates a uniform field within the bore. The particular advantage of a uniform field is that, together with uniform current density, it produces uniform thrust density that is, at least theoretically, free of shear and turbulence. When several dipoles are clustered, however, the uniformity of the field is degraded. The field is increased most at the extreme corners closest to the neighboring ducts and at the ends. These deviations from uniformity can be addressed by relocating some of the windings.

Because the winding and associated structure extends around the whole circumference, the individual duct regions of a clustered array are separated. A substantial volume of the field therefore cannot generate thrust.

The forces in the cos θ dipole are supported partly by the longitudinal tensile structure (including the winding, as in the racetrack coil), partly by the circumferential compression in the winding, and by bending in a circumscribing ring structure.

Shielded Solenoid

The general arrangement of this configuration is shown in Fig. 4. The current flow in the shielded solenoid duct is radial between concentric cylindrical electrodes. This current interacts with the axial component of field to generate an azimuthal thrust. Vanes must be used to convert this thrust to the helical direction so that an axial component is produced.

The radial current density in the duct varies as $1/r$. This leads to varying thrust density and shear, with serious problems of shear and turbulence. A series of concentric electrodes will reduce the radial variation in current density by allowing additional current to be introduced at increasing radial increments. But, in practice, this is difficult and defeats the simplicity of the SCEMT. The designer must also recognize the limits imposed by practical considerations of thickness of the cylindrical electrodes and thickness of any helical vanes.

To maximize the magnet confinement, a relationship between the number of coils, the number of turns, the current, and the mean radius of each coil must be satisfied.

Figures 5-7 show the calculated results for the racetrack toroid, the clustered dipole, and the shielded solenoid, respectively. The parameters plotted against field are the integrated Lorentz force in the duct, the stored energy in the magnet and the fraction associated with the useful field in the duct, the tesla-ampere meters of conductor, and the power dissipated in the duct assuming 0.25 Ω-m resistivity and 500 A m^{-2} of electrode current density.

It must be noted that in the toroid and dipoles the Lorentz force vector is parallel to the system axis, whereas in the solenoid it is in the azimuthal curvilinear vector direction. The force vector varies for the three configurations, with the solenoid the lowest, the clustered dipole next higher, and the torus the highest. This is a direct consequence of more area in the torus (ignoring limitations of entrance and exit constructions).

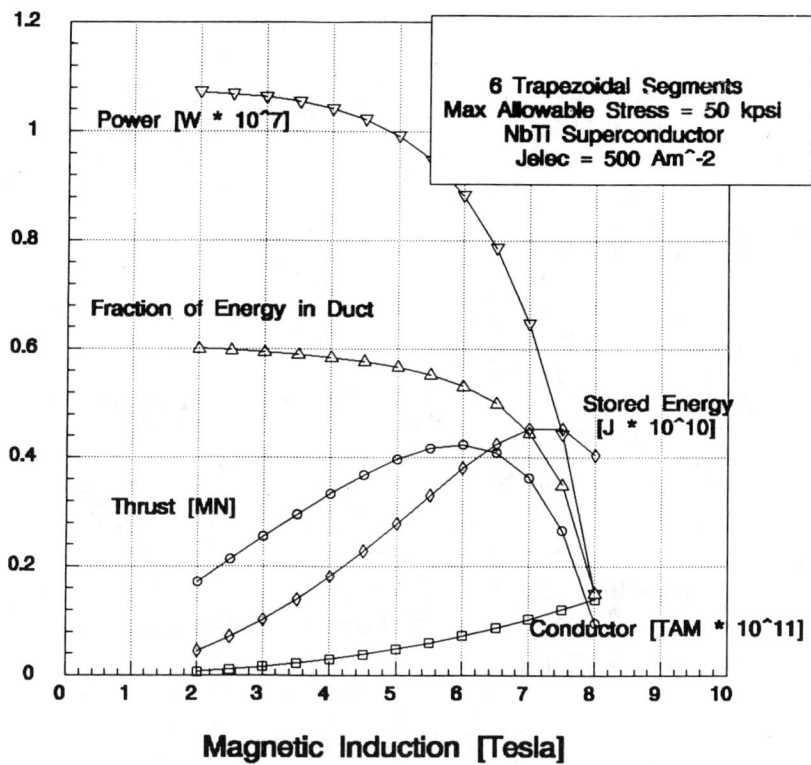

Fig. 5 Performance characteristics for racetrack toroid.

For the racetrack toroid configuration both field and current density vary as 1/r, and the resulting Lorentz force will be a function of the chord factor, which is defined as the ratio of the outer radius of the channels to the inner radius. The performance curves of Fig. 5 assume no field nonuniformity, which can be accomplished by either shading the current density or by tailoring the winding to produce a B field that increases with r. However, it yields optimistic results because the implied hydraulic configuration may be difficult or impractical.

For the solenoid case the field is approximately uniform as a function of the radius; however, the current density varies as 1/r. A duct in which the current density is uniform between electrodes can be

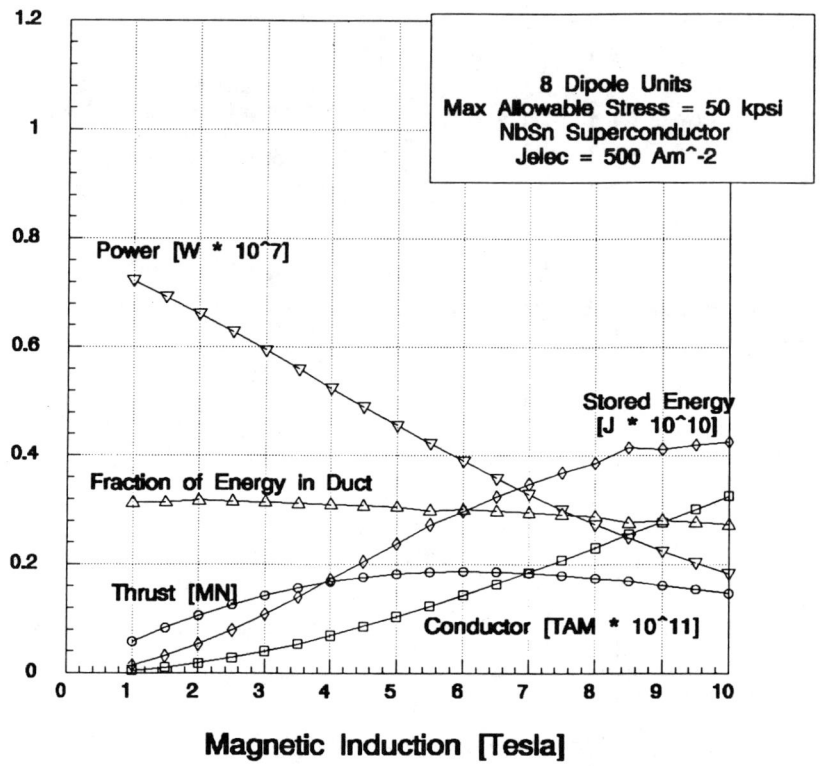

Fig. 6 Performance characteristics for clustered dipole.

achieved by the use of vanes with trapezoidal cross sections. The thrust will be reduced by the thickness of the electrodes and by the thickness of the vane. It should be noted that, in a practical sense, there is less reduction in thrust by use of the uniform channel than at first appears, because accommodation for power cables is required in any case and is provided by the trapezoidal vane. However, the power dissipation in the channel will significantly increase by the layers of electrodes. Most of the thrust of the shielded solenoid is generated in the shielding (low field) region; furthermore, the center field of the solenoid does not uniquely define the configuration, unlike the dipole and toroid. Therefore, the variation of

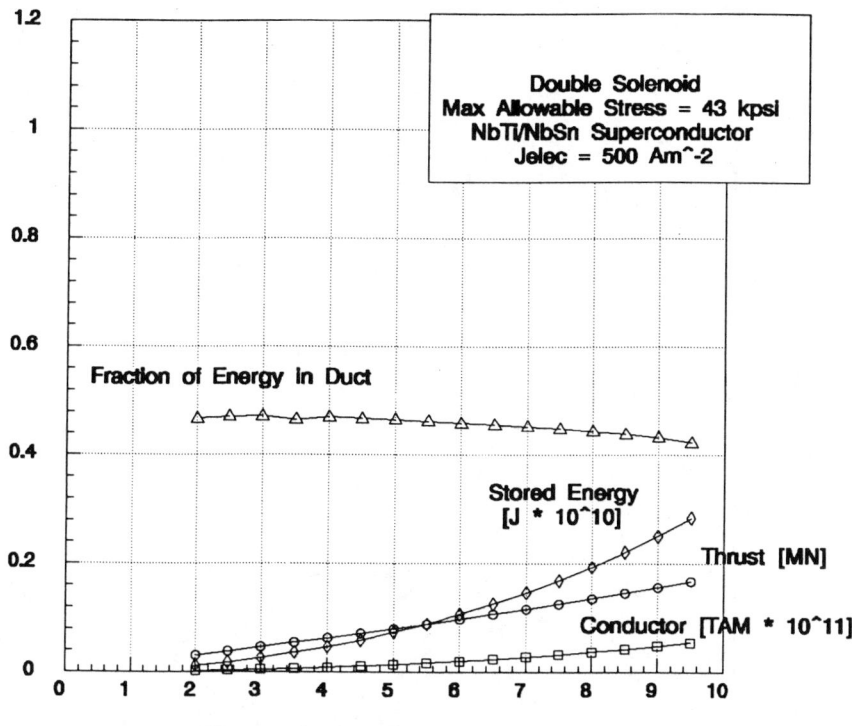

Fig. 7 Performance characteristics for shielded solenoid. (Thrust is the axial component of total static thrust.)

thrust as a function of maximum field shown may be misleading.

The comparisons among the three configurations implied by Figs. 5-7 are not (indeed, cannot be) based on identical assumptions. Because of the small innermost diameter of the solenoid ducts, two additional electrodes have been assumed for that case. In the dipole and toroid no extra electrodes have been assumed. This leads to a relatively larger solenoid thrust.

Figures 8-10 show the comparison of three FOMs. The relative ranking does not change with the particular FOM; the racetrack toroid ranks the highest. However, the decrease in thrust at high fields in the toroid is sharp. The duct volume

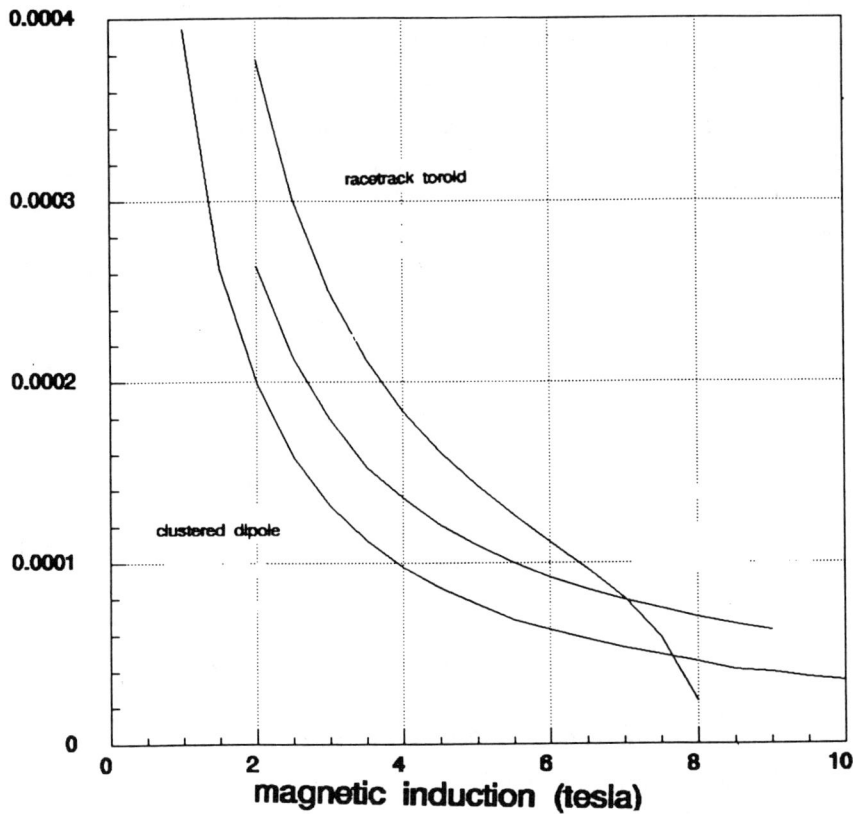

Fig. 8 Lorentz force per unit of stored magnetic energy.

decreases rapidly as the winding volume increases. In the other two configurations the decrease is slower. In the solenoid, little thrust is developed in the high field duct so that, even though duct volume is decreasing with field, it does not rapidly affect the total thrust as it does in the toroid. If the field parameter for the solenoid had been in the outer duct instead of the center, the cutoff of the solenoid thrust would have been at least as rapid as that for the toroid.

Size and Weight

This section presents estimates of size and weight based on a simple model of the SCEMT

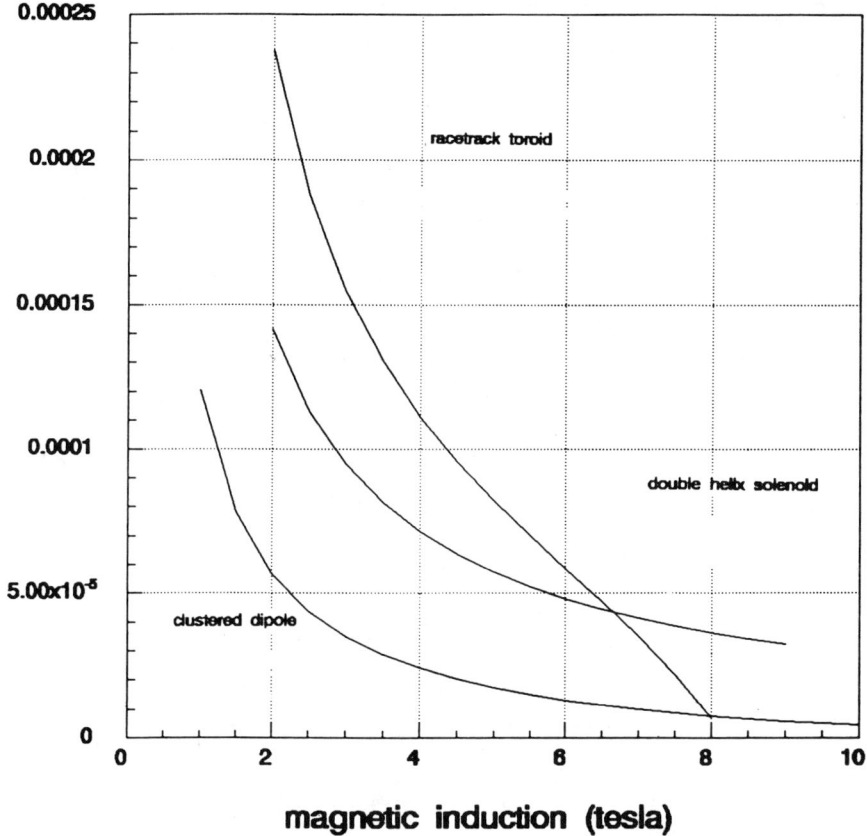

Fig. 9 Lorentz force per unit tesla-ampere-meter.

propulsor. The model assumes that the propulsion system is composed of a number of thruster cells, each contributing an equal amount of thrust. The total thrust from the group of cells is balanced by the drag from the vehicle and the thrusters.

The dimensions of the magnet primarily depend on the magnetic induction, the electric current density in the magnet winding, and the working stress of the structural support system and the cryostat. Once the initial values for the physical dimensions for the thruster, including the winding build, structural support, and the cryostat, are specified, the size and weight of the thrusters are calculated based on

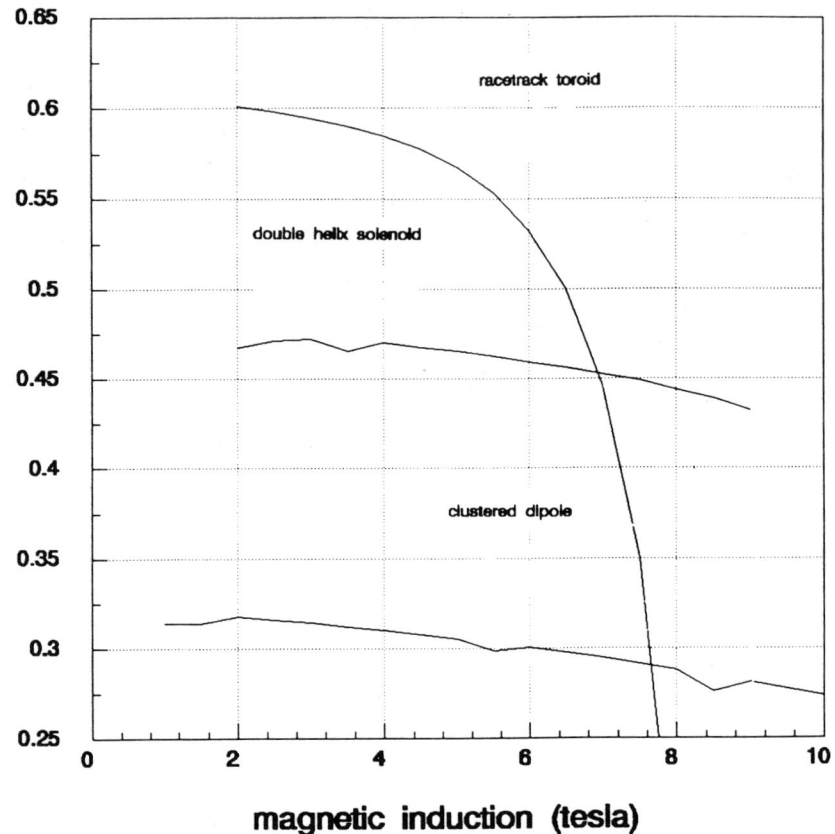

Fig. 10 Fraction of magnetic energy in the MHD duct.

these parameters and other geometric factors. The resultant size of the thruster does have an effect on the overall drag of the system and hence the thrust requirement. This suggests an iterative procedure; such a scheme has been implemented, and the solution converges rapidly.

Winding Build

Calculations of the magnet winding, structural support, and cryostat are based on formulas derived in two papers by Stekly.[6,7] Each magnet is assumed to be a dipole or saddle magnet with the conductor running along the sides of the duct parallel to the flow. The cross section of the winding build is modeled as a cosine distribution. The thickness of the winding varies sinusoidally from zero to a maximum at a position 90 deg from zero and back to zero at the position diametrically opposed from the first zero. The cosine distribution has been chosen for the model because it is a good approximation to most practical winding builds, it is easy to analyze, and it leads to a uniform field. End effects have been neglected, although the saddle geometry requires that the conductor wrap around the duct at each end of the winding.

The amount of material in the winding, its thickness, and its mass depend on the current density flowing in the conductor. The higher the current density, the lighter and more compact the winding. However, conductors with higher current densities have less copper for protection, and the magnet design is less robust and more susceptible to damage from quench. Compact windings also are desirable because the windings closer to the duct are more effective in generating a field in the duct.

According to this model, the winding build is linearly proportional to the magnetic field and inversely proportional to the current density.

The current density in any particular application is a compromise between the need for a compact lightweight winding and questions relating to the stability of operation, and a safe nondestructive shutdown should the superconductor, for any reason, become normally resistive. The lower current densities are generally associated with magnets having large magnetic energy because of the larger mass in the magnetic field. This stored energy must be removed or dissipated when the magnet becomes normal.

Magnet Structural Support

High field magnets are subjected to large forces. The local force exerted per unit volume of winding or conductor is equal to the current density in the winding times the local magnetic field. These forces act directly on the conductors carrying the current so that the windings in the transverse field magnets are compressed and also must be supported by a structure.

The simplest structural element is a cylinder, which just fits over the outer winding layers. This applied load must be supported by the structure. Based on these considerations, the formulas for the loads, bending moments, stresses, and thickness for a cylindrical support structure can be obtained.

The main factors that need to be considered are the bearing stress at the winding-structure interface, the stress in the structure for a given radial thickness, or, alternatively, the required thickness to keep the structural stresses to an allowable level. The required radial structural build is a function of a single parameter b, which depends on the ratio $B^2/(2 \mu_0 \sigma_{max})$ (in addition to some geometrical parameters). This parameter represents the ratio of the maximum magnetic load to the operating stress of the structural material used.

Cryostat Model

The winding and structure are placed inside a cryostat, which has a room temperature bore. This model assumes that each thruster cell has its own inner and outer cryostat shell, even though the individual cells may be clustered.

Each of the cryostat walls is composed of two vacuum vessel walls. The thickness of these walls depends on their diameter and the operating depth or pressure to which the vehicle is exposed.

The design assumes two shells in each wall, one for the helium vacuum vessel and one for the room temperature vacuum vessel. A vacuum assembly is required both at the warm bore and at the outer wall of the thruster. In addition to the thickness of the structural elements, clearance is required for the possible irregularities in the components.

Finally, the high-temperature side of the vacuum vessel uses super insulation, which requires a thickness of 15 mm.

Combining the three elements previously given, some calculations have been carried out on the dependence of the weight on the two key parameters, the peak magnetic field, and the thruster duct diameter. Figure 11 shows an interesting result of the ratio of efficiency to the thruster system weight

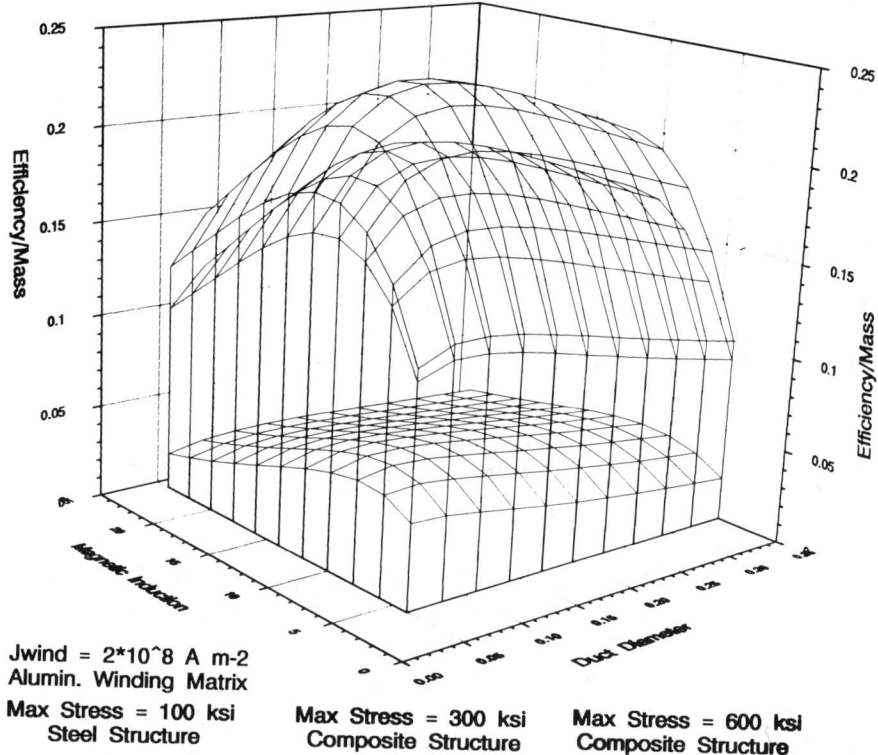

Fig. 11 Ratio of efficiency to system weight vs the magnetic induction and the duct diameter.

vs magnetic induction in the duct and the duct diameter. Maximum efficiency and minimum system weights do exist as a function of both variables and represent a significant factor for optimization analysis in the future.

Acknowledgments

The author acknowledges the contributions of Dr. Q. Huynh for the hydropropulsor model; Dr. L. E. Lema and G. S. Khoury for the electrode model; the MIT team, led by Prof. J. E. C. Williams, for the superconducting magnet model; and Dr. P. J. Hendricks for the size and weight model.

References

[1] Phillips, O. W., "The Prospects for Magnetohydrodynamic Ship Propulsion," *Journal of Ship Research*, Mar. 1962.

[2] Way, S., "Electromagnetic Propulsion for Cargo Submarines," *Journal of Hydronautics*, Vol. 2, No. 2, Apr. 1968.

[3] Cott, D. W., Daniel, V. W. and Carrington, R. A., "Annular DC MHD Thrusters for Submarines," 27th Symposium on Engineering Aspects of MHD, Reno, NV, Jun. 27-29, 1989.

[4] Hummert, G. T., "An Evaluation of Direct Current Electromagnetic Propulsion in Sea Water," Office of Naval Research, Arlington, VA, ONR Rept. CR168-007-1, Jul. 27, 1979.

[5] Tada, E., Saji, Y., Kuroishi, K., and Fujinaga, T., "Fundamental Design of a Superconducting EMT Icebreaker," *Transactions of the International Marine Engineering Conference*, Vol. 97, 1978.

[6] Stekly, Z. J. J., "Transverse Magnets for EMT Applications," Intermagnetics General Corp., Acton, MA, Final Rept. submitted to BST Systems, Plainfield, CT, Aug. 1988.

[7] Stekly, Z. J. J., "Mechanical and Thermal Model of Cryostat," Intermagnetics General Corp., Acton MA, Draft Rept. submitted to the Naval Underwater Systems Center, Newport, RI, Feb. 1989.

Liquid-Metal MHD Research and Development in Israel

H. Branover*
Ben-Gurion University of the Negev, Beer-Sheva, Israel

Abstract

The history and the proposed status of the Israeli liquid metal MHD energy conversion program are presented. The most advanced part of this program relates to ETGAR-type gravitational liquid metal MHD systems which are already in the commercialization stage. Thorough technical and economic analyses and evaluation of the ETGAR technology was done by a number of internationally recognized institutions and it was proven convincingly that this technology is especially advantageous in the cogeneration mode of operation. ETGAR systems have been considered for industrial cogeneration, at electrical power level of up to 25 MW. Under those conditions, they have substantially lower installation and electrical energy costs than that of steam-turbine engines.

History

The liquid metal MHD energy conversion program was started in Israel in 1978. This program was elaborated mainly at the Center for MHD Studies of Ben-Gurion University in Beer-Sheva, with the participation of specialists from the Technion in Haifa, Hebrew University in Jerusalem, Israel Atomic Energy Commission, and others. The program was sponsored initially by the Israel Ministry of Energy, and later by the Ministry of Industry and Trade. Since 1980 Solmecs, a private commercial company, became a major factor in elaborating the liquid metal MHD program of Israel. The work was from the very beginning based on broad international cooperation. A number of overseas institutions and individuals became participants in the program, particularly the Argonne National Laboratory, Purdue University at Calumet, Indiana, Energy Technology Engineering Center of Rockwell International at Canoga Park, California, Westinghouse R&D Center in Pittsburgh, Pennsylvania, Colorado School of Mines in Denver, Colorado, Nottingham University in England, the Institut de Mecanique in Grenoble, France; and more recently the Institute for High Temperatures of the Russian Academy of Science (IVTAN) and some other Russian research institutes, as well as the Institute of Physics of the Latvian Academy of Sciences.

Copyright © 1992 by the American Institute of Aeronautics and Astronautics, Inc. All rights reserved.
*Professor, Head, Center of MHD Studies.

It was established that the most promising concept for industrial application demanding the relatively shortest period of development is the gravitational system using heavy metals (lead, lead alloys, and others) as the magnetohydrodynamic fluid and steam or gases as the thermodynamic fluids.[1-3] This system was chosen for further development and industrial application and the program related to those systems was named the "ETGAR Program" (ETGAR stands for "challenge" in Hebrew).

The main directions of research and development activities have been defined as follows: 1) investigation of physical phenomena; 2) development of a universal numerical code for parametric studies, optimization, and design of the system; 3) material studies; 4) development of component engineering; 5) building and testing of integrated small-scale ETGAR-type systems; 6) economic evaluation of the system and comparison with conventional technologies; and 7) development of a moderate-scale industrial demonstration plant.

At present, items 1-6 have already been fully implemented and activities on item 7 commenced.

ETGAR Concept and Advantages

The ETGAR concept in its specific embodiments have been already described in a number of papers (see, for example Refs. 2 and 3). However, for the convenience of the reader of the present paper we will repeat here a short description of the concept and of the operation of a system based on this concept.

A schematic of the ETGAR liquid metal MHD energy conversion system concept is illustrated in Fig 1. The basic system consists of two pipes (an upcomer and downcomer) connected at the bottom with a crossover pipe and with a separator joining them at the top. A mixer is located at the bottom of the upcomer and a single-phase MHD generator, from which electrical power is extracted, is located in the downcomer or lower crossover pipe. Operation of the system is as follows. A thermodynamic fluid — vapor, steam, gas (or volatile liquid boiled in direct contact with the hot liquid metal) is introduced into the mixer at the bottom of the upcomer at an appropriate temperature and pressure. A two-phase fluid with density lower than the liquid metal density is created. As the two-phase fluid flows to the separator, the gaseous phase undergoes an expansion from the high pressure in the mixer to the low pressure in the separator, lifting liquid metal. The gaseous phase (thermodynamic working fluid) is removed in the separator (gravitational or cyclone type) and single-phase flow of the liquid metal (LM) returns into the downcomer. The separator has to be designed in such a way that a possibly high portion of the kinetic energy of the LM could be preserved.

The pressure differential that exists between the upcomer and the downcomer due to the density difference causes the LM to circulate in the system; as the single-phase LM downflow crosses the magnetic field in the MHD generator, an electrical potential is generated and power is extracted. The flow rate in the loop self-adjusts to balance the density differential between the upcomer and downcomer with the frictional and acceleration flow losses and the electromagnetic forces acting in the MHD generator. Heavy LM, such as lead, is used for the electrodynamic fluid so as to maximize the pressure differential (or the expansion ratio of the thermodynamic working fluid) for a given loop height.

Several advantages of the described system stem from the simplicity of its design and ease of control. As it will be shown below the installation cost is much lower than with conventional turbine systems. Moreover, the LMMHD system components can be manufactured by any reasonably equipped modern mechanical workshop.

However, the major advantage of the ETGAR concept is related to the fact that LMMHD energy conversion system (ECS) performs a very special type of a thermodynamic cycle. The LMMHD cycles differ from the conventional turbine cycle mainly by the expansion process: in the LMMHD system the expansion of the thermodynamic working fluid is nearly isothermal. This unique feature of LMMHD power conversion is due to the intimate contact between the thermodynamic working fluid (e.g., steam) and the LM, leading to a continuous heat transfer from the LM (which has a very high heat capacity relative to the heat capacity of the vapor) to the vapor or gas throughout the expansion process. The direct contact heat transfer in the MHD expansion process leads to two important results: 1) it increases the average temperature of energy delivery from the heat source to the cycle, thus increasing the cycle efficiency, and 2) the isothermal expansion can be considered to be a process with infinite number of reheat stages, while the additional heat delivery occurs without the need for extra reheaters.

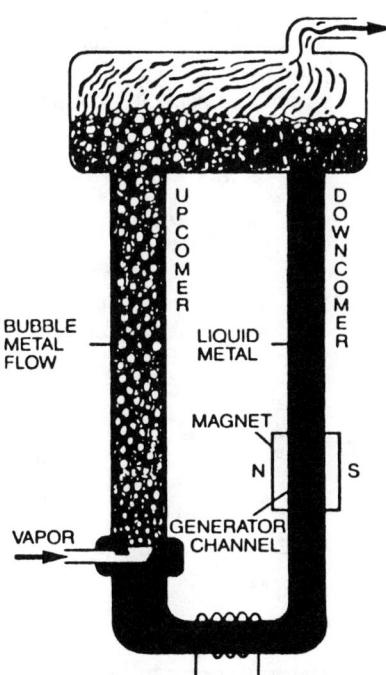

Fig. 1 Schematic of the ETGAR type liquid metal MHD energy conversion system.

The LMMHD EC technology appears to be especially attractive for cogeneration applications (production of electricity and heat simultaneously) and in particular for industrial cogeneration applications associated with heat sources which can deliver a large fraction of their energy at or above the cycle high temperature (constant temperature heat sources, e.g., solar collectors, fluidized bed combustors, boiling water, nuclear fission reactors, etc.).

One attribute of the LMMHD EC technology which contributes to its attractiveness for cogeneration is the already discussed isothermal expansion process. Comparing this cycle with a turbine Rankine cycle (without reheating) commonly used in small power plants, one can find that, for the same amount of equal quality steam delivered to the customer, the LMMHD ECS can generate significantly more (up to 30%) electricity. Thus, the LMMHD EC technology can offer a higher electricity-to-thermal (steam and/or heat) energy ratio than conventional cogeneration plants, while using a relatively simple system (without reheaters). The higher electricity generation ability of LMMHD cogeneration plants for a given rate of thermal energy supply can be a significant economic advance of the plant if the utility is committed to purchase the supply electricity at a reasonable price.

By bypassing the regenerator fully or partially (recovering the high temperature heat back to the boiler feed water) the electric to thermal ratio can be changed by up to 33% (depending on the cycle pressure difference) for a given power system and capacity, not like in steam turbines. This advantage is very important since the industrial need for process heat changes during the seasons and even during the day. The present conventional technologies can offer a solution to this basic requirement by designing the systems to maximum capacity of process steam and reduce it by reducing the electricity production. However, a turbine which works on partial load has a significantly reduced cycle efficiency. This is not the case in the LMMHD technology where the electricity production capacity remains constant while changing the electric to thermal ratio.

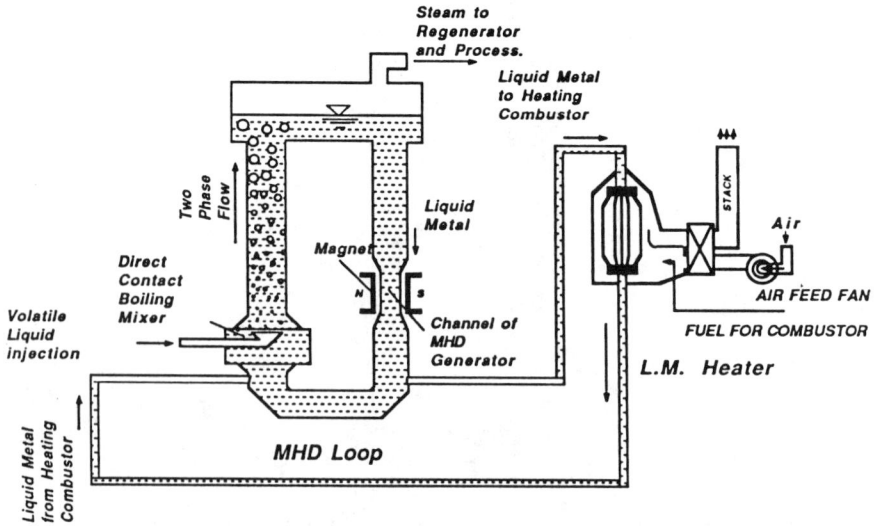

Fig. 2 ETGAR facility with direct contact boiling mixer.

Moreover, the possibility of reducing simultaneously the electrical power capacity and process heat exists also in the LMMHD technology, but here the cycle efficiency at partial load increases due to the reduction of the slip ratio in the two-phase flow and hydraulic friction losses at reduced two-phase and single-phase flow rates.

Figure 2 presents a schematic of an ETGAR system in which the volatile liquid is boiled in direct contact with the hot liquid metal and no separate boiler is needed.

ETGAR Program

As already mentioned, the development of LMMHD ECSs was started in Israel in 1978 and was elaborated mainly by the Center for MHD Studies of Ben Gurion University.[2,3] The first few years have been dedicated to experimental and theoretical studies of phenomena and to parametric studies of systems. Since 1983 several integrated pilot-type facilities have been built and operated for the purpose of characterizing and developing the natural circulation system and its key components (separator, mixer, single-phase generator and two-phase riser pipe, as well as solving the problems of confinement material compatibility with the two-phase LM flow).

Experiments have been conducted in the two following fields: 1) Studies of basic phenomena (two-phase flows, mixing, separation, carry-under of gas, etc.) and 2) building and testing of complete LMMHD power conversion facilities.

One of the most important outcomes of experimental studies of phenomena was the development of a new general dimensionless correlation for predicting the characteristics of the vertical two-phase LM flow in the upcomer pipe, which plays a major role in determining the conversion efficiency. Mixers, separators, and single-phase flow Faraday-type MHD generators have also been carefully studied.

Results of all these extensive experimental studies created an empirical data base, which together with data on thermodynamic properties of liquid metals and thermodynamic fluid, enabled the development of universal computer code for the design and optimization of ETGAR-type LMMHD systems (as well as of LMMHD system embodiments based on concepts different from the ETGAR concept).

Two complete integrated power conversion systems have been built and tested, together with laboratory-scale experimental facilities. The ETGAR-3 plant, commissioned in 1985, is the most advanced integrated pilot plant built to date. It employs lead-bismuth alloy and steam and works at temperatures up to 170°C. The smaller ER-4 system works with mercury and steam at the same temperatures as ETGAR 3.

Several years of testing of the ETGAR 3 facility verified a number of predictions regarding the performance of the ETGAR-type LMMHD energy conversion systems. There have not been any major failures or any material corrosion problems in the ETGAR 3 facility which was operated for more than 3000h (accumulatively). As it will be shown below, detailed measurements of performance characteristics of ETGAR 3 (as well as of the smaller mercury-steam integrated system ER-4) demonstrated that the previously mentioned universal computer code can be used with confidence for the optimization and design of LMMHD energy conversion systems. Over the years a number of scientific and engineering institutions in Israel and overseas became involved in the ETGAR research development program (see History Section).

Analysis of results obtained from all phenomena and materials studies, and engineering development of system components indicated that the data base and

Table 1 - ETGAR-5 cogeneration demonstration system parameters

Cycle high temperature, °C		480
Steam temperature at boiler exit, °C		390
Steam generator (boiler) mass flow rate, ton/hr		4.9
Cycle high pressure, bar		28
Cycle low pressure, bar		5
Total heat input, MWth		3.7
Electric plant efficiency, %		8.5
Net electric power, kW		315
Steam delivery to process:		
Saturated steam conditions, bar/163°C		5
Mass flow rate, ton/hr		5
Structure configuration:		

Stage no.	Riser diameter, in.	Pipe schedule	Riser height, m
1	10	40	17.74
2	8	40S	17.72

practical experience necessary for entering the commercial stage of the LMMHD power system development has already been accumulated. This statement does not mean that no further studies are necessary, especially for future generations of ETGAR systems and for the development of concepts other than ETGAR. What it does mean, however, is that the immediate next stage of the ETGAR program must be the development of a commercial multimegawatt LMMHD power facility. The practical implementation of this commercialization program began in 1988.

The first moderate-scale commercial-scale demonstration plant to be built in the framework of the ETGAR commercialization program will be the ETGAR-5 plant. It will be working with lead as the electrodynamic fluid and steam as the thermodynamic fluid. Steam will be generated initially by a conventional boiler. For a later stage of operation it is envisaged that the steam may be supplied by a fluidized bed oil shale combustor. Provisions will also be made for the possibility to inject water directly into the hot lead and thus to eliminate the boiler.

Although the specific site on which the ETGAR-5 plant will be erected has yet to be identified, it is a strategic decision to have ETGAR-5 operating continuously in a real industrial environment. The main technical parameters of the ETGAR-5 plant are presented in Table 1.

Electricity generated by ETGAR-5 will undergo the to inversion and will be sold to the power grid. Connection to the grid will take place after half a year of continuous operation of the plant. Steam exhausted from the ETGAR-5 system and enriched by heat through the isothermal expansion process will be supplied to a low-pressure steam network of the industry which will host the plant. The ETGAR-5 plant is expected to commence operations 24 months after the beginning of this stage of the program.

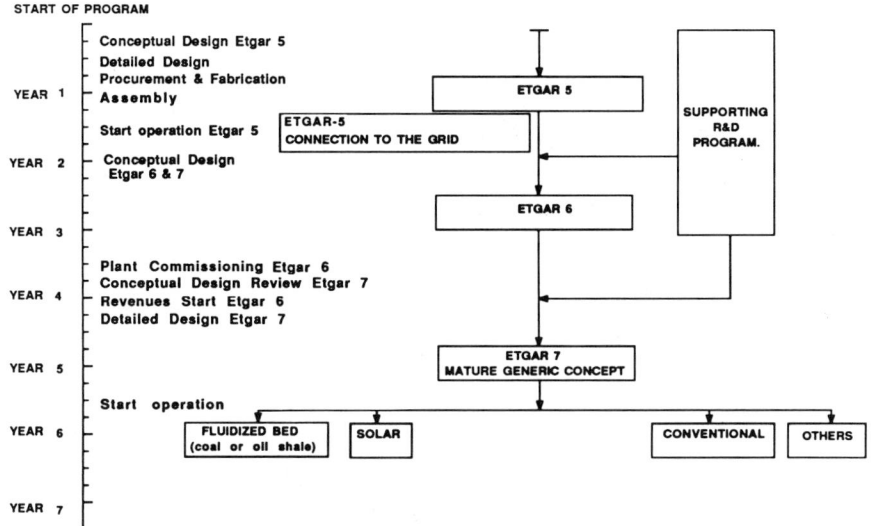

Fig. 3 ETGAR Program Timetable.

To reduce as much as possible uncertainties and risk in the development of this plant, it was decided to design ETGAR-5 using only fully tested processes, materials and engineering solutions, so that no further research and development would be necessary for accomplishing the ETGAR-5 stage. It is believed that the fact that ETGAR-5 is based completely on the current state of the technology will help to ensure that it will be put into operation according to schedule and will perform as designed.

The ETGAR-5 system has been designed to verify the performance predictions for an upscaled plant, to demonstrate the applicability of the presently developed LMMHD technology to the conditions of an industry with continuous operation. Its continuous operation will also give the opportunity to acquire all the necessary knowledge in running and controlling these type of plants.

An extensive research and development program will be implemented in parallel with the construction and testing of the ETGAR-5 plant. This program will be related first of all to the development of the next plant - ETGAR-6 in the 3-5 MW_e power range, and the problem of further upscaling the technology. The research and development program will also consider next generations of ETGAR plants as well as plants based on concepts other than ETGAR. The timetable and the work plan is presented in Fig. 3.

The development beyond the ETGAR-5 stage, namely, the design and construction of ETGAR-6 plant, will incorporate all the information obtained from operating ETGAR-5 and the inputs from the supporting research and development program. This program, which will be conducted in parallel with ETGAR-5 design and construction, concentrates on further investigation of physical processes of liquid metal MHD power facilities (mainly involving the ability to control two-phase flow characteristics), further parametric studies, component optimization, and material engineering.

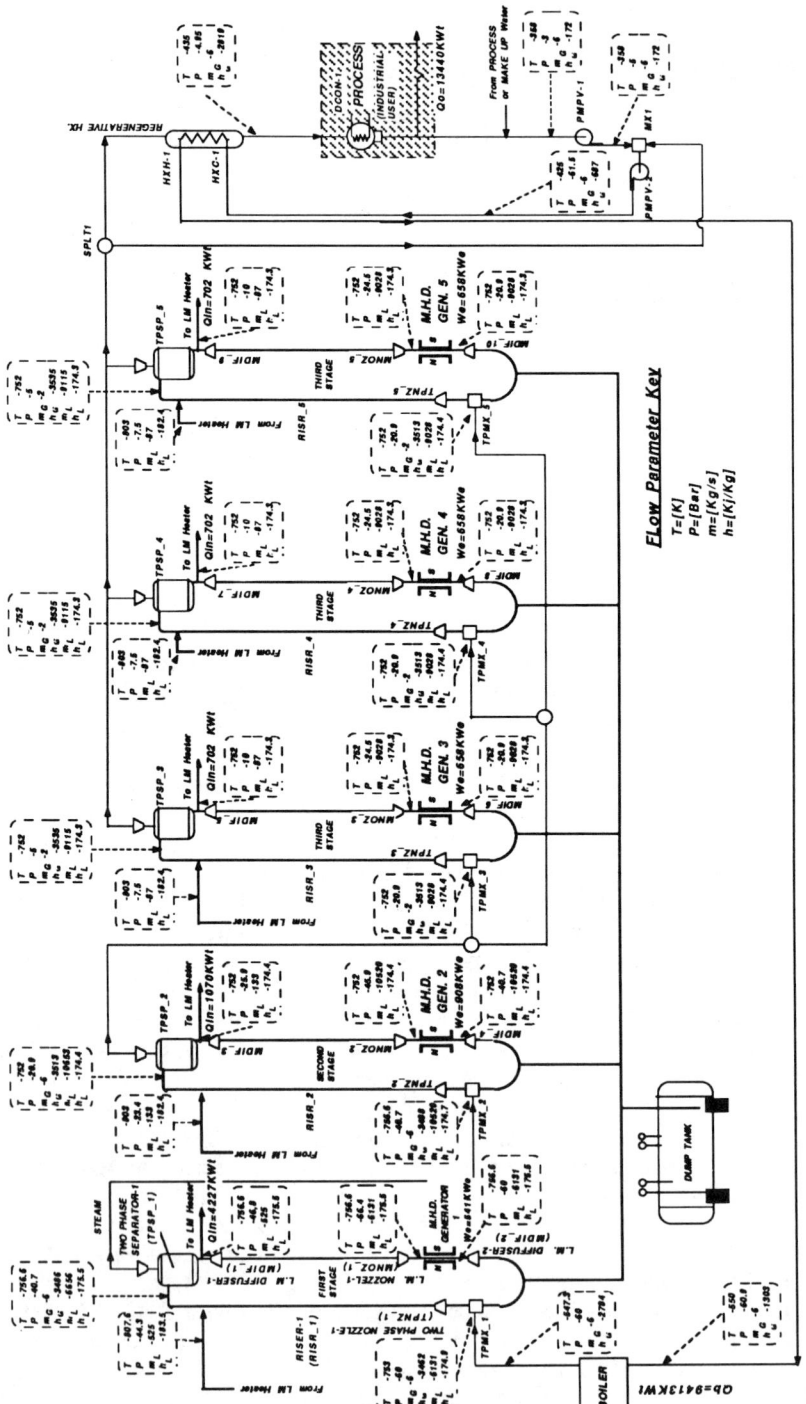

Fig. 4 ETGAR-7 steam/lead Rankine cycle cogeneration power plant (preliminary version).

Table 2 - ETGAR-7 Mature Plant Main Parameters as Compared With Steam Turbines

	ETGAR-7	Steam Turbine isentropic efficiency		
		70%	74%	80%
Pressure, bar	60	63	60	60
Temperature, °C	277	408	480	480
Gross electrical output, kW$_e$	3668	3315	2960	3200
Net electrical output, kW$_e$	3429	3000	2715	2950
Heat to industrial processes, kJ/h	56395			
Heat to industrial processes, MWth	15.66	17.8	14.2	14.0
Heat from fuel, MWth	21.1	23.1	18.6	18.6
Steam flow, t/h	24.3	31.7	24.6	24.6
Steam to industrial processes				
236°C, t/h	21.7			
160°C, t/h	23.4	26.6	21.1	20.7
Heat to electrical power ratio, MWth/MW$_e$	4.6	5.9	5.2	4.7
Fuel flow, kg/h	1912	2096	1688	1688
Fuel flow allocated to, kg/h				
process heat	1422			
Fuel flow allocated to				
electric power, kg/h	468	481	400	398
Fuel consumption, g/kWh	134	145	135	131
Total efficiency, %	90.5	90	91.4	91.0
Electrical efficiency, %	17.4	14.3	15.9	17.2
Installation cost $/kW$_e$	1339	1793	1920	1905
Electricity cost cents/kW$_e$h	3.497	4.647	4.985	4.936

The ETGAR-6 cogeneration integrated system (net electric power of 5 MW) is the final development stage and is to operate under the same optimal parameters as that of the mature ETGAR-7 system. Its development will begin 24 months after program commencement. ETGAR-6 will be commercial and start to generate revenues 48 months from the program startup. In parallel with the design, construction, and operation of the ETGAR-6 system, research and development will be conducted for adjusting the technology to applications other than cogeneration, coupling of the power block to different kinds of heat sources, further optimization and reduction of the cost of system components, as well as the development of the second generation technology.

The ETGAR-7 mature plant will be designed as a generic concept applicable to the utilization of different heat sources and operation over a wide spectrum of applications in the 1-20 MW$_e$ size range. The actual implementation of ETGAR-7 will depend on specific orders for such a plant. The first mature commercial ETGAR-7 plant is expected to be completed in five years from the commencement of development activities.

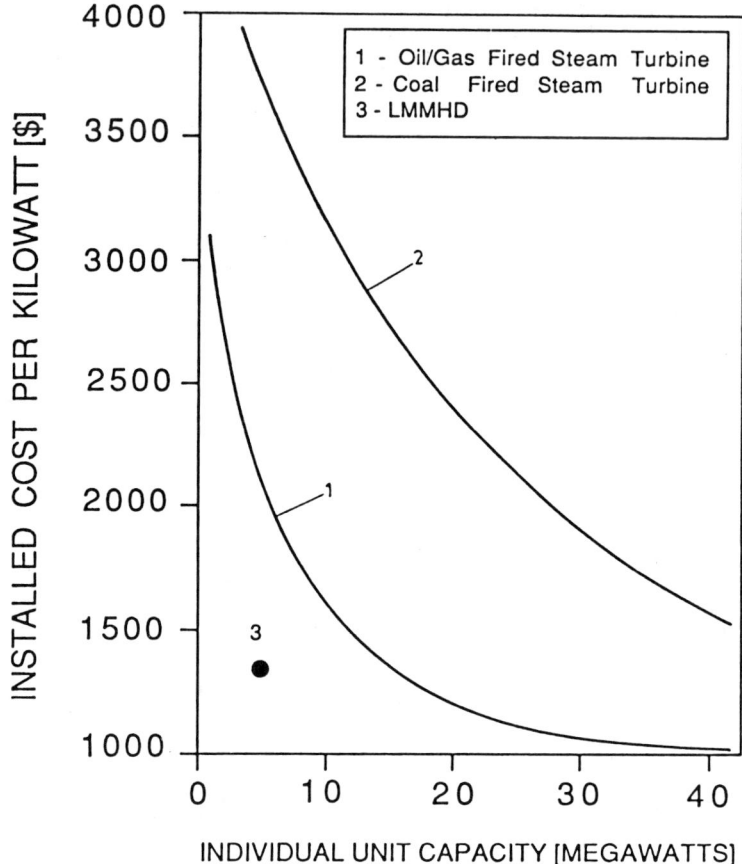

Fig. 5 Installed cost of steam turbine plants
(data for turbines from Dunn and Bradstreet[6]).

Main parameters of ETGAR-7 plant stations for 3.67 MW$_e$ gross electrical output appears in Fig. 4 and Table 2 (which also contains data for competing turbines). The scheme shown in Fig. 4 is one of the versions; not the one reflected in Table 2. One parameter not mentioned in the table is the magnetic field strength for the MHD generator. It should be stressed that one of the basic features of the ETGAR-type MHD facility is that it demands a very moderate magnetic field — usually about 0.5T or even less. Thus the system can be equipped with a conventional electromagnet consuming very little power. If superconductive magnetics will become substantially advanced and much cheaper than they are now, the use of a superconductive magnet can become desirable.

Table 3 - ETGAR-7 versus Steam Turbo-Generator: Economic Comparison

	ETGAR 7	Steam turbine Eff. 70%	Steam turbine Eff. 74%	Steam turbine Eff. 80%
Life of station, y	30	25	25	25
Availability, h/y	7884	7000	7000	7000
Gross electricity production, kW_e	3668	3315	2960	3200
Net electricity production, kW_e	3429	3000	2715	2950
Installed cost				
Total cost of system, $000	4911	5944	5684	6096
Cost per kW_e, $/$kW_e$	1339	1793	1920	1905
Energy cost				
When fully allocated to electricity, cents/kW_h	8.046	10.368	10.067	9.548
Partial allocation to electricity after cleaning steam cost, cents/kWh	3.497	4.647	4.985	4.936

Table 4 - ETGAR-7 Annual Economic Projections

	US($000)
Income from steam 23.4 ton/h x 7880 h x $8.2/ton	1513
Income from electricity 3668 kWe/h x 7880 h x $0.06/kWe	1735
Total expected income	3248
Operational expenses	(1849)
Operational profit	1399
Levelized capital cost (at 9% interest rate)	(478)
Expected profit	921

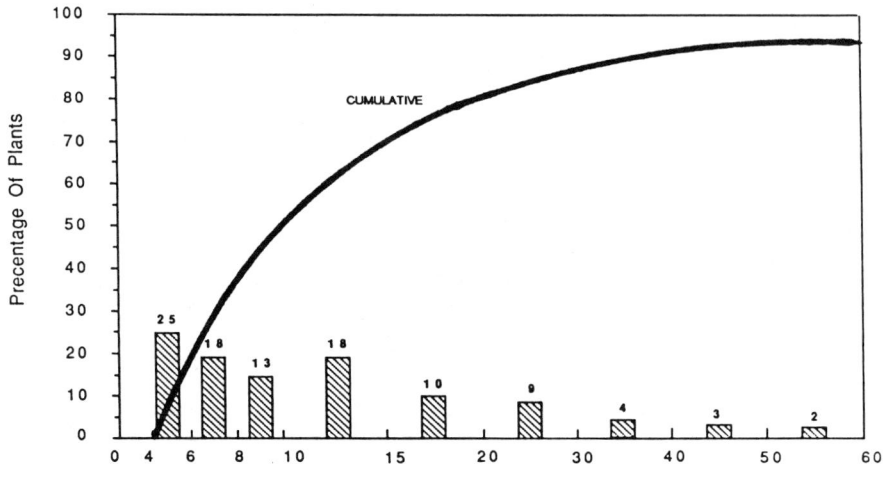

Fig. 6 Demand for cogeneration plants in U.S. (from Dunn and Bradstreet[6]).

Economic Analysis of the Cogeneration ETGAR Technology and market surveys

Several economic studies of the ETGAR technology as well as surveys of its market potential have been performed. A thorough investigation of the installed cost, performance characteristics, and cost of electricity for a mature ETGAR cogeneration system was performed by the United Project Services (UPS), a leading Israeli engineering company.[4] This investigation also compared all the parameters with corresponding parameters of steam turbines working in the cogeneration mode. The main results of this investigation performed for a 3.68 MW_e plant are given in Table 2 and in Table 3. It should be stressed that UPS established all costs through obtaining real price proposals for each component of the system from actual manufacturers (and not on the basis of some averaged data given in the literature). The numbers presented in Tables 2 and 3 show very convincingly the superiority of the ETGAR system over turbines in terms of substantially lower installation and electricity costs. Annual economic projections for a 3.68 MW_e ETGAR plant are given in Table 4. These results lead to the conclusion that the payback period for such a plant is about 4y.

Using the economic data for ETGAR cogeneration systems, a market study was performed by Arthur D. Little Inc. in Boston.[5] This study takes into consideration the substantially lower installed cost of an ETGAR system with power up to 20 MW_e as compared to steam turbine systems. It also reflects the fact that the ETGAR technology has a very flexible thermal to electrical power ratio and other unique features. Figure 5 gives a comparison of the installed cost of the ETGAR system with the cost of turbine systems, the latter being taken from a Dunn and Bradstreet study[6] with adjustment of the data to the present value of the dollar. The same Dunn and Bradstreet study also states that about 80% of all cogeneration power installed in the U.S.A. is in plants below 20 MW_e (Fig. 6).

The principal conclusions of Arthur D. Little's study are as follows:

LMMHD EC is best matched (in scale and thermal/electric power ratio) to the industrial/commercial (I/C) cogeneration market. It substantially supercedes competing technologies both in performance and in economic characteristics.

Because of lower installation costs, reliability, and simplicity, the LMMHD technology is highly competitive and can expect market penetration in industrial cogeneration markets for plants up to 20 MW_e. The forecasted penetration by the end of the decade is between 25-33% of the market for plants up to 20 MW_e. This forecast leads to annual installations of LMMHD plants below 20 MW_e in the range from $125 million to $165 million for the U.S. market alone. The worldwide market is estimated to be at least three times that of the United States, i.e., from $375 million to $495 million, and it is anticipated that due to the simplicity of the technology much higher penetration rates will be achieved in countries which are not producers of turbines.

Conclusions

The LMMHD technology reached the stage of commercialization and according to a very thorough technical and economic evaluation and comparison is highly advantageous and competitive. Its penetration into the market is feasible by the middle of the 1990's. The first system to be commercialized will be the ETGAR type cogeneration system. It is expected that this will be followed by modification of the technology which will broaden substantially the areas and scale of its application.

Acknowledgment

This work was sponsored by Solmecs (Israel) Ltd. The author is thankful to all researchers of Solmecs and of the Center for MHD Studies of Ben Gurion University who participated in the studies presented in this paper.

References

[1] Petrick, M., and Branover, H., "Liquid Metal MHD Power Generation — Its Evolution and Status," *Single- and multi-phase flow in an electromagnetic field, Energy, Metallurgical and solar Applications Seminar*, Ben Gurion University (Beer-Sheva, Israel) 1984, p. 371.

[2] Branover, H., "MHD Power Generation Program of the State of Israel," *Proceedings of the 23rd Symposium on Engineering Aspects of Magnetohydrodynamics*, (Somerset, PA) 1986, pp. 44-60.

[3] Branover, H. and El-Boher, A., "The "ETGAR" Program for the Development of Liquid Metal MHD Conversion Technology," *Proceedings of the 10th International Conference on Magnetohydrodynamic Electrical Power Generation* (Tiruchirapalli, India) Vol. 1, 1989, pp. 1.19-1.26.

[4] United Project Services, Ltd., "MHD Cogeneration Power Plant - ETGAR-7," Techno-Economic Study Report, Beer Sheva, Israel, Sept. 1990.

[5] Cook, E.J. and Kleinschmidt, D.E., "The United States Cogeneration Market Relative to Solmecs LMMHD Technology," prepared for Solmecs (Israel) Ltd., Arthur D. Little, Inc., 1989.

[6] Dunn and Bradstreet Technical Economic Services, "Comprehensive 1984 Plant Profiles", prepared for US Department of Energy, 1984.

OMACON Technology for Seawater Desalination

A. Barak* and E. Greenspan†

Israel Atomic Energy Commission, Tel-Aviv, Israel

Abstract

A preliminary feasibility assessment of a couple of novel seawater evaporative desalination processes is undertaken. One of these processes involves the injection of preheated seawater into the hot liquid metal of an OMACON-type liquid metal MHD energy conversion system. The other process involves the injection of the superheated steam coming out from an OMACON system into a seawater containing vessel. Both processes are characterized by direct-contact heat transfer, implying the need for relatively simple and inexpensive equipment and, practically, no problem of scale formation. The figures of merit considered in this assessment include the rate of desalination per MW_e of electricity-generating capacity, the extra energy that needs to be invested in the plant due to the use of seawater rather than regular water, the corresponding extra capital that needs to be invested, and the cost for water desalination. It is found that both desalination processes are likely to be economically viable. The desalination by seawater injection into hot liquid metal appears to be the more attractive process from the point of view of the desalted water-to-electricity production rates ratio as well as of the cost of the desalted water. A thorough study of the OMACON-related desalination processes is therefore recommended.

1.0   Introduction

Of the many processes developed for seawater desalination, evaporation are the dominant processes so far. Economic analysis shows that the largest cost components for sea water desalination via evaporation are the energy for the process, the cost of the heat-transfer surfaces, and the cost of the vessels in which the evaporation takes place. The energy for the evaporation process is preferably taken from a powerplant in the form of a low temperature steam

---

Copyright © 1992 by the American Institute of Aeronautics and Astronautics, Inc. All rights reserved.

*Senior Staff Engineer, Division of Nuclear Engineering.

†Visiting Scientist; also at Department of Nuclear Engineering, University of California at Berkeley, Berkeley, California.

exhausted or extracted from the turbine. From the energy flow point of view, the desalination plant serves as the heat sink of the energy conversion (EC) system; it is "tied-up" to a backpressure turbine.

The liquid metal (LM) magnetohydrodynamic (MHD) energy conversion technology opens new possibilities for seawater desalination in dual-purpose plants. It appears that a number of these possibilities could benefit from reduced energy investment and/or reduced heat-transfer surface or special vessels requirements. One purpose of this work is to describe these possibilities. Another purpose of the work is to perform a preliminary evaluation of the economic viability of a couple of these approaches, both involving an evaporation process.

The LMMHD EC technology, which features natural circulation appears to be particularly suitable for the desalination applications. Section 2.0 provides a brief review of the technology called OMACON. Also discussed in Section 2.0 are the unique features of the OMACON technology that make it attractive for desalination applications.

Sections 3.0 and 4.0 describe the novel evaporative seawater desalination processes and evaluate their viability. This viability evaluation is only the first phase of a study that needs to be undertaken before the feasibility of the novel processes proposed can be ascertained with a high degree of confidence. Such an undertaking, which needs to involve quite expensive material compatibility tests, will be justified if, and only if, the outcome of the first phase viability evaluation turns out to be positive.

2.0 OMACON Technology

2.1 General Review

A review of the LMMHD principles for energy conversion, the LMMHD EC cycles conceived and developed, and the unique features of this technology and of potential applications can be found in Refs. 1-4. Briefly, the LMMHD approach to energy conversion involves the conversion of the thermal energy provided by the heat source to mechanical energy of a liquid metal that is converted (by the Faraday effect), possibly simultaneously, to dc electricity. The conversion of thermal to mechanical energy is done by direct contact between the thermodynamic working fluid (such as water or gas) and the liquid metal.

There is much flexibility in the design of LMMHD EC systems.[1-4] In the following we shall restrict our consideration to the so-called OMACON (Optimized Magnetohydrodynamic energy Conversion system) technology. This approach to LMMHD EC features natural (or gravitational) circulation. Figure 1 gives a schematic layout of the basic system; it consists of two vertical pipes (referred to as a "riser" and a "downcomer") connected at the bottom and top by a crossover pipe and a separator element, respectively. A mixer is located at the bottom of the riser, and a single-phase MHD generator is located in the downcomer or lower crossover pipe. Liquid metal that is chemically compatible with water is used as the electrically conducting fluid. In the following we shall assume that this LM is lead. Water (which could be steam or other volatile liquid or gas, in principle) is introduced into the lead in the form of small droplets (or bubbles) via the mixer. A two-phase fluid consisting of steam and lead (with, possibly, some liquid water at the beginning) is created. Since the average fluid density in the riser is lower than that in the downcomer, the fluid circulates in the loop; the driving pressure difference is balanced by the Lorenz forces in the MHD channel and by friction. As the steam bubbles move up (along with the lead) in the riser, they experience lower pressure and thus expand. The steam is separated at the top of the loop, from where it can be introduced into another LMMHD loop (albeit, at a lower pressure), a turbine, a regenerative heat exchanger (HX), or a steam customer.

Notice that most of the LM driving force in the loop is caused by the steam expansion in the riser. Since the steam is in direct contact with the hot lead, heat is transferred from the lead to the steam throughout the expansion process. Since the heat capacity of lead is higher than that of steam by orders of magnitude (on per unit volume basis), the steam expansion is nearly isothermal (vs. nearly isentropic expansion in turbines). Assuming that the average void fraction in the riser is 0.5, the steam can expand approximately 1 bar per 2 m of the loop heights. Since it may not be desirable to design loops that are too high, OMACON systems featuring relatively large steam expansion are likely to be designed to have two or more loops in a series, as illustrated in Fig. 2.

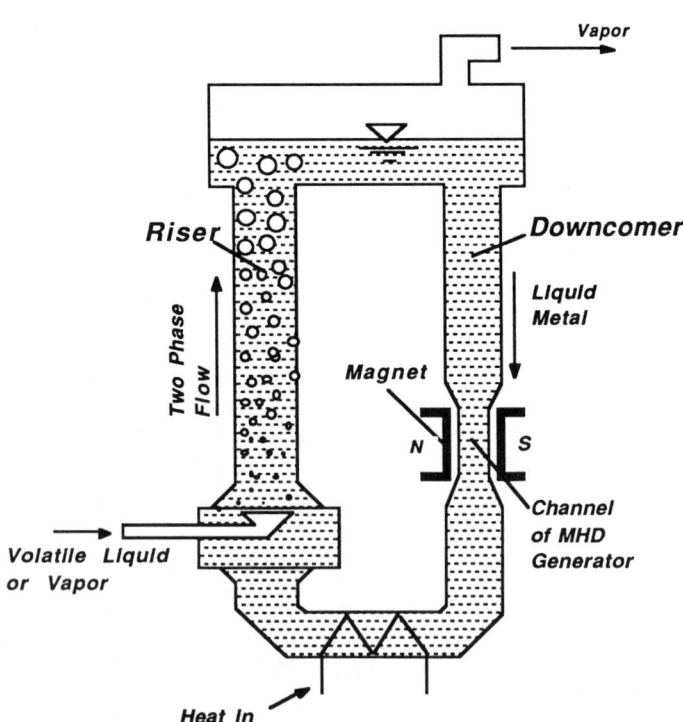

Fig. 1 Schematic layout of the Optimized MAgnetohydrodynamic energy CONversion system (OMACON).

SEAWATER DESALINATION

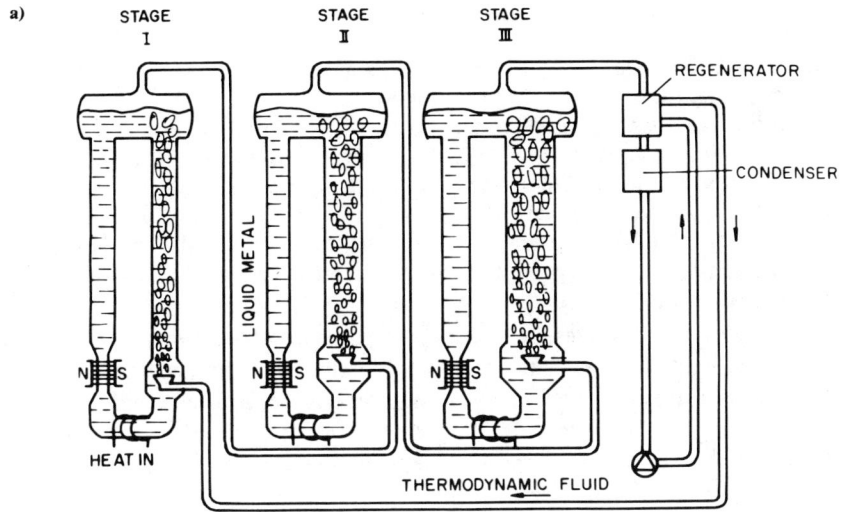

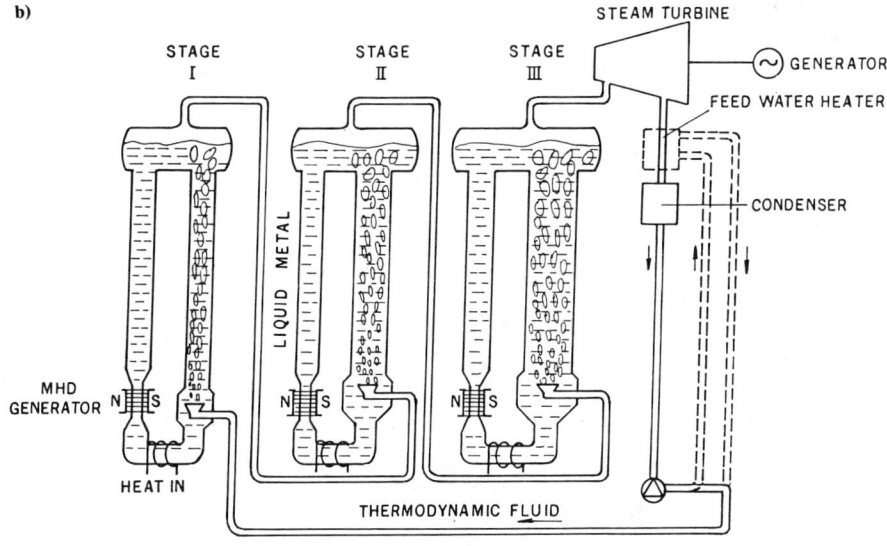

SCHEMATIC OF A MULTISTAGE OMACON SYSTEM

Fig. 2 A schematic illustration of a multistage OMACON system featuring (a) regeneration, and (b) low pressure steam turbine.

2.2 Approaches to Desalination

The OMACON technology opens unique and promising new approaches to seawater desalination by virtue of the following features:

1) Direct contact evaporation of water. If seawater were to be injected into the first stage of an OMACON system, the water would vaporize (and cause the LM to circulate) while the salts would tend to float in the separator (from which they can be skimmed off). This approach to evaporation eliminates the need for metallic heat-transfer surfaces and for special vessels for the desalination process. It is also free from concerns about scale deposition and its adverse implications.

2) Discharging the low-pressure steam at superheated conditions. Because of the nearly isothermal expansion, optimal LMMHD steam cycles can feature superheated steam at the condensation temperature, as illustrated in Fig. 3. By bringing the low-pressure superheated steam in direct contact with seawater, it is possible to practice desalination without the need for solid heat-transfer surfaces.

3) Reverse osmosis (RO) of feedwater. Consider approach "1" in which seawater is being used for the feedwater. By installing a RO facility between the feedwater heaters and the first OMACON loop, it is possible to desalinate part of the seawater (or saline water) without the need to make a significant investment in pumping power and heating [the higher the water temperature (up to a limit), the more efficient the RO process can be].

4) Direct current (dc) electricity. This type of electricity is required for the electrodialysis (ED) desalination process. The RO and vapor compression (VC) processes could use dc electricity as well, if they were to use dc motors to drive the pumps/compressors.

A preliminary evaluation of approaches 1) and 2) is undertaken in Sections 3.0 and 4.0, respectively.

2.3 Reference OMACON System

Table 1 summarizes selected characteristics of the OMACON system considered as the reference system for the purpose of the present analysis. It makes a relatively small extrapolation from Etgar-7 -- the first OMACON system expected to be commercial.[5] The main difference in the design parameters is, perhaps, the system high pressure -- 135 bars vs. the 60 bars of Etgar-7. The higher pressure was chosen for the desalination plant to enable only partial evaporation of the seawater. The transition pressure between the MHD and the turbine systems was determined to maximize the electricity generation efficiency; the lower the transition pressure the higher the average temperature for heat supply to the cycle becomes, but the higher the temperature of the steam exhausted from the turbine becomes. It is assumed that the sensible heat of the steam exhausted from the turbine is used for feedwater heating. It should be emphasized that the OMACON of Table 1 was chosen somewhat arbitrarily; thus, it is not necessarily the optimal system for seawater desalination.

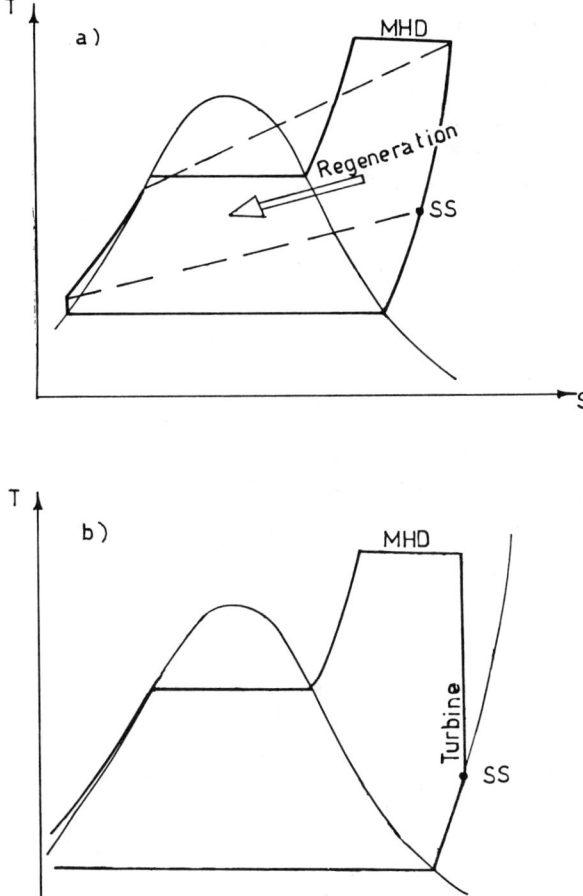

Fig. 3 A schematic illustration of (a) an all-LMMHD, and (b) of a combined LMMHD-Turbine Rankine cycles indicating the availability of low pressure superheated steam (SS) for desalination.

Table 1 - Selected characteristics of the OMACON system used as the reference for desalination evaluation

Characteristic	Value
Cycle high pressure, bars	135
Cycle high temperature, °C	530
Turbine inlet pressure, bars	2
Condensation pressure, bars	0.08
Total heat input, MW_{th}	30.0
Electricity generated, MW_e	
MHD generators (dc) [a]	6.5
Turbo-generator (ac)	3.8
Total (ac) [a]	10.0
Auxiliary power needs, MW_e	0.16
Net efficiency, %	
Without converting dc to ac	33.8
DC converted to AC	32.8
Riser height, m	60
No. of loops	4 [b]

[a] Assuming a dc-to-ac conversion efficiency of 95%
[b] The high pressure of the loops is 135, 125, 84, and 43 bars.

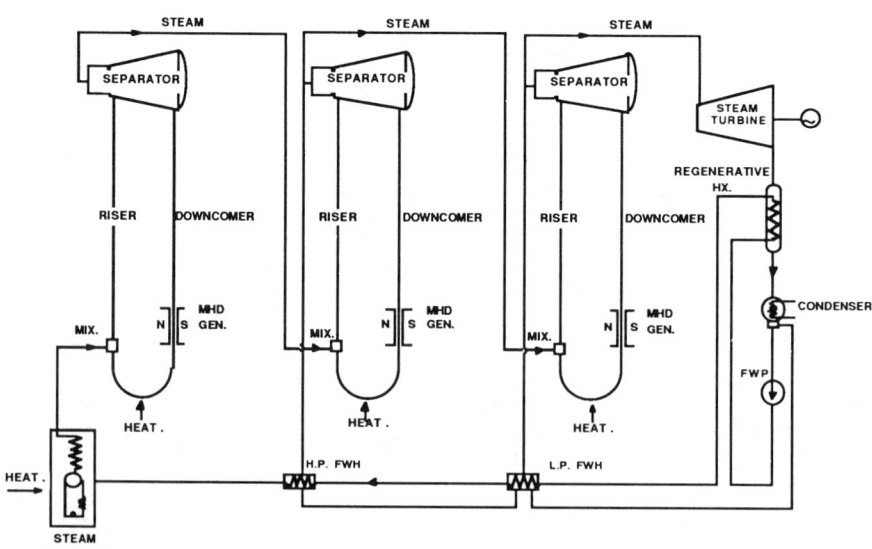

Fig. 4 Schematics of the combined OMACON-turbine steam cycle of the reference single purpose plant.

The reference OMACON system, illustrated in Fig. 4, is assumed to have two feedwater heaters. The hotter one gets 1.35 kg/s, 43 bars steam extracted from between the last two OMACON expansion stages. The feedwater is heated almost up to the saturation temperature of the extracted steam (254°C). The colder feedwater heater is heated by steam extracted from between the outlet from the MHD system and inlet to the turbine. The flow rate is 0.29 kg/s at 2 bars, heating the feedwater to 120°C.

3.0 Desalination via Seawater Injection into an OMACON System

3.1 General Description

Figure 5 shows a schematic of an OMACON EC system in which the desalination is done by injecting preheated seawater into the first, relatively small loop, the temperature of which is likely to be lower than the cycle high temperature. The system operates as an open loop -- receiving seawater and delivering desalted water (in addition to electricity and salt). No extra equipment is needed for the desalination except for the removal and handling of the salts left over in the first loop and for the recovery of the sensible heat carried away by the salt. It is hoped that being immiscible in lead and having significantly lower density than the lead, the salt will float on top of the lead and could be "skimmed off" without undue complications. Needless to say, an experimental demonstration of the salt removal possibility, as well as of the chemical compatibility of the seawater with the OMACON system, is essential for assessing the practical feasibility of the proposed process.

In practice it may be desirable to operate the first OMACON loop at the lowest feasible temperature (at the cycle high pressure) -- about 350°C (the lead melting occurs at 327°C). A significantly lower temperature might be feasible if a lead-bismuth alloy is to be used for the LM instead of lead. (The melting temperature of the Pb-Bi eutectic is 125°C.) Even though the bismuth is significantly more expensive than the lead, the benefits from the low-temperature evaporation may outweigh the extra cost of the LM; the volume of the first (high-pressure) loop is relatively small. The lower the evaporation temperature the lower becomes the probability (or rate) of undesirable chemical reactions between the seawater and the lead (and bismuth) and structural materials becomes, the lower the need for waste heat recovery from the removed salts becomes, and the lower the probability of scale formation in the components preceding the OMACON first loop becomes, in case seawater preheating is to be carried out.

Another potential advantage of the low-temperature evaporation is the feasibility of partial evaporation. This may simplify the waste-disposal problem: Rather than dealing with solid waste, one will be dealing with a liquid waste (i.e., brine).

3.2 Desalination: Ability

Consider the reference OMACON EC system described in Section 2.3. Heat and mass balance calculations show that the seawater feeding rate needs to be 6.0

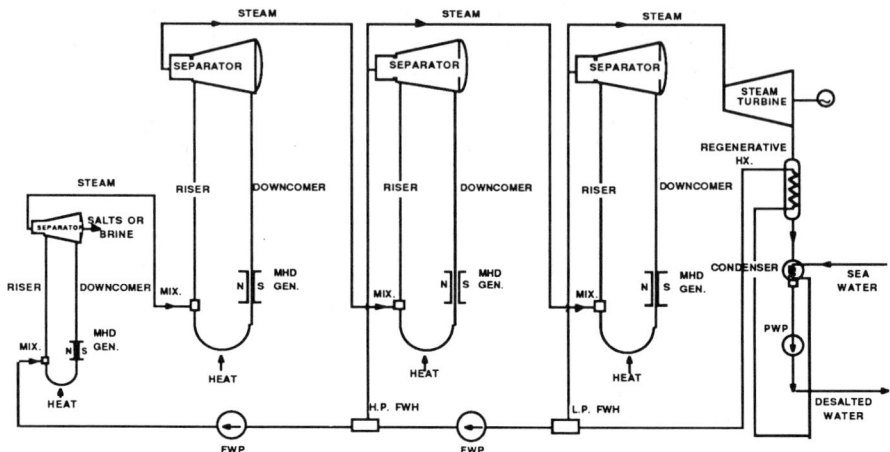

Fig. 5 Schematics of a combined OMACON-turbine steam cycle of a dual purpose plant featuring desalination by sea water injection.

or 7.2 kg/s if the water is to evaporate completely, or to become a concentrated brine with a salinity of about 28%, respectively. The corresponding desalination rate is about 5.8 and 6.3 kg/s, i.e., 500/550 m^3/day or ~55 (m^3/day/MW$_e$). This is to be compared with a specific desalination rate of 300-00 m^3/day/MW$_e$ attainable in dual-purpose plants based on the commonly used multi-effect distillation (MED) method. The difference in the specific desalination rates is because in the OMACON approach under consideration the desalination rate is actually the flow rate of steam in the power plant, whereas in the MED approach the latent heat of the steam coming out of the turbine can be used to evaporate (and hence desalinate), in multiple effects, many times that amount of steam.

Suppose that the first loop of this OMACON system is to evaporate the seawater at 350°C. If the water is to vaporize only partially, the lowest pressure in the first loop needs to be approximately 125 bars. (A 15°C boiling point elevation was assumed, to account for the effect of the salt.) Assuming that the water flow velocity in the riser is 4.2 m/s, and that the average void fraction in the riser is 50%, the diameter of the riser of this evaporating loop needs to be less than 30 cm.

It should be recalled that, if the seawater is to be completely vaporized, the evaporating loop pressure can be significantly lower than 125 bars. Alternatively, if Pb-Bi is to be used for the LM of the first loop, the system can be designed to have partial evaporation at a significantly lower pressure. In the following we shall consider the OMACON system designed for partial evaporation of the seawater at maximum pressure of 135 bars as the reference dual-purpose plants; it is assumed that 87% of the seawater is being vaporized. The waste heat carried by the brine is assumed to be recovered (for seawater preheating), using a couple of feedwater heaters (as in the reference single-purpose plant).

3.3 Energy Consumption

The efficiency of the dual-purpose plant is expected to be somewhat lower than that of the corresponding single-purpose plant due to three factors: 1) the boiling point elevation due to salt dissolved in the seawater (for the same steam pressure saline water needs to be heated to a higher temperature); 2) the enthalpy carried away by the salt or brine removed from the evaporating loop, which is not recovered, and 3) the extra pumping power requirements (for pumping the salt or brine along with the water to be desalinated). Generally speaking, the lower the evaporation temperature and the better the waste heat recovery the lower the extra energy that needs to be invested in the desalination and the higher the plant efficiency.

Table 2 compares our preliminary estimates of the energy balance of the reference dual- and single-purpose OMACON plants. "Charging" the desalted water throughput (≤ 550 m^3/days or ~22.5 m^3/h) by the difference in the electrical energy generated by the two plants gives an energy cost component of about 11.5 kWh/m^3. This turns out to be comparable to the specific energy for desalination in similar capacity plants based on conventional desalination methods (such as mechanical vapor compression and reverse osmosis). However, in order-of-magnitude bigger plants the specific energy for desalination can be significantly lower - between 10 to 5 kWh/m^3.

It should be realized, though, that an OMACON-type dual-purpose plant can be designed to have a specific energy for desalination that is significantly lower than 11.5 kWh/m^3. This could be done by 1) concentrating the brine to the maximum practical salinity, 2) lowering of the evaporation temperature, and 3) using the sensible heat of the extracted brine for countercurrent preheating of the feed seawater.

Consider the reference dual-purpose plant modified as follows:
1) The brine is concentrated to constitute only about 17% of the seawater feed.
2) The sensible heat of the brine is used for preheating one-eighth of the seawater feed using a countercurrent heat exchanger that features a 15°C

Table 2 - Energy Balance of desalination via seawater injection into an OMACON EC system

Characteristic	Plant type	
	Dual purpose	Single purpose
Heat input, kW$_{th}$	30,000	30,000
Electricity (ac) generated, kW$_e$		
MHD generator	6,096	6,200
Turbo generator	3,666	3,800
Total	9,762	10,000
Auxiliary power needs, kW$_e$	178	160
Net electricity (AC) output, kW$_e$	9584	9840
Net efficiency, %	31.95	32.80

Table 3 - Preliminary estimates of desalted water cost from the OMACON system

Case no.	Description	Cost component, $/m^3$			
		Energy	Capital	O&M	Total
1	Reference case	0.69	0.13	0.05	0.87
2	Improved case	0.23	0.14	0.05	0.42

temperature difference (compared with approximately 5°C common in desalination applications where scale formation is not a problem).
3) The evaporation temperature is 290°C, corresponding to a high cycle pressure of 60 bar.

The specific energy for desalination for this improved case is estimated to be less than 4 kWh/m^3. This is significantly lower than the specific energy for desalination of even the most energy efficient of the common desalination method.

3.4 Equipment Cost

Very little additional equipment that is specific for desalination is needed. Some design modifications are, nevertheless, to be expected. For example, it will be necessary to add the seawater intake system (unless it is used as the heat sink for the condenser), the seawater treatment system, the desalted water containment system, and the salt disposal system. On the other hand, the water treatment system of the conventional (i.e., single-purpose) plant could probably be eliminated.

A reliable evaluation of the overall effect of the desalination on the OMACON plant capital cost necessitates a conceptual design study and is therefore beyond the scope of this work. A very rough estimate is $200,000. Fortunately, the cost of the desalted water in this method is less sensitive to the capital cost component than in conventional desalination methods; a 100% change in the extra equipment cost will have only about 15-35% change in the cost of the product (see Section 3.5).

3.5 Product Cost

Table 3 provides a summary of the preliminary estimates of the cost of production of desalted water in the couple of OMACON systems considered. In obtaining these estimates it was assumed that the plant capacity factor is 80%, the investment needs to be recovered within 20 years, the cost of electricity is 6¢/kWh; and the operation and maintenance cost is 5¢/m^3.

For comparison, the cost of desalted water from dual-purpose plants of a similar capacity that are based on a conventional desalination technology is,

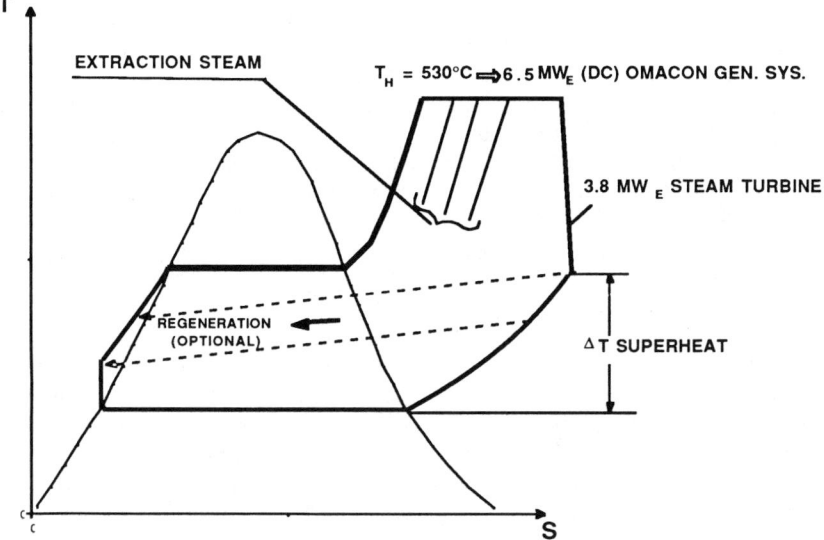

Fig. 6 T-S diagram of combined OMACON-turbine steam cycle with desalination by evaporative desuperheating.

typically, in the range of 1 - 2 \$/m^3. Large capacity plants can produce desalinated water at as low as about 0.70 \$/m^3. It is therefore concluded that the OMACON technology considered has a potential of being economically attractive for seawater desalination. However, more thorough study is necessary before the promise of this technology could be reliably assessed.

4.0 Desalination via Desuperheating Evaporation
4.1 General Description

Consider a combined OMACON-turbine system of the type illustrated in Fig. 2b, which has a T-S diagram like the type shown in Fig. 3b. For given inlet steam conditions and condensation temperature, there exists an optimal pressure for transition from expansion in the OMACON loops to expansion in the steam turbine. This optimal transition pressure, measured in terms of the maximum attainable efficiency, also depends on the efficiency of the system components as well as on the system mode of operation. In the systems considered in the present work the transition pressure is in the range of 0.5-5 bars. For a transition pressure of 1.4 bars, for example, the steam outlet temperature from the turbine is approximately 225°C; that is about 185°C above the condensation temperature.

By bringing the superheated steam exhausted from the LP turbine in direct contact with seawater, it is possible to make use of the sensible heat of the steam for evaporating part of the seawater. This desuperheating-evaporation process (the thermodynamics of which is illustrated in Fig. 6) can take place in an empty vessel. It might be desirable to increase the contact area between the steam and the seawater in order to make the desalination plant more compact and, hopefully, of a lower cost. This can be attained by the use of spray nozzles, packing material (such as Pall Rings), rotating disks, etc. The flow diagram of the process is shown in Fig. 7.

Relative to a single-purpose plant, the dual-purpose plant described earlier might need several design modifications. These include the following:

1) Changes in the condenser. Because of the increase in the amount of the steam to be condensed (with the extra steam coming from the seawater evaporation), the condenser volume, heat transfer surface and venting systems will have to be increased.
2) Addition of seawater preheaters. This will maximize their evaporation rate for a given turbine discharge rate.
3) Increase in the amount of steam extracted at pressures exceeding the condensation pressure. This is to compensate for the diversion of the sensible heat of the superheated steam coming out of the turbine from feedwater heating to seawater evaporation, in case the turbine exhaust steam is used for feedwater heating in the reference plant.
4) Change in the transition pressure from isothermal (MHD) to adiabatic (turbine) expansion. This may be necessary in order to maximize the cycle efficiency in view of the changes done in the energy balance.

4.2 Desalination Ability

Table 4 gives our estimates of the desalination ability of the desuperheating-evaporation approach described earlier, for three cases. These cases differ in the

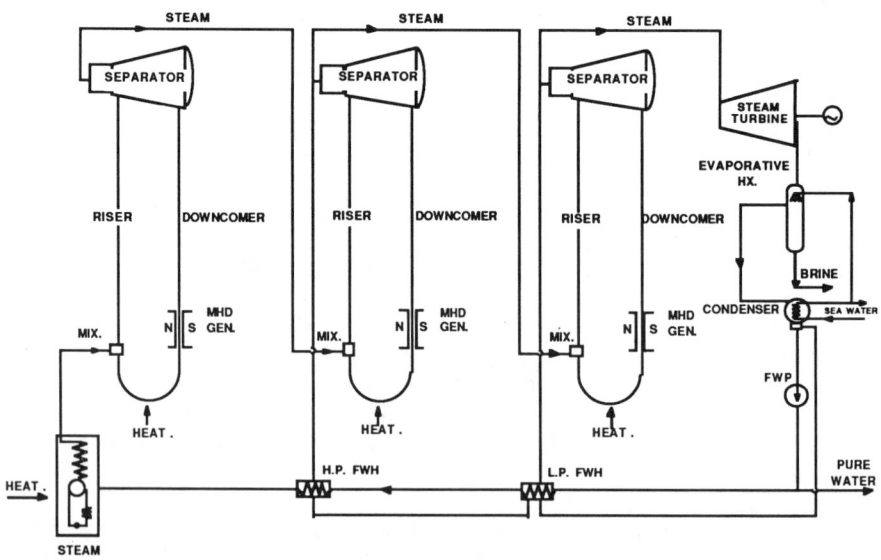

Fig. 7 Schematics of a combined OMACON-turbine steam cycle of a dual purpose plant featuring desalination by desuperheating of the turbine exhaust steam.

Table 4 - Desalination ability of a combined OMACON-turbine plant using the desuperheating-evaporation approach

Case	Transition pressure, bars	Turbine efficiency, 1st stage/overall	Exhaust temperature, °C	Product ratio, kg/kg[a]	Specific desalination, (m^3/day)/MW$_e$[b]
A	2	0.72/0.76	208	0.118	8.8
B	5	0.72/0.78	135	0.068	5.0
C	5	0.69/0.75	160	0.085	6.3

[a] Kilogram of desalted water produced per kilogram of steam getting out from the turbine.
[b] Rate of desalted water that can be produced per MW$_e$ of installed generation capacity.

transition (from isothermal to adiabatic) pressure and in the assumed turbine efficiency. In arriving at the estimates of Table 4, it was assumed that the condensation pressure is 0.08 bar, that 83% of the saline water is vaporized, and that the seawater is preheated up to 37°C.

The specific desalination ability of this process is almost an order of magnitude smaller than that of the seawater injection process described in Section 3.0. However, the desuperheating-evaporation process is simpler to implement, since the saline water is maintained at relatively low temperatures and does not get in touch with the LM. It should be realized, nevertheless, that a dual-purpose plant designed to generate 100 MW$_e$ will be considered a medium-sized desalination plant, even when the desuperheating-evaporation process is used.

4.3 Energy Consumption

A couple of scenarios are considered concerning the use of the sensible heat of the superheated steam coming out of the turbine of the reference single purpose plant:

Scenario I: The superheated steam energy is not recovered. This might be the case when the cost of the heat exchangers required for this energy recovery is prohibitively expensive.

Scenario II: The superheated steam energy is recovered; it is used for feedwater heating. In this case the optimal turbine inlet pressure is reduced to about 0.5 bar.

Table 5 compares preliminary estimates of the energy balance of the dual- and single-purpose plants for the two scenarios considered, both referring to case A of Table 4. The resulting specific energy for desalination is 3.2 and 62 kWh/m^3 for scenarios I and II, respectively. Whereas scenario I yields attractive desalination energetics, scenario II appears to be prohibitively expensive, energetically.

Table 5 - Energy balance of desalination via turbine exhaust steam desuperheating

Characteristic	Scenario I		Scenario II	
	Dual	Single	Dual	Single
Heat input, kW_{th}	30,000	30,000	30,000	30,000
Electricity (AC) generated, kW_e				
MHD generator	6,200	6,200	6,200	6,200
Turbo generator	3,640	3648	3640	3800
Total	9,840	9,848	9,840	10,000
Auxiliary power needs, kW_e	161	160	161	160
Net electricity (AC), kW_e	9,679	9,688	9,679	9,840
Net efficiency, %	32.26	32.29	32.26	32.80

It should be realized, though, that by changing the OMACON-to-turbine transition pressure it is possible to significantly effect the specific energy consumption (as well as the desalination ability; see Table 4). This is illustrated in Table 6.

4.4 Equipment Cost

Consider case A in scenario I. A very preliminary rough estimate of the extra investment in equipment gives $70,000; it consists of, primarily, an evaporator ($50,000), extra pumps, pipes, electrical systems and controls ($15,000); and miscellaneous items ($5000). The corresponding specific investment is $1150/($m^3$ desalted water/day). The specific investment for cases B and C of scenario I are inferred (assuming a -0.2 power factor for scale up) to be approximately $1300/($m^3$/day) and $1230/($m^3$/day), respectively.

A completely different situation exists in scenario II; the conversion of the single- purpose to dual-purpose plant is associated with a reduction in the overall investment due to the elimination of the desuperheated steam heat recovery heat exchangers (feedwater heaters). This reduction is estimated to be approximately $400,000 and $250,000 for case A and cases B and C, respectively.

4.5 Product Cost

Table 7 summarizes the preliminary estimates of the cost of desalted water obtained in the different cases examined for desalination by exhaust steam desuperheating. In deriving these cost figures it was assumed, as in Section 3.5, that the real interest rate is 8%, the plant lifetime and capacity factor are 20 years and 80% respectively and the cost of electricity is 6¢/kWh.

The negative cost of the desalted water obtained in cases B and C of scenario II implies that it does not pay to invest in feedwater heaters-turbine exhaust steam desuperheaters (i.e., that scenario II is probably not likely to be of practical interest).

Table 6 - Effect of system design (primarily transition pressure) on the desalination energetics in kWh/m³

Energy consumption	Scenario I			Scenario II		
	A	B	C	A	B	C
"Taken" from cycle	2.9	5.1	4.1	61.9	22.7	23.1
Pumping and miscellaneous	0.3	0.3	0.3	0.3	0.3	0.3
Total	3.2	5.4	4.4	62.2	23.0	23.4

Table 7 Preliminary estimates of the cost (in $/m³) of desalted water using the exhaust steam desuperheating approach

Cost component	Scenario I, Case			Scenario II, Case		
	A	B	C	A	B	C
Energy	0.19	0.32	0.26	3.73	1.38	1.40
Capital	0.40	0.46	0.43	-1.88	-2.02	-1.63
O&M	0.05	0.05	0.05	0.05	0.05	0.05
Total	0.64	0.83	0.74	1.90	-0.59	-0.18

Considering scenario I, it is observed that the cost of the product water is quite reasonable - it is significantly lower than the cost of water desalination in similar capacity dual purpose plants that use one of the conventional desalination technologies.

5.0 Concluding Remarks

The analysis carried out in this work indicates that the OMACON technology offers new and interesting possibilities for seawater desalination. Of the couple of approaches examined, the direct contact evaporation of seawater in the first OMACON loop appears to be the more promising: It offers a significantly higher water-to-electricity ratio and a lower water cost. In fact, the estimated cost of the desalted water is highly competitive with the cost of water desalinated in even large-capacity dual-purpose plants that use the most economical desalination processes available.

However, at our present state of knowledge there are many uncertainties concerning the feasibility of the desalination processes hereby proposed in connection with the OMACON technology. Among these uncertainties are the

following:
1) the compatibility of seawater with lead or another liquid metal (such as mercury) and with structural materials;
2) the feasibility of extracting the salt or brine from the OMACON loop;
3) the handling of the seawater and the waste brine and, particularly, of preheating the seawater and waste heat recovery from the brine; and
4) the extra energy and capital cost investment requirements.

Consequently, a more reliable assessment of the promise of the OMACON technology for seawater desalination should await the outcome of R&D programs addressing certain key issues and should be based on a thorough conceptual design study.

At the same time it should be realized that the dual purpose OMACON systems considered in this work are not necessarily the optimal. It is probable that detailed optimization studies, based on economic rather than thermodynamic criteria, will identify even more promising schemes and system designs for seawater desalination using the OMACON technology.

In view of the encouraging results on the viability of seawater desalination determined in the preliminary assessment, it is highly recommended that the previously mentioned R&D and conceptual design studies will be undertaken.

Acknowledgment

This work was supported by Solmecs (Israel).

References

[1] Petrick, M., and Branover, H., "Liquid Metal MHD Power Generation -- Its Evolution and Status," *Single and Multiphase Flows in an Electromagnetic Field Energy Metallurgical and Solar Application,* Progress in Astronautics and Aeronautics Series, Vol. 100, AIAA, New York, 1985, pp. 371-400.

[2] Branover, H., "Liquid Metal MHD," *Proceedings of the 9th International Conference on MHD Electrical Power Generation,* Society for MHD Electrical Power Generation of Japan, Japan, Vol. V, 1986, pp. 1735-1743.

[3] Blumenau, L., Branover, H., El-Boher, A., Spero, E., Sukoriansky, S., Talmage, G., and Greenspan, E., "Liquid Metal MHD Power Conversion Systems with Conventional and Nuclear Heat Sources," *Proceedings of the 24th Symposium on Engineering Aspects of MHD,* Symposia for Engineering Aspects of Magnetohydrodynamics, Inc., USA, 1986, pp. 33-41.

[4] Greenspan, E., Barak, A.Z., Blumenau, L., Branover, H., El-Boher, A., Spero, E., and Sukoriansky, S., "Liquid Metal MHD Conversion of Nuclear Energy to Electricity: Possibilities and Implications," *Liquid Metal Flows; Magnetohydrodynamics and Applications,* Progress in Astronautics and Aeronautics Series, AIAA, Washington, DC, Vol. 111, 1988, pp. 129-157.

[5] Branover, H., and El-Boher, A., "The "Etgar" Program for the Development of Liquid Metal MHD Energy Conversion Technology," *Proceedings of the 10th International Conference on MHD Electrical Power Generation,* Bharat Heavy Electricals Ltd., MHD Center, Tiruchirapalli, India, Vol. 1, 1989, pp. I.19-I.26.

MHD Generator for Waste Heat Utilization in Northern Conditions

I. M. Kirko*
Perm Institute of Technical Physics, Perm, Russia

Abstract

The purpose of this note is to introduce a novel concept in MHD electrical power generation applicable for small units utilizing low grade heat. The Institute of Physical Problems of Technology in Perm, Russia is presently investigating and developing this concept. Since this development is still in its very early stages, this note deals only with general assessments and does not provide any firmly established quantitative data.

Introduction

The problem of using renewable energy sources is one of the most important problems in contemporary energy engineering.

The international seminars on MHD flows and turbulence take place in Israel — a country which is particularly rich in solar energy and, hence, its utilization is of particular importance. As for northern countries, they also possess rich low grade heat sources which (as will be shown) can be used by means of devices similar to those developed for solar energy. Both northern and southern hemispheres can cooperate in this field with mutual benefit.

In this connection, the contribution of the Center for MHD Studies at Ben Gurion University, Israel, should be emphasized. Investigations done there have given rise to a principally new direction in applied magnetohydrodynamics: that of the utilization of low-temperature energy sources.

To confirm the importance of the problem in question, let us consider two examples of relatively large waste heat sources which as yet do not find application in the Perm region (in the northeast of European Russia) and which in some cases are ecologically adverse (see Table 1). The electrical power which could be generated is also given in Table 1. Using MHD-installations of the type suggested below, we assume solar energy density to be roughly 1.06 kW/m^2, and determine that the area occupied by solar collectors which would absorb the same amount of heat equal to waste heat in our first example approximates 1670 m^2. Thus the problem of utilizing heat emissions in

---

Copyright © 1992 by the American Institute of Aeronautics and Astronautics, Inc. All rights reserved.
*Director.

Table 1 Typical waste heat sources in the Perm region

	Values names	Perm hydroelectric	Berezniky metallurgical plant
1	Kind of waste heat	Water cooling electrical capacitors	Steam generated through evaporation of cooling water
2	Temperature of effluent, °C	15	100
3	Average temperature of the environment (October-April), °C	-20	-25
4	Average value of Carnot efficiency	0.121	0.335
5	Usable thermal power, MW	1.67	2.24
6	Electric power output at conversion efficiency attainable with the suggested conversion system, kW	180	241

northern climates is comparable with that of building small size solar power stations in the south.

What are the possible ways of connecting these two directions of new power engineering? The magnetohydrodynamic methods of converting heat energy into electrical energy utilizing low grade heat are, we believe, this joining link.

Let us consider the problem of a low-temperature liquid metal MHD-generator in the retrospective of the history of its development. Single phase flow generators with high velocity of metal as well as two-phase flow liquid metal-gas MHD generators would have extremely low efficiencies at working conditions considered here. In 1984 Petrick and Branover[1] suggested an ingenious scheme in which a liquid metal circuit is driven by steam airlift (see Fig. 1).

This scheme overcomes a number of deficiencies of the previously mentioned concepts, however, for conditions considered here the Branover-Petrick concept would still have too low efficiency because of high slip ratio (ratio of gas to liquid velocities).

Therefore, we introduced a number of changes into this scheme which leads to higher efficiency and makes it applicable to low grade heat.

Let us consider some possible electrodynamically active media. We suggest use of a medium[2] which we called "magnetostabilized" which later a number of authors called magnetorheologic. Imagine a "porridge" made of a liquid metal, for example, a sodium-potassium eutectic alloy (potassium 78%,

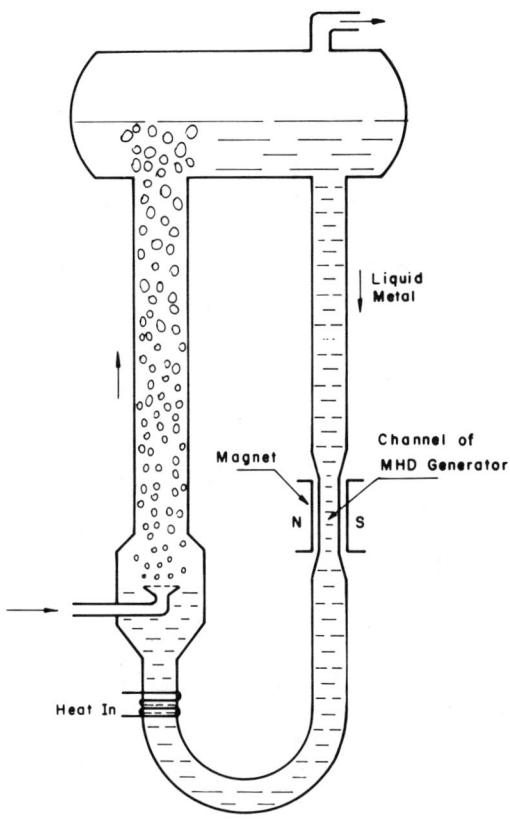

Fig. 1 Schematic of optimized magnetohydrodynamic conversion system (from Ref. 1).

sodium 22%, melting temperature 12°C) and ferromagnetic particles in the form of long needles of iron or strips of amorphous steel. An even better version would be to use an easily boiling liquid instead of sodium-potassium. The most suitable one in this case, in our opinion, is methylamine (CH_3NH_2) used previously as a fuel for large rockets. The freezing temperature is 93°C, the latent heat of evaporation $r = 2.0 \cdot 10^6$ J/kg. The advantages of methylamine include its weak corrosion properties, low toxicity, and inexpensiveness. Thus, let us imagine such a medium consisting of methylamine with suspended ferromagnetic particles. In moving such media in channels with longitudinal or transverse magnetic fields one can easily implement a "plug" operation. The idea of using the plug operation in MHD-generators is not new. There were many works dealing with the possibility of pushing liquid metal plugs, separated from each other by gas or steam layers, through an MHD generator channel. However, the instability of liquid metal plugs in the magnetic field made this approach impractical, and these works gradually faded. The physical

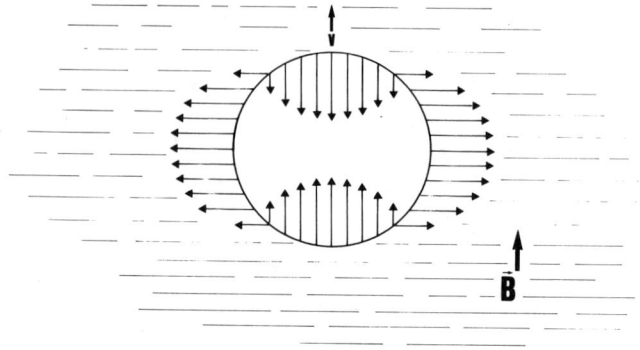

Fig. 2a   Gas bubble in magnetic field with longitudinal magnetic field.

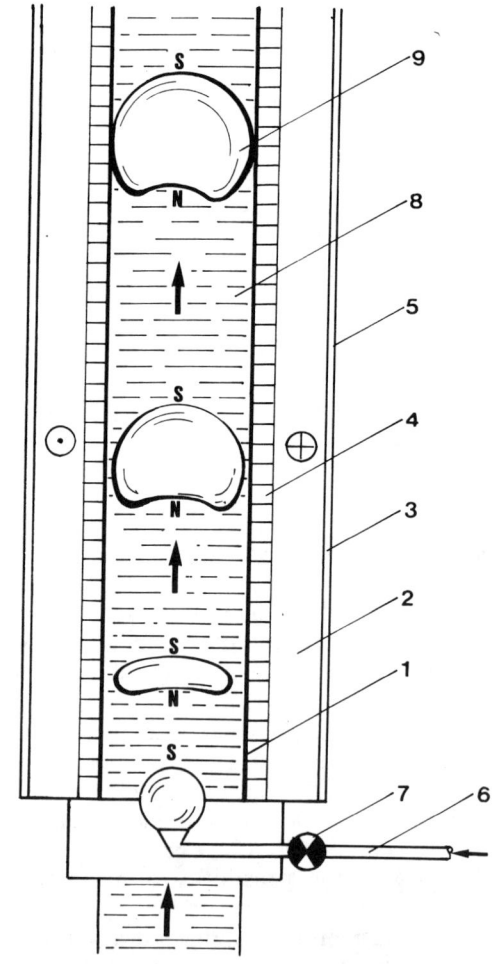

Fig. 2b   New MHD Generator with magnetostabilized liquid pistons (see the explanation in the text).

cause of this instability is simple: in moving a liquid metal piston with free boundaries in a tube, induction currents are generated in the cross magnetic field, the ponderomotive interactions of these currents with the magnetic field are such that they tend to destroy the piston, and this leads to the transition of the plug into two longitudinal streamlines. The general electrodynamic rule is in force: induction interactions in a mobile system are always such that they tend to break the resulting induction currents.

Quite a different situation can be observed in a magnetostabilized medium moving, for example, in a longitudinal magnetic field (see Fig. 2a). Gas bubbles being blown into the magnetic "porridge" would contract in the direction of the magnetic field and expand in the cross direction until they turn into a cross gas layer. The ferromagnetic inserts in an induction winding would promote this process.

A schematic diagram of a magnetodynamic machine having magnetostabilized pistons as its working medium is shown in Fig. 2b where 1 is a stainless steel tube, 2 windings creating a uniform constant longitudinal magnetic field, 3 thermal insulation, 4 induction elements (windings) connected to the load circuit, 5 ferromagnet inserts, 6 injection of gas or volatile liquid, 7 valve setting time intervals between introducing bubbles, 8 magnetostabilized liquid (suspension of ferromagnetic particles), 9 bubble in the process of expansion. One can view the layers of a ferromagnetic liquid between the lifting bubbles as constant magnets moving through solenoids - i.e., Faraday's famous experiment.

It is interesting to note that Faraday's apparatus consisting of a constant magnet moving through a solenoid is an electric machine with the highest efficiency as compared with all the other ways of converting mechanical energy into electric known at present. In this way it is possible to eliminate slip (i.e., to achieve a slip ratio equal to unity) and achieve efficiency even higher than that achieved by using the nonmodified device (Fig. 1).

In conclusion it should be mentioned that although much more theoretical work has to be done and practical tests are of vital importance, the scheme of utilizing the heat of industrial effluents with relatively low temperature in the northern conditions, developed as a modification of Branover-Petrick's concept appear to be technically and economically viable. The use of modern materials such as ceramic superconductors, magnetic rheologic liquids, and methylamine as a working medium will make it possible to bring up the efficiency of converting low grade waste heat into electric energy to the highest attainable value. Facilities of the described type will probably have a very long lifetime with no maintenance needed.

References

[1] Petrick, M. and Branover, H., "Liquid Metal MHD Power Generation - its Evolution and Status," *Single- and Multi-Phase Flow in an Electromagnetic Field, Progress in Astronautics and Aeronautics,* Vol. 100, edited H. Branover, P. S. Lykoudis and M. Mond, AIAA, Washington, DC., 1985, pp. 371-400.

[2] Kirko, J. M., and Filipov, M. V., "Particularities of the weighted layer of ferromagnetic particles in the magnetic field," *Journal of Theoretical Physics* (in Russian), Vol. 30, No. 9, 1960, pp. 1081-1084.

Recent Results on LMMHD Induction Generators

Ph. Marty,* L. Leboucher,* A. Alemany,* and A. Pilaud*
Institut de Mécanique de Grenoble, Grenoble, France
Ph. Massé†
*Madylam-Institut National Polytechnique de Grenoble,
St. Martin d'Héres, France*
and
H. Branover,‡ A. El Boher,‡ and Y. Kaplan‡
Ben-Gurion University of the Negev, Beer-Sheva, Israel

Abstract

This paper presents recent results on liquid metal magnetohydrodynamic (LMMHD) flat induction generators working under self-excitation. A numerical modelization of an induction generator with a finite element software is described and the main parameters of an experimental prototype are given as well as its predicted performances. Experimental results of this prototype working with NaK are presented and compared with predictions for a magnetic Reynolds number below approximatively 2 and a frequency range between 7 and 25 Hz. The evolution of the critical velocity to obtain self-excitation and of the electrical efficiency is described. The discrepancies with the numerical predictions are discussed as well as some possible explanations to justify these differences. Finally, an analytical model describing the influence of the lateral side bars is presented and found eventually able to explain these discrepancies.

Nomenclature

ν = kinematic viscosity of the fluid ($=10^{-7}$ $m^2.s^{-1}$ for NaK)
σ_f = electrical conductivity of the fluid = $2.5.10^6$ $\Omega^{-1}.m^{-1}$ for NaK
σ_w = electrical conductivity of the walls
σ_y = electrical conductivity of the magnetic iron yoke
μ_r = magnetic permeability

Copyright © 1991 by the American Institute of Aeronautics and Astronautics, Inc. All rights reserved.
*Fluid Mechanics Department, B.P. 53 X.
†MADYLAM, P.O. Box 95, 38402 St. Martin D'Hères.
‡Center for MHD Studies, P.O. Box 653.

ρ	=	density of the fluid (= 850 kg.m^{-3} for NaK)
λ	=	wavelength of the winding
k	=	wave number ($k = 2\pi/\lambda$)
ω	=	electrical pulsation ($\omega = 2\pi f$)
V_s	=	synchronism velocity of the magnetic field ($V_s = \lambda f = \omega/k$)
Vc	=	critical velocity of the fluid to achieve self-excitation.
a	=	thickness of the liquid metal flow
ε	=	thickness of each wall
e	=	resulting magnetic gap ($e = a + 2\varepsilon$)
l	=	width of the liquid metal flow
L	=	length of the generator
r	=	electrical resistance of one phase of the stator winding
L	=	electrical inductance of one phase of the stator winding
C	=	value of each capacitor
R	=	electrical resistance of the load
j	=	complex electrical density in the fluid
b	=	complex magnetic field
J	=	lineic source density of the primary winding (A/m)
P_L	=	power transmitted to the load
P_{Jf}	=	Joule losses in the fluid
P_{Jw}	=	Joule losses in the channel walls
P_{Js}	=	Joule losses in the primary copper stator windings
P_{MHD}	=	mechanical power drawn from the fluid by Lorentz forces
I	=	electrical current density in the primary winding
U	=	output voltage across one phase
p	=	pressure
Q	=	volumic flow rate
M	=	Hartmann number : $M = Ba\sqrt{\sigma f/\rho\nu}$
R_m	=	magnetic Reynolds number = $\mu_0 \sigma_f (V-V_s)/k$
R_e	=	hydrodynamic Reynolds number = Va/ν
N	=	interaction parameter = M^2/R_e
s	=	slip ratio between the fluid and the magnetic field: $s = 1 - V/V_s$
Λ	=	hydraulic friction coefficient
η_{el}	=	air gap efficiency = power transmitted to the stator / P_{MHD}
		$= 1 - (P_{Jf} + P_{Jw})/P_{MHD}$

I. Introduction

The capability of LMMHD linear induction generators of working under self-excitation conditions has been clearly illustrated through the works at the French Mechanical Engineering Laboratory in Grenoble during recent years (Fig. 1). On the theoretical point of view[1] an infinitely long machine has been studied and an expression for the critical velocity of the fluid to achieve self-excitation has been proposed for the case of small values of the magnetic Reynolds number. In this model, the actual distribution of the primary currents has been replaced by an ideal sinusoidal current sheet and a one-dimensional bulk flow hypothesis has been used. The theoretical expression of the critical velocity has been checked on a mercury facility [2]. The experimental results have confirmed the theoretical works but have led us to introduce additional coefficients, α_1 and α_2, which did not appear in the analytical formula and which are defined as follows:

$$V_c = V_s \left(1 + \frac{2\varepsilon}{a} \frac{\sigma_w}{\sigma_f}\right) + \alpha_1 \frac{e}{a \mu_0 \sigma_f} \frac{r}{V_s L} + \alpha_2 \frac{e}{a \mu_0 \sigma_f V_s RC} \quad (1)$$

On the one hand, these coefficients α_1 and α_2 have to be considered as the contribution of the various three-dimensionnal effects, such as finite width or finite length. On the other hand, they express the influence of the real velocity profile, which differs from a flat profile, and the influence of the real primary currents distribution, which is not a pure spatially sinusoidal sheet. Indeed for an ideal infinite generator with a flat velocity profile, $\alpha_1 = \alpha_2 = 1$.

Nevertheless, the technical performances of the experimental facility[2] did not allow a complete study of all of the parameters that influence the value of the critical self-excitation velocity as well as the energetic efficiency of a liquid metal induction generator. Furthermore, the density of mercury has been responsible for experimental difficulties (cavitation problems have been lowered by an extremely high pressure level in the loop) which have prevented us from achieving a significant amount of power at the generator terminals. Consequently, a fluid with a smaller density than that of mercury and with a good electrical conductivity has appeared suitable in order to complete this described works. This study will present experimental results obtained at the Ben Gurion University on a NaK (22% Na-78% K) loop and the numerical modelling that has been undertaken in order to predict the critical self-excitation velocity and the mechanical into electrical energy conversion efficiency. It is worth noting that this work is more concerned with the capability of our numerical program in predicting the performances of an MHD generator than with the achievement of a high electrical efficiency. As a matter of fact, when starting this work, we already were aware that our technological choices (rectangular

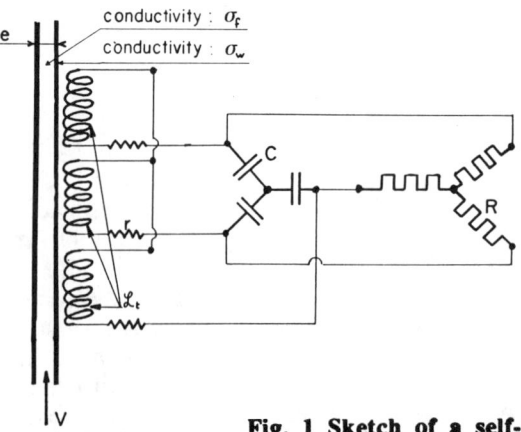

Fig. 1 Sketch of a self-excited induction generator.

channel instead of annular, for example) as well as the restrictions imposed by an already existing NaK loop could not permit us to achieve a high efficiency.

II. Numerical Modelling of the Experimental Generator

A. Generalities

The choice of the main parameters of the experimental prototype has been widely dictated by the hydraulic caracteristics of the Ben Gurion University NaK loop. In a second stage, the exact performances have been numerically calculated with the Flux-Expert finite elements software. For technological purposes, a flat induction generator with lateral copper side bars has been chosen ; consequently, a two-dimensional modelisation in the (x, y) plane, itself perpendicular to the primary currents, has been possible (Fig. 2). The modelling of an annular induction machine should be possible by choosing cylindrical coordinates as well. Then, a scalar equation on the A_z component of the vector potential of the magnetic field has to be solved :

$$\sigma \frac{\partial A_z}{\partial t} + \sigma V \frac{\partial A_z}{\partial x} - \frac{1}{\mu_0 \mu_r} \nabla^2 A_z = 0 \qquad (2)$$

At first, the velocity profile has been assumed to be constant although this hypothesis could be untrue at the entry and exit of the generator where the fluid suddenly enters (or leaves) the magnetic field. On the other hand the high values of the hydraulic Reynolds number in the test section allow us to neglect the channel walls boundary layers influence. Nevertheless, an analytical study of the small weakening effect of these layers on the electrical efficiency of a conduction generator is to be found in a paper from Elliott[3] where a 1/7 - power velocity profile has been used. Equation (2) is then solved by the numerical program in a 2-D domain, where all of the geometrical and physical characteristics have been described for each region (electrical conductivity, magnetic permeability, electrical primary current source density and frequency, etc...). More detailed information on this part will be found in Ref. 2. From the knowledge of the complex quantity A_z everywhere in the domain, one can obtain the current

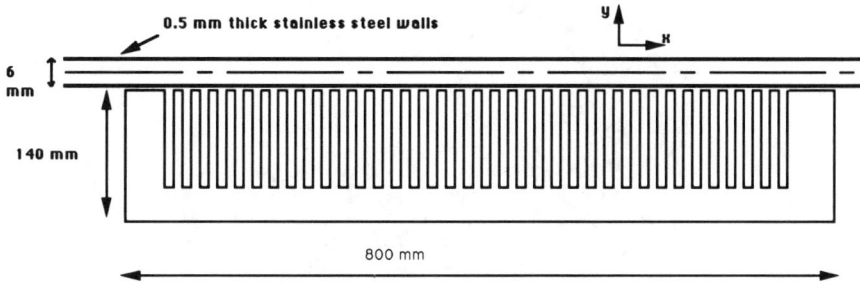

Fig. 2 Experimental generator.

density and the magnetic field as follows:

$$j_z = -\sigma_f (i\omega A_z + V \frac{\partial A_z}{\partial x}) \tag{3}$$

$$B_x = \frac{\partial A_z}{\partial y} \; ; \; B_y = -\frac{\partial A_z}{\partial x} \tag{4}$$

One can then compute the mechanical power extracted from the fluid, P_{MHD}, as well as the Joule losses in the fluid, P_{Jf}, or in the channel walls, P_{Jw}, (when conducting):

$$P_{MHD} = \frac{1}{2} \text{Re} \{ \iint_{fluid} j_z \cdot B_y^* \, V \, dS \} \tag{5}$$

$$P_{Jf} = \frac{1}{2\sigma_f} \text{Re} \{ \iint_{fluid} j_z \cdot j_z^* \, dS \} \tag{6}$$

$$P_{Jw} = \frac{1}{2\sigma_w} \cdot \text{Re} \{ \iint_{walls} j_z \cdot j_z^* \, dS \} \tag{7}$$

where the * notation denotes the conjugated complex quantity. By assuming the electrical conductivity of the magnetic iron yoke σ_y to be zero, one can neglect the induced currents in the magnetic circuit and express the air gap efficiency η_{el}

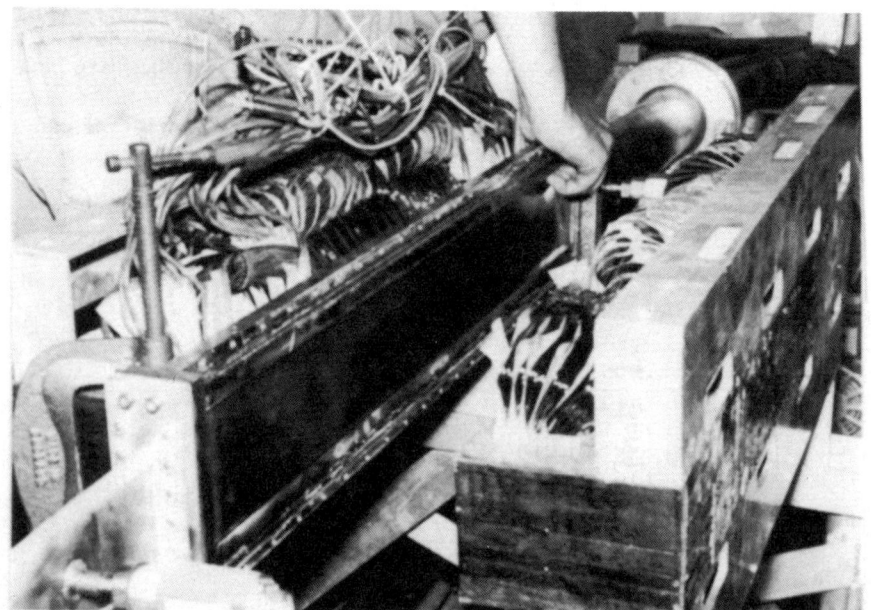

Photo 1 Experimental generator.

as follows :

$$\eta_{el} = 1 - \frac{P_{Jf} + P_{Jw}}{P_{MHD}} \tag{8}$$

On the other hand, it has also been possible to make a numerical prediction of the critical velocity V_c for self-excitation. As a matter of fact, this phenomenon can only arise when the power P_{MHD} given by the fluid is exactly equal to the sum of the various Joule losses (in fluid, walls, primary windings and magnetic circuit if σ_y is not equal to zero) and of the power P_L transmitted to the resistive load R. Then the numerical prediction of V_c is quite simple to undertake.

B. Technical Description

In the first stage, mainly due to hydraulic considerations, the cross-sectional area and the length of the channel have been determined. In turn, this choice has imposed a typical value of the wavelength, λ. The order of magnitude of the fluid velocity being given by the hydraulic circuit, one can deduce what range of electrical frequency will give a reasonable slip ratio, s. According to these restrictions, the following parameters have been expected to give the best energy performances :

1) Double-sided magnetic yoke with an electrical copper winding identical to that described in Ref. 2.
2) Cross-section area of the rectangular flow : a x l = 5 x 100 mm^2.
3) Cross-section area of the lateral copper side bars : 5 x 20 mm^2.
4) Active length of the channel L = 800 mm.
5) Thickness of the stainless steel walls : ε = 0.5 mm.
6) Resulting magnetic gap : e = a + 2ε = 6 mm.
7) Wavelength : λ = 36 cm.
8) Total number of slots per block = 36 cm.
9) Number of slots/pole/phase = 3.
10) Number of phases : 3 or 6 (adjustable).
11) Resistance/inductance ratio : r/L = 4.34 Ω/H for serie connexion
    = 4.62 Ω/H for parallel connexion.

The numerical calculations have clearly confirmed the negative influence of the local decrease of the intensity of the magnetic field in front of each slot; consequently, the two magnetic blocks and the windings (Photo 1) have been designed in such a way that the so called "tooth opposite tooth" or "tooth opposite slot" configurations (see Pierson[4]) could be tested. These two different geometrical situations will be hereafter refered to as A and B geometries, respectively. On the other hand, in order to work in a frequency range as large as possible, the inductance of the windings has been varied according to the type of connexion of the two primary blocks (serie or parallel). Figure 3 shows the numerical prediction of the electrical efficiency η_{el} vs the slip s for the A geometry and the very weak influence of the frequency can be observed. Figure 4 shows the numerical calculation of the expected critical velocity V_c for an infinite value of the load (R = ∞) and for the same A geometry.

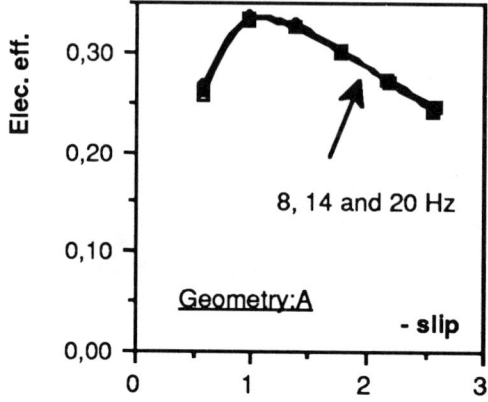

Fig. 3 Numerical calculation of the electrical efficiency vs slip at 8, 14, and 20 Hz (Results concerning geometry B would show the same evolution with a maximum value approximately 5% higher than A).

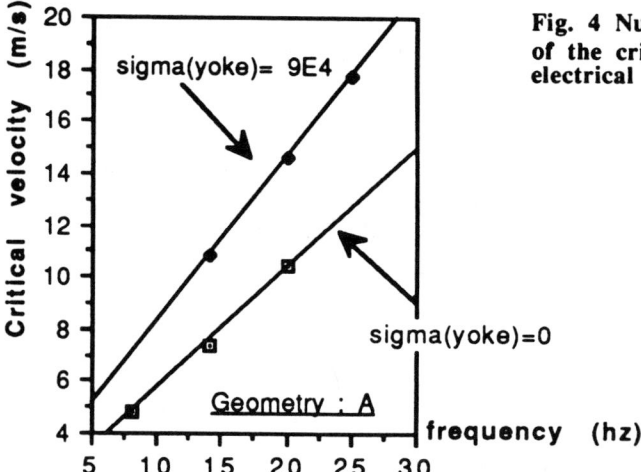

Fig. 4 Numerical calculation of the critical velocity (no electrical load).

III. Experimental Facility

A. NaK Loop Description

The Ben Gurion University NaK loop (Fig. 5a) has an overall volume of about 90 liters, and the flow is powered by a seal-less centrifugal CRANE pump supplied with a variable frequency (0 to 60 Hz). The flow rate adjustment can be accomplished by using a by-pass valve, as well, and the flow rate is measured via a Venturi nozzle. The various running operations of the loop (NaK fill, run, test, drain, emergency, standby, etc.) as well as a part of the data acquisitions are made by a computer. Finally, all of the NaK free surfaces (sump tank, expansion tank, etc.) are covered with a high purity nitrogen atmosphere. Figure 5b shows the dimensions of the nozzle and the diffusor at the inlet and outlet of the generator, respectively, as well as the location of the pressure taps used for the measurement of P_{MHD}.

INDUCTION GENERATORS

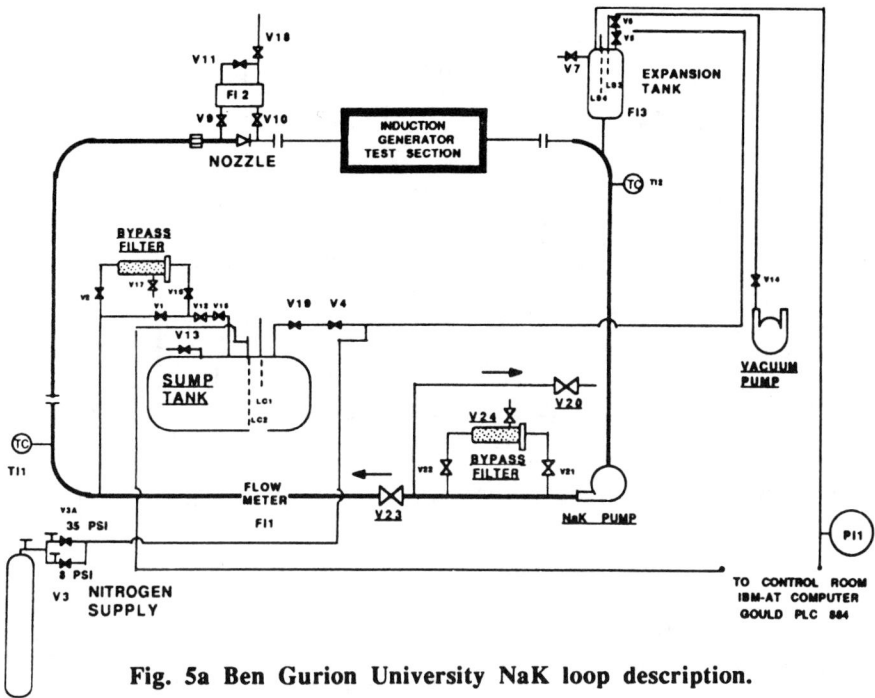

Fig. 5a Ben Gurion University NaK loop description.

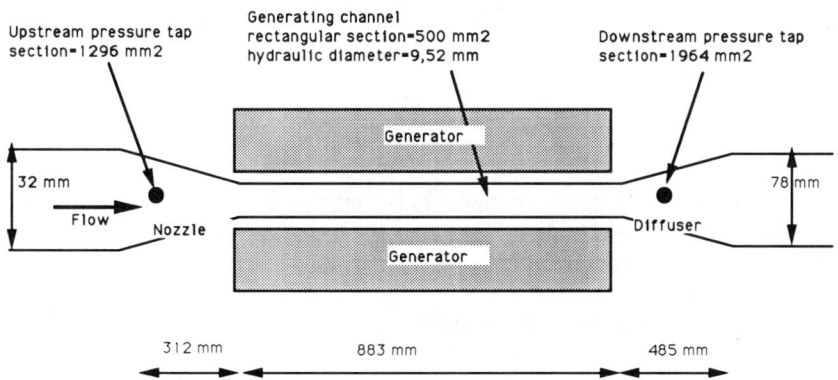

Fig. 5b Hydraulic description of the test section.

B. Experimental Procedure

The measurement of P_{MHD} requires the knowledge of the hydraulic friction coefficient Λ between the upstream nozzle and the downstream diffusor. This coefficient can be measured when no magnetic field is applied:

$$\Delta (p + \frac{1}{2} \rho V^2) = \Lambda \frac{L_{gen}}{D_H} \frac{1}{2} \rho V_{gen}^2 \qquad (9)$$

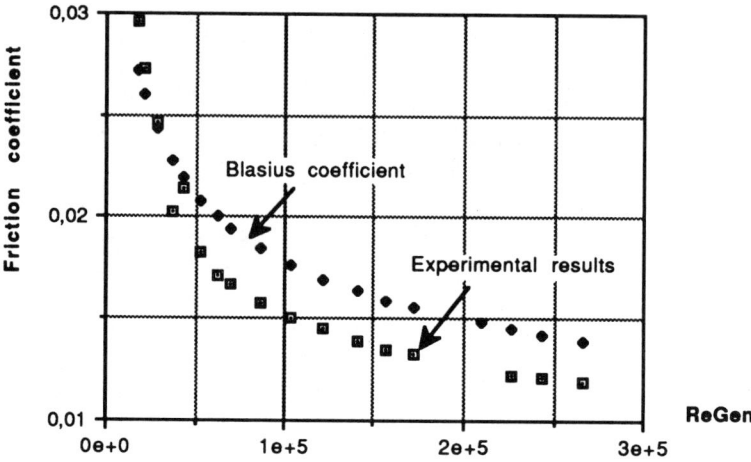

Fig. 6 Experimental friction coefficient compared with Blasius coefficient.

where L_{gen} denotes the distance between the two pressure taps (≈ 1 m) and D_H stands for the hydraulic diameter of the rectangular channel ($D_H = \frac{2al}{a+l} = 9.52$ mm).

Figure 6 shows the evolution of Λ with the Reynolds number in the generator. Its approximate expression can be written as:

$$\Lambda = \frac{0.7353}{Re^{0,3344}} \qquad (10)$$

It is worth comparing this experimental result with the Blasius formula: $\Lambda = 0.3164/Re^{1/4}$. The knowledge of Λ and of the upstream and downstream pressure measurements then allow the calculation of P_{MHD} under the assumption that Λ stays unchanged by the magnetic field. Unfortunately, few results reporting the influence of a traveling magnetic field wave on Λ are available. On the other hand, the results obtained for the case of a constant magnetic field by Brouillette and Lykoudis[5] show that for Re ranging from 1 to 2.10^5 and $M/Re < 2.10^{-3}$ (which correspond to our experimental situation) the increase in Λ with the magnetic field applied was less than 20%. Consequently, the value of Λ expressed in Eq. (10) for B = 0 has been (more or less arbitrarily) increased by 10% in the order to measure P_{MHD}. Finally, the measurement of the power transmitted to the load, P_L, and of the primary Joule losses in the stator windings P_{Jst} lead to the value of η_{el} previously defined in Eq. (8):

$$\eta_{el} = \frac{P_L + P_{Jst}}{P_{MHD}} \qquad (11)$$

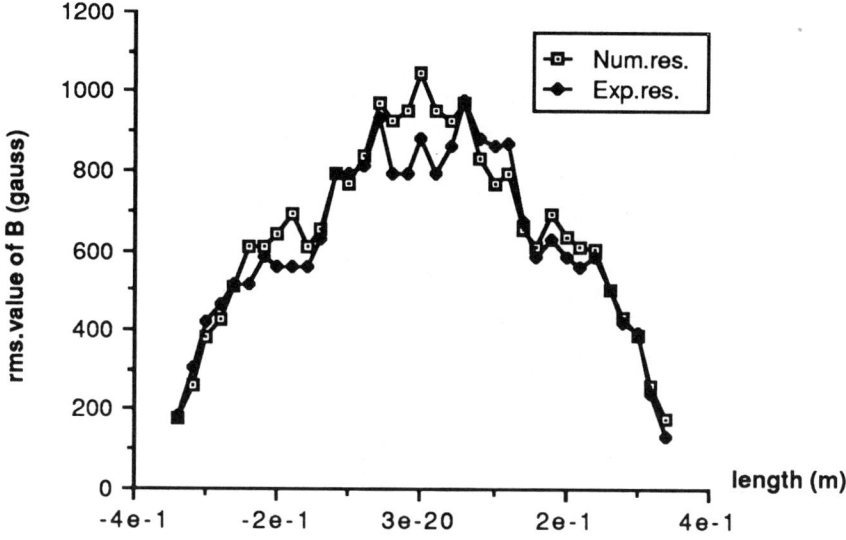

Fig. 7 Empty air gap magnetic field distribution.

IV. Experimental Results

A. Preliminary Tests and Measurements

After removing the channel from the generator, the magnetic field distribution in the air gap has been measured with a Hall effect probe. Because of the size of the probe, the measurements have been made with a 16-mm air gap and at a frequency of 50 Hz (industrial frequency). Figure 7 shows a comparison between experimental results and numerical predictions. A good agreement can be observed despite a small discrepancy at the center of the machine, which could be possibly explained by an unbalancing of intensity between the 3 phases during the measurements. This unbalancing, which had an average value of 10% has not been modelised in the numerical calculation, where a 120-deg phase difference was assumed. During the NaK tests, this unbalancing has been almost canceled by additional turns in the slots located near the inlet and outlet of the machine. This technique has been previously described in Ref. 2.

B. Study of the Self-excitation Critical Velocity

In contrast to mercury experiments described in Ref. 2 where self-excitation had to be initiated, it appeared spontaneously with NaK. For the particular case of a 6-phase connection, Fig. 8 shows the time evolution of the output voltage during the transitory self-excitation stage. During this short period (a few hundredths of ms) the flow slows down under the effect of the j x B retarding forces until it reaches the critical velocity V_c solely depending upon the frequency f and the load R.

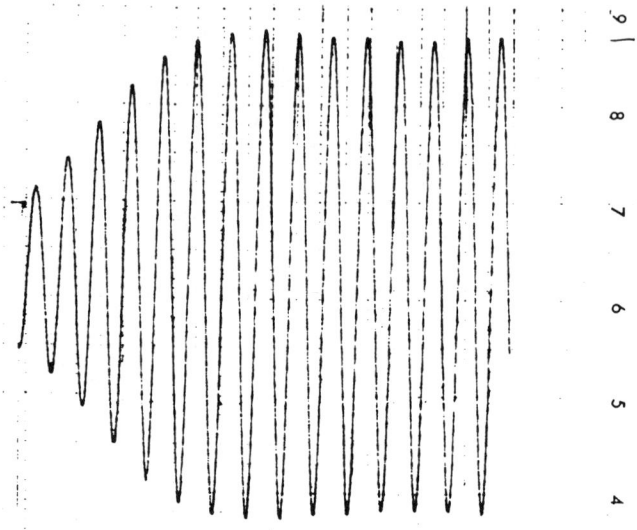

Fig. 8 Generator output voltage during the transient stage of self-excitation for a 6-phase connection (typical duration = approximately 1s).

Figures 9a and 9b show the critical velocity dependence with respect to 1/R for A and B geometries, respectively. Several frequencies have been experimented with. The linear dependence with 1/R is perfectly satisfied, as it was foreseeable from Eq. (1), despite the fact that the magnetic Reynolds number was not small compared to unity (R_m had a maximum value of about 1.8). The slope of each line allows the calculation of the α_2 coefficient defined in Eq. (1). The average value of α_2 that was drawn from these results is approximatively $\alpha_2 = 3$ compared to the results previously obtained with mercury[2]. Furthermore, for a given load (1/R) and frequency, V_c is approximatively 0.5 m/s smaller for B than for A geometry (see Fig. 10).

C. Study of the Electrical Efficiency η_{el}

For various frequencies, Fig. 11 shows the evolution of η_{el} with respect to the slip ratio s. As explained in Sec. III, the friction coefficient Λ without magnetic field has been increased by 10%. One can observe that the experimental points are below the numerical predictions, as it will be discuss in the next section. Moreover, the maximum efficiency experimentally occurs at s=-1.8 instead of -1 numerically (see also Fig. 3). Nevertheless, it appeared that geometry B had a overall maximum efficiency of about 25% instead of 23% for geometry A.

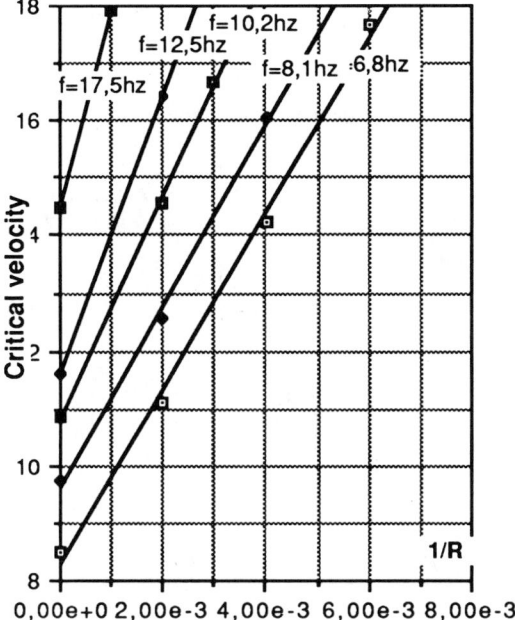

Fig. 9a Critical velocity vs 1/R for the A geometry.

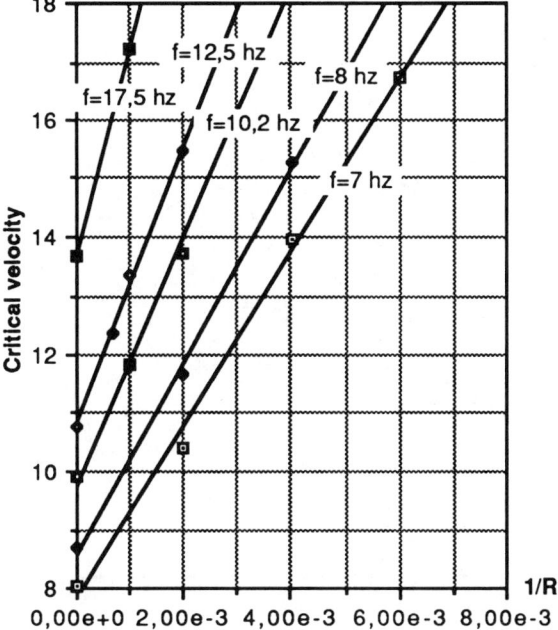

Fig. 9b Critical velocity vs 1/R for the B geometry.

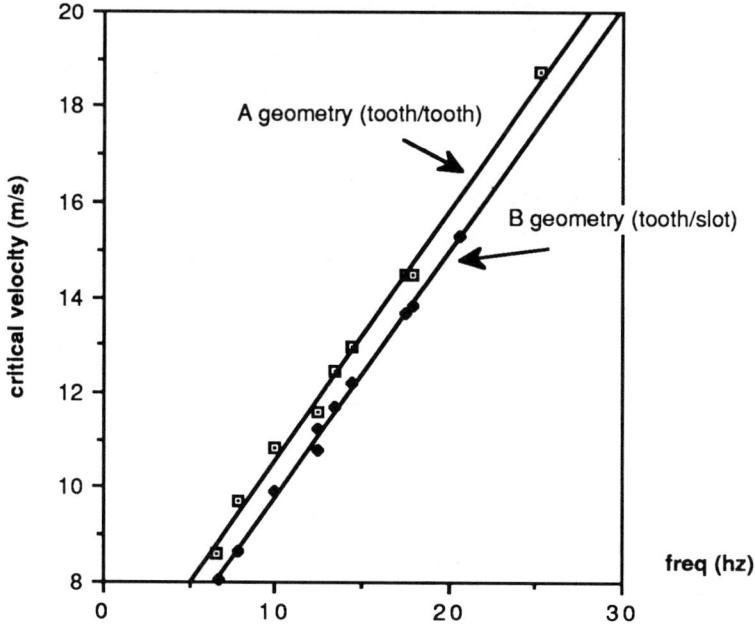

Fig. 10 No load (infinite R) experimental critical velocity vs frequency.

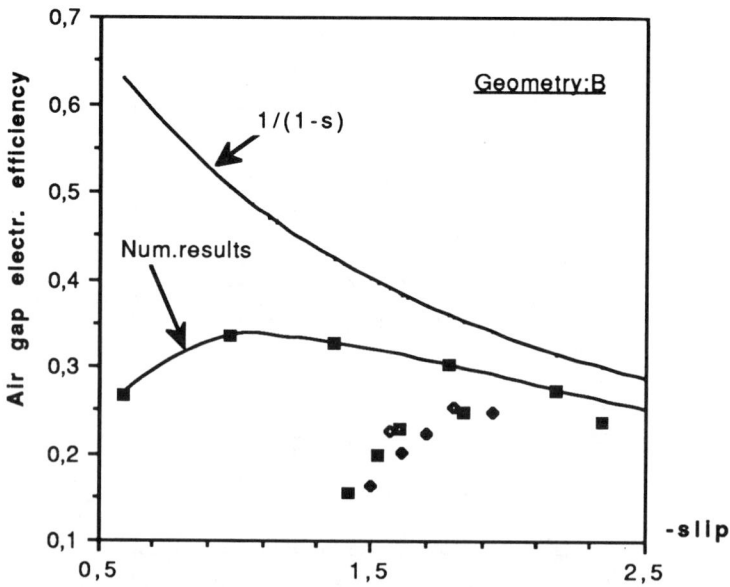

Fig. 11 Experimental efficiency compared to numerical predictions (10% increase on the friction coefficient).

V. Discussion

Two quantities are mainly involved in the discrepancies between numerical calculations and experimental results, namely, critical velocity and electrical efficiency. We currently think that the same reasons could be valid for both parameters, i.e.:

1) uncertainty in velocity and pressure measurements, which are of capital importance in determining η_{el}. Nevertheless, the weak discrepancy between the experimentally measured Λ coefficient and the one from Blasius law (Fig. 6) indicates that this problem should be of limited importance.

2) The iron sheets that we used to build the iron stack were laminated in a direction perpendicular to the teeth. Consequently, the flux density in the teeth was perpendicular to the lamination direction and could have suffered some magnetic saturation.

3) The velocity profile could have been badly distorted by the magnetic field at the entrance of the machine, more especially as the interaction parameter N (built with the thickness a of the flow) is greater than unity. Furthermore, a numerical test introducing a parabolic profile has shown the dramatic influence of a distorted velocity profile on η_{el}.

4) Because of the self-excitation conditions, the phase displacement between the currents was not exactly 120 deg (for the 3 phases connection) as it was assumed in the numerical model.

5) The tightening of the iron stack with metallic bolts as well as the tool-machining of the slots could be responsible for additional power losses in the magnetic circuit. With the aim of modelizing these short-circuit effects, an imaginary electrical conductivity σ_y has been numerically introduced in the magnetic iron circuit. The upper curve on Fig. 4 shows that for $\sigma_y = 9.10^4 \ \Omega^{-1}.m^{-1}$, the numerical results for critical velocity prediction are much closer to the experimental measurements presented on Fig. 10.

6) Finally, and this is probably the most important point, we strongly suspect a bad electrical contact between the fluid and the lateral copper side bars. This problem could be easily explained by the large number of draining and filling operations on the loop; it could be responsible for a two-dimensional distribution of the induced electrical current paths in the fluid, which in turn, weakens the electrical efficiency. In order to quantify the importance of such a phenomenon we have studied an analytical model with totally insulating side bars as shown in Fig. 12 (see also Ref. 6). The main hypotheses are :

1) Infinitely long machine: all quantities have a (x, t) dependance in $\exp[i(\omega t - kx)]$.

2) The walls have been removed ($\varepsilon = 0$).

3) The flow is uniform.

4) The primary current sources, $\overline{J}$, which lay along the z axis, are assumed to have a lineic density (in A/m) as follows:

$$\overline{J} = J_m \exp[i(\omega t - kx)] \qquad (12)$$

By choosing $\bar{j}_x$ and $\bar{j}_z$ as unknowns, the boundary conditions are:

$$\bar{j}_z (z = \pm \frac{1}{2}) = 0 \tag{13}$$

By taking the curl of induction equation, we obtain the electric current distribution in the fluid:

$$\bar{j}_x = - \frac{J_m \cdot k^2 R_m}{e\bar{\gamma} \, ch \frac{\bar{\gamma} l}{2}} \tag{14}$$

$$\bar{j}_z = \frac{i\, J_m\, k^2\, R_m}{e\bar{\gamma}^2} (1 - \frac{ch\, \bar{\gamma} z}{ch\, \bar{\gamma} \frac{1}{2}})$$

where $\bar{\gamma}^2 = k^2 (1 - i R_m)$.

Using Ampere's theorem leads to the magnetic field expression:

$$\bar{b}_y = \frac{-i.\mu_0.J_m}{ke(1 - iR_m)} \left(1 - i\, R_m \frac{ch\, \bar{\gamma} z}{ch\bar{\gamma} \frac{1}{2}} \right) \tag{15}$$

A calculation similar to that described in equation (5) allows the knowledge of the retarding force on the fluid as well as P_{MHD}. For one wavelength, the Joule losses in the fluid are:

$$P_{Jf} = \frac{\lambda e}{2\, \sigma_f} \int_{-1/2}^{+1/2} (\bar{j}_x . \bar{j}^*_x + \bar{j}_z \bar{j}^*_z)\, dz \tag{16}$$

Writing the necessary energy balance between P_{MHD} on the one hand and the losses in the primary windings, the fluid, and the load on the other hand, yields the critical velocity V_c, via the magnetic Reynolds number R_m.

For arbitrary values of R_m, the calculations appear too heavy, although rather simple; then the solution has only been performed for the case $R_m \ll 1$ where the solution for V_c is quite similar to Eq. (1):

$$V_c = V_s + \frac{r}{\mu_0\, \sigma_f^*\, LV_s} + \frac{1}{\mu_0\, \sigma_f^*\, V_s\, R\, C} \tag{17}$$

INDUCTION GENERATORS

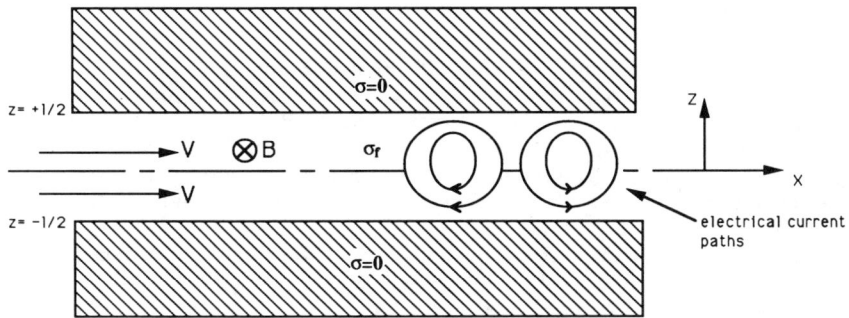

Fig. 12 Model used to study the influence of insulating lateral side bars.

where σ_f^* is an imaginary equivalent electrical conductivity of the fluid:

$$\sigma_f^* = \sigma_f \left(1 - \frac{\text{th}(k\,l/2)}{k\,l/2}\right) \qquad (18)$$

In the case of our experiment, this conductivity would be only less than 20% of the real NaK conductivity (with $l = 10$ cm and $\lambda = 36$ cm, then $\sigma_f^* = \sigma_f/5.13$).

The solution to this problem for $R_m \gg 1$ is not achieved yet, but first estimations drawn from these calculations show that the possibility of a bad contact between the fluid and the lateral side bars is not to be rejected.

VI. Conclusion

Both numerical modelling and NaK tests of a LMMHD induction generator have been presented. The evolution with frequency of the electrical efficiency and of the critical self-excitation velocity has been studied.

The maximum efficiency experimentally achieved has been about 25% instead of the 35% numerically predicted. Furthermore, the critical velocity V_c has been found higher than the theoretical predictions. These two phenomena are currently expected to have the same physical origin. Among the large number of possible sources of discrepancies, the dramatic influence of a bad electrical contact between the fluid and the lateral side bars has been analytically estimated.

Beside the framework of MHD energy conversion, this work can be also connected with metallurgical applications. As a matter of fact, the self-excitation phenomenon can be examined as a liquid metal flowrate regulation process and further investigations will be undertaken in this respect in the near future.

Acknowledgments

The authors are grateful to the PIRSEM Foundation for financial support (Programme Interdisciplinaire de Recherches sur les Sciences pour l'Energie et les Matières Premières) and to the Israelian National Council for Research and Development for supporting a part of the travel expenses of the French team.

References

[1] Joussellin, F., Alemany, A., Werkoff, F. and Marty, Ph., "MHD Induction Generator at Weak Magnetic Reynolds Number. Part 1 : Self-excitation Criterion and Efficiency". *European Journal of Mechanics*, B/Fluids, Vol. 8, N° 1, pp.23-29, 1989.

[2] Joussellin, F., Marty, Ph., Alemany, A. and Werkoff, F., "MHD Induction Generator at Weak Magnetic Reynolds Number - Part 2 : Numerical Modelisation and Experimental Study ", *European Journal of Mechanics*, B/Fluids, Vol. 8, N° 4, pp. 327-350, 1989.

[3] Elliott, D.G., "Direct Current Liquid Metal Magnetohydro-dynamic Power Generation". *AIAA Journal*, Vol. 4, N° 4, April 1966.

[4] Pierson, E.S., "Experimental Study of a One-wavelength Uncompensated MHD Induction Generator", 5th International Conference on MHD electric power generation, Munich, Germany, 1971, p. 81.

[5] Brouillette, E.C. and Lykoudis, P.S., "Magneto-Fluid-Mechanic Channel Flow. I : Experiment". *Physics of Fluids*, Vol. 10, N° 5, May 1967.

[6] Poloujadoff, M., "The Theory of Linear Induction Machinery". Oxford University Press, Oxford, England, UK, 1980.

Analytical and Experimental Studies of End Effects in an LMMHD Generator

T. K. Thiyagarajan,* P. Satyamurthy,* N. S. Dixit,*
and N. Venkatramani†
Bhabha Atomic Research Centre, Bombay 400085, India
and
V. K. Rohatgi‡
University of Poona, Pune 411007, India

Abstract

The end effect that arises due to the magnetic field overhang at the entrance and exit of the liquid metal MHD generator was analyzed. The increase of power due to the introduction of the insulating vane was estimated by solving the required electrodynamic equations. The calculations were performed for a single-phase generator by considering the typical velocity variation across the generator. The calculations were also performed for a two-phase generator by considering typical axial and transverse variations of electical conductivity in addition to the velocity variation. These results were compared with an earlier model that assumes only uniform values of fluid velocity and electrical conductivity. It was

---
Copyright © 1992 by the American Institute of Aeronautics and Astronautics, Inc. All rights reserved.
 * Scientific Officer, Laser & Plasma Technology Division.
 † Section Head, Thermal Plasma, Laser & Plasma Technology Division.
 ‡ Professor of Physics Emeritus, Physics Department.

found that, for the two-phase generator, there is an 18% increase of power for the short magnetic field decay and a 9% increase of power for the long magnetic field decay, values that are larger than the predicted values of the earlier model. Experiments were conducted by use of a single-phase generator with a single movable insulating vane at the entrance of the generator. The experimental results were compared with the predicted values of the theoretical calculation, which assumed a constant velocity. The deviation between the experimental results and the theoretical values was 10% for most of the cases.

Introduction

The liquid-metal magnetohydrodynamic (LMMHD) power conversion system is one of the promising methods of converting thermal energy from low-temperature heat sources (e.g., solar, geothermal, industrial waste heat) to electrical energy.[1] Different stages of development from laboratory scale to commercial demonstration of this power conversion systems are occurring at various places.[2,9] The performance of the MHD generator in which the electrical power is generated is one of the important factors for deciding the overall efficiency of the LMMHD power conversion system. An important phenomenon that gives rise to the loss of useful power in the LMMHD generator are the end losses. This phenomenon arises due to the magnetic field overhang that leads to the reversal of current at entrance and exit of the generator.[4] The losses due to these reversal end currents can be reduced by 1) increasing the aspect ratio of the channel, 2) properly shaping the magnetic field at end regions, and 3) introducing an insulator vane in order to increase the resistance for the end currents. But each of these methods has limitations. The aspect ratio is mainly decided by the available pressure drop and the terminal voltage. The magnetic field shaping involves the complex design. Introduction of the insulator vane can cause an additional pressure drop. Thus a systematic analysis of all of these methods is required.

The end loss analysis was carried out by Sutton et al.[5] for the exponential-type magnetic field distribution and without insulator vane using the conformal mapping technique. Gherson et al.[6] analyzed the end losses with and without the vane by numerically solving the required electrodynamic equations. But all of these analyses assumed uniform velocity and uniform electrical conductivity throughout the generator and end regions (exit and entrance of the generator). Very little experimental or reliable empirical data are available regarding velocity and electrical conductivity profiles in two-phase flow in the presence of an electric and magnetic field. In this paper some typical power law profiles with variable profile parameters for veocity and electrical conductivity are considered to realistically estimate the end losses by theoretical calculation. To the author's knowledge few experiments have been performed to analyze the end losses. In view of this some preliminary experiments were performed in a single-phase room temperature mercury LMMHD generator with a movable single insulator vane at the entrance of the channel. Since experiments were conducted in a single-phase LMMHD generator and the velocity profiles could not be measured, the experimental data are compared with Gherson's constant velocity model[6] for different vane locations.

Mathematical Model

Basic Assumptions

Our analysis is based on a two-dimensional approach. A long duct of uniform rectangular crosssection is considered. The channel configuration and coordinate system is shown in Fig.1. The required equations are developed with the following assumptions:

1) Variations of parameters (both fluid and electrical) along the magnetic field direction are negligible.
2) The duct wall (except for the electrode) and vane are perfect insulators. The electrodes are perfect conductors.
3) The presence of the vane does not change the flow and disturbance at the edge of the vane is neglected.
4) The variation of the flow parameters due to the $\vec{J} \times \vec{B}$ force is neglected and flow velocity is predominently along the x direction.

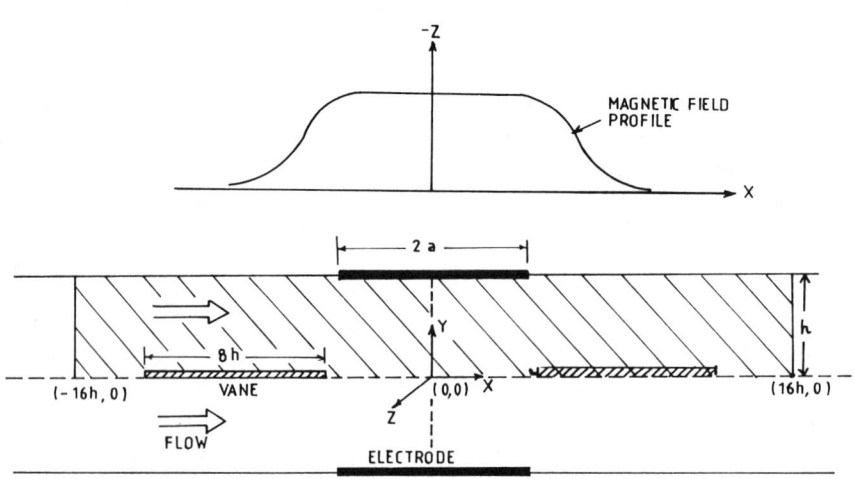

Fig. 1 Coordinate system for theoretical model.

Mathematical Equations

The generalized Ohm's law with the above assumptions is

$$j_x = \sigma E_x, \quad j_y = \sigma (E_y + uB) \quad (1)$$

where j, E and B are the current density, electric field, and magnetic induction, respectively. Also, u and σ are the velocity and electrical conductivity of the working fuid. For two-phase flow the variation of conductivity is due to the variation of the void fraction. The following profiles are assumed for the velocity and the void fraction:

$$u(y) = u_o \{1 - (y/h)\}^{1/n} \quad (2)$$

$$\alpha(x,y) = \alpha_o(x)\{1-(y/h)\}^{1/m} \quad (3)$$

where n is varied from 2 to 5 and m is taken as 0.5.[6] The n variation accounts for wide range of flow conditions. For the case of two-phase flows the experimental data are more scarce regarding the void fraction profile. Hence, we have assumed a typical m value. The u_o and $\alpha_o(x)$ are respective values at the center of the channel. The assumed axial variation of the central void fraction α_o is given in Fig.2. For the electrical conductivity the following equation, which is generally used in LMMHD flows[7] is considered

$$\sigma = \sigma_l \exp\{-3.8\alpha(x,y)\} \quad (4)$$

where σ_l is the liquid-metal electrical conductivity. The nondimensionalized electrical conductivity ($\xi = \sigma/\sigma_l$) distribution is shown in Fig.3.

The magnetic field distribution is given by

$$B = B_o, \quad |x| \leq a \quad (5a)$$

$$B = B_o \exp\{-(x - a)/Le\}, \quad |x| > a \quad (5b)$$

where B_o is the magnetic field at the center of the generator. This type of magnetic field is suggested by Sutton[5] and Le is the decay length. The current conservation is given by

$$\frac{\partial j_x}{\partial x} + \frac{\partial j_y}{\partial y} = 0, \quad |x| < x_{max}, \quad h > y > 0 \qquad (6)$$

with $E = -\nabla\phi$, where ϕ is the electrical potential and by considering the nature of σ, u, and B as described earlier and from Eqs.(1) and (6) we obtain the following equation:

$$\nabla^2\phi + \left(\frac{1}{\sigma}\right)\left(\frac{\partial\sigma}{\partial y}\right)\left(\frac{\partial\phi}{\partial y} - uB\right) + \left(\frac{1}{\sigma}\right)\left(\frac{\partial\sigma}{\partial x}\right)\left(\frac{\partial\phi}{\partial x}\right) = \frac{\partial}{\partial y}(uB) \qquad (7)$$

The third term on the left side arises due to the transverse conductivity variation and the fourth term is due to the axial variation. The term on the right side of the equation is due to the transverse velocity variation. The area marked in Fig.1 is taken as working area. The required boundary conditions are as follows

$$\phi = \eta \int_0^h u(y)B_o \, dy, \quad |x| \leq a, \quad y = h \qquad (8a)$$

$$\frac{\partial\phi}{\partial y} = u(h) \cdot B(x), \quad |x| > a, \quad y = h \qquad (8b)$$

$$\phi = 0, \quad y = 0 \qquad (8c)$$

$$\phi = 0, \quad x = \pm x_{max}, \quad h > y > 0 \qquad (8d)$$

Equation (8a) follows from the defintion of load voltage, and η is the load factor. The absence of a normal current at the wall other than the electrode region leads to Eq.(8b).

At the central line, Eq.(8c) is taken if there is no vane. Eq.(8d) follows from the consideration that the electrical potential vanishes at a distance x_{max}, which is far away from the generator. The distance x_{max} is chosen as 16h. At the central line, if vanes are there, the required boundary condition at the vane region is

$$\frac{\partial \phi}{\partial y} = u_o \cdot B(x) \qquad (8e)$$

The elliptic equation (7) is solved by a finite difference technique with the mixed-type boundary conditions (8). The steepness of the u,σ profile near the wall requires a denser grid spacing near the wall than at the center. The load current I_L is calculated from the following relation:

$$I_L = \int_{-x_{max}}^{+x_{max}} (-\frac{\partial \phi}{\partial y} + uB) \, dx \qquad (9)$$

The integration is performed by trapezoidal rule. Introduction of the vane increases the load current as well as the

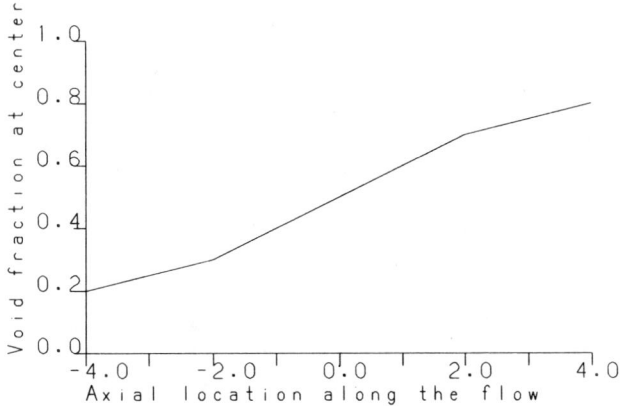

Fig. 2 Void fraction variation at center.

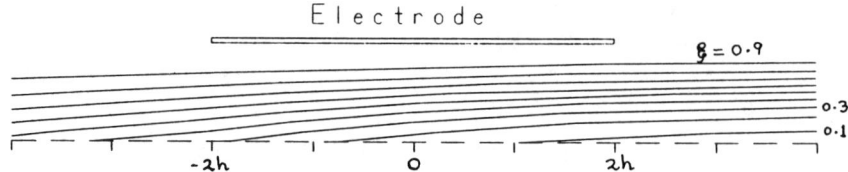

Fig. 3   Void fraction distribution.

load factor. In the calculation the increased load current and the load factor are estimated for the fixed load resistance. The increase in power by the introduction of the vane is considered in order to analyze the end losses. The computer code takes ~8000 iterations to converge with an error limit of 10^{-6}, and the typical CPU time is 90 min in a Norskdata computer.

Theoretical Results and Comparision with the Standard Model

In order to study the effect of variable velocity and conductivity on end losses the following cases have been computed.
1) case a: constant velocity, constant conductivity;
2) case b: variable velocity (no slip at boundary), constant conductivity; and
3) case c: variable velocity (no slip at boundary), variable conductivity.

Case a is the same as the Gherson's model, in which the conductivity and velocity are assumed constant everywhere. Case b and c are applicable for single-phase LMMHD generators and two-phase generators, respectively. Gherson's model (case a) will hereafter be referred to as the standard model. The results of cases b and c are compared with this standard model in this analysis.

Current Distributions and Effect of Electrical Conductivity Variation

A detailed analysis of cases a and b was reported earlier.[a] For case c in addition to the variation of the electrical conductivity across the channel, the axial variation of electrical conductivity is also considered. Since the end losses are greater in the two-phase LMMHD generator, the two-dimensional functional variation of electrical conductivity helps one understand the end effect more realistically. For the purpose of illustration typical current density distribution plots for the two magnetic field decays (Le=0.5, 1) with and without the vane are given in Figs. 4 and 5. The nondimensionalized normal current density j_y at the electrode is also plotted for the aforementioned cases and is shown in Figs. 6 and 7.

It can be seen from Figs. 4 and 5 that the current density at the central region of the generator shows no deviation from the axis to the electrode. Because of the absence of the j_x component, the current conservation equation makes the j_y component constant. Hence of all the current generated at the axis reaches the electrode without any loss. The normal current at the electrode in the central region of the generator decreases along the flow direction. This is due to the decrease of overall electrical conductivity along the flow direction. But at the entrance and exit region of the generator, the deviation of the current can be seen from the figures. In the region slightly away from the entrance and exit of the generator the current reversal can be seen clearly. As can be seen at the exit region, most of the current flows near the wall and the reversal current takes a longer path. This is due to the lower electrical conductivity at the exit region compared to the entrance region. The

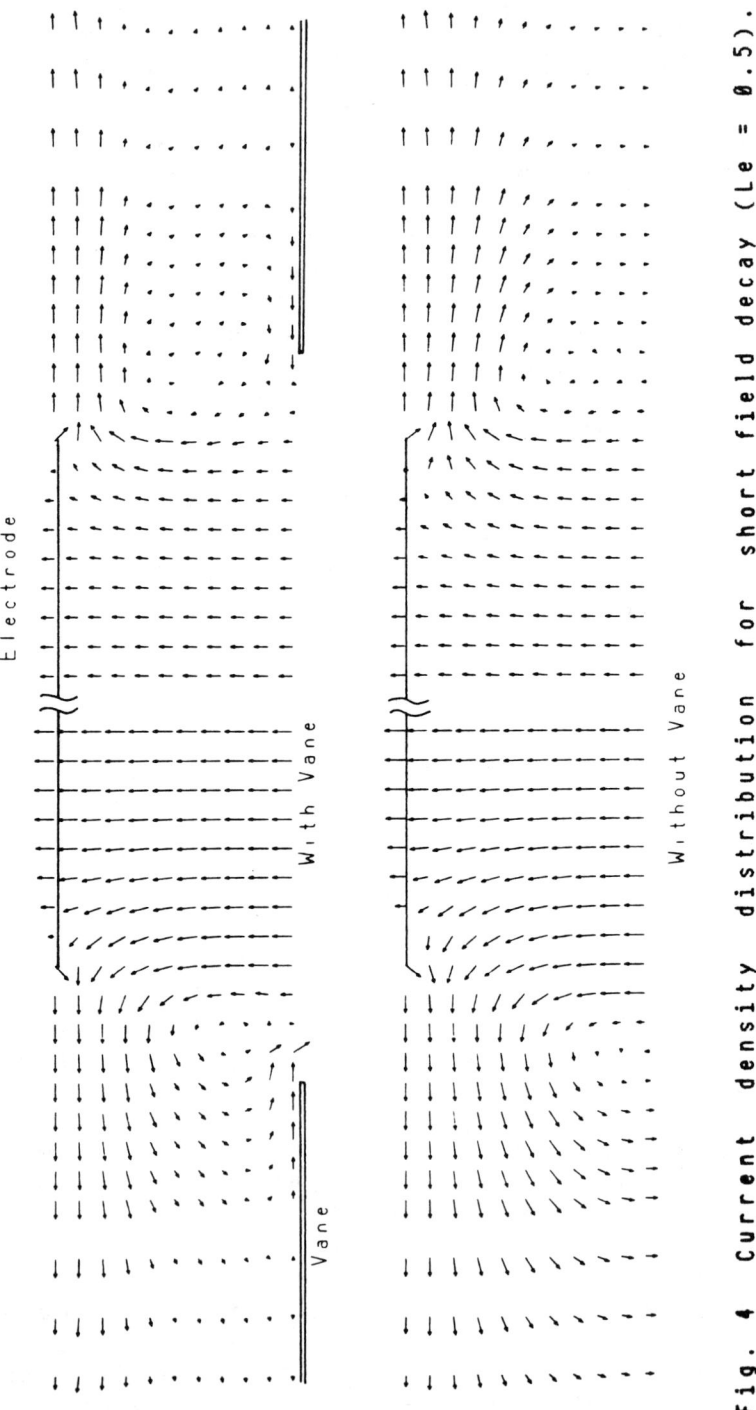

Fig. 4 Current density distribution for short field decay (Le = 0.5).

END EFFECTS

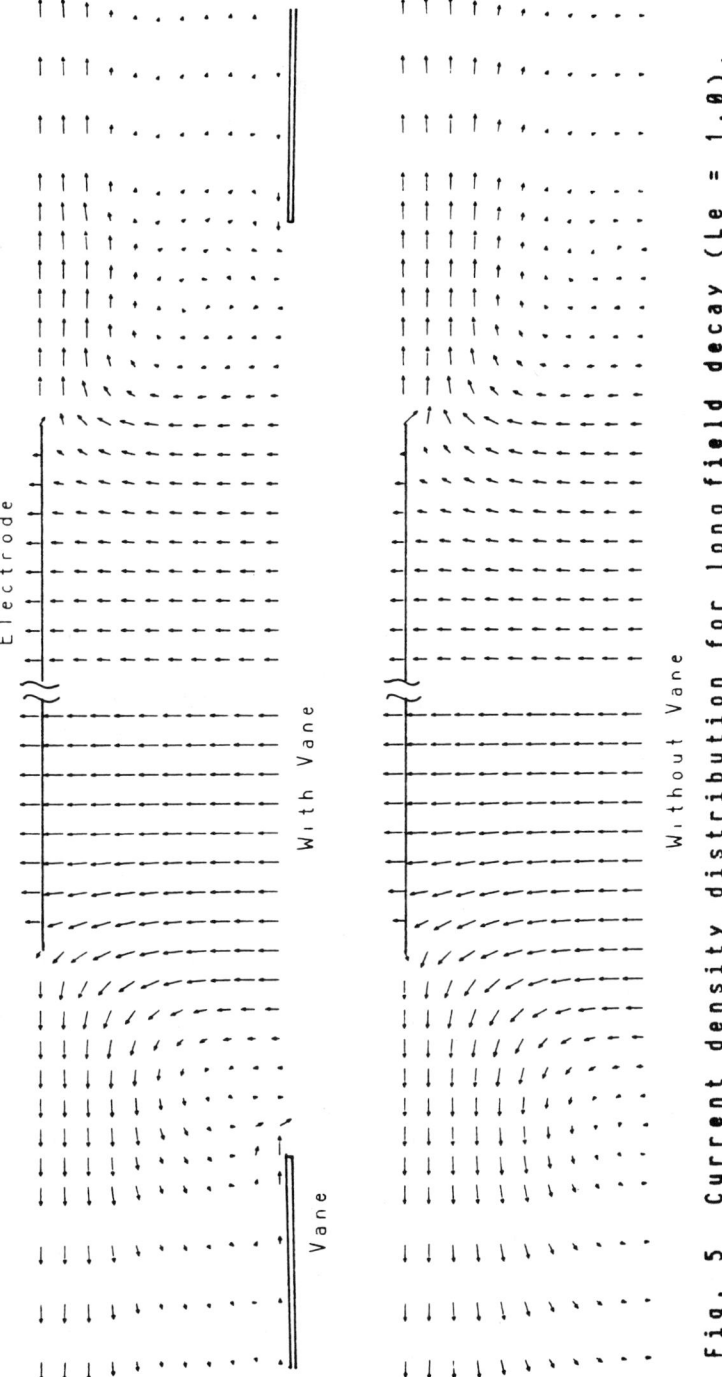

Fig. 5 Current density distribution for long field decay (Le = 1.0).

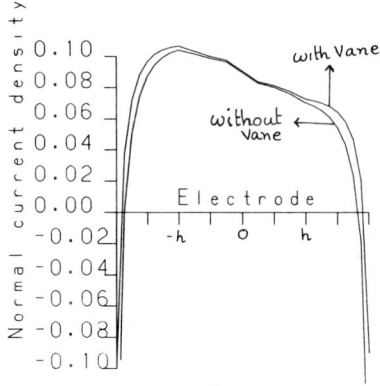

Fig. 6 Normal current density at electrode for short field decay (Le = 0.5).

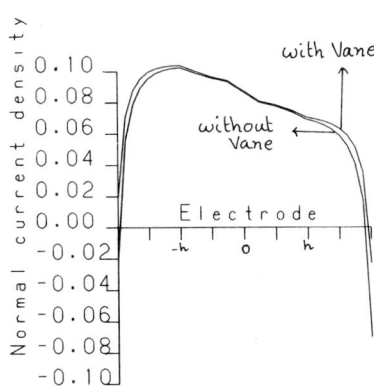

Fig. 7 Normal current density at electrode for long field decay (Le = 1.0)

Table 1 Increase of power due to the insulator vane (n = 2, Le=0.5)

Theoretical model type	η	$\bar{I}_L$	$\eta \bar{I}_L$	η	$\bar{I}_L$	$\eta \bar{I}_L$	Increase in power
Constant u Constant σ case a	0.7	0.327	0.229	0.72	0.336	0.242	5.7%
Variable u Constant σ case b	0.7	0.312	0.218	0.73	0.325	0.237	8.7%
Variable u Variable σ case c	0.7	0.276	0.193	0.76	0.300	0.228	18.13%

Table 2 Increase of power due to the insulator vane
(n = 2, Le = 1.0)

Theoretical model type	η	I_L	$\overline{\eta I_L}$	$\overline{\eta}$	$\overline{I_L}$	$\overline{\eta}\,\overline{I_L}$	Increase in power
Constant u Constant σ case a	0.7	0.364	0.255	0.71	0.368	0.261	2.35%
Variable u Constant σ case b	0.7	0.348	0.244	0.71	0.353	0.251	2.87%
Variable u Variable σ case c	0.7	0.356	0.249	0.73	0.372	0.272	9.24%

blocking of the reversal current by the vane at exit and end regions can be seen from the Figs. 4 and 5. The computed nondimensionalized load current and the load factor with and without the vane for all the three cases are given in Tables 1 and 2.

It is found that the standard model estimates only a 5.7% increase of power for short field decay (Le=0.5) case and a 2% increase of power for long field decay; these values are less compared to the single-phase and two-phase generator values. This is due to the assumption of constant velocity and constant conductivity. The two-phase generator shows more of an increase of power due to the vane than the single-phase generator. The normal current density for the MHD generator is given by

$$j_y = \sigma \left(-\frac{\partial \phi}{\partial y} + uB \right) \qquad (10)$$

The normal current density consists of two components: 1) the electrostatic component $-\sigma(\partial\phi/\partial y)$ and 2) the induced component σuB. The potential gradient $\partial\phi/\partial y$ increases at the electrode edge and makes the overall normal current density decrease and sometimes reverse the normal current. In the standard model this potential gradient is calculated with constant σ, u; hence, it estimates less end loss and the increase of power is less. In the two-phase generator the potential gradient is enhanced by the electrical conductivity, which is maximum at the electrode. Hence, for the two-phase generator the end losses are more and show more increase in power due to the introduction of the vane. The maximum increase of 18% in power is found for the short field decay (Le=0.5) case in the two-phase generator due to the insulator vane. But for the long field decay (Le=1.0) only a 9% increase of power is obtained.

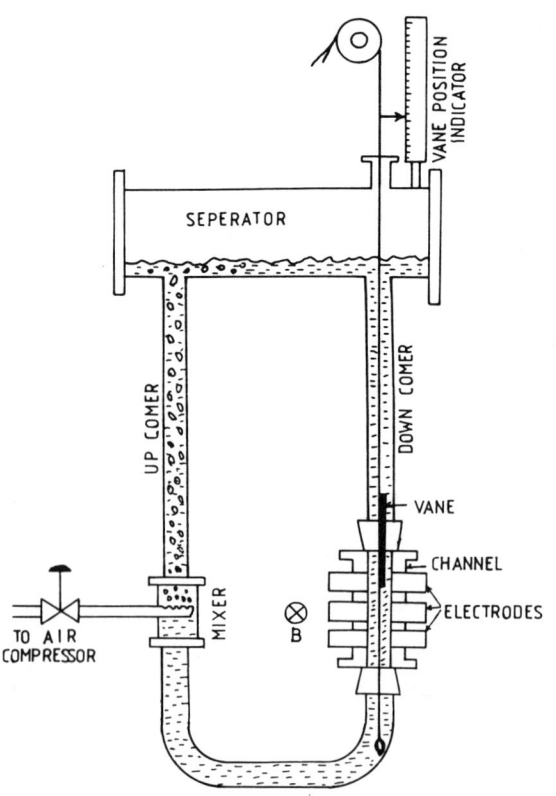

Fig. 8 Schematic of the experimental facility.

Description of the Experimental System

Some limited experiments have been conducted in a single-phase LMMHD generator with mercury as a electrodynamic fluid. The experiments were performed with a movable single insulator vane at the entrance of the channel. The schematic of the experimental system is shown in Fig.8. It consists of the mixer, the upcomer, the separator, the downcomer, and the MHD channel. Compressed air at room temperature, introduced through the mixer, drives the mercury through the upcomer. The two-phases become separated in the separator. Air is released to the atmosphere through a liquid nitrogen trap. Separated mercury falls through the downcomer due to gravity and passes through the MHD generator. The MHD channel is composed of Perspex with copper electrodes. It has a cross section of 40 mm (width) x 6 mm (height) and 175 mm length. A bakelite vane 1 mm thick, 6 mm wide, and 160 mm long is

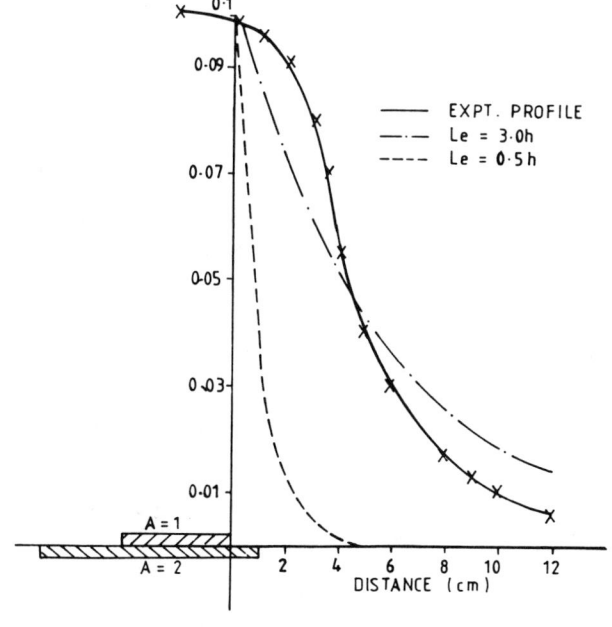

Fig. 9   MAGNETIC FIELD PROFILE.

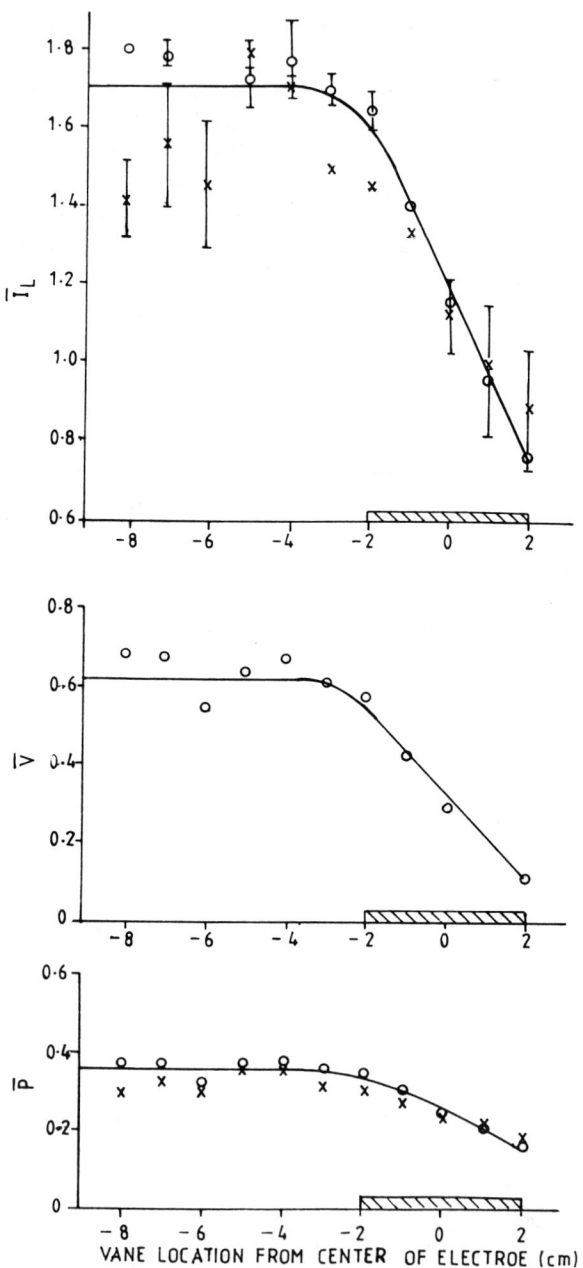

Fig. 18 Aspect ratio 1 measurements for 67 μΩ load resistance.

used. The electrodes are segmented (40 mm each and three in number) so that the aspect ratio can be varied from 1 to 3. The load resistances were made of copper plates of various thicknesses, and the contact points were gold plated to minimize the contact resistance. A C-shaped permanent magnet of 0.1 T field was used. The magnet has a pole

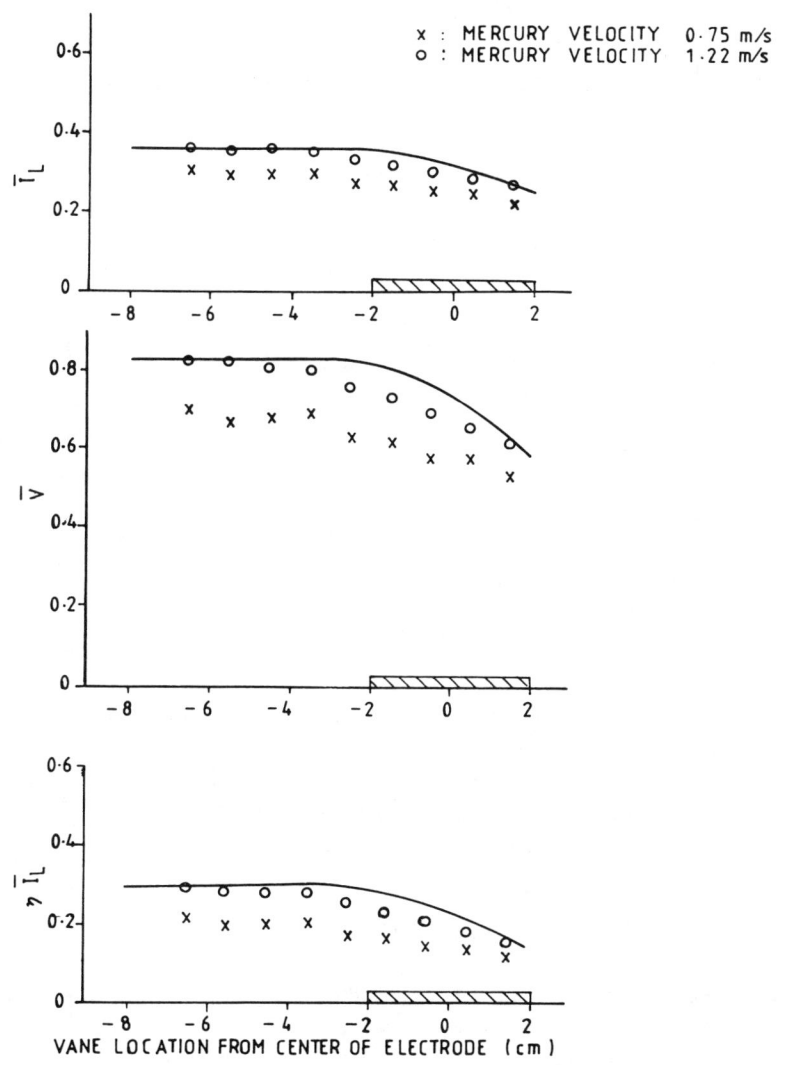

Fig. 11 Aspect ratio 1 measurements for 728 $\mu\Omega$ load resistance.

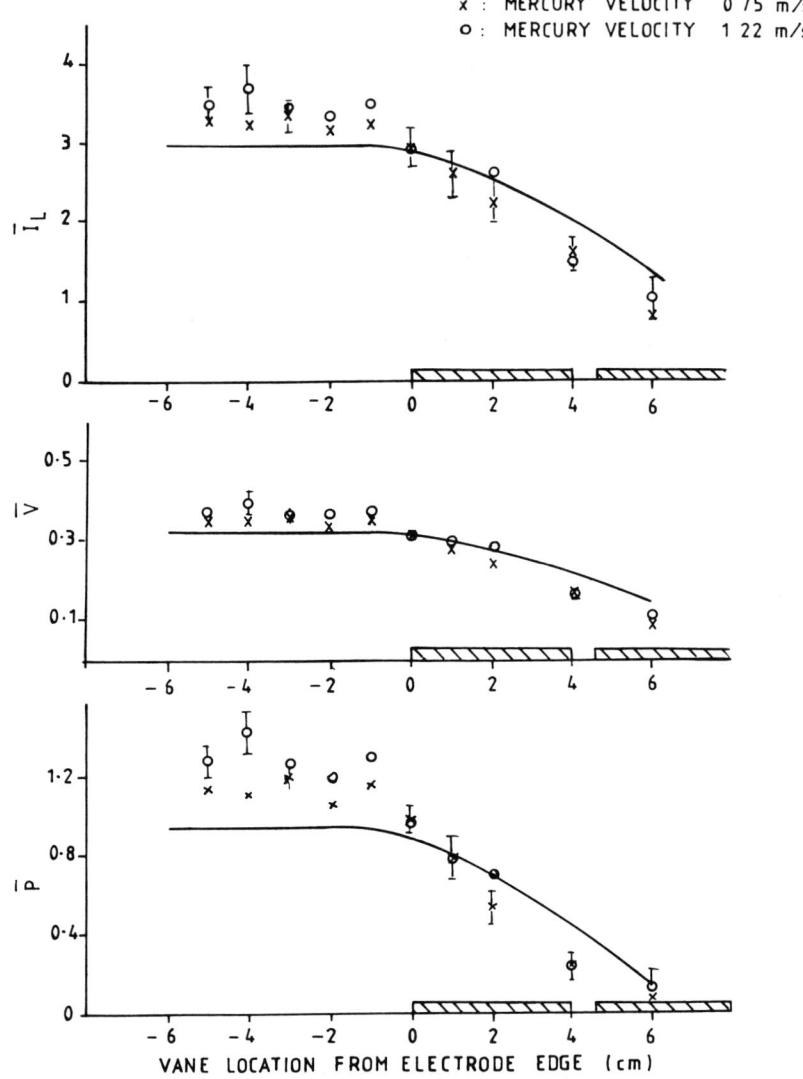

Fig. 12 Aspect ratio 2 measurements for 67 $\mu\Omega$ load resistance.

gap of 60 mm, a width of 50 mm, and a length of 80 mm.

By varying the air mass flow rate, the mercury mass flow rate was varied from 0 to 4 kg/s corresponding to a velocity of 0-1.2 m/s in the generator. The load resistances varied from 67 $\mu\Omega$ to 728 $\mu\Omega$. With these resistances

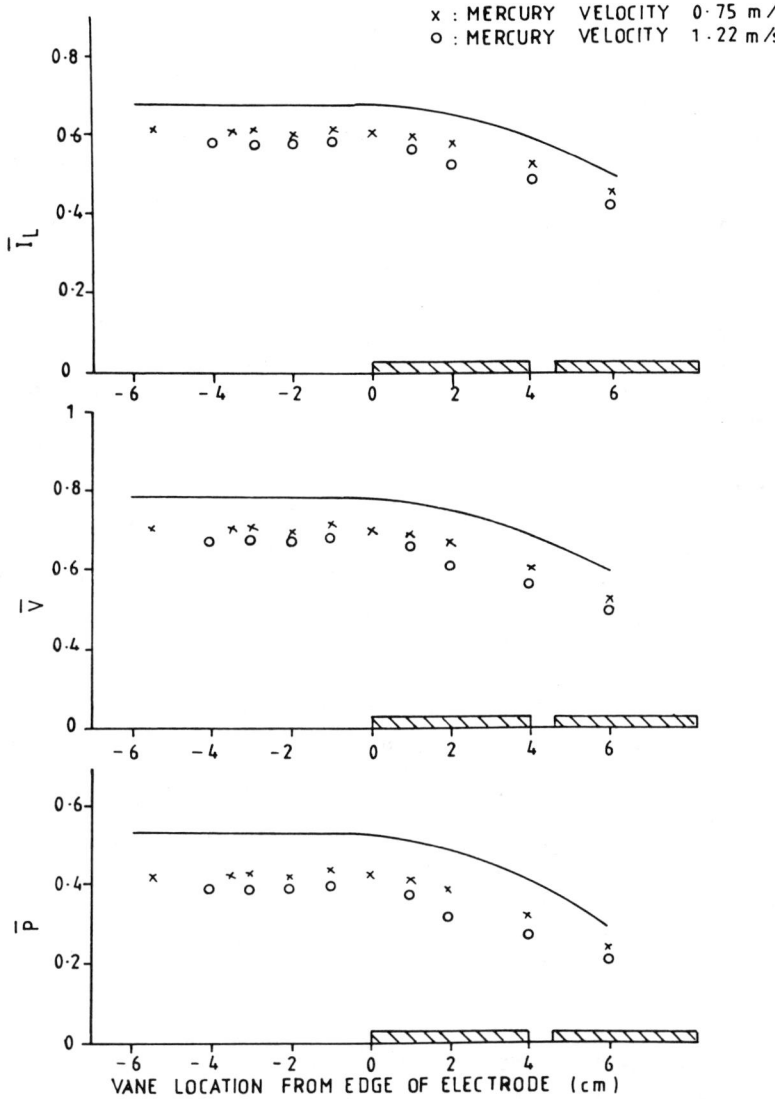

Fig. 13 Aspect ratio 2 measurements for 728 μΩ load resistance.

the load factor could be varied from 0.3 to 1.0. The mass flow rate of the mercury as a function of the air flow rate was determined using another electromagnet of large pole length (so that the end loss is completely negligible).

Experimental Results

For the aspect ratios of 1 and 2, the load currents were measured for different vane locations. These measurements were performed for three different load resistances (67, 264, and 728 $\mu\Omega$) and two different mercury velocities at the channel (0.75 and 1.22 m/s). A typical open-circuit voltage was around 6.5 mV (E =0.15 V/m) for the magnetic field of 0.1 T. The maximum current obtained was ~10 A (J = 40000 A/m^2) when the mercury velocity was 1.22 m/s. The field distribution of the magnet used is given in Fig.9. In the same figure the type of magnetic field distribution is suggested by Sutton and is used for theoretical calculations are shown. It can be seen that the experimental magnetic field has a large field decay length (>Le=3). Because of the large field decay length, it is expected that the magnitude of reversal currents is small and hence voltages/currents are likely to be insensitive to the locations of the vane when positioned outside the channel. In order to make the vane location sensitive, it was introduced in to the geneator for better comparisions with the model. Based on this comparision, the overall accuracy of the model is determined. The nondimensionalized load current $\overline{I_L}$ ($I_L/\sigma uBhd$), voltage $\overline{V_L}$ (V_L/uBh), and power $\overline{P_L}$ ($V_L \cdot I_L$) are given in Figs. 10-13. Here d is the height of the electrode. In the same figures the values predicted by the standard model are also shown.

From the experimental results obtained, the following conclusions can be made:
1) In the results of aspect ratio 1, for the higher load factor (η = 0.8, R_L = 728 $\mu\Omega$) the theoretical values matched well with the experimental values of 1.22 m/s velocity case, whereas at a low load factor(η = 0.3, R_L = 67 $\mu\Omega$), the theoretical data are between two flow rate cases.

2) For aspect ratio 2 it is found that the theory consistently predicts smaller values than the experimental data for both the flow cases. However, for a larger load factor (η = 0.8) with the 0.75 m/s case the data matched well with the theory. The maximum deviation of 15% is found between theoretical and experimental values.

Conclusion

The following conclusions can be made from this work:
1) The usefulness of the vane to reduce the end losses is analyzed by numerically solving the required electrodynamic equations with the velocity and electrical conduvctivity variation.
2) The increase in power due to the vane for the two-phase LMMHD generator is greater than that for the single-phase LMMHD generator.
3) The estimations of end losses are greater if variations of electrical conductivity and velocity are considered. The increase of conductivity near the wall increases the end losses.
4) Some preliminary experiments were conducted to study the effect of the vane on the load current in a room temperature single-phase mercury LMMHD system with aspect ratios of 1 and 2
5) The experimental results with an aspect ratio of 1 agree with the predicted values of Gherson's constant velocity model. It is found that the deviation is within 10% for all loads (η = 0.3-1.0). However, for low velocity flow (0.75 m/s), around 15% deviation is found for the large load factor (η = 0.8)
6) The experimental results with aspect ratio 2 are close to the values predicted by the constant velocity model except for the high velocity (1.22 m/s) case. The theoretical values are larger by 15% for the high load factor (η = 0.8)

7) Since velocity profiles could not be measured experimentally, no comparision could be made with the variable velocity model. Further experiments should be conducted in two-phase flow systems with wide parametric range to verify the model presented in this paper.

References

[1] Branover, H., Borde, I., El-Boher, A., and Leitner, A., "On the Possible Use of MHD Generator in Solar Energy Systems," 2nd Bat-sheva International Seminar on MHD Flows and Turbulence, Beer-sheva, Israel, March 28-31, 1978.

[2] Branover, H., and El-Boher, A., "The 'ETGAR' Program for the Development of Liquid Metal MHD Energy Conversion technology," 10th International Conference on MHD Electrical Power Generation, Tiruchirapalli, India, Dec.4-8, 1989.

[3] Satyamurthy, P., Thiyagarajan, T. K., and Venkatramani, N.,"Design of 500 Watt Electrical Steam-Mercury Liquid Metal MHD Experimental System," 10thInternational Conference on MHD Electrical Power Generation, Tiruchirappalli, India, Dec.4-8, 1989.

[4] Petrick, M., Fabris, G., Pierson, E. S., Fischer, A. K., Johnson, C. E., Gherson, P., Lykoudis, P. S., and Lynch, R. E., "Experimental Two-Phase Liquid Metal Magnetohydrodynamic GeneratorProgram," Argonne National Laboratory Report/MHD-79-1, April 1979.

[5] Sutton, G. W., Hurwitz, H., Jr., and Poritsky, H., "Electrical and Pressure Losses in a Magnetohydrodynamic Channel due to End Current Loops," Transactions of AIEE Pt.I, Communications and Electronics, 1961. pp.687-695.

[6]Gherson, P., Lykoudis, P. S., and Lynch, R. E., "Analytical Study of End Effects in Liquid Metal MHD Generators," Proceedings of the 7th International Conference on MHD Electrical Power Generation, Boston, MA, June 1980.

[7]Walis, G. B., One-Dimensional Two-Phase Flow, McGraw-Hill, New York, 1969.

[8]Thiyagarajan, T. K., Satyamurthy, P., and Rohatgi, V. K., "Effect of Non-Uniform Velocity and Electrical Conductivity on End Effects in LMMHD Generator with Insulating Vane," International Forum on Mathematical Modelling and Computer Simulation Process in Energy Systems, Sarejevo, Yugoslavia, March 20-24, 1989.

Theoretical Magnetic Field Distributions Eliminating End Losses in Linear High Magnetic Reynolds Number MHD Channels

L. Blumenau*
Ben-Gurion University of the Negev, Beer-Sheva, Israel

Abstract

To date, so-called "end losses" in liquid metal dc Faraday MHD-channels have been evaluated while assuming the magnetic Reynolds number, Rem, is negligibly small. For reasonably large MHD-channels, Rem is, in reality, quite large, leading to considerable magnetic field compression in the end regions, where cancellation busbars cannot be applied. A new class of end field decay functions is derived based on the complete MHD-equation for electric potential, including Rem, and while assuming boundary conditions with an ideal device in the central electrode region. These end field decay functions are validated in a general 2-D model as yielding zero current leakage from the central electrode region and, thus, zero end losses for large Rem and certain ranges of other channel parameters. In addition, arbitrary aspect ratio MHD-channels can be composed with zero current leakage channel modules separated by insulated vanes in the end regions.

Nomenclature

A	=	c/nh = MHD channel aspect ratio
A_{MOD}	=	c/h = module aspect ratio
b_z	=	B_z/B_o dimensionless magnetic field
B_z	=	vertical field distribution
B_o	=	$B_z(x; -c \leq x \leq 0)$ = vertical field in central electrode region
c	=	length of electrodes
C	=	c/L_1 = dimensionless length of electrodes
E	=	electric intensity
f_i	=	function
h	=	distance between electrodes (or between insulating vanes with zero current leakage modules)

Copyright © 1992 by L. Blumenau. Published by the American Institute of Aeronautics and Astronautics, Inc., with permission.
*Center for MHD Studies, P.O. Box 653.

H	=	h/L_1 = dimensionless distance between electrodes
Ha	=	$B_z w \sqrt{\sigma/\mu}$ = Hartmann number
I	=	current
J	=	current density
K	=	load factor
K_F	=	friction factor
L_i	=	length of end regions; i = 1 $\triangleq$ outflow end region, i = 2 $\triangleq$ inflow end region
N	=	interaction parameter
n	=	number of zero current leakage modules
p	=	pressure
P	=	power
Re	=	$\rho c U_x/\mu$ = conventional Reynolds number
Rem_i	=	$\sigma \mu_o U_x L_i$ = magnetic Reynolds number
Q	=	volume flow rate
U_x	=	velocity in x-direction
w	=	height of channel out-of-plane
x,y,z	=	coordinates
$X = x/L_1$, $Y = y/L_1$, $Z = z/L_1$ = dimensionless coordinates		
Δ	=	LaPlace operator or difference
$\emptyset$	=	electric potential
$\emptyset^*$	=	$\emptyset/(K U_x B_o h)$ = dimensionless electric potential
σ	=	electric conductivity
η_{PUMP}	=	pump efficiency
η_{GEN}	=	generator efficiency
$\eta_{E.L.}$	=	efficiency for elimination of leakage currents (end loss efficiency)
ρ	=	density
μ	=	viscosity
μ_o	=	magnetic permeability in vacuum

All dimensional units are according to MKSA.

Introduction

The electrical efficiencies of dc Faraday MHD pumps and generators may be significantly reduced by the effects associated with the passage of the working fluid through the end regions. So called end losses are present due to:
1) A short circuiting effect of the electric current loop passing between the electrodes and through the external circuit, manifest in some current bypassing, in an opposite sense, the electrodes in the end regions (shunt currents) and
2) an additional pressure drop due to extra ponderomotive force, determined by the shunt currents.

End losses have been studied for four different cases of end field distributions, e.g., constant extension, abrupt termination at the end of the electrodes, linear decay, and exponential decay.[1,2] Among these, exponential decay was found to yield the smallest end losses, and the effect of an employed e-fold distance was determined. End losses were seen to be a function of the channel aspect ratio (length of electrodes to distance between electrodes). This should be reasonably large to produce small end losses, e.g., greater than about four. Additionally, the channel aspect ratio could be artificially increased beyond the electrode aspect ratio by providing insulating vanes in the end regions serving as obstacles for shunt currents. Zheng Linjing[3] went a step further by actually determining the optimal magnetic field-decay, with and without insulating vanes, for obtaining maximum efficiency. All of these studies presupposed that the magnetic Reynolds number be close or equal to zero. Even so, the results and conclusions, especially those reached by Sutton et al,[1] have been freely applied to single and two-phase flow liquid metal MHD channels where the magnetic Reynolds number clearly is not zero but rather between 1 and 10.[4-6]

This is so much more surprising since Sutton et al explicitly state that "the analysis is limited to cases where the Magnetic Reynolds Number ... is small compared with unity."[1] Directly resulting from the mistreatment of high magnetic Reynolds numbers liquid metal MHD channels, aside from displaying only modest electrical efficiencies, also tended to become long and slender and associated with low magnetic fields. This again reduced the efficiency and led to characteristically small electrical potential difference across the electrodes, which again produced high currents and associated Joule losses in busbars and conversion equipment.

The effort by Zheng Linjing,[3] although not applying to liquid metal MHD channels, is interesting from another point of view. As mentioned, his concern is to derive the optimal magnetic field distribution yielding maximum channel efficiency. This objective differs from the entire state of the art summed up in Ref. 4, where the objective rather has been maximizing the efficiency, subject to the condition that the magnetic field distribution is dictated by that produced by electromagnets having parallel-faced flat pole shoes (commonly configured laboratory magnets). To achieve this, it was attempted to fit the exponential end field decay function, having a prescribed e-fold distance, with the "natural" field decay of such laboratory magnets. In some respects, inadequate design rules had to be applied with respect to location of the edge of the electrodes relative to the start of the curve-fitted exponential function, location and length of insulating vanes, etc. This has led to a great deal of inconsistency in the design of MHD devices and, probably as a consequence, their performance. Indeed, in order to take these out of the laboratory, the optimal field distribution should, ideally speaking, be specified first, and the magnet system should then be designed to produce, at least approximately, this very distribution, not the other way around. Magnetic field shaping is done in a variety of areas ranging from nuclear magnetic resonance to, notably, magnetic confinement fusion. The time has also come to apply this philosophy to the design of MHD devices.

The approach taken here for the elimination of end losses proceeds as follows:

1. The 2-D MHD-equation for electric potential is derived while not ignoring the effect of magnetic field compression due to the presence of fairly large magnetic Reynolds numbers, $Rem_i = \sigma\mu_o U_x L_i$.
2. A class of magnetic field decay functions, one for each end region, is derived, if possible, by solving the MHD-equation in the respective end regions while satisfying boundary conditions to an "ideal device" postulated to exist in the central electrode region. If this is successful and assuming Rem_i is large, this approach leads to zero or negligible current leakage out of the central electrode region and, thus, by implication, approaching a condition of zero end losses. The other immediate implication is that one can compose MHD-channels with arbitrary channel aspect ratios by stacking such zero current leakage modules side-by-side separated by insulating vanes in the end regions.
3. The near zero current leakage end field decay functions are validated in a general 2-D model applying these MHD-equation and pertaining boundary conditions to the entire domain comprising the central electrode region and inflow and outflow end regions.
4. Subsequently, parametric studies are performed with the purpose to render, if possible, the end losses due to leakage currents to vanish with finite Rem_i. If this is successful, this approach would lead, theoretically speaking, to ideal device modules that can be used to compose ideal MHD-channels of practically any aspect ratio.

With this, there is already hinted the possibility that, assuming appropriate end field decay functions can be provided, high magnetic Reynolds number MHD channels can achieve high efficiency and electric potential-difference while using small channel aspect ratio and high magnetic field.

Problem Formulation

The treated channel (see Fig. 1) has a constant rectangular cross section. Equidistant electrodes face each other mounted in two opposite sides of the channel. A rectangular coordinate system is placed with its origin at the right-hand edge of the electrodes and in the middle of the channel. Flow of conductive fluid is assumed in the positive x-direction. A magnetic field B_z is applied in the z-direction, cross-wise to the fluid flow as well as the current flowing in the plane between the electrodes.

The channel is divided into three distinct regions: 1) the central electrode region ($-c \leq x \leq 0$), 2) the inflow end region ($-c - L_2 \leq x \leq -c$), and 3) the outflow end region ($0 \leq x \leq L_1$). With the application of a cancellation busbar (see Fig. 2) it is assumed that the central field in the region between the electrodes $B_z = B_o$ is unperturbed by the induced current J_y (since an equal but opposite direction

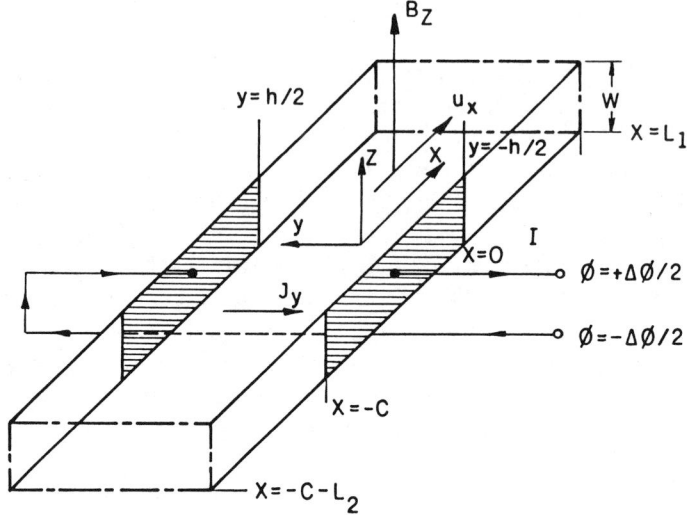

Fig. 1   Linear MHD Channel

current returns nearby through the cancellation busbar). The end field decay functions will be derived herein.

The basic model assumes: 1) no variations in properties out of the xy-plane (2-D model), 2) steady uniform fluid velocity U_x in all regions with viscous effects ignored initially, 3) electrodes are perfect electrical conductors, whereas walls are perfect insulators; and 4) high magnetic Reynolds number $Rem_i = \sigma\mu_o U_x L_i$, where L_i is the length of the respective end regions.

This model should be fairly accurate and consistent assuming the Hartmann number $Ha = B_z w \sqrt{\sigma/\mu}$ is also large. Then, in accordance with Hartmann flow, the boundary layers in the z-direction would be thin and the velocity profile quite uniform.[7] This, naturally, becomes less than a good assumption at the very entrance and exit of the end regions, where the magnetic field B_z approaches zero. Here there may be a redistribution of velocity in accordance with viscous flow. The important thing is, however, that the model be accurate in and around the central electrode region.

In accordance with these assumptions the MHD-equation for electric potential (satisfying Kirchhoff's first law, $\nabla \cdot \overline{J} = 0$) has been derived in Appendix A:

$$\Delta\emptyset - \frac{Rem_i}{L_i}\frac{\partial\emptyset}{\partial x} = 0 \; ; -c - L_2 \leq x \leq L_1 \tag{1}$$

where

$$\Delta\emptyset = \frac{\partial^2\emptyset}{\partial y^2} + \frac{\partial^2\emptyset}{\partial x^2} \tag{2}$$

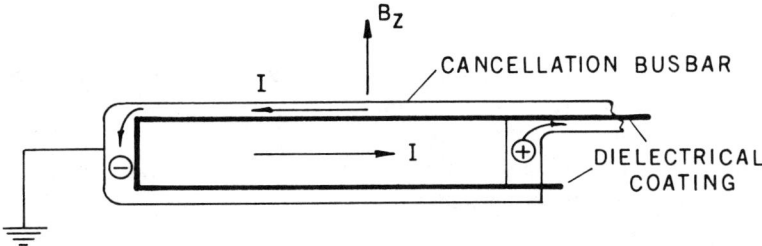

Fig. 2 Schematic Cross-Section of MHD Channel with Cancellation Bussbar

It is observed that an extra term must be added to the Laplace equation to take care of magnetic field compression due to induced current. Assuming the magnetic Reynolds number is large, this term may be substantial even should the gradient $\partial \emptyset / \partial x$ be small. One also realizes that Eq. (1) is not symmetrical like the Laplace equation, $\Delta \emptyset = 0$. Thus, it is to be expected that the end field decay functions for the respective end regions will not be symmetrical. Thus, the distinction is made between inflow and outflow end regions. This is already a major deviation from conventional wisdom, where the Laplace equation is applied to the entire domain.

The MHD-equation is, generally speaking, satisfied by the auxiliary relationships:

$$\frac{\partial B_z}{\partial x} = \mu_o \sigma \left(\frac{\partial \emptyset}{\partial y} + U_x B_z \right) \tag{3}$$

$$\frac{\partial B_z}{\partial y} = \mu_o \sigma \frac{\partial \emptyset}{\partial x} \tag{4}$$

Once the electrical potential gradients are known, these relationships can be used for evaluation of the magnetic field distributions. Notice, however, in the central electrode region $\partial B_z / \partial x$ can be set approximately equal to zero due to the presence of cancellation busbars (see Appendix A on this issue).

Hypotheses for Elimination of Current Leakage

A hypothesis will be established whereby current leakage from the central electrode region into the end regions will approach asymptotically zero with increasing magnetic Reynold number, Rem_i. Fundamentally, it is being postulated that the central electrode region is constituted by an ideal device (to be defined below). It is then assumed that one can find magnetic field decay functions in the inflow and outflow end regions that implicitly have the following characteristics:

Postulate #1: Electric potential in the interfaces between the central electrode region and the respective end regions is continuous, assuming Rem_i is large.

Postulate #2: Electric potential in the end regions is decaying to zero within a finite distance.

The mathematical equivalence to these postulates is derived below.
The current density in an ideal device is according to Ohm's law [see Eq. (A.10)]:

$$J_x = -\sigma \frac{\partial \emptyset}{\partial x} = 0 \qquad (5)$$

$$J_y = -\sigma \left(\frac{\partial \emptyset}{\partial y} + U_x B_z\right) \qquad (6)$$

A load factor is introduced as:

$$\frac{\partial \emptyset}{\partial y} = K \left(\frac{\partial \emptyset}{\partial y}\right)_{J_y = 0} \qquad (7)$$

where $K = \leq 1$ for generator, ≥ 1 for pump.
Using Eq. (7) yields:

$$\frac{\partial \emptyset}{\partial y} = -KU_x B_z \quad \text{for } B_z = B_o; \ -c < x < 0 \qquad (8)$$

It has now been established the boundary conditions with the end regions according to Postulate #1:

$$\frac{\partial \emptyset (0;y)}{\partial y} = \frac{\partial \emptyset (-c;y)}{\partial y} = -KU_x B_o \qquad (9)$$

$$\lim_{Rem_1 \to \infty} \frac{\partial \emptyset (0;y)}{\partial x} = \lim_{Rem_2 \to \infty} \frac{\partial \emptyset (-c;y)}{\partial x} = 0 \qquad (10)$$

Postulate #2 is trivially expressed as:

$$\emptyset(L_1, y) = \emptyset(-L_2 - c, y) = 0 \qquad (11)$$

Derivation of Zero Current Leakage End Field Decay Functions

Applying Eq. (1) to the outflow end region together with boundary conditions from Eqs. (9) and (11) yields the electrical potential distribution (see

Appendix B):

$$\emptyset(x,y) = - KU_xB_oy\ [1 - \exp[Rem(x/L_1-1)]\ /\ [1-\exp(-Rem)];$$

$$0 \le x \le L_1\ ;\ -h/2 \le y \le +h/2 \tag{12}$$

where $Rem = Rem_1$

Performing a limit analysis where current goes to zero in the end region, together with the auxiliary relationships in Eqs. (3) and (4), further yields the field distribution:

$$B_z(x) = KB_o\ [1-\exp[Rem(x/L_1-1)]]\ /\ [1-\exp(-Rem)] \tag{13}$$

The dimensionless outflow region magnetic field distribution $b_z = B_z/(KB_o)$ is shown plotted in Fig. 3.

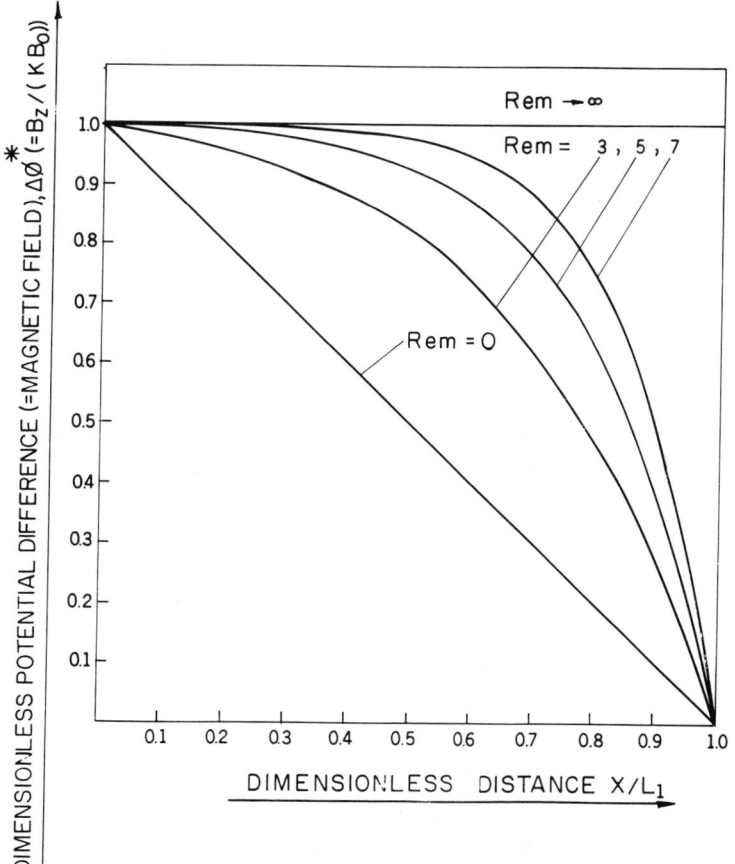

Fig. 3 Dimensionless Potential Difference
$\Delta\emptyset^* = 2\emptyset(x,h/2)/KU_xB_oh = B_z/KB_o$ vs. Dimensionless Distance from the Edge of the Electrodes in the Out-flow End Region

As indicated in Appendix C, the inflow end region cannot be treated in the same way as the outflow end region since it would lead to the conclusion that $\partial\emptyset(-c,y)/\partial x \to \infty$ for $\mathrm{Rem}_2 \to \infty$. Instead, there is applied a general algorithm that by implication [via the auxiliary relationships Eqs. (3) and (4)] characteristically satisfies all of the boundary conditions according to Eqs. (9), (10) and (11) as well as complying with the MHD-equation, at least in the vicinity of the electrodes, assuming $\mathrm{Rem}_2 = \mathrm{Rem}\frac{L_2}{L_1}$ is very large. It is elected here to use an end field decay function in the inflow region that is symmetrical to the outflow,

$$B_z(x) = KB_o \left[1 - \exp[-\mathrm{Rem}\,(\frac{L_2}{L_1} + \frac{c+x}{L_1})]\right] / \left[1 - \exp(-\mathrm{Rem}\,\frac{L_2}{L_1})\right] ;$$
$$- c - L_2 \leq x \leq -c \tag{14}$$

The corresponding potential function is:

$$\emptyset(x,y) = - U_x y\, B_z(x) \text{ for Rem} \to \infty \tag{15}$$

As it turns out, this decay function satisfies the MHD-equation not only at $x = -c$ but in the entire regime x; $-c - L_2 < x \leq -c$, again assuming Rem is very large.

Figure 4 shows schematic representations of magnetic field and electric potential distributions. The first things to note are the characteristic discontinuities in magnetic field on either side of the central electrode region, represented by the load factor K. This could be around 0.97 to yield maximum efficiency (considering friction losses). To produce a theoretical step is, of course, an impossible demand with real magnets, but some form of S-transition should be realizable. Figure 5 shows a suggested method for shaping of the pole shoes using central focusing and surrounding notches for attainment of this transition. The second thing to note is that the gradient $\partial\emptyset/\partial x$ is practically zero at the inflow and outflow edges of the central electrode region, which is a prerequisite to avoiding current leakage. In effect, the gradients rapidly approach zero with increasing magnetic Reynolds number. The third thing to note is that, in reality, both magnetic field and electric potential must decay towards zero only gradually. The fact that the present model does not predict this derives from the artifact that the velocity distribution is uniform everywhere. Natural redistribution due to viscous effects will cause the characteristic tails. It is emphasized that what is important here is the presumed accurate behavior in and in the immediate vicinity of the central electrode region. Far from the electrodes, errors in the model should have only a minor influence. This may be seen in contrast to the conventional exponential decay functions, which may be fairly accurate far away from the electrodes but provide arbitrarily large gradients in electric potential close to the electrodes.

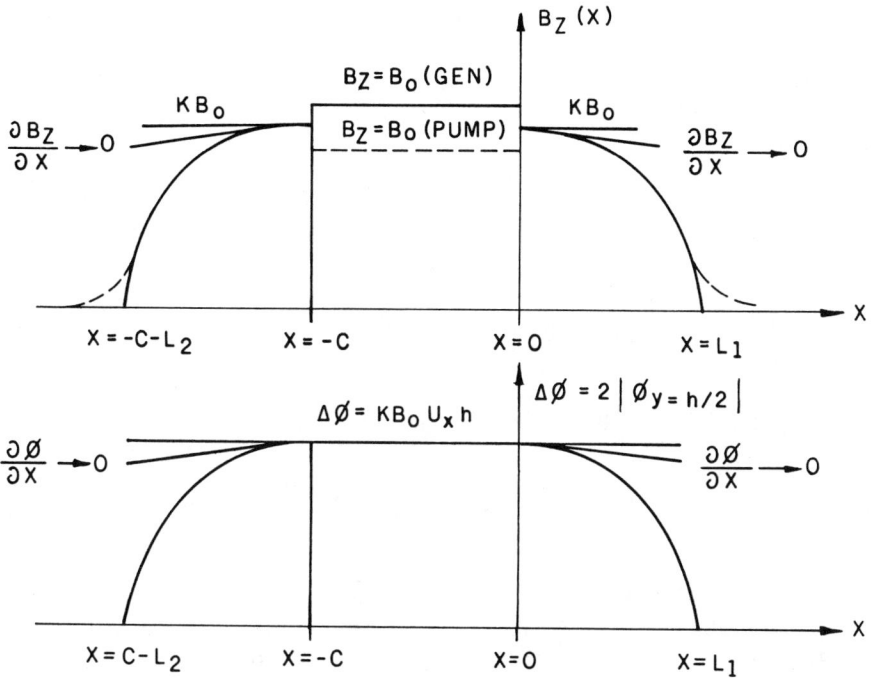

Fig. 4   Schematic Representation of Magnetic Field and Electric Potential Distributions

General 2-D Model and Its Solution

The 2-D model formulated in Appendix D was analyzed using the numerical method of finite-differences. Referring to Figs. 6 and 7, the domain ($-c - L \leq X \leq 1$; $0 \leq Y \leq H/2$) has been filled by the uniform finite-difference grid:

$$X_i = (i - 1) \Delta X + 0.5 \Delta X \qquad (16)$$
$$Y_j = (j - 1) \Delta Y + 0.5 \Delta X \qquad (17)$$

where $i = 1, 2, 3, ..., i_{MAX}$; $j = 1, 2, 3, ..., j_{MAX}$, and

$$\Delta X = (1 + L + C) / (i_{MAX} - 2) \qquad (18)$$
$$\Delta Y = 0.5H / (j_{MAX} - 2) \qquad (19)$$

The grid dimensions were usually selected to be $j_{MAX} = 102$ and $i_{MAX} = 102$. The boundary conditions stated in Appendix D, namely Eqs. (D.3 -

D.5), were approximated with the second order of accuracy. The finite-difference equations were solved by the line-by-line method using the standard TDMA algorithm.[8] The iterations were terminated when the maximum residual of the finite-difference equations became less than 10^{-5}.

Device Efficiencies and Parametric Studies

A generator efficiency is derived in Appendix E as:

$$\eta_{GEN} = K \frac{1 - K/\eta_{E.L.}}{1 - K/\eta_{E.L.} + K_F/2N}, \qquad K \leq 1$$

In pump operation the device efficiency would be:

$$\eta_{PUMP} = \frac{1}{K} \frac{K/\eta_{E.L.} - 1 - K_F/2N}{K/\eta_{E.L.} - 1}, \qquad K \geq 1$$

Here $\eta_{E.L.}$ is an efficiency relating to current leakage out of the central electrode region or actually an end loss efficiency. For slot like Hartmann flow[7] the friction term $K_F/2N \cong 1/Ha$. This number decreases with increasing magnetic field strength and increasing width w of the channel (however, for validity w/h must by the same token be small). As shown in Appendix E, the load factor K can be readily optimized at the outset. The optimal efficiency is proportional to the end loss efficiency and also an expression containing only the Hartmann number. Consequently, it makes sense to study only the end loss efficiency in terms of how it is affected by the independent parameters = (Rem, $A_{MOD} = C/H$, $L = L_2/L_1$, and $C = c/L_1$). Parametric studies using the general 2-D model for generator application yielded the following. The end loss efficiency $\eta_{E.L.}$ increases monotonically with L, at first quite slowly around $L = 1$ and then gradually much faster around $L = 2$. Since it may not be practical and may possibly be uneconomical to produce magnetic field coverage twice as long in

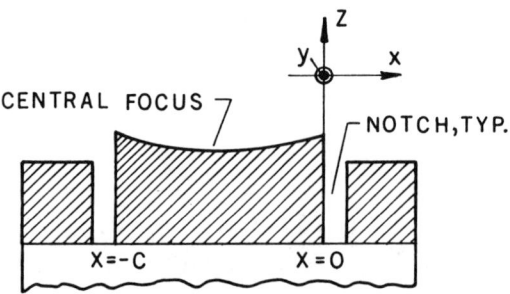

Fig. 5   Conceptual Shaping of Pole Shoes of Electro Magnet for Achieving Step in Field

the inflow region compared with the outflow region, one can just as well settle for L = 1. This has the benefit of yielding symmetrical end field decay functions, thus facilitating the magnet design. Figures 8 and 9 show the end loss efficiency vs magnetic Reynolds number and parametrically with module aspect ratio for the two cases C = 2.0 and 1.0, respectively.

In all cases studied, the end loss efficiency increases monotonically with A_{MOD} but tends to saturate towards a given value with increasing A_{MOD}. Generally, this becomes higher the smaller the value for C. So for C = 2.0 the end loss efficiency saturates at $\eta_{E.L.} \cong 0.980$. With C = 1.0 end losses can be made to disappear altogether, even for relatively small values of A_{MOD}. The ability to produce an end loss efficiency of 1 occurs for C smaller than about 1.5. The fact that the end regions need to be quite long relative to the length of the central electrode region is quite analogous to the demand of exponential field decay function to have a long e-fold distance.

Once C is in the useful domain, there appear local maxima with respect to the magnetic Reynolds number. The locus for these is indicated in Fig. 9. As the end loss efficiency approaches 1 the optimal Rem seems to approach between 4 and 5. At zero end loss saturation the range for useful magnetic Reynolds numbers widens considerably with increasing A_{MOD}. For C = 1.0 and A_{MOD} = 2.0 the end loss efficiency is 1 for Rem between 4 and 7. For C = 1.0 and $A_{MOD} \cong 10$ practically any Rem can be used towards eliminating end losses.

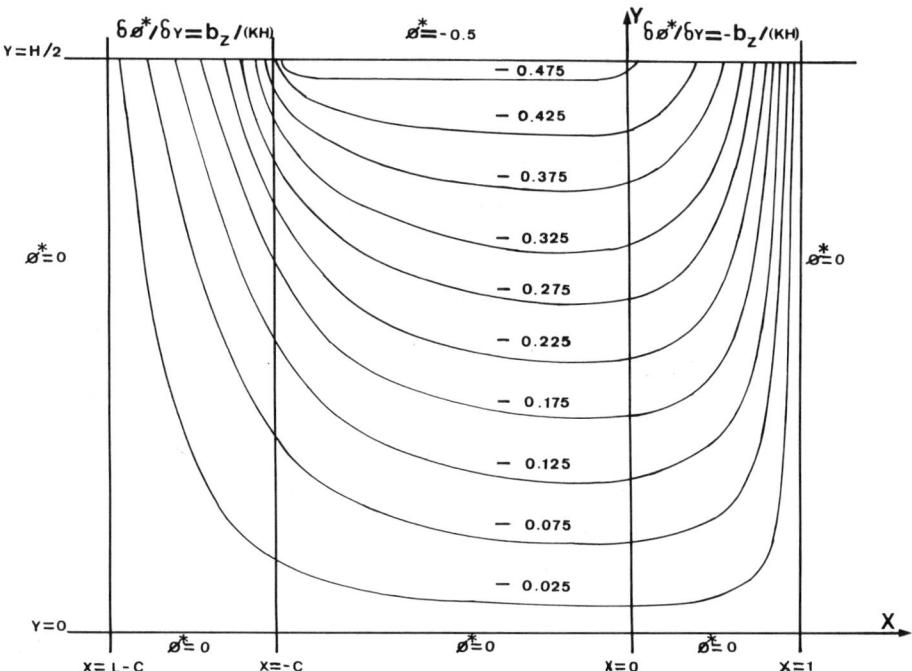

Fig. 6   Electrical   Potential   Map   (C = 2.0, L = 1.0, H = 2.0)

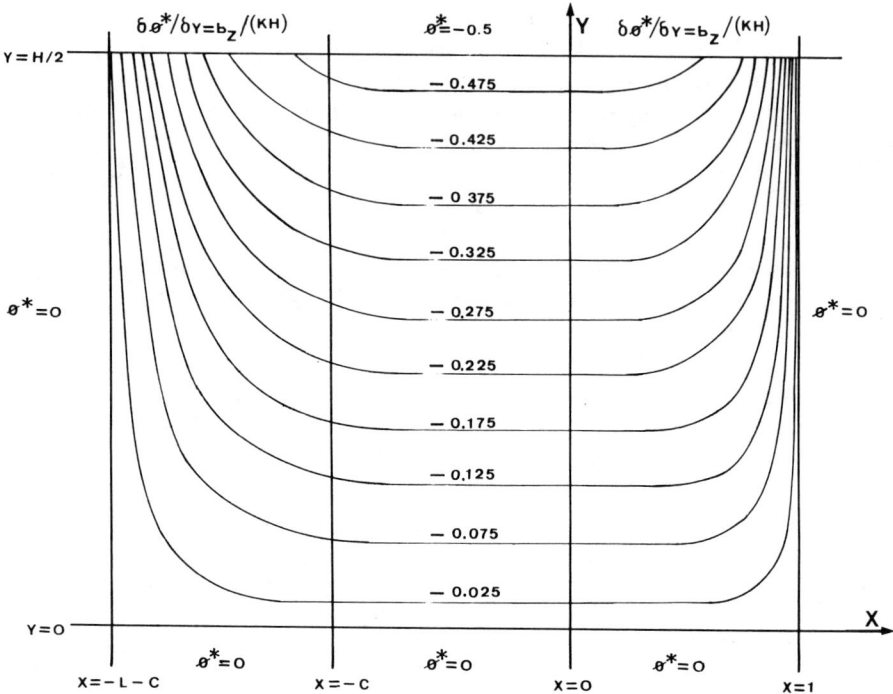

Fig. 7   Electrical Potential Map   (C = 1.0, L = 1.0, H = 0.5)

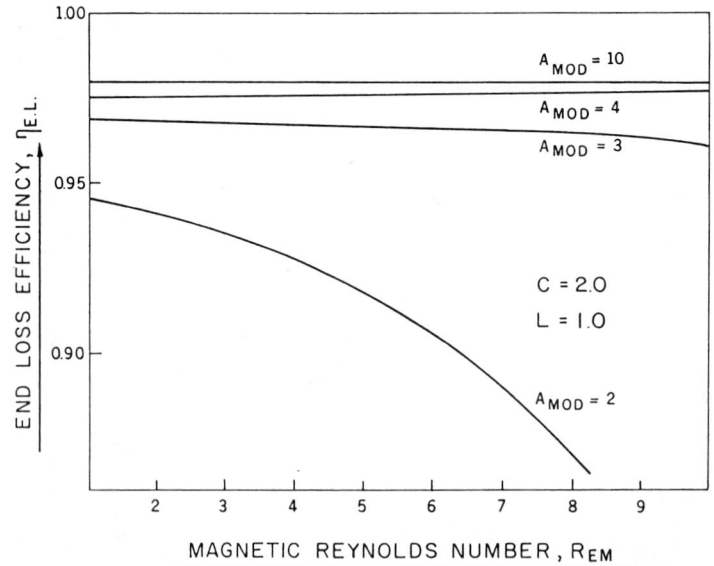

Fig. 8   End Loss Efficiency, $h_{E.L.}$, vs. Magnetic Reynolds Number, Rem, and Parametrically with Module Aspect Ratio, A_{MOD} (C = 2.0, L = 1.0)

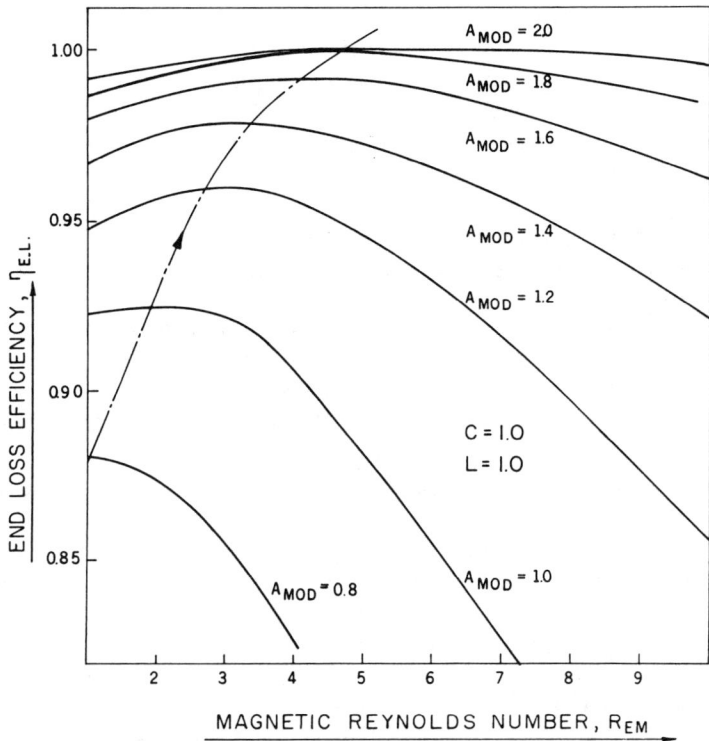

Fig. 9 End Loss Efficiency, $h_{E.L.}$, vs. Magnetic Reynolds Number, Rem, and Parametrically with Module Aspect Ratio, A_{MOD} (C = 1.0, L = 1.0)

This is, of course, of great practical value since it would permit the device to be operated at various flow rates with constant efficiency.

Electrical potential plots are provided in Figs. 6 and 7, one for a low efficiency case and the other for a high efficiency case. Comparing the two plots, it becomes apparent how the potential tends to become linear and uniform in the central electrode region for the high efficiency case. The effect of magnetic field compression is also apparent as the electric potential tends to balloon in the flow direction.

Conclusions

End field decay functions aiming at zero current leakage out of the central electrode region for a large magnetic Reynold's number, Rem, have been derived based on assuming the existence of an ideal device in the central electrode region. These decay functions are, generally speaking, different from one another applying to the inflow and outflow end regions. They have in common, however, a theoretical step change in the magnetic field at the edges convex, as opposed to the concave exponential field decay function commonly used for benchmarking liquid metal MHD channels.

The end field decay functions were validated in a generally formulated 2-D model for electrical potential, including the effect of a nontrivial magnetic Reynolds number. From parametric studies it was concluded that to obtain zero current leakage from the central electrode region (implying zero end losses) the following prevails:

1. The maximum length of the electrodes must be smaller than 1.5 the lengths of the respective end regions (defined as the distance required to obtain zero potential).
2. The smaller the ratio of the above lengths the smaller is the required channel aspect ratio (length of electrodes/distance between electrodes).
3. Although it is easier to obtain zero end losses where the inflow end region is longer than the outflow end region equal decay distance and, thus, symmetrical end field decay functions can be used.
4. Where the length ratio according to item 1 is sufficiently small and the channel aspect ratio is sufficiently large, a great range of magnetic Reynolds numbers would achieve zero end losses.

There has been derived a maximum device efficiency that is proportional to an end loss efficiency and a function of the Hartmann number. As has been demonstrated, the end loss efficiency can, theoretically speaking, be made equal to 1. The Hartmann number function approaches 1 as the Hartmann Number goes to infinity. The Hartmann number is in reality in the order of 1000 with large conductive liquid metal MHD-channels and reasonably high magnetic fields. This would give an optimal load factor of 97% and a maximum efficiency of about 94%. It is also implied from item 4 previously that one can have constant high efficiency operation for a wide range of flow rates using one and the same channel.

In addition, since it would be possible to achieve zero current leakage out of the central electrode region, practically any effective channel aspect ratio can be created by stacking zero current leakage modules side by side, separated by insulated vanes in the end regions. This implies that, in theory, the electrical potential difference can be made orders of magnitude greater than presently experienced with liquid metal dc MHD-channels for a given pressure difference (the only limitations being minimum lengths of electrodes and maximum economic magnetic field strength).

The present approach suffers from the use of a simple 2-D model and the simplifying assumptions, notably the assumption for slug flow. As an artifact of this model, there are no characteristic exponential decay tails associated with the present end field decay functions. It is implied, however, that such can be added artificially without affecting much the validity of the present results. This is so since the correction is relatively far away from the edge of the electrodes. A more severe issue relates to whether the characteristic step, albeit small, is even approximately reproducible with an applied magnetic field. This warrants a study in itself. Complementing this study one should find to what degree the present good results are dependent on the distinctness of the step. That the step can, in actuality, be smoothed over is borne out by the employed general 2-D numerical analysis, where the finite differences naturally are not infinitely small and grid points rarely coincide with the location of the step.

Appendix A
Derivation of the MHD-Equations

The treated channel (see Fig. 1) has a constant rectangular cross section. Equidistant electrodes face each other mounted in two opposite sides of the channel. A rectangular coordinate system is placed with its origin at the right-hand edge of the electrodes and in the middle of the channel. Flow of conductive liquid is assumed in the positive x-direction. A magnetic field B_z is applied cross-wise to the flow as well as the plane of the electrodes. It will be assumed for the purpose of developing this model that the velocity is uniform in all regions and that there are no variations in the z-direction (2-D model).

A current is generated following Ohm's law:

$$\vec{J} = \sigma(\vec{E} + \vec{U} \times \vec{B}) \tag{A.1}$$

Since $\vec{B} = (0,0,B_z)$ is steady it follows from Faraday's law:

$$\nabla \times \vec{E} = 0 \tag{A.2}$$

and the electrical field intensity may be expressed in terms of electric potential:

$$\vec{E} = -\nabla \varnothing \tag{A.3}$$

Upon taking the divergence of Eq. (A.1), one obtains:

$$\nabla \cdot \vec{J} = \sigma(-\nabla^2 \varnothing + \nabla \cdot (\vec{U} \times \vec{B})) = 0 \tag{A.4}$$

Now, from vector analysis:

$$\nabla \cdot (\vec{U} \times \vec{B}) = \vec{B} \cdot (\nabla \times \vec{U}) - \vec{U} \cdot (\nabla \times \vec{B}) \tag{A.5}$$

For potential flow:

$$\nabla \times \vec{U} = 0 \tag{A.6}$$

From Ampere's law:

$$\nabla \times \vec{B} = \mu_o \vec{J} \tag{A.7}$$

Combining Eqs. (A.4), through (A.7) yields:

$$\Delta \varnothing + \mu_o \vec{U} \cdot \vec{J} = 0 \tag{A.8}$$

There follows from Eq. (A.3):

$$\vec{E} = \left(-\frac{\partial \varnothing}{\partial x} ; -\frac{\partial \varnothing}{\partial y} ; 0 \right) \tag{A.9}$$

Substituting into Ohm's law yields the current distribution:

$$\vec{J} = \sigma \left(-\frac{\partial \varnothing}{\partial x} ; -\frac{\partial \varnothing}{\partial y} - U_x B_z ; 0 \right) \tag{A.10}$$

Using this result and considering that $U = (U_x, 0, 0)$, one obtains from Eq. (A.8):

$$\Delta \varnothing - \sigma \mu_o U_x \frac{\partial \varnothing}{\partial x} = 0 \tag{A.11}$$

or with a magnetic Reynolds number defined as $Rem = \sigma \mu_o U_x L$:

$$\frac{\partial^2 \varnothing}{\partial y^2} + \frac{\partial^2 \varnothing}{\partial x^2} - \frac{Rem}{L} \frac{\partial \varnothing}{\partial x} = 0 \tag{A.12}$$

This is the MHD-equation for electric potential.

From Ampere's law is obtained:

$$\vec{J} = \frac{1}{\mu_o} \left(\frac{\partial B_z}{\partial y} ; -\frac{\partial B_z}{\partial x} ; 0 \right) \tag{A.13}$$

Equating this result with Eq. (A.10) yields:

$$\frac{\partial B_z}{\partial x} = \mu_o \sigma \left(\frac{\partial \varnothing}{\partial y} + U_x B_z \right) \quad (= -\mu_o J_y) \tag{A.14}$$

$$\frac{\partial B_z}{\partial y} = -\mu_o \sigma \frac{\partial \varnothing}{\partial x} \quad (= \mu_o J_x) \tag{A.15}$$

Once the voltage gradients are known, these auxiliary relationships that, naturally, satisfy Eq. (A.12), may be used to evaluate the magnetic field distributions.

Note here that the gradient $\partial B_z/\partial x$ is approximately equal to zero in the central electrode region due to the presence of cancellation busbars (see Fig. 2). It can be shown that the z-component of magnetic field produced by the Ohm's law current J_y [see Eq. (A.10)] is effectively canceled by the field generated by the same current returning through the nearby cancellation busbar.

Appendix B
Derivation of the Zero End Loss Condition for the Outflow End Region

Consider the outflow end region see Fig. 1. Applied to this region is the MHD-equation derived in Appendix A, which is repeated here:

$$\frac{\partial^2 \emptyset}{\partial y^2} + \frac{\partial^2 \emptyset}{\partial x^2} - \frac{\text{Rem}}{L_1} \frac{\partial \emptyset}{\partial x} = 0; \; 0 < x \le L_1; \; -\frac{h}{2} < y < +\frac{h}{2} \qquad (B.1)$$

where $\text{Rem} = \text{Rem}_1 = \sigma \mu_o U_x L_1$.

Hypothetically speaking, the outflow end region interfaces with an ideal device in the central electrode region. Thus, at the mutual boundary the following must be satisfied:

$$\frac{\partial \emptyset (0;y)}{\partial y} = - K U_x B_o \qquad (B.2)$$

and

$$\lim_{\text{Rem} \to \infty} \frac{\partial \emptyset (0;y)}{\partial x} = 0 \qquad (B.3)$$

Note that the latter condition is not, mathematically speaking, a boundary condition, but must follow from the nature of the solution. In addition, it is assumed that the electrical potential is broken down to zero within a finite distance L_1. Note that this characteristic distance is included in the magnetic Reynolds number. Thus,

$$\emptyset(L_1, y) = 0 \qquad (B.4)$$

The electrical potential distribution is antisymmetric around $y = 0$. Thus, considering the linear varition according to Eq. (B.2), one can arbitrarily set:

$$\emptyset(x, 0) = 0 \qquad (B.5)$$

The partial differential equation is solved by the method of separation of variables.

Following this, a product solution is postulated as:

$$\emptyset(x, y) = H(x) \, G(y) \qquad (B.6)$$

After differentiation and insertion into the main equation, two ordinary differential equations are obtained:

$$\frac{d^2H}{dx^2} - \frac{Rem}{L_1} \frac{dH}{dx} - k^2H = 0 \tag{B.7}$$

$$\frac{d^2G}{dy^2} + k^2G = 0 \tag{B.8}$$

where k^2 is the separation variable.
Equations (B.4) and (B.5) are satisfied for:

$$H(L_1) = 0 \tag{B.9}$$

$$G(0) = 0 \tag{B.10}$$

Equation (B.2) implies that:

$$H(0) \frac{dG(y)}{dy} = -KU_xB_o \tag{B.11}$$

Equation (B.8) can have a solution on this form only if the separation variable $k^2 \equiv 0$. Thus, it follows that:

$$G(y) = A_1 + A_2y \tag{B.12}$$

Equation (B.10) yields $A_1 = 0$ and, thus, $dG/dy = A_2$. With $k^2 \equiv 0$, the solution for Eq. (B.7) is:

$$H(x) = C_1 + C_2 \exp[Rem(x/L_1)] \tag{B.13}$$

Substituting Eqs. (B.12) and (B.13) into (B.11), yields:

$$(C_1 + C_2) A_2 = -KU_xB_o \tag{B.14}$$

Making use of Eq. (B.9) yields:

$$C_2 = -C_1 \exp(-Rem) \tag{B.15}$$

Combining the last two expressions yields:

$$C_1 A_2 = -KU_xB_o / [1 - \exp(-Rem)] \tag{B.16}$$

and from Eq. (B.6):

$$\emptyset(x,y) = -KU_xB_o y \, [1 - \exp[Rem(x/L_1)]] / [1 - \exp(-Rem)];$$
$$0 \leq x \leq L_1 \; ; \; -h/2 \leq y \leq +h/2 \tag{B.17}$$

Indeed, this solution does satisfy Eq. (B.1). There remains to satisfy the condition of Eq. (B.3).

Taking the first derivative of Eq. (B.17) gives:

$$\frac{\partial \phi(o;y)}{\partial x} = KU_x B_o (y/h)(h/L) f_o (Rem) \tag{B.18}$$

where
$$f_o(Rem) = Rem \exp(-Rem)/[1-\exp(-Rem)] \tag{B.19}$$

As can be seen, the gradient of electrical potential in the direction of flow can reach zero for large values in the magnetic Reynolds number and also, alternatively, for small ratios h/L_1. In actuality, the gradient approaches zero rapidly with increasing Rem. In other words, regardless of the electrical conductivity of the liquid and its velocity, the gradient approaches zero as the decay distance L_1, increases. The corollary is that if the conductivity and velocity are fairly high the decay distance need not be very long in order to achieve the condition of near zero gradient. It is concluded that it is quite feasible to provide such conditions, especially with high-conductivity liquid metals, leading to zero current leakage into the outflow end region. The question is now what should be the corresponding magnetic field distribution in the end region. Ohm's law, see Eq. (A.10), indicates that the zero current condition is for:

$$\frac{\partial \phi}{\partial x} = 0 \tag{B.20}$$

and

$$\frac{\partial \phi}{\partial y} = - U_x B_z \tag{B.21}$$

Should this prevail, then according to the auxiliary relationships of Eqs. (A.14) and (A.15):

$$\frac{\partial B_z}{\partial x} \to 0 \text{ and } \frac{\partial B_z}{\partial y} \to 0 \tag{B.22}$$

Using Eq. (B.21), yields:

$$B_z(x) = KB_o [1-\exp[Rem(x/L_1-1)]] / [1-\exp(-Rem)] \tag{B.23}$$

According to this (after differentiation):

$$\frac{\partial B_z}{\partial x} = - \frac{KB_o}{L} Rem \exp[Rem(x/L_1-1)] / [1-\exp(-Rem)] \tag{B.24}$$

and

$$\frac{\partial B_z}{\partial y} = 0 \tag{B.25}$$

As can be seen, Eq. (B.25) is consistent with (B.22), and also Eq. (B.24), as long as L_1 and, thus, Rem are sufficiently large. In the vicinity of $x/L_1 = 1$, strictly speaking, L_1 and Rem must approach infinity. This is, however, not required in reality.

In the solution for B_z it should be further noted the required discontinuity in magnetic field between the central electrode region and the outflow end region. The step change is given by the load factor K.

Appendix C
Conditioning of Algorithm for Magnetic Field Distribution in Inflow End Region

The approach taken in Appendix B, where magnetic field compression is utilized for sealing off the central electrode region from the outflow end region, is not useful with regard to the inflow end region. Following that approach the electric potential distribution would be:

$$\emptyset(x,y) = -KU_xB_oy \ [\ 1\text{-exp}\ [\text{Rem}_2(1+\frac{x+c}{L_2})]\ ] \ /[1\text{-exp}(\text{Rem}_2)\ ];$$
$$-L_2 - c \leq x \leq -c; \ -h/2 < y < +h/2 \qquad (C.1)$$

Consequently, the gradient at the edge of the electrodes is given by:

$$\frac{d\emptyset(-c;y)}{dx} = KB_oU_x\left(\frac{y}{h}\right)\frac{h}{L_2}\text{Rem}_2\frac{e^{\text{Rem}2}}{1-e^{\text{Rem}2}} \qquad (C.2)$$

It cannot approach zero with conductive liquids. Thus, one could conclude from this that it is impossible for the inflow end region to interface properly with an ideal device central electrode region. Not yielding to this conclusion, it becomes necessary to introduce a new idea. This will be developed below.

It is suggested to study the influence of a general algorithm for magnetic field distribution that is of the following form:

$$B_z(x) = KB_o\ [1\text{-exp}\ [-\text{Rem}_2(1+\frac{c+x}{L_2})]]/[1\text{-exp}(-\text{Rem}_2)]; \ -c - L_2 \leq x \leq -c \qquad (C.3)$$

The first thing to note, is that this function is symmetrical to the outflow region end field decay function according to Eq. (B.23) (see also Fig. 3). However, the algorithm can be manipulated via the decay distance L_2 which does not necessarily coincide with the outflow region decay distance L_1.

The behavior of Eq. (C.3) is inspected in more depth here. Taking the derivations of Eq. (C.3) yields:

$$\frac{\partial B_z}{\partial y} = 0 \qquad (C.4)$$

$$\frac{\partial B_z}{\partial x} = f_1(Rem_2) = \frac{Rem_2}{L_2} \exp\left[-Rem_2\left(1+\frac{c+x}{L_2}\right)\right]/[1-\exp(-Rem_2)] \qquad (C.5)$$

Here $\partial B_z/\partial x \to 0$ for $Rem_2 \to \infty$; thus, according to Eq. (A.14) the current $J_y \to 0$, if also:

$$\emptyset = -\int_0^y U_x B_z(x)\,dy = -KU_x B_o\, y\left[1-\exp\left[-Rem_2\left(1+\frac{c+x}{L_2}\right)\right]\right]/[1-\exp(-Rem_2)] \qquad (C.6)$$

It is recognized, at once, that Eq. (C.6) satisfies the boundary conditions:

$$\partial\emptyset(-c,y)/\partial y = -KU_x B_o, \quad \text{as well as } \emptyset(-L_2-c,y) = 0.$$

It is also required that Eq. (C.6) satisfies the MHD-equation at least at the boundary $x = -c$. Differentiating Eq. (C.6) and inserting into Eq. (A.12), yields:

$$\frac{\partial\emptyset}{\partial x} = -KU_x B_o y\, f_1(Rem_2) \qquad (C.7)$$

$$\frac{\partial^2\emptyset}{\partial x^2} KU_x B_o y\, \frac{Rem_2}{L_2} f_1(Rem_2) \qquad (C.8)$$

$$\frac{\partial^2\emptyset}{\partial y^2} = 0 \qquad (C.9)$$

and

$$\frac{\partial^2\emptyset}{\partial y^2} + \frac{\partial^2\emptyset}{\partial x^2} - \frac{Rem_2}{L_2}\frac{\partial\emptyset}{\partial x} \equiv 2KU_x B_o \frac{Rem_2}{L_2} f_1(Rem_2) \qquad (C.10)$$

This approaches zero for large Rem_2, for all x; $-L_2 -c < x \le -c$. By the same token, the gradient at the edge of the electrodes $\partial\emptyset(-c,y)/\partial x \to 0$ for $Rem_2 \to \infty$.

To conclude, Eq. (C.3) seems to be uniquely suited as end field decay function in the inflow end region.

Appendix D
Formulation of General 2-D Model for Linear MHD-Channel

Consider the MHD-channel shown in Fig. 1. Applied to the entire 2-D region are MHD-equations according to Eqs. (A.12) and (A.13). Introducing dimensionless variables, e.g.:

$$b_z = B_z/B_o$$
$$\emptyset^* = \emptyset/KB_oU_xh$$
$$X = x/L_1$$
$$Y = y/L_1$$
$$C = c/L_1$$
$$L = L_2/L_1$$
$$H = h/L_1$$

yields the dimensionless MHD-equation:

$$\frac{\partial^2 \emptyset^*}{\partial Y^2} + \frac{\partial^2 \emptyset^*}{\partial X^2} - \text{Rem} \frac{\partial \emptyset^*}{\partial X} = 0 \; ; \; -C-L \leq X \leq 1 \; ; \; -H/2 \leq Y \leq H/2 \quad (D.1)$$

where $\text{Rem} = \sigma\mu_o U_x L_1$.

Since the problem is antisymmetric around $Y = 0$, it is sufficient to study the region $Y \geq 0$. Also referring to Fig. 6, the boundary conditions are as follows:

$$\emptyset^*(X,0) = \emptyset^*(1,Y) = \emptyset^*(-C-L, Y) = 0 \quad (D.2)$$

$$\emptyset^*(X,H/2) = -1/2 \text{ for } -C < X < 0 \quad (D.3)$$

$$\frac{\partial \emptyset^*(X;H/2)}{\partial Y} = -\frac{1}{H}\left(\frac{b_z}{K}\right) \text{ for } 0 \leq X \leq 1 \text{ and } -C-L \leq X \leq -C \quad (D.4)$$

where according to Eq. (B.26):

$$\frac{b_z}{K} = \frac{1 - \exp(\text{Rem}(X-1))}{1 - \exp(-\text{Rem})} \text{ for } 0 \leq X \leq 1 \quad (D.5)$$

and according to Eq. (C.3):

$$\frac{b_z}{K} = \frac{1 - \exp(-\text{Rem}(L+X+C))}{1 - \exp(-\text{Rem}L)} \text{ for } -C-L \leq X \leq -C \quad (D.6)$$

As can be seen, the dimensionless electrical potential in any point (X,Y) is a function

$$\emptyset^* = \emptyset^*(\text{Rem}, H, L, C)$$

It must be observed that the load factor K is not involved in a direct way in this dimensionless formulation.

Appendix E
Derivation of Electrical and End Loss Efficiencies

The electrical power is given by:

$$P_E = I \Delta \phi \tag{E.1}$$

where the potential difference:

$$\Delta \phi = K B_o U_x h \quad (\Delta \phi^* = 1) \tag{E.2}$$

and the current passing though the electrodes:

$$I = -\int_{-L_2}^{0} (j_y)_{y=h/2} \, w \, dx \tag{E.3}$$

Using Ohm's law [Eq. (A.10)] yields the current (consistent with the assumption for slug flow where $U_x \neq 0$ even on the boundary):

$$I = \sigma \int_{-L_2}^{0} \left(\frac{\partial \phi}{\partial y} + U_x B_z\right)_{y=h/2} w \, dx = B_o U_x c w \sigma \left[\frac{1}{B_o U_x c} \int_{-L_2}^{0} \left(\frac{\partial \phi}{\partial y}\right)_{y=h/2} dx + 1\right] \tag{E.4}$$

The hydraulic power is given by:

$$P_H = Q (\Delta p + \Delta p_F) \tag{E.5}$$

Here the flow rate is:

$$Q = U_x w h \tag{E.6}$$

and the friction drop:

$$\Delta p_F = K_F \rho U_x^2 / 2 \tag{E.7}$$

The hydraulic pressure drop due to ponderomotive force can be found by taking the pressure drop along the wall (as any stream line path is valid with the assumption of slug flow), thus:

$$\Delta p = \int_{-c-L_2}^{L_1} (\vec{J} \times \vec{B})_{y=h/2} \, dx = \sigma \int_{-c}^{0} \left[\left(\frac{\partial \phi}{\partial y} + U_x B_z\right) B_z\right]_{y=h/2} dx \tag{E.8}$$

or

$$\Delta p = U_x B_o^2 \sigma c \left[\frac{1}{B_o U_x c} \int_{-c}^{0} \left(\frac{\partial \phi}{\partial y}\right)_{y=h/2} dx + 1\right] \tag{E.9}$$

Other investigators using the present assumption for slug flow prefer, for some unknown reason, to take the integrated average in the entire domain. In this case, this would require evaluating the magnetic field compression in the entire domain.

The electrical efficiency is:

$$\eta_{GEN} = P_E/P_H \text{ for generator}$$
$$\eta_{PUMP} = P_H/P_E \text{ for pump} \qquad (E.10)$$

Evaluating the efficiency from these equations, yields:

$$\eta_{GEN} = K \frac{1 - K/\eta_{E.L.}}{1 - K/\eta_{E.L.} + K_F/2N} \qquad (E.11)$$

$$\eta_{PUMP} = \frac{1}{K} \frac{K/\eta_{E.L.} - 1 - K_F/2N}{K/\eta_{E.L.} - 1} \qquad (E.12)$$

where after identifying terms and introducing dimensionless variables (see Appendix D):

$$\eta_{E.L.} = -\frac{C}{H} / \int_{-c}^{0} \left(\frac{d\emptyset^*}{dY}\right)_{y=h/2} dx \qquad (E.13)$$

and N is the so-called interaction parameter:

$$N = \frac{\sigma B_o^2 c}{\rho U_x} \qquad (E.14)$$

In an ideal generator: $K_F / 2N \rightarrow 0$
Thus (regardless of the value for $\eta_{E.L.}$), the generator efficiency in this case approaches the value of the load factor K; $K \leq 1$.
Similarly, the efficiency for an ideal MHD pump approaches the inverse of the load factor, 1/K; $K \geq 1$.
Under certain circumstances, the friction factor is approximately (from study of Hartmann flow[7]):

$$K_F \cong \frac{2c}{w} \sqrt{\frac{N}{Re}} \qquad (E.15)$$

where $Re = \rho U_x c/\mu$ is the regular fluid flow Reynolds number.
Thus,

$$\frac{K_F}{2N} \cong \frac{c}{w} \frac{1}{\sqrt{NRe}} = \frac{1}{B_o w \sqrt{\sigma/\mu}} = \frac{1}{Ha} \qquad (E.16)$$

In an ideal generator, one also has that

$$\frac{\partial \emptyset^*}{\partial y} = -\frac{1}{H} \text{ for } C \leq X \leq 0 \qquad (E.17)$$

This gives the upper value for $\eta_{E.L.} = 1$. Clearly, $\eta_{E.L}$ represents an end loss efficiency. For $\eta_{E.L} = 1$ there would be no current leakage from the central electrode region. One can readily optimize, lets say, the generator efficiency with respect to the load factor. Setting $d\eta_{GEN}/dK = 0$ gives:

$$K_{OPT} = \eta_{E.L} \left[1 + \frac{1}{Ha} - \sqrt{\frac{1}{Ha} + \frac{1}{Ha^2}} \right] \qquad (E.18)$$

If the friction loss approaches zero or equivalently $Ha \to \infty$, then $K_{OPT} \to \eta_{E.L}$. From this it is implied that the maximum attainable theoretical efficiency equals the end loss efficiency.

Acknowledgment

The author wishes to thank Dr. B. Abramzon from the Center for MHD Studies, Ben-Gurion University, Beer-Sheva, Israel, for his help in the numerical analysis and programming of the problem.

References

[1] Sutton, G.W., et al., "Electrical and Pressure Losses in a Magnetohydrodynamic Channel Due to End Current Loops," *Trans. AIEE*, Pt 1, Vol. 80, 1961, pp. 687-695.

[2] Gherson, P. et al., *Proceedings of 7th International Conference on MHD Electric Power Generators* (A.M. Dawson and D. Overlan, Massachusetts Institute of Technology, Cambridge, MA, 1980, p. 590.

[3] Zheng Linjing, "A Method to Optimize the Configuration of the End Magnetic Field of a D-C MHD Generator," *Proceedings of the 9th International Conference on MHD Electrical Power Generation*, Vol. 4, p. 1630, Tsukubu, Japan, 1986.

[4] Gherson, P. and Lykoudis, P.S., "Analytical Study of End Effects in Liquid Metal MHD generators," Final Rept. PNE-79-144 prepared for ANL under Contract No. NTP 0176, School of Nuclear Engineering, Purdue Univ., West Lafayette, IN, 1979.

[5] Roberts, J.J., et al., "Direct Current MHD Generators with Variable Conductivity, Velocity and Magnetic Field," *Journal of Spacecraft and Rockets*, Vol. 6, No. 6, June 1964, pp. 729-734.

[6] Elliot, D.G., "DC Liquid-Metal Magnetohydrodynamic Power Generation," JPL prepared for NASA under contract No. NAS 7-100.

[7] Branover, H., *Magnetohydrodynamic Flow in Ducts*, Wiley, Israel Universities Press, 1978.

[8] Patankar, S.V., *Numerical Heat Transfer and Fluid Flow*, Hemisphere, 1980.

Embrittlement of Steels by Lead

Rangarajan Venkataraman,* Michael D. Baldwin,†
and Glen R. Edwards‡
Colorado School of Mines, Golden, Colorado 80401

Abstract

This work reviews the current transport and embrittling mechanisms for metal-induced embrittlement, with the objective to promote a better fundamental understanding of the causes for and nature of embrittlement of steels by lead. Embrittlement of steels can be caused by external lead or internal lead, as in leaded steels, in either the liquid or solid state. Most lead embrittlement happens in the liquid state and is known as liquid-metal-induced embrittlement (LMIE). This paper gives general prerequisites, influencing factors, and characteristics of LMIE and the specific facts about lead embrittlement. Mechanisms for embrittlement are reviewed, with emphasis on the "adsorption-induced decohesion" model. This model is believed to best describe lead LMIE of steel by the authors. Additionally, solid-metal-induced embrittlement (SMIE) is briefly considered. SMIE is not as severe as LMIE, because of a slower rate of transport, but occurs below the melting point of lead. Finally, specific results are reviewed to emphasize the significant factors and recommendations for avoiding metal-induced embrittlement for liquid lead/steel structures.

Introduction

Metal-induced embrittlement (MIE) is a phenomenon wherein a loss in ductility and fracture toughness occurs in a metal subjected to tensile stresses while in contact with a lower-melting-point embrittler. The

---

Copyright © 1991 by the American Institute of Aeronautics and Astronautics, Inc. All rights reserved.
* Graduate Research Assistant, Center for Welding and Joining Research.
† Research Associate, Center for Welding and Joining Research.
‡ Director, Center for Welding and Joining Research.

lower-melting-point metals can cause embrittlement either in the liquid state (LMIE) or the solid state (SMIE). Intimate contact on an atomic scale between base metal and embrittler is necessary for these phenomena to occur. Another prerequisite for MIE is the presence of tensile stresses sufficient to cause microplastic deformation at crack tips. The onset of LMIE usually occurs at the melting temperature of the liquid embrittler. A trough in ductility then exists over a range of temperatures which varies in magnitude, depending on severity of embrittlement (Fig. 1).[1] Severity of the embrittlement, as measured by the height and width of the trough, is a function of the factors that influence LMIE. Fracture occurs by the nucleation of a single crack that usually propagates intergranularly at rapid rates (several m·s^{-1}).[2] Growth of the crack can be stopped by interrupting the supply of the liquid embrittler.

Material factors that influence LMIE include the vapor pressure, viscosity, and wetting tendency of the liquid embrittler, and the microstructure, strength level, and metallurgical integrity of the embrittled solid. For steels the type of matrix is an important factor; i.e., ferritic, austenitic, bainitic, and martensitic steels exhibit different susceptibilities to embrittlement. Temperature, strain rate, and stress state are the external or imposed factors that affect MIE.

An empirical understanding of the effects of these factors on MIE currently exists, but a clear understanding of the fundamental principles involved in the phenomena is lacking. Several mechanisms have been proposed for MIE. However, each mechanism adequately explains only a fraction of all of the observed parent metal/embrittler systems, with several exceptions found for each proposed mechanism.

This paper introduces the problem of lead embrittlement in steels, then briefly describes previously proposed mechanisms. Emphasis is placed

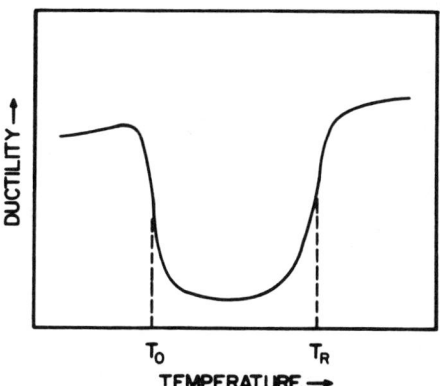

Fig. 1 Schematic of the characteristic ductility trough that illustrates the temperature dependence of liquid-metal-induced embrittlement, where T_O and T_R represent the onset and recovery temperatures, respectively.[1]

on the "adsorption-induced decohesion" model for MIE, which seems most adequate for describing embrittlement in the steel-lead system. Experimental results of lead embrittlement in steels are provided, followed by suggestions to avoid MIE in the steel-lead system.

Lead Induced Embrittlement in Steels

In many technological applications lead embrittles different types of steels. A summary of the available literature on lead embrittlement is presented in Ref. 2. Armco iron, and low alloy steels AISI 4145, 4335, and 4140 steels have been reported to be embrittled. The embrittling potency of lead is increased by additions of antimony, bismuth, copper, tin, or zinc.[2] Lead-lithium liquids have been shown to embrittle pressure vessel steels.[3,4] LMIE by lead is sensitive to both composition and metallurgical effects. For example, lead embrittlement of AISI 4145 steel was eliminated by cold work.[5] Embrittlement of both l2Cr-lMo and 2.25Cr-lMo pressure vessel steels by lead-lithium liquids was ameliorated by strength-reducing tempers.[3,4] In AISI 3340 steels segregation of phosphorus to grain boundaries reduced the susceptibility to LMIE by lead, whereas segregation of tin and antimony had the opposite effect.[6] Leaded steels are susceptible to LMIE by internal lead. For example, leaded AISI 4145 steel has been shown to be embrittled in the temperature range of 200-480°C (Ref. 7) (melting temperature of lead is 328°C). In such cases the compositions of the lead inclusions were shown to be very critical for determining the extent of embrittlement.[7] Lead embrittlement can occur in a variety of leaded steels, ranging from low to high carbon and plain carbon to high-alloy steels.[7]

Before we consider the specific nature of embrittlement by lead, let us review the mechanistic explanations that have been proposed for MIE.

Brittle-Ductile Fracture Criteria

Because LMIE is typically a brittle fracture phenomena, a convenient thinking tool that has been frequently invoked[8] to help explain MIE is the brittle-ductile fracture criteria:

$$\frac{\sigma_{app}}{\tau_{app}} > \frac{\sigma_c}{\tau_f} \quad \dashrightarrow \quad \text{Brittle Fracture} \qquad (1)$$

This inequality quantifies the observation that a crack propagates in a brittle manner only when the ratio of applied tensile stress, σ_{app} to applied shear stress, τ_{app}, is greater than the ratio of fracture stress, σ_c, to plastic flow stress τ_f.[9,10] When the ratio of applied tensile stress to applied shear stress is <u>less</u>

than the ratio of fracture stress to flow stress, plastic blunting of the crack tip prevents brittle crack extension, and a ductile rupture mode intervenes. Clearly, then, any externally imposed factor that reduces the fracture stress, σ_c, will promote brittle fracture. Also, any factor that increases the flow stress of the material, τ_f, will also decrease the ratio σ_c/τ_f and increase the MIE susceptibility.

The application of Eq. (1) is illustrated schematically in Fig. 2, where a temperature-insensitive σ_c line and two τ_f lines, which monotonically decrease with temperature, have been shown for a hypothetical metallic substrate. Each separate τ_f line represents a specific microstructural condition, and therefore strength, of the given material. A discontinuity in σ_c at a specific temperature, T_o, is indicated. In this illustration the temperature T_o can be taken as the embrittling metal melting temperature

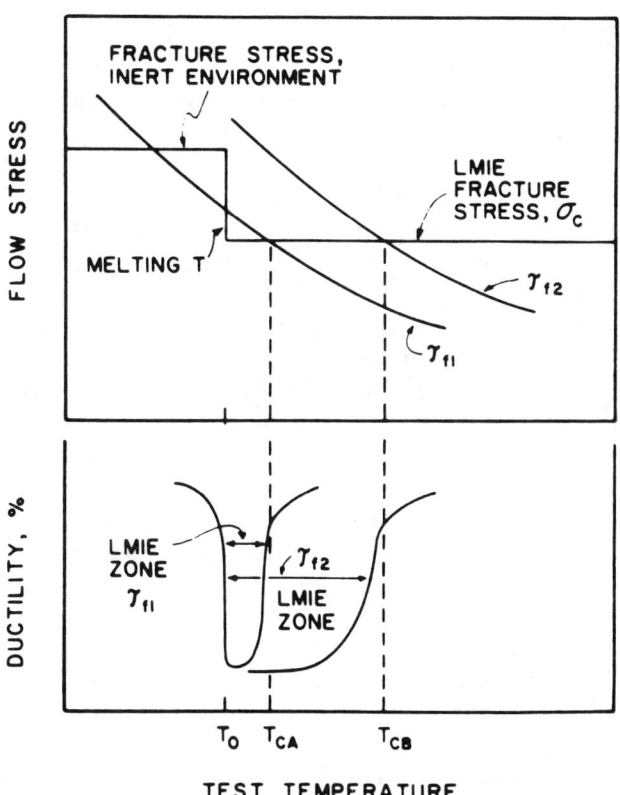

Fig. 2 Schematic illustration of the temperature dependence of both LMIE fracture stress and flow stress for a metal exposed to an embrittling metallic liquid. Note that the flow stress is assumed to be monotonically decreasing with increasing temperature. Corresponding ductilities are also shown.[9]

T_m. Figure 2 then suggests that, for all temperatures between T_o and T_c (the intersection of the σ_c line and either of the τ_f lines), LMIE is observed. Clearly, if the substrate material is strengthened (τ_{f2} curve vs the τ_{f1} curve), the LMIE zone is expanded (temperature T_{CB} vs temperature T_{CA}). Also clear is the fact that further reducing σ_c (lowering the LMIE σ_c line in Fig. 2) would expand the LMIE temperature range, regardless of the strength level of the substrate.

Mechanisms in MIE

It is helpful to consider separately the means by which the embrittling species is transported to the crack tip and then address the mechanism by which the embrittling atom affects the cracking susceptibility of the substrate.

Transport Mechanisms

One of the requirements for MIE is the transport of embrittler atoms to the advancing crack tip. Most MIE failures at high stress levels are instantaneous, therefore, transport times of 1 s. or less are required.[11] Based on this knowledge, predictions can be made concerning a viable transport mechanism for the embrittler. For LMIE this is either bulk liquid flow or a vapor transport mechanism. In the case of liquid lead the vapor pressure is too low to provide sufficiently rapid vapor transport rates. Therefore, bulk liquid flow must be the primary embrittling transport mechanism.[11] For SMIE in general, surface diffusion has been suggested as the operative transport mechanism.[11,12] This is true for SMIE of steels by lead.

Embrittling Mechanisms

Once the embrittler atoms reach the crack tip, the actual embrittlement has been hypothesized to occur by several different mechanisms. These include the following[5]:

 1) stress-assisted dissolution;
 2) enhanced plasticity at the crack tip;
 3) formation of a weakly bonded alloy zone ahead of the crack tip; and
 4) weakening of interatomic bonds by liquid metal at the crack tip.

In the "stress-assisted dissolution" model[13] a crack propagates because the highly stressed atoms at the crack tip are dissolved into and carried away by diffusion through the liquid embrittler, as illustrated schematically in Fig.

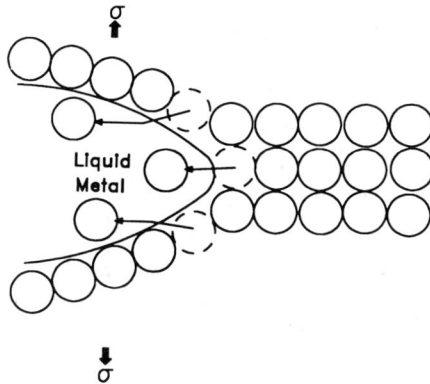

Fig. 3 Schematic illustration of the "stress-assisted dissolution" model. Crack propagation by the highly stressed atoms at the crack tip dissolve into and are carried away by diffusion through the liquid embrittler.[13]

3. This model suggests that the severity of embrittlement will increase with the solubility of the solid in the liquid. This has not been practically observed.[5] A similar mechanism was suggested for the Cu/BiPb melt system by Glikman and Goryunov.[14] They suggested that nucleation and subcritical growth of cracks takes place along the grain boundaries because of selective dissolution of copper atoms in these regions at the crack tip. A higher chemical potential at the grain boundaries contributed to the selective dissolution. The dissolution was proposed to be followed by rapid diffusion and subsequent precipitation on stress-free walls of the surface.[14] Glikman and Goryunov contend that the principal observation during LMIE is an increase in the rate of fracture initiation, not a decrease in fracture stress. Thus, the liquid enhances the subcritical crack growth rate, but plays no role in decreasing fracture toughness.[14]

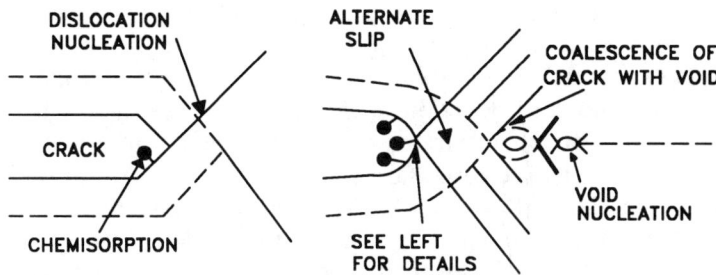

Fig. 4 Schematic illustration of Lynch's "enhanced plasticity" model. Adsorbed embrittler atoms decrease the stress for dislocation nucleation, leading to intense localized slip and microvoid nucleation.[16]

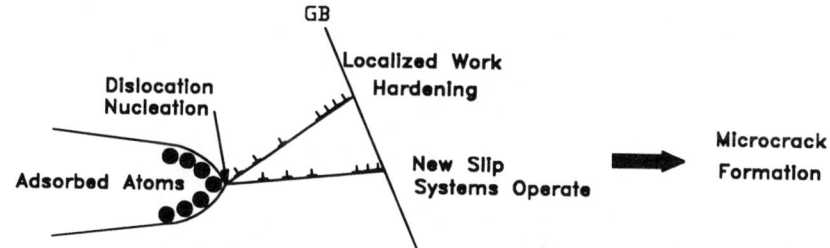

Fig. 5 Schematic illustration of Popovich's "enhanced plasticity" model. Adsorbed embrittler atoms decrease the stress for dislocation nucleation, leading to localized work hardening, operation of new slip systems, and finally, microcrack formation.[21]

The "enhanced plasticity at the crack tip" model suggests that adsorbed liquid metal atoms facilitate dislocation nucleation and movement.[15,16] Two similar but significantly different interpretations have been given for the enhanced plasticity model. Lynch[17-19] proposed that absorbed embrittler atoms decreased the stress necessary to nucleate dislocations, and the pileup stresses caused fine microvoid nucleation ahead of the crack tip. His explanation is one of intense localized slip, illustrated in Fig. 4, and is very similar to what others[20] have suggested for hydrogen embrittlement. Popvich[16] claims that adsorption of embrittler atoms causes more than just dislocation nucleation. He proposes that reduction in the liquid/solid interfacial energy, caused by adsorption of the embrittling species with no significant solution formation or diffusion of atoms, results in ease of dislocation outlet to the surface, operation of new slip systems, and localized work hardening. Strengthening of the surface layers promotes the formation of microcracks.[16] Surface cracks propagate through a hardened surface layer into a ductile matrix and get blunted. The new surfaces opened up by the crack are coated by adsorbed atoms of the embrittling medium, and the crack propagates. Popovich's concept is depicted in Fig. 5 (Ref. 21).

Weakening of interatomic bonds by adsorbed liquid metal atoms at the crack tip is a widely accepted conceptual model for LMIE. The model, which does not permit a quantitative assessment of embrittlement, has been given two different interpretations.[5] One is the fracture of a weakly bonded alloyed region, formed in a limited volume at the crack tip, as illustrated in Fig. 6. The second is a reduction in the fracture stress simply because of atomic interaction of the substrate with the adsorbed embrittler atoms, as shown in Fig. 7. This interpretation has been called "adsorption-induced reduction in cohesion.[22] Reduction of fracture stress can also be interpreted as a reduction in the solid-liquid interfacial energy. Most aspects of LMIE can be understood on the basis of this model.

Kelly and Stoloff[23] predicted embrittling tendencies based on thermodynamic calculations of interaction energies between the liquid embrittler atom and the solid embrittled atom and the corresponding

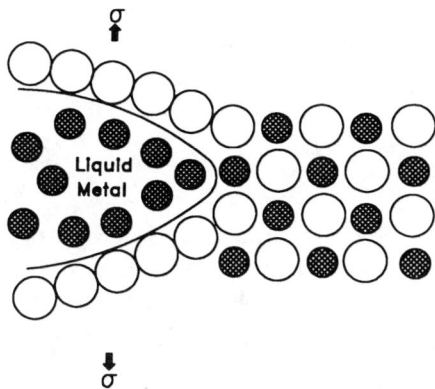

Fig. 6 Schematic illustration of the "formation of a weakly bonded alloy zone" model. An alloyed zone forms in a limited volume at the crack tip, thus decreasing the required stress for crack propagation.[5]

reductions in fracture surface energy. Their approach gives a thermodynamic tendency for embrittlement in a given system. However, even when there is a tendency for LMIE to occur, the kinetics of embrittler atom-induced crack growth is important.[14] Macroscopic manifestations of LMIE occur only over a certain range of experimental conditions, within which the subcritical crack growth rate is fast enough to compete with other routes of failure, such as necking, due to plastic instability in tension tests.[14]

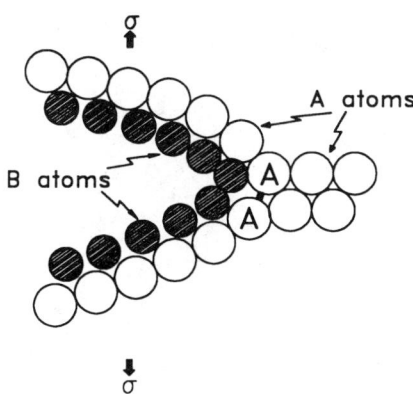

Fig. 7 Schematic illustration of the "weakening of interatomic bonds by liquid metal," ("adsorption-induced reduction in cohesion") model. The atoms of the substrate interact with the adsorbed embrittler atoms, thus reducing the fracture stress.[22]

The mechanisms discussed thus far suggest that LMIE results from one of the following potential phenomena:

1) The presence of a liquid metal causes selective dissolution, leading to formation of cracks at grain boundaries.

2) The adsorption of embrittler atoms causes enhanced plasticity. The result may be either intense and localized slip, which nucleates microvoids at the crack tip, or work-hardened surface layers, which then crack.

3) The formation, by alloying, of a weakly bonded volume of metal at the crack tip reduces the fracture stress.

4) The adsorption of liquid atoms onto the solid surface at the crack tip promotes brittle failure because electronic interaction weakens atom bonds or reduces the surface energy.

Adsorption-Induced Reduction in Cohesion

Druschitz and Gordon[24] consider MIE as occurring in three stages. The first stage is the embrittler-induced crack initiation, which is characterized by an incubation period. This is followed by an embrittler-controlled crack propagation stage and a final overload stage. This discussion will first consider the aspects of crack initiation, and then return to the more thoroughly studied phenomenon of crack propagation.

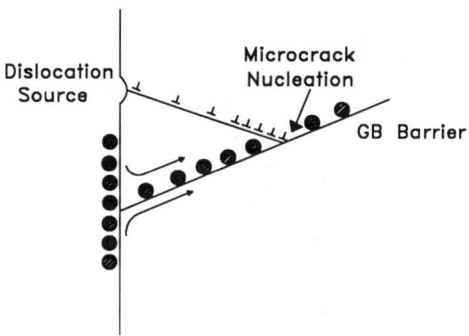

Fig. 8 Schematic illustration of a model for LMIE crack initiation. Embrittler atoms diffuse into the parent metal grain boundary. Once a sufficient number of embrittler atoms is attained, microcracks are nucleated in accordance with Zener's microcrack nucleation model.[26]

Crack Initiation

Most MIE failures initiate intergranularly in polycrystalline material[24], and a well-defined incubation time for initiation has been observed.[25] Because of the incubation period for LMIE crack initiation, delayed failure under constant load has also been observed. The delay time in failure was found to be both stress and temperature dependent.[25] Gordon and An[25] have proposed that the actual crack nucleation event is not the rate-controlling step in crack initiation. They suggest that, during the incubation period, a preparation process for crack nucleation takes place. The preparation process involves the penetration of embrittler atoms into the grain boundaries of the parent metal, driven by stress-assisted grain boundary diffusion. The embrittler atoms go from the adsorbed state to the dissolved state in the surface, and subsequently diffuse into the parent metal grain boundaries. The process is depicted in Fig. 8. The incubation times associated with this process depend on the activation energies for localized incorporation and grain boundary diffusion and thus are stress and temperature dependent.[25] The presence of the embrittler atoms was assumed to lower the crack resistance in the parent metal. Once a sufficient concentration of embrittler atoms was attained in one of the penetration zones, a crack was assumed to nucleate at the head of a dislocation pile-up, in accordance with the Zener[26] concept of microcrack nucleation.

Crack Propagation

For the embrittlement of steels by lead, the "adsorption-induced reduction in cohesion" model of crack propagation, depicted in Fig. 7, has been proposed by Westwood and Kamdar[22] and Kamdar[27]. In this model adsorbed embrittler atoms at the crack tip reduce the principal stress required to break the bonds between atoms in the parent metal at the crack tip. Consider an elastic crack that propagates by the breaking of atomic bonds (A-A bonds in Fig. 7). The potential energy of the system as a function of the separation distance is shown in Fig. 9. The corresponding stress required at the crack tip to break the bonds is obtained by differentiating the potential energy function with respect to distance of separation. The plot of stress vs distance is also given in Fig. 9. Now consider an embrittler atom (B) arriving at the crack tip. If the B atom can cause a reduction in density of the cohesive electron cloud between A atoms, then both the potential energy and stress functions are altered as shown in Fig. 9. The stress required to break the bonds is reduced from σ to σ_B. If the critical stress necessary for crack propagation is calculated on the basis of the rupture of atomic bonds at the crack tip, the presence of the embrittler atoms causes the critical stress to be reached at a lower applied stress. Stoloff and Johnston[28] originally proposed a similar mechanism for crack propagation in a liquid metal environment. Since the embrittler atom

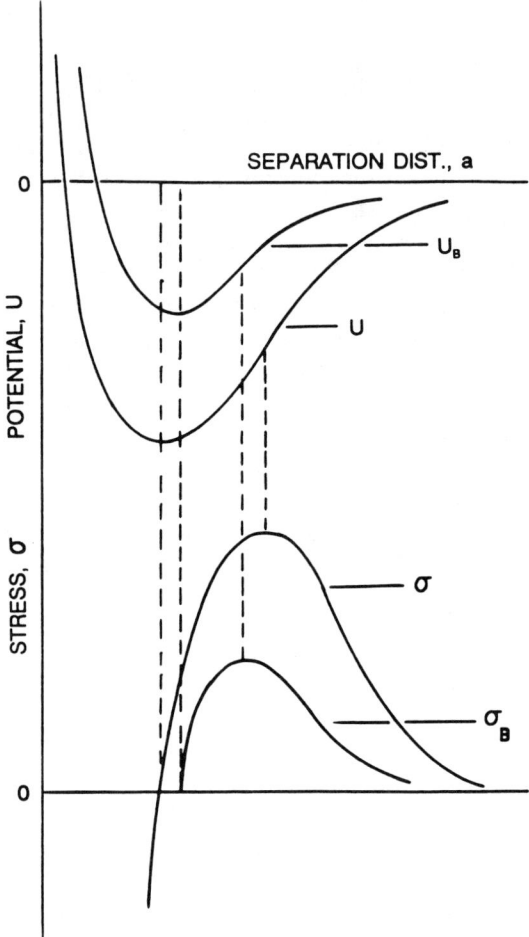

Fig. 9 Potential energy, U and U_B, and the resulting maximum stress, σ and σ_B, vs separation distance for bonds of the type A-A, both in the absence and in the presence of an embrittling chemisorbed atom B.[22]

has to interact with the electron cloud surrounding the bonds in the parent metal, a chemical interaction is required; i.e., atom B has to chemisorb onto the surface. Such chemisorption can lead to A_xB complex formation and weakening of A-A bonds.[22] The mechanism corresponds to a decreased fracture stress [σ_c in Eq. (1)] in the zone ahead of the crack tip.

An alternative and equally useful interpretation of the decohesion model for crack propagation focuses on the reduction in the free energy of the fracture surfaces in the presence of the embrittler.[5] The classic Griffith criteria[29] for fracture equates the fracture stress σ_c [see Eq. (1)] to the free energy of the fracture surface in the following way:

$$\sigma_c = \alpha \sqrt{\frac{2(\gamma+U)E}{a}} \quad (2)$$

In this expression a is a proportionality constant dependent on crack geometry, γ is the fracture surface energy, U quantifies any additional energy/unit area absorbed during crack propagation, E is the elastic modulus, and a is the half-crack length. The term γ for an intergranular failure is given by[5]

$$\gamma = 0.5(2\gamma_{sv} - \gamma_{gb}) \quad (3)$$

where γ_{sv} and γ_{gb} are surface and grain boundary energies, respectively. In the presence of the embrittler this term is modified to

$$\gamma = 0.5(2\gamma_{sl} - \gamma_{gb}) \quad (4)$$

where γ_{sl} is the solid-liquid interfacial energy for either a transgranular or an intergranular failure. If γ_{sl} is less than γ_{sv}, crack propagation is promoted. This is the case for a wetting liquid. Wetting is a prerequisite for LMIE, for intimate atomic contact between the embrittler and the parent metal must be established. Wetting may also promote chemical interaction. For example, Eberhard et al.[30] found that, for short times, 2.25Cr-1Mo steel was not embrittled by pure liquid lead, since it did not wet the steel. However, prior exposure of the steel surface to a fluxing agent such as $ZnCl_2$ promoted wetting by lead and caused embrittlement.

Temperature Dependence of Crack Propagation

Chemisorption can be considered as strain activated in the vicinity of the crack tip.[22] Because the chemisorption is thermally activated, the embrittling phenomena should be a thermally activated process as well. Preece and Westwood[31] have suggested that the adsorption coverage of the embrittler atoms is a function of the free energy decrease resulting from embrittler-parent metal atomic interaction. If the free energy change is ΔG, then the fractional surface coverage by adsorbed atoms, θ, is given by[31]

$$\theta = [1 - \exp(-\Delta G/kT)] \quad (5)$$

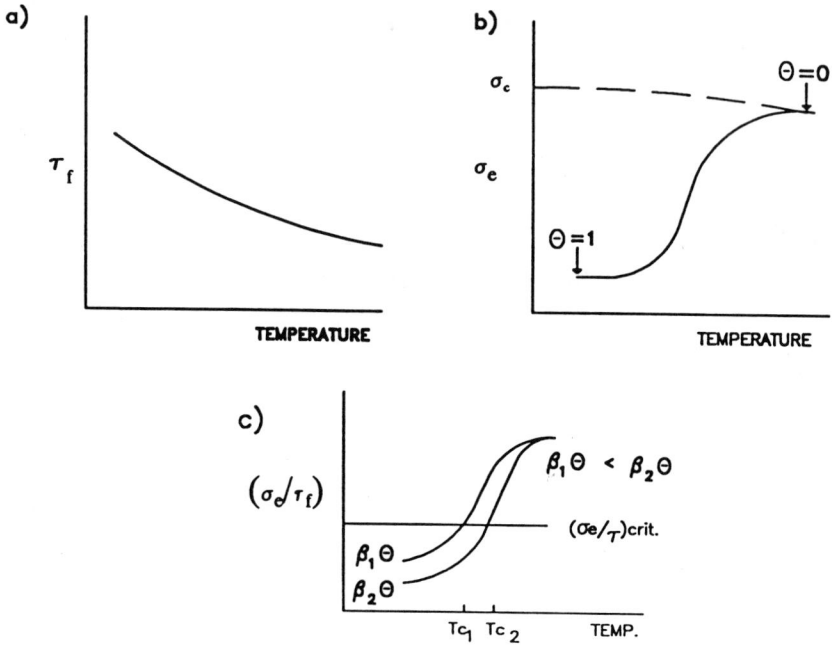

Fig. 10 Schematic illustration of the possible variations with temperature of a) τ_f, the flow stress, b) σ_e, the fracture stress, and c) the ratio, σ_e/τ_f (Ref. 31).

Equation (5) expresses the exponential decrease in chemisorbed atoms at the crack tip as the temperature increases. If the adsorbed embrittler atoms weaken the atomic bonds of the substrate at the crack tip, then the fracture stress σ_e, will increase as θ decreases, or the fracture stress should increase with increasing temperature:

$$\sigma_e = \sigma_c (1-\beta\theta) \tag{6}$$

where σ_c is the unembrittled fracture stress, β is a constant for the particular system, and θ was defined earlier.

Preece and Westwood[31] have used the temperature dependence of σ_e suggested in Eq. (6) to explain observations concerning increased concentrations of embrittler atoms in binary liquids such as mercury-gallium in contact with solid aluminum. Consider the flow stress of the solid to decrease uniformly with increasing temperature as shown in Fig. 10a. Equations (5) and (6) suggest that the fracture stress σ_e will increase with increasing temperature, as the concentration of adsorbed embrittler atoms is reduced at the crack tip by thermodynamic disordering (see Fig. 10b). The ratio σ_e/τ must likewise increase with temperature (Fig. 10c), so that a

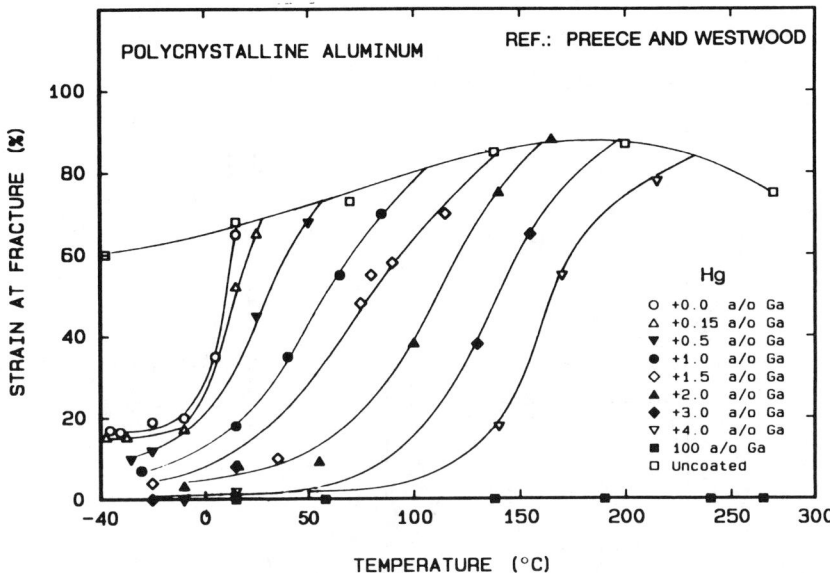

Fig. 11 Influence of mercury, gallium, or mercury-gallium environments on the strain at fracture of aluminum (grain diam ~1 mm) as a function of temperature.[31]

temperature T_c will be reached, above which no LMIE of the material is observed. If the concentration of embrittling atoms is increased in the liquid metal, the fraction of adsorbed embrittler atoms at a given temperature should increase, thereby decreasing σ_e or σ_e/τ at that temperature. Higher embrittler concentrations should therefore require that a higher temperature be reached before ductility is restored, as indicated in Fig. 10c. The experimental data of Preece and Westwood[31] for the embrittlement of aluminum by mercury-gallium liquids are shown in Fig. 11. The increased range of embrittling temperature with increasing concentration of embrittling gallium is qualitatively predicted by their proposed temperature dependence of the fracture stress.

Edwards et al.[9] studied the LMIE of pressure vessel steels by lithium and lead-lithium liquids and suggested an entirely different type of temperature dependence. These studies concentrated on different strength levels of the same steel composition exposed to a constant composition of embrittling liquid (refer back to Fig. 2). The stronger the steel, the wider the temperature range of embrittlement, as shown in Fig. 2.

The behavior illustrated in Fig. 2 did not explain their observation that the maximum embrittlement of 2.25Cr-1Mo steel by lead-lithium liquids occurred at temperatures greater than the embrittling liquid melting temperature.[32] A logical interpretation of this phenomena is shown in Fig.

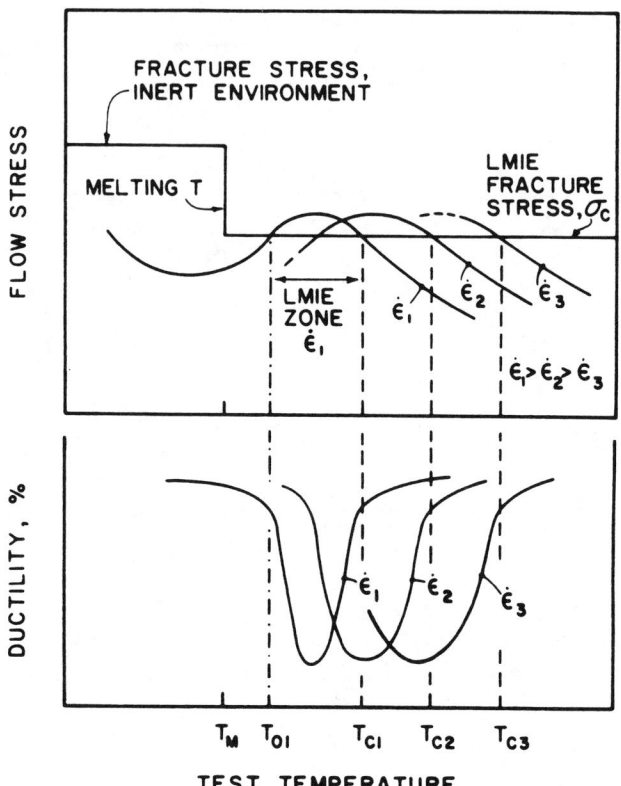

Fig. 12 Schematic illustration of the temperature dependence of both LMIE fracture stress and ductility for a metal exposed to an embrittling metallic liquid. Note that the flow stress is assumed to maximize at a temperature above the melting temperature of the embrittling liquid. Corresponding ductilities are also shown.[9]

12, where τ_f is assumed to exhibit a maximum. Flow stress curves for three different strain rates are illustrated. Specific metallurgical phenomena that create flow stress maxima in hot tensile testing, such as dynamic strain aging and secondary hardening, are relatively common and do occur during hot deformation of 2.25Cr-1Mo steel. A flow stress maximum, occurring at a temperature above the melting point of the embrittling liquid, could result in LMIE, which begins at a temperature well above the embrittling liquid melting temperature, as indicated in Fig. 12. Furthermore, material strengthening or increased strain rate, both factors that increase the flow stress, should move the entire LMIE regime to higher embrittling temperatures as shown.

Specific Results: Steel LMIE by Lead

Many studies[6] could be reviewed that consider LMIE of steels by lead. This review will be kept brief and limited to examples that, in the view of the authors, teach important concepts concerning control of LMIE in this system.

Effect of Impurities

Common impurities in lead include tin, arsenic, zinc, antimony, and bismuth. Trace additions of antimony, tin, zinc, or arsenic to the liquid lead have been reported[6,33] to increase the LMIE of steel by lead. This effect is exemplified by the data of Breyer and Johnson[34] as reported by Old[35] and shown in Fig. 13. The common trace elements listed earlier affect the LMIE of steel as suggested by Eqs. (5) and (6) and as depicted in Figs. 10 and 11. These more aggressive embrittlers must be eliminated from molten lead in contact with steel whenever doing so is technically feasible.

Common trace impurities in steel include antimony, tin, phosphorus, and arsenic. Dinda and Warke[6] studied the effect of grain boundary segregation of these trace elements on the LMIE by lead and tin. Segregation of tin or antimony caused increased LMIE susceptibility, whereas segregation of phosphorus or arsenic caused a slight decrease. These data are exemplified by Fig. 14, which shows the effect of antimony in the steel on the LMIE susceptibility. The effects were explained on the basis of whether the segregated element should or should not facilitate the adsorption of lead. Steels subjected to molten lead should contain low concentrations of such trace elements.

Metallurgical Factors

The strength level of the steel exposed to molten lead is perhaps the most significant concern. This point is exemplified by the data of Edwards

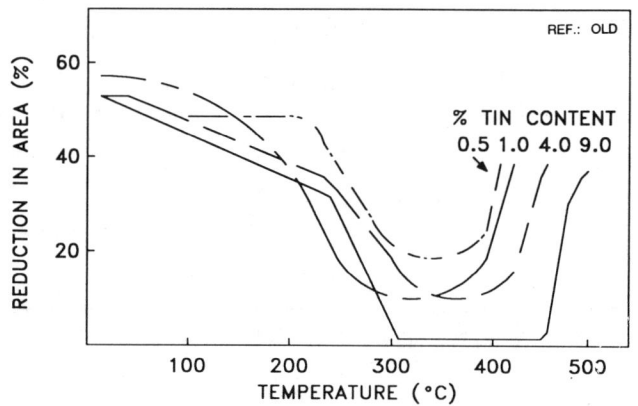

Fig. 13 Embrittlement of AISI 4145 steel by lead-tin alloys.[35]

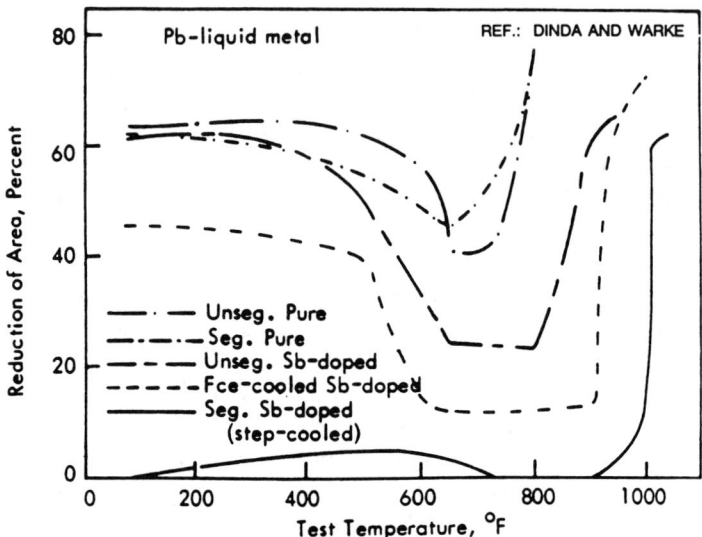

Fig. 14 LMIE susceptibility of both "pure" steel and antimony-doped steel subjected to the heat treatments specified.[6]

Table 1 Tempering and LMIE susceptibility heat treatments for 2.25Cr-1Mo steel[36]

Heat Treatment	Condition	Comments
HT-1	Austenitized at 1300°C (10 min), oil quenched	Simulated as-welded HAZ
HT-2	HT-1, plus 690°C temper for 25 h, oil quenched	Represents a normally extensive PWHT* for 2.25Cr-1Mo Steel
HT-3	HT-1, plus 690°C temper for 1.6 h, oil quenched	Represents a minimal PWHT* Condition

*PWHT = Post Weld Heat Treatment

Table 2 Tempering and LMIE susceptibility heat treatments for HT-9 steel[36]

Heat Treatment	Condition	Comments
HT-1	Austenitized at 1300°C (3 min), oil quenched	Simulated as-welded HAZ
HT-2	HT-1, plus 760°C temper for 2.5 h, oil quenched	Represents a normally extensive PWHT for HT-9 steel
HT-3	HT-1, plus 600°C temper for 1.0 h, oil quenched	Represents a minimal PWHT condition

et al.[36] who studied the effects of tempering 2.25Cr-1Mo and 12Cr-1Mo (HT-9) pressure vessel steels prior to exposure to a 17%Li-83%Pb liquid. Each steel was heat treated to simulate

1) an as-welded heat affected zone of a weld;
2) a minimally code-acceptable post weld heat treatment of the weld in point 1; and
3) an extensive postweld heat treatment of the weld in point 1.

Details of the heat treatments are given in Tables 1 and 2. The measured tensile ductilities for each steel, heat treated according to points 1, 2, and 3, are shown in Figs. 15, 16, and 17, respectively. A minimal postweld heat treatment was found to be insufficient to remove LMIE susceptibility under the imposed test condition. The extensive postweld heat treatment greatly reduced the LMIE susceptibility, but did not completely remove it, as shown by fractography and reduction of area measurements.[36]

The dynamic flow stress characteristic of the steel and not just the strength level may be important. Reference has already been made to observations in the LMIE of 2.25Cr-1Mo steel by lead-lithium liquids[9] which suggested the presence of maxima in the steels flow stress that depended on deformation rate and temperature. Edwards et al.[9] found that the LMIE temperature range for 2.25Cr-1Mo steel heat treated to several different strength levels and exposed to a 1%Pb-99%Li liquid correlated extremely well with the temperature of maximum dynamic strain aging at a fixed strain rate.[9] Specific heat treatments are given in Table 3. These data are shown in Fig. 18, where the amplitudes of the dynamic strain aging serrations and ductilities are both plotted against test temperature.

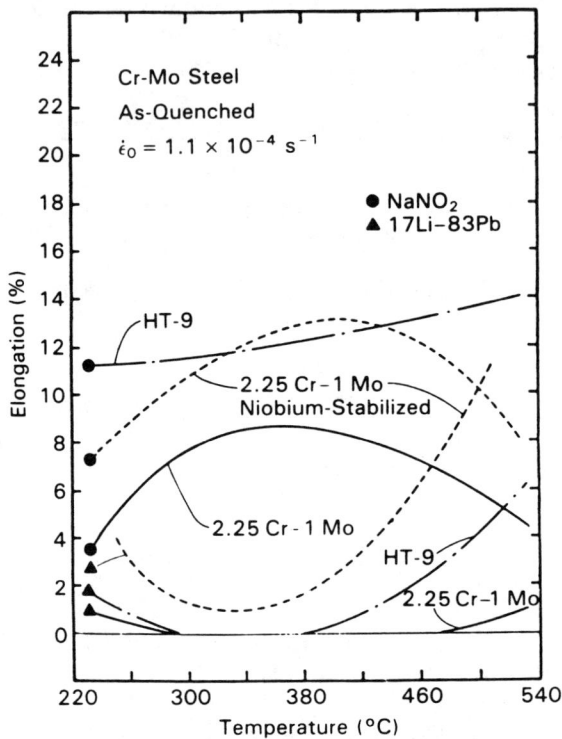

Fig. 15 Elongation at fracture vs test temperature for the Cr-Mo steels given a simulated heat affected zone (HAZ) heat treatment (HT-1).[36]

Spencer[37] previously established a critical value of the Zener-Holloman parameter:

$$\tau = g(Z) = g(\dot{\epsilon} \exp\Delta H/RT) \tag{7}$$

to track the maximum crack propagation during LMIE-enhanced fatigue of 2.25Cr-1Mo steel by lithium. Use of this critical value ($Z = 7.6 \times 10^5 s^{-1}$), even when extrapolated to the much slower strain rate ($\dot{\epsilon}_o = 1.1 \times 10^{-4} s^{-1}$) for the tensile testing shown in Fig. 18, accurately predicted the temperature of maximum embrittlement ($T_{calc} = 279°C$ $T_{meas} \approx 250°C$; see Fig. 18).

Grain size can be a complicated factor in LMIE. Grain size strengthening, as expressed by the Hall-Petch relation ($\tau \propto d^{-1/2}$), would ordinarily be expected to enhance the LMIE susceptibility by decreasing σ_c/τ. However, microcrack nucleation, and hence the fracture stress, is also known to depend on grain size in a similar way:

$$\sigma_c \propto d^{-1/2} \tag{8}$$

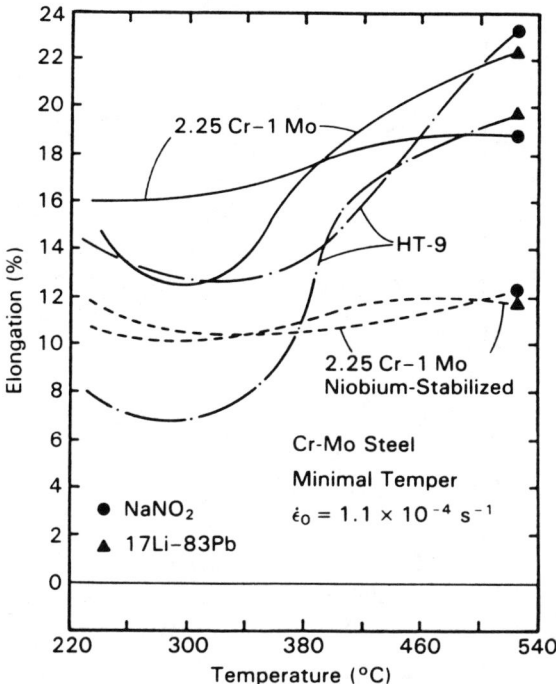

Fig. 16 Elongation at fracture vs test temperature for Cr-Mo steels given a minimal PWHT (HT-3).[36]

The slip dispersal resulting from grain size refinement is generally more effective in increasing the fracture stress σ_c than in increasing the flow stress τ. Consequently, grain refinement is usually considered beneficial in reducing LMIE susceptibility.[2]

Summary and Recommendations

Lead is a known embrittler for ferrous alloys, and steel structures exposed to lead near its melting point can be expected to manifest this LMIE susceptibility. The phenomenon is most conveniently viewed as a form of brittle fracture, so that all of the factors used in design to reduce brittle fracture susceptibility are appropriately applied. Strength levels must be kept low, and design factors should be conservative. Discontinuities that create high localized tensile stresses should be avoided, and cyclic loads should be avoided or minimized.

Other trace elements common to lead are, in general, more aggressive than lead itself, and less LMIE of steel will be experienced if the

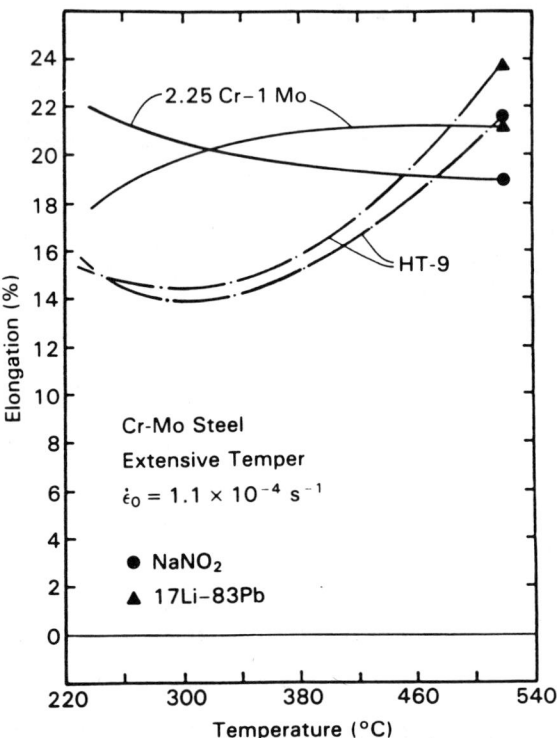

Fig. 17 Elongation at fracture vs test temperature for the Cr-Mo steels given an extensive PWHT (HT-2).[36]

Table 3 Dynamic strain aging and LMIE susceptibility

Heat treatment	Condition	Comments
HT-1	Austenitized at 1300°C (20 min), oil quenched	Simulated as-welded HAZ
HT-2	HT-1, plus 740°C temper for 10 h	Represents extensive PWHT at maximum allowable temperature
HT-3	HT-1, plus 695°C temper for 15.7 h	Represents a minimal PWHT condition
HT-4	HT-1, plus 695°C temper for 121 h	Represents extensive PWHT at minimum allowable temperature

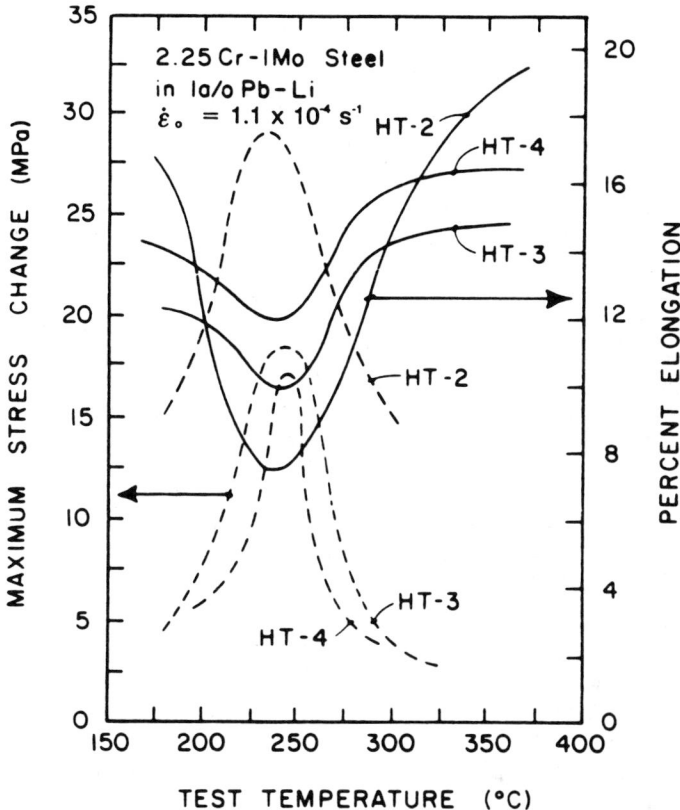

Fig. 18 Ductilities observed for three heat treatments of 2.25Cr-1Mo steel. Specimens were fractured in tension while submerged in a 1 atomic percent Pb, 99 atomic percent Li liquid at various temperatures. Also plotted are the amplitudes of the dynamic strain aging serrations observed during the tensile tests.[9]

molten lead is kept pure. Metallurgical phenomena such as secondary hardening or dynamic strain aging can accentuate LMIE tendencies, and a good knowledge of the dynamic deformation characteristics of the steel can be important. Trace impurities common to steel, such as tin and antimony, should be minimized, and in general, those metallurgical factors known to promote good toughness, such as low inclusion content, fine grain size, and uniform microstructure, should be optimized.

The phenomenon of LMIE is not sufficiently understood so that engineering schemes to avoid embrittlement can be applied with confidence. Current engineering structures must be constructed in order to minimize the impact of embrittlement, and diligent inspection and maintenance will be necessary for utilization of steel in contact with molten lead.

References

[1] Wilkinson, B. D., "The Effect of Lead Concentration on the Corrosion Susceptibility of 2.25Cr-1Mo Steel in Lead-Lithium Solutions," MS Thesis, No. T-2442, Colorado School of Mines, Golden, CO, July 1982.

[2] Nicholas, M. G., "A Survey of Liquid Metal Embrittlement of Metals and Alloys," Proceedings of the Symposium on Embrittlement by Liquid and Solid Metals, edited by M. H. Kamdar, Metallurgical Society of American Institute of Mining and Metallurgical Engineers, 1984, pp. 27-50.

[3] Jones, K. A., "The Liquid Metal Induced Embrittlement Susceptibility of 2.25Cr-1Mo Steel Subjected to an 83Pb-17Li Liquid," MS Thesis, No. T-3105, Colorado School of Mines, Golden, CO, Aug. 1985.

[4] Halvorson, S., "The Liquid Metal Induced Embrittlement Susceptibility of 12Cr-1Mo Steel Subjected to Lithium and Lead-Lithium Liquids," MS Thesis, No. T-3167, Colorado School of Mines, Golden, CO, April 1986.

[5] Nicholas, M. G., and Old, C. F., "Review-Liquid Metal Embrittlement," Journal of Materials Science, Vol. 14, Jan. 1979, pp. 1-18.

[6] Dinda, K., and Warke, W. R., "The Effect of Grain Boundary Segregation on Liquid Metal Induced Embrittlement of Steel," Materials Science and Engineering, Vol. 24, Aug. 1976, pp. 199-208.

[7] Warke, W. R., and Breyer, N. N., "Effect of Steel Composition on Lead Embrittlement," Journal of Iron and Steel Institute, Vol. 209, Oct. 1971, pp. 779-784.

[8] Kelly, A., Tyson, W. R., and Cottrell, A. H., "Ductile and Brittle Crystals," Philosophical Magazine, Vol. 15, Mar. 1967, p. 567.

[9] Edwards, G. R., Matlock, D. K., and Eberhard, B. A., "Embrittlement of 2.25Cr-1Mo Steel by Lithium and a Lead-Lithium Liquid," Fusion Technology, Vol. 8, July 1985, pp. 937-943.

[10] Kamdar, M. H., Embrittlement by Liquid Metals, Pergamon, Oxford, UK, 1973.

[11] Gordon, P., "Metal-Induced Embrittlement of Metals-An Evaluation of Embrittler Transport Mechanisms," Metallurgical Transactions A, Vol. 9A, Feb. 1978, pp. 267-273.

[12] Lynn, J. C., Warke, W. R., and Gordon, P., "Solid Metal-Induced Embrittlement of Steel," Materials Science and Engineering, Vol. 18, Jan. 1975, pp. 51-62.

[13] Robertson, W. M., "Propogation of a Crack Filled with Liquid Metal," Metallurgical Society of AIME Transactions, Vol. 236, Oct. 1966, pp. 1478-1482.

[14] Glikman, E. E., and Goryunov, Y. V., "Mechanism of Embrittlement by Liquid Metals and Other Manifestations of the Rebinder Effect," Fiziko-Khimicheskaya Mekhanika Materialov, Vol. 14, No. 4, July-Aug. 1978, pp. 20-30.

[15] Lynch, S. P., "A Fractrographic Study of Gaseous Hydrogen Embrittlement and Liquid-Metal Embrittlement in a Tempered-Martensitic Steel," Acta Metallurgica, Vol. 32, Jan. 1984, pp. 79-90.

[16] Popovich, V. V., "Mechanisms of Liquid-Metal Embrittlement," Fiziko-Khimicheskaya Mekhanika Materialov, Vol. 15, No. 5, Sept.-Oct. 1979, pp. 11-20.

[17]Lynch, S. P., Proceedings of the 34th International Conference on Hydrogen Effects Met., edited by J. M. Bernstein and A. W. Thompson, American Institute of Mining and Metallurgical Engineers, 1981, pp. 863-871.

[18]Lynch, S. P., "Liquid-Metal Embrittlement in an Al 6%Zn 3%Mg Alloy," Acta Metallurgica, Vol. 29, No. 2, Feb. 1981, pp. 325-340.

[19]Lynch, S. P., Proceedings of the 4th International Conference on Fracture, Vol. 2, University of Waterloo Press, Waterloo, Ontario, 1977, p. 859.

[20]Birnbaum, H. K., "Hydrogen Related Failure Mechanisms in Metals," Proceedings of the Symposium on Environment-Sensitive Fracture of Engineering Materials, edited by Z. A. Foroulis, American Institute of Mining and Metallurgical Engineers, 1979, pp. 326-360.

[21]Pickens, J. R., and Westwood, A. R. C., "Recent Soviet Contributions to the Understanding and Application of Liquid Metal Embrittlement Phenomena," Proceedings of the Symposium on Embrittlement by Liquid and Solid Metals, edited by M. H. Kamdar, Metallurgical Society of American Institute of Mining and Metallurgical Engineers, 1984, pp. 51-64.

[22]Westwood, A. R. C., Kamdar, M. H., "Concerning Liquid Metal Embrittlement, Particularly of Zinc Monocrystals by Mercury," Philosophical Magazine, Vol. 8, May 1963, pp. 787-804.

[23]Kelly, M. J., and Stoloff, N. S., "Analysis of Liquid Metal Embrittlement from a Bond Energy Viewpoint," Metallurgical Transactions A, Vol. 6A, Jan. 1975, pp. 159-166.

[24]Druschitz, A. P., and Gordon, P., "Solid Metal-Induced Embrittlement of Metals," Proceedings of the Symposium on Embrittlement by Liquid and Solid Metals, edited by M.H. Kamdar, Metallurgical Society of American Institute of Mining and Metallurgical Engineers, 1984, pp. 285-316.

[25]Gordon, P., and An, H. H., "Mechanisms of Crack Initiation and Crack Propagation in Metal-Induced Embrittlement of Metals," Metallurgical Transactions A, Vol. 13A, March 1982, pp. 457-472.

[26]Zener, C., Fracturing of Metals, American Society of Metals, Metals Park, OH, 1943, p. 3.

[27]Kamdar, M. H., "Embrittlement of 4340 Type Steel by Liquid Lead and Antimony and Lead-Antimony Solutions," Proceedings of the Symposium on Embrittlement by Liquid and Solid Metals, edited by M.H. Kamdar, Metallurgical Society of American Institute of Mining and Metallurgical Engineers, 1984, pp. 149-159.

[28]Stoloff, N. S., and Johnston, T. L., "Crack Propogation in a Liquid Metal Environment," Acta Metallurgica, Vol. 11, March 1963, pp. 251-256.

[29]Griffith, A. A., "The Phenomena of Rupture and Flow in Solids," Philosophical Transactions of the Royal Society of London, Vol. 221A, Oct. 1921, pp. 163-198.

[30]Eberhard, B. A., Mullinaux, B. A., and Edwards, G. R., "The Susceptibility of 2.25Cr-1Mo Steel to Liquid Metal Embrittlement by Lithium-Lead Solutions," Liquid Metal Engineering and Technology, Vol. 1, British Nuclear Energy Society, London, 1984, pp. 337-345.

[31]Preece, C. M., and Westwood, A. R. C., "Temperature-Sensitive Embrittlement of FCC Metals by Liquid Metal Solutions," Transactions of ASM, Vol. 62, No. 2, June 1969, pp. 418-425.

[32] Eberhard, B. A., Mullinaux, B. A., and Edwards, G. R., "The Susceptibility of 2 1/4Cr-1Mo Steel to Liquid Metal Embrittlement by Lithium-Lead Solutions," Liquid Metal Engineering and Technology, British Nuclear Energy Society, London, 1984, p. 337.

[33] Warke, W. R., Johnson, K. L., and Breyer, N. M., Corrosion by Liquid Metals, Plenum, New York, 1970, p. 417.

[34] Breyer, N. N., and Johnson, K. L., Journal of Testing Evaluation, Vol. 2, 1974, p. 471.

[35] Old, C. F., "Liquid Metal Embrittlement of Nuclear Materials," Journal of Nuclear Materials, Vol. 92, 1980, pp. 2-25.

[36] Edwards, G. R., Jones, K. A., and Halvorson, S. F., "Tempering of 2.25Cr-1Mo Steel and HT-9 Steel to Reduce Liquid-Metal-Induced Embrittlement Susceptibility in 17Li-83Pb Liquid," Fusion Technology, Vol. 10, Sept. 1986, pp. 243-252.

[37] Spencer, R. E., Matlock, D. K., and Olson, D. L., "The Effects of Liquid Metal Embrittlement on High Temperature Fatigue of 2.25Cr-1Mo Steel in Liquid Lithium," Journal of Material Energy Systems, Vol. 4, No. 4, March 1983, p. 187.

Materials Compatibility of Mercury for Practical Applications at Elevated Temperatures

Noel W. Kirshenbaum*

Placer Dome U.S. Inc., San Francisco, California 94111

Abstract

The compatibility of mercury with structural or containment materials at elevated temperatures is assessed by review of the literature on heat pipes and mercury topping cycles for electric power generation. The dissolution of iron in mercury at elevated temperatures is inhibited by minute additions of titanium to the mercury; mercury boiler tubes operated up to temperatures ranging from $1050°$ to $1180°$ F ($565°$ to $635°$ C) in long term utility applications. Enhanced wettability of mercury is achieved by addition of magnesium. Concentrations of Ti and Mg used are about 0.001 and 0.01%, respectively. Mercury may thereby be considered as a suitable working fluid for use in liquid metal magnetohydrodynamics (MHD) systems, even at elevated temperatures.

Introduction

Liquid metal magnetohydrodynamic (LMMHD) energy-conversion systems can utilize various liquid metals and alloys for their electrodynamic working fluids. In many cases the selection of the specific liquid metal will be determined by the desirable temperature range of operation and by materials compatibility considerations.

In the research and development of LMMHD systems, mercury is often used to simulate--or to be a surrogate for--metals having considerable higher melting points.[1-3] But it is with less frequency that mercury or mercury alloys are used or considered for the actual working fluid in operation, even when conditions are appropriate for their use in terms of the temperature and resulting vapor

---

Copyright © 1991 by Noel W. Kirschenbaum. Published by the American Institute of Aeronautics and Astronautics, Inc., with permission.
*Manager, Mineral Projects Development.

pressure. Mercury may be overlooked for use in the ultimate applications because of concerns regarding wettability and corrosion, especially at elevated temperatures. However, review of the literature shows that these two concerns were addressed years ago in the process of solving problems in designing and building electric generating plants using mercury as a thermodynamic working fluid.

It is well known that mercury forms amalgams or low melting eutectics with such metals as tin, copper, lead, and aluminum (once its protective oxide film is penetrated).[4] However, the logical and preferred materials for containing mercury in operating systems are ferrous metals, and the fact that the solubility of iron in mercury is very low suggests that mercury could be satisfactorily used with steels even at elevated temperatures. For example, the solubility of mercury in iron at room temperature is cited at 0.15×10^{-5} wt %, increasing to 1.1×10^{-5} at $400°C$.[5]

Mercury Power Plants

Mercury boilers were first tested in 1912 by the General Electric Company. Over the years a number of installations of such units were made (Table 1), and a 1942 paper[6] reported that a total of 17 mercury boilers had been built to that date. Most were used in a binary cycle for generation of electric power, although the Sun Oil Company employed seven mercury boilers for extremely close control of large-scale distillation of oils at its Marcus Hook, PA, refinery. Other units were built until at least 1950, including units for marine vessel propulsion.[7] The advent of efficient combined-cycle generating plants, higher temperature and pressure steam units, and very large utility-size power plants, with their economies of scale, probably combined to end the building of these mercury-steam-generating plants.

In these binary cycle plants mercury that vaporized in a boiler generated power in a turbine and then passed to a condenser-boiler. There, mercury's latent heat of vaporization served to boil water into steam at any desired pressure, for operating a steam cycle turbine. Since the mercury condenser-steam boiler is a nonfired unit, water of poor quality can be tolerated because it does not come in contact with flame-heated surfaces. The units generated electrical energy with greater efficiency than those with the steam cycle alone.[8] For example, the unit at Kearny, NJ, produced 21,000 kW from the mercury turbine and 30,000 kW from by-product steam, operating on 0.5 lb of fuel oil per kWh. The heat rate was 9200 Btu/kWh, then the lowest rate in history, although mercury units subsequently reported 8500-8900 Btu/kWh.[9] Before the era of combined-cycle plants, these thermal efficiencies of over 37% were certainly attractive. The 1942 paper cited stated that power units could be built with then-existing materials to produce power at 41% efficiency.

It is not the purpose here to describe the detailed functioning of mercury boilers, but to relate what metallurgical developments they achieved that may be useful in the context of applications of interest today. In the operation of the

Table 1 Mercury-steam binary cycle power plants

1912	First mercury boiler operated by W. L. R. Emmet of General Electric Co.
1922	Installation of Dutch Point Station, Hartford Electric Light Co. 1.8-MW generator driven by mercury turbine, boiler replaced in 1925.
1928	Installation of South Meadow Station, Hartford Electric Light Co. 10-MW mercury turbine output. Operated 19 yrs.
1932	Installation of General Electric Schenectady Works Station. 20-MW mercury turbine output. Dismantled 1950.
1932	Installation of Kearny Generating Station of Public Service Electric and Gas Co. of NJ. 20-MW mercury turbine. First operated in 1933.
1937	Installation of General Electric River Works Plant, Lynn MA. 1-MW mercury turbine generator. Successful operation for 6000 h. Shut down in Aug. 1939.
1937	Installation of Small Test Boiler at General Electric, Pittsfield, MA Works. Advances in treatment of mercury were made here. Generator, mercury turbine, valves, and condenser-boiler were originally used in Dutch Point Station.
1940	New mercury boiler installed at Kearny Station, 20-MW from mercury turbine.
1947	Original South Meadow Station removed to make way for new, larger mercury boiler unit.
1949	New South Meadow Station put into operation. 15-MW mercury plant.
1950	Schiller Station at Portsmouth, NH Installed. Public Service Co. of New Hampshire.

mercury boiler, General Electric found the major problem to be in eliminating mercury attack on the steel tubes, which in time became plugged with an iron crystalline deposit. An extensive research program was conducted by the G.E. Laboratory in which metallurgical and chemical properties of mercury, combined with other metals, were thoroughly studied. Test pieces of polished and weighed steel were placed in hundreds of small boilers that were operated at temperatures of 1200-1300°F to accelerate the results. The mercury in each of the test boilers contained small quantities of such metals as tin, lead, sodium, magnesium, silver, titanium, or zirconium, singly or in combination.

After more than 300 tests, it was determined that titanium and sodium additions improved wetting characteristics of the mercury and prevented solution of the iron in the boiler mercury. "The steel maintained its original weight and size with no loss of metal. The mercury-filled tube became a suitable and reliable absorber of heat at unbelievably high rates. Plugging and overheating of test boilers had been eliminated."[8] An extremely dilute mercury sodium-titanium amalgam "perfectly" wetted the steel, with the liquid spreading over the tube surface to form a tenacious layer difficult to wipe away. Sodium was believed to be the effective wetting agent, causing a breakdown and removal of oxides on the steel's surface and in the mercury. Titanium was found as a thin microscopic layer deposited on the steel, as well as a eutectic compound with iron in the mercury. The surface deposit of the Fe-Ti alloy may have inhibited dissolving of the iron.

It was subsequently found that a reaction between sodium, iron oxide, and oxygen led to plugging of tubes by a ball of hard sodium ferrite, $Na_2OFe_2O_3$, and that oxygen was the principal offender. Better shaft seals were built that greatly reduced air infiltration, and further laboratory and field tests showed that magnesium and titanium had all of the desirable wetting features of sodium and titanium and none of the undesirable ones; hence, magnesium replaced sodium in the boiler mercury. Hard inclusions did not form when magnesium in the boiler mercury reacted with iron oxide or free oxygen. The final products of such a reaction were an easily removed very fine, dry, harmless powder consisting of free iron and magnesium oxide.

The amount of titanium required in the mercury to prevent solubility of the steel is a function of the temperature, ranging from 1 ppm at 850°F up to 10 ppm at 1000°F.[6] To scavenge any oxygen and assist in the wetting of the steel by the mercury, it was found desirable to maintain up to 0.01% metallic magnesium in solution in the mercury.[9]

The furnace and slag-screen tubes were made of Sicromo 5S, or Croloy 5 MSi, a ductile chrome-alloy steel. Sicromo 5S (formerly known as Sicromo 5) contains 0.12% carbon, 0.50% molybdenum, 5.0% chromium, and 1.5% silicon. These tubes operated at temperatures of 1050-1180°F (565-635°C) and pressures as high as 395 psig at the lower header. Because of lower temperatures in the convection tubes, Croloy 3 was used in the lower bank and Croloy 2 in the upper two banks of tubes. Wear and erosion of turbine parts were negligible; the turbine at Hartford was said to have operated some 80,000 h from the time it had been opened for inspection. Data in Table 2 (see Ref. 6) show how iron solubility in mercury was retarded by use of Sicromo 5S steel. These data do not reflect the inhibition of solubility by means of minute additions of titanium to the mercury.

Mercury as Working Fluid for Heat Pipes

Although mercury is mentioned in a number of texts written about heat pipes, perhaps the most comprehensive information in the literature on this application of mercury appeared in a 1970 paper by Deveral.[10] In his tests at Los Alamos, magnesium and titanium additions were made to the mercury to achieve wetting and inhibit corrosion in heat pipes, one of which was life-tested for 10,000 h at 330°C. The tube was type 347 stainless steel, and the wick was a 100-mesh type 304 stainless steel screen. Good wetting was achieved in the test, and at the conclusion of 10,000 h, heat-pipe operation was still essentially isothermal. When the unit was sectioned, the wick structure and tube wall appeared to be in good condition. However, microscopic examination and chemical analysis showed that some corrosion and mass transfer had occurred. Changes in tube-wall thickness were negligible; the major corrosion effect evidently was the stress corrosion resulting from the manner of fabricating the

Table 2 Solubility of steel in mercury (mils/yr)

Temperature	Carbon Steel[a]	5% Cr steel[b]	Sicromo 5S[c]
900°F	2	2	0.2
1000°F	9	4	0.5
1100°F	22	10	1.1
1200°F	53	25	2.5

Data were obtained from tests conducted over 5-year period.
[a]Steel used in boilers at Hartford, Schenectady, Pittsfield, and original unit at Kearny.
[b]5% Cr steel used in boiler at Lynn, MA.
[c]Sicromo 5S steel, used in 1940 boiler at Kearny.
Note that data do not reflect inhibition of iron solubility, achievable with minute additions of Ti to the mercury.
Reference 6 was used as a source.

wick structure. Appreciable corrosion was detected only at the crossover points of the screen wires.

Results of the Los Alamos life test indicate that, at the temperature and heat-transfer conditions under which the pipe operated, no catastrophic corrosion occurred in 10,000 h. A "worst case" corrosion assumption for the screen wires would suggest at least a 4-yr life expectancy. Deverall states that the materials used were not necessarily the most corrosion-resistant to mercury, and suggests that use of different steels for the container and wick structure could have caused an electrolytic effect. Moreover, annealing of the wick after assembly could lessen the possibility of stress corrosion occurring.

Significantly, Deverall's paper concludes that it "appears that mercury is a suitable heat-pipe fluid from the standpoint of wettability and corrosion" and that it should be an excellent working fluid for space application, despite its high density: "The properties of mercury (e.g., vapor pressure, surface tension, vapor density, viscosity, and latent heat of vaporization) make it a suitable fluid for use in heat pipes operating above $200^{\circ}C$. Its operational characteristics as a working fluid compare favorably with those of other liquid metals. Although there are still some problems to be solved with regard to the wetting-in of fine-pore wick structures, the mercury heat pipe has excellent heat-transfer capability and appears to be the most promising system for the temperature range from 200 to $400^{\circ}C$." As depicted in Fig. 1 (see Ref. 11), over this $200^{\circ}C$ range of temperature mercury has an excellent potential to be the working fluid of choice because water has too high a pressure for containment and the vapor pressure of alkali metals is too low to sustain operation.

Dunn and Reay[12] have noted that a diphenyl/diphenyl ether eutectic fluid (Dowtherm A in the United States, Thermex in the United Kingdom) may be useful in a temperature range starting at $150^{\circ}C$. However, they caution that breakdown of this fluid occurs progressively above $160^{\circ}C$. Moreover, this liquid's low surface tension and its low latent heat of vaporization make it less useful than mercury as a working fluid. Dunn and Reay also noted that the attractive thermodynamic properties of mercury, and its being a liquid at room temperature, facilitates handling, filling, and startup of the heat pipe.

One U.S. firm has sold "about 50" mercury heat pipes for use as isothermal furnace liners where flat temperature profiles are required. With no mechanical components and a single heater and controller, the units provide better temperature uniformity than is possible with other techniques. These mercury heat pipes, working successfully after years of service, are used primarily for crystal growing in the electronics industry.

A number of applications for mercury heat pipes in spacecraft installations and for a plasma MHD project are currently being evaluated. A potentially significant use has been proposed in the nuclear power industry, to transfer heat safely and economically from sodium-cooled reactors.[13] In the latter application, heat pipes, with mercury as the working fluid, would enable the entire nuclear steam supply system to be placed in the same compact vessel as the core, without the intermediary of a secondary (nonradioactive) sodium

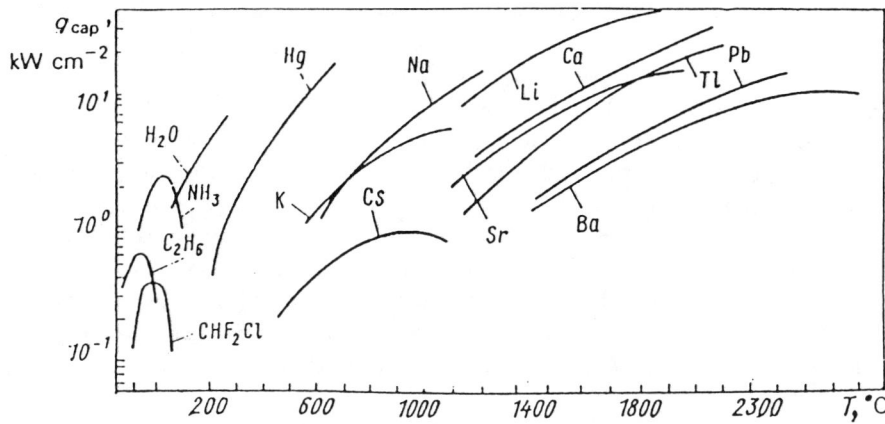

Fig. 1 Heat flux for heat pipes with various working fluids. Figure shows temperature range between 200 and 400°C, which is filled by mercury. It also shows that mercury provides the highest heat flux until much higher temperatures are reached with sodium and lithium. (By permission of Oxford University Press.)

system that is spread through many buildings with interconnected piping and having extensive components.

Application to Liquid Metal MHD Systems

Inasmuch as the working fluid in liquid metal MHD systems need not be at temperatures elevated much above the melting point of the metal or alloy (unlike the situation that pertains to heat pipes or even more particularly to mercury boilers), mercury offers some special advantages in terms of materials behavior. Even where there is limited materials compatibility, the life of components can usually be substantially extended by operating at lower temperatures. Mercury's low melting temperature, -39°C, enhances the opportunity of operating at low temperatures. As for elevated temperature operation, the test work performed by General Electric, their experience with commercial boilers, and the results of the Los Alamos heat pipe project strongly indicate that mercury can be satisfactorily contained by commercial steels at temperatures of at least 400-500°C, depending on the application.

Conclusion

Extensive test work and operating experience with mercury boilers for electric power generation provide assurance of good compatibility of mercury with steels at elevated temperatures. Certain chromium steels were favored for the mercury boilers. More recently, nickel-alloy steels have been used successfully in mercury heat pipes. Excellent resistance to high-temperature solubility of iron is achieved by minute additions of titanium to the mercury. Enhanced wettability is achieved by addition of magnesium. Concentrations of titanium and magnesium used are about 0.001 and 0.01%, respectively. Moreover, mercury heat pipes, operating at 200-400°C, have performed well, with excellent operating results for extended service at these temperatures which happen to be inappropriate for other working fluids having good heat-transfer characteristics. It is concluded that past experience provides the materials data base for the use of mercury as a working fluid for power generation, heat transfer, and other elevated temperature applications.

References

[1] Lesin, S.; El-Boher, A., and Branover, H., "Mercury Natural Circulation Systems with Internal and External Boiling of Thermodynamic Fluid," Sixth International Beer Sheva Seminar on MHD Flows and Turbulence, 1990.

[2] Lykoudis, P. S., and Takahashi, M., "Double Conductivity Measurements in Nucleate Boiling of Mercury," Sixth International Beer Sheva Seminar on MHD Flows and Turbulence, 1990.

[3] Pigny, S., and Moreau, R. J., "Stability of Fluid Interfaces in the Presence of a Transverse Electric Current," Sixth International Beer Sheva Seminar on MHD Flows and Turbulence, 1990.

[4] Dillon, C. P., Corrosion Control in the Chemical Process Industries, McGraw-Hill, New York, 1986.

[5] Hansen, M., Constitution of Binary Alloys, 2nd ed., McGraw-Hill, 1958.

[6] Smith, A. R., and Thompson, E.S., "The Mercury-Vapor Process," Transactions of ASME, Oct. 1942, pp. 625-646.

[7] Marine propulsion systems using mercury were built subsequent to the paper by Emmet, W. L. R., "A Mercury Propelled Cargo Ship," Transactions of the Society of Naval Architects and Marine Engineers, Vol. 48, 1940, pp. 371-381.

[8] Hackett, H. N., "Mercury for the Generation of Light, Heat and Power," Transactions of ASME, Oct. 1942, pp. 647-656.

[9] Shields, C. D., Boilers: Types, Characteristics, and Functions, Dodge, New York, 1961.

[10] Deverall, J. E., "Mercury as a Heat Pipe Fluid," Los Alamos Scientific Laboratory of the University of California, Los Alamos, NM, Rept. LA-4300 MS, UC-34, Physics, TID-4500, 1970.

[11] Ivanoskii, M. N., Sorokin, V. P., and Yagodkin, I. V., The Physical Principles of Heat Pipes, Clarendon, Oxford, UK, 1982, (Oxford Studies in Physics).

[12] Dunn, P. D., and Reay, D. A., Heat Pipes, 3rd ed., Pergamon, Oxford, UK, 1982.

[13] "Argonne Lab's New Heat Pipe Will Cut Nuclear Costs 25 Per Cent," Electric Light and Power, May 1986, pp. 21.

Status of MHD Energy Conversion Research in Poland

B. Zaporowski*
Technical University of Posnan, Posnan, Poland

Abstract

A review of the status of the MHD energy conversion research in Poland, starting from 1960, is presented. Nowadays, the primary activities are at the nuclear research center in Swierk and at the Technical University in Poznan. The work at Swierk concentrates on coal driven PMHD systems; it includes the investigation of a rotational high temperature pressurized combustion chamber and of thermodynamic processes in the MHD channel. The work at Poznan includes the study of thermodynamic and electrical properties of plasmas of combustion products, modelling of the combustion chamber and MHD generator and energetic analysis of combined PMHD-steam power plants integrated with coal gasification, as well as of thyristors based inverters from DC to AC. In addition, the economic viability of combined cycles featuring a PMHD topping cycle is being studied at the Technical University of Lublin while insulation and electrode materials for PMHD channels are being studied at the Academy of Mining and Metallurgy. Combined coal gasification - MHD plants appear to be economically viable in Poland; the combined cycle efficiency is predicted to be 46-48%. Presently there are no plans to construct MHD plants in Poland.

Introduction

Investigations in the field of MHD energy conversion have been conducted in Poland since 1960. First they started at the Institute of Nuclear Research in Swierk (INR Swierk) under the management of P. Nowacki and W.S. Brzozowski, and since 1961 they have been conducted at the Technical University of Poznan (TU Poznan) under the management of Z. Jasicki and at the Institute of Fluid-Flow Machines of the Polish Academy of Sciences in Gdansk (IFFM PAS Gdansk).

From the beginning, research work in the field of MHD energy conversion at INR Swierk and at TU Poznan has been directed at the construction and

---

Copyright © 1992 by the American Institute of Aeronautics and Astronautics, Inc. All rights reserved.
*Professor and Dean, Faculty of Engineering.

building of MHD generators. The first laboratory MHD generators, of thermal power 150-200 kW, were built at INR Swierk and at TU Poznan in 1962.[1] In this period, intensive theoretical research work was also undertaken in the field of MHD generators, particularly in the areas of mathematical models of electrical phenomena in the MHD channel,[2] mathematical models of thermodynamic processes in MHD generators,[3] and methods of plasma diagnosis.[4,5]

In this period, at IFFM PAS Gdansk, research work was developed in the area of synchronous inductive MHD generators[6] as well as of electrogasdynamic generators (EGD).

From 1964 to 1970, two experimental MHD generators were built in Poland, one at INR Swierk and one at TU Poznan. The one INR Swierk had a thermal power of 3 MW, expanded later to a thermal power of 4.5 MW. The one at TU Poznan has a thermal power of 4 MW. They work in an open cycle.

The MHD generator built in INR Swierk worked initially on liquid fuel. After reconstruction (since 1976), it has been coal fired. The generator works on an experimental stand composed of: a mixture heater, a rotational combustion chamber, a slag receiver, a channel of Faraday's type, and a channel of diagonal type as well as of an electromagnet (1.4 T). The stand is equipped with a carburizing installation with a pneumatic batch meter. In the combustion chamber, plasma with a pressure of 0.3 MPa and a temperature of 2800-3000 K can be generated. The degree of slag separation amounts to 0.3 - 0.87. The results of investigations obtained on a research stand with an MHD generator in INR Swierk, were presented at several international conferences.[7,8,9]

The experimental MHD generator built at TU Poznan, works on liquid fuel. The combustion chamber of the generator is composed of two basic elements: the proper cylindrical chamber and the head. In the combustion chamber, plasma with a pressure of 0.3 MPa and a temperature of 2800-3300K can be generated. Enrichment of the oxidizer in oxygen is applied to regulate the temperature of the plasma. The channel of the MHD generator has a ceramic metal construction and is water-cooled. The electromagnet of the generator allows one to generate induction of a magnetic field of the value of 2.7 T in the volume of the channel. The construction and some results of investigations of the generator were presented in Ref. 10.

Currently, in Poland, there are two principal centers of research in the domain of MHD energy conversion at INR Swierk and at TU Poznan, that have experimental MHD generators at their disposal.

First, in INR Swierk, research is conducted of high-temperature coal and semicoke combustion in a rotational pressure combustion chamber. The influence of basic parameters of the combustion process in tested on the separation of liquid slag and investigations are conducted on thermodynamic processes in the MHD channel and other devices of the test stand.

In the field of MHD generators at TU Poznan, investigations are performed in several areas:
1) Analysis and calculations of the composition and thermodynamic and electrical properties of plasmas of combustion products.[3]

2) Mathematical models of thermodynamic and electrical phenomena in the combustion chamber and in the channel of the generator in stabilized and transient states.[11]
3) Energy analysis of thermal systems of combined MHD-steam power plants, integrated with the process of coal gasification.[12,13]
4) Thyristor inverter systems, allowing mating of dc MHD generators with the power system.[14]

Apart from INF Swierk and TU Poznan, as the principal centers of research in the field of MHD generators in Poland, there are some smaller research groups that conduct investigations in this domain but in a more restricted area. Such groups exist at the Technical University of Lublin and at the Academy of Mining and Metallurgy in Krakow, as well as the Technical University of Gdansk.

The Technical University of Lublin conducts research evaluating of economic impacts resulting from coal-fired MHD electrical power plant operation (topping cycle) and MHD-steam power plant (bottoming cycle) operation in a power system.[15] The influence of plasma thermodynamic parameters pulsation on the electrical output and efficiency of MHD generators has also been examined.

From 1972 to 1980 at the Academy of Mining and Metallurgy, a wide program of research in the field of MHD generators was realized, particularly in the areas of investigating and improving fuels for MHD generators as well as testing plasma in the channel of the MHD generator.[16] Recently, research in this center has concentrated on insulation and electrode materials for the construction of channels of MHD generators.[17]

From 1972 to 1985 at the Technical University of Gdansk, work was performed on the technology of coal gasification in thermal systems of power plants with MHD generators, particularly on making use of outlet gases from the MHD generator.[18]

In Poland, research work in the field of MHD generators is directed at the construction of MHD generators in which coal is used as fuel, because the country has large resources of this fuel. Moreover, there is a shortage of gaseous and liquid fuels in Poland.

Polish researchers in the field of MHD generators have always exhibited great activity at international conferences, on magnetohydrodynamic electrical power generation. During all of the conferences of this cycle, from the second conference in Paris, France, (1964) to the 10th in Tiruchirapolli, India (1989), 54 papers were presented. Some of them are represented in the references.

References

[1]Brzozowski, W.S., Dul, J., Fuksiewicz, E., Mikos, M. and Wang, R. "Experimental Direct Current MHD Generator of the Open Cycle," *Proceedings of the 2nd International Symposium on Magnetohydrodynamic Electrical Power Generation*, Vol. I, Paris, France, 1964.

[2] Celinski, Z.N., "Electrical Equivalent Circuits of DC MHD Generators," *Proceedings of the 3rd International Symposium on Magnetohydrodynamic Electrical Power Generation*, Salzburg, 1966, Vol. I, International Atomic Energy Agency, Vienna, Austria, 1966, pp. 323-332.

[3] Zaporowski, B., "Temperature and Physical Properties of Combustion Gases in MHD Generators," *Proceedings of 4th International Symposium on Magnetohydrodynamic Electrical Power Generation*, Warsaw, 1968, Vol. IV, International Atomic Energy Agency, Vienna, Austria, 1968, pp. 2195-2209.

[4] Kordus, A., "Mesure et Analyse de al Conductivite Tensoriele des Gaz le Long de L'axe D'une Tuyere MHD de Petite Puissance," *Proceedings of 4th International Symposium on Magnetohydrodynamic Electrical Power Generation*, Warsaw, 1968, Vol. IV, International Atomic Energy Agency, Vienna, Austria, 1968, pp. 2367-2378.

[5] Suckewer, S., Licki, J., Mizera, B. and Zelazny, P., "Study of Various Methods of Determining the Temperature of a Plasma in an MHD Duct," *Proceedings of the 4th International Symposium on Magnetohydrodynamic Electrical Power Generation*, Warsaw, 1968, Vol. IV, International Atomic Energy Agency, Vienna, Austria, 1968, pp. 2179-2194.

[6] Milewski, J., "Striated Flow in the Synchronous Induction Magnetohydrodynamic Generator," *Proceedings of the 3rd International Symposium on Magnetohydrodynamic Electrical Power Generation*, Salzburg, 1966, Vol. III, International Atomic Energy Agency, Vienna, Austria, 1966, pp. 469-483.

[7] Brzozowski, W.S., Celinski, Z.N. and Kozlowski, T., "Operating Experience with an Open-Cycle MHD Test Facility," Proceedings of the 3rd International Symposium on Magnetohydrodynamic Electrical Power Generation, Salzburg, 1966, Vol. III, International Atomic Energy Agency, Vienna, Austria, 1966, pp. 221-232.

[8] Brzozowski, W.S., Celinski, Z., Dul, J., Fuksiewicz, E., Kozlowski, T., Mokwinski, A., Plata, M., and Rybacki, Z., "Le Resultats des Premiers Experiments sur L'installation MHD — 3 MW a Swierk, *Proceedings of the 5th International Conference on Magnetohydrodynamic Electrical Power Generation*, Vol. I, Munich, Germany, 1971, pp. 151-170.

[9] Brzozowski, W.S., Kozlowski, T., Rybacki, Z., Adamus, A., Licki, J., "Results of the Experiments from the 4 MWth Coal Fired MHD Test Facility," *Proceedings of 7th International Conference on Magnetohydrodynamic Electrical Power Generation*, Vol. I, Cambridge, MA, USA, 1980, pp. 79-87.

[10] Zaporowski, B., Pawlaczyk, H., Roszkiewicz, J. and Switala, B., "Investigations of the 4 MWth Experimental MHD Generator," *Proceedings of the 6th International Conference on Magnetohydrodynamic Electrical Power Generation*, Vol. I, Washington, D.C., 1975, pp. 419-434.

[11] Zaporowski, B. and Sroka, K., "Transient Processes in the Technological System of the MHD Generator at Load Changes," *Proceedings of 9th International Conference on Magnetohydrodynamic Electrical Power Generation*, Vol. III, Tsukuba, Japan, 1986, pp. 1395-1404.

[12] Zaporowski, B., Roskiewicz, J. and Sroka, K., "Parameters Analysis of the MHD/Steam Power Plant Thermal System," *Proceedings of 8th International*

Conference on Magnetohydrodynamic Electrical Power Generation, Vol. VI, Moscow, USSR, 1983, pp. 75-78.

[13]Zaporowski, B., Roszkiewicz, J., and Sroka, K., "Analysis of the Energy Conversion System of a Combined MHD-Steam Power Plant Integrated with Coal Gasification," *Proceedings of 10th International Conference on Magnetohydrodynamic Electrical Power Generation*, Vol. I, Tiruchirapalli, India, 1989, pp. III, 47-54.

[14]Mitkowski, E., Grzybowski, A. and Stiller, J., "Chosen Problems of Cooperation of Great Power MHD Generator with Power System," *Proceedings of 9th International Conference on Magnetohydrodynamic Electrical Power Generation*, Vol. III, Tsukuba, Japan, 1986, pp. 1272-1281.

[15]Gora, St. and Kapron, H., "Economic Aspects of Operation of MHD Electrical Power in Power System," *Proceedings of 9th International Conference on Magnetohydrodynamic Electrical Power Generation*, Vol. I, Tsukuba, Japan, 1986, pp. 78-86.

[16]Jasicki, Z., Sakalus, E., Kalinski, A., Siwek, A., and Wojtaszek, G., "The Diffuser and Heat Exchanger as Elements of an MHD Generator System in a Combined Cycle Power Plant of: 720-257 MWE," *Proceedings of 6th International Conference on Magnetohydrodynamic Electrical Power Generation*, Vol. I, Washington, D.C., 1975, pp. 91-104.

[17]Kielski, A.M., "Calcium Oxide as a New Electro-Insulating Material for an MHD Generator Channel," *Proceedings of 9th International Conference on Magnetohydrodynamic Electrical Power Generation*, Vol. III, Tsukuba, Japan, 1986, pp. 1012-1018.

[18]Pudlik, W., Stasiek, J., Rogowski, M., and Cieslinski, J., "Experiments with Gasification of Ground Coal by Outlet Gases from the MHD Generator, *Proceedings of 8th International Conference on Magnetohydrodynamic Electrical Power Generation*, Vol. V, Moscow, USSR, 1983, pp. 44-49.

Open Cycle Disk Generator Operating Conditions

H. K. Messerle*
University of Sydney, Sydney, Australia

Abstract

MHD disk generators might be a feasible and economic alternative to linear MHD generators, which have been favored in the past. Experiments have been carried out with shock tube driven disks, and these have demonstrated that significantly large enthalpy extraction is feasible. Performance studies indicate that outflow generators should operate in the supersonic regime, and inflow generators in the subsonic regime. Performance is affected by the development of a swirl motion, which can be offset by providing counterswirl at the inlet. Recent studies on outflow and inflow geometries have shown that each have specific advantages. The outflow has lower thermal wall losses, whereas the inflow offers operation with high swirl without the need for guide vanes, and the counterswirl is preserved along the flow path in the channel by conservation of angular momentum. The inflow can operate more effectively at subsonic speed. This means, however, that the outflow does not need the high magnetic fields, which leads to cheaper field structures. There is also the possibility of operating a disk in a closed-cycle as well as open-cycle system. Again, both offer specific advantages, and a great deal of work is still required to assess and compare the performance of the outflow and inflow disk geometries under various operating conditions. Other features discussed in the paper are operation at subatmospheric pressures and the effect of the diffuser on overall performance.

Introduction

In a fossil-fuel-fired disk generator the combustion gases are flowing radially outward or inward, and this flow induces a tangential Faraday current in an applied axial magnetic field. The Faraday current

---
Copyright © 1992 by the American Institute of Aeronautics and Astronautics, Inc. All rights reserved.
*Professor, School of Electrical Engineering.

rotates freely around the flow axis between the two parallel disks confining the radial flow.[1,6]

The Faraday current generates a Hall voltage that depends critically on the Hall factor γ and the electrical conductivity γ. A Hall current can then be extracted via ring electrodes placed at different radii along the disk walls. The radial current will impose a tangential force on the gas and cause rotation; i.e., a swirl may develop that has to be accounted for.

A schematic diagram of an outflow disk generator is shown in Fig. 1, and a schematic cross section of the experimental disk generator at the University of Sydney is shown in Fig. 2. The disk walls have to be insulated to prevent Hall current flow and this is achieved by using pegged wall construction. The pegs are water cooled with copper surfaces brazed on stainless steel.[4]

An experimental set of current-voltage characteristics is shown in Fig. 3. Voltage V is plotted along radius I_r for a set of values of radial current Iv up to short circuit. Here the cathode drop is compensated for on the diagram. Results compare favorably with calculated expected values. The simulation program has therefore proved reliable and has been used both to predict performance of large-scale facilities and to design our proposed 30-MW thermal facility.

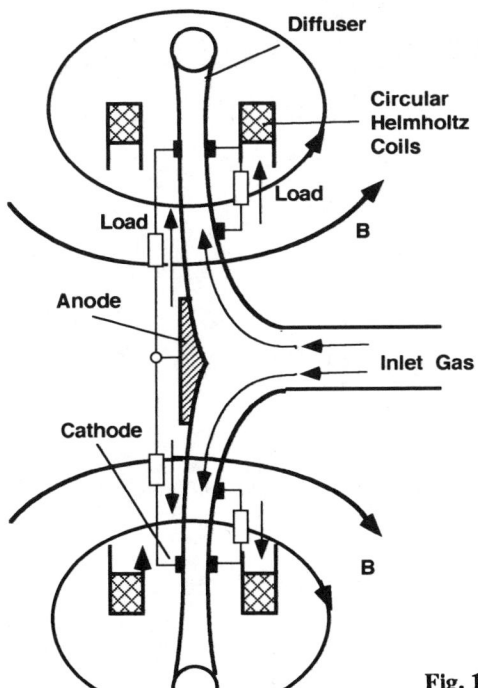

Fig. 1 Schematic diagram of an outflow disk generator.

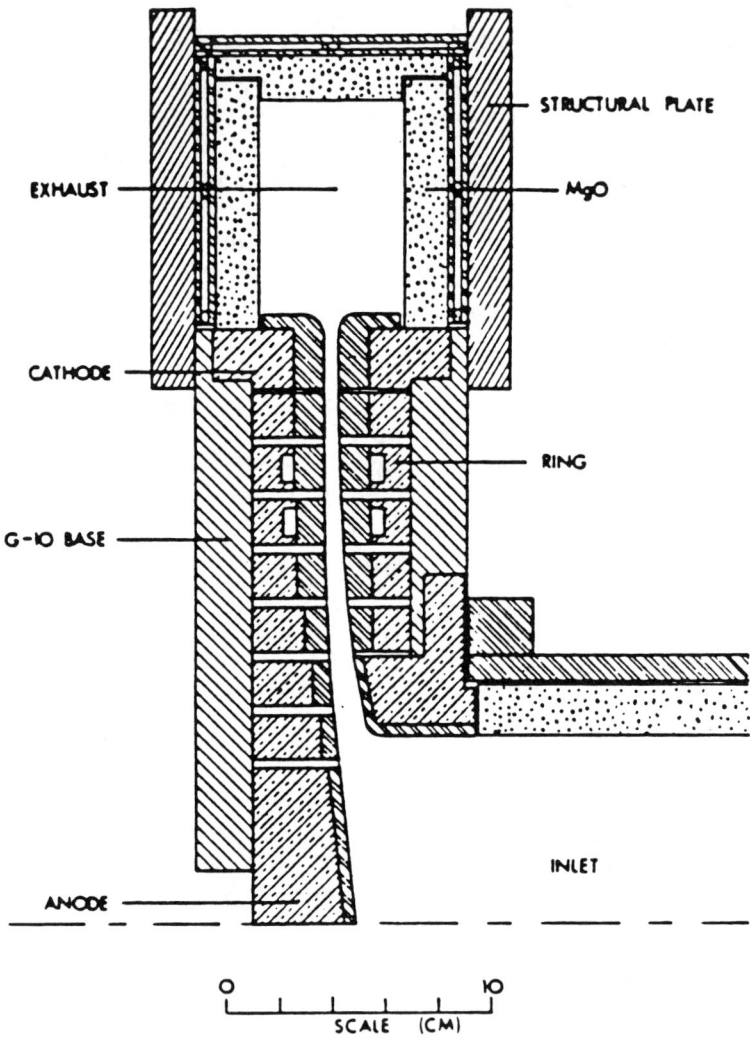

Fig. 2 Schematic layout of experimental disk generator at the University of Sydney.

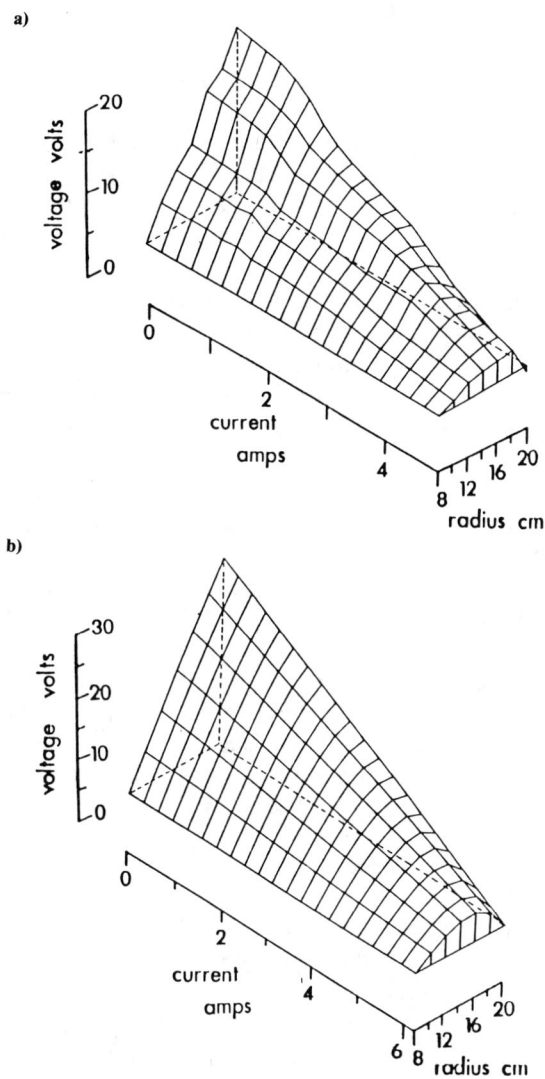

Fig. 3 Set of a) experimental and b) theoretical current-voltage characteristic of University of Sydney disk generator.

Inflow Versus Outflow Disk Generators

For an inflow disk geometry optimal performance occurs at subsonic conditions for Mach 0.8 to 0.9. The Hall parameter varies over a wide range along the radius in such a design and this leads to several load take off ring electrodes. This is necessary for reasonable load matching.

For an outflow disk generator optimal performance develops for supersonic flow. Counterswirl may be introduced at the inlet using guide vanes, but the temperatures are very high in that region. Hall parameters are very high, and this may lead to a greater tendency for instabilities to develop.

The boundary layers on the insulating walls will prove very critical in the disk performance since the wall surface area is relatively large. This is particularly pronounced because of the high magnetohydrodynamic (MHD) interaction in the disk. The effects are in opposite direction for a subsonic inflow and a supersonic outflow geometry.

In the outflow design the viscous dissipation at the wall can lead to a rise in temperature; i.e., the boundary-layer temperature can overshoot and can be higher than the core temperature. This results in a rise in the electrical condutivity, and the azimuthal current can increase. As a result, the plasma in the layer is decelerated, which causes boundary layer growth.

In the subsonic inflow disk the boundary-layer temperature is lower than that in the core. This results in a decrease in electrical conductivity and hence smaller deceleration and an overshoot in velocity. As a result, the exit diffuser in the flow design will exhibit a higher diffuser efficiency than that for a diffuser in a supersonic outflow generator.

Instabilities

Because of the high interaction conditions, in particular the very high value of the Hall parameter, the disk generator exhibits an extreme current sensitivity. Operation is approaching a constant current characteristic. At high magnetic fields we find that magnetoacoustic waves become unstable. This fact has been demonstrated by both linear perturbation analysis and nonlinear time-dependent growth studies.[7,8]

Figure 4 shows the radial electric field along the channel and how it changes with time for an outflow disk channel. The initial conditions are set by the steady-state conditions, as determined by the design operating at 7 T. As shown, steady state does not last long, and instability causes the shock to move rapidly upstream from the diffuser, the time involved being within 6 ms.

If the field is reduced from 7 T to 5 T, the flow is stable for some values of load current. In general the flow becomes more stable for

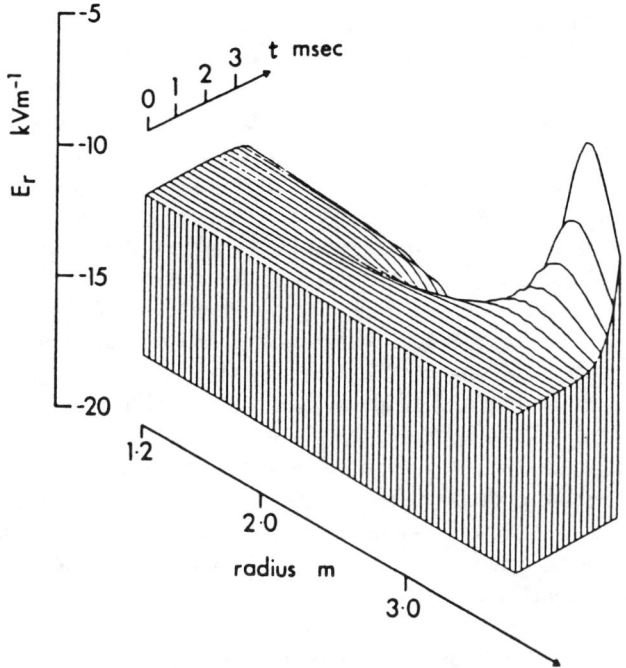

Fig. 4 Time development of fluctuations in electric field (curves are shown at intervals of 0.2 seconds; B = 7 Tesla; initial constant electric field strength E_r = -12 kV/m).

lower fields, shorter channels, or if control electrodes are introduced that hold the interelectrode voltage constant. The instability develops close to the exit where the flow reaches Mach 1 and conditions are favorable.

More work is necessary to determine the effect of chemical kinetics, boundary layers, and azimuthal currents have on the development of instability.

Low Pressure Operation

In an MHD disk generator operating with a subatmospheric diffuser,[9] the pressure of the combustion gases at the channel exit would be reduced through the use of a flue gas extractor placed downstream of the heat and seed recovery subsystem. At low pressures, both the electrical conductivity and the Hall parameter increase significantly, resulting in increased enthalpy extraction and a possible reduction in the gas temperature for feasible operation. As yet, however, there are no data for open-cycle generator performance at high Hall parameters, and the actual physical limit on this property is not known. Both ion slip and nonuniformities in the flow would be expected to reduce the effective Hall parameter.[10,11]

When operating at a low pressure, the capacity and power of the air compressor for the combustor can be much smaller than that required for conventional normal pressure operation. The losses in the flue gas extractor must, of course, be taken into account. The efficiency of the extractor would be expected to be lower than that of a compressor because of particulate matter in the exhaust gases. Nonetheless, for realistic values of extractor efficiency, the net power generated could still be greater than the power obtained by a comparable conventional design. Performance is also enhanced as a result of the reduction of heat losses and friction with the increased disk channel height in the low-pressure design. Increased channel height would mean a larger warm-bore volume of the superconducting magnet. However, the cost of the magnet would still decrease, as the required field strength can be much lower since the Hall parameter should be kept at a reasonable value.

A substantial improvement in performance is predicted for the case of low-pressure operation. However, there are some factors that may cause problems. At low pressure, chemical nonequilibrium in the gas flow would be expected to worsen. Also, problems associated with stability may become more serious.

Effects of Pressure on Performance

Both the electrical conductivity σ and the Hall factor β increase with decreasing pressure, rapidly at pressures lower than half an atmosphere. This means that, by operating the disk generator at low pressures, increased current can be obtained at reduced magnetic field strengths, and the gas enthalpy at the exit of the generator can be lower than it would need to be for effective power generation at normal pressure. As a result, the enthalpy extraction can be increased.

The convective heat transfer coefficient and the ratio of wall area to channel volume decrease almost in direct proportion to the decrease in pressure, and for the same static enthalpy the gas temperature decreases with decreasing pressure. Thus, the heat loss in a disk generator operating at low pressure would drop considerably.

Since operation at low pressure improves the power output per unit length, the diameter of the channel is reduced. In addition, the required magnetic induction is lower so that the cost of channel and magnet would be expected to be reduced.

In a low-pressure MHD power generation system, the pressure drop in the system is maintained partly (if both extractor and compressor are used) or totally (if only a flue gas extractor is used) by the pressure difference across a flue gas extractor.

In a system employing both a combustion air compressor and a flue gas extractor, the pressures at the inlet of the compressor and at the exit of the extractor are both 1 atm, thus, the total power consumption of the compressor and extractor is approximately proportional to (P_{co}/P_{ei}), where P_{co} is the pressure at the exit of the compressor, and P_{ei} is the pressure at the inlet of the extractor. The pressure difference $\Delta P = P_{co} - P_{ei}$ is equal to the sum of the pressure drops of each component along the gas flow path from the air preheater to the seed recovery equipment, with the pressure drop in the MHD generator being the most important.

In a low-pressure MHD generator, as the gas density decreases significantly, the pressure drop caused by the gas acceleration in the nozzle and the friction along the flow path would be smaller than those for high-pressure operation. In full-scale disk generators the largest part of the pressure drop is due to the electromagnetic braking force. For the same power output the electromagnetic braking force in a low-pressure disk generator would be smaller than that for a high-pressure generator. However, with the decreasing pressure the increase in efficiency is not as fast as the increase in power required for the extractor,[4] therefore, a low-pressure disk generator operating with high enthalpy extraction would have a higher overall power loss in the compressor and extractor than a high-pressure disk generator operating only with a compressor.

Effect of Magnetic Field on the Diffuser

For an MHD disk generator the diffuser must reduce the velocity of the gas flow at the exit of the channel in order to match the requirements of downstream components. It must also recover a portion of potential pressure rise. The gas flow of high velocity at the diffuser inlet possesses considerable kinetic energy, which can be determined by the difference between the total and static pressures, i.e., the potential pressure rise ΔP_p. The enthalpy extraction of disk MHD generators is sensitive to the pressure in the channel since both the electrical conductivity σ and Hall parameter β of the gas flow increase with decreasing pressure. The pressure at the diffuser exit, P_{d2}, is set by downstream requirements. The recovery of this potential pressure rise implies a reduced pressure at channel exit and thus an enhanced enthalpy extraction. Unfortunately, the diffuser can only recover a portion of the ideal value of ΔP_p. The pressure recovery coefficient C_{pr} is defined by the ratio of achievable static pressure rise [i.e., the difference of static pressures at the inlet and exit of the diffuser, $(P_{d1} - P_{d2})$ to ΔP_p]. C_{pr} is an important measure of the performance of a diffuser.

Subsonic diffuser performance used for linear MHD generators has been investigated experimentally and theoretically.[13,14] The ideal recovery coefficient is 0.88 for no boundary-layer blockage. However, in large MHD generators the blockage at the diffuser inlet is usually unavoidable. The recovery coefficient achieved is about 0.5 or less without the effect of magnetic fields being considered.

For an outflow disk generator the leakage magnetic flux could pass across the diffuser and is maximum in the region near the diffuser inlet, where the main portion of pressure rise is recovered. The effect of this magnetic field[1,2] on the pressure recovery must be considered. The diffusers used for disk generators are usually shorter than those for linear generators. The effects of the fringe magnetic field of the magnet on the performance of diffusers are thus more severe, and the performance of the diffusers becomes worse. For the disk generator geometry, so far the design and development of diffusers has not received the attention it deserves. In practice, the diffuser may even be used as an effective section for power generation, and the resulting benefits, due to improved pressure recovery and increased power output, can be demonstrated using sample calculations. A one-dimensional model is used here, in which heat loss, friction, current leakage, and electrode voltage drops are taken into account.

The magnet coils are not normally designed to cover the subsonic diffuser section. However, there still exists a considerable fringe magnetic field, though its strength, B, is not as high as that in the channel. The fringe field interacting with the gas flow of a radial velocity u_r will induce a tangential current in the subsonic diffuser. The tangential current density J_θ is given by

$$J_\theta = \sigma u_r B - \beta J_r$$

where J_r is the radial current density. Interaction with the magnetic field, in turn, will produce a radial electromagnetic braking force $J_\theta B$ on the gas flow that decreases the pressure recovery of the diffuser and raises the pressure in the disk channel, resulting in reduced power output. At the same time, J_θ, interacting with the magnetic field, will also induce a radial E-field determined by

$$E = u_\theta B - \frac{1}{\sigma}(\beta J_\theta - J_r)$$

In the region where the return magnetic flux passes through the diffuser, B becomes negative. The direction of radial electromagnetic braking force $J_\theta B$ and the E-field induced by J_θ remain the same, but the direction of the E-field induced by the tangential velocity u_θ changes. In the diffuser, u_θ is usually negative, which, interacting with a negative B, will increase the value of E.

In the diffuser the decrease in gas velocity will cause an increase in the gas temperature and gas conductivity. Consequently, the tangential current density J_θ would be large if there is no radial current (i.e., $J_r = 0$). This tangential current introduces an electromagnetic braking force, and the diffuser performance will be seriously affected by reducing the effective pressure recovery. The decrease in enthalpy extraction of the channel because of the poor performance of the diffuser will increase the total enthalpy at the diffuser inlet. This will increase the tangential current, and the performance of the diffuser will become worse. In addition, the radial electrical intensity E at the diffuser inlet can reach an unacceptable value that will lead to excessive wall leakage currents and wall breakdown. Therefore, the diffuser must not be left electrically open circuited.

An open circuit may be avoided if a radial current can be generated in the diffuser by providing conducting walls or by extracting the current from the diffuser externally using an additional electrode at its exit. A radial current will effectively reduce the E-field, J_θ and ΔP_b, and thus enhance the pressure recovery. A radial short circuit across the diffuser can be accomplished either externally or internally by providing conducting walls in the diffuser. When conducting walls are provided the wall temperature should be high. Cold boundary layers and layers of slag attached on the cold wall will reduce the radial current.

Conclusions

Full-scale commercial system studies indicate several operational features as well as problems:

1) Electrode number: In a linear generator hundreds of electrode pairs have to be aligned along the channel, with the associated problems of inter electrode breakdown due to Hall voltages. In addition, there is a need to consolidate the electrode interconnection to the output circuits. In a disk generator only a few output ring electrodes are required, with a few more possibly for control purposes. This offers a considerable simplification in channel design and in the output circuitry.

2) Size of generator: A disk generator is more compact than a linear generator at equivalent power load and should offer a simpler assembly arrangement, which should prove important for maintenance purposes.

3) Magnet coils: The magnet structure should be much simpler, and a smaller magnetic volume means a substantial reduction in costs. No detailed disk channel design for a commercial size facility and no study of performance due to non axial field effects have been carried out so far.

4) Enthalpy extraction: The enthalpy extraction for both linear and disk generators depends critically on the operating conditions set and also the assumptions made. Calculations indicate similar performance for both, although heat losses in a disk may be slightly higher for a comparable heat input.

5) Instabilities: Disk generators have to be designed for a high energy density and operate at high Hall parameters. Under these conditions problems of instabilities can arise. In fact, this may give the

inflow disk design an advantage since its optimal performance requires subsonic operation.

6) Low pressure operation: For a disk generator operating into a subatmospheric diffuser, the performance can be improved, allowing an increase in conductivity and Hall parameter. Good performance can be achieved with zero swirl; thus, no guide vanes are needed for an outflow geometry.

At low pressure, a larger volume is required for the combustor. Low pressure simplifies design, but slag rejection is reduced in a cyclone combustor. In a combined cycle MHD/steam plant, the steam plant must be sealed on the gas side to maintain the low pressure. The temperature at the inlet of the boiler section can be lower, and low density leads to reduced heat stress in the high-temperature heat exchanger but implies an increase in heat transfer areas. Little experience has been gained in the past with low-pressure boiler operation. The size and cost of heat and seed recovery subsystems would increase, and the economies of low-pressure operation must be assessed in more detail.

7) Magnet costs: It can be expected that the cost of the superconducting magnet could be reduced significantly since a lower magnetic induction can be used. The channel volume increases somewhat; however, the diameter can be reduced.

8) Small-scale generators: For small-scale generators, calculations indicate that considerable performance improvement is feasible at field strength produced with conventional iron-cored magnets. Therefore, low-pressure operation may find application for low-power systems and in space power systems that operate in a low-pressure environment.

9) Diffuser: The magnetic field in the diffuser can induce a tangential current and this would generate a radial electromagnetic braking force against the gas flow. Thus, the pressure recovery of the diffuser and the power output of the generator would be reduced. When the diffuser is open circuited, such a tangential current reaches its maximum value. The E-field in the region near the inlet of the diffuser will increase to an unacceptable value, which will lead to breakdown. In practice, the diffuser should not be open circuited. Guide vanes may be used to block the tangential current, but these vanes would have to withstand very high temperatures.

10) Combined channel-diffuser power: The use of the subsonic diffuser for power generation enhances the total output, and the diffuser can be used as a means for pressure control. In this case the extra electrode is needed at the exit of the diffuser, and the walls of the diffuser would have to be insulated radially as in the channel.

Acknowledgments

This study has been carried out by the MHD group at the University of Sydney, and the author wishes to express his appreciation to S. W. Simpson, S. Marty, and Y. Fang.

References

[1] Jenkins, M. K., Nakamura, T., Vilas, T. R., and Eustis, R. H., "Experimental Results of a High Magnetic Field Combustion Disk Generator," SEVENTH INTERNATIONAL CONFERENCE ON MHD ELECTRICAL POWER GENERATION, 1980, pp. 495-502.

[2] Nakamura, T., Lear, W. E., and Eustis, R. H., "Feasibility Study of the Inflow Disk Generator for Open-Cycle MHD Power Generation," 19TH SYMPOSIUM ON ENGINEERING ASPECTS OF MHD, 1981, pp. 3.1.1-3.1.14.

[3] Nakamura, T., and Lear, W. E., "Results of Combustion Driven Inflow Disk Generator Experiments," 20TH SYMPOSIUM ON ENGINEERING ASPECTS OF MAGNETOHYDRODYNAMICS, 1982, pp. 6.2.1-6.2.5 and addendum.

[4] Simpson, S. W., Marty, S. M. and Messerle, H. K., "Radial Outflow Disk Generator Experiments," 25TH SYMPOSIUM ON ENGINEERING ASPECTS OF MHD, 1987, pp. 2.5.1-2.5.8.

[5] Simpson, S. W., Marty, S. M., Rankin, R. R. and Messerle, H. K. "Disk Generator Project at Sydney University," 20TH SYMPOSIUM ON ENGINEERING ASPECTS OF MHD, 1982, pp. 6.4.1-6.4.5.

[6] Simpson, S. W. and Messerle, H. K., "Performance of Open-Cycle Disk MHD Generators," EIGHTH INTERNATIONAL CONFERENCE ON MHD ELECTRICAL POWER GENERATION, Vol. 4, 1983, pp. 201-206.

[7] Marty, S. M., Fishman, F. J., Simpson, S. W. and Messerle, H. K., "Operating Characteristics of MHD Disk Generators," 23RD SYMPOSIUM ON ENGINEERING ASPECTS OF MAGNETOHYDRODYNAMICS, 1985, pp. 605-623.

[8] Marty, S. M., Simpson, S. W. and Messerle, H. K., "Stability of Open-Cycle Disk Generators," 26TH SYMPOSIUM ON ENGINEERING ASPECTS OF MAGNETOHYDRODYNAMICS, 1988, pp. 7.4.1-7.4.11.

[9] Messerle, H. K., Fang, Y., Simpson, S. W. and Marty, S. M., "Coal-Fired MHD/Steam Combined Cycle Power Plant Using Low Pressure Disk Generators," 26TH SYMPOSIUM ON ENGINEERING ASPECTS OF MAGNETOHYDRODYNAMICS, 1988, pp. 7.3.1-7.3.5.

[10] Loubsky, W. J., Lytle, J. K., Teare, J. D. and Louis, J. F., "Molecular Gas Performance of a Disk Generator with Swirl," 16TH SYMPOSIUM ON ENGINEERING ASPECTS OF MHD, 1977.

[11] Loubsky, W. J., Teare, J. D., Lytle, J. K. and Cohen, H. D., "Kinetic Effects on the Performance of a Disk Generator with Swirl Driven by Combustion Gases," 17TH SYMPOSIUM ON ENGINEERING ASPECTS OF MHD, 1978.

[12] Messerle, H. K., "MHD Disk Generator Operation at Low Pressure," INTERNATIONAL SCIENTIFIC SESSION ON MHD, 1988, pp. 18-32.

[13] Reneau, L. R., Johnston, J. P., and Kline, S. J., "Performance and Design of Straight Two-Dimensional Diffuser," JOURNAL OF BASIC ENGINEERING, Vol 89, March 1967, pp. 141-160.

[14] Doss, E. D., "Subsonic MHD-Diffuser Performance with High Blockage," 16TH SYMPOSIUM ON ENGINEERING ASPECTS OF MHD, 1977.

Conceptual Design of an MHD Retrofit of the Corette Plant in Billings, Montana

R. Labrie* and N. Egan†
MHD Development Corporation, Butte, Montana 59703
and
F. Walter†
Montana Power Company, Butte, Montana 59703

Abstract
This study presents the results of the conceptual design of an MHD retrofit for the Corette plant in Billings, Montana.

Introduction
Magnetohydrodymanics (MHD) burns coal with reduced emissions, at lower cost, and with efficiencies in excess of 50 percent. The U.S. will need this new technology in about 10 years to provide a clean coal alternative.

The Department of Energy (DOE) proof-of-concept (POC) program provides commercial prototype hardware testing and evaluation by 1993. This technology can then be demonstrated and commercialized before the year 2000.

DOE proposed the conceptual design of a MHD retrofit plant as a way to ensure that the POC program would produce practical results. It will help guide the POC program to a successful conclusion.

MHD Development
Scale up from the test units in Montana and Tennessee was initially planned to be an engineering test facility (ETF) of approximately 200 - 500 MWe in size. ETF was to be a completely new power plant incorporating the MHD generating cycle and the steam generating cycle. As technical risk and capital costs were further defined, it became apparent that a different tack had to be taken. Analyses showed that the MHD cycle could be added or

---
Copyright © 1992 by Robert J. Labrie. Published by the American Institute of Aeronautics and Astronautics, Inc., with permission.
* President.
† Project Manager.

retrofitted to an existing steam generating plant and could be built on a smaller scale and lower cost than a new stand-alone facility. The objective of showing the power generator the benefits of MHD could still be achieved with this retrofit. Consequently, the design steps that would result in an MHD retrofit were initiated by DOE's request to industry for a conceptual design of a coal-fired MHD retrofit. The Corette retrofit design reports, contained in three volumes, are one of industry's responses to DOE. A schematic showing DOE's planned progression through hardware testing to the retrofit is shown in Fig. 1.

The DOE advanced power train studies have shown that a scale up of about five times from present hardware was generally acceptable, that a steam connection to an existing plant was feasible, and that the retrofit has to perform reliably in a utility environment to be of value to utilities. A subsequent study sponsored by the state of Montana reinforced these points, particularly the size and type of power plant to be retrofitted. Canvassing more than 100 utilities resulted in a general response that the retrofitted power plant should be more than 150 MWe, have a steam reheat cycle, be base-loaded, and be operated by a utility. The power plant selected from the Montana study was the Corette plant, which is 180 MWe, has reheat, is base-loaded, and is operated by the Montana Power Company (MPC), a privately owned, regulated utility. The consensus was that these design features would produce believable and usable results upon which utilities may base future generating plans. These elements, particularly the utilities, guided the design presented here by the MHD Development Corporation (MDC).

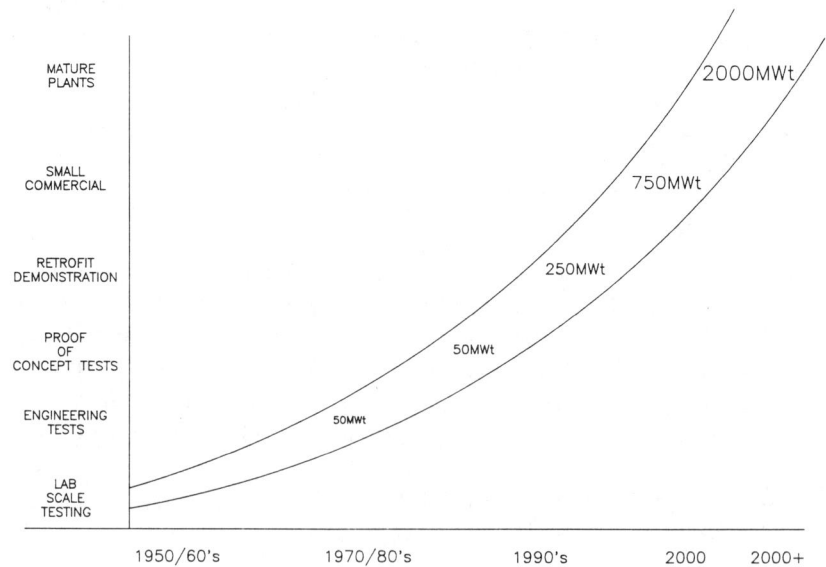

Fig. 1 Path to MHD commercialization.

TABLE 1 Corporate Entities and Technical Advisory Committee

Corporate Entities

    MDC
    AVCO Research Laboratory, Inc.
    Babcock & Wilcox, Inc.
    General Dynamics
    Montana Energy Research and Development Institute (MERDI)
    MPC
    MSE, Inc.
    TRW, Inc.
    Westinghouse Electric Corporation

Technical Advisory Committee

    Paul Probert, Chairman, Babcock & Wilcox
    Harold Bell, Arizona Public Service Company
    Dr. Robert Kessler, AVCO Research Laboratory
    Dr. Arthur Cohn, EPRI
    R. A. Johnson, General Dynamics
    Dr. B.R. Arora, Houston Lighting & Power
    John C. Orth, MERDI
    Robert Labrie, MDC
    Fred Walter, MPC
    Neal Egan, MSE, Inc.
    Clinton P. Ashworth, Pacific Gas & Electric Company
    Bob Johnson, Pennsylvania Power & Light Company
    Bob Cunningham, TRW, Inc.
    Norm R. Johnson, P.E., UTSI
    Larry E. Van Bibber, Westinghouse Electric Corporation

MHD Development Corporation

As the DOE MHD program progressed to the consideration of a retrofit project and the opportunity for MHD within the clean coal technology program became evident, the MDC was formed with the express purpose of moving MHD into commercialization. The MDC is a private corporation comprising the corporate entities most involved with the development of MHD. Table 1 lists those companies and the Technical Advisory Committee to MDC. The MDC is the contractual entity that developed this retrofit conceptual design in response to DOE's Pittsburgh Energy Technology Center request. MDC's members were subcontracted to provide the appropriate areas within their expertise. The project design team and their areas of responsibility are shown in Table 2. The project was to take approximately 18 months starting in October 1987 and cost about $1 million,

TABLE 2 Project Design Team Responsibilities

MDC	MHD Development Corporation: Overall program management — Bob Labrie, Neal Egan, Randy Senf, Dennis Moore
MPC	Montana Power Company: Owner of the J. E. Corette Station, host utility for the MHD retrofit, utility system economics, and planning support — Fred Walter
MSE	MSE, Inc.: Overall technical project management and coordination, operations assessment, environmental impact, oxygen plant design coordination — Gene Ashby
B&W	Babcock and Wilcox, Inc.: Coal feed, heat and seed recovery systems, balance of plant, coordination of cost estimates, scheduling and A/E services — Paul Robert, B&W, and Jerry Martin, Hudson Engineering
TRW	TRW, Inc.: Applied Technology Division: Slagging combustor and nozzle — Gil Ogle
AVCO	AVCO Research Laboratory: MHD channel and (in cooperation with W) current collection and consolidation electronics — Stan Petty
W	Westinghouse Electric Corporation: MHD current collections and consolidation electronics (in cooperation with AVCO), and dc-to-ac conversion and associated filtering, reactive compensation and protective circuitry — Rudy Putkovich
GDSS	General Dynamics Space Systems Division: Superconducting magnet and cryogenics — Husam Gurol
UTSI	University of Tennessee Space Institute: Recommendations for environmental controls and bottoming cycle integration — Dick Attig
	Advisory Committees
EEI	Edison Electric Institute — MHD Task Force
EPRI	Electric Power Research Institute — Arthur Cohn
MDC	MHD Development Corporation — Technical Advisory Committee

of which approximately one-third was to be contributed by MDC through cash and supplied services.

Design Results

The conceptual design of a coal-fired MHD retrofit by the MDC and its project team has shown that the Corette plant in Billings, Montana, meets the criteria for a meaningful scale up of existing MHD technology. The Corette retrofit design exhibits these features:
1) Enhances retrofit reliability through a conservative design approach.
2) Operates in a utility environment.
3) Has as the ultimate design goal the demonstration of sufficient effectiveness to be included by the utility (MPC) and the regulatory agency (public service commission) into the rate base.
4) Integrates the steam, reheat, and feedwater systems of the present steam generating facility and the MHD unit.
5) Is five times larger than existing test facilities.
6) Allows the existing facility to operate normally if the MHD unit is off-line.
7) Increases utilization of the existing turbine.
8) Attains an MHD cycle efficiency of 37.5 percent.
9) Increases plant capacity by 19 percent.
10) Improves plant efficiency by approximately 10 percent.
11) Reduces environmental impacts, particularly particulates and SO_x.
12) Delivery of long-lead items (superconducting magnet, channel, and combustor) could take five years.
13) Six- to seven-year design, construction, startup time.
14) Costs to design, construct, and startup this first-of-a-kind plant would be $330 million—about double a conventional coal-fired steam plant of comparable size.

Conceptual Design Project

The conceptual design project was separated into six tasks that would furnish the design information required to describe the scope of the work to accomplish a retrofit and provide an estimate of the cost and schedule of the project through acceptance testing. The tasks are as follows:
1) Component identification and description.
2) Plant layout and integration.
3) Reporting and design review.
4) Preparation of cost estimates.
5) Preparation of design and construction schedules.
6) Interfacing with the national MHD program.

The output of these tasks is reported in Vol. I—Executive Summary, Vol. II—Technical Discussion, and Vol. III—Appendices (technical backup

data). General guidelines for the design were chosen as follows:
1) Sixty-five percent plant capacity factor.
2) Eighty-five percent plant operational availability.
3) Plant operational considerations in the range of 75 to 100 percent of design.
4) Each of the MHD components is to be design for a continuous operation of 4,000 hours.
5) Rosebud coal is the reference fuel.
6) Dry potassium carbonate is the seed material.
7) Component description shall include appropriate sections for inspection, maintenance, and operational considerations.

Retrofit Description

The power plant to be retrofitted is the J. E. Corette plant. Specific information is shown in Table 3.

The Corette is restricted to less than full boiler capacity of 180 MWe environmental concerns, particularly particulates, and less than full turbine capacity (182 MWe) by an undersized boiler. The MHD retrofit would enable full capacity to be attained while realizing a reduction in emissions.

The type of retrofit envisioned at the Corette is a steam connection between the MHD unit and the Corette. This implies that steam generated in the heat recovery seed recovery (HRSR) system of the MHD unit would be

TABLE 3 Corette Description

Owner	Montana Power Co.
Built	1969
Location	70 acres on the Yellowstone River near Billings, Montana
Nominal capacity	180 MWe
Operating capacity	166 MWe
Fuel	Montana Rosebud
Moisture	25.3%
Sulfur	9.1%
Energy	8,600 Btu/lb
Thermal input	506.5 MWt
Steam generation	
Main steam	1,800 psi/1000 °F
Reheat steam	1,000 °F
Net output	157.0 MWe
Heat rate	11,010 Btu
Efficiency	31%

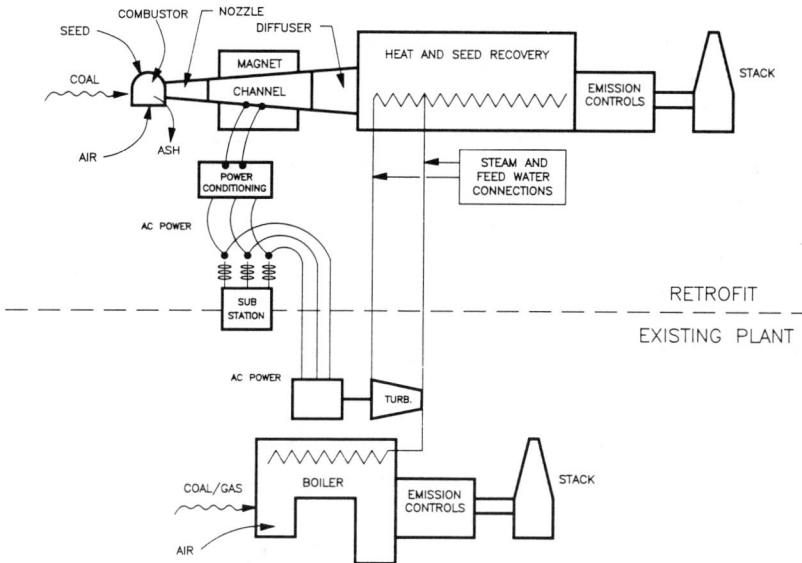

Fig. 2 Steam connection concept.

sent to the existing turbine to generate electricity, and part of the boiler feedwater from the Corette would be heated by heat rejected from the MHD power train components. A simplified schematic of this is shown in Fig. 2.

The guiding philosophy in the Corette retrofit conceptual design was to make it simple and reliable. The design team believed that the primary result that would impress potential users was an MHD unit that could run when called upon and stay running. Consequently, all component designs are within the current state of the art for MHD.

Concurrent with designing is reliability, and the design team realized that the MHD unit must be large enough to show a sizable scale up from current test units and must demonstrate cycle efficiency and environmental benefits. Consequently, the MHD design parameters shown in Table 4 were determined by present-day hardware capabilities, calculations of performance, performance optimization within conservative design constraints, and heat and mass balancing of the MHD system and the Corette plant.

Heat and Mass Balance, Power Output, and Plant Efficiency

Starting with the Component Development and Integration Facility (CDIF) 50-MWt channel and the scale up decision of 5:1, the size of the MHD topping cycle was fixed at 250 MWt input. The coal was assumed to be Rosebud dried to 5 percent moisture content, and the combustor was assumed to reject 70 percent of the slag, leaving 30 percent to be carried through the channel.

Next, a supersonic rather than subsonic channel was selected. Performance at 1.2 Mach is about the same as a subsonic or a higher supersonic channel; however, more test data exist at 1.2 Mach, thereby increasing design reliability.

High-temperature air heaters have not yet been developed, so oxygen enrichment was chosen with 1200 °F preheated air to get the high combustion temperatures required.

A 4.5-tesla superconducting magnet was selected as a conservative design approach to maximize system reliability. Design challenges at 4.5 tesla were considered to be less than those at higher field strengths.

A computer model was used to evaluate the topping cycle parameters in Table 4. Oxygen enrichment and preheat, seed content, and Mach number were varied to determine performance effects, and the values shown in Table 4 were selected.

The HRSR system was designed to recover the heat from the exit gases of the MHD power train. The combustion gases entering the HRSR system are fuel rich, so it is equipped with an afterburner at an appropriate location to bring the stoichiomentery up to 1.05. Oxidant air is also preheated in the HRSR System. The result is that the HRSR will produce about 183 MWt of superheated and reheated steam at the Corette turbine conditions. At full load, the HRSR System will produce about 40 percent of the turbine steam, whereas the Corette boiler will produce 60 percent.

Cooling for the combustor/channel provides heat that is transferred to the feedwater system through heat exchangers. The diffuser cooling is connected in parallel with the HRSR radiant boiler so boiling water (20 percent steam) circulates to the drum.

The electrical output from the MHD channel-inverter is 27.2 MWe. The HRSR system supplies 183 MWt of steam to the Corette turbine. Cooling for the combustor, nozzle, and channel add an additional 32.5 MWt to the

TABLE 4 MHD Design Parameters

Thermal input	250 MWt
Fuel type	Montana Rosebud from Corette storage piles
Moisture maximum	5%
Combustion stoichiometry	0.9 (fuel rich)
HRSR afterburner stoichiometry	1.05
Ash rejection	70% in combustor
Oxygen enrichment	38%
Oxidant preheat	1200 °F
Seed content	1% by weight potassium
Mach number	1.2 at channel inlet

Corette feedwater cycle. When 12.6 MWt is taken out for coal drying, as well as the MHD power train losses and 15.5 MWe for the air separation unit, the MHD facility operates at an efficiency of 37.5 percent, and the combined retrofit facility operates at 33.8 percent. The Corette plant without MHD operates at 31 percent. Operation at 75 percent load for the topping cycle was also studied and appears to be practical. Fig. 3 shows the combined energy flow.

Description of MHD Operations

The MHD retrofit will verify technical and commercial feasibility and provide the information necessary to proceed with commercializing MHD. It will utilize the information developed from the POC testing at the CDIF and Coal-Fired Flow Facility. This new facility will provide for a fully integrated unit within a commercial utility environment.

A two-year acceptance program will be needed to perform component tests, debug, tune the controls, and establish that the system operates as specified. Conformance with environmental standards will also be verified. After the two-year acceptance tests, any required modifications will be performed. The objective is a facility acceptable to the MPC and the regulatory agency.

The retrofit will be designed such that the MPC can operate the Corette plant with the MHD system shut down. The Corette unit will have to be on-line, however, for the MHD unit to supply steam to the turbine. Tests can be run on the topping cycle using a startup flash tank system to dissipate the HRSR steam that is produced. In the final design, controls can be installed to allow the MHD unit to supply the steam turbine without the boiler in operation.

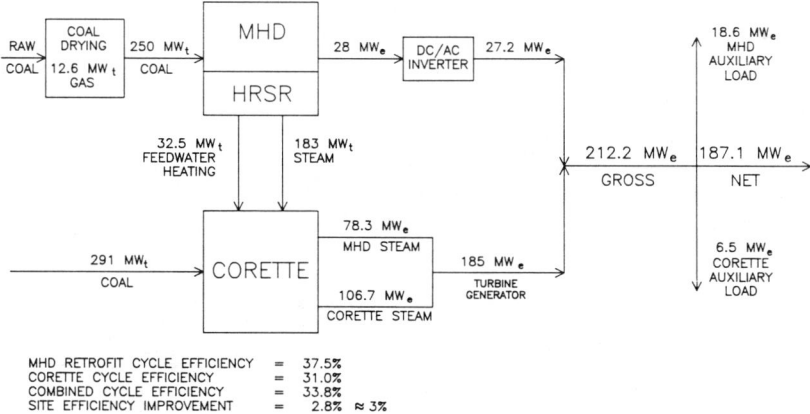

Fig. 3 MHD/Corette retrofit energy input-output diagram.

The initial steam in the HRSR system will be diverted to a flash tank until the steam conditions nearly match those of the Corette boiler, then the steam will be redirected to the turbine so the two boilers are operating in parallel. There will be a detailed startup procedure on the MHD side so that all of the components are integrated into the system; e.g., the superconducting magnet will need to be cooled down to 4.1 K before the unit can start up.

The MHD facility will be controlled from the Corette control room by a supervisory control system. The control systems for the two plants will be coordinated and optimized for steady-state, dynamic, and fault conditions. This can be done with a fairly conventional control system utilizing the electric demand signal and proportional controls. Time constants can be developed using a dynamic model that has been developed.

Dynamic simulations to date indicate controls can be designed so the Corette plant can survive the loss of the MHD combustor without a turbine trip.

Careful attention will have to be paid to coordination of the support systems such as the raw coal supply, the oxygen-enriched air supply, and the seed supply during the detailed design.

Project Schedule

The overall project schedule for the MHD retrofit project is seven years from the beginning of project mobilization through startup (see Fig. 4).

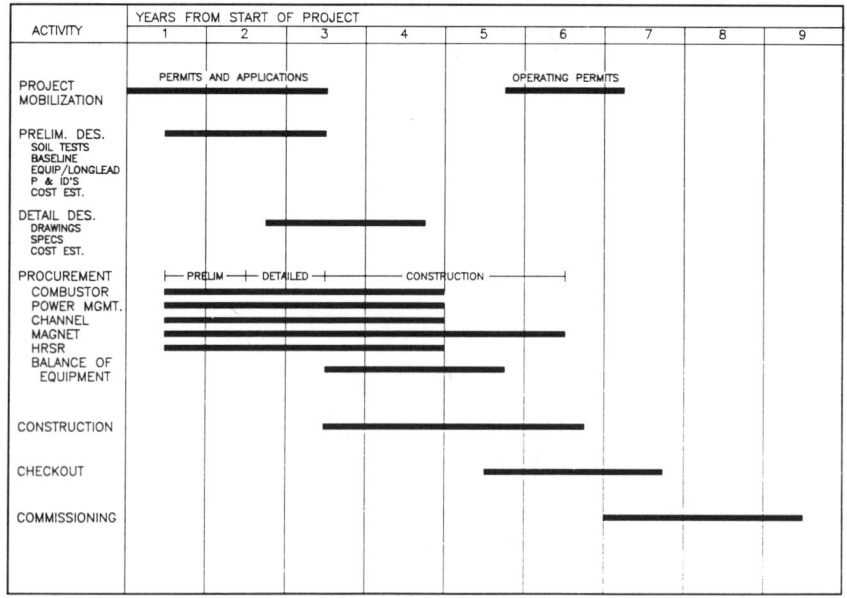

Fig. 4 Project schedule.

The long lead-time items are five years for the channel, combustor, and magnet and four years for the HRSR system. These items will have to be ordered as early as possible in the preliminary design.

The summary level schedule of activities is broken down into the following major categories: project mobilization, preliminary design, detailed design, construction, checkout, and commercial operation.

Cost Estimate

The costs of the Corette retrofit facility were compiled by estimating the costs of each major component (e.g., combustor, channel, etc.), then applying factors for associated bulk commodities, labor, indirect costs, project contingency, and engineering. This resulted in a retrofit plant cost of approximately $330 million in 1989 dollars (see Table 5). This facility produces a capacity of 98.5 MWe, so the cost is about $3350/KW—about twice the cost of a 100-MWe commercial pulverized coal steam plant installed in 1995. For a first-of-a-kind demonstration unit, this seems reasonable. However, approximately $30 million is needed to complete the development of the new technologies involved in the retrofit facilities beyond the effort planned for the current POC work. Then the MHD retrofit power

TABLE 5 Cost Estimate Summary

Title I — Preliminary engineering		$ 1,196 million
Title II — Detail engineering		6,777 million
Equipment		201,686 million
Construction		
Installation	$22,017 million	
Indirects	8,382 million	30,399 million
Miscellaneous		
Project integration	9,975 million	
Spare parts	9,082 million	
Environmental engineering	1,500 million	20,557 million
Contingency		
Process	25,468 million	
Project	42,710 million	68,178 million
Total estimated cost		$328,793 million

train should be equipped with special diagnostic instrumentation that will not be required on commercial MHD facilities. This cost could range from $5-$20 million, depending on the instrumentation selected. The basic equipment costs were produced by the individual manufacturers (e.g., combustor—TRW, etc.) from data that they are developing for the POC designs for the CDIF; therefore, costs are preliminary and probably conservative. Hopefully, they will come down somewhat as the POC design progresses.

Operating costs during the testing phase were estimated at about $35 million per year. These costs will be offset a little by the value of the power produced during the testing phase. This value is based upon nonfirm prices since the facility cannot be scheduled reliably during this period.

After testing and debugging are completed, the power produced by the facility should be valued at firm prices, and the cost of operation should be reduced.

Recommendations for Support Test Programs

To clarify some of the design considerations and provide a firm basis for the retrofit, certain items need to be investigated prior to final design. Following is a list of these items.
1) Reliability and availability.
2) Dynamic modeling and control simulation of combined boiler systems.
3) Electrical and magnetic isolation.
4) Channel/magnet interfaces.
5) K_2/S molar ratio.
6) Alternate seed materials—potassium, formate.
7) Combustor second-stage oxidant levels.
8) MHD channel additives.
9) Coal flow measuring technologies.

Developing a finished package to implement the MHD retrofit is an economic challenge of equal importance as the technical issues. For the retrofit, there are two funding sources that must be successfully worked together. The first is the federal government:
1) The exiting fossil energy R&D MHD line item.
2) Clean coal technology.
3) Congressional line item.
4) Environmental legislation—acid rain/greenhouse.

The second major funding source is the nongovernment sector:
1) The sponsoring utilities.
2) Electric Power Research Institute (EPRI).
3) Industrial suppliers.
4) State governments.
5) Foreign partners.

Constricted Discharges in Ar-Cs MHD Generators

W. F. H. Merck,* A. P. C. Holten,† A. Veefkind,*
and E. M. van Veldhuizen*
Eindhoven University of Technology, 5600 MB, Eindhoven, The Netherlands
V. A. Bityurin‡ and A. P. Likhachev§
USSR Academy of Sciences, Moscow, Russia
and
B. Stefanov¶ and L. Zarkova**
Institute of Electronics, Sofia 1784, Bulgaria

Abstract

In this paper the constricted discharges, usually called streamers, that appear in alkali-seeded noble gas magnetohydrodynamic (MHD)-generators are treated from different points of view. The plasma physical approach led to the conclusion that the current is flowing through a bunch of several hundreds thin arcs, called filaments. This was confirmed by CO_2 laser scattering experiments. The thin filaments suffer from high energy loss through electron heat conduction to the surrounding gas. The electrodynamical approach was used to calculate Hall field distributions around a streamer. The calculations for insulator walls and electrode walls correspond well with measurements in the Eindhoven Blow-Down Facility. The gasdynamical approach of the interaction of the streamer with the supersonic flow by means of a two-dimensional time dependent model shows that a shock is created in the streamer which moves upstream and that the gas around the streamer is accelerated in the dynamic nozzles between streamer and insulator walls. Pressure variations calculated with this model were confirmed by the experi-

---

Copyright © 1991 by the American Institute of Aeronautics and Astronautics, Inc. All rights reserved.
* Senior Lecturer, Department of Electrical Engineering.
† Electronic Engineer, Department of Electrical Engineering.
‡Senior Scientist, Group Theoretical Physics, Institute of High Temperature.
§Scientist, Group Theoretical Physics, Institute of High Temperature.
¶ Professor, Plasma Division.
** Senior Scientist, Plasma Division.

ments. A semi-empirical model of the electrical performance of the generator leads to a relationship between the conductance per unit length of the streamers, the streamer current, the gasdensity and the magnetic induction which enables us to predict the performance of large size linear Ar-Cs MHD generators. From this it is concluded that under the given conditions the linear Ar-Cs MHD generator operating in the streamer mode is not feasible for commercial application. This is mainly due to the high joule dissipation in the aforementioned thin filaments.

Introduction

It has long been known that in argon-alkali driven linear magnetohydrodynamic (MHD) generators nonuniformities in the current distribution appear. These current concentrations, called "streamers", behave like stable entities and move downstream in the channel with a velocity approximately equal to the average gas velocity. This has been observed by fast photography with an electronic camera; another feature has also been observed by stereoscopic measurement: the helical structure of the streamers (Fig. 1). Helical structures of discharges in a magnetic

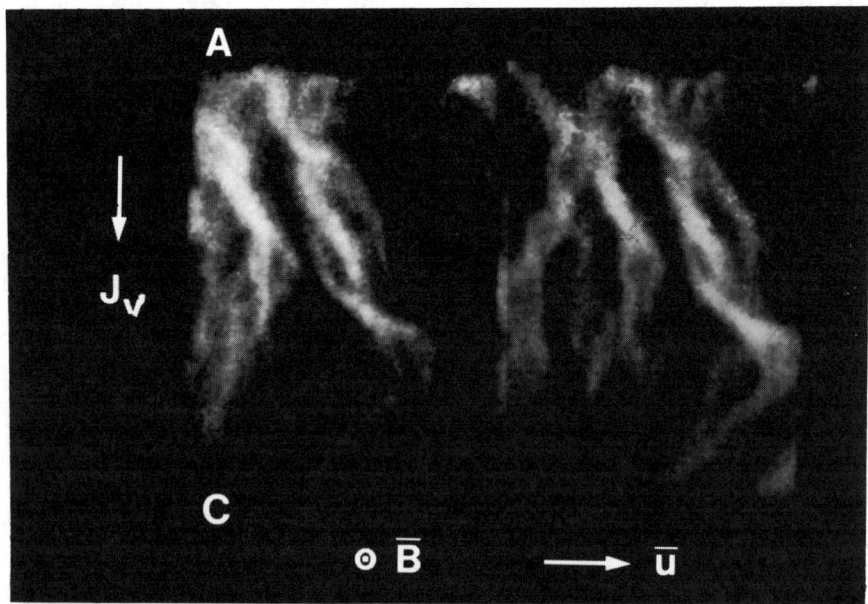

Fig. 1   Stereoscopic picture of a helical discharge structure in the Ar-Cs-driven shock-tube MHD generator.

field have also been found by Montgomery et al.[1] through numerical-calculations using the principle of minimum energy dissipation.

Another detail in the streamers was found by de Haas et al.[2] by means of CO_2 laser scattering experiments. He found a fine structure of several hundreds of threadlike arcs ("filaments") with diameters smaller than 1 mm. This observation was supported by numerical solutions of the plasma conservation equations by Stefanov et al.[3].

From the aforementioned phenomena it is clear that the electric field in the vicinity of the streamers will be strongly non uniform. Electrodynamic calculations of the fields inside of and around the moving streamers by Bityurin[4] led to a better understanding of the time-resolved measurements of Hall voltages performed in the Eindhoven Blow-Down Facility (EBDF). A geometrical layout is presented in Fig. 2.

The existence of streamers as stable current-carrying entities interacting with a supersonic flow causes a severe disturbance of the re gular gas flow. Numerical calculations of the gasdynamic performance were made by Bityurin et al.[5] and were qualitatively confirmed by time-resolved pressure measurements in the EBDF.

When the complicated three-dimensional structure of the streamers and their time-dependent interaction with a supersonic flow are examined, the question arises whether a full-fledged theory and computational method can be formulated. In order to circumvent these problems a semi empirical approach was used by Merck[6] for the prediction of the electrical performance of Ar-Cs-driven linear MHD generators, based on the EBDF experiments.

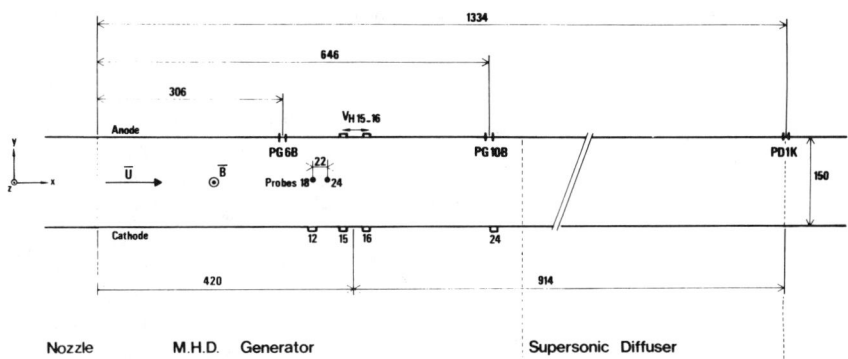

Fig. 2 Geometrical layout of the Eindhoven blow down MHD generator with locations of sensors: pressure probes PG6B, PG10B, and PD1K; electrostatic probes 18-24; electrodes 15-16.

Filaments

It appears from experimental observations that the streamers in an alkali seeded inert gas MHD generator have a filament substructure with typical cross-sectional dimensions of the filaments between 10^{-4} and 10^{-3} m.[2]

Each filament can be considered as a thin arc, and its physical properties can be derived from the solution of conservation equations. Here cesium-seeded argon will be considered. In order to simplify the model the following assumptions are made: 1) constant pressure, 2) steady state, 3) circular cross section of the arcs, 4) infinitely long arcs and 5) no axial gradients.

In order to be able to account for moderate deviations from local thermo-dynamic-equilibrium (LTE), a partial local thermodynamic equilibrium (PLTE) description[2] has been employed. It is used in its simplest form where only the population of the ground state of the cesium neutrals is allowed to deviate from the Saha value with respect to the electron density. The deviation from LTE is then expressed by the difference D between collisional excitations and de-excitations of the first excited level:

$$\Delta = nN_1 K_{12} - nN_2 K_{21} \qquad (1)$$

where n is the electron density: N and N are the populations of the ground state and the first excited state, respectively; and K_{12} and K_{21} are the rate coefficients of the excitations and de-excitations, respectively. (It follows from estimations that electron collisions are the dominant process.)

The governing equations read as follows:

$$\frac{1}{\rho}\frac{d}{d\rho}\left(\rho\lambda_e \frac{dT_e}{d\rho}\right) + \sigma E_*^2 - 3mnk[\bar{v}_{ea}/M_a$$
$$+ (\bar{v}_{es} + \bar{v}_{el})/M_s](T_e - T_\bullet) - \varepsilon_{12}\Delta = 0 \qquad (2)$$

$$\frac{1}{\rho}\frac{d}{d\rho}\left(\rho\lambda_e \frac{dT_e}{d\rho}\right) - 3mnk[\bar{v}_{ea}/M_a$$
$$+ (\bar{v}_{es} + \bar{v}_{el})/M_s](T_e - T_\bullet) = 0 \qquad (3)$$

$$\frac{1}{\rho}\frac{d}{d\rho}\left(\rho D_A \frac{dn}{d\rho}\right) + \Delta = 0 \qquad (4)$$

In these equations ρ is the radial coordinate; T_e and T_a are the electron and heavy particle temperatures, respectively; λ_e is the electron thermal conductivity; λ_a is the argon thermal conductivity: ε_{12} is the excitation energy of the first excited level; and E_* is the component of the total electric field $(E + u \times B)$ along the filament. In Eqs. (2) and (3) $\bar{v}_{ea}$, $\bar{v}_{es}$, and $\bar{v}_{ei}$ are averaged electron collision frequencies with noble gas atoms, seed atoms, and ions, respectively. M_a is the noble gas atom mass, and M_s is the seed atom mass. In Eq. (4), D_a represents the ambipolar diffusion coefficient. The steady-state assumption requires that the outward flow of electrons and ions is equal to the inward flow of seed neutrals. Furthermore, it is assumed that the flow of ground-state seed neutrals can be approximated by the total flow of seed neutrals. As a result, Δ as defined in Eq. (1) can also be interpreted as the net ionization rate and thus appears as the source term in the electron continuity equation [Eq.(4)].

The equations are solved for the input parameter values presented in Table 1.

The solution of Eqs. (2-4) for the standard reference case of Table 1 is presented in Fig. 3. The shapes of the profiles are to a large extent determined by the electron energy transport. It has been demonstrated earlier (ref. 2) that in the central part of the arc the Joule heating and the thermal conduction are the most important terms in Eq. (2) and that, consequently, the characteristic cross-sectional dimension is related to the central value of the electron temperature (see Ref. 3). The sharp edge that appears in the electron temperature profile reflects the fact that the thermal conductivity decreases with the electron temperature.

The electron density follows from the electron temperature by applying LTE relationships in very good approximation, except for radii larger than 1.75×10^{-4} m, where the deviation of n from its LTE value becomes more than 5%. The real situation in the outer part of the arc is much more complicated than is represented by the model. Because of the large gradients, the nonequilibrium will be so strong that the PLTE description will no longer hold. Furthermore, it has to be expected that the electron velocity distribution function deviates from the Maxwellian and that, therefore, the hydrodynamic description of the arc [Eqs. (2-4)] is no longer adequate.

Fast electrons produced by recombinations and de-excitations travel over considerable distances when moving outward without being thermalized. It can be estimated that is is likely that most of the heat is removed from the arc by fast electrons produced in the outer region of the arc and able to escape from the arc volume. If, indeed, in the outer layer of the arc the electron thermal flow is replaced by the energy transport through escaping fast electrons, it can be expected that the slope of the electron temperature profile is less steep than the slope in the solution. (See the dashed line in Fig. 1 obtained with $\Delta = K_{12}nN_1$, in

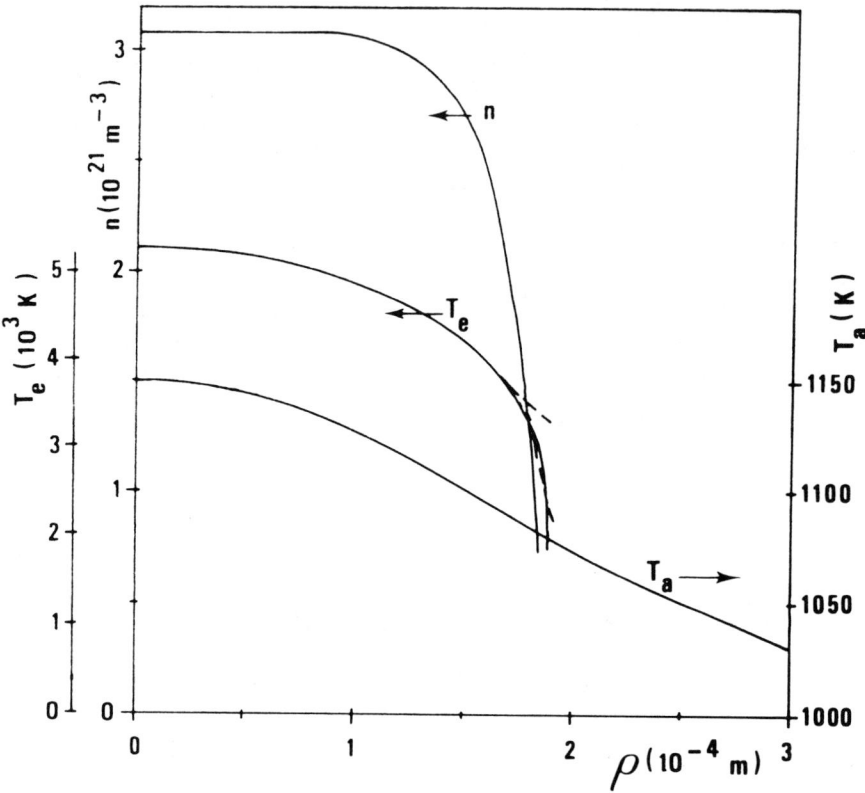

Fig. 3 Radial distributions of electron density n, electron temperature T_e, and gas temperature T_a in a filament of a streamer, for cesium-seeded argon. The seed fraction is 5×10^{-4}, and the total electric field is 4000 V/m. For $\rho > 1.7 \times 10^{-4}$ m fast electrons are assumed to leave the arc, resulting in changed T_e (ρ) and u (ρ) (dashed lines).

Table 1 Input parameter values

Parameter	Standard reference value	Range of variation
p	0.5 bar	0.125 ÷ 2 bar
sf	1×10^{-3}	$(0.125 \div 2) \times 10^{-3}$
$T_{a\infty}$	1000 K	500 ÷ 1500 K
E_*	2000 V·m^{-1}	1000 ÷ 3000 V·m^{-1}
B	3 T	1 ÷ 5 T
T_{eo} [a]	5280 K	3800 ÷ 6220 K

[a]The electron temperature at the center of the are T_{eo} is chosen from a stability analysis[3] and represents a function of the seed fraction sf only.

Eq. (2) for $\rho > 1.65 \times 10^{-4}$ m; i.e., from this point outward the fast electrons were assumed to leave the arc.)

From the calculations (including the input parameter variations indicated in Table 1), the following conclusions are drawn:

1. The arc characteristics are to a large extent determined by the electron energy transport.
2. The electron concentration profile exhibits a sharp edge.
3. The deviation from LTE does not affect the parameters in the arc except in a thin layer near the edge.
4. Near the edge of the arc the situation can no longer be described by the model. Estimations show that it is likely that there the electron thermal flux is replaced by energy transport through fast electrons that, once produced in the outer part of the arc, may travel considerable distances in the outward direction.
5. By increasing the electric field E_* the arc dimension is decreased. Therefore, the nonuniformity of the discharge under the conditions of closed-cycle MHD generators follows from the high value of the electric field E_* as observed in the reference frame moving with the gas.
6. For input data chosen in accordance with noble gas MHD generator experiments, especially electron temperatures between 4000 and 6000 K, the calculated arc radius is $1 \div 3 \times 10^{-4}$ m, in agreement with the result of laser-scattering experiments.[2]
7. The behavior of the radial dimension with respect to variation of plasma parameters and electrical and magnetic characteristics can be explained in good approximation by the following simple expression:

$$\rho_{max} = 1.1 \times 10^{-4} \frac{T_{eo}}{E_* (1 + \beta^2)^{0.15}} \quad (5)$$

where ρ_{max} is the radial dimension of the filament, and β is the Hall parameter.

Equation (5) follows from a parabolic electron temperature profile and the application of Spitzer values for the electrical conductivity and electron thermal conductivity.

Electrodynamics of a Streamer

In both shock-tube and blow-down experiments it has been observed that at any time a limited amount of streamers is present in the channel, so that each streamer is connected with two or more electrodes. On the other hand, it has been found that the overall Hall voltage is always smaller than that predicted by the existing theories. In this section two electrodynamic models of the field distribution around a streamer will be developed in order to compare the results with measured Hall voltages.

Theory and Calculations

The first model treats the electric field distribution in the u-B midplane of the channel (y = 0.5 h). We consider a simplified two-dimensional model. The streamer is a long elliptical cylinder with small axis 2a (Fig. 4). End effects of the streamer at electrode walls are neglected: hence $\partial\varphi/\partial y = 0$. Inside the streamer the Hall parameter β_s and conductivity σ_s are considered constant, and outside the streamer β and σ are constants, with $\sigma \ll \sigma_s$. The longitudinal current density within the streamer, J_{ys}, is assumed to be a given constant. From the x component of Ohm's law we can now derive the following:

$$E_{xs} = \frac{\beta_s J_{ys}}{\sigma_s} + \frac{J_{xs}}{\sigma_s} + E_{xs}^0 + \frac{J_{xs}}{\sigma_s} \tag{6}$$

where $E^°_{xs}$ is the ideal value of the streamer Hall field under zero current conditions, $J_{xs} = 0$. For the region outside the streamer we find

$$\nabla \cdot \bar{J} = \frac{\partial}{\partial x} \{ \frac{\sigma}{1+\beta^2} [E_x - \beta (E_y - u B)] \} + \frac{\partial}{\partial z} (\sigma E_z) = 0 \tag{7}$$

With

$$\zeta = z (1 + \beta^2)^{-1/2}, \quad \bar{E} = -\nabla\varphi$$

and

$$\phi = \varphi - \beta (u B - E_y) x$$

Equation (7) can be transferred to

$$\Delta\phi = \frac{\partial^2\phi}{\partial x^2} + \frac{\partial^2\phi}{\partial \zeta^2} = 0 \tag{8}$$

This equation will be solved with the boundary conditions at the streamer perimeter:

$$\phi_p = \varphi_p - \beta (uB - E_y) x_p, \quad \varphi_p = -(E_{xs} + \frac{J_{xs}}{\sigma_s}) x_p \tag{9}$$

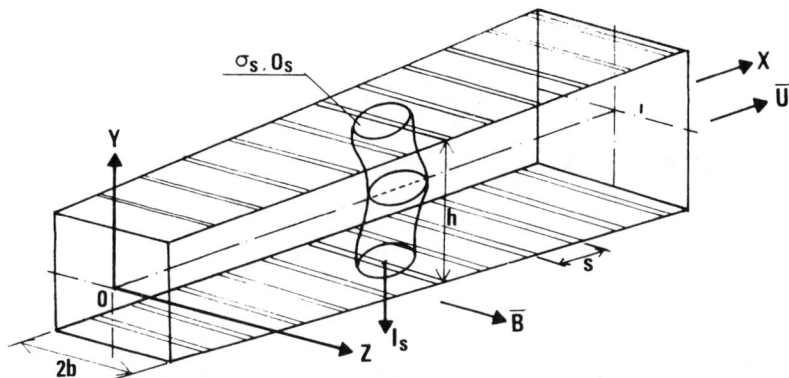

a) Configuration of the channel with width 2b, height h, and segment pitch s. I_s is streamer current, σ_s is streamer conductivity, and O_s is streamer cross section.

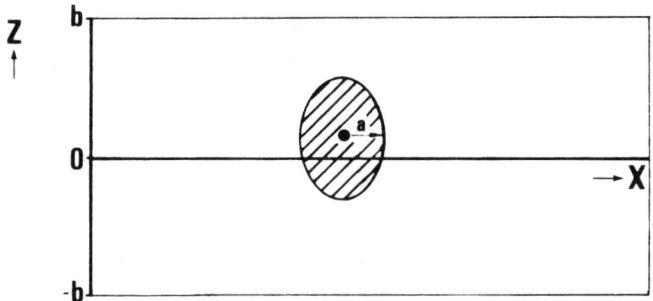

b) Streamer with elliptical cross section in the x-z-midplane of the generator (y = 0.5 h).

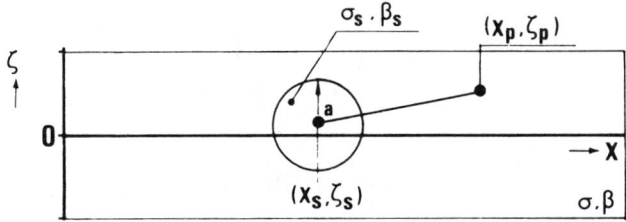

c) Streamer with circular cylindrical cross section after transformation to the x-ς plane.

Fig. 4   Schematic view of a streamer in an MHD generator channel.

where x_p = x coordinate at the streamer perimeter (Fig. 4). The boundary conditions at the insulator walls $z = \pm b$ yield

$$J_z = 0 \quad or \quad \frac{\partial \phi}{\partial z} = \frac{\partial \varphi}{\partial z} = 0 \tag{10}$$

The boundary conditions for $x \to \pm \infty$ are

$$\left(\frac{\partial \phi}{\partial x}\right)_{\pm \infty} = \frac{\partial \varphi}{\partial x} - \beta(uB - E_{x\infty}) = \varepsilon_x = const \tag{11}$$

It is assumed that $\varepsilon_x = 0$, which means that only the Hall voltage generated by the streamer is considered. Further defining

$$V_p = -\left(E_{xs}^0 + \frac{J_{xs}}{\sigma_s}\right)2a \tag{12}$$

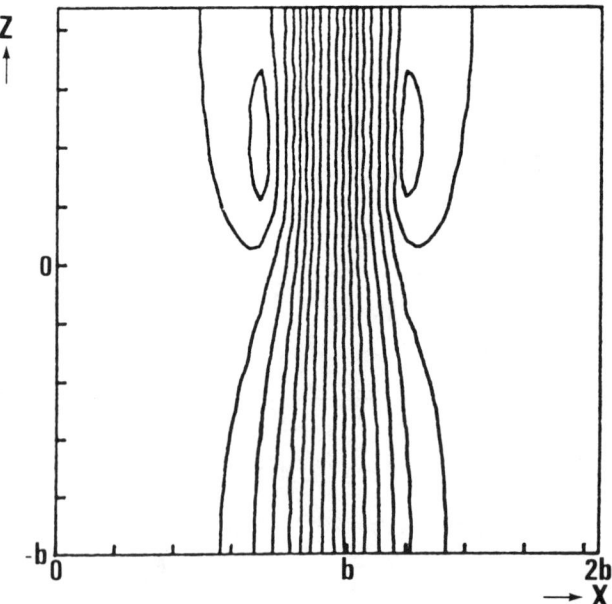

Fig. 5 Two-dimensional equipotential line distribution in the x-z plane with the streamer center at $x_c = 0$ and $z_c = 0.5$ and with radius $a = 0.25\ b$. Numerical solution for ß = 5.

we obtain

$$\phi_p = \frac{V_p}{2a} x_p \qquad (13)$$

Numerical solutions of the set of equations in a 41 x 41 mesh are presented in Fig. 5 as the equipotential line distribution around a circular streamer, centered at the point $x_c = 0$, $z_c = 0.5b$ with radius $a = 0.25b$.

From Fig. 5 we notice that the Hall voltage generated within the streamer is partially shorted in upstream and downstream regions close to the streamer. Furthermore we see that a substantial part of the Hall field of the streamer interior is transmitted to the insulator walls, which can be measured by electrostatic probes. Finally, we see that the streamer potential distribution is similar to the potential distribution of a double layer with a slightly lower potential difference. These numerical results were compared with results from a simplified analytical approach that gave close agreement.

Fig. 6 shows the time-dependent Hall voltage that is calculated from the electric field distribution in Fig. 5 for two probes in the insulator

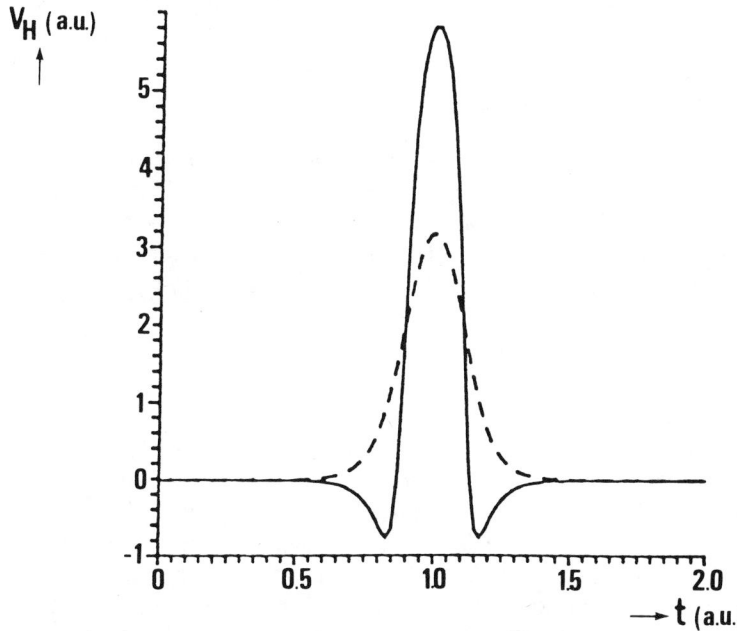

Fig. 6 Hall field distributions along the upper insulator wall (solid line) and the lower insulator wall (dashed line) for the conditions given at Fig. 5.

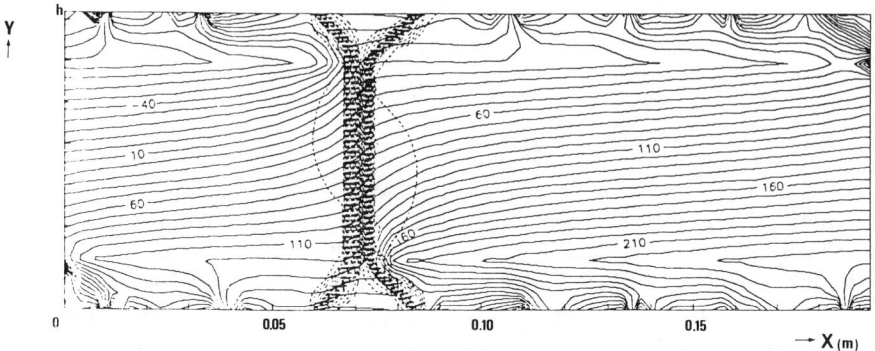

Fig. 7 Schematic streamer configuration in the x-y midplane between the insulator walls (z = 0). The streamer is connected to two electrodes and is indicated by the dashed current stream lines. The solid lines represent the equipotential lines. Case: $\sigma_s = 100\ \sigma$; $\beta_s = 3$; $\beta = 5$.

wall of the channel when the streamer is moving downstream. The solid curve is the Hall voltage measured at the upper insulator wall, close to the streamer, and the dashed curve is the one at the lower insulator wall, away from the streamer.

The second model treats the two-dimensional electric field distribution around a streamer in the x-y midplane between the insulator walls as illustrated in Fig. 7. The shape of the streamer has been derived from the high-speed photography data obtained with the Eindhoven shock-tube facility (Fig. 7 is an example).

In order to simulate the time-dependent behavior of Hall voltage between two adjacent anodes, a number of similar two-dimensional electric field distributions have been calculated by moving the streamer downstream step by step, whereas the Hall voltage between two adjacent electrodes was solved as a function of the streamer location (or time). This time-dependent Hall voltage is shown in Fig. 8.

Experiments

Hall voltage measurements have been performed in the EBDF using electrostatic probes in the insulator wall near electrode 12 at 22 mm streamwise distance (Fig. 2) and using anodes 15 and 16 in the electrode wall, 28 mm apart. An example of the probe Hall voltage is shown in Fig. 9. Outstanding is the sharp single peak during about 15 µs. Comparing Figs. 6 and 9 for the theoretical, respectively, experimental probe Hall voltage, we can state there is a qualitatively good agreement between calculated and measured probe voltage curves. The measured curve in Fig. 9 shows a small remaining Hall voltage that declines slowly after the streamer has passed by. This may be due to

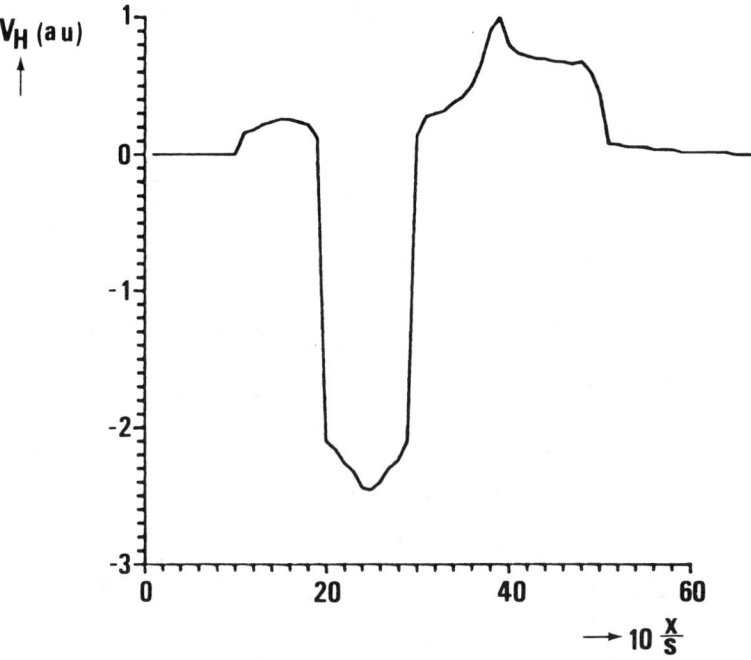

Fig. 8 Hall voltage between adjacent anodes as a function of time for a streamer passing by in the positive x direction (Fig. 7).

the fact that, in practice, the external Hall field is not equal to zero, as was assumed in the theory [Eq. (11)].

The Hall voltage measured between anodes 15 and 16 in the EBDF, run 810, is shown in Fig. 10 for the case where a single streamer is passing by. Compared with the calculated curve in Fig. 8, we notice a striking resemblance again. The explanation of the special features of the Hall voltage is as follows: The first negative peak is created by the first electrical connection of the streamer moving downstream with the first (upstream) electrode pair, i.e., 15. Then a Faraday load current I_F passes through the load resistor R_L and causes a Faraday voltage $V_F = I_L R_L$. In the next electrode pair (16) no current is flowing yet, and the Faraday voltage remains zero. Thus, the upstream anode is suddenly negative related to the down stream anode. After the streamer is connected to the downstream electrode pair (16), the situation is completely changed. Now Faraday currents of approximately the same values pass through both load resistors; hence only a small portion of the Hall voltage at that stage can be defined by a difference in Faraday voltages, and the main portion of the Hall voltage is contributed by the Hall field in the streamer center part (Fig. 5).

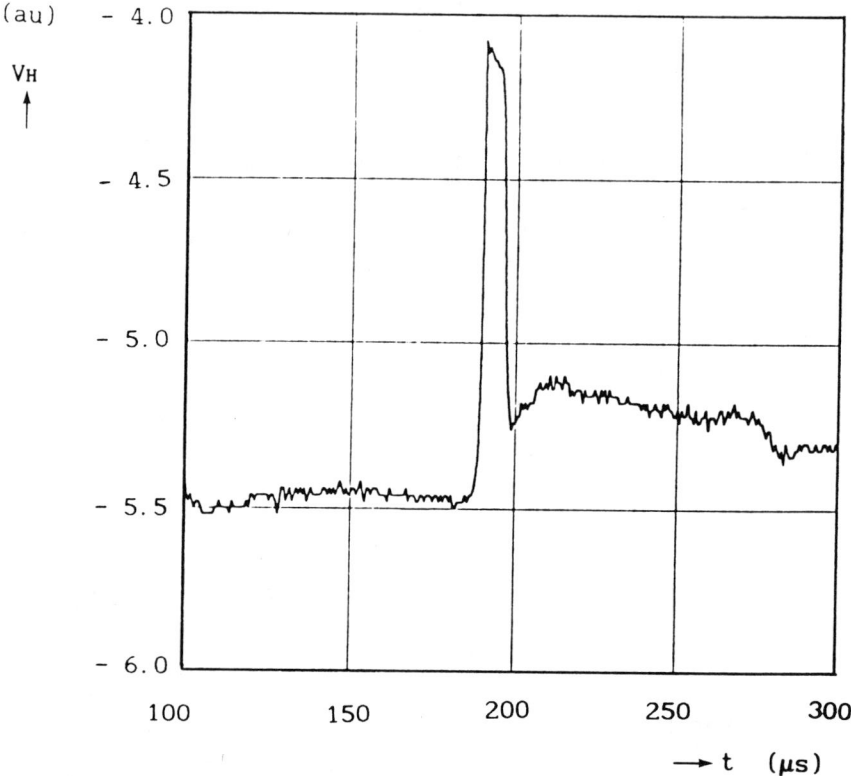

Fig. 9 Hall voltage measured at probes 18-24 in the insulator wall of the EBDF for a single streamer passing by. The sharp peak of 15 µs is comparable with the Hall field peaks in Fig. 6.

Gasdynamics of a Streamer

Theory and Calculations

In this chapter the two-dimensional flow development in the u-B plane of the supersonic flow of low-conducting gas (main flow) around a highly conducting streamer[4] is considered. Hot highly conducting rectangular plasma clots are introduced periodically in the main flow at the entrance section of the duct with parallel insulator walls. The conductivity of the gas is given by

$$\sigma = \sigma_o \, T^{3/4} \, \exp(-I/T) \, p^{1/2} \qquad (14)$$

where T and p are gas temperature and pressure, respectively, and σ_o and I are constants. The calculations are carried out for a combustion

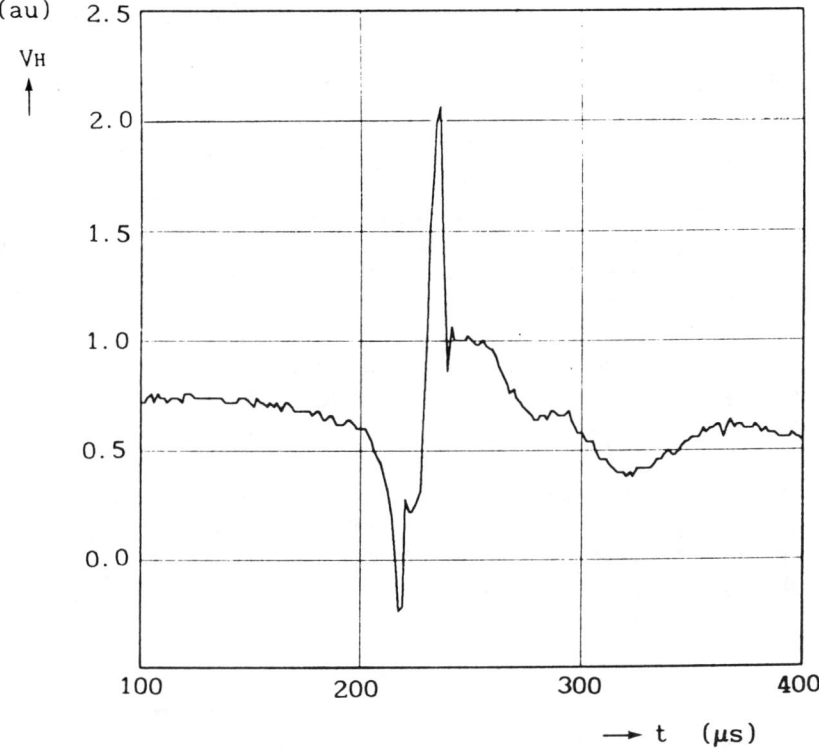

Fig. 10 Hall voltage measured between anodes 15 and 16 in the EBDF. The negative and positive peaks are also found in Fig. 8. For both figures the attachment time of the streamer is about 4 s/u.

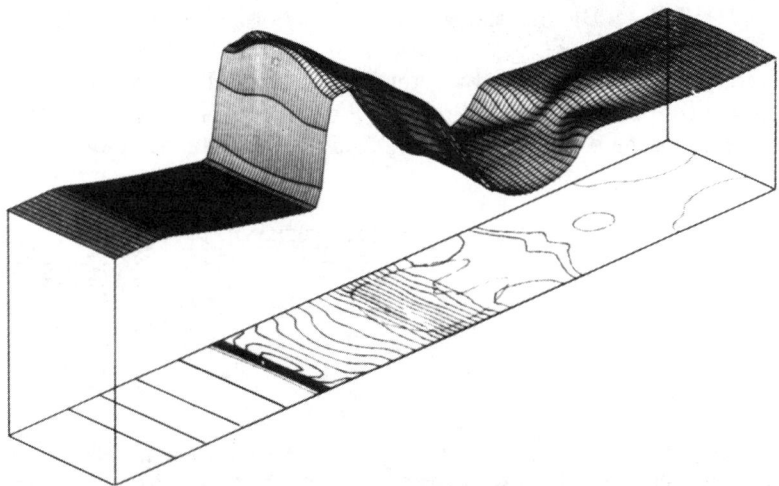

Fig. 11 Pressure distribution p(x,z) around a current-carrying hot plasma clot immersed in a cold main flow. In the x-z plane the isobars are shown. The shock wave upstream of the plasma clot is clearly visible.

gas, but the conclusions apply qualitatively to an Ar-alkali plasma as well. Furthermore, the standard gasdynamic equations are used; thus, viscosity and heat conduction are neglected. The current is assumed to be directed along the y direction. One extra equation for the hot clot location is introduced. It is assumed that plasma clot interacts with the main flow as a mobile impermeable and deformable body. Fig. 11 shows the pressure distribution p(x,z) and corresponding level lines p = const, and Fig. 12 gives the velocity vector distribution in the coordinate system, which refers to the clot mass center (only the half-plane is shown), both for the same moment. Within the hot clot a shock wave builds up that enters upstream into the cold main flow. The retarded deformed hot clot now acts as a dynamic nozzle, and the main flow is accelerated in the gaps between the clot and the insulating walls. Further oblique shocks are created by the interaction of the supersonic jets from the dynamic nozzles with the downstream flow.

The time-dependent two-dimensional solution obtained in Ref. 4 allows one to plot the pressure as a function of time, p(t), at any point of the area considered. These pressure plots are compared with the experimental results in the next section.

Experiments

During the EBDF experiments the pressure was measured as a function of time at different locations in the generator channel and in the supersonic diffuser, as indicated in Fig. 2. The frequency response was limited to 20 kHz.

Figures 13 shows the p(t) signal as calculated on the channel axis, point (x=4, z=0) (Fig. 13a) and measured at electrode 24, (Fig. 13b); see Fig. 2 (PG 10 B). Six stages can be distinguished in the calculated curve. Stage 1 corresponds to the main gas flow passing by the selected point, undisturbed by interactions with clots. Stage 2 corresponds to the zone downstream of the first plasma clot where the pressure drops to a minimum. Stage 3 is the first plasma clot zone with

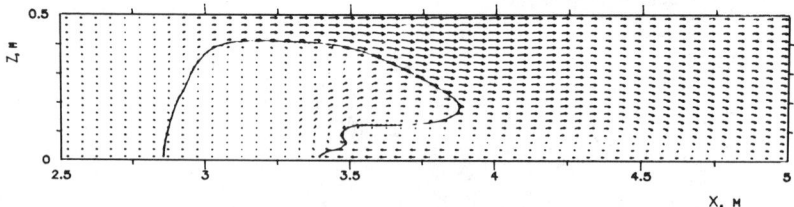

Fig. 12 Velocity vector distribution in the channel region around the hot plasma clot. The solid line indicates the perimeter of the hot clot. The effect of the dynamic nozzle between hot clot and upper wall is clearly shown by the increased length of velocity vectors.

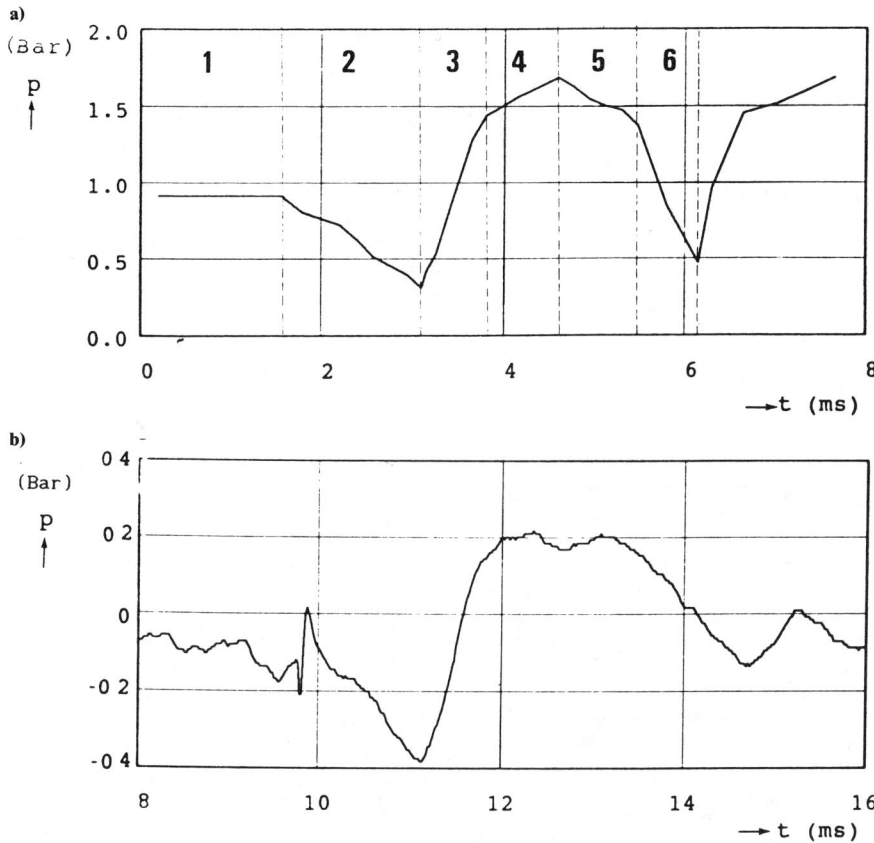

Fig. 13 Pressure variations p(t): a) calculated with the two-dimensional time-dependent model for a fixed position in the channel; b) as measured at pressure probe PG10B (Fig. 2) in the EBDF. See text for a description of stages 1 to 6.

strong interaction and thus increasing pressure. Stage 4 is the shock compressed low conducting gas. Stages 5 and 6 correspond to the expansion and compression waves generated by the interaction of the shock wave propagating upstream of the first clot with the disturbed zone downstream of the second clot. A very simular pressurecurve was measured as shown in the lower track (Fig. 13b). The agreement is even more striking when we realize that the calculations were performed for an open cycle generator with inlet conditions (stagnation temperature T_{st} = 2800 K, static pressure p = 2 bar, cold gas density ρ_c = 0.3 kg/m^3, and u = 1800 m/s, with load factor K = 0.8 and induction B = 5 T), whereas the measurements were performed at a closed-cycle generator (Ar+Cs) (T_{st} = 2000 K, p_{st} = 7 bar, M = 1.6, load factor < 0.5, and induction 5 T). Nevertheless, the main features as mentioned under stages 1-6 apply to both types of generators.

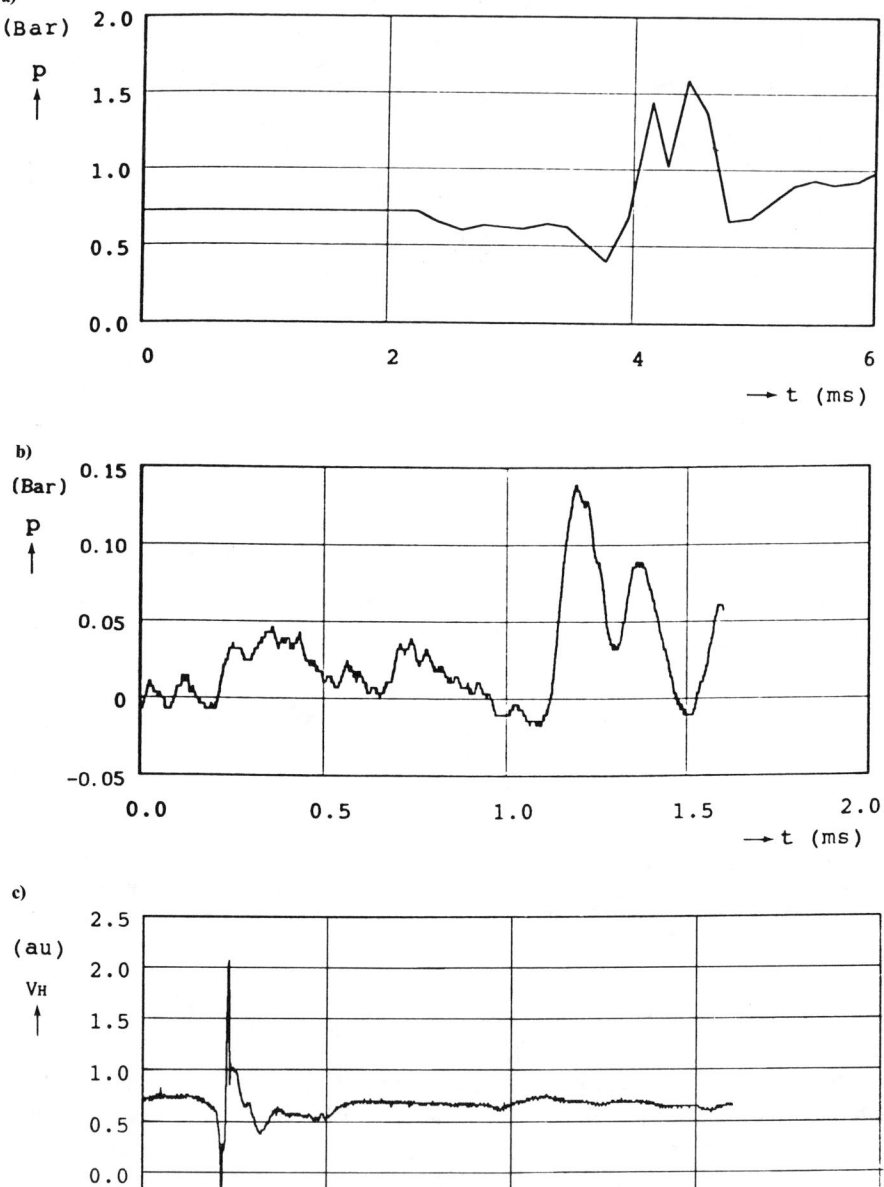

Fig. 14 Calculated pressure variation p(t): a) at the channel exit; b) the measured pressure variation in the supersonic diffuser (PD1K) of the EBDF 0.3 m downstream of the channel exit. c) Signal b has been produced by a single streamer passing by anodes 15-16 about 0.9 ms earlier.

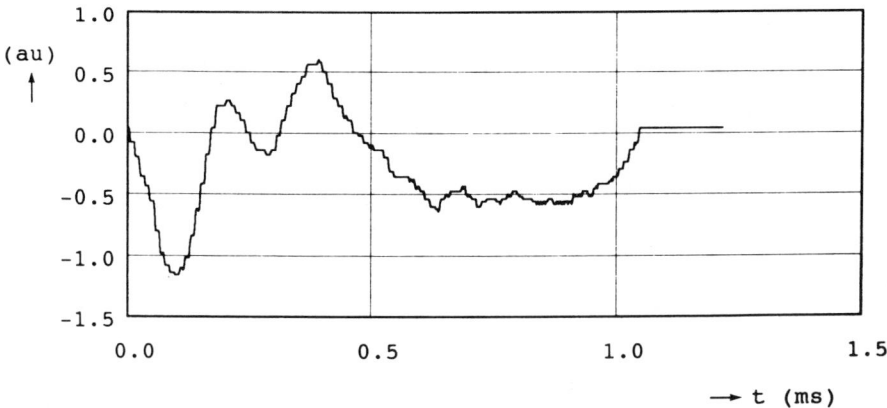

Fig. 15 Cross-correlation signal of Hall voltage at electrodes 15-16 and pressure signal at PG6B where PG6B is located 0.11 m upstream of the electrodes. The pressure signal had to be retarded in relation with the Hall voltage, which means that the shock wave appears at PG6B after the streamer has passed by electrodes 15-16.

Fig. 14 shows the calculated pressure curve at the channel exit (Fig. 14a) and the measured pressure signal in the supersonic diffuser of the EBDF 0.3 m downstream of the channel exit (Fig. 14b) as produced by a single streamer (Fig. 14c). The double peak (Fig. 14a) corresponds to the entrance and exit of the counter pressure shock wave at the first plasma clot passing by the reference point. Curve b shows a double peak between 1.0 and 1.5 ms about 1 ms after the streamer has passed electrodes 15-16 located 0.91 m upstream, which corresponds to the mean gas velocity of ~ 1000 m/s.

Furthermore with many streamers present in the generator channel, cross correlations were made of the Hall voltage at electrodes 15-16 and pressure signals at PG6B (see Fig. 2), located 0.11 m upstream of the electrodes. It was found (Fig. 15), that there is only a significant cross-correlation signal if the pressure signal is retarded in relation with the Hall voltage, which means that the main pressure phenomenon, i.e., the shock wave, appears at PG6B after the streamer has passed by electrodes 15-16.

Again we notice a good qualitative agreement between calculations and experiments. Moreover it is evident that the streamers mainly act as a impermeable body and that the filamentary substructure has no substantial effect on the interaction with the main flow.

Empirical Relations and Power Production

In order to derive the empirical relations the streamer in a generator segment (Fig. 4a) is considered as a part of a Kirchhoff current loop

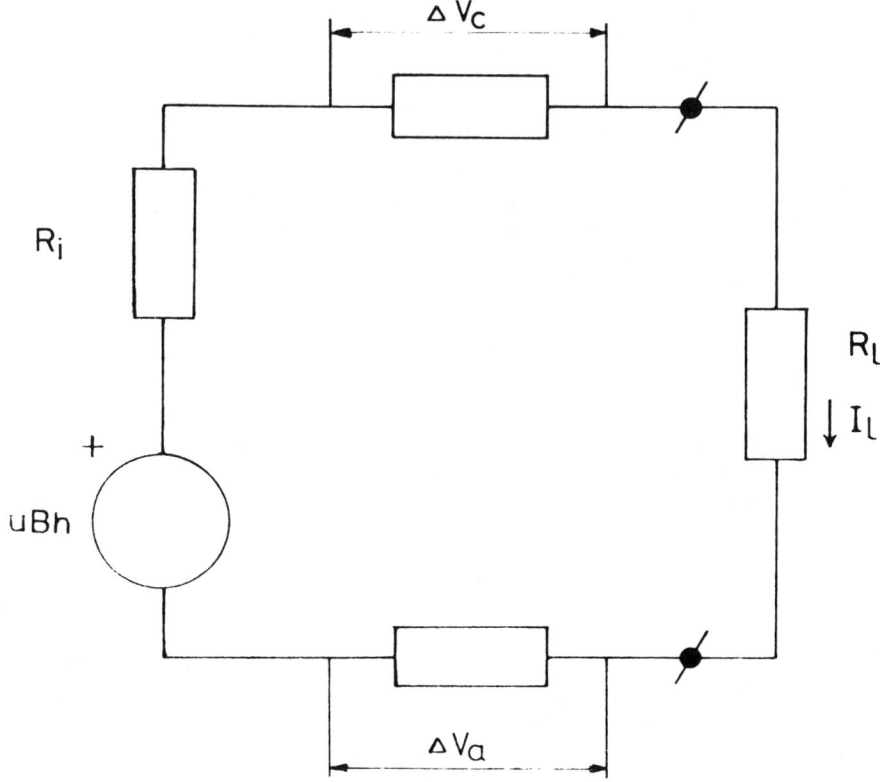

Fig. 16 Kirchhoff current loop of a generator segment with load resistance R_L. uBh = induced voltage; R_i = internal resistance; ΔV_a, ΔV_c = voltage drop at anode resp. cathode.

as indicated in Fig. 16. The internal resistance R_i of the current loop can be calculated from measured quantities:

$$R_i = (UBh - \Delta V_a - \Delta V_c - I_L R_L) / I_L \qquad (15)$$

It should be realized that the internal resistance differs from the streamer resistance R_s. Streamers pass by with a frequency f of about 10^4 Hz, whereas the measured data are mean values over a 0.25-s time interval. With streamer velocity u and segment pitch s the relationships between streamer quantities and segment quantities are found to be[6]

$$I_s = \frac{u}{sf} I_L \qquad (16)$$

$$R_i = \frac{u}{sf} \quad R_s = \frac{u}{sf} \cdot \frac{h}{\sigma_s O_s} \tag{17}$$

The streamer conductance per unit length $\Sigma_s = \sigma_s O_s$ can be found from R_i by

$$\Sigma_s = \frac{h}{R_i} \frac{u}{sf} \tag{18}$$

From the measured load current distributions along the generator channel, it was clear[6] that streamers develop gradually with time in the so-called "relaxation region" and after that remain stable, i.e., "fully developed". Now at given gasdynamic conditions the following is assumed:

1) The ignition frequency f of the streamers is mainly governed by the magnetic induction B.

2) The streamer development in the relaxation region is dependent on external loading and magnetic induction.

3) In the fully developed region the streamer conductance per unit length Σ_s can be expressed as a function of streamer current I_s, gas density ρ, and magnetic induction.

For the streamer conductance per unit length the following expression has been introduced:

$$\Sigma_s = I_s^{a_1} e^{-I_s/I_{so}} \rho^{a_2} (1 + a_3 B^2)^{-1} a_4 \tag{19}$$

where I_{so} is a reference current. The parameters a_1-a_4 were found by a least-squares method applied to a set of 300 measured data points, where I_{so} = 600 A, yielding

$$a_1 = 1.57 \pm 0.04 \tag{20a}$$

$$a_2 = 0.8 \pm 0.1 \tag{20b}$$

$$a_3 = -3.1 \ 10^{-3} \pm 0.9 \ 10^{-3} \tag{20c}$$

$$a_4 = 9 \ 10^{-6} \pm 2 \ 10^{-6} \tag{20d}$$

It should be noticed from Eq. (19) that it does not contain any geometrical size of the channel used. Applying Eqs. (20) to the EBDF experimental results, an agreement within 5% of measured and calculated electric power output was achieved. Thus the empirical relations were considered to be applicable for scaling up, and a case study of a 200-MW thermal linear MHD generator has been worked out. This size of MHD generator is considered for first type retrofit installations. Table 2 shows

Table 2 Inlet conditions of an argon-cesium linear MHD generator with 200-MW thermal input power

Height	1.25 - 1.79	m
Width	0.42 - 1.13	m
Length	8.00	m
Segment pitch	0.20	m
Stagnation temperature	2000.00	K
Stagnation pressure	7.00	bar
Inlet Mach no.	2.50	
Mass flow	192.00	kg/s
Seed fraction	0.07	%
Magnetic induction	5.00	T
Load resistance	4-20.00	Ω

the inlet conditions of this generator. The generator loading has been chosen so that an optimum was obtained for power extraction, isentropic efficiency, local efficiency, and gasdynamic performance.

Figures 17a-17f show the distributions along the generator channel of load current, electric conductivity averaged over a segment, Mach number, static pressure, incompressible shape factor (ISF), and the efficiencies. The averaged electric conductivity (Fig. 17b) is very low because of the small section of the segment occupied by the streamer. The high magnetic interaction causes a sharp decline of the Mach number (Fig. 17c). Downstream of the 30th generator segment the static pressure (Fig. 17d) increases, which is accompanied by the sharp increase of the ISF (Fig. 17e). This means that the generator is operating close to boundary-layer separation.

In Fig. 17f the efficiencies are presented. Compared with wanted values, we find the following:
1) Enthalpic efficiency = 24%, ≥30% wanted;
2) Isentropic efficiency = 47%, ≥70% wanted; and
3) Local efficiency = 47-64%, ≥75% wanted.

From these values it is clear that this generator under the prevailing conditions and assumptions does not meet the wanted terms, and it may be concluded that the Ar-Cs linear MHD generator operating in the streamer mode is not feasible for commercial electric power generation. Only if the streamer conductance per unit length Σ_s can be increased by a factor 2 or more can the desired terms be met, as was proved by computer experiments.

The main cause of the aforementioned problems is found in the filamentation of the streamers. The thin filaments are responsible for high energy losses due to electron heat conduction (see the section on filaments). Only when the diameter of the filaments can be increased can the losses in the streamers be reduced and the performance of the generator improved. This point will be thoroughly investigated.

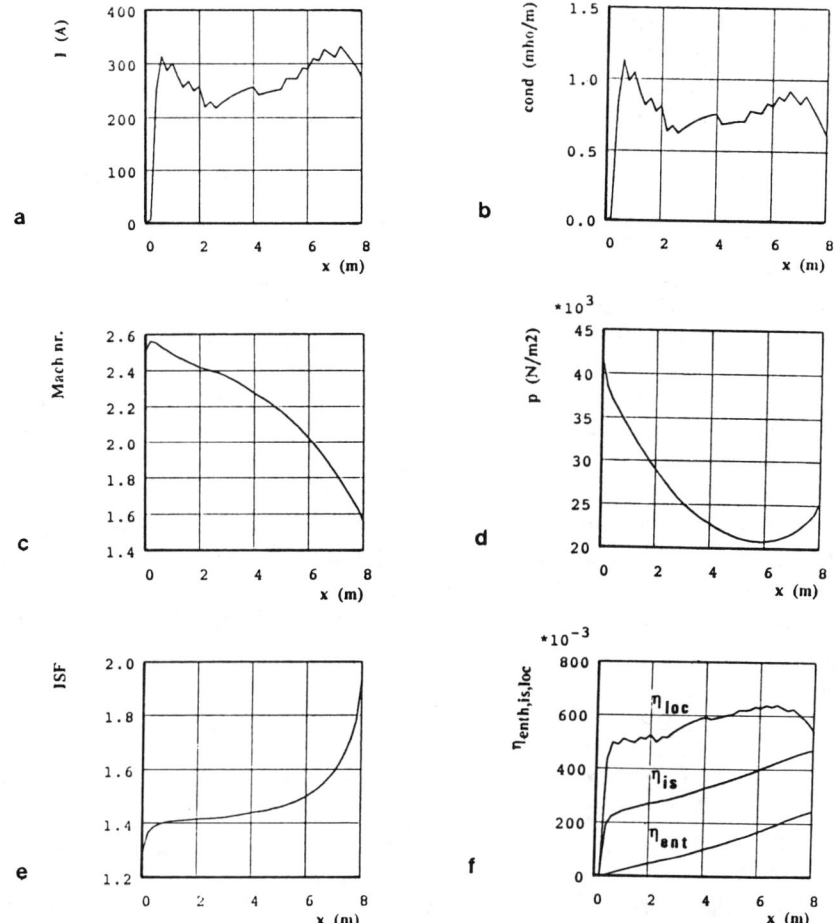

Fig. 17 Distribution along the generator channel of a) Load current; b) Averaged electrical conductivity; c) Mach number; d) Static pressure; e) Incompressible shape factor; f) enthalpic, isentropic, and local efficiency. 200-MW thermal input MHD generator with inlet conditions of Table 2 and according Eqs. (19) and (20).

Conclusions

In the preceding sections the constricted discharges as they appear in linear alkali seeded noble gas MHD generators were considered from different points of view, leading to a series of conclusions, of which the most important ones are reformulated as follows:

1) Streamers are composed of a bunch of thin filamentary arcs, of which the radius is defined by a sharp decrease in the electron concentration.

2) In the filaments the heat of joule dissipation is transported by electron heat conduction and at the edge by the loss of fast electrons.

3) The radius of a filament is proportional to the central electron temperature and inversely proportional to the electric field (E_y-uB).

4) A two-dimensional electrodynamic approach to the cross-sectional plane of the streamer gives a good prediction of the Hall fields to be measured at the insulator walls.

5) A two-dimensional electrodynamic approach to the longitudinal sectional plane of the streamer gives a good prediction of Hall voltages measured along the electrode walls.

6) The overall Hall voltage of the generator is mainly contributed to by the Hall field inside the streamer.

7) The interaction of a current-carrying streamer in a supersonic flow induces a shock wave moving upstream of that streamer.

8) Around the streamer the gas is being accelerated by a dynamic nozzle,

9) The two-dimensional time-dependent gasdynamic approach gives a good prediction of the time-dependent pressure changes at different locations in the MHD generator.

10) The electrical performance of the Ar-Cs MHD generator can be described by an empirical method using the streamer conductance per unit length Σ_s, which is a function of streamer current, gas density, and magnetic induction.

11) Computer experiments using the streamer conductance per unit length show that a 200-MW_{th} Ar-Cs-driven MHD generator does not work satisfactorily because of too low of an isentropic efficiency.

12) The main loss factor in linear Ar-Cs generators operating in the streamer mode is the joule dissipation in the filaments.

From the aforementioned conclusions it may be clear that there is now a good understanding of the performance of MHD generators working in the streamer mode. In order to improve the performance a reduction of the losses in the filaments is necessary. An increase of the averaged electrical conductivity by a factor of 2 has been calculated to be sufficient. At Eindhoven University of Technology this problem will be further investigated. Another approach to the improvement of Ar-Cs MHD generators is to operate them in the fully ionized seed mode, work which is being conducted at the Tokyo Institute of Technology. It has been shown[5] that a highly conducting hot clot in an open-cycle generator can substantially improve its power production.

References

[1]Montgomery, D., Phillips, L., and Theobald, M. L., *"Helical, Dissipative, Magnetohydrodynamic States with Flow,"* <u>Pysical Review, Section A</u>, Vol. 40, No. 3, Aug. 1989, pp. 1515-1523.

[2]De Haas, J. C. M., Schenkelaars, H. J. W., Van de Mortel, P. J., Schram, D. C., and Veefkind, A., "Collective CO_2 Laser Scattering on Moving Discharge Structures in the Submillimeter Range in a Magnetohydrodynamic Generator," Physics of Fluids, Vol. 29, No. 5, May 1986, pp. 1725-1730.

[3]Stefanov, B., Veefkind, A., and Zarkova, L., "Thin Free Arcs in Alkali-Seeded Noble Gases," IEEE Transactions on Plasma Sciences, Vol. 17, No. 1, Feb. 1989, pp. 51-59.

[4] Bityurin, V. A., Holten, A. P. C., and Veefkind, A., "Notes on Gasdynamic Aspects and Streamer Characteristics in Noble Gas MHD Generators II," Eindhoven University of Technology, Eindhoven, The Netherlands, Internal Rept., EG/89/464, 1989.

[5]Bityurin, V. A., Likhachev, A. P., Lyubimov, G. A., and Median, S. A., "On the Dynamics of a Non-Uniform Conducting Flow in an MHD Generator Channel," Magnetohydrodynamics International Journal, Vol. 2, Nos. 2-3, 1989, pp. 173-184.

[6]Merck, W. F. H. and Rietjens, L. H. Th., "Empirical Description of Discharge Phenomena in Ar-Cs MHD Generators," 10th International Conference on MHD Electrical Power Generation, 1989, Vol. II, pp. XII.38-44.

Pseudo Two-Phase Flow in an Open-Cycle MHD Generator

V. A. Bityurin* and A. P. Likhachev†

USSR Academy of Sciences, Moscow, Russia

Abstract

Several types of magnetohydrodynamic (MHD) generators that operate with a strongly non uniform non-steady-state flows are known to date. In an open cycle-combustion-driven MHD generator, such a flow can be created artificially or forms spontaneously. As a result of an interaction of non uniform flow with an externally applied magnetic field, this flow develops into a two-phase flow, consisting of a main low-conducting portion (phase 1) and high-conducting clots overheated by Joule dissipation (phase 2). The integral and local characteristics of such an MHD generator are discussed in this paper. Particularly, it is shown that the enthalpy extraction coefficient can be as high as 35%, isentropic efficiency is about 80%. The interaction between main flow hot plasma clots and an external electromagnetic field is also discussed on the basis of two-dimensional-time-dependent numerical modeling of the flow.

Introduction

Analysis of a linear open-cycle combustion-driven magneto-hyrodynamic (MHD) generator has shown that the prospect of improvement of its characteristics is rather limited for two main reasons. First, the inlet temperature for baseload application can hardly be higher than 3000 K for fossil fuel combustion gases; second, the outlet temperature should be higher than 2300 K because the electrical conductivity becomes too low. This temperature range defines the enthalpy extraction coefficient at the maximal level as 20-25% which is inadequate to design a competitive combined power station.

Copyright © 1992 by the American Institute of Aeronautics and Astronautics, Inc. All rights reserved.
*Head of Laboratory, Institute of High Temperature.
†Institute of High Temperature.

One of the attractive ways of increasing the enthalpy extraction coefficient of a combustion driven MHD generator is to use a non-steady-state "pseudo-two-phase" flow in an MHD generator channel.[1,2] This non uniform flow consists of two different working media or phases: the main flow of fossil fuel combustion products with low conductivity and 5-10% of the volume portion of the second "hot" high-conducting phase. The second phase can differ from the main flow media by the significant higher conductivity provided by higher temperatures or by some other sources. The hot phase portion can be created artificially or can appear spontaneously due to some non linear mechanism.

The most well-known type of such an MHD generator is the so-called MHD generator with a T-layer.[3,4] In this MHD generator, the hot-high conducting portions are created by electrical discharge at the MHD generator inlet which overheats the plasma up to 10^4 K with conductivity above 10^3 mho/m.

The main idea of such pseudo-two-phase open-cycle MHD generators is to divide the functions of each phase. The "hot" high conducting phase is responsible for MHD interaction with the external magnetic field, and the low-conducting phase works by pushing the hot plasma clot as a piston. The Joule dissipation heat which is now located in a relatively small volume of the hot plasma clot maintains the conductivity level during expansion of working gas in the MHD generator channel. The disadvantages of the MHD generator with a T-layer are as follows: at first, the energy balance in the T-layer is very sensitive to operation conditions and shock waves created by usually strong MHD interaction with the magnetic field.[4,5]

Recently, some other MHD generator types have been considered.[6,7] In contrast to the T-layer MHD generator, we assume that at least hot plasma clots are seeded by alkali metal to reach a high conductivity level even at relatively low temperatures of about 3300-3500 K. There are two advantages in this case: first, it is a much more economical way of hot plasma clot creation; second, it is a markedly lower interaction of a single clot with the magnetic field that allows us to avoid practically shock-wave processes.[8] In frames of this approach, a very high enthalpy extraction coefficient (above 30%) was demonstrated initially for an open-cycle MHD generator.[7]

In this paper, we continue to study the overall performance of such an MHD generator to understand better the field of application and investigations of interaction mechanisms between two phases, especially for non-regular cases.

Integral Characteristics

In Ref. 7 the rather high performance of an MHD generator with current-carrying nonuniformities was demonstrated. The MHD generator flow was simulated numerically with a quasi–one-dimensional time-dependent model developed for this study. The peculiarities of the numerical procedure is the "quasi-Lagrange" approach in combination with the modified Godunov's technique, which allowed a high resolution of the flow non uniformities.

The full and detailed problem formulation is also presented in Ref. 7. In this paper, new loading conditions at the inlet of the MHD generator channel and a modification of outlet pressure conditions allowed output characteristics to improve significantly.

The MHD generator considered here is characterized by the following parameters: the length of the channel is about 20 m, the magnetic induction (maximum) is 4 T (see also Fig.1), the cross-sectional area changes from ~0.1 to 2 m^2 at the outlet. The main flow parameters are as follows: inlet velocity, 1200 m/s, static pressure at the inlet 1.3 MPa, and at the diffusor outlet, 0.1 MPa; and inlet static temperature, 2195 K. The working media is assumed to be an ideal gas with a specific heat ratio $\gamma = 1.2$ and the gas constant R = 300 J/kg·K.

It was assumed also that the hot plasma clot was a slab blocking all of the channel cross-section. These slabs entered the MHD channel periodically during the time τ_h = 0.2 ms, interrupted by the main flow portion entering during τ_c = 1.4 ms. The temperature of plasma clots at the inlet was 3332 K. The initial conditions were related to an adiabatic steady-state flow of the main gas.

The most important results are presented in Figs. 2 and 3. Figure 2 shows that the enthalpy extraction coefficient reached a very high level at 35%. It is important to outline, that under these conditions, eight hot plasma clots are situated in the MHD channel simultaneously. Figure 3 depicts the static pressure distribution. One can see that the pressure

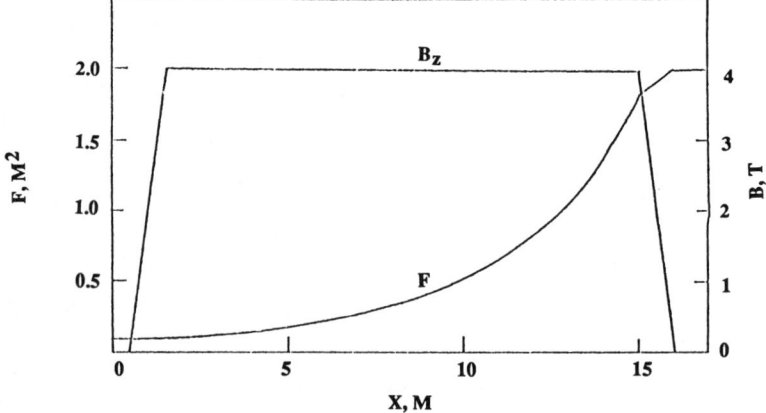

Fig. 1 Magnetic induction and cross-sectional area vs distance in quasi-one-dimensinal simulation.

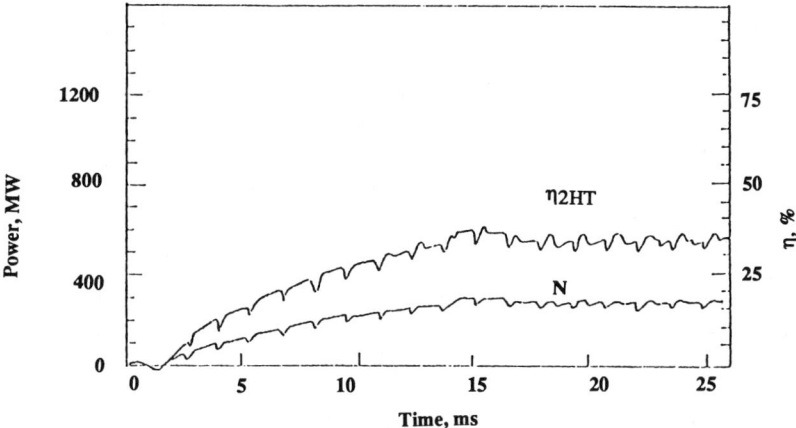

Fig. 2 Time variations of integral power N and enthalpy extraction ratio η in quasi-one-dimensional simulation.

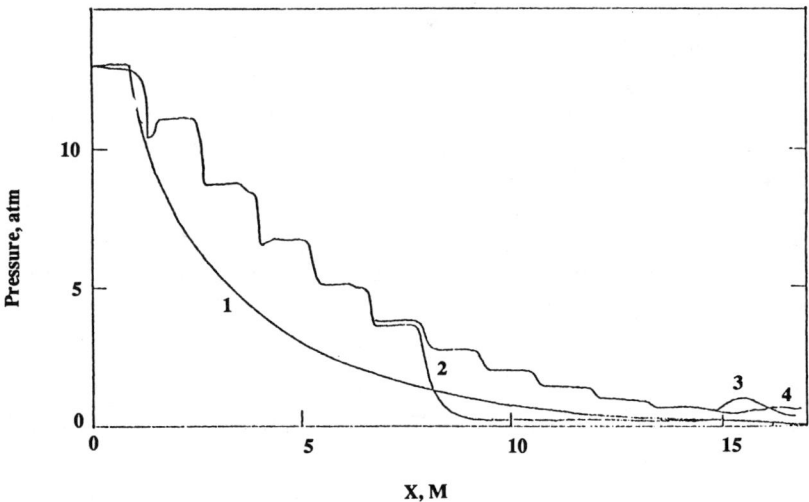

Fig. 3 Static pressure vs distance in quasi-one-dimensional simulation for the moments of time: 1-t = 0; 2-t = 8.4 ms; 3-t = 16.8 ms; 4 - t= 25.5 ms.

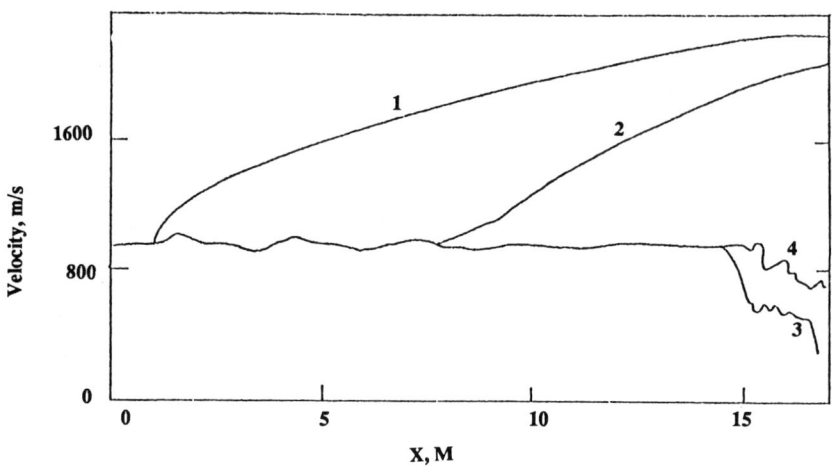

Fig. 4 Velocity vs distance in quasi-one-dimensional simulation for the different moments of time (for legends see Fig. 3).

distribution has a specific "staircase" shape, which corresponds to the flow with current carrying non uniformities. It can be shown that plate regions correspond to practically adiabatic flow of the main gas portion and the steps correspond to the high-conducting plasma clots. This figure clearly shows the energy conversion mechanism in the MHD generator of such a kind: the main gas portion expands and pushes the plasma portions in the externally applied magnetic field. As a result of motion, some current induced in the plasma clot creates an electromagnetic drag force working against which the main flow converts its heat energy to the electrical power.

In Fig.4 the development of velocity distribution along the channel is also presented. One can see that the velocity at the on-design conditions is practically constant. It is also important to mention that the value of the isentropic efficiency

$$\eta_{oi} = \Delta H_{MHD} / \Delta H_{ise}$$

in the case discussed was as high as 84% which corresponds to a very effective MHD generator.

Two-Dimensional Effects

In practice, it is very difficult to expect that an ideal structure of the flow can be maintained automatically. The shape of the plasma clot can be disturbed, for example, by interaction with an insulator wall or by Rayleigh-Taylor instability.

In Ref. 6, the mechanical interaction between the main flow and that limited in a B-direction plasma clot was studied. It was shown that during the channel flow evolution the interphase interaction passes two stages. The first stage relates to a significantly non-steady state interaction when entering the magnetic field. At this stage the plasma clot is decelerated by electromagnetic forces $\bar{J} \cdot \bar{B}$, which results in the main flow pattern transformation accompanied by effective gasdynamic force appearance.

The second stage is a quasi-stationary critical flow in a relative motion. In this case, the first-stage clot deceleration results in "dynamic nozzle" construction, which is very similar to a subsonic-supersonic nozzle formed by the insulator wall and the plasma clot. The static pressure in the nozzle changes monotonically, which results in creation of the total force acting on the clot from the main flow and then directed downstream.

The typical pressure distribution in the B-u plane (perpendicular to the Faraday current) calculated in Ref. 6 is presented in Fig. 5. It is shown that initially the brick-shaped plasma clot transformed into a kidney-shape clot which practically fully blocked the channel cross section. A small portion of the main flow can pass through the gap between the clot and the insulator wall. The flow pattern in motion in the relation of the clot mass center is presented in Fig. 6 and obviously supports the explanation.

It is very interesting and important to examine the influence of the initial location of the hot plasma clot at the channel inlet. In contrast to Ref. 6, where the symmetrical case was studied, in this paper the results of non symmetrical entrances to the channel are considered. Under the same initial and loading conditions, the plasma clot entering was shifted in the B-direction at one tenth of the channel width. The solution has shown that, in the first order magnitudes, the process develops similar to the symmetrical case. At the same time, because the geometry of the flowfield was locally significantly changed, the more or less markedly discrepancies of symmetrical and asymmetrical cases could be observed. At the beginning, the flowfield pattern develops practically independently at the two sides of the clots. However, later the interaction of shocks, compression and expansion waves coming from different sides occurs, creating a very complicated structure (see Figs. 7 and 8). It is also important that the clot was deformed weakly, not to change the process in principle.

The second series of two-dimensional process simulation deals with the plasma clot stability study in respect to finite disturbances of the clot shape or border. It is well known that one of the main reasons for stopping the theoretical and experimental study of such an MHD generator in the early 1960s was the conclusion based on an absolute instability in the normal contact border in respect to the Rayleigh-Taylor

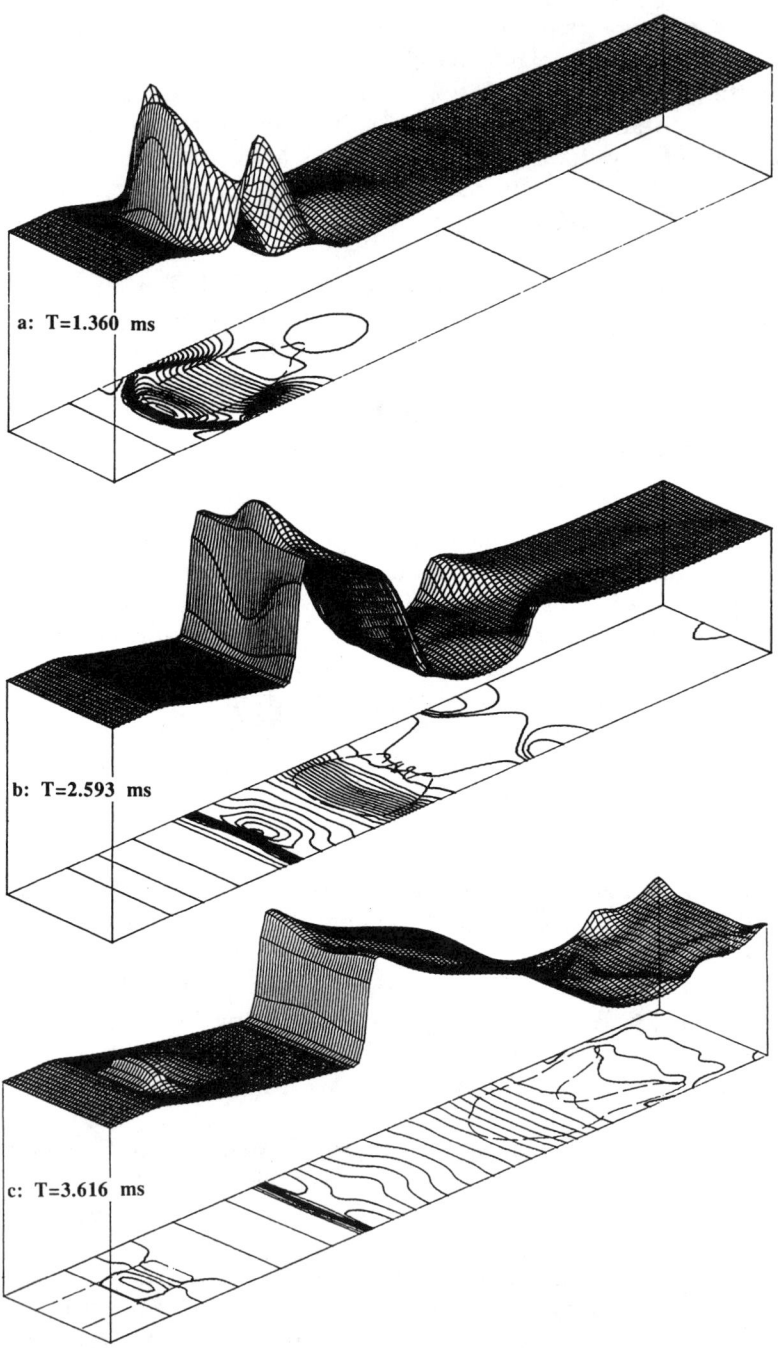

Fig. 5 Static pressure distribution p(x,z) and corresponding level lines p = const for the different moments of time in quasi-two-dimensional simulation (symmetrical entrance case). At the channel inlet (left-hand boundary of the picture base), p = 0.2 MPa, line interval Δp = 15 kPa is reckoned from zero.

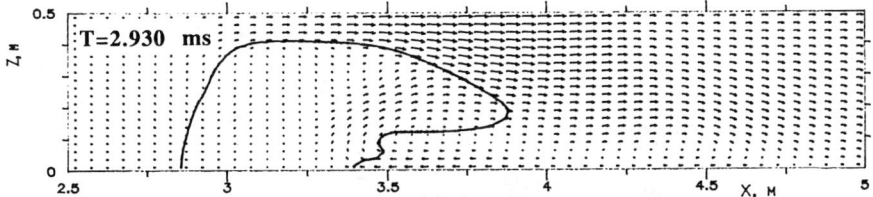

Fig. 6 Vector diagram of velocity determined in the coordinate system moving with the clot mass center (quasi-two-dimensional simulation, symmetrical entrance case).

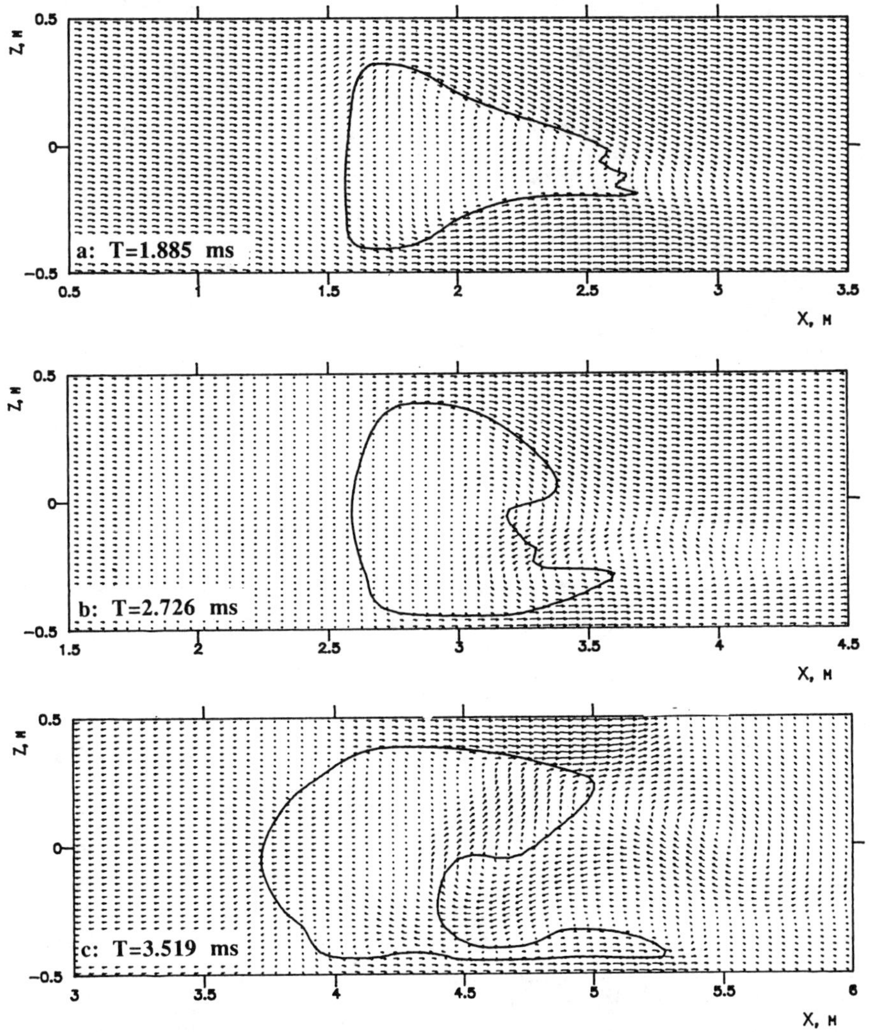

Fig. 7 Vector diagram of relative velocity (quasi-two-dimensional simulation, nonsymmetrical entrance case).

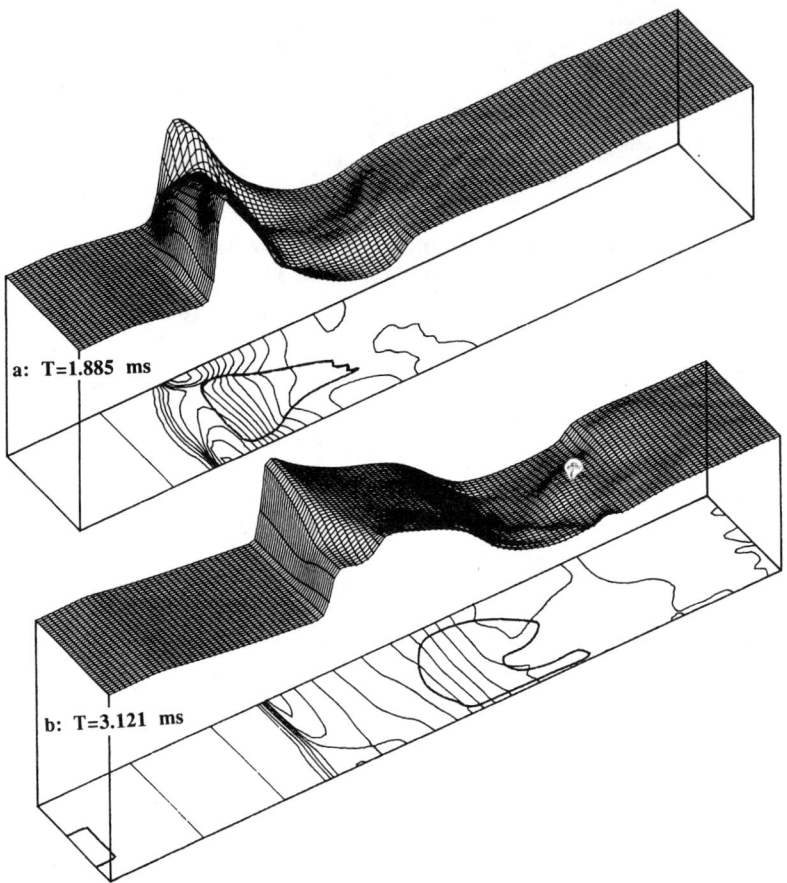

Fig. 8 Static pressure distribution p(x,z) and corresponding level lines p = const for the different moments of time (quasi-two-dimensional simulation, nonsymmetrical entrance case). Line interval Δp = 30 kPa: other conditions are the same as in Fig. 5.

modes.[8] Nevertheless, in last decade, experimental data were obtained for different types of quasi-periodically nonuniform flow which showed less increment than from linear analysis. The recent theoretical study has also shown an increase of the characteristic time of the disturbance growth about 50 times due to strong non linear effects.[9]

Therefore, the numerical simulation of the development of upstream clot border disturbances was carried out. Under the same initial and boundary conditions discussed earlier, the upstream border of the clot was disturbed after the plasma clot had entered the nominal magnetic field. The shape of the disturbance can be seen in Fig. 9. This disturbance was a shutdown of the electrical conductivity at the moment t_0 = 2.33 ms,

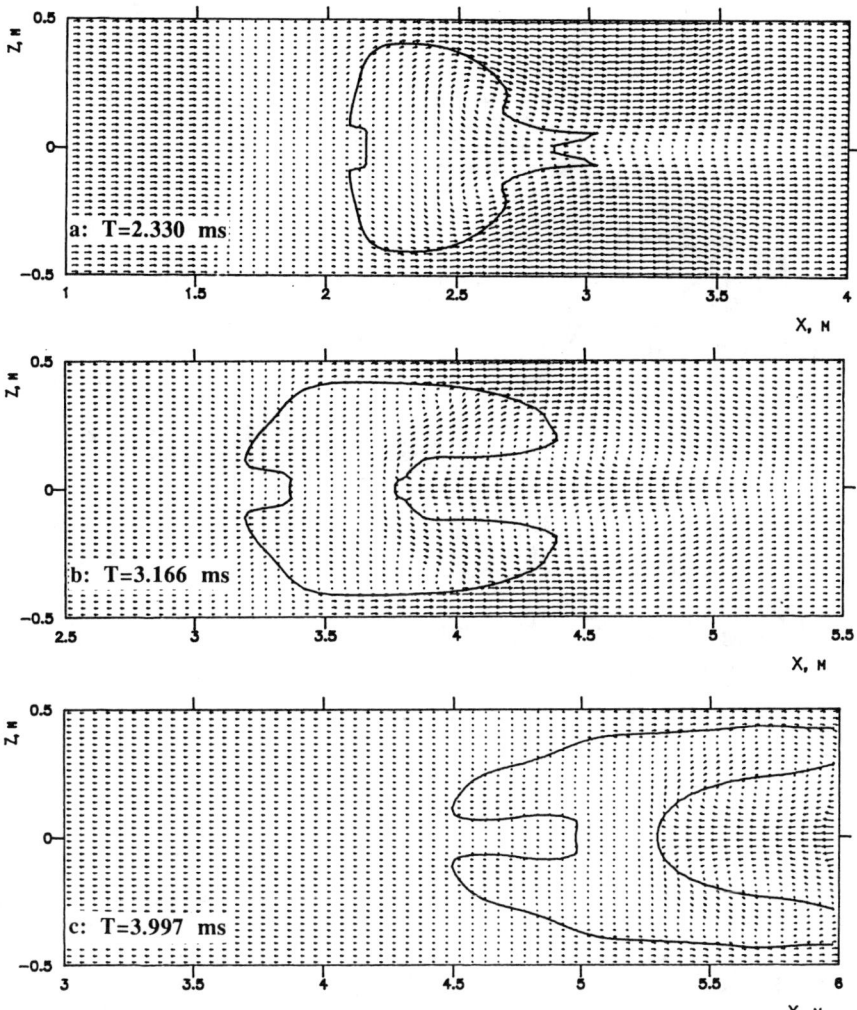

Fig. 9 Vector diagram of relative velocity (quasi-two-dimensional simulation, development of the artificial disturbance of the clot surface).

so that some portion of the clot became unbalanced because of the electromagnetic body force disappeared but the pressure gradient still existed. The development of the disturbances is presented in Fig. 9 a–c (velocity diagrams) and in Figs. 10a and 10b. (static pressure distribution). From these figures, one can conclude that the growth of the amplitude has a rather regular character with a characteristic time of about several milliseconds. During this process the interaction efficiency does not change. The details of the process depend sufficiently on the size, shape, and creation time of the disturbances. To make a final conclusion on this matter, more studies are needed.

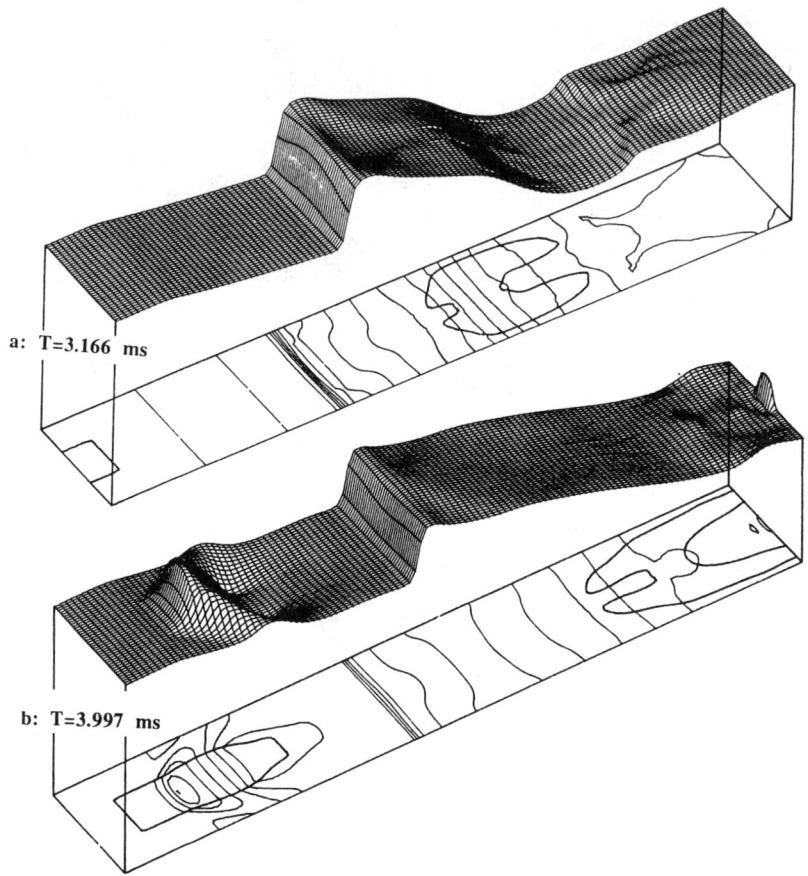

Fig. 10 Static pressure distribution p(x,z) and corresponding level lines p = const for the different moments of time (quasi-two-dimensional simulation, development of the artificial disturbance of the clot surface). Conditions are the same as in Fig. 8.

Conclusion

The results of two types of numerical simulation of a pseudo–two-phase flow in an open-cycle combustion-driven MHD generator are presented and discussed.

It is shown that, under rather low initial gasdynamic parameters and magnetic induction, a very high enthalpy extraction coefficient with high isentropic efficiency (35% and 80%, respectively) can be reached, in principle.

Asymmetrical location of the plasma clot at the channel inlet weakly influences the integral characteristics.

The finite plasma clot disturbance develops with low velocity. During the disturbed clot evolution the MHD interaction efficiency is practically stable.

References

[1]Fraidenraioh, N., Medin, S. A., and Tring, M. V., "Vozmoghnosti MGD generatora so 'sloyevym' potokom rabochego tela," *Magnetogidrodinamichekoye preobrazovaniye energii*, edited by V. A. Popov, VINITI, Moscow, 1966, pt. 1, pp. 271-284.

[2]Ricato, P. and Zettvoog, P., "MHD generators neodnorodnym potokom rabochego tela," *Prikladnaya magnitnaya gidrodinamika*, edited by A. V. Gubarev, Mir Publishers, Moscow, 1965, pp. 93-109.

[3]Gridnev, N.P., Katsnelson, S.S., and Fomichev, V.P., *Neodnorodnye MGD techeniya T-sloyem*, Nauka Publishers, Novosibirsk, 1984.

[4]Derevyanko, V. A., Slavin, V. S., and Sokolov, V. S., "Vysokoeffektivny MGD generator ispolzuyushiy gazoplazmennye potoki s krupnomasshtabnymi neodnorodnostyami," *Teplotehnicheskiye problemy preobrazovaniya energii*, Naukova Dumka, Kiev, 1979, pp. 54-58.

[5]Slavin, V. S., "Final Results of the Theoretical Study of the Development Problems of an MHD Generator with Self-Maintained Current Layers," *Magnetohydrodynamics: The International Journal*, Vol. 2, Nos. 2-3, 1989, pp. 127-140.

[6]Bityurin, V. A., Likhachev, A. P.,Medin, S. A., and Lyubimov, G. A., "On the Dynamics of a Non-Uniform Conducting Flow in an MHD Generator Channel," *Magnetohydrodynamics: The International Journal*, Vol. 2, Nos. 2-3, 1989, pp.173-184.

[7]Bityurin, V. A.and Likhachev, A. P., "High-Efficient MHD-Generator with Space and Time Dependent Current Carrying Non-Uniformities," *Tenth International Conference on MHD Electrical Power Generation*, Vol. 2, 1989, pp. X.191-X.197.

[8]Nedospassov A.V. and Khait, V. D., *Kolebania i neustoichivosti niskotemperaturnoi plasmy*, Nauka Publishers, Moscow, 1979

[9]Gasilov, V. A., Zakharov, S. V., and Tkaohenko S. I., "Modelirovanie dinamiki tokovoso sloja," Preprint No. 66, Keldysh Institute of Applied Mathematics, USSR Academy of Science, Moscow, USSR, 1989.

Analysis of Flow Parameters in MHD Channel at Various Load Conditions

B. Zaporowski* and K. Sroka†
Technical University of Posnan, Posnan, Poland

Abstract

This study shows the analysis of the parameters of flow in the channel of an MHD generator, which works in an open cycle at various loads. A mathematical model was formulated of the flow of plasma in the channel. This formed the basis, after elaboration, for the computer program. The analysis constitutes the results of calculations of multivariant values of electric power, generated in the MHD channel, as a function of plasma pressure in the combustion chamber of the MHD generator, Mach's number, and the load factor of the MHD channel. The calculations were performed for the MHD channels in which the electrodes are connected in Faraday's, Hall's, and Montardy's systems. The changes in parameters of plasma are also determined, such as temperature, pressure, and velocity along the channels of Faraday's and Montardy's types, as well as the exchange in the cross sections of the channels of Faraday's and Montardy's types along their length.

Introduction

The channel of an MHD generator is a device that combines the functions of a turbine and generator. Electric energy is generated in it due to breaking of the flow of the partly ionized plasma flux in the transverse magnetic field.

The electromotive force, induced in the plasma, produces the flow of electric current through the electrodes, connected with the external receivers. Thus, the electric current flowing in the receivers is the source of useful electric energy. The value of the electromotive force induced in plasma and the value of electric current generated by it depends first of all on the parameters of plasma such as temperature, pressure, velocity, and electrical conductance, as well as on the induction of the magnetic field in which it moves and on the load factor of the channel. The parameters of plasma first change along the length of the channel.

Copyright © 1992 by the American Institute of Aeronautics and Astronautics, Inc. All rights reserved.
*Professor and Dean, Faculty of Electrical Engineering.
†Professor, Faculty of Electrical Engineering.

In order to have the possibility of determining the basic parameters of the flow of plasma along the channel of the MHD generator, it is necessary to formulate a suitable mathematical model describing the chemical, thermal, flow, and electrical phenomena that occur in plasma during its flow in the magnetic field. This study aims in principle at formulating such a model and elaborating a method of its solution, as well as at performing calculations by means of a special computer program for different load factors of the channel.

Mathematical Model of Plasma Flow in the Channel of an MHD Generator

Phenomena, occurring in the flux of plasma flowing in the channel of an MHD generator, can be described by a system of equations, including basic equations of gasdynamics and electrodynamics as well as by the equation of Ohm's law for the ionized gas. Because of the relationships among these equations, we call them equations of magnetohydrodynamics.[1]

In the equations of magnetohydrodynamics, regard for all of the chemical, thermal, flow and electrical phenomena occurring during the flow of plasma in the channel of the MHD generator is very difficult.[2-4] Moreover, by making allowances for all of these phenomena, the system of magnetohydrodynamics equations, would be difficult to solve. Therefore, in this study, the following simplifying assumptions have been used to describe the phenomena in the channel of the MHD generator:
1) The flow of plasma is steady.
2) The parameters of plasma change only along the channel (one-dimensional flow).
3) The working medium constitutes a mixture of molecules, atoms, ions, and electrons of averaged parameters.
4) The magnetic field has a constant value in the whole volume of the channel.

With these simplified assumptions, the mathematical model of plasma flow in the channel of the MHD generator has been based on four basic equations: continuity of flow, conservation of momentum, conservation of energy, and generalized Ohm's law, complemented by the equation of gaseous state.

In the equation of momentum conservation, friction losses have been taken into account. In the equation of energy conservation, heat losses of cooling the channel of the MHD generator have been taken into account. The mathematical model of the one-dimensional flow of plasma in the MHD channel, with these assumptions, assumes the following form:

$$\frac{d(\rho u A)}{dx} = 0 \qquad (1)$$

$$\frac{d(\rho u^2 A)}{dx} = -A(\vec{I} \times \vec{B})_x + A\frac{dp}{dx} + A \cdot f = 0 \qquad (2)$$

$$\frac{d[\rho u(i + \frac{u^2}{2})A]}{dx} - A(\vec{I}\ \vec{E}) + Aq = 0 \qquad (3)$$

$$\vec{I} = \sigma(\vec{E} - u\vec{B}) - \beta(\vec{I} \times \frac{\vec{B}}{B}) \qquad (4)$$

$$p = \rho RT \qquad (5)$$

where:

$\vec{B}$ - vector of the magnetic field induction
$\vec{I}$ - vector of electric current density
$\vec{E}$ - vector of electric field intensity
u - plasma velocity
p - plasma pressure
T - plasma temperature
ρ - plasma density
σ - plasma conductivity
R - gas constant of plasma
i - plasma specific enthalpy
f - force of friction in relation to the unit of the channel volume,
q - heat given up to the walls of the channel per unit of plasma mass and unit of the channel length
A - cross section of the channel (A = y·z).

Specific enthalpy of plasma can be determined approximately by means of the following relation:

$$i = \frac{\gamma}{\gamma - 1} RT \qquad (6)$$

where:
γ - the adiabatic exponent.

Heat given up to the walls of the channel has been defined by means of dependence

$$q = 4\alpha_h (T - T_w) \frac{1}{\rho u D} \qquad (7)$$

where:
α_h - coefficient of heat exchange
T_w - temperature of the channel walls
D = 2yz/(y+z) - hydraulic diameter of the channel.

The force of friction has been defined by means of relation

$$f = \frac{\xi \rho u^2}{2D} \qquad (8)$$

where:
ξ - friction coefficient

On performing transformations of Eqs. (1-5) and on introducing Eqs. (6-8) into them, differential equations, describing the one-dimensional steady flow of plasma in the channel of the MHD generator, obtain the form:

$$\frac{dp}{dx} = \frac{1+(\gamma-1)M^2}{1-M^2} F - \frac{(\gamma-1)M^2}{1-M^2} \frac{1}{u} G + \frac{\gamma M^2}{1-M^2} \frac{p}{A} \frac{dA}{dx} \qquad (9)$$

$$\frac{du}{dx} = \frac{\gamma-1}{1-M^2} \frac{u}{p} F + \frac{\gamma-1}{1-M^2} \frac{1}{\gamma p} G - \frac{1}{1-M^2} \frac{u}{A} \frac{dA}{dx} \qquad (10)$$

$$\frac{dT}{dx} = \frac{(\gamma-1)M^2}{1-M^2} \frac{T}{p} F + \frac{(\gamma-1)(1-\gamma M^2)}{1-M^2} \frac{T}{pu\gamma} G + \frac{(\gamma-1)M^2}{1-M^2} \frac{T}{A} \frac{dA}{dx} \qquad (11)$$

$$\frac{dM^2}{dx} = \frac{-[2+(\gamma-1)M^2]}{1-M^2} \frac{M^2}{p} F + \frac{(\gamma-1)(1+\gamma M^2)}{1-M^2} \frac{M^2}{pu\gamma} G$$

$$+ \frac{2+(\gamma-1)M^2}{1-M^2} \frac{M^2}{A} \frac{dA}{dx} \qquad (12)$$

where:
F and G   - quantities, the values of which depend on the connection system of electrodes

$M = \dfrac{u}{\sqrt{\gamma RT}}$ - Mach number

For the system of connection of Faraday with electrodes ideally sectionalized, it is advantageous to introduce load factor

$$K_y = \frac{E_y}{uB} \qquad (13)$$

Quantities F and G, occurring in the system of Eqs. (9-12), for the ideally sectionalized MHD channel in which the electrodes are connected in Faraday's system, have the form:

$$F = (K_y - 1)\sigma u B^2 - \xi \frac{\gamma(y-z)}{4yz} M^2 p \qquad (14)$$

$$G = (K_y - 1)K_y \sigma u^2 B^2 - 2\alpha_h \frac{T - T_w}{yz}(y + z) \tag{15}$$

For the channel of the MHD generator in which the electrodes are connected in Hall's system, these quantities have the form:

$$F = -\frac{K_x \beta^2 + 1}{1 + \beta^2} \sigma u B^2 - \xi \frac{\gamma(y+z)}{4yz} M^2 p \tag{16}$$

$$G = \beta^2 K_x \frac{K_x - 1}{1 + \beta^2} \sigma u^2 B^2 - 2\alpha_h \frac{T - T_w}{yz}(y + z) \tag{17}$$

where:
β - Hall's parameter
$K_x = \frac{E_x}{uB}$ - axial load factor of the channel.

For the channel of the MHD generator in which the electrodes are connected in Montardy's system, F and G have the form:

$$F = \frac{K_y[(1 + \alpha\beta) - 1]\sigma u B^2}{1 + \beta^2} - \xi \frac{\gamma(y+z)}{4yz} M^2 p \tag{18}$$

$$G = \frac{K_y}{1 + \beta^2}[K_y(\alpha^2 + 1) + \alpha\beta - 1]\sigma u^2 B^2 - 2\alpha_h \frac{T - T_w}{yz}(y + z) \tag{19}$$

where:
$\alpha = \frac{E_y}{E_x}$ - quantity defining the direction of the lines of electric field forces.

It is assumed in the presented mathematical model that the following physical quantities will have the character of assumed parameters:

1) Induction of the magnetic field (B).
2) Load factors (K_y) and (K_x).
3) Temperature of the channel walls (T_w).
4) Friction coefficient (ξ).
5) Heat exchange coefficient (α_h).

Moreover, the system of Eqs. (9-12) was completed by the relation, describing conductivity of plasma, as a function of its pressure and temperature:

$$\sigma = \sigma(p, T) \tag{20}$$

On thus formulating the mathematical model, the number of variables has been limited to five (T, p, u, M, and A) and this allows one to realize a wide program of investigations.

The first task consists of searching the geometry of the channel of the MHD generator in defining one of the following quantities:

$$u = u(x), \qquad p = p(x), \qquad T = T(x) \text{ or } M^2 = M^2(x)$$

The second group of tasks consists of finding the distribution of velocity, pressure, and temperature along the channel at its defined geometry.

The formulated mathematical model of the one-dimensional steady flow of plasma in the channel of the MHD generator has been utilized after elaboration, to construct a computer program of numerical calculations, which allows one to determine the distribution of parameters of plasma and changes in the cross section of the channel along its length as well as the value of generated electric power.

Calculation of Parameters of Plasma Flow in the Channel of the MHD Generator

On utilizing the elaborated mathematical model of plasma flow in the channel of the MHD generator, multivariant calculations of different channels were carried out. The analysis of results of calculations of various geometry channels showed that the highest efficiency of generating electric energy is obtained in channels of geometry that ensure maintenance of constant Mach's number along the channel. That is why further calculations have been performed for the channels with constant Mach's number.

Calculations were performed for channels in which the electrodes were connected in Faraday's, Hall's and Montardy's systems, assuming the following quantities as constant:

$$B = 6 [T] \qquad \alpha = 450 \ [W \cdot K^{-1} \cdot m^{-1}] \qquad \xi = 0.005 \qquad \gamma = 1.25$$

In the calculations, combustion products of fuel oil, on adding 1% of ionizing seeding in the form of potassium, were assumed as a working medium in the MHD channel.

In the calculations, temperature T_0 and pressure p_0 of plasma in the combustion chamber, as well as the value of Mach's number in the MHD channel, were assumed as reference parameters.

In all of the variants of the calculations, amounting to $p_e = 0.072$ MPa, were assumed. The program of calculations was thus planned to allow an analysis of the influence of basic flow parameters on the efficiency of generating electric energy in the channels of the MHD generators of different systems of electrode connections, namely:
1) Parameters of plasma in combustion chamber T_0 and p_0.
2) Intensity of plasma flow D.
3) Mach's number M.
4) Load factors K_y and K_x.

Analysis of Results of Calculations and Conclusions

The quantity of generated electric power in the channel of the MHD generator is directly proportional to the temperature of plasma in the combustion chamber T_0 and to the intensity of plasma flow in channel D. The quantity of electric power generated in the channel of the MHD generator depends in a more complex way on pressure in the combustion chamber p_0. Figure 1 shows the determined dependence of the value of the generated electric power in the MHD channel of Faraday's type of the pressure of plasma in the combustion chamber p_0 for the constant temperature T_0. The results of calculations show that the length of the MHD channel increases proportionally to the pressure in the combustion chamber p_0. However, the choice of the optimum value of pressure in the combustion chamber must result from analysis of the entire power plant in which the MHD generator operates.

Figure 2 shows the influence of the value of the assumed Mach's number, on the power operation of the MHD channel. The increase in the assumed Mach's number causes a shortening of the MHD channel length and a decrease in the total power generated in the channel. However, in this case an increase in the density of power generated in a unit of channel volume is obtained.

Figure 2 presents the result of calculations that show that the preferred Mach numbers should be values approximating 1.

Figure 3 shows the influence of the load factor of Faraday's type MHD channel on the value of the generated electric power. The maximum electric

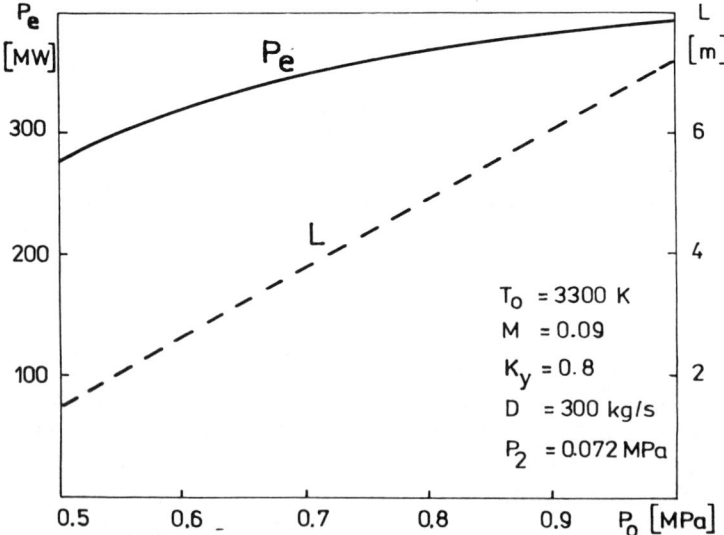

Fig. 1 Dependence of the value of the generated electric power P_e and the length of the MHD channel L on the pressure in the combustion chamber p_0 for Faraday's connections of electrodes.

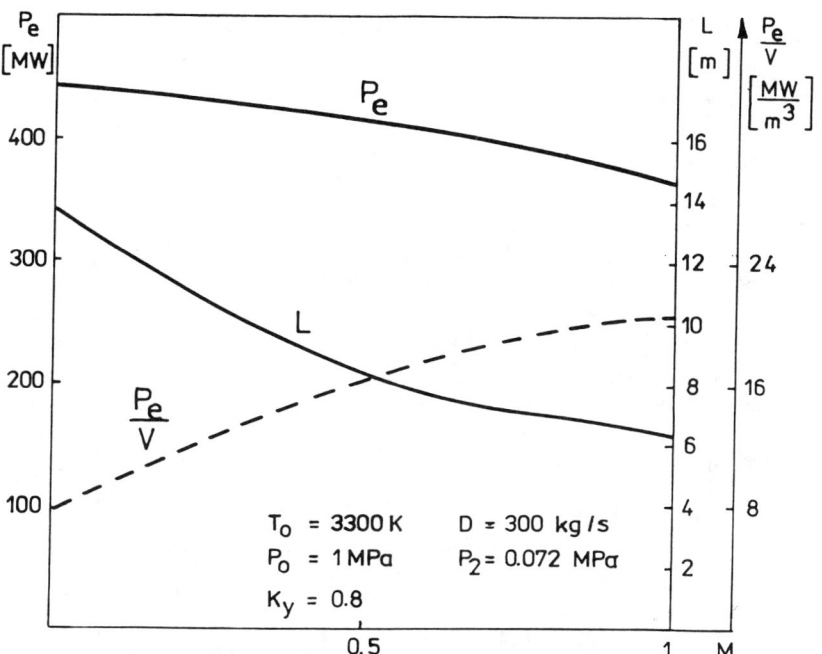

Fig. 2 Dependence of the value of the generated electric power P_e and its density P_e/V as well as of the length of the MHD channel on Mach's number of the channel of Faraday's type connections of electrodes.

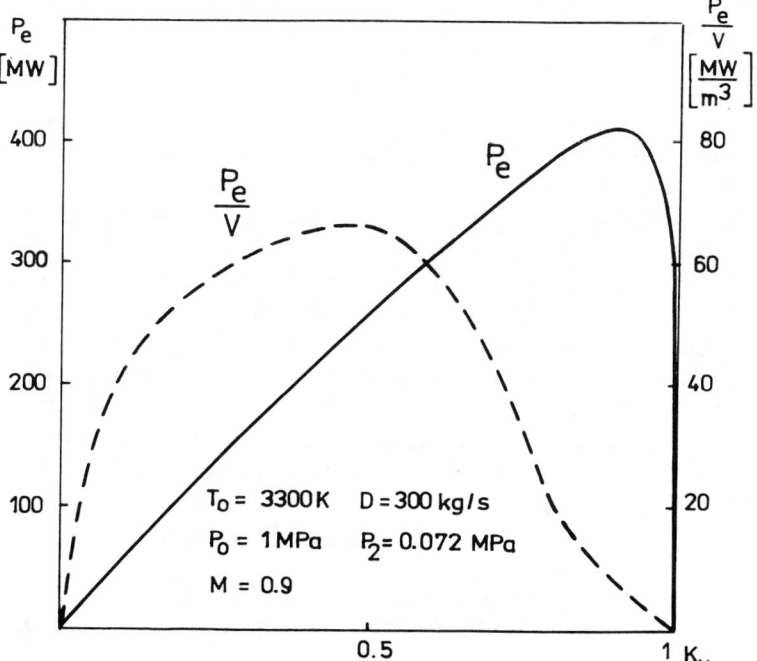

Fig. 3 Dependence of value of the generated electric power Pe and its density (P_e/V) on the load factor of the MHD K_y for the system of connections of Faraday's type.

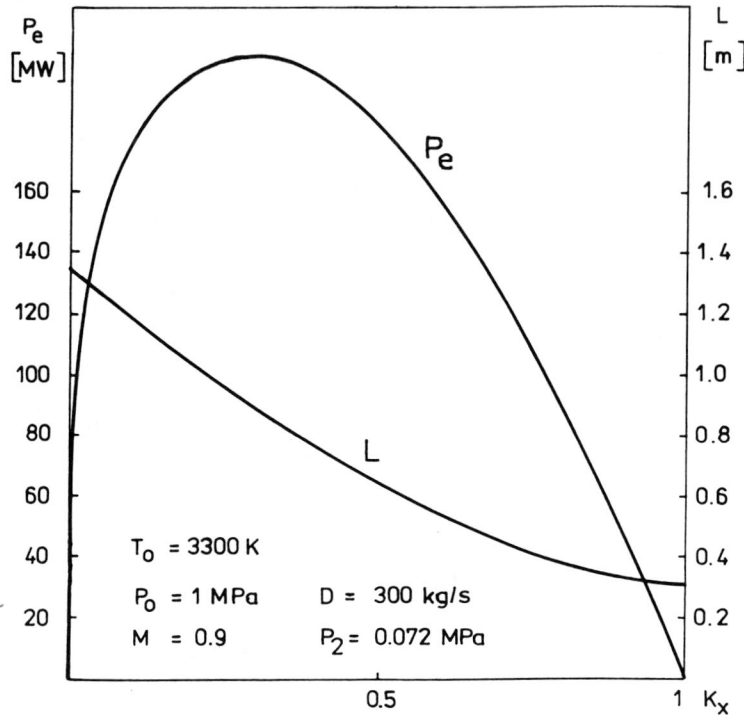

Fig. 4 Dependence of the value of the generated electric power P_e and the length of the MHD channel L on the load factor of the MHD channel K_x for the system of connections of Hall's type.

power is obtained in such a channel for $K_y = 0.85$, whereas the greatest density of power occurs at $K_y = 0.5$.

Figure 4 shows the influence of the longitudinal load factor K_x on the value of the generated electric power in the MHD channel of Hall's type. The maximum of the generated electric power is obtained for K_x, which has a low value of about 0.3. It should be emphasized that the maximum value of electric power generated in the MHD channel of Hall's type is about two times lower than the value of electric power generated in the MHD channel of Faraday's type (Fig. 3) for the same parameters p_0 and T_0. It must be added, however, that this concerns the fuel products of plasma, which have a relatively low value of Hall's parameters. The calculations point out that the channel of Hall's type is shorter than that of Faraday's type, but it has a higher coefficient of expansion along its length. Therefore, the channel of Hall's type can be used for MHD generators of disk type in closed cycles.

Figure 5 presents the results of the MHD channel calculations in which the electrodes are connected in Montardy's system. The influence of load factor K_y, on the value of the generated electric power in the channel depends on the tan of the angle of inclination of commutative frames. The channel of Montardy's type

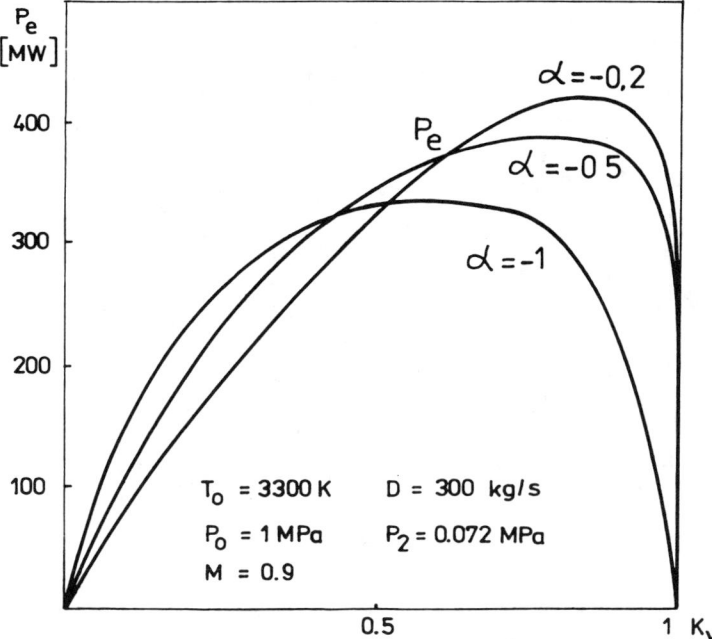

Fig. 5 Dependence of the value of the generated electric power P_e on the load factor of the MHD channel K_y for the system of connections of Montardy's type.

will reach higher electric power at small angles of inclination of the commutative frames, because the applied plasma has low values of Hall's parameters. The most advantageous is the work of the MHD channel of Montardy's type with the load factor $K_y = 0.85$ and the tan of the angle $\alpha = -0.5$. For the value of tan δ approaching zero, the obtained power has high values, but the length of the channel is very great (tens of meters).

The channels of the MHD generators in which plasma of fuel products is the working medium ought to work in the system of connections of electrodes of Faraday's or Montardy's type with the value of $\alpha = \tan \delta = -0.5$.

Figure 6 shows the change in the width of the MHD channel and the change in parameters of plasma, such as temperature, pressure, and velocity along the MHD channel of Faraday's type, for the following parameters: $T_0 = 3300$ K, $p_0 = 1$ MPa, $M = 0.9$, $D = 300$ kg/s, $K_y = 0.8$. For the same parameters, the results of calculations for the MHD channel of Montardy's type, for $\alpha = \tan \delta = -0.5$, are presented in Fig. 7. The values of electric power obtained in these two generators amount to:

1) In the MHD channel of Faraday's type, $P_e = 392$ MW.
2) In the MHD channel of Montardy's type, $P_e = 387$ MW.

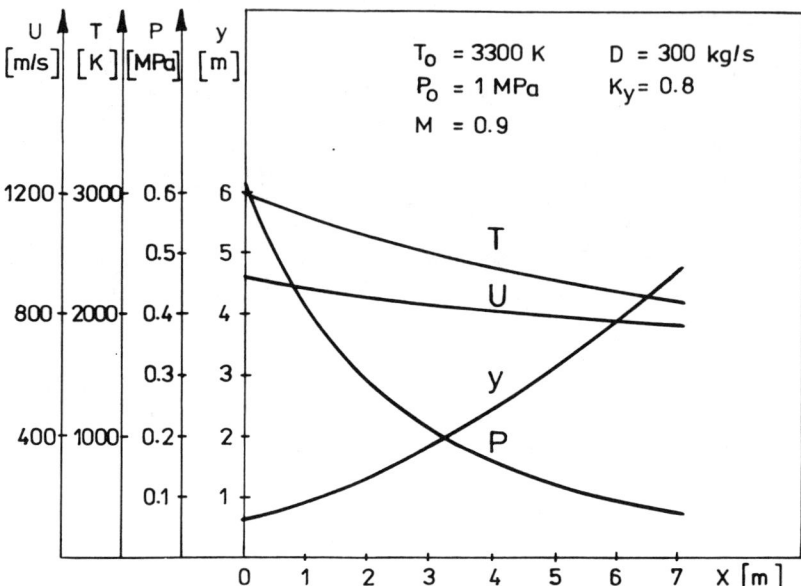

Fig. 6 Change in temperature T, pressure p and velocity u of plasma as well as the change in width y of the MHD channel along its length for the system of connections of Faraday's type.

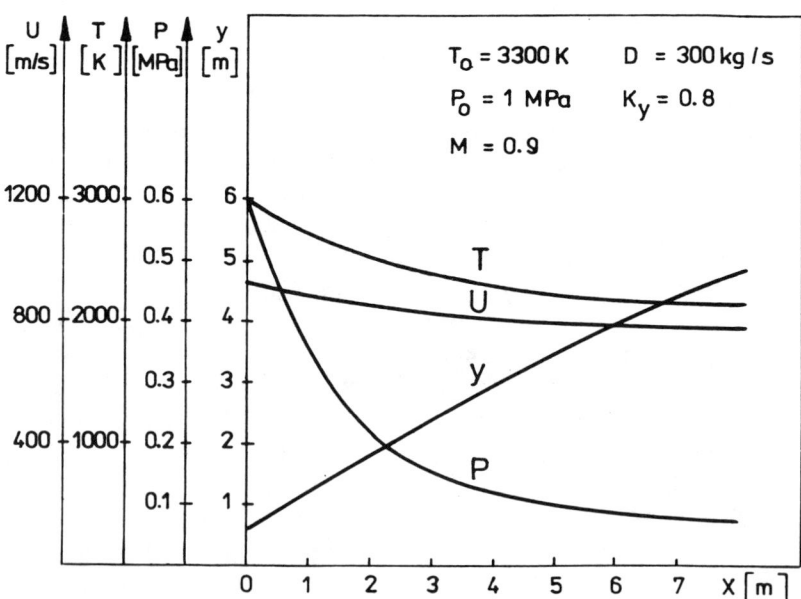

Fig. 7 Change in temperature T, pressure p and velocity u of plasma as well as the change in width y of the MHD channel along its length for the system of connections of Montardy's type.

References

[1] Wulis, L.A., Genkin, A.L., and Fomenko, W.A., *Theory and Calculation of Magnetohydrodynamic Flow*, Moscow, 1971.

[2] Zaporowski, B., "Temperature and Physical Properties of Combustion Gases in MHD Generators," *Proceedings of the 4th International Symposium on Magnetohydrodynamic Electrical Power Generation*, Warsaw, 1968, Vol. IV, International Atomic Energy Agency, Vienna, 1968, pp. 2195-2209.

[3] Zaporowski, B. and Sroka, K. "Analysis of Parameters of the MHD Generators Operation at Chanding Loads," *24th Symposium on the Engineering Aspects of Magnetohydrodynamics*, Butte, MT, 1986, pp. 100-108.

[4] Zaporowski, B., Roskiewicz, J. and Sorka, K., "Analysis of the Energy Conversion System of a Combined MHD-Steam Power Plant Integrated with Coal Gasification," *Proceedings of 10th International Conference on Magnetohydrodynamic Electrical Power Generation*, Vol. I, Tiruchirapalli, India, 1989, pp. III 47-54.

Simulation and Comparison with the Experiment: The Dynamic Processes in an MHD Facility Flow Train

A. M. Levints,* V. R. Satanovsky,† and V. N. Zatelepin‡
Russian Academy of Sciences, Moscow, Russia

Abstract

The main working fluid flow train of an MHD facility consists of not only the MHD generator but also such units as the compressor or pump, the oxidizer heater, diffusers, the heat exchangers, the electrostatic precipitator, and the stack. The main features of this apparatus are a very low Mach number and a large volume in comparison with MHD channel. The numerical model of dynamic processes in such a flow train for a time range more than 10s has a number of assumptions that have to be approved in comparison with the experiments. The experimental facility U25BM in the Institute of High Temperatures includes the main units of the MHD flow train. The comparison of experimental data obtained on U25BM with the results of numerical simulation gives good agreement for time dependence of such parameters as the pressure, temperature, and mass flow rate.

Introduction

The study of dynamic characteristics of an MHD powerplant is a new area for numerical simulation in MHD technology. An MHD powerplant incorporates more than a dozen various subsystems. For example, only the main working fluid flow train of an open-cycle MHD powerplant includes such units as a compressor, a system of oxidizer heaters, a combustor, an MHD channel and diffusers, a steam generator, an electrostatic precipitator, a low-temperature heat exchanger, smoke exhausters, and a stack. The characteristic time scale of this unit changes in wide range, and this obstacle imposes the serious restrictions on the numerical model of such a flow train.

Copyright © 1992 by the American Institute of Aeronautics and Astronautics, Inc. All rights reserved.
*Scientific Researcher, Design Bureau, Institute for High Temperature.
†Head of Laboratory, Experimental Division, Institute for High Temperature.
‡Senior Researcher, Science Division, Institute for High Temperature.

One way to solve this problem is to develop the different numerical model for a different time scale. Reference 1 presents three numerical approaches in cases of characteristic time: 1) less than 10s, 2) more than 10s, and 3) more than 1 h.

Processes with a characteristic time more or of order of 10s are important ones for both industrial and experimental facilities, because this is a typical time for filling or dumping by working media the flow train. Processes of this type arise at any control influence on flow train in that facility. Maximum amplitudes take place on shutdown and start-up of the facility, but in some cases time-dependent processes of this nature periodically occur in the normal mode operation. Particularly, these are the periodical disturbances of flow parameters in an open-cycle MHD powerplant with the heat exchangers of cauper type. These caupers are turned off and on periodically to the main flow train, and at this moment the pressure and temperature in the flow train are disturbed.

To describe the processes of the type described in Ref. 1 a numerical model is developed that takes into account that the characteristic time of the unit of the MHD flow train alone is much less than the transient time of the total flow train. The second important point of this model is the relationship of mass and enthalpy flow rates as functions of pressure and density. These assumptions do not take into account the momentum equation.

In this paper the verification of this model on the basis of experimental data that are acquired on the U25BM facility is presented. Such a conventional MHD powerplant of open-cycle units as MHD channel and steam generator are absent in this facility, but there is a system of caupers and many adiabatic ducts for air supplying with large values of mass and heat capacity.

Model Description

A numerical model that is used for simulating time-dependent processes in U25BM takes into account that convective time (t > 100 s) for the main flow train of this facility is much more than the time for the sound wave. All flow trains of the gaseous main working media are divided on the series of volumes.

Let us suppose that $Sh = X^*/U^*t \ll 1$, where U^* and X^* are the typical values of velocity and dimensions of these volumes. We may present all gasdynamic values in time-dependent processes as an asymptotic set of number Sh. For example, for pressure we can write

$$P(x,t) = P_0(x,t) + Sh P_1(x,t) + ...$$

The system of gasdynamic equations in this case can be divided into systems for different powers of Sh. For values with a subscript 0 the system is a stationary one. The solution of this system gives a distribution of gasdynamic parameters inside the elementary volume of the flow train. The time dependence of the values with the subscript 0 result from the boundary conditions. If we average the system of the first approximation on Sh by the elementary volume, we obtain the following system for the averaged mass and energy:

$$V_i dR_{i0}/dt = (G_{il} - G_{ir} + B_{ri})$$

$$V_i dP_{i0}/dt = (\gamma-1)(GH_{il} + GH_{ir} + B_{hi})$$

(1)

where R_{i0} and P_{i0} are the averaged by space of zero-order approximation density and pressure in the elementary volume V_i, i is number of this volume, G_{il} and G_{ir} are the mass flow rate on the entrance and exit bounds of the ith volume, GH_{il} and GH_{ir} are the enthalpy flow rates of this one, B_{ri} and B_{hi} are the sources of mass and enthalpy, and γ is an isentropic coefficient. We suppose that the Mach number in the elementary volume V_i is much less than 1. Let us suppose that

$$G_{ir} - G_{il} \cong G_{0ir} - G_{0il} \neq 0$$

where the mass flow rate with the subscript 0 is the value that is founded on the basis of boundary values of pressure and density of zero-order approximation. This assumption shows that the unsteady process in elementary volume arises as a result of the disturbances in the neighboring elementary volume. We also suppose that the mass and enthalpy flow rate on entrance and exit bounds of the elementary volume V_i are determined by the following formula:

$$G_{0ir} = G(P_{0ir}, R_{0ir}, R_{0i+1l}), \quad G_{0il} = G(P_{0i-1r}, R_{0i-1r}, P_{0il})$$

(2)

where the arguments of formula (2) are the values on an appropriate bound (entrance or exit) of the elementary volume.

The scheme of the main working fluid of U25BM is shown on Fig. 1:

1. Oxygen plant and atmosphere. In the model we assume that this is an elementary volume with constant parameters.
2. Roll filters.
3. Compressor. A special procedure controls the mass and enthalpy flow rate values, which is based upon pressure and density of compressor inlet and exit.
4. Cauper of the oxidizer heating system.
5. Special distributor of the hot oxidizer, which has the three exit ducts. This enables the hot oxidizer to be supplied to three different facilities. One of them is U25BM.
6. Section of measurements of the hot oxidizer mass flow rate.
7. Combustor and nozzle.
8-10. Experimental and diagnostic section with a high Mach number. In the model there is a procedure that is analogous to the procedure for the compressor. This procedure determines the mass and enthalpy flow rate in

this section as functions of the pressure and density in the combustor and cooling system.
11. Cooling system and stack with fixed values of pressure and temperature.

Different units of this flow train are connected to each other with ducts. Some of these ducts are lined by the refractory materials. In this case in model the equation for ceramic wall temperature T_w is solved by

$$R*C*r * dT_w/dt = d(\lambda r\, dT_w/dr)/dr \qquad (3)$$

where R, C, and λ are the density, specific heat, and heat conductance of wall, respectively; and r is distance from the surface of wall. Surface wall temperature is used for calculating the heat sources B_h in Eqs. (2).

Results of Numerical and Physical Experiments

As was mentioned earlier, the numerical model has some important assumptions that have to be approved by comparison with experiments. Comparison of numerical simulation was carried out with the experimental data obtained in run 29 of the U25BM facility. In the course of this run the mass flow rate, pressure, and temperature in different points of the flow train are denoted by small letters on Fig. 1 (points a-g).

It was chosen for comparison the quasistationary mode with cauper B as a heater of oxidizer. This mode is characterized by the following main parameters:

1. the mass flow rate of the hot air enriched up 40% by oxygen, 4.2 kg/s;
2. stoichiometry coefficient in the combustor, 0.98;

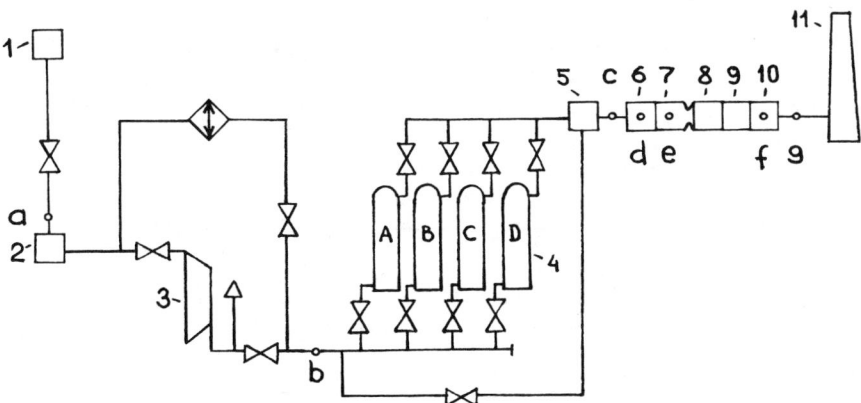

Fig. 1 U-25BM main flow train scheme: 1-oxygen plant; 2-filter; 3-compressor; 4-cauper; 5-distributor of hot oxidizer; 6-mass flow rate measurement section; 7-combustor and nozzle; 8 and 9-experimental and diagnostic section; 10-cooler; 11-stack.

3. the pressure after the compressor, 3.7 bar;
4. the hot oxidizer temperature at the combustor entrance, 1040K;
5. the mass flow rate of the ionizing seed, 0.09 kg/s;
6. the pressure in the nozzle, 2.57 bar; and
7. the total combustion products mass flow rate, 4.8 kg/s.

Several correction coefficients are included in the numerical model that gave a possibility to tune the numerical results on experimental data. The model tuning is carried out by selection of correction coefficients for the mass flow rate K_g in formula (2) and this one for enthalpy sources K_n in Eq. (1). The correction coefficients are included in the model as follows:

$$K_g = -G_n/G_t$$
$$K_n = B_n/B_t \qquad (4)$$

where values with the subscripts n and t are the ones used in the numerical model and recommended by theory for the stationary mode of different units of flow train. The data of Ref. 2 are used particularly for the calculation of G_t. These correction coefficients were chosen for one of the transition modes of the flow train to minimize the deviation of the experimental and numerical data.

Figure 2 shows the cyclogram of the operation on the oxidizer heater valves when the cauper C is substituted by the cauper D. On horizontal axes the time from the beginning of this action is shown. From 40s the filling valve on the cold side of cauper D is opened. The valve on hot side of cauper D is opened at some moment when the pressure in this cauper will be equal to that in the hot oxidizer. From 600s to 623s cauper C is turned off, and after that moment only cauper D is working.

The substitution of the cauper by another one generates an intensive dynamical process in flow train. On Figs. 3-7 experimental and calculated time-dependent variations of different values are shown. The time dependence of the pressure at two points of the flow train is shown in Fig. 3. The upper and lower

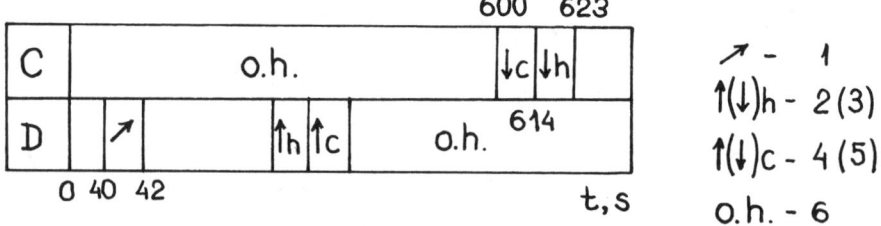

Fig. 2 Cyclic operation of valves on hot and cold sides of caupers C and D of heat exchanger system: 1-opening of valve for cauper filling; 2 and 3-hot valve opening and closing; 4 and 5-cool valve opening and closing.

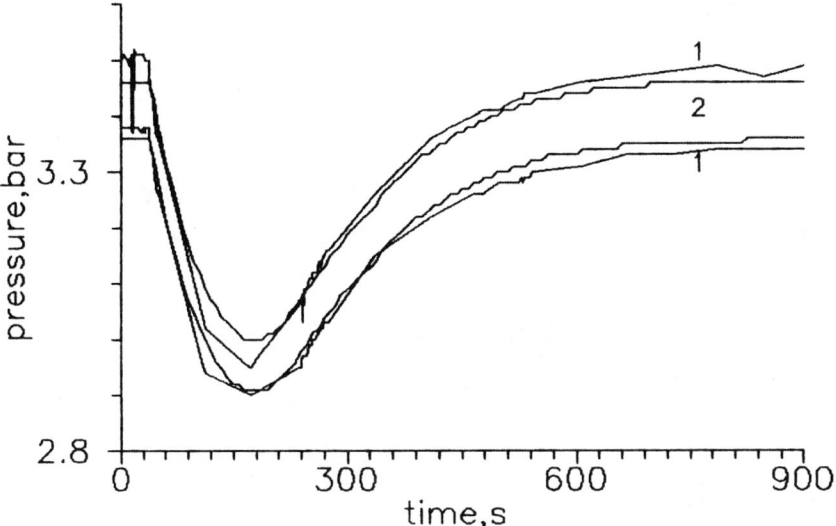

Fig. 3 Experimental (1) and calculated (2) pressure time dependence in two train points: upper curves, point c on Fig. 1; lower curves, point e.

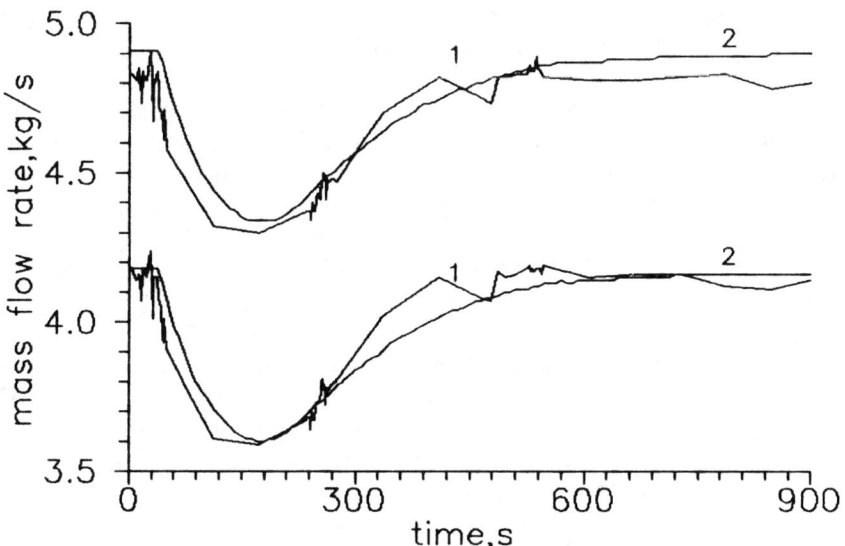

Fig. 4 Experimental (1) and calculated (2) time dependence of mass flow rate: upper curves, point e on Fig. 1; lower curves, point c.

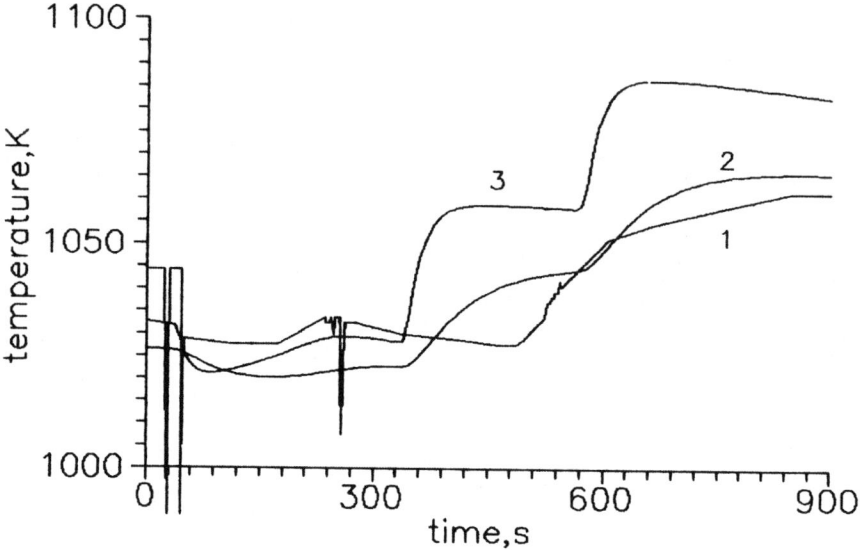

Fig. 5 Time dependence of temperature: 1-experimental data for thermocouple in point c on Fig. 1; 2-calculated ceramic wall temperature in this point; 3-working media temperature.

curves are the experimental and numerical results for points c and e (see Fig. 1). The decreasing of pressure is the result of filling cauper D. Some part of the compressor mass flow rate is used for this filling, and this causes to dump pressure in the flow train. In Fig. 4 the mass flow rates at the same points are shown.

In Fig. 5 the temperature time dependence in point c is shown. The thermocouple for temperature measurements in the hot oxidizer duct was used. This thermocouple was placed inside the ceramic wall at some distance from the surface of the duct. Equation (3) gives the time-dependent temperature distribution through the wall thickness. Curves 1 and 2 show the calculated and experimental results for temperature in this point inside of wall. Curve 3 presents the oxidizer temperature under the position where the thermocouple is placed.

From a comparison of the experimental and computing data we can derive the maximum deviations of different values: 3% for the pressure in the duct of the hot oxidizer, 3% of the pressure at the entrance of the combustor, 3% for the mass flow rate at the combustor entrance, 3% for the mass flow rate after the combustor, and 4% for the temperature of the hot oxidizer.

The numerical model gives much more data than can be measured by experiment. Figures 6 and 7 show the unmeasured values of the mass flow rate on the cold and hot side of caupers C and D in this time-dependent process. It can be seen that very intensive time variations of these parameters take place.

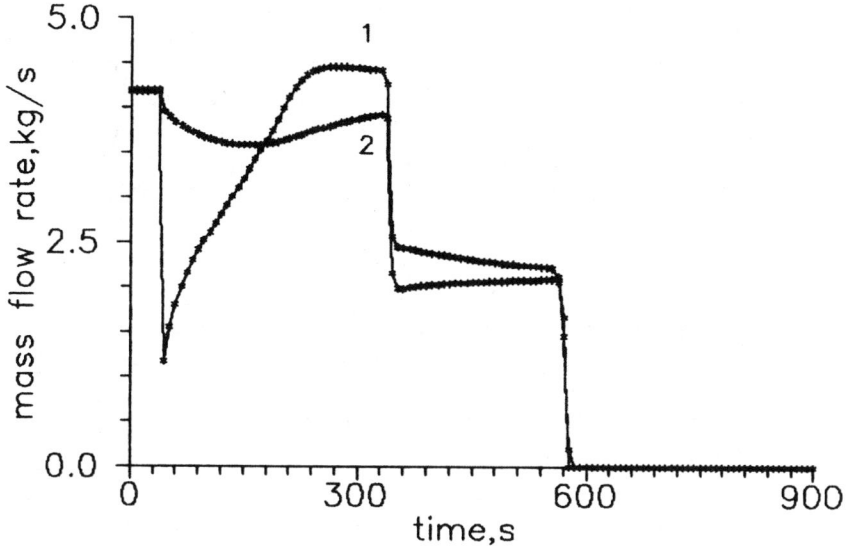

Fig. 6 Time dependence of mass flow rate: 1-entrance of heat exchanger C; 2-exit of heat exchanger C.

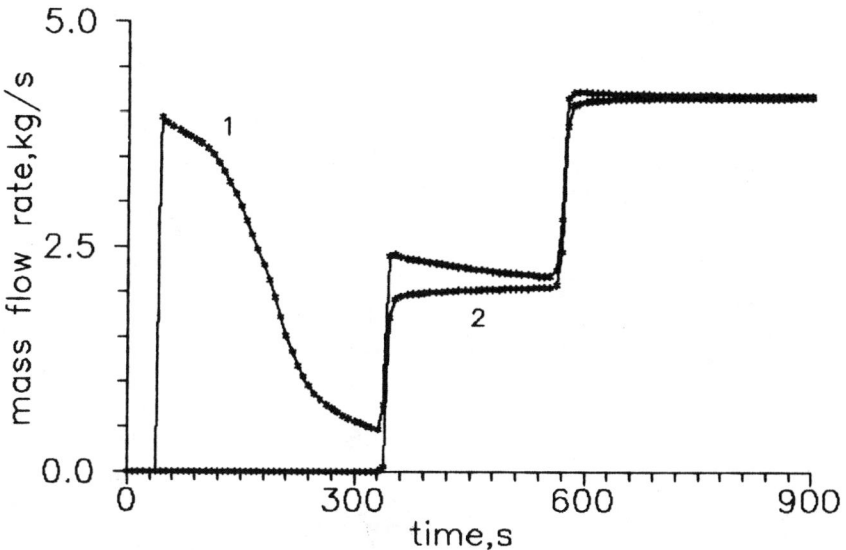

Fig. 7 Time dependence of mass flow rate: 1-entrance of heat exchanger D; 2-exit of heat exchanger D.

Conclusion

A comparison of the developed numerical model for simulation of unsteady processes in low Mach flows of main working media of MHD facility with experimental data was carried out. This model has the correction coefficients for mass flow rate and enthalpy sources in the flow train. The selection of these coefficients on some operation mode results in good agreement between numerical and experimental data in the various time-dependent processes.

References

[1] "Mathematical Simulation of Slow Dynamical Processes in Flow Train of MHD Power Station," *Thermal Physics of High Temperature*, Vol. 1, 1990, Moscow (in Russian).

[2] Idelchik, G.N. *Handbook of Hydraulic Resistances*, Heavy Industry Publishing, Moscow, 1973 (in Russian).

Acceleration of Gas-Liquid Piston Flows for Molten-Metal MHD Generators

A. Kolesnichencko* and V. Malakhov†
Ukraine Academy of Sciences, Kiev, Ukraine

Abstract

A concept of a gas driven liquid metal (LM) piston for LM magnetohydrodynamic (MHD) energy conversion system is described. It features forced circulation of slugs (or pistons) of LM brought about by alternating the introduction of liquid metal and gas (or vapor) into an acceleration channel. It is being claimed that instabilities formed at the back side of the liquid metal piston (which is being "pushed" by the gas) are developed slowly enought to enable the bulk of the liquid metal piston to pass through the MHD generator (of a constant cross-section) before breaking down and becoming dispersed in the driving gas. The concept was first demonstrated experimentally in 1974. Liquid metal velocities of up to 94 m/s were attained with molten tin piston initial mass of 8 kgs. A number of pilot plants were tested.

Introduction

The acceleration device[1] is an integral part of any heatpower magnetohydrodynamic (MHD) installation with the Rankine cycle.[2,3] (see Fig. 1). The thermodynamic work of the steam phase expansion is transformed here to the kinetic one and then in the diffuser partially into potential energy of the molten-metal flow.

The accelerating gas flow entrains usually a dispersed liquid phase. The efficiency of the momentum exchange between phases is the higher, the smaller is drop size. However, for dispersion of molten metal, it is

---

Copyright © 1992 by the American Institute of Aeronautics and Astronautics, Inc. All rights reserved.
*Chief of Division, Institute of Electrodynamics.
†Institute of Electrodynamics.

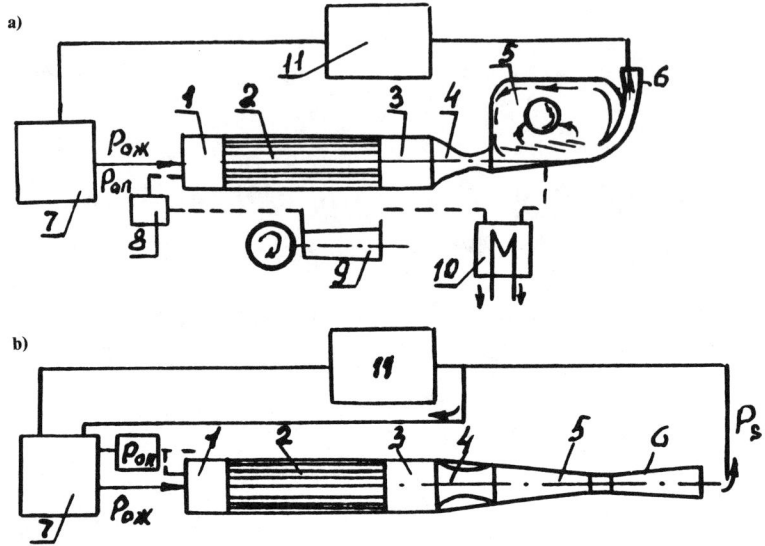

Fig. 1 Heat energy accelerating device (1, piston former; 2, accelerating channels; 3, receiver; 4, two-phase nozzle; 5, separator; 6, diffuser; 7, heat source; 8, gas distributor; 9, feed pump).

necessary to have a sufficiently large difference of the phase velocities. Thus, a considerable part of the enthalpy is wasted on creation of such velocities and the liquid dispersion.

In piston gas-liquid flows, these losses may be practically eliminated. Gas-liquid metal flows in the channels are formed by alternating doses of gas and molten metal. Liquid doses set as pistons in the piston heat engines.

Stability and Failure of Liquid Pistons

The front face (edge) of the piston moving with acceleration is stable. Its surface is orthogonal to the velocity vector. The Rayleigh-Taylor instability develops at the back end. Gas penetrates into the piston at one or more cavities. Their quantity is determined by the largest length of the perturbed wave $L_{max} = \frac{2\pi}{W_m}$ of the free surface, where

$$W_{max} = \frac{D}{2\pi} \sqrt{\frac{(\rho_L - \rho_G)a}{3\sigma}}$$

$$a = \frac{d^2 x}{dT^2}$$

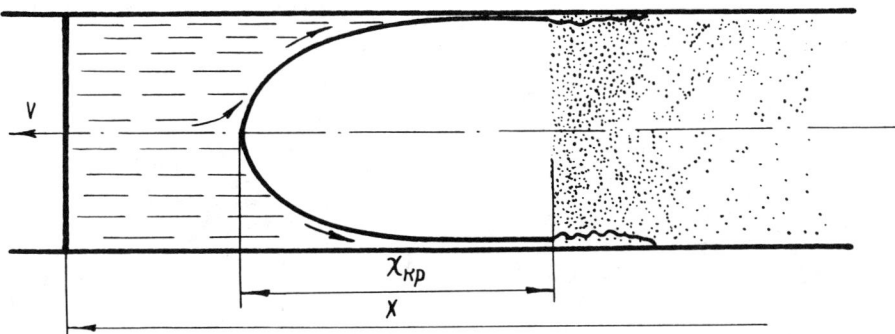

Fig. 2 Piston flows include alternating doses, such as gas and liquid metal.

The liquid phase is expelled to the channel walls, its velocity decreases sharply here.

Penetration intensity of usually one cavity is determined by a ration of the face convergence. This rate is equal to

$$w = 0.204 \sqrt{\left(a + \frac{f_M}{\rho_L}\right)D}$$

or

$$w = 0.334 \sqrt{\left(a + \frac{f_M}{f_L}\right)D} + 0.049v$$

where f_M is the electromagnetic forces acting in the piston if it moves in the channel of the synchronous generator.[4]

Formation of Piston Flows

Piston flows (see Fig. 2) are formed by means of alternating introduction of liquid and gaseous doses into the channel. It proceeds in the formers of the piston flows. Their design may vary greatly. In Fig. 3a, one of the formers with a moving rotor is presented. The rotor with holes measures gas and liquid of different pressures. It favorably compares the previously discussed acceleration system with the known ones.

A former, where there are no moving or rotating parts, is presented in Figs. 3b and 3c. Here the elements of hydropneumoautomatic gas-jet and liquid switches of the flow are employed. In the gas switch, the effect of the free gas adhesion to the wall is used; in the liquid switch a dynamic stability of the eddy flow. Formation of a pair of doses (liquid and gaseous) is performed for one period. In the acceleration channel, a dose of gas expands and accelerates the piston.

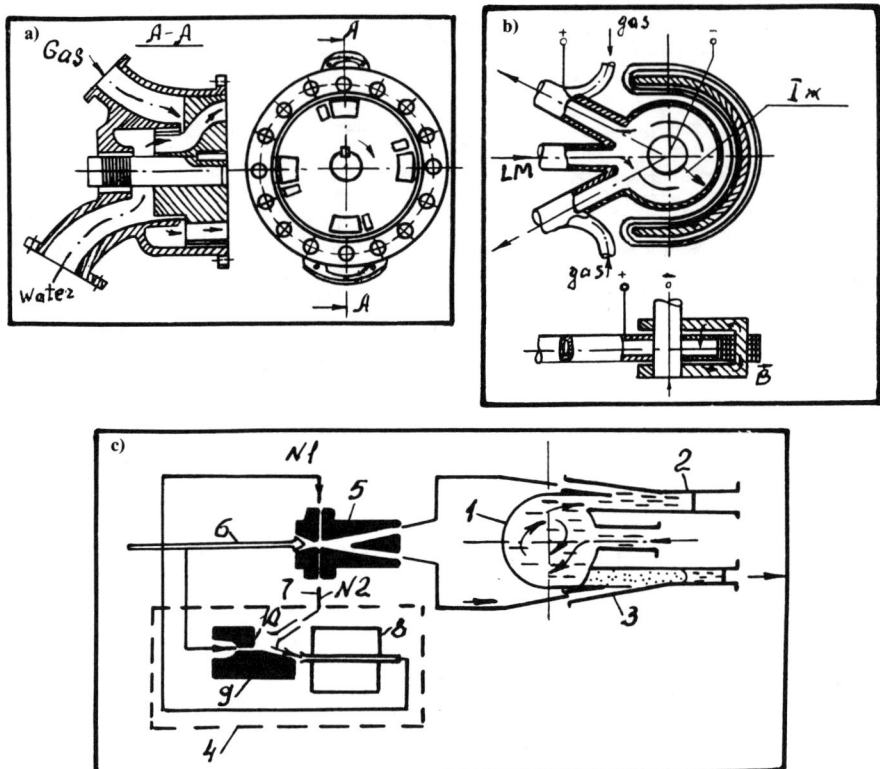

Fig. 3 Piston flow formers a) rotating former; b) MHD former; c) gas-hydroformer.

A dimensionless parameter that determines the work of the piston flow former is obtained from a system of equations, including the equations of motion of the liquid and gas doses and continuities of the gas flow.

$$K^2 = \left(\frac{G_G}{G_L}\right)^2 \frac{\Delta \overline{P}}{P_{G,0} - P_{G,1}} \cdot \frac{\rho_L \cdot \rho_{G,1}}{\rho_{G,0}}$$

This complex relates the parameters of the former and acceleration channels. Values of the complex K as dependent on the relative length of the piston L/D.

The longer is the piston, the larger is the excess of the partial gas pressure $P_{G,o}$ over the pressure of the molten metal $P_{L,0}$ at the beginning of acceleration (Fig. 4) $P_{G,0}/P_{L,0} > 1$ only for piston flows.

Operation of Acceleration Channels

The acceleration destroys both as a result of the penetration of a gas cavity and as a result of viscous friction. This process was studied by means of static and impact pressure pickups. The impact pressure (a sum of the static and velocity head) permits determining the velocity of both solid and dispersed parts of the piston. The impact pressure was measured by means of the pneumoelectric pickup created especially for this purpose. It is base on comparison of two pressures—the impact pressure of the piston flow and the preset return pressure. The pickup (see Fig. 4) consists of a mobile small piston with a pair of electric contacts, set-point device, and the return pressure meter. Shorting of the contacts proceeds in the moment when the impact pressure exceeds the measured return pressure, the contacts are broken when the return

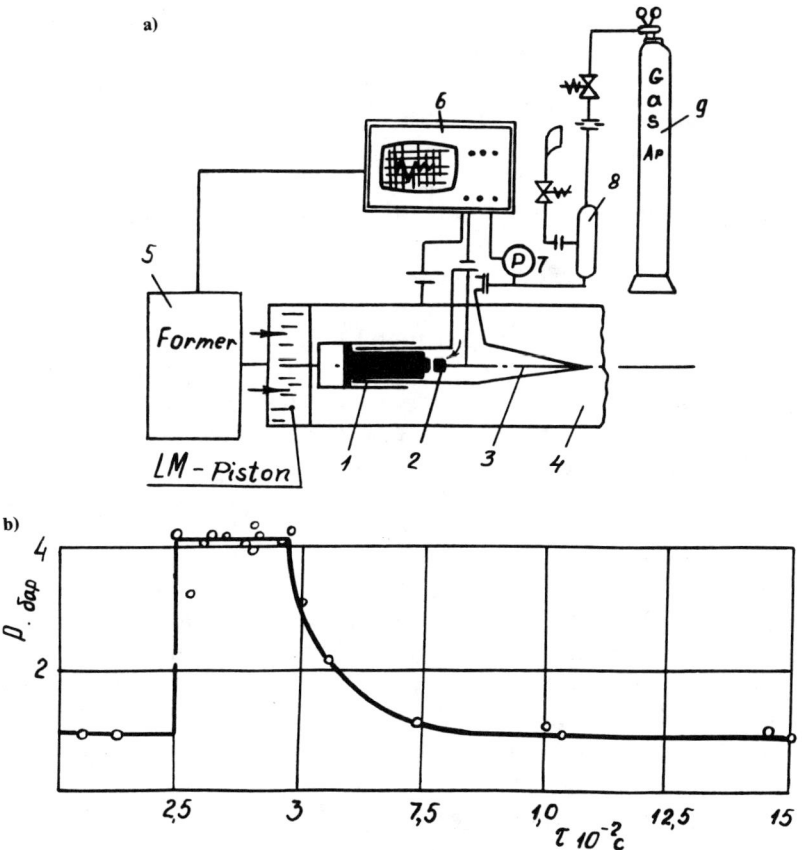

Fig. 4 Measuring of pressure in liquid pistons (a) the pressure in liquid piston (b).

pressure is higher than the impact pressure of the piston. Value of the pressure and the shorting-breaking moments are registered by the oscillograph. Time function of the preset pressure may be different in principles. A diagram of the impact pressure is given in Fig. 4.

Velocities up to 94 m/s were attained in the experiments on acceleration of pistons of the molten tin with their initial mass to 8 kg (steam-tin).

To the moment of penetration of the gas cavity apex to the front face of the liquid piston (the moment of complete failure) the gas pressure remains sufficiently high. Multifunctional heat drop achieves 30% of the entropy heat drop. These losses are eliminated in the additional two-phase nozzle located behind the acceleration channel. Between the channels (their number may be 10 or more) there is a receiver. It equalizes pressure before the nozzle. A dispersed liquid phase formed from the liquid film behind the piston comes into the nozzle. If the values $Re = [(w_G - w_L)D] / v > 500$ are exceeded, the film is swelled with gas. Dispersion liquid formed after complete failure of pistons is accelerated by gas in the two-phase nozzle. In the acceleration device of the pilot plant with heat capacity of 75 KWt (tin vapor-steam) efficiency calculated according to the formula $\eta = (G_L) / (G_G)[(v_1^2 - v_0^2) / (i_0 - i_1)]$ achieves 0.65. Indices "0" and "1" here indicate velocity and enthalpy at the beginning and end of the acceleration channels. If the acceleration device is equipped with a two-phase nozzle, this value increases to 0.78.

Separation of Gas and Liquid Phase

It is necessary for the molten metal to find its way to the MHD generator without gas. Our experiments and the experiment of our colleagues from the Institute of Thermal Physics[5] with curvilinear separators have shown the efficiency of the separator with idealized (nondestroyed) piston flows attains 0.75 at

$$\beta = x_L / (x_{G,0} + x_L)$$

where x is the length of the dose. With an increase of gaps between liquid doses to $\beta = 5$ or under, complete failure of the liquid doses ejected into the separator from the two-channel acceleration device efficiency falls to 0.1 but then increases to 0.6 in the conical diffusor with 12 channels working on it (Fig. 5). In this case, our separator in may way resembles the Elliot separator.

Mathematical Model

The mathematical model of acceleration of piston flows is plotted proceeding from the following assumptions.

GAS-LIQUID PISTON FLOWS

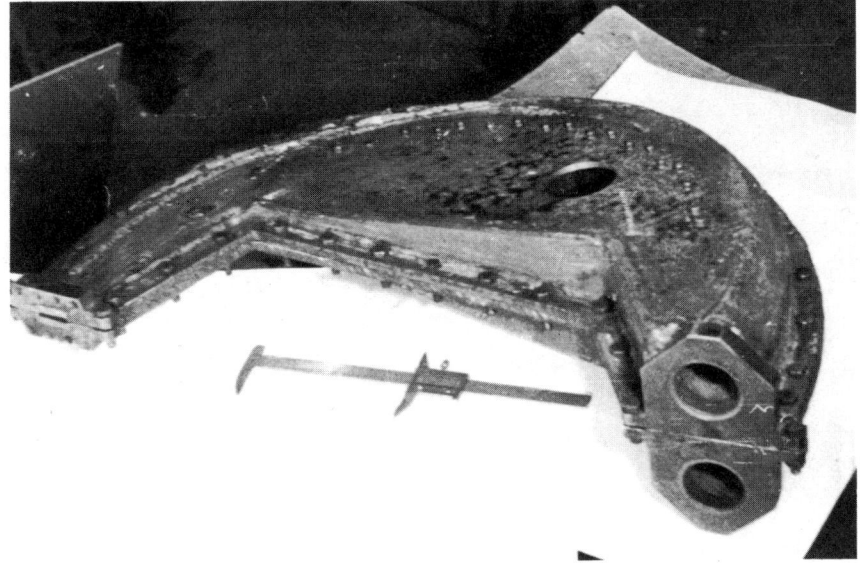

Fig. 5 The separator.

1) Failure of the piston is specified by the initiation of one cavity at its back face.

2) All liquid that leaves the piston is dispersed.

3) Drops of dispersed liquid are sphere of one diameter. Its value is determined by the Weber critical number under which drops are broken.

4) The Stokes law of viscous cohesion of gas and drops is true.

5) The medium which forms the piston flow is at three temperatures (solid part of the piston, drops, and gas).

6) State of the gas behind the piston is at equilibrium,

7) Gravitational forces are small vs other forces.

8) The channel length is of constant section.

A system of equation, which relates velocities of the piston v, drops v_d, cavity w_e, dispersed front w, and L is the length of the piston, S the section, DC the diameter of the channel, j the density of the current in the piston, σ_e the conductivity, μ_G, μ_L the dynamic viscosities, α the convective heat coefficient, a, b the coefficients in the Van der Waals equation state, R the gas constant, T the temperature, Nu = $\alpha D/\lambda$ the Nussel number, is presented as

$$\frac{L}{L_0}v\frac{dv}{d\tau}+\frac{L_0-L}{L_0}v_d\frac{dv_d}{d\tau}(c_{p,L}T_d)\pm\frac{w\psi}{\rho_L SL}+\frac{L}{L_0}\frac{d}{d\tau}(c_{p,L}T_L)\pm vF_e/\rho_L S=0$$

$$\frac{L}{L_0}\frac{d}{d\tau}(c_{p,L}T_L)=\frac{j^2}{\sigma_e}+\frac{c_f\rho_L v^3}{2L_0 S}$$

$$(P + a\rho_G^2)\left(\frac{1}{\rho_G} - b\right) = RT_G$$

$$L = L_0 - \frac{D}{6}\int_0^\tau w\,d\tau \qquad w = 0.472\sqrt{\frac{Ddv}{2d\tau} + 0.049v}$$

$$\frac{dv_d}{d\tau} = 0.163\frac{\rho_G^2}{\rho_L}\frac{\mu_G}{\sigma^2}(v - v_d)^5$$

$$\bar{v}_d = 0.5(v_d + v + w + w_k)$$

$$w_k = 4.17w\sqrt{\frac{x_k}{D}} \qquad x_k = \frac{D}{(4.17 - \frac{\varphi}{w})^2}$$

$$\frac{(T_G - T_L)\sigma c_{P,L}Wew\rho_L}{3(T_G - T_d)\alpha_d(v - v_d)^2\rho_G(L_0 - L)} + \frac{8\pi\sigma^3(T_G - T_s)}{\rho_G^3 S(L_0 - L)(T_G - T_d)(v - v_d)^6} = 1$$

$$\alpha_d = \frac{6Nu_v(L_0 - L)\lambda_g}{\Delta x_k D^2}$$

$$T_d = 0.5(T_L + T_S), \qquad T_L/T_S \geq 0.7$$

$$T_d = \frac{T_L - T_S}{\ln\frac{T_L}{T_S}}, \qquad T_L/T_S \leq 0.7$$

The initial conditions are $\tau = 0$, $v = v_0$, $dv/dT = [(P_0 - P)/\rho_L L_0] S, w = 0$, $T_G = T_{G,0}$, $p = p_0$, $T_L = T_d = T_{L,0}$.

The system comprises equation of movement of the piston with its mass decreasing, the equation of energy, the equation of the liquid phase, mass conservation of the solid part of the piston, the equations of dispersion intensity, and the equation of the convective hear exchange between phases.

Figure 6 presents the results of the solution of a system of equations for the acceleration devices, where the following working bodies are employed: alloy Sn-Bi and steam $D = 0.05$, $p_{0,L} = 100$ kPa, $p_{0,g} = 400$ kPa, $L_0/D = 5$, $v_0 = 5$ m/s. η_a, the adiabatic efficiency, and η_p the polytropic efficiency of acceleration device, are shown in Fig. 6.

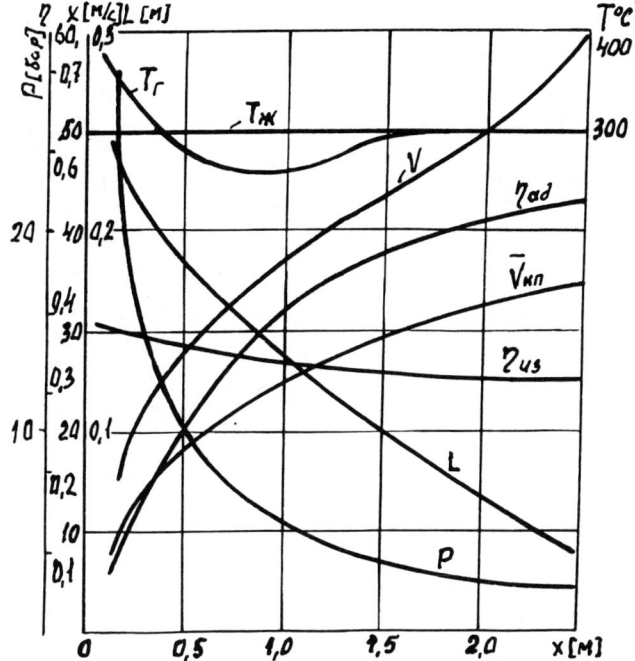

Fig. 6 Results of solving complete system of equations.

Pilot Plants

A number of liquid metal MHD installations with one- and multipistons were tested.

Pumping of the molten metal in the circuit at the expense of the thermodynamic acceleration device was, for the first time, realized in 1974. At the same time, current in the liquid-metal MHD generator was obtained. Parameters of the installation are as follows: working substances, steam-tin; initial parameters of steam, 400 kPa, 773 K; steam yield, 0.1 kg/s; consumption of the molten metal, 8×10^{-3} m^3/s; highest metal velocity, 63 m/s; polytropic efficiency of the accelerator device (without separator, 0.72; with separator, 0.51); efficiency of the molten-metal circuit, 0.12; and electrical efficiency, 0.031.

References

[1] Kolesnichenko, A. F., Technological MHD Installations and Processes, *Naukova Dumka*, USSR, 1980.

[2] Kalafati, D. D., and Kozlov, V. B., Thermodynamics of Molten-Metal MHD Converters, *Atomizdat*, Moscow, 1972.

[3]Kim, K. I., Kolsenichenko, A. F., and Gorislavets, Yu. M., Analysis of Certain Circuits of Binary Units of Electric Power Plants, *Naukova Dumka*, USSR, 1974, pp. 72-79.

[4]Dezusy, L. G., Kim, K. I., Scheglov, G. M., et al., "Molten-Metal MHD Systems with complex Flow and Selection of Power by the Synchronous Principle," *Symposium on Production of Electric Energy by Means of MHD Generators*, 1968, pp. 107-179.

[5]Kolesnichenko, A. F., Malakhov, V. V., and Gorislavets, Yu. M., "Pumping of Liquid-Metal Heat Carrier with Application of Acceleration Devices on Piston Gas-Liquid Flows," *Teplifika Cysokikj Temperatur*, Vol. 15, No. 1, 1977, pp. 172-178.

Recent Developments in Liquid-Metal MHD Thermoacoustic Engines

D. Hamann*
Dresden University of Technology, Dresden, Germany
and
G. Gerbeth†
Central Institute for Nuclear Research, Rossendorf, Germany

Abstract

A literature review on thermoacoustic engines (TAEs), with particular emphasis on liquid metal MHD TAE's is presented. The main aim of this paper is to draw the attention and the interest of the international MHD community to these new developments since it has only been discussed in the literature on acoustics.

TAEs provide a new way to convert heat to mechanical energy, or more strictly speaking, to acoustic power. They have an efficiency comparable to existing techniques but with the possibility of increasing reliability because there are no moving parts. TAEs utilize heat flow from a high-tempeature source to a low-temperature sink to generate acoustic power in the form of high-amplitude sound waves in liquid sodium. Since acoustic power is inconvenient in most situations, a power transducer is required to convert acoustic power into an electric one. Though there are a number of converter mechanisms, the magnetohydrodynamic one is particularly suited for sound waves in liquid metals. A magnetic field perpendicular to the direction of sound propagation is applied to the center of the resonator, in which the sound has been generated. There are electrodes in the sodium that form an electric current path perpendicular to both magnetic field and sound velocity.

Introduction

Small, compact, safe, reliable and if possible cheap electric power supply systems are necessary for space and deep ocean research programs as well as for

---
Copyright © 1992 by the American Institute of Aeronautics and Astronautics, Inc. All rights reserved.
*Professor, Department of Acoustics.
†Professor, Department of Sodium Technology.

other purposes. Though there are several well-developed power generators, TAEs could be competitive due to their extreme simplicity.

TAEs belong to the so-called heat engines. As shown in Fig. 1, there are two classees of these engines: prime movers and heat pumps. In a prime mover, heat flows through the engine from a high-temperature source to a low-temprature sink, and the engine generates mechanical work. In heat pumps, work is absorbed by the engine, resulting in the pumping of heat from a point with low temperature to a high one. In this study, only prime movers are reviewed. Consequently, the term TAE is used as a synonym for them, although acoustically excited heat pumps are TAEs too.

The main feature of TAEs is the generation of acoustic power in the form of high-amplitude sound waves in gas or liquid-filled resonators using heat flow from a high-temperature source to a low-temepature sink. In the engines under consideration the generated acoustic power is converted into an electric one magnetohydrodynamically. Although technological development is only beginning, it seems useful to discuss the ideas in the international community.

The aim of this study is
1) to draw attention to this interesting phenomenon, which has been discussed mainly in the acoustic literature only.
2) to describe some unsolved physical and technological problems.
3) to propose an idea for international cooperation in this promising field.

In the following section, the history of TAEs as well as a review of converter mechanisms is briefly given. Then, the main components of these engines are described. After that, the most advanced TAEs using liquid sodium as the working substance, as well as the MHD conversion principle are considered. Finally, concluding remarks as well as a review of future perspectives complete the study.

It should be emphasized that this review rests mainly on the remarkable series of papers by Swift et al.[1-5] (see acknowledgments).

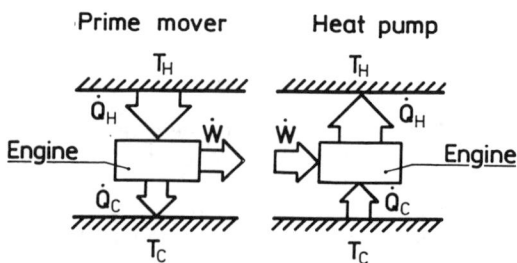

Fig. 1 Two types of heat engines: T_H, T_C = temperature of hot and cold reservoir; $\dot{Q}_H$, $\dot{Q}_C$ = associated heat flows; W = work flow.

History
Sound Generation by Heat

The history of thermally driven sound generators is rather long. In 1777, Byron Higgins observed acoustic oscillations in a large pipe excited by a suitable location of a hydrogen flame inside (see Fig. 2). Higgins work was continued and extended by Rijke in 1859. He placed a heated wire screen in the lower half of an open-ended vertical pipe, as shown in Fig. 3.

In 1850 Sondhauss developed the earliest TAE with a direct relationship to the prime mover of this reivew. He discovered that when a hot glass bulb was attached to a cold glass tubular stem sometimes sound emission by the stem tip occurred (see Fig. 4). Soundhauss derived a relation between the pitch of sound and the dimensions of the apparatus. Low-temperature physicists have been very cautious as to the effect described by Taconis in 1949 (relevant references can be found in Ref. 1) since it could pose a severe danger for cryogenic apparatus. He observed extremely high-amplitude oscillations when a gas-filled tube is cooled from room temperature to a cryogenic one.

The most important advance in modern thermo-acoustics was made by Carter (cf. Ref. 8) in 1962. He found empirically that the performance of Sondhauss tubes can be considerably improved by placing suitable structures

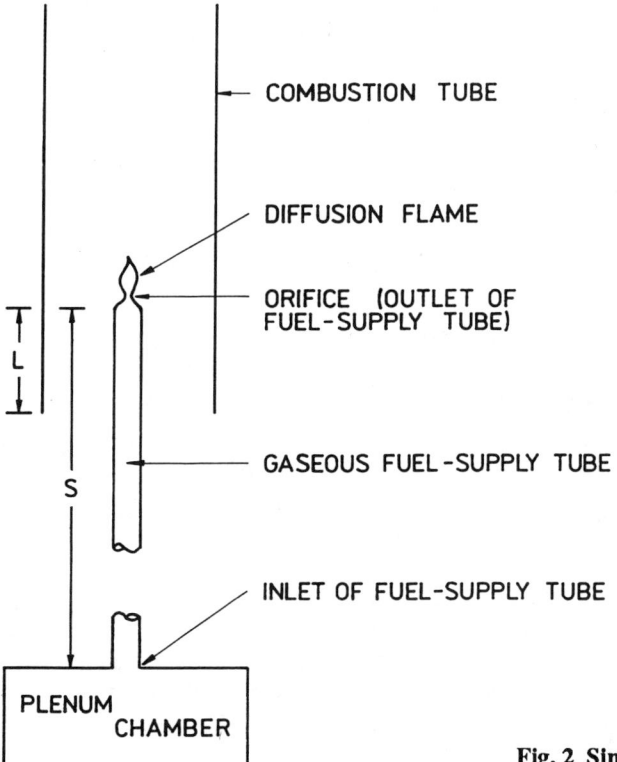

Fig. 2 Singing-flame apparatus.[6]

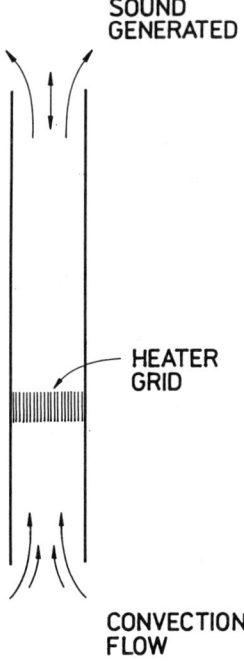

Fig. 3 Rijke tube.[7]

inside the tube. Examples for such insertion (large concentric tube, bundles of small tubes, wire screen gauze, etc.) are shown in Figs. 5 and 6. Developing Carter's ideas, Feldman[8] produced 27W of acoustic powr from 600 W heat. It should be noted that all of the recently built TAEs are based on the pioneering papers of Carter.

The thermo-acoustic effect that is reponsible for the operation of these modern TAEs consist of the two following subeffects:

Convective heat flow. Stacks of plates with a sufficiently high temperature gradient are commonly used as insertions in the TAE. A significant convective heat flow from the hot to the cold part of the plates takes place.

Sound generation. Simultaneously, the fluid near each plate produces acoustic power.

It is noteworthy that in an ordinary acoustic standing wave only adiabatic temperature oscillations accompanying the pressure oscillations can be found. The addition of a stationary plate, which simply imposes a temporarily isothermal boundary condition, leads to a dramatic change of the situation. The thermal interaction of the wave with the plate generates the effects described previously.

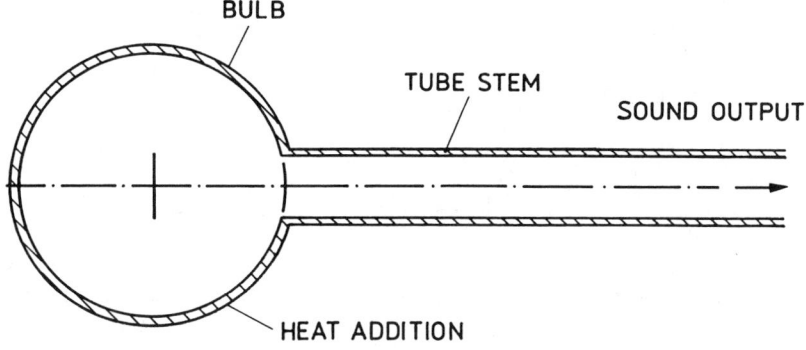

Fig. 4 Sondhauss tube.[8]

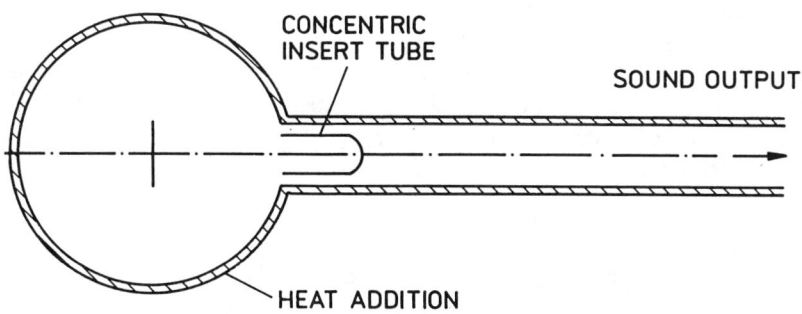

Fig. 5 Sondhauss tube with concentric tube insert.[8]

Conversion of Acoustic Energy

Since acoustic power is inconvenient in most situations, a power transducer is required to convert acoustic power into some other form of work, mostly into electric power.

Up to the beginning of the 1950's, the thermo-acoustic machines described in the previous section were usually meant as demonstration devices, serving no useful purpose other than to produce a loud noise. However, in 1951 and 1958, Hartley[9] and Marrison[10] respectively, of Bell Telephone Laboratories received patents on coupling "singing" pipes (comparable to that shown in Fig. 2) to acoustical-to-electrical transducers, which should generate usable electrical power. In 1962 Carter[8] suggested using a special kind of transducer, namely the electrostatic one.

Acoustic MHD Transduction

Even at the beginning of the 1960's Carter (cf. Ref. 8) started a very promising development — the generation of an alternating electrical current by using oscillations of a suitable conducting gas in a magnetic field (MHD effect).

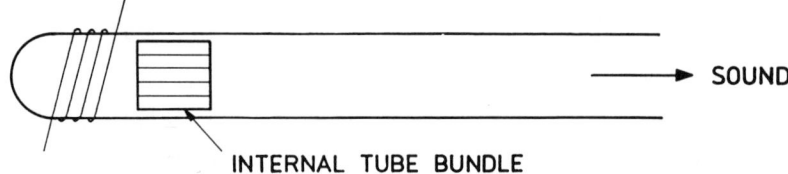

Fig. 6 Sondhauss tube with internal tube bundle externally heat.[8]

The first experiment in the history of TAEs in which electric power was generated by an acoustic wave field was performed by Moose[11] in 1979. As can be seen from Fig. 7, he used the MHD transduction effect. Whereas Moose conducted his investigations at 10 kHz in a simple laboratory-scale tank filled with salt water, Bogorodskij[12] initiated similar experiments in the open ocean at 10 Hz in the Earth's magnetic field.

Since 1985 a new approach in TAE research was taken. Liquid metal was used as a conducting fluid. In that year Swift[2] proposed his liquid metal MHD TAE in which, first, a sound field in the liquid metal is generated thermally, and second, the acoustic power is converted into electric power magnetohydrodynamically.

Components of TAEs

From the statements of the previous section the scheme of a complete TAE, presented in Fig. 8 can be derived. It consists of the following

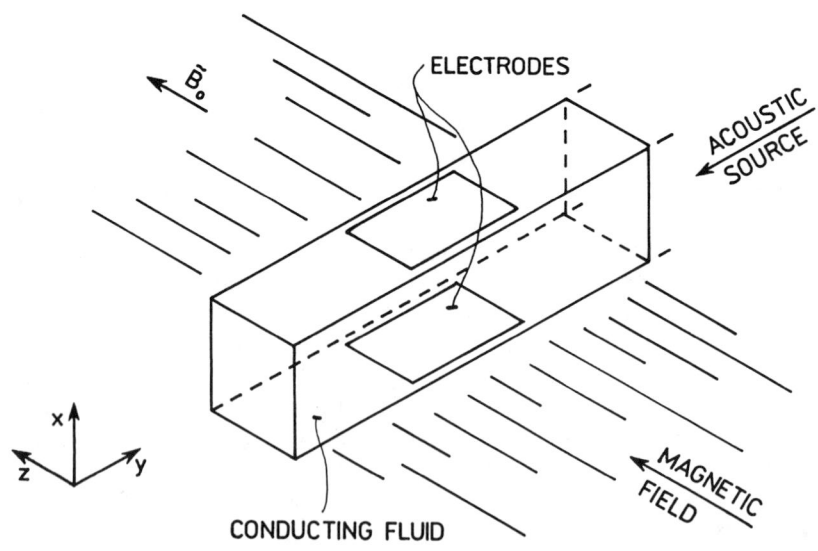

Fig. 7 Experiment of Moose demonstrating the conversion of acoustic energy into electricity by MHD effect.[11]

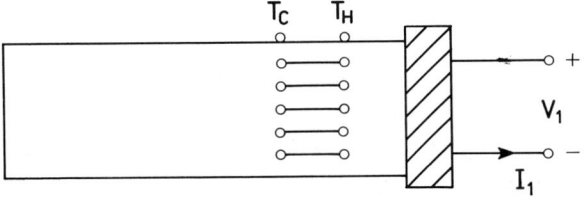

Fig. 8 Schematics of a complete TAE consisting of a stack of plates, hot and cold heat exchanger, resonator, and the transducer producing oscillating voltage V_1 and current I_1.

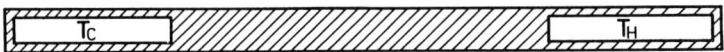

Fig. 9 Heat exchangers at the ends of a plate.[1]

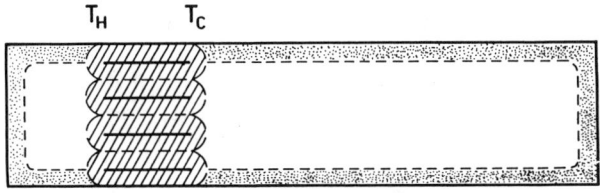

Fig. 10 Three fluid regions inside a TAE.[1]

components:

1) <u>Stack of plates</u>. As already mentioned, this component is the "core" of the TAE. Here the presence of a standing wave stimulates a time-averaged convective heat flow along with the temperature gradient, accompanied by the generation of acoustic power.

2) <u>Heat exchanger.</u> The heat exchanger is required to supply and to extract heat at the ends of the stack, as shown in Fig. 9. From the physical picture that heat flow along the plate is performed by fluid parcels like a bucket brigade, it follows that the heat exchanger should have an appropriate length, strictly speaking a length of about the peak-to-peak displacement amplitude (for details see Ref. 1). If it is longer, some parcels exist that contact only the heat exchanger at both extreme points of their positions. They perform no useful function, and their presence is purely dissipative, e.g., they absorb work. If the heat exchanger is shorter, some parcels "jump" past the heat exchanger (never contacting the heat exchanger at all). They and their bucket brigade partners are ineffective in carrying heat.

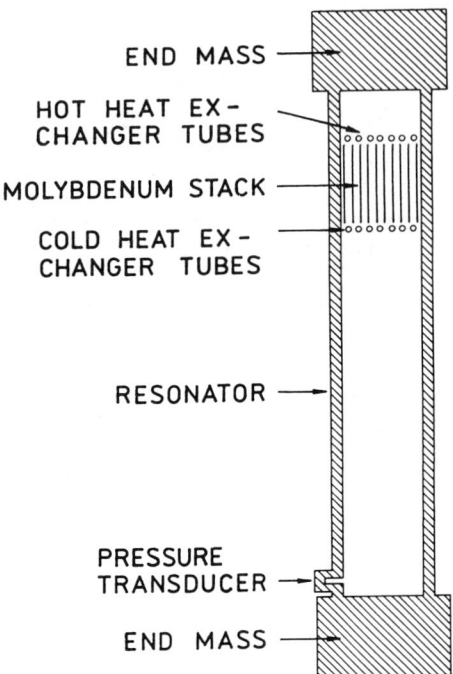

Fig. 11 Schematics of the liquid sodium TAE.[5]

3) <u>Resonator.</u> It is necessary that the resonator contains the stack and the heat exchanger. Moreover, its presence is one prerequisite for the generation of a standing acoustic wave. The resonator should have a high acoustic quality factor Q to minimize the dissipation of acoustic power into heat. Generally, the volume of a TAE résonator can be divided into three parts as shown in Fig. 10. The fluid in the near neighborhood of the plates, i.e., in the so-called penetration depth, is productive since it generates the sound. The fluid near the resonator surface is completely dissipative, whereas the rest is reactive, i.e., it stores energy and determines the frequency of the acoustic resonance.

The most frequently used resonator shape is that of a cylinder. However, many other forms are possible.

4) <u>Transducer.</u> It is required to convert acoustic power into electric power. Several basic conversion mechanisms are applicable, such as the electrodynamic, electrostatic, piezoelectric, MHD, etc. The preferable transducer should be simple, reliable, and efficient. Thus, the inherent virtues of the TAE are not compromised. The MHD transducer used in the liquid metal TAE meets all of the requirements.

Liquid-Sodium MHD TAE

In 1931 Malone (cf. Ref. 1) built several reciprocating heat engines using water as the working fluid. Moreover, he realized that from a theoretical point of view other liquids, including sodium, could be good working substances, too. However, he rejected sodium as being too dangerous.

Now, materials technology is sufficiently advanced to permit safe, routine handling of liquid sodium. At the beginning of the 1980s the Los Alamos group initiated its liquid sodium MHD TAE program.

As the first step Swift[5] built an engine shown in Fig. 11. It has no power transducer, so that all acoustic power produced by the stack is absorbed by resonator losses. The stack plates were made of molybdenum because it is not soluble in liquid sodium. They were 0.021 cm thick, spaced 0.040 cm apart and 5.21 cm long, 2nd mounted with their hot ends 12.65 cm from the hot end of the resonator. The resonator itself is 122 cm long with 4.89 cm o.d. and 3.8 cm i.d. In order to provide nearly fixed acoustic boundary conditions, the ends of the resonator were welded to masses of 32 kg. With the help of a flow and ballast volume, a constant mean pressure p_m = 9.7MPa was maintained. The two heat exchangers consisted each of 18 stainless steel hypodermic tubes, with 1.1 mm o.d. and a 0.2 mm thick wall. The hot heat exchanger was connected to a liquid sodium heat exchange loop and the cold one to a pressurized-water heat exchange loop.

In Fig. 12 the observed performance is presented. The cold temperature T_c was fixed at 388 + 10K, and the hot one T_H could be varied. It can be seen that below $\dot{Q}_H$ = 630 W ($\dot{Q}_H$ is hot heat flux) there were no acoustic oscillations and,

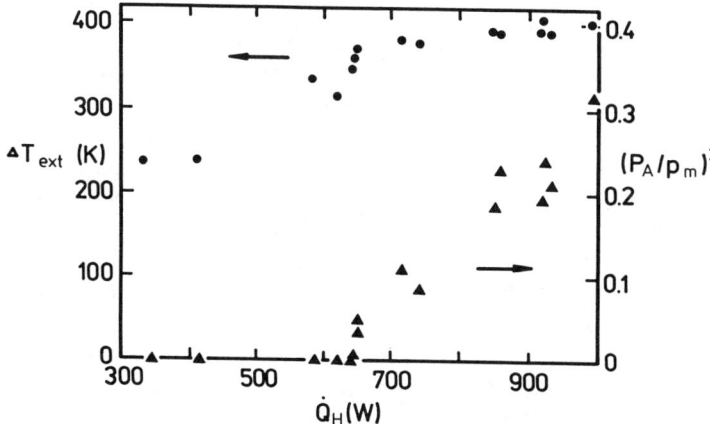

Fig. 12 Observed performance of the sodium TAE; acoustic pressure amplitude P_A and $\Delta T_{ext} = T_H - T_C$ are plotted against hot heat flux $\dot{Q}_H$.

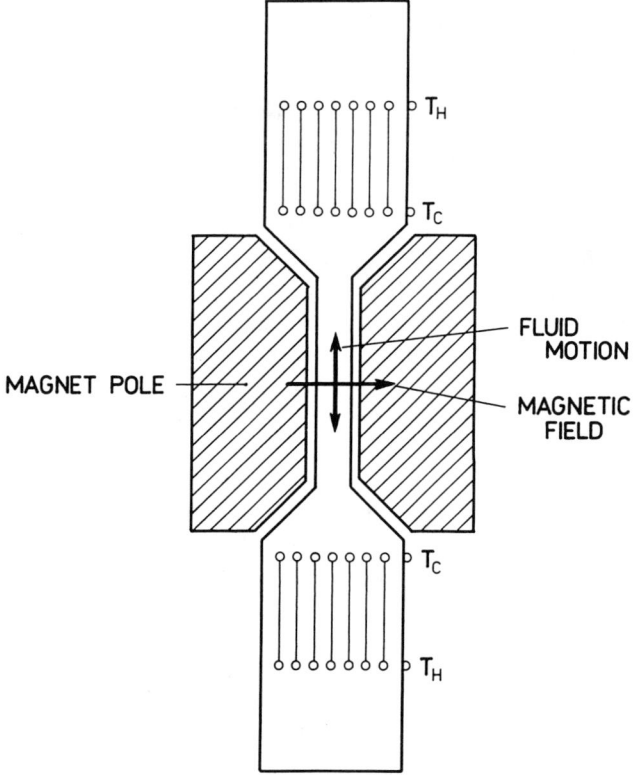

Fig. 13 Advanced liquid metal MHD TAE.

consequently, the acoustic pressure amplitude P_A is zero. Above $\dot{Q}_H = 630$ W the sodium oscillated at 910 Hz, with P_A^2 increasing rapidly and linearly with $\dot{Q}_H$. Above $\dot{Q}_H = 900$ W, Swift did not obtain data becaus P_A was large enough to cause cavitation in one of the resonator filling tubes, which was acting like another small, driven resonator. In that situation, the amplitude rose exponentially until cavitation occured, violently killing the oscillations in a few ms and restarting the exponential buildup. Fearing catastrophic failure of the whole facility, the researchers avoided such high $\dot{Q}_H$. At the highest acoustic amplitude that could be safely achieved, the engine produced 18 W of acoustic power from 990 W of heat. Summarizing these results, one can state that the first experiment was successful.

As the second step, the MHD transducer has been developed separately.[4] It consists of a rectangular channel for the sodium of 1.2 cm thickness in the direction of the 2.3T magnetic field, 7.6 cm width in the direction of electric current flow, and 31cm length in the direction of acoustic fluid flow, with the middle 20 cm of that length actually in the magnetic field and in contact with the electrodes. The transducer was welded into the center of a 1 m long, 10 cm^2

cross section, high Q resonator filled with sodium at 130°C. Now, this transducer is well checked both experimentally and theoretically. Up to now the highest observed efficiency was 45%.

Thirdly, the experience gained during both previously described periods were joined together.[1] Figure 13 shows the most advanced liquid metal MHD TAE developed by Swift and his group. It is a half-wave acoustic resonator with two symmetrically located stacks of plates, as well as hot and cold heat exchangers. There is a sound velocity antinode at the center, where electrical power is extracted magnetohydrodynamically. Unfortunately, neither detailed geometrical nor performance data are available from the open literature. Preliminary estimates promise 60 W/cm² acoustic power at T_H = 1000K and ΔT = 600K with an efficiency of about 1/3 of the Carnot efficiency. The performance is expected to be substantially better at higher operation temperatures.

Compared with the existing concepts of liquid metal MHD power generation[13] the main advantage of the described MHD transducer is lack of a global circulation of the liquid metal. Therefore, there is no need to use vapor or gas in order to drive the liquid metal through the MHD generator. In the liquid sodium MHD TAE, the sodium is used as an electrodynamic as well as a thermodynamic working fluid.

Concluding Remarks

The electric generators described in this study are distinguished by their extremely simple construction and by the complete lack of moving parts and, due to the fact that there are no circulating fluids, of pumps. They could be used in all of the cases where reliability is paramount, such as in space and in deep ocean applications. The core of the TAEs is the stack of plates inside the resonator where the acoustic wave is generated by a sufficiently large temperature gradient. The heat flow necessary for this purpose can be taken from nuclear reactors, conventional power stations, and strongly exogeneous chemical reactions, as well as from solar energy sources TAEs have the ability to work as a single generator as well as jointly with other power facilities.

Though first tests are promising, it is obvious that the main disadvantage of the existing engines is their low efficiency. Therefore, future activities must be focused on increasing efficiency. In this context, the solution of the following problem is desirable:

1) Improvement of the theory. The recent state of TAE theory can be found in Rott's papers (for a review, see Ref. 14). However, by using this model it is impossible to predict the onset of acoustic oscillations. Rott's theory describes the physical phenomena quite well, assuming that a standing acoustic wave field already exists.
2) Decrease of acoustic losses.
3) Increase of the maximum acoustic energy density.
4) Development of better heat exchangers.

5) Exclusion of cavitation.
6) Improvement of the MHD transducer.
7) Investigations on the possibility of using MHD self-excitation to improve TAEs performance.

These points should be considered as a joint research program of both of the institutes to which the authors belong. It is expected that a certain progress can be achieved combining the experiences gained at the Dresden Institute of Acoustics with those of the Department of Sodium Technology.

However, in the authors' opinion full success in the form of liquid metal MHD TAEs having a significantly better efficiency is only possible if this project is performed with international cooperation. Therefore, this article should be understood as a call for taking part in this program. Contributions from laboratories (MHD, acoustics, etc.) as well as sponsorships from organizations and companies are welcome.

Acknowledgments

The authors wish to thank G.W. Swift at Los Alamos National Laboratory, Los Alamos, New Mexico, for sending a number of his and his group's papers. D. Hamann is also grateful for the support and encouragmeent of A. Alemany's group at the Grenoble Institute of Mechanics, Grenoble, France.

References

[1] Swift, G.W., "Thermoacoustic Engines," *Journal of the Acoustic Society of America*, Vol. 84, Oct. 1988, pp. 1145-1180.

[2] Swift, G.W., Migliori, A., Hofler, T., and Wheatley, J., "Theory and Calculations for an Intrinsically Irreversible Acoustic Prime Mover Using Liquid Sodium as Primary Working Fluid," *Journal of the Acoustic Society of America*, Vol. 78, Aug. 1985, pp. 767-781.

[3] Wheatley, J., Swift, G.W., and Migliori, A., "The Natural Heat Engine," *Los Alamos Science*, No. 14, fall, 1986, pp. 1-33.

[4] Swift, G.W., "A Liquid-metal Magnetohydrodynamic Acoustic Transducer," *Journal of the Acoustic Society of America*, Vol. 83, Jan., 1988, pp. 350-361.

[5] Migliori, A., and Swift, G.W., "Liquid-sodium Thermoacoustics Engine," *Applied Physics Letters*, Vol. 53, Aug., 1988, pp. 355-357.

[6] Puntnam, A.A., and Dennis, W.R., "Survey of Organ-pipe Oscillations in Combustion Systems," *Journal of the Acoustic Society of America*, Vol. 28, March 1956, pp. 246-259.

[7] Feldman, K.T., "Review of the Literature on Rijke Thermoacoustic Phenomena," *Journal of Sound and Vibration*, Vol. 7, No. 1, 1968, pp. 83-89.

[8] Feldman, K.T., "Review of the Literature on Soundhauss Thermoacoustic Phenomena," *Journal of Sound and Vibration*, Vol. 7, No. 1, 1968, pp. 71-82.

[9] Hartley, R.V.L., U.S. Patent No. 2, 549, 464 (1951).

[10] Marrison, W.A., U.S. Patent No. 2, 836, 033 (1958).

[11]Moose, P.H., and Klaus, R.F., "Experimental Observations of Magnetoacoustic Fields," *Journal of the Acoustic Society of America*, Vol. 74, Sept. 1983, pp. 1066-1068.

[12]Bogorodskij, V.V., Gusev, A.V., Zufrin, A.M., Polyakov, A.P., Ponomarev, A.N., and Stepanov, B.M., "Synchronous Measurements of Infralow-frequency Acoustic and Electromagnetic Waves in a Marine Medium," Arktic Antarktic, Nauchno-Issledovatelnoi Institute, Vol. 374, 1980, pp. 80-86.

[13]Petrick, M. and Branover, H., "Liquid Metal MHD Power Generation. Its Evolution and Status," *Single-and Multiphase Flows in an Electromagnetic Field*, Vol. 100, edited by H. Branover, P.S., Lykoudis, and M. Mond, Vol. 100, Progress in Astronautics and Aeronautics, AIAA, New York, 1985, pp. 371-400.

[14]Rott, N., "Thermoacoustics," *Advances in Applied Mechanics*, Vol. 20, 1980, pp. 135-175.

Chapter 3. Magnetohydrodynamic Flows

Interfacial Instabilities in the Presence of Electric Current and Magnetic Field

Sylvain Pigny* and René Moreau*
*Madylam-Institut National Polytechnique de Grenoble,
St. Martin d' Héres, France*

ABSTRACT: The stability of an interface between two liquid layers, sandwiched by solid electrodes, is studied by a perturbation method. The horizontal directions of the four media are considered to be infinite. Thicknesses and electric conductivities of each medium are finite, so the influence on the interface position of electromagnetic phenomena inside the electrodes can be taken into account. The importance of non uniformity of electric currents is underlined, as its interaction with a non uniform magnetic field. It appears necessary to lay an emphasis on large wavelengths phenomena.

Introduction

In metal processing there are a number of situations where two liquid layers, separated by an interface, are crossed by an electric current. Even though the interface is usually almost flat and horizontal, addition of electric current and magnetic field may drive stirring motion in each fluid and generate curved shape and instability of the interface. Electric current lines usually come from a solid anode. Before being collected by a cathode, they generally cross the interface. But in particular situations, as in Robinson's experiment,[1] the whole electric current does not cross the interface, and some current lines are contained in only one of the two fluids.

Some magnetic field is always present and contains two parts: the near field (induced by the internal current distribution), which varies according to curl$\mathbf{B} = \mu \mathbf{J}$, and the far field, (induced by the external conductors), whose curl is zero. The two fluids are therefore submitted to Lorentz forces $\mathbf{J} \times \mathbf{B}$. In simple situations where the forces are irrotational, they may augment or reduce the effective gravity and thus influence the stability of the interface. But in more complex situations, where curl$\mathbf{F}$

Copyright © 1991 by the American Institute of Aeronautics and Astronautics, Inc. All rights reserved.
* Laboratoire Madylam, B.P. 95.

is not zero, some stirring motion takes place and may significantly differ from one fluid to the other. Then the shearing of the interface may also result in instability.

We limit our attention to situations where the relevant magnetic Reynolds number is supposed to be much smaller than 1 and where the disturbances of the electromagnetic quantities are independent of the velocity field. They depend only on the geometry of the interface and instantaneously adjust themselves to any change in that geometry, thus reacting to its disturbance. Another situation was studied by Descloux and Romerio,[2] who take into account the electric field $\mathbf{u} \times \mathbf{B}$. Our philosophy is that, in most situations in the industry, the relevant magnetic Reynolds number and the far magnetic field are both small enough for $\mathbf{u} \times \mathbf{B}$ to be much smaller than $\mathbf{J}/\sigma$.

In this paper the thicknesses of the two layers are assumed to be small; thus the stirring motion in each fluid is modeled by a friction coefficient κ, as introduced in the stationnary model of aluminium reduction cells by Moreau and Evans.[3] The relative difference of density may vary from 0.1 to 10, and gravity has some stabilizing influence, as has surface tension. The stability of an interface has been studied using several approaches. Shercliff[4] emphasizes the extremely anisotropic character of the waves at the interface. The waves influence the electric current and hence the magnetic force to an extent that depends on the direction of the propagation. Sneyd[5] shows that the non uniformity of the magnetic field created by external conductors can destabilize the interface. Moreau and Ziegler[6] underline the importance of a horizontal component of the electric current and show that it can also destabilize the interface.

The purpose of this paper is to develop a linear stability analysis, in which both the motion of the fluids and the non-uniformity of the electric current density and magnetic field are taken into account. It appears necessary to estimate the perturbations of current and magnetic field in the electrodes, on the one hand to study their influence on the electric quantities in the two fluid layers, on the other hand to ensure the continuity of the magnetic field. A striking result is the anisotropy and spacial dependence of the stability criteria, which we will explain by some elementary mechanisms. Another point is that the instability appears above all for the large wavelengths.

Linear Stability Analysis

Undisturbed State

In this stability study the four media (see Fig. 1) are considered to be infinite in the horizontal directions x and y. In the z direction the anode is limited by $h < z < h_a$, the cathode by $-h_c < z < -H$, and the two fluid layers by $-H < z < h$, and the interface is the $z = 0$ plane when not perturbed.

In each medium the electric potential is developed in a polynomial pattern, which satisfies the Laplace equation and the conditions of continuity of Φ and $\sigma \partial \Phi / \partial z$ at the interfaces. The four media are referred to as a for the anode, 2 for the upper fluid, 1 for the lower, and c for the cathode; their electric conductivities are $\sigma_a, \sigma_2, \sigma_1, \sigma_c$:

$$\begin{cases} \Phi_a = \dfrac{J}{\sigma_2}\left[\epsilon_a z + \delta_a + ax + by + \dfrac{c}{H}\left(x^2 - z^2\right)\right] \\ \Phi_2 = \dfrac{J}{\sigma_2}\left[z + ax + by + \dfrac{c}{H}\left(x^2 - z^2\right)\right] \\ \Phi_1 = \dfrac{J}{\sigma_2}\left[\dfrac{z}{\Sigma} + ax + by + \dfrac{c}{H}\left(x^2 - z^2\right)\right] \\ \Phi_c = \dfrac{J}{\sigma_2}\left[\epsilon_c z + \delta_c + ax + by + \dfrac{c}{H}\left(x^2 - z^2\right)\right] \end{cases} \quad (1)$$

The following non-dimensionnal quantities are introduced:

$$\Sigma = \frac{\sigma_1}{\sigma_2}, \ \Sigma_a = \frac{\sigma_2}{\sigma_a}, \ \Sigma_c = \frac{\sigma_1}{\sigma_c}$$

$$\alpha_a = \frac{h_a}{H}, \ \alpha_c = \frac{h_c}{H}, \ \alpha = \frac{h}{H}, \ e_a = \alpha_a - \alpha, \ e_c = \alpha_c - 1$$

$$\epsilon_a = \Sigma_a + 2a\alpha\left(1 - \Sigma_a\right), \ \delta_a = h\left(1 - \Sigma_a\right)\left(1 - 2a\alpha\right)$$

$$\epsilon_c = \frac{\Sigma_c}{\Sigma} + 2a\left(\Sigma_c - 1\right), \ \delta_c = H\left(\Sigma_c - 1\right)\left(\frac{1}{\Sigma} + 2a\right)$$

The average current density crossing the interface is denoted as J. The first term proportional to z represents a uniform vertical current crossing the horizontal interface. Additional electric currents, which are commonly observed, are modeled via

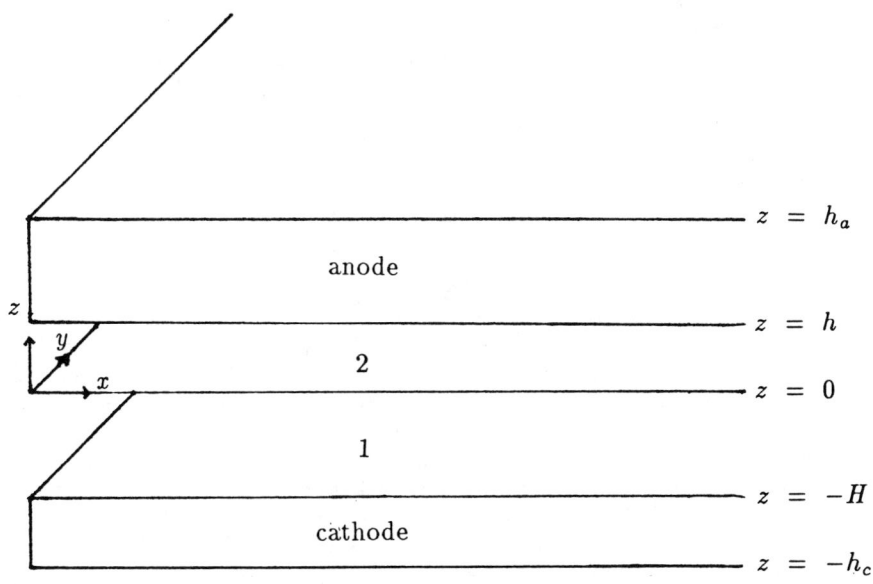

Fig. 1 The four media.

three other terms, characterized by the non dimensional coefficients a, b and c, which may be supposed to be much smaller than 1. The factors a and b are responsible for a pure horizontal current in each fluid. They can be used to oppose gravity in an experiment such as Robinson's.[1] The current associated with c entirely crosses the interface and allows variation in the shape of the electric current lines. Because of b, the pure horizontal current can have any orientation with the current associated with c.

We follow Sneyd[5] and model the far field by a linear relation, $B_i = B_{0i} + \alpha_{ij} x_j$. The matrix of coefficients α_{ij} is symmetric because curl$\mathbf{B} = 0$ in each medium, and $\alpha_{ii} = 0$ because $\nabla \cdot \mathbf{B} = 0$. Then the actual field in each fluid layer may be expressed:

$$B_{1,2} = \begin{cases} B_{0x} + \gamma_1 x + \beta_3 y + \beta_2 z + \mu J \left(\frac{y}{2} - b_{1,2} z\right) \\ B_{0y} + \beta_3 x + \gamma_2 y + \beta_1 z + \mu J \left(-\frac{x}{2} + a_{1,2} z + \frac{2c_{1,2} xz}{H}\right) \\ B_{0z} + \beta_2 x + \beta_1 y + \gamma_3 z \end{cases} \quad (2)$$

with $a_2 = a$, $a_1 = a\Sigma$, $b_2 = b$, $b_1 = b\Sigma$, $c_2 = c$, $c_1 = c\Sigma$, $\gamma_1 + \gamma_2 + \gamma_3 = 0$. One should remark that our expression for the far magnetic field is a first order development in x and y, when for Φ a second-order development in x is written. The reason is that, in order to fit with observations in some electrolysis cells, the second-order term $c(x^2 - y^2)/H$ may be significantly larger than the first-order terms ax and by. If L_n is a scale for the spatial dependence of Φ on x and y, L_r is a scale for the distance between the fluid layers and the conductors that generate the far magnetic field, the ratio L_n/L_r is generally much smaller than 1. Consequently, the second-order terms for the far magnetic field may be neglected. The order of magnitude of L_n is given by comparing the importance of the terms $c(x^2 - z^2)/H$ and z in the expression of Φ. Therefore, L_n is given by $cL_n^2/H \simeq H$.

Generally the $\mathbf{J} \times \mathbf{B}$ force field may be rotational and generate some deformation of the interface as well as some motion in each liquid layer. In the particular case of aluminium reduction cells this undisturbed regime has been analysed by Moreau and Evans.[3] They model friction with a linear drag, such that the motion equations are simply:

$$\begin{cases} \nabla \cdot \mathbf{U} = 0, \\ \kappa \mathbf{U} = \nabla P + \mathbf{J} \times \mathbf{B} \end{cases} \quad (3)$$

where the friction coefficient κ may be supposed to be constant and of the order of 0.1 to 0.3 s^{-1}. It follows from their analysis that the first order solution of Eq. (3) is the purely static equilibrium $\mathbf{U} = 0$, because $\mathbf{J} \times \mathbf{B}$ is almost irrotationnal, and that the maximum elevation of the interface is so small that its slope may be considered to be zero. The same general tendancy may be obtained with the expression of $\mathbf{J} \times \mathbf{B}$ which follows from (1) and (2) when the coefficient a, b, c, β_i and γ_i are supposed small enough. Therefore, rather than solving Eq. (3) to get the undisturbed pressure and velocities which result from the applied $\mathbf{J} \times \mathbf{B}$ force field, we suppose that the interface is plane and the motion is modelled introducing two horizontal velocities $\mathbf{U}_1$ and $\mathbf{U}_2$, which are supposed to be constant and given a priori. Indeed the main justification of that assumption is to allow this theory to compare the instability due to the electromagnetic effect with the classical Kelvin-Helmholtz instability of sheared interfaces.

Disturbances

Generally, with Φ and B varying with x and y the perturbation procedure should be particularly difficult, requiring first to find the correct spatial eigen-functions, and then to decompose any perturbation into these normal modes. However the analysis of some particular cases (electric current purely vertical as in Sneyd,[5] or admitting an uniform horizontal component in each liquid when one of them is a much better electrical conductor than the other) reveals that the variation with x and y of the pressure disturbance remains compatible with a disturbance of the interface of the form

$$\eta = Ae^{st+ilx+imy} \tag{4}$$

where (l, m) are the components of the wave vector $\mathbf{k}$. In these particular cases the characteristic equation remains homogeneous after cancellation of all the terms varying with x and y which come from the polynomial expressions (1) and (2).

Also one may consider that, if the typical length scale of the horizontal variation of Φ and B (say L) is much larger than the wave length of the disturbance, then Φ and B are almost constant at the scale of the disturbance.

These two arguments suggest that the stability of this complex situation may be studied with the simplified form (4). This is certainly justified when $kL \gg 1$, and for any value of k in all particular cases where the dispersion relation remains homogeneous. It is hoped that the results keep at least some qualitative validity in other circumstances.

Notice that the expression for Φ does not depend symmetrically on x and y. A dependence on x would be enough to get a horizontal current. But since the wave vector has two components (l, m), situations where horizontal currents (uniform or not) have any orientation with respect to $\mathbf{k}$ are taken into account. Consequently, electromagnetic and mechanical quantities do not depend symmetrically on x and y. In the following θ denotes the angle between $\mathbf{k}$ and x direction.

The perturbation of the electric potential ϕ is, of course, proportional to η. It satisfies Laplace's equation, $\Delta \phi = 0$ in the four media and obeys the conditions that $\Phi + \phi$ and $\sigma \partial (\Phi + \phi)/\partial n$ be continuous when crossing the three interfaces. We neglect the discontinuity in $\Phi + \phi$ due to the electrode polarisation, a small effect studied by Grjotheim et al.,[7] usually stabilizing, present in electrolysis cells. We impose $\partial \phi/\partial z = 0$ at the planes $z = h_a$ and $z = -h_c$, so the perturbation of current cannot exist beyond the four media. Eight relations can be written. At the fluid-fluid interface, the first order expression for ϕ satisfies

$$\Phi_1(\eta) + \phi_1(0) = \Phi_2(\eta) + \phi_2(0) \tag{5}$$

$$\sigma_1 \left[\frac{\partial \Phi_1}{\partial n}(\eta) + \frac{\partial \phi_1}{\partial z}(0) \right] = \sigma_2 \left[\frac{\partial \Phi_2}{\partial n}(\eta) + \frac{\partial \phi_2}{\partial z}(0) \right] \tag{6}$$

where n is the normal to the interface. The jumps of Φ and of $\partial \Phi/\partial n$ at $z = \eta$ are of the form $u + vx$, where u and v are functions that do not depend on the space variables, and where v is proportional to the c coefficient. Therefore, an expression for ϕ is sought in each medium of the form

$$\phi = \frac{J}{\sigma_2}\left(f_0(z) + \frac{x}{H}f_1(z)\right)\eta \qquad (7)$$

Since $\Delta\phi = 0$, f_0 and f_1 obey the following equations

$$\begin{cases} f_1'' - k^2 f_1 = 0 \\ f_0'' - k^2 f_0 = -\frac{2il}{H}f_1 \end{cases} \qquad (8)$$

Their solutions are of the following type:

$$\begin{cases} f_1 = A_1\left(\mu_1 \sinh kz + (1-\mu_1)\cosh kz\right) \\ f_0 = A_0\left(\mu_0 \sinh kz + (1-\mu_0)\cosh kz\right) + \text{term}(f_1) \end{cases} \qquad (9)$$

where A_1 and A_0 are amplitude coefficients, and where μ_1 and μ_0 can be considered as orientation coefficients, connected with the ratio of electric conductivities and thicknesses of two consecutive media. In each medium four unknown quantities have to be determined, thus 16 must be determined for the complete system. Separating the eight relations in x^0 and x^1, 16 equations are obtained, and every coefficient may be determined. The A_1 coefficient is proportional to c, which involves the non uniformity of Φ and ϕ in each medium. The μ_1 coefficient is a real number. The A_0 coefficient obeys a non linear equation:

$$A_0^2\left[cX_c + X + i(aX_a + bX_b)\right]$$

$$+ A_0\left[c^2 Y_{cc} + cY_c + Y + a^2 Y_{aa} + abY_{ab} + b^2 Y_{bb} + i(acY_{ac} + bcY_{bc} + aY_a + bY_b)\right]$$

$$+ c^3 Z_{ccc} + c^2 Z_{cc} + cZ_c + a^2 c Z_{aac} + abc Z_{abc} + b^2 c Z_{bbc}$$

$$+ i\left(ac^2 Z_{acc} + bc^2 Z_{bcc} + acZ_{ac} + bcZ_{bc}\right) = 0 \qquad (10)$$

The orientation coefficient μ_0 is given by

$$\mu_0 = \frac{cX_\mu}{A_0} + \mu_1 \qquad (11)$$

where X, Y, Z, X_μ are dimensionless functions of electric conductivities and thicknesses of the four media, and of the components l, m of the wave vector.

Eq. (10) leads to two distinct solutions. When c is zero, it becomes

$$A_0^2\left[X + i(aX_a + bX_b)\right]$$

$$+ A_0\left[Y + a^2 Y_{aa} + abY_{ab} + b^2 Y_{bb} + i(aY_a + bY_b)\right] = 0 \qquad (12)$$

Obviously, one of the two solutions for Eq. (10) is zero as c is zero, and therefore is not taken into account. The non-linear character of (10) disappears when the electric current is uniform ($c = 0$). In this case the A_0 coefficient is given by

$$A_0 = V + i(aV_a + bV_b), \qquad (13)$$

where V, V_a and V_b are dimensionless functions of electric conductivities and thicknesses of the four media, and of the components (l, m) of the wave vector. In the simple case where the electric conductivity of an electrode is much more (much less) than those of the nearest fluid, the perturbed current line is normal (parallel) to the interface and μ_1 and μ_0 become 1 (0) in the fluid. The result is that the non linearity disappears in Eq. (10). Generally, the amplitude of the perturbation of current is a non linear function of the a, b, c coefficients; therefore A_0 and μ_0 are complex numbers because the orientation of the perturbed current lines does not depend on a, b, c.

The perturbation of current is obtained by Ohm's law, $\mathbf{j} = -\sigma \nabla \phi$, in each medium. Since the perturbed potential is of the type

$$\phi = \frac{J}{\sigma} F(x, z) \eta \qquad (14)$$

the disturbed current density writes

$$\mathbf{j} = -J\eta \begin{cases} ilF + \frac{\partial F}{\partial x} \\ imF \\ \frac{\partial F}{\partial z} \end{cases} \qquad (15)$$

A noticeable point is that, in the case where c is zero, A_1 is zero, f_1 is zero, and so is $\partial F/\partial x$. In this case the perturbation of current is therefore in the $(\mathbf{k}, z)$ plane.

The perturbation of magnetic field is the sum of two terms

$$\mathbf{b} = \mathbf{b_A} + \mathbf{b_P} \qquad (16)$$

where $\mathbf{b_A}$ is obtained by Ampère's law $curl \mathbf{b_A} = \mu \mathbf{j}$. Its expression in each medium is

$$\mathbf{b_A} = \mu J \eta \begin{cases} -im \int F \, dz \\ il \int F \, dz + \int \frac{\partial F}{\partial x} dz \\ 0 \end{cases} \qquad (17)$$

When c is zero, b_A is normal to the $(\mathbf{k}, z)$ plane. In addition to this mathematically simple expression, an irrotational term $\mathbf{b_P}$ is considered in order to ensure the continuity of the actual magnetic field $\mathbf{B_T} = \mathbf{B} + \mathbf{b}$. Across the three interfaces, the tangential and normal components of $\mathbf{B_T}$ are continuous; thus six relations are obtained. The perturbation of current is assumed not to exist beyond the electrodes. Since $\mathbf{b_A}$ is horizontal everywhere, on $z = h_a$ and $z = -h_c$, the nullity of the horizontal component of $\mathbf{b}$ only is imposed. Two other relations can be written. The vertical component of $\mathbf{b_P}$ is not necessarily zero on $z = h_a, -h_c$. It would be possible to consider two other media, beyond the electrodes, where b_{P_z} can decay, in order to obtain b_{P_z} zero at $\pm \infty$.

The jumps of $\mathbf{B} + \mathbf{b_A}$ across every interface are of the form $u + vx$, where v is proportional to the c coefficient. Therefore an expression for $\mathbf{b_P}$ is sought in each medium of the form

$$\mathbf{b_P} = \mu J \nabla (G\eta) \quad (18)$$

where

$$G = g_0(z) + \frac{x}{H} g_1(z) \quad (19)$$

Since $\nabla \bullet \mathbf{b_P} = 0$, $\Delta(G\eta) = 0$, g_0 and g_1 obey the following equations

$$\begin{cases} g_1'' - k^2 g_1 = 0 \\ g_0'' - k^2 g_0 = -\frac{2il}{H} g_1 \end{cases} \quad (20)$$

Solutions for these equations are of the following type

$$\begin{cases} g_1 = C_1 (\nu_1 \sinh kz + (1-\nu_1) \cosh kz) \\ g_0 = C_0 (\nu_0 \sinh kz + (1-\nu_0) \cosh kz) + \text{term}(g_1) \end{cases} \quad (21)$$

where C_1 and C_0 are amplitude coefficients, and ν_1 and ν_0 can be considered as orientation coefficients, connected with the ratio of electric conductivities and thicknesses of two consecutive media. In each medium appear 4 unknown values, thus there are 16 for the complete system. Separating the 8 relations in x^0 and x^1, 16 equations are obtained, and every coefficient is determined.

The C_1 coefficient is proportional to c. Since the far magnetic field is continuous everywhere, C_1 involves the non uniformity of $\mathbf{b}$ in each medium. The ν_1 coefficient is a real number. The C_0 coefficient obeys a non linear equation:

$$C_0^2 [cL_c + i(aL_a + bL_b)]$$

$$+ C_0 \left[c^2 M_{cc} + a^2 M_{aa} + ab M_{ab} + b^2 M_{bb} + i(ac M_{ac} + bc M_{bc}) \right]$$

$$+ c^3 N_{ccc} + a^2 c N_{aac} + abc N_{abc} + b^2 c N_{bbc} + i \left(ac^2 N_{acc} + bc^2 N_{bcc} \right) = 0 \quad (22)$$

The orientation coefficient ν_0 is given by

$$\nu_0 = \frac{cL_\nu}{C_0} + \nu_1 \quad (23)$$

where L, M, N and L_ν are dimensionless functions of electric conductivities and thicknesses of the four media and of the components (l, m) of the wave vector.

Equation (22) leads to two distinct solutions. When c is zero, it becomes

$$C_0^2 [i(aL_a + bL_b)] + c_0 \left(a^2 M_{aa} + ab M_{ab} + b^2 M_{bb} \right) = 0 \quad (24)$$

Obviously, one of the two solutions for Eq. (22) is zero as c is zero, and therefore is not taken into account. The non linear character of (22) disappears when the electric current is uniform ($c = 0$). In this case the C_0 coefficient is given by

$$C_0 = i(aW_a + bW_b) \quad (25)$$

where W_a, W_b are dimensionless functions of electric conductivities and thicknesses of the four media, and of the components l, m of the wave vector. In the simple case, where the electric conductivity of an electrode is much more (much less) than those of the nearest fluid, ν_1 and ν_0 become 1 (0) in the fluid. The result is that the non linearity disappears in Eq. (22). Generally, the amplitude of b_P is a non linear function of the a, b, c coefficients and C_0 and ν_0 are then complex numbers. One should notice the similarity between the Eqs. (2,10) and (2,22): G and ϕ are two harmonic functions, with the same type of boundary conditions.

The disturbances of velocity $\mathbf{u}$ and pressure p are also proportional to η. The perturbation of the body force $\mathbf{f}$ is obtained by

$$\mathbf{f} = \mathbf{j} \times \mathbf{B} + \mathbf{J} \times \mathbf{b} \tag{26}$$

and in each medium $\mathbf{u}$ and p obey the following equations

$$\begin{cases} \rho \left(\dfrac{\partial}{\partial t} + \kappa \right) \mathbf{u} = -\nabla p + \mathbf{f} \\ \nabla \cdot \mathbf{u} = 0 \end{cases} \tag{27}$$

As noticed by Sneyd[5], the disturbance of Lorentz forces satisfies

$$\nabla \cdot \mathbf{f} = -2\mu \mathbf{J} \cdot \mathbf{j}$$

so that

$$\nabla^2 p = -2\mu \mathbf{J} \cdot \mathbf{j} \tag{28}$$

Equation (28) has to be solved with the following boundary conditions on the vertical component of the velocity

$$\begin{cases} w(z = h) & = 0 \\ w(z = \eta) & = A(s + i\mathbf{k} \cdot \mathbf{U}) \\ w(z = -H) & = 0 \end{cases} \tag{29}$$

which are easily transformed into conditions on pressure

$$\begin{cases} \dfrac{\partial p}{\partial z}(h) = f_z^{(2)}(h) \\ \dfrac{\partial p}{\partial z}(0) = -(s + i\mathbf{k} \cdot \mathbf{U_2} + \kappa)(s + i\mathbf{k} \cdot \mathbf{U_2})\rho_2 + f_z^{(2)}(0) \end{cases} \tag{30a}$$

$$\begin{cases} \dfrac{\partial p}{\partial z}(0) = -(s + i\mathbf{k} \cdot \mathbf{U_1} + \kappa)(s + i\mathbf{k} \cdot \mathbf{U_1})\rho_1 + f_z^{(1)}(0) \\ \dfrac{\partial p}{\partial z}(-H) = f_z^{(1)}(-H) \end{cases} \tag{30b}$$

It is noticeable that, since $f_z = j_x B_y - j_y B_x + J_x b_y - J_y b_x$, the far magnetic field only appears in the boundary conditions for p.

It is also noticeable that B_{oz} is not present in the expression for p. The vertical magnetic field can therefore influence the stability of the interface only through its effects on the horizontal motion and small-scale turbulence in each fluid, represented by the independent variables $\mathbf{U}$ and κ in the present analysis.

In each medium the vertical component of $\mathbf{f}$ is of the form

$$\left(F_{00} + F_{10}x + F_{01}y + F_{20}x^2 + F_{11}xy\right)\eta \tag{31}$$

where the F_{ab} functions do not depend on the space variables x and y, but only on z. This suggests for the disturbance of pressure an expression in each medium of the following type:

$$p = -2\mu J^2 H \left(P_{00} + \frac{x}{H}P_{10} + \frac{y}{H}P_{01} + \frac{x^2}{H^2}P_{20} + \frac{xy}{H^2}P_{11}\right)\eta \tag{32}$$

where P_{ab} are non dimensional functions depending only on the space variable z, and whose expressions are obtained by solving Eq. (28), with the boundary conditions given by Eq. (30).

Characteristic Equation

The characteristic equation may be deduced from the condition of continuity of the pressure at the interface

$$P^{(2)}(\eta) + p^{(2)}(0) - P^{(1)}(\eta) - p^{(1)}(0) = 0 \tag{33}$$

The disturbance of pressure is obtained from Eq. (28). The non-perturbed pressure difference contains a term that represents the irrotational part of the non perturbed electric force $\mathbf{J} \times \mathbf{B}$, the gravity term $g\delta\rho$, and the surface tension term Γk^2. At the interface, it is written as

$$\left[P^{(2)} - P^{(1)}\right](\eta) = \eta\left[g\delta\rho + \Gamma k^2 - \mu J^2 H(\Sigma - 1)\left(\frac{ax}{H} + \frac{2cx^2}{H^2}\right)\right] \tag{34}$$

Finally, the characteristic equation has the following classical form

$$\frac{(s + i\mathbf{k}\cdot\mathbf{U_1} + \kappa)(s + i\mathbf{k}\cdot\mathbf{U_1})\rho_1}{k} \cdot \frac{1}{t_1} + \frac{(s + i\mathbf{k}\cdot\mathbf{U_2} + \kappa)(s + i\mathbf{k}\cdot\mathbf{U_2})\rho_2}{k} \cdot \frac{1}{t_2}$$

$$+ g\delta\rho + \Gamma k^2 + \mu J^2 H[n + im] = 0 \tag{35}$$

where n and m are real numbers, which depend on x and y. One should notice the presence of the imaginary term im, which is zero when curl$\mathbf{f}$ is zero.

Neutral Curve

The marginal condition which ensures that the real part of s be zero follows from Eq. (35). Dimensionless quantities are introduced, based on the characteristic length H and the scale of magnetic field $\mu J H$. Let us define the following non dimensional numbers:

Froude number: $\quad F = \dfrac{|\mathbf{k}\cdot(\mathbf{U_2}-\mathbf{U_1})|}{k\sqrt{gH}}$

Ratio of electromagnetic forces to gravity forces: $\quad Q = \dfrac{\mu J^2 H}{g\delta\rho}$

Ratio of square of electromagnetic forces
on the product of gravity and friction forces: $\quad P = \dfrac{\mu J^4 H}{g\delta\rho^2 \kappa^2}$

Ratio of surface tension on gravity forces: $\quad T = \dfrac{\Gamma}{g\delta\rho H^2}$

The marginal condition is written as

$$\frac{\alpha_1 r_1 r_2}{t_1 t_2} F^2 = -P\alpha_1 m'^2 + \left(\frac{r_1}{t_1} + \frac{r_2}{t_2}\right)(Qn' + 1 + T\alpha_1^2) \tag{36}$$

where $\alpha_1 = kH$, $\alpha_2 = kh$, $r_1 = \frac{\rho_1}{\delta\rho}$, $r_2 = \frac{\rho_2}{\delta\rho}$, $t_1 = \tanh(\alpha_1)$, and $t_2 = \tanh(\alpha_2)$. The two functions m' and n' depend on the far magnetic field components in a linear way.

Typical neutral curves are presented in Fig. 2 for a particular situation where a layer of liquid gallium ($\rho = 6000$ kg.m^{-3}), and a layer of mercury are sandwiched between an upper carbon anode and a lower copper cathode, with a zero far magnetic field. These curves are obtained for $x = y = 0$ and for a wave vector parallel to the x direction ($\theta = 0$). The influence of electromagnetic effects on the shape of the neutral curve, which is negligible for large wave numbers ($\alpha_1 > 10^2$), dominated by the surface tension effect, becomes predominant for small wave numbers. It may be stabilizing or destabilizing in the range of medium wave numbers ($10^{-1} \leq \alpha_1 \leq 10^2$) depending on the values of the two electromagnetic parameters P and Q and of the coefficients a, b, c. Then electromagnetic effects become systematically stabilizing in some range of larger wave lengths ($10^{-2} \leq \alpha_1 \leq 10^{-1}$) where the neutral curve exhibits a maximum. For extremely large wave lengths, ($\alpha_1 \leq 10^{-2}$), electromagnetic effects are responsible for an asymptotic branch which may be either negative (as in the conditions of Figure 2), or positive, depending on the sign of the difference $\Sigma - 1$, and the relative thicknesses. The most striking result is certainly this possible strong destabilization of extremely large wave lengths. Its understanding requires an asymptotic analysis of expression (36) when α_1 tends to zero, which is easy to deduce from the asymptotic values of m and n:

$$m \sim \frac{c(\Sigma - 1)^2 M}{\alpha_1^3}, \quad n \sim \frac{c^2 \sin^2\theta (\Sigma - 1) N}{\alpha_1^4}$$

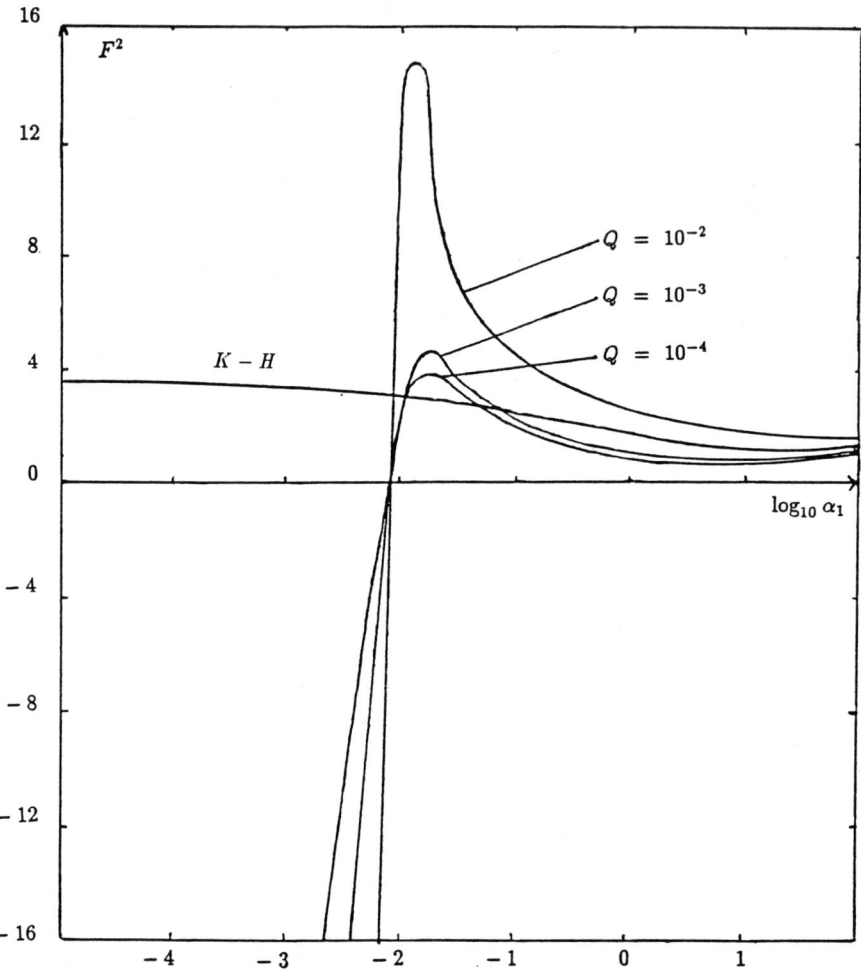

Fig. 2 Neutral curves: electric parameters $c = 10^{-7}$, $a = 0$, $b = 0$; dimensionless parameters $P = 10^{-3}$, $T = 6.10^{-3}$; $K - H$ (classical Kelvin-Helmholtz curve) $P = Q = 0$.

When the following notations are introduced

$$\zeta = \frac{(\alpha_a - \alpha)}{\Sigma_a} + \alpha + \Sigma + \frac{\Sigma}{\Sigma_c}(\alpha_c - 1), \quad \xi = \frac{\Sigma^2}{\Sigma_c}(\alpha_c - 1) - \frac{1}{\alpha}\frac{(\alpha_a - \alpha)}{\Sigma_a},$$

$$M = \frac{8\cos\theta \sin\theta \,(a\sin\theta - b\cos\theta)}{\alpha_a + \alpha_c} + \frac{2b\sin\theta}{\alpha_a + \alpha_c}$$

$$+ \frac{2\left(1 + 2\cos^2\theta\right)}{\zeta}(\beta_2 \sin\theta - \beta_1 \cos\theta) + \frac{2\beta_1 \cos\theta}{\zeta} + \frac{24c\sin^2\theta \cos\theta}{\alpha_a + \alpha_c}\frac{x}{H}$$

$$N = -\frac{2\left(\Sigma^2 - 1\right)}{\zeta}\left(\cos^2\theta - \sin^2\theta\right) + \frac{2\xi\sin^2\theta}{\zeta} - \frac{8\left(5\cos^2\theta - \sin^2\theta\right)(\Sigma - 1)}{\alpha_a + \alpha_c}$$

we obtain

$$F^2 \sim \frac{c^2(\Sigma - 1)}{\alpha_1^4}\left[-\frac{P\alpha(\Sigma - 1)^3 M^2}{r_1 r_2} + \left(\frac{\alpha}{r_2} + \frac{1}{r_1}\right)QN\sin^2\theta\right]. \quad (37)$$

This expression (37) shows that the asymptotic behaviour when $\alpha_1 \to 0$ is essentially dependent on the c coefficient, which characterizes the second order term $c(x^2 - y^2)/H$ in the Φ polynomial expression (1). Depending on the sign of the difference $(\Sigma - 1)$ the effect may be strongly destabilizing $(\Sigma - 1 < 0)$, or stabilizing $(\Sigma - 1 > 0)$. It is however necessary to remember that these results are only justified when the wave length of the perturbation is significantly smaller than the horizontal length scale L_n of the Φ variations. This condition, which requires that $\alpha_1 > 2\pi c^{\frac{1}{2}}$, suggests that the asymptotic values of F^2 when $\alpha_1 \to 0$ have no realistic significance. Nevertheless, the tendency of F^2 to decrease (or to increase) seems to keep some validity. It suggests that, when F^2 decreases, as in the case of Fig. 2, it would not be possible to keep an extended layer of heavy liquid below the other. In such a case the heavy liquid might gather on one side of the cell as soon as the electric current is switched on, possibly without any stable position. On the contrary, when F^2 increases, a flat interface with instability only in some circumstances and for values of α_1 above some limit should be expected.

In the oral presentation of this paper, different elementary mechanisms were discussed. One, called differential pinch effect, is due to the interaction between the vertical electric current and its self-perturbed magnetic field. Another one, due to the interaction between vertical and horizontal electric current, generates a propagation of the interface in the direction opposite to the horizontal current.

References

[1] Robinson, I. S., "A Novel Form of the MHD Rayleigh-Taylor Instability," Journal of Fluid Mechanics, Vol. 72, 1975, pp. 135-143.

[2] Descloux, J., and Romerio, M. V., "On the Analysis by Perturbation Method of the Anodic Current Fluctuations in an Electrolytic Cell for Aluminium," Light Metals, 1989, Vol. 1, pp. 237-243.

[3] Moreau, R., and Evans, J. W., "An Analysis of the Hydrodynamics of Aluminium Reduction Cells," Journal of the Electrochemical Society: Electrochemical Science and Technology, Vol. 131, No 10, 1984, pp. 2251-2259.

[4] Shercliff, J. A., "Anisotropic Surface Waves under a Vertical Magnetic Force," Journal of Fluid Mechanics, Vol. 38, Pt. 2, 1969, pp. 353-364.

[5] Sneyd, A. D., "Stability of Fluid Layers carrying a Normal Electric Current," Journal of Fluid Mechanics, Vol. 156, 1985, pp. 223-236.

[6] Moreau, R., and Ziegler, D., "Stability of Aluminium Cells. A New Approach," Light Metals, 1986, Vol. 1, pp. 359-364.

Survey of Liquid-Metal MHD Activities in Dresden

G. Gerbeth* and G. Uhlmann†
Central Institute for Nuclear Research, Rossendorf, Germany
and
D. Hamann‡
Dresden University of Technology, Dresden, Germany

Abstract

This study briefly summarizes the activities of **our group** in the field of liquid metal MHD. It shows in which way our **present investigations** on basic problems in liquid metal MHD followed from the fast-breeder research. Special interest is focused on liquid metal two-phase flow and MHD flow around obstacles, as well as the laminar-turbulent transition in two-dimensional MHD flows. Most of the investigations are theoretical, but partly connected to experiments performed in Riga, Latvia or Grenoble, France. Our own experiments at a sodium loop are described. Finally, the most promising directions of our future research are presented.

Introduction

The activities of our group in the field of liquid metals started at the beginning of the 1970's. They were devoted to the technical diagnosis of the sodium loops of a liquid metal fast breeder reactor (LMFBR). Interest was focused on processes affecting nuclear safety and plant reliability, in particular, sodium boiling due to local coolant flow blockages and the occurrence of leaks in the sodium-heated steam generator. The main goal of the investigations was to detect and, if possible, to locate these processes. This was done mainly by acoustic methods. On the other hand, both sodium boiling as well as leaks lead to a sodium two-phase flow. In this way our research was connected with liquid metal two-phase flow and acoustics from the beginning. Though the LMFBR

Copyright © 1992 by the American Institute of Aeronautics and Astronautics, Inc. All rights reserved.
*Professor, Department of Sodium Technology.
†Research Head, Department of Sodium Technology.
‡Professor, Department of Acoustics.

program was finished five years ago, these two topics are still characteristic of our present research. Since the beginning of the 1980's, the research subject has shifted more and more from LMFBR to basic research in liquid metal MHD.

The present study summarizes all of these activities briefly. The main goal is to point out the promising directions for further research and to stimulate international cooperation in this field.

Investigations on Sodium Loops, Test Facilities and LMFBR

Our sodium loop was built in 1973.[1] One purpose was to perform simulations of water microleaks in order to study the emission of sound during the sodium-water reaction that is typical for a leak in a sodium-heated steam generator. Piezoelectric pressure transducers applicable up to temperatures of 600°C were developed. Analysis of the acoustic signals resulted in a leak detection conception based on statistical decision theory. In the next step, a prototype system for leak detection was developed, using six different measuring positions at the steam generator.[2] Now this system is part of a methodically diverse surveillance system that uses further information resulting from a leak (e.g., impurity content and absolute pressure increase).[3]

In order to study possibilities of boiling detection, an experiment under reactor conditions was performed in cooperation with the Research Institute for Nuclear Reactors, Dimitroffgrad, USSR, at the LMFBR Bor-60. Sodium boiling was generated in the core by heating a tungsten rod bundle in a special test assembly. The possibility of boiling detection by analysis of acoustic and neutronic noise was shown.[4] The results are summarized and a variety of signal processing techniques are compared in a coordinated research program on Sodium Boiling Noise Detection initiated by the International Working Group on Fast Reactors of the IAEA.[5]

Moreover, a method of detecting inhomogeneities in liquid metals by inductive flow meters was developed. Void fractions of at least 1% are detectable by analyzing the random fluctuations of the flow meter output signal.[6]

Basic Research

The study of the previously described LMFBR failures was directly connected with the investigation of the arising two-phase flow. For instance, a microleak in a sodium-heated steam generator represents a source of small hydrogen bubbles in the vicinity of the wall. In order to study the local distribution of the gas phase, a diffusion model was developed.[7] The main prerequisite of this approach is to compare small bubble radii to the characteristic length scale of the turbulent flow. Using this diffusion model, the following problems were considered:
1) Optimization of the location of acoustical and chemical sensors in sodium-heated steam generators on the basis of local void fraction calculations for potential leak positions.
2) Prediction of local void fraction in a boiling channel with inlet subcooling.[8]

3) Investigation of local void fraction in bubble flow through rod bundles.

Recently, the model was extended to a two-phase flow in an external magnetic field. This was undertaken since the study of two-phase flows inside and outside magnetic fields is presently one of the most important directions in liquid metal MHD research.[9] The magnetic field is taken into account by its influence on the diffusion coefficient tensor and the relative bubble velocity. The diffusion tensor is predicted, assuming the fluid correlation tensor in the absence of the magnetic field to be known. The tensor of the diffusion coefficients is anisotropic due to the two preferred directions of the system, namely, the direction of the magnetic field and the direction of the deterministic part of the bubble velocity. In a broad range of real situations the deterministic particle velocity has a strong influence on the diffusion coefficients. This so-called "crossing trajectory effect" is even more pronounced in the presence of a magnetic field. In a certain range of parameters, increasing diffusion coefficients are obtained for an increasing magnetic field.[10]

Now an experimental program is in preparation to measure local void fractions in a vertical sodium-argon flow with and without magnetic field. Global information on the flow is obtained by electromagnetic flow meters and acoustic transducers. However, it appeared to be very difficult to operate with local sensors in liquid sodium. A hot-wire sensor was developed for measurements at about 250°C, producing stable results over a period of three hours at most.[11] Local resistance probes, developed in our group and extensively used in water-steam systems, have been tested in sodium. Wettability of the probe tip with sodium appeared as a serious problem. Much effort has been expended to find a suitable insulator-conductor pair for the probe. This problem seems to be solved now since recent tests with two tungsten-glass probes yield reproducible results over a period of 12 hours with several interruptions concerning the sodium contact.

In the context of two-phase flow measuring techniques, a scheme was developed that allows the determination of bubble-size distribution from the chord length distribution obtained by local double-wire probes.[12] First, investigations were performed concerning the sound propagation in a liquid-metal two-phase flow.[13] Quantitative estimations were obtained for phase velocity and damping of sound propagation through liquid sodium-gas mixtures on the basis of experimentally obtained bubble-size distributions.

One starting point in two-phase flow modelling is always the drag coefficient of a single particle. Therefore, a variety of analytical and numerical calculations were carried out on MHD flow around obstacles. In particular, the following problems were considered.

1) Stokes flow around spheres and cylinders in a strong magnetic field (high Hartmann number).
2) High Reynolds number flow around a cylinder in an external magnetic field.
3) Electromagnetically induced flow along a cylinder.

Please refer to Refs. 14-16 for more information.

The laminar-turbulent transition is one of the most serious problems in fluid dynamics, and it is generally accepted that two-dimensional flows provide an efficient tool to facilitate theoretical and experimental studies. Liquid metal flows maintained by the interaction of a homogeneous magnetic field with an inhomogeneous electric current present the opportunity to produce and study two-dimensional flows. The instability of such flows was investigated by Kolesnikov[17] (starting with a Kolmogorov-flow) and by Sommeria[18] (starting with a square array of alternating vortices). A linear stability analysis of the governing equations (which differ, in fact, by the forcing term only) was performed in our group[19] taking into account the effects of viscosity, bottom friction, and confining lateral walls. The obtained critical values of the characteristic non-dimensional parameters are in good agreement with that of the experiments. Analysis of the most unstable perturbations leads to a good prediction of the structure of the weakly turbulent flow above the instability threshold.

The application of liquid metals in a magnetically confined fusion reactor has found growing interest in recent years. Most of the research is focused on blanket flow, but there is also the proposal of using liquid metals inside the torus[20]. The main aim of this idea is to replace the solid plate limiters or diverters (which are necessary to protect the first wall from high heat and ion fluxes) by a liquid metal flow. Estimations indicate that a gravity-driven flow of a liquid metal having a free surface will not move fast enough due to strong MHD forces. In order to circumvent these MHD penalties, the novel system of a liquid metal droplet rain was proposed. A first experiment, performed at the Moscow experimental reactor T-10 showed the feasibility of this approach.[20] Because of fruitful cooperation between the Riga MHD Laboratories and our group, we are partly involved in this program. Distortion of a free falling drop in an inhomogeneous magnetic field was determined, and good agreement with an experiment performed in Riga was found.[21] In the entrance region of the magnetic field the drop is well approximated by an ellipsoid, with the large half axis along the direction of the magnetic field. Heating of a drop from one side leads to a convective flow inside it. The structure of this flow in the presence of a strong magnetic field was analyzed recently. The magnetic field suppresses the convective flow so strongly that even under fusion reactor conditions the heat conduction predominates over the convective heat flow.

Outlook

In continuation of the described investigations, the following activities will be at the center of our interest in the near future:

1) Measurement of void fraction distribution in a vertical sodium-argon flow with and without a magnetic field.
2) Numerical simulation of the two-dimensional MHD flow around a cylinder.
3) Investigation on the transition to turbulence in two-dimensional MHD flows.

On the basis of our liquid metal MHD experiences, we plan to shift partly to more advanced studies, and, if possible, to industrial applications. We see potential possibilities in the following areas:
1) Electromagnetic separation of small particles in melts.
2) Influence of a magnetic field on the flux of a containment into the bulk of the melt in Czochralski crystal growth.
3) Liquid metal flows under fusion reactor conditions.

A new approach to liquid metal MHD power generation was created by Swift[22] using the principle of a "thermo-acoustic engine." Compared to the existing concepts of liquid metal MHD power generation, the main advantage of this approach is the lack of a global circulation of the liquid metal. There is one resonator of liquid metal only. The primary heat energy is used to generate high-amplitude sound waves in the liquid metal. This acoustic energy is converted into electricity magnetohydrodynamically. Details of this concept are presented in a literature review.[23] This approach is very attractive for us since it combines the two main research activities of our group, namely, liquid metal MHD and acoustics.

It would be highly desirable if the present study would stimulate a creative exchange of ideas and could possibly be the starting point of international research cooperation in one of the described topics.

Acknowledgment

The authors are grateful to Dr. D. Ziegenbein, Head of the Reactor Physics Department, for encouraging these investigations. Fruitful research cooperation with the groups of Dr. O.A. Lielausis, Riga-Salaspils Latvia and Dr. A. Alemany, Grenoble, France, benefited and supported our work.

References

[1]Langenbrunner, H., Grunwald, G., and May, R., "Die Natriumversuchsanlage NAVA," *Kernenergie*, Vol. 19, No. 1, 1976, pp. 9-13.

[2]Pridöhl, E. and Höhnel, E., The Acoustic Leak Detection System ALDES," *Kernenergie*, Vol. 27, No. 3, 1984, pp. 107-110.

[3]Mauersberger, H., Froehlich, K. J., Teske, K., Clauss, V., Matal, O., Banovec, J., and Sirek, J., "A Methodically Diverse Steam Generator Surveillance System," *Kernenergie*, Vol. 31, No. 1, 1988, pp. 15-21.

[4]Afanasiev, V.A., et al., "Sodium Boiling Experiment in the Reactor Bor-60," Internal Report Zentralinstitut für Kernforschung-344, Central Institute for Nuclear Research, Rossendorf, Germany, 1977.

[5]Mauersberger, H., Fröhlich, K. J., and Clauss, V., "Sodium Boiling Noise Detection," *Kernenergie*, Vol. 31, No. 7, 1988, pp. 307-316.

[6]Langenbrunner, H., "Detektion von Inhomgenitöten in Flüssigmetallen mittels induktiver Strömungsmessung," Internal Report, Zentralinstitut für Kernforschung-387, Central Institute for Nuclear Research, Rossendorf, Germany, 1979.

[7]Katona, T., Kozma, R., and Uhlmann, G., "Ein Modell für die Phasenverteilung in Blasenströmungen," Internal Report, Zentralinstitut für Kernforschung-432, Central Institute for Nuclear Research, Rossendorf, Germany, 1981.

[8]Hamann, D., and Uhlmann, G., "Prediction of Subcooled Boiling Void Fraction," *Kernenergie*, Vol. 28, No. 8, 1985, pp. 338-347.

[9]Branover, H., "Experimental Studies in Liquid Metal Magnetohydrodynamics," *Proceedings of the International Symposium on Experimental Heat Transfer, Fluid Mechanics and Thermodynamics*, edited by R.K. Shah, E.N. Ganic, and K.T. Yang, Dubrovnik, Yugoslavia, Sept. 1988, pp. 737-748.

[10]Gerbeth, G. and Hamann, D., "Dispersion of Small Particles in MHD Flows," IUTAM-Symposium on Liquid Metal MHD, Riga, Latvia, May 1988.

[11]Platnieks, I., and Uhlmann, G., "Hot-wire Sensor for Liquid Sodium," Journal *of Physics E: Scientific Instruments*," Vol. 17, 1984, pp. 862-863.

[12]Hamann, D., Lotzmann, R., and Uhlmann, G., "A Method for the Determination of the Bubble Size Distribution in Liquid-Metal Two-Phase Flows," *Kernenergie*, Vol. 28, No. 7, 1985, pp. 297-301.

[13]Hamann, D., and Thess, A., "Sound Propagation in Liquid Sodium-gas Mixtures," *Kernenergie*, Vol. 31, No. 6, 1988, pp. 246-250.

[14]Gerbeth, G., "New Results on MHD Drag Coefficients," 6th Beer Sheva International Seminar, Jerusalem, Israel, Feb. 1990.

[15]Alemany, A., Josserand, J., Marty, Ph., and Gerbeth, G., "MHD Flow Around a Cylinder in an Aligned Magnetic Field," 6th Beer Sheva International Seminar, Jerusalem, Israel, Feb. 1990.

[16]Thess, A., Gerbeth, G., and Marty, P., "Electromagnetically Induced Flow Around a Cylinder," 6th Beer Sheva International Seminar, Jerusalem, Israel, Feb. 1990.

[17]Kolesnikov, Y.B., "Investigation of Flat Shear Flow Instability in a Magnetic Field," *Magnetohydrodynamics*, No. 1, 1985, pp. 60-66.

[18]Sommeria, J., "Experimental Study of the Two-Dimensional Inverse Energy Cascade," *Journal of Fluid Mechanics*, Vol. 170, 1986, pp. 139-168.

[19]Thess, A., "Stability of a Class of Two-Dimensional Wall-Bounded Magnetohydrodynamic Flows," Internal Report, Zentralinstitut für Kernforschung-674, Central Institute for Nuclear Research, Rossendorf, Germany 1989.

[20]Demjanenko, B.N., Karasev, B. G., Kolesnitschenkov, A. F., Lavrentiev, I. B., Lielausis, O. A., Muravjov, E. B. and Tananejev, A. B., "Liquid Metal in the Magnetic Field of a TOKAMAK Reactor," *Magnetohydrodynamics*, No. 1, 1988, pp. 104-124.

[21]Gerbeth, G., Galititis, A. and Kasudze, M., "Deformation of an Electrically Conducting Drop in a Magnetic Field," IUTAM Symposium on Liquid Metal MHD, Riga, Latvia, May 1988.

[22]Swift, G.W., "A Liquid Metal Magnetohydrodynamic Acoustic Transducer," *Journal of the Acoustic Society of America*, Vol. 83, 1988, pp. 350-361.

[23]Hamann, D. and Gerbeth, G., "Recent Developments of Liquid Metal MHD Thermoacoustic Engines," 6th Beer Sheva International Seminar, Jerusalem, Israel, Feb. 1990.

Experiments with a Superconducting Magnet on an InGaSn Loop

O. Lielausis,* E. Platacis,† I. Platnieks,† M. Pukis,‡ and A. Shishko‡
Latvian Academy of Sciences, Riga-Salaspils, Latvia

Abstract

A superconducting solenoid is introduced mainly for liquid-metal MHD experiments. The maximum field in the center is 5.06 T at a current of 816 A and a stored energy of 1.33 MJ. Taking into account specific features of MHD experiments, the following design has been chosen. The device is mobile. The diameter of the available working volume is 0.32m. A 0.64 m long specially shaped coil ensures an acceptable 5% homogeneity of the field in an 0.3 m long central region. Thus, the possibility is provided to place channels with complex geometry into the field, to perform experiments at higher temperatures, etc. A nonstandard composition of main elements of the construction (feeder tank + cryostat) enables the orientation of the field lines to be changed, which is important for the experiments influenced by gravity (e.g., in the case of free surface flows). The first run of experiments was performed on a 50 liter loop with a low-temperature (melting point of 10.5°C) eutectic melt (20% In + 67.5% Ga + 12.5% Sn). In a set of 180° bends, pressure drop and potential distribution were measured. The influence of the relative wall conductivity on the pressure losses and velocity distribution was demonstrated. A comparison with another experiment where a smaller superconducting solenoid was installed on the Na loop is presented.

Introduction

The growing interest in MHD processes in strong magnetic fields is caused by mainly two reasons. First, one should ask if it is not high time to make the MHD ready for the golden age of high-temperature superconductivity. But, of course, the main reason is connected with the proposed applications of liquid metals in various systems of thermonuclear reactors. Convincing experiments under conditions as similar to the expected natural ones as possible are necessary here. It is worth mentioning that interesting positive results in this field were obtained more than 15 years ago. Thus, in Ref. 1, the possibility of a conducting liquid metal pumping in the presence of a 4-T high field was

---

Copyright © 1992 by the American Institute of Aeronautics and Astronautics, Inc. All rights reserved.
*Head of Laboratory, Institute of Physics.
†Head of Group, Institute of Physics.
‡Senior Researcher, Institute of Physics.

demonstrated. In Ref. 2 it was shown that a field of equal intensity allows a nucleate boiling of liquid metal.

The values of dimensionless MHD parameters characterizing the flows under discussion are very high: $Re \cong 10^4$, $Ha \cong 10^4$, and $N \cong 10^4$. Only experiments in very strong fields, caused mainly by superconducting coils, give the opportunity for such high values for the whole system of criteria to be preserved.

Description of the Magnet

The design was aimed at carrying out, if possible, a number of various experiments after the magnet is put into operation. The relation of consumption typical of superconducting devices (getting ready vs. the experiments themselves) compels one to search for such a possibility. Hence, in the case under consideration 19 h were spent, 500 liters of nitrogen and 600 liters of liquid helium were expended to cool the device down to the temperature of liquid helium. After that, 3 h were used to fill the device with 200 liters of He. Let us compare: To keep the coil in a superconducting state throughout the experiments, only 5-9 liters/h were consumed.

The magnet consists of two main elements (Fig. 1a): a feeder tank (1) with N_2 and He coolants and a cryostat (2) with a superconducting solenoidal coil. The feeder tank is inclined 45 deg to the axis of the solenoid. Such a

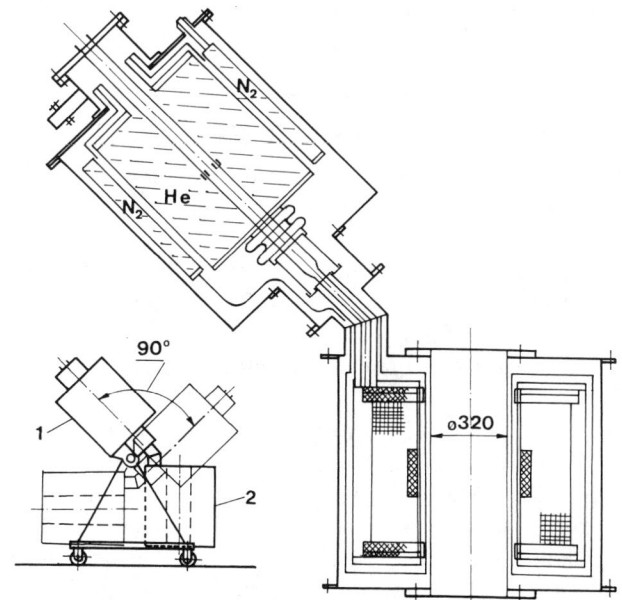

Fig. 1 Diagram of the superconducting magnet:
  1. feeder tank
  2. cryostat

nonstandard composition enables the device to be turned by a definite angle around the horizontal axis. The device is fixed up on a stand, permitting it to be swung by 0-90 deg around the equilibrium axis. In such a way it is possible to set the field liens oriented in all of the positions between vertical and horizontal ones. The stand, together with the magnet, is mobile.

The main characteristics of the construction are presented in Fig. 1b. The diameter of the working volume is 0.32 m. A 0.64 m long specially shaped solenoidal coil ensures an acceptable 5% homogeneity of the field in an approximately 0.3 m long central region of the working volume. The inductance of the coil is 4 He. The maximum intensity of the field in the center is 5.06 T at a current of 816 A and a stored energy of 1.33 MJ. A special absorbent is introduced in the vacuum chamber that allows vacuum pumping to be interrupted during experiments. The coil is supplied with a protecting system that detects the emergence of a normal zone in the coil, cuts the current, and discharges the stored energy on an outer resistance.

The first run of experiments was performed on a 50 liter loop with a low-temperature nonaggressive, nontoxic eutectic melt 20%In + 67.5% Ga + 12.5% Sn. Four local MHD resistances were tested. From the physical point of view, the experiments were mainly aimed at evaluating resistance coefficients and indicating the influence of the wall conductivity.

The loop based on a 32-mm stainless steel pipe is linearly arranged. The experimental objects are taken into the center of the magnet by means of relatively long legs connected with the loop through flanges and short bellows. The bellows make some declination of the legs possible.

Comparison Between Experiments with Two Similar Elements on Two Different Superconducting Solenoids

One of the elements, a rectangular 180-deg bend of a round pipe, was chosen so that the results could be compared with the results of another experiment carried out a few years ago — in Ref. 4, a shorter solenoid was installed on a Na loop, and very high values of the MHD parameters were achieved. The relative conductivity of the walls for the element under discussion reached a far as the value of $\alpha = 0.012$, which was caused by the ideal contact of sodium with stainless steel. Here $\alpha = t_w \sigma_w / d\sigma$, where t_w is the thickness of the wall, σ_w is the conductivity of the walls, d is the diameter of the pipe, and σ is the conductivity of sodium. The use of InGsSn provides realization of the wall isolation ($\alpha = 0$) in a very simple way, i.e., by means of a nitrocellulose enamel coating. The pressure losses in the elements under consideration are depicted in Fig. 2. The figure shows that the transition from $\alpha = 0.012$ to $\alpha = 0$ is connected with a noticeable reduction of the losses. It should be noted that the interest to flows in channels with practically isolated walls grows simultaneously with the development of the techniques for creating the corresponding coatings. Thus, in Ref. 5 the losses in a straight round pipe placed in a homogeneous transversal field were reduced by an order by means of

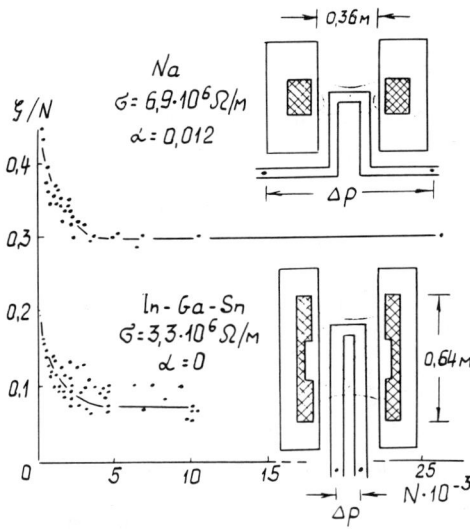

Fig. 2 Experiments on two similar MHD resistances in two superconducting solenoids.

special isolating multilayer inserts. Both in the presence of the insert and without it a good agreement with the well known theoretical dependences was achieved. It should be noted that in the case of local resistances the scope for a quantitative theoretical treatment is very limited.

To bring the MHD experiments on superconducting devices to a quantitative level, new problems have been taken into account. Hence, the field in superconducting solenoids used for the MHD experiments will practically always be inhomogeneous because of the limited length of the coils. The dashed region in Fig. 2 corresponds to the calculated 5% homogeneity of the field for each of the solenoids. One may say that, in respect of homogeneity, both elements were placed under nearly identical conditions. However, more detailed differences in the picture of the field are able to considerably influence the result of comparison, and this is worth keeping in mind. Furthermore, there are serious problems regarding the removal of the pressure readings from the pints placed in strong fields. In order to eliminate these difficulties the pressure differences were measured between the points placed outside the field. As a result, losses in a relatively long supplying legs were included in the readings. In measurements in InGaSn the possibility was provided to incline these supplying legs by some angle, i.e., to change the distribution of the field along the legs. Thus, it has been confirmed that, beginning with some definite meaning of the field, the losses in the supplying legs can be ignored. Practically all of the losses become localized on the very bend.

The dependence of the pressure losses on a gradually increasing field was similar for both of the cases ($\alpha = 0.012$ and $\alpha = 0$), as described in Ref. 5. In a weak field dependence $\xi = f(Re, Ha)$ turns out to be rather complicated. The losses at $B = 0$ still have to be taken into account, and the complex processes connected with the suppression of nonlinear effects come out, etc. But, at $\alpha = 0.012$, as well as at $\alpha = 0$, beginning with the determined value of $Ha^2/Re = N \cong 3 \cdot 10^3$, the coefficient of pressure losses $\xi = 2\Delta P/rV_o^2$ simply become proportional to the parameter of the MHD interaction N: $\xi = kN$ (see Fig. 2). Here ΔP is the pressure drop on the bend, ρ is the density, and V_o is the average velocity.

Comparison Between Experiments on Two Identical Elements Different only in Relative Conductivity of the Hartman Walls

Measurements on two elements (Fig. 3) were compared representing the changeover from a round pipe directed along the field to a rectangular channel directed perpendicular to the field. The ratio of the sides for the rectangular channel is 1:10. The long side lies along the field. It is a typical situation aimed at restricting the losses. In one case all of the walls were non-conducting (made of plexiglass); in the other case the walls perpendicular to the field were covered with a 0.05 mm copper foil. The corresponding relative conductivity of these walls based on dimension a is $\alpha_1 = 0.012$, as in the previous case. The pressure drop upon the whole element ΔP as well as potential differences $\Delta \varphi$ in the middle plane shown in Fig. 3 were measured. These readings of potential differences allow one to reason indirectly about the velocity field in the most interesting flat part of the element. In the case I ($\alpha = 0$) the distribution of $\Delta \varphi$ along the direction a appears to be practically constant. In Fig. 3, the measured values of $\Delta \varphi$ are displayed by the height of the dashed rectangles.

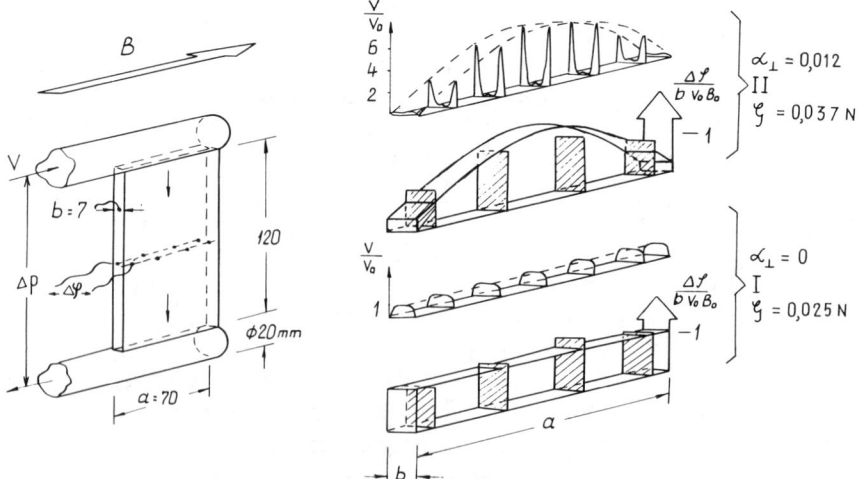

Fig. 3 Experiments on two MHD resistances differing in the Hartman wall conductivity.

For the flow under discussion the classical direct solution allows to introduce a picture of a fully developed velocity field characteristic of a long enough channel. In Fig. 3 these fields are represented by a set of equidistant profiles. The value of Ha based on dimension a is 104. The corresponding calculated distributions of potential differences are also shown. In the case of $\alpha = 0$, the calculated fields are practically homogeneous. A good agreement between the calculated and measured values of $\Delta\varphi/bv_oB_o$ allows one to conclude that the flat part of the element was long enough for the evolution of a fully developed MHD flow.

The calculations, as well as measurements of potential differences, attest that the introduction of the two conducting Hartman walls considerably changes the field of velocities (Fig. 3, case II). The distribution along a becomes symmetric but particularly nonhomogeneous. The changes along b are the most drastic.

The reliability of the statements just presented is limited because the measurements of the potential differences illustrate only the distribution of the velocities averaged over b. However, the general conclusion indicates that in local MHD resistances the velocity fields may be heavily influenced by the conductivity of the walls. That is related to such processes as generation of perturbations, corrosion, cavitation, etc.

In view of the aforementioned statements. it is interesting to note that after the second measured parameter (pressure drop) the influence of the α-value is less noticeable. Again, beginning with some definite value of N, we simply get ξ = kN. If N is based on dimension a, then k = 0.025 at $\alpha_1 = 0$, and k = 0.037 at α_1 = 0.012. It turns out that the determination of these coefficients can be treated as practically important a task for experiments on local MHD resistances in strong fields.

References

[1]Frass, A. P., Young, F. J. and Holcomb, R. S. "Magnetohydrodynamic Test of a One-Sixth Scale Model of a CTR Recirculation Lithium Blanket," Annual Meeting of the American Nuclear Society, New Orleans, LA, 1975.

[2]Frass, A. P., Lloid, D. B., and MacPherson, R. E., "Effects of a Strong Magnetic Field on Boiling Potassium," Oak Ridge National Lab., Oak Ridge, TN, TM-4218, 1974.

[3]Keilin, V.Y., et al. "Superconducting Magnetic Systems With Increased Uniformity of Magnetic Field," *Voprosy Atomnoy Nauki i Techniki, Seria Obschey Zadernoy Fiziki*, Vol. 2, No. 42, 1988, pp. 142-143.

[4]Grinberg, G., Lielausis, O., Kaudze, N., "Research of Local MHD-Resistances on a Na-Loop with Superconducting Magnet." Magnitnaya Hidrodinamika, Vol. 1, 1985, pp. 121-126.

[5]Barleon, L., et al., "MHD-Stromung in Flüssigmetallgekülten Blankets," Statusbericht KFK 1989. Kernforschungszentrum Karlsruhe, Germany, 1989.

Comparision of the Core Flow Solution and the Full Solution for MHD Flow

Lutz Lenhart*
Kernforschungszentrum Karlsruhe GmbH, Germany
and
Kathy McCarthy†
Idaho National Engineering Laboratory, Idaho Falls, Idaho 83415

Abstract

The self-cooled liquid metal blanket is a prime candidate for use in a fusion reactor. The presence of a magnetic field affects heat transfer, mass transfer, and pressure drop. Therefor, it is important to analyze general features of a blanket to assess the viability of a particular design.

The most detailed information can be obtained by solving the full set of three-dimensional nonlinear MHD equations, but for other than fully developed flow this can take extensive computer time and storage. An alternative method to predict the flow variables is desirable. One such method is based on the fact that viscous and inertial terms are negligible if the Hartmann number M and the interaction parameter N are high. Then certain characteristics of the resulting linear equations allow the use of a two-dimensional code without losing the three-dimensional information. To establish the range of M and N in which this core flow solution (CFS) is valid, the results are compared to the full solution (FS) of a flow in a rectangular duct with a varying magnetic field. Both codes assume steady-state conditions and neglect the induced magnetic field. Only the FS includes inertial and viscous terms.

Comparisons are made for different M and N. Computer storage and thereby the mesh Reynolds number determine the combination of M and N in the FS code. The results show similar trends in velocities, potentials, and pressures. The quantitative agreement is within acceptable error limits, and the core velocities differ less than 1 percent for $M \geq 200$. Differences in the velocity profile near the walls are evident, because the CFS does not yield exact values in boundary layers and neglects the jump of the potentials across the side layers. The FS calculates the typical M-shaped profile for fully developed MHD flow and an ordinary hydrodynamic distribution where the magnetic field is low. Since the CFS provides a good estimation of the volume flux, it can be used to predict heat transfer, whereas the FS also can be used for corrosion analysis. All interaction parameters available for the FS, $N \geq 100$ for $M=50$ and $N \geq 2000$ for $M=300$, do not effect the results significantly.

Copyright © 1992 by Lutz Lenhart. Published by the American Institute of Aeronautics and Astronautics, Inc., with permission.
*Research Engineer, Institut für Angewandte Thermo- und Fluiddynamic.
†Engineering Specialist, **Fusion Safety** Program.

The CFS is much faster than the FS, and, if the equations are reduced to two dimensions, the numerical solution needs less computer storage. Therefor, this method should be preferred when applicable. Further comparisons should be done at lower N and in more complex geometries, such as bends or expansions, where inertial forces and viscosity may have a greater effect.

Nomenclature

The typical numbers set in parentheses are from Refs. 1 and 2.

a	= half width of the channel (~ 0.3m)
B	= magnetic field (3-5 tesla)
c	= wall conductance ratio (10^{-3}-10^{-2})
j	= current density
M	= Hartmann number (10^3-10^4)
N	= interaction parameter (10^2-10^4)
p	= pressure
Re_m	= magnetic Reynolds number ($\ll 1$)
t_w	= wall thickness (~ 0.001m)
v	= velocity (average velocity ~ 0.5m/s)
x,y,z	= Cartesian coordinates
η	= viscosity (10^{-4}-10^{-3} Ns/m^2)
ρ	= density ($\rho_{\text{fluid}} \sim 10^3$kg/m^3)
σ	= electrical conductivity ($\sigma_{\text{fluid}} \sim 10^6$ 1/Ωm)
ϕ	= electrical potential

Introduction

Examining the flow of a conducting fluid under the influence of a strong magnetic field is essential for the design of a self-cooled liquid metal blanket. In a fusion reactor, this blanket is responsible for both heat transfer and tritium breeding simplifying the structure, which is one of the reasons that it is a prime candidate for use. In order to determine the viability of the blanket, the velocity distribution and the pressure losses of the fluid must be predictable. Experimental and theoretical studies are desirable (see, e.g., Refs. 3-5). Because analytical methods such as asymptotic expansions are not always efficient or possible for all reactor relevant applications, the theoretical analysis is done through development of numerical schemes.

This paper compares two methods used to predict the three-dimensional magnetohydrodynamic (MHD) flow of a conducting fluid in a rectangular duct under the influence of a varying magnetic field. The main difference between them is, in general terms, that one neglects viscous and inertial effects to simplify the equations and thereby obtains solutions much faster. As a consequence of these approximations, the range of application is restricted and a check of the validity of results is required.

Different Sets of Equations

The two different models studied here, the full solution (FS) and the core flow solution (CFS), refer to different sets of equations. The FS is more general, but much more programming effort, computational time, and storage is required. Because the storage of modern vector computers is still limited, it is desirable to use an easier alternative method, such as the CFS, for predicting flow variables.

Equations of the Full Solution

Magnetic fields in fusion reactors are slowly varying, and the magnetic Reynolds number Re_m, the ratio of induced to applied magnetic field, is small. The induced field is often negligible. A typical MHD flow in this case can be assumed inductionless, incompressible, and isothermal. The latter approximation makes buoyancy forces negligible and decouples the Navier-Stokes equation from the conservation of energy. Thus the flow variables can be computed without solving the energy equation simultaneously. Then the full set of dimensionless equations describing nonrelativistic MHD flow consists of the Navier-Stokes equation, Maxwell's equations, conservation off mass, conservation of electric charge, and Ohm's law. They can be written in operational form[6]:

$$\frac{1}{N}\left[\partial_t \mathbf{v} + (\mathbf{v} \cdot \nabla)\mathbf{v}\right] = -\nabla p + \mathbf{j} \times \mathbf{B} + \frac{1}{M^2}\Delta \mathbf{v} \quad \text{Navier-Stokes (1)}$$

$$\nabla \cdot \mathbf{v} = 0 \quad \text{conservation of mass (2)}$$

$$\Delta \phi = \nabla \cdot (\mathbf{v} \times \mathbf{B}) = \mathbf{B} \cdot \nabla \times \mathbf{v} \quad \text{conservation of el. charge (3)}$$

$$\mathbf{j} = -\nabla \phi + \mathbf{v} \times \mathbf{B} \quad \text{Ohm's law (4)}$$

where t, $\mathbf{v}$, p, $\mathbf{j}$, $\mathbf{B}$, and ϕ are the dimensionless time, velocity, pressure, current density, magnetic field, and electric potential, nondimensionalized by $a/v_0, \rho_0 v_0^2 B$, $\sigma v_0 B_0$, B_0, $a v_0 B_0$, respectively. B_0 is the applied magnetic field, a is a characteristic length (the half width of the rectangular channel), v_0 is the average velocity, σ and ρ_0 are the fluid's conductivity and density, Δ is the Laplacian operator, and ∂_t is the time derivative. The Hartmann number $M = aB_0(\sigma/\eta)^{1/2}$ gives the square root of the ratio of electromagnetic forces to viscous forces, with η being the viscosity of the fluid. $N = \sigma a B_0^2 / (\rho_0 v_0)$ is the interaction parameter, giving the ratio of electromagnetic forces to inertial forces.

The boundary conditions are the no-slip condition at each wall of the channel

$$\mathbf{v}\big|_{wall} = 0 \quad (5)$$

and the thin wall approximation, as introduced by Shercliff[7] and later made more general by Walker[8]:

$$\partial_n \phi \big|_{wall} = c\Delta n_t \phi \big|_{wall} \quad (6)$$

Here $c = \sigma_w t_w / \sigma a$ is the wall conductance ratio, giving the ratio of wall conductance to fluid conductance, σ_w and t_w being the conductivity of the wall and it's thickness. ∂_n is the normal derivative pointing into the wall, and Δ_t is the tangential part of the Laplacian operator. Applying the thin wall approximation, the flow can be simulated without solving Maxwell's equations in the wall. The Hartmann number, interaction parameter, and conductance ratio are the dimensionless numbers characteristic of MHD flow. Typical values for fusion blankets are $10^3 \leq M \leq 10^4$, $10^2 \leq N \leq 10^4$, and $10^{-3} \leq c \leq 10^{-2}$, (Refs. 1 and 2). Since M and M are large, inertial effects and friction are generally confined to thin layers with steep velocity gradients. Numerical simulation of these shear layers is difficult, because they must be resolved properly. On the other hand, the greater part of the flow (i.e., the whole core) is determined by a dominating Lorentz force $j \times B$. This leads to the CFS, which calculates the variables in the core and provides the volume flux carried by the side layers. Exact description of the flow in these boundary layers is assumed to be of minor importance.

Equations of the Core Flow Solution

The idea of the CFS is to neglect nonlinear inertial terms and viscous effects which make the numerical scheme of the FS difficult. This is reasonable if M and N are high enough, as can be seen from Eq. (1). In this case only the Lorentz force balances the pressure gradient, and the governing equations, representing the core flow, read[8]:

$$\nabla p = j \times B \qquad \text{Navier-Stokes} \quad (7)$$

$$\nabla \bullet v = 0 \qquad \text{conservation of mass} \quad (8)$$

$$\nabla \bullet j = 0 \qquad \text{conservation of el. charge} \quad (9)$$

$$j = -\nabla \phi + v \times B \qquad \text{Ohm's law} \quad (10)$$

Notice that neither the Hartmann number nor the interaction parameter appear in this set of linear equations, but still the same number of variables are involved. Because viscosity is assumed to be negligible, the no-slip condition cannot be satisfied here. Instead of this the normal component of the velocity vanishes at each wall:

$$v_n \big|_{wall} = 0 \quad (11)$$

The thin wall approximation of FS and CFS are identical:

$$\partial_n \phi\big|_{wall} = c\Delta_t \phi\big|_{wall} \qquad (12)$$

Algorithm

A rectangular channel as shown in Fig. 1 is examined. Cartesian coordinates are used with x pointing in main stream direction. The cross section extends from $-1 \le y \le 1$, $-1 \le z \le 1$, and the axial domain from $-15 \le x \le 15$. For c a typical value for fusion reactors of 0.07 is chosen. A variable magnetic field

$$B_y(x) = \frac{0.95}{1 + e^{0.75x}} + 0.05 \qquad (13)$$

is applied with the subscript indicating its direction. This particular form models the exit of a magnet.

Full Solution

Equations (1) to (6) are evaluated with a vectorized version of the numerical method described in Ref. 6. A staggered grid discretizes the computational domain defining the scalars p and ϕ in the middle of the cell, and the flux variables j and v in the middle of the sides. This grid generation is chosen to ensure mass and current conservation. A linear interpolation of variables is employed on points where variables are not defined. Equidistant central differences with second order accuracy are applied. After an initial guess for the three velocity components, the equations are advanced forward in time until the steady-state solution is reached. A time-splitting procedure is used for the time integration. The integration of the Navier-Stokes equation is performed by an ADI method. Two Poisson-type equations for electric potential and pressure, respectively, have to be evaluated for each time step with a fast Poisson solver.

The storage available on the VP50 computer at KfK limits the discretization to about $70 \times 62 \times 62$ nodes. The gradients of the velocity profile in the layers close to the wall increase at higher Hartmann number. In a rectangular duct the thickness of side layers at walls parallel to the magnetic field lines is proportional to $M^{-1/2}$. The thickness of Hartmann layers at walls perpendicular to B is proportional to M^{-1}. Since velocity variations have to be resolved properly, there is an upper limit of the Hartmann number depending on the number of grid points and the magnetic field. Steeper gradients of B produce steeper gradient of the velocity. For the variable magnetic field chosen here and the storage available on a VP50, the maximum of M is restricted to 300.

As shown, for example, in Ref. 9, there is another restriction for numerical calculations with the FS. If the relation for the mesh Reynolds number

$$\frac{M^2}{N} \Delta x U \le 2 \qquad (14)$$

is not satisfied, the validity of the solution is not guaranteed. Here Δx is the dimensionless distance between two neighbouring grid points, and U is the local velocity. This relation yields a minimum N, due to a maximum number of grid points determined by computer storage and the applied magnetic field. A faster variation of B enhances the local U in side layers. If N is too low, the calculation can become inconsistent and yields results alternating at every computational time step. The minimum interaction parameter for this magnetic field and a Hartmann number of 50 is about 100. For $M=300$ the minimum N increases to 2000.

The typical computational time for the FS depends on the number of nodes and the parameters chosen. With $34 \times 34 \times 34$ grid points, the CPU-time on a VP50 for $M=50$ and $N=2000$ is a few minutes, rising to more than 15 hours for $M=300$ with $62 \times 62 \times 62$ points. This time can be reduced by a factor of five to ten if a "good" guess for the initial velocities is done. For example, the results at a lower Hartmann number can be taken as initial values for a run with a higher M. If B and N remain unchanged convergence is reached much faster.

Core Flow Solution (CFS)

Equations (7) to (12) can be solved in the present form with a three-dimensional computer code. They are linear and therefor easier to solve than the set of nonlinear equations of the FS. But MHD flows at high Hartmann numbers, and interaction parameters have special characteristics that can be taken advantage of. It can be shown that the pressure and all components of the current density perpendicular to the applied field do not vary along field lines. After the equations are integrated along these lines, they can be solved on the two-dimensional surface of the duct without losing the three-dimensional information of the entire domain.[3] The core variables of the duct flow can be derived from the values on the surface, because their variation along field lies is known. This reduces greatly the computer time needed, the code complexity, and the storage needed, whereas the analytical preliminaries imply more effort.

After an initial guess for the core flow velocity, SOR is employed to calculate the potential distribution in the fluid and the wall. As long as the potential does not vary along field lines in the prescribed manner, the velocities are adjusted and normalized, and the procedure is repeated. If $M^{-1/2} \leq c$ is not satisfied, then the current returning through the side layers parallel to the magnetic field is not negligible and is included in the analysis with respect to c and M.[3] In this case the variables in the core must be adjusted considering the higher volume flux through the side layers.

This method only holds for $B_y \neq 0$, and if a significant portion of B is perpendicular to the duct. As mentioned above, the Hartmann number and interaction parameter need to be high. Abrupt expansions or contractions require higher M and N than a smooth geometry. For any more complex geometry, like bends and smooth variations of the duct's cross section, the analytical integration along field lines becomes more and more expanded. Due to the fact that viscosity is neglected, the velocities in the thin shear layers are not evaluated exactly, and corrosion analysis is not possible with CFS.

Computer storage and CPU-time is not a strong restriction for the CFS. The typical time for a rectangular duct, the magnetic field defined in Eq. (13), and 90 ×30 grid points is less than 10 minutes.

Numerical Results

One restriction of the CFS is that high Hartmann numbers and interaction parameters are presumed. It is important to know how high these numbers must be in order to estimate the range of validity of this approximate method. As an example, a straight rectangular duct (see Fig. 1) with a constant wall conductance ratio of $c=0.07$ under a magnetic field specified in Eq. (13) is investigated. Results of the FS and CFS are compared at $M=50$, 100, 200, 250, and 300 and at several N. To find out the influence of M, being characteristic of the magnitude of viscous effects, N is kept constant at 2000. This value satisfies condition (14) for all M checked here. In a second part of computations, flow variables at a constant M and various Hartmann numbers are investigated.

Figures 1 to 4 show curves of the dimensionless pressure, velocity in x-direction, and electrical potential against the axial x-direction of the duct for FS and CFS. The applied field is plotted for reference. Values for p, ϕ, and the velocity are given at the axis of the duct, at $y=0$ and $z=0$, and close to the middle of the side wall, at $y=0$ and $z=0.92$. These are positions of the cross section where variables differ greatly. There is a slight difference between the coordinates Of FS and CFS (~0.02), because the discretization in the schemes is not exactly the same. This has little effect on the comparison.

The MHD flows in all cases presented here have three distinct regions. In the first, $B_y=1$ is constant, in the second it slowly decreases; and in the third $B_y=0.005$ is constant again. This minimum field is applied, because $B_y=0$ is not allowed in the CFS. Areas one and three have some similarities. The velocity, the potential, and the pressure gradient are constant in the x-direction of these fully developed regions (see Fig.2 to 4). The pressures in the middle and the side are exactly the same. The whole flow is two-dimensional. A difference between the results of the

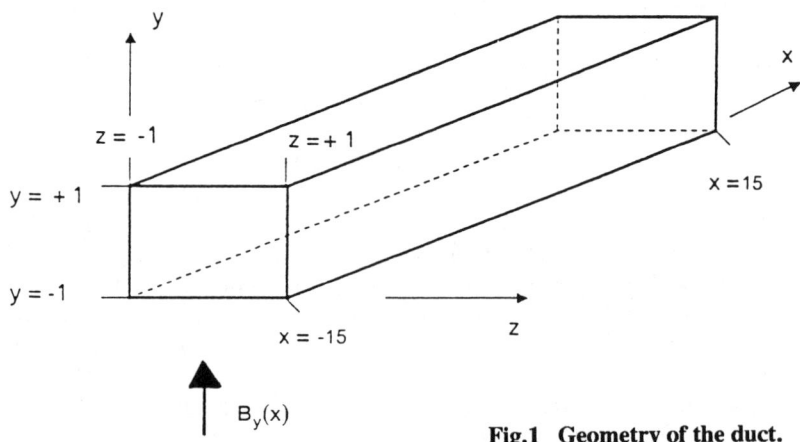

Fig.1 Geometry of the duct.

CORE AND FULL SOLUTIONS

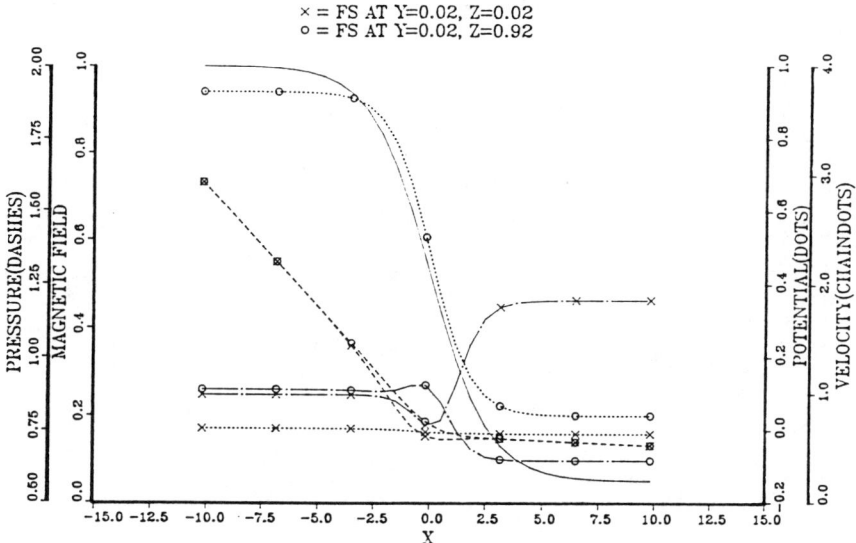

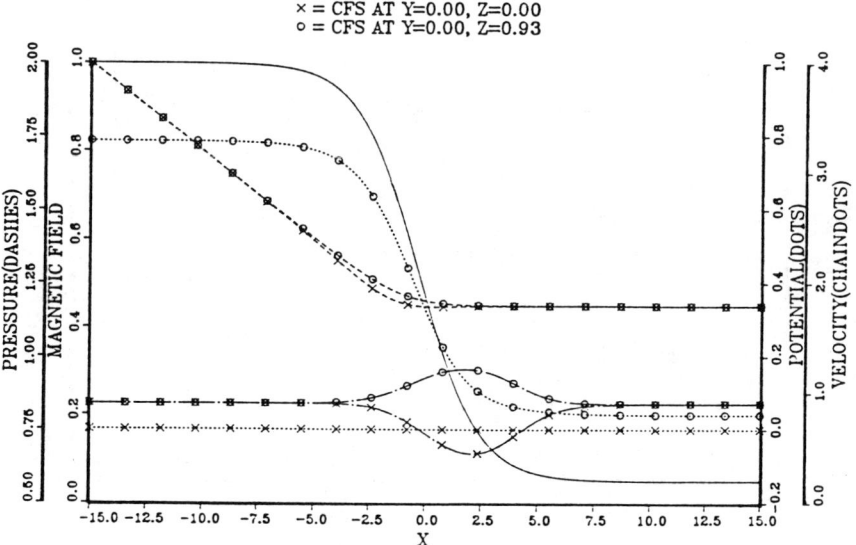

Fig. 2 Results for $M=50$, $N=2000$, $c=0$; FS (above) and CFS (below).

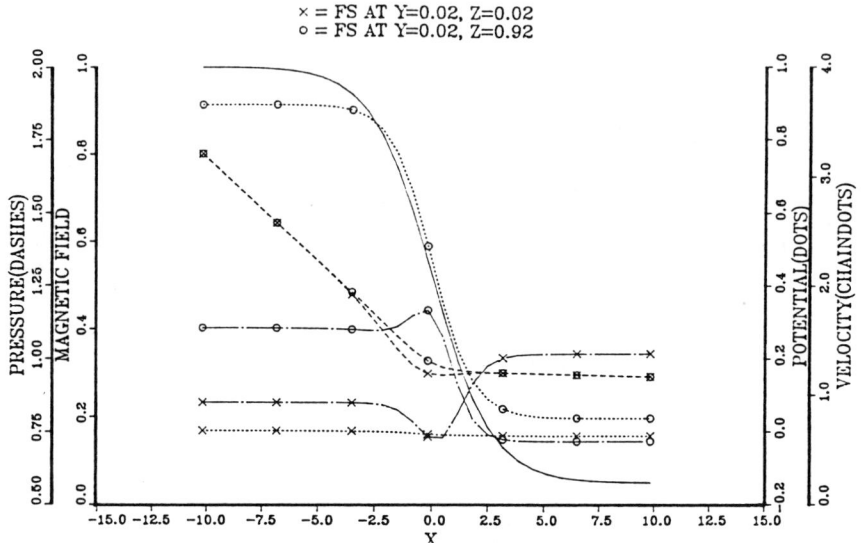

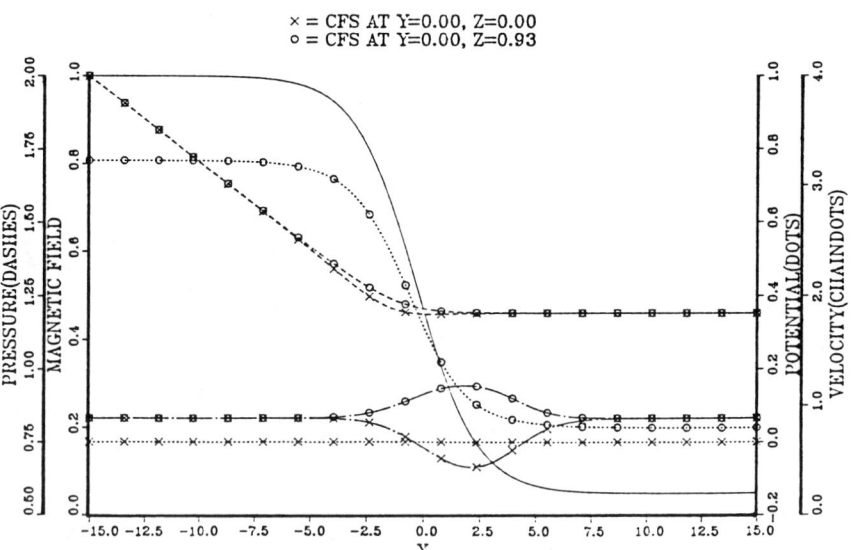

Fig. 3 **Results for $M=100$, $N=2000$, $c=0$;** FS (above) and CFS (below).

CORE AND FULL SOLUTIONS

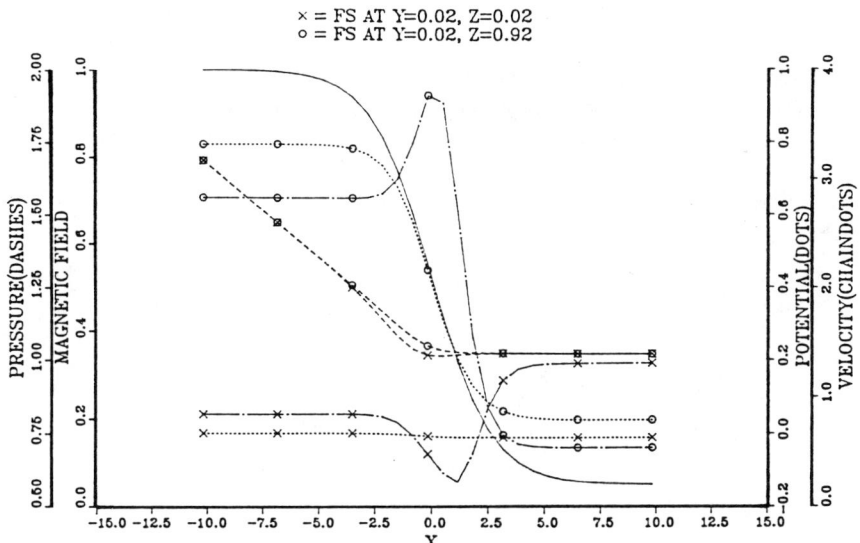

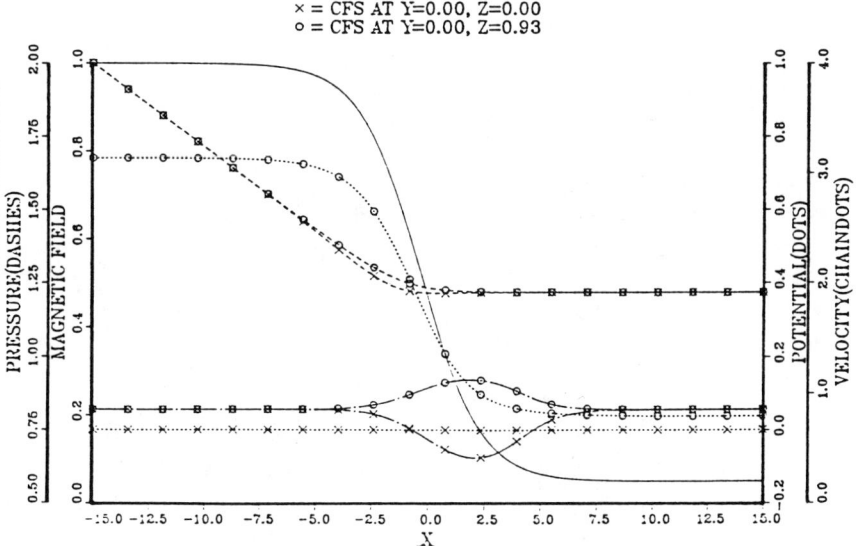

Fig. 4 Results for $M=300$, $N=2000$, $c=0$.
FS (above) and CFS (below).

two methods can be detected at the duct outlet at $x>5$. For the CFS, the pressure gradient vanishes, when B_y is nearly zero, whereas the FS still predicts a small pressure gradient, because this method accounts for the viscous effects. Initial effects do not appear in the fully developed flow. There is no time-dependence and no y- or z-component of the velocity under steady state conditions. Surprisingly, it seems as if the viscous pressure gradient is higher for a lower Hartmann number (compare Fig. 2 and 4 for example). But these nondimensionalized numbers have to be multiplied by M^2 for a comparison at equal volume fluxes.

The velocities obtained from the CFS turn out to be the same in the middle of the duct and at $z=0.93$, as long as B does not change Figure 5 displays velocity

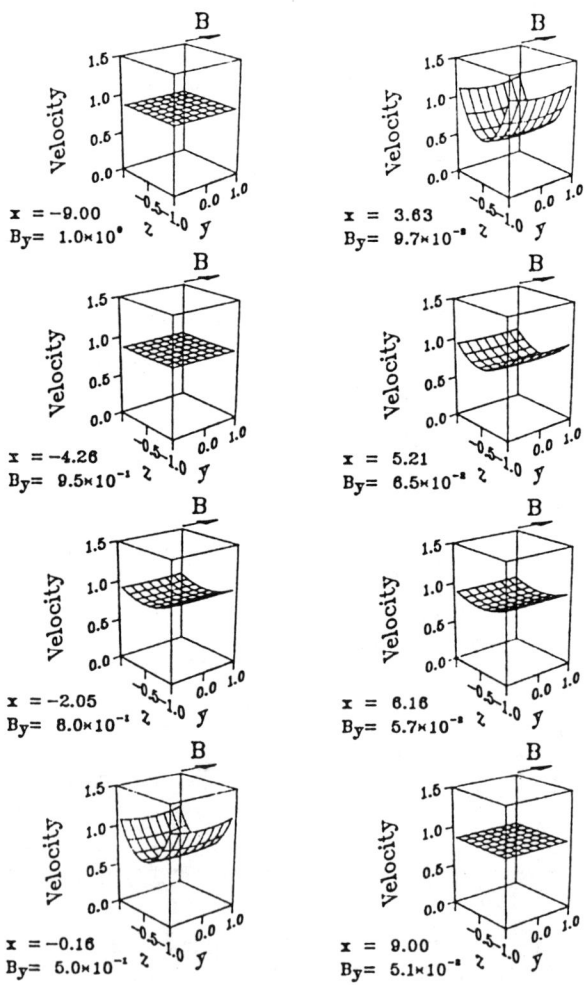

Fig. 5 Development of velocity distribution for the CFS; $M=200$, $N=2000$, $c=0$.

distributions for several cross sections along the duct axis. The magnitude of **B** and the corresponding coordinate is given below each subplot. Lorentz forces are uniform if **B** is constant. The CFS calculates the core variables only, and as a result the velocity profile is flat.

Figure 6 shows results for the FS. With $B_y=1$, the typical M-shaped velocity profile of fully developed viscous MHD-flow can be seen. The no-slip condition is satisfied at the walls, in contrast to the CFS. An ordinary hydrodynamic profile results downstream for $x>3$, where **B** is low. Figures 2 to 4 again demonstrate the differences between the two solutions. In the FS the velocity changes from a higher value in the side layer at $x<-3$ to a lower one at $x>3$, in accordance with

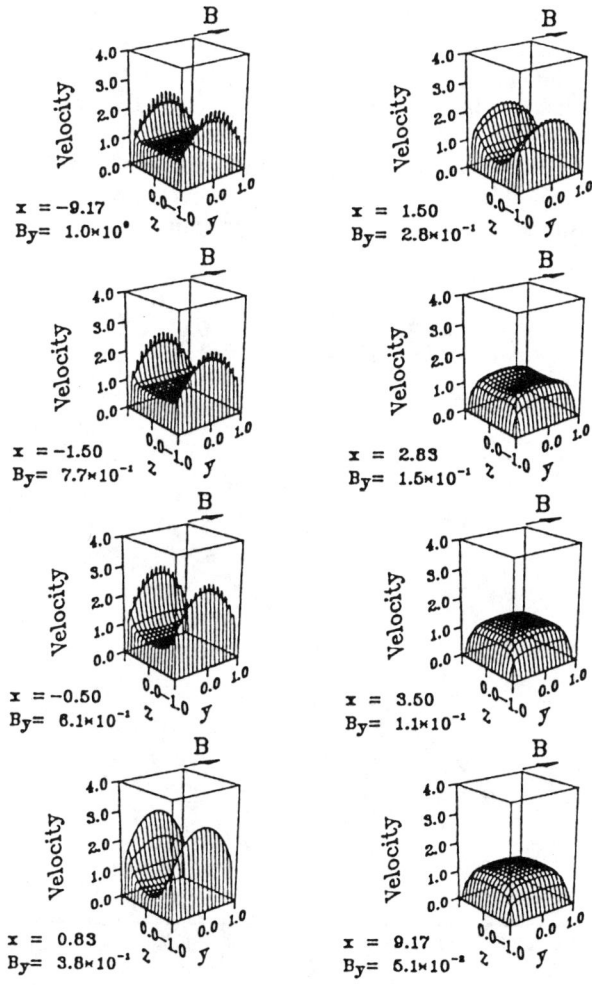

Fig. 6   Development of velocity distribution for the FS; $M=200$, $N=2000$, $c=0$.

the M-shaped profile and the nearly hydrodynamic profile, respectively. Once more the CFS shows uniform velocities for $x<-3$ and $x>3$. For high M and N, these differences only occur at the boundary layers. As mentioned before, these layers are thin and decrease with M. A good estimation of the volume flux in the side layers is possible with respect to the wall conductance ratio and the Hartmann number.[3] For this reason the CFS yields good predictions of the heat transfer, but cannot be taken for corrosion analysis, where exact velocities are required.

The development of the potential in the axial direction show similar behaviour for FS and CFS. As a symmetry condition it can be adjusted to zero in the axis of the duct, at $z=0$. Potential differences between the middle and the side are reduced as the magnetic field decays. he potential difference at $x<-5$ where the flow is fully developed is about 17 percent higher for the FS at $M=50$, falling down to 9 percent at $M=300$. This is due to the fact that the CFS does not include the potential jumps across the side layer. The potential distributions given in Figs. 2 to 4 show the different result for these side layers, whereas the potential in the core indicates a very good agreement of both methods (see Fig. 7). For higher M, a lower difference in he potentials is expected, and Figs. 2 to 4 show this trend.

Three-dimensional effects can be seen in region two, where the magnetic field decreases. the lower downstream magnetic field induces lower potential differ-

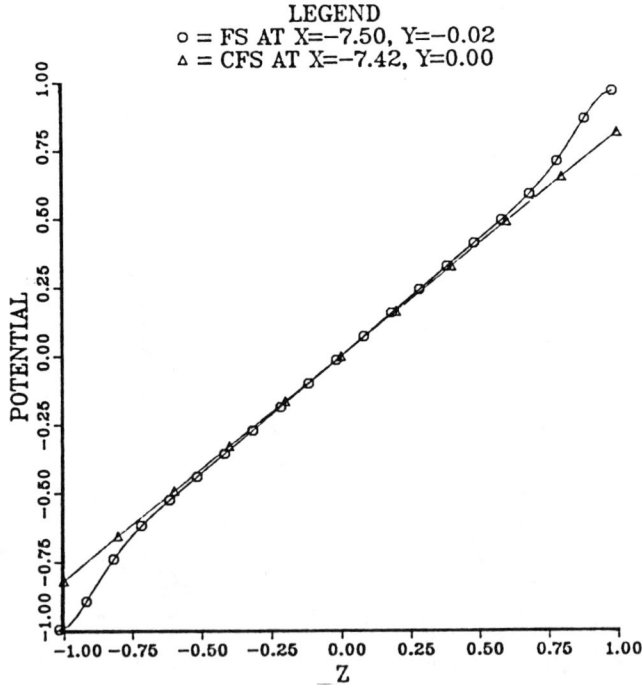

Fig. 7 Potential distribution in the y-z-plane at $y=0$ and $x=-7.5$; $M=200$, $N=2000$, $c=0$.

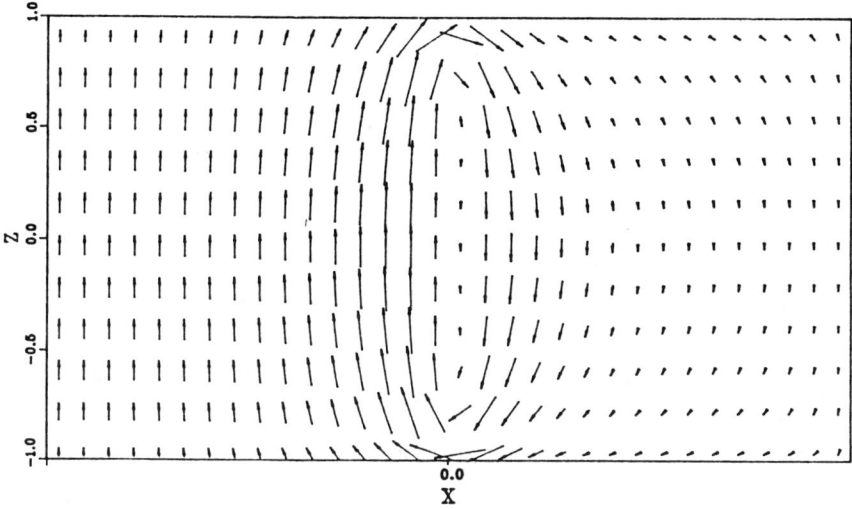

Fig. 8 Current in the x-z-plane at $y=0$; $M=50$, schematic.

ences between the side and the center of the channel. Due to the resulting axial potential differences, axial currents are induced. They are responsible for Lorentz forces, which have to be balanced by a pressure gradient. This results in a pressure difference seen in Figs. 2 to 4, where the pressure is higher close to the side wall. Here FS and CFS have very similar results. These axial currents have a short circuit in the x-z-plane due to conservation of electrical charge (see Fig. 8). The additional z-component of the current is higher in the core than close to the side walls. Subsequently there is a velocity jet in the side layers. Again FS and CFS show the same trend, but for the reasons mentioned above the magnitude of these jets is much higher in the FS. To satisfy conservation of mass, the velocity must slow down in the middle of the channel. The whole flow is driven toward the side layers in the decreasing field. Figure 6 shows the effects in more detail for the FS. The flow has the expected M-shape at $x=-9.17$, is driven to the side layers in the three-dimensional region, and forms an ordinary hydrodynamic profile at $x>3$. Figure 6 indicates an additional volume flux through the side layers in the varying field for the FS.

Comparison of the axial pressure gradient in the fully developed region with $B_y=1$ provides another possibility to check the quality of the calculations. Figure 9 demonstrates these gradients for FS and CFS with Hartmann numbers between 50 and 300 and $N=2000$. The third curve in this plot shows the difference between FS and CFS in percent. Viscous effects considered in the FS require a higher pressure difference between the inlet and the outlet of the duct, ensuring the same volume flux as the CFS. The difference between FS and CFS is reduced at higher M, as expected, because viscosity becomes less important. For $M>100$ the agreement is within acceptable error limits.

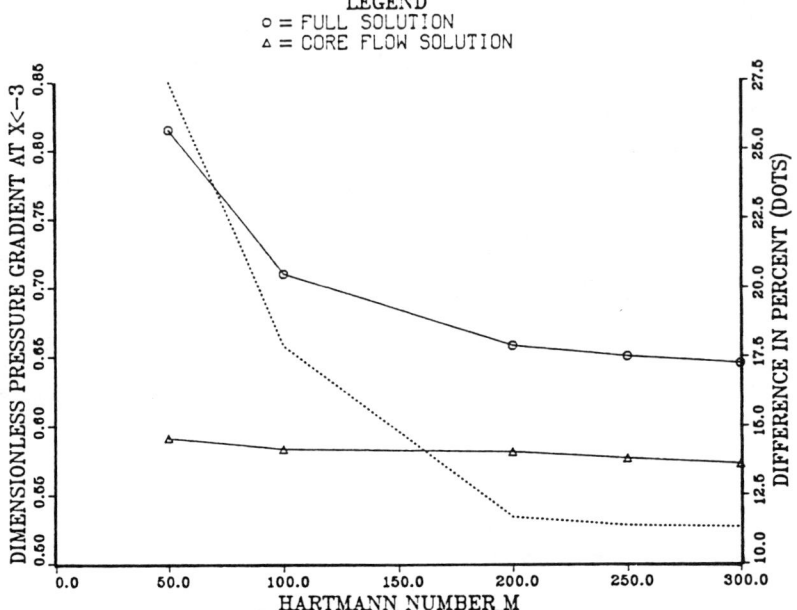

Fig. 9 Comparison of pressure gradient at $B_y=1$.

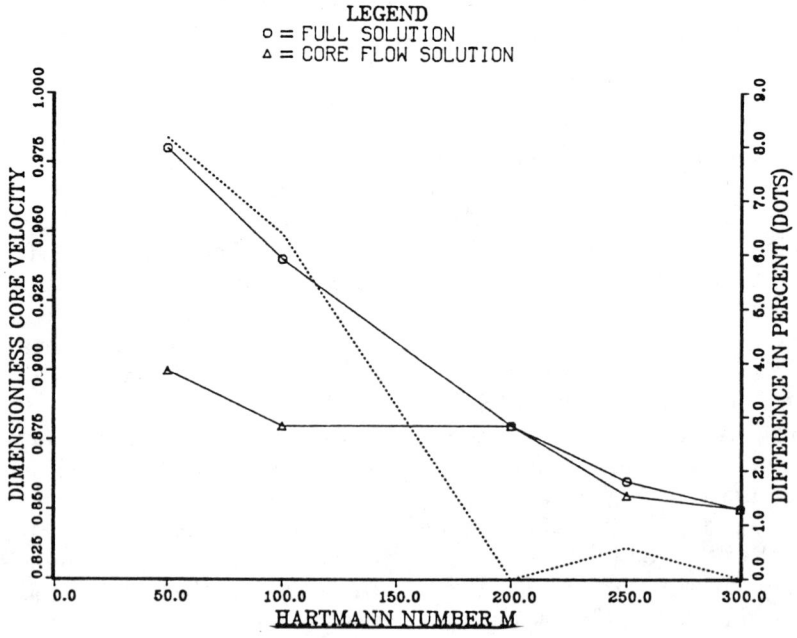

Fig. 10 Comparison of core velocities at $B_y=1$.

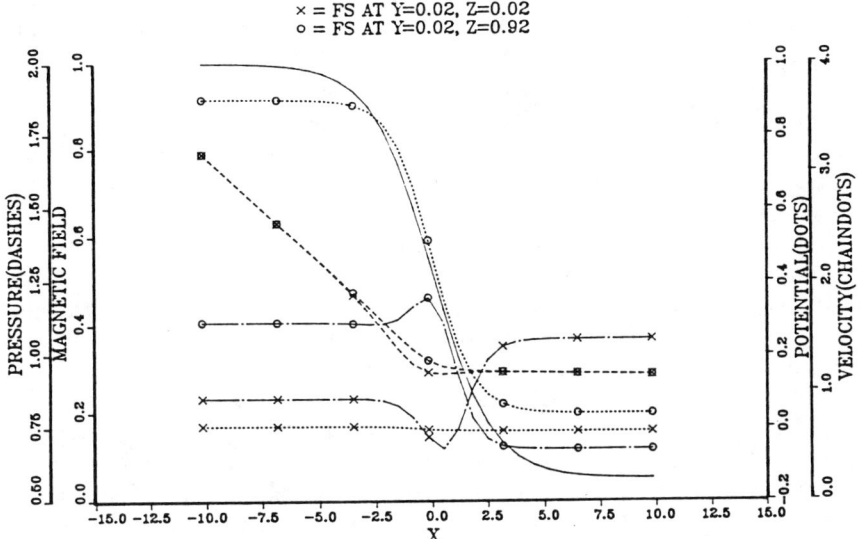

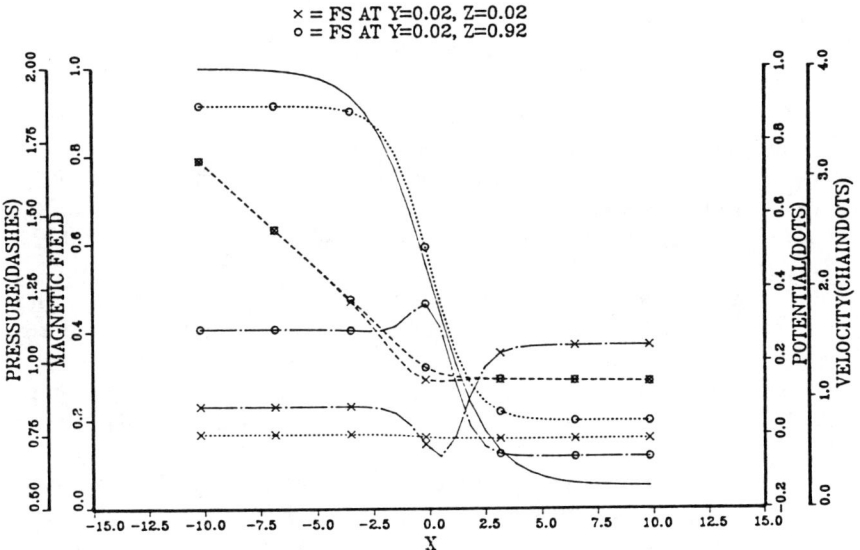

Fig. 11 FS with $M=100$, $c=0.07$, various N;
$N=1000$ (above), $N=200$ (below).

The core velocity also is compared in the same region. Figure 10 illustrates results for the FS and the CFS with $50 \leq M \leq 300$ with $N=2000$. The numbers obtained from the CFS match very well to Walker's asymptotic solution valid for $M^{-2} \ll c \ll M^{-1/2}$.[8] This relation does not hold for Hartmann numbers less than 100, if the wall conductance ratio is $c=0.07$. In this case the currents closing through the side layers are not negligible. The same can be obtained from Fig. 10. For $M \geq 200$ the agreement between FS and CFS is within 1 percent. Results for $M \leq 100$ show the expected deviation, but even for $M=50$ the difference is less than 9 percent.

In the three-dimensional region and at the transition from three- to two-dimensional flow, inertia might influences the redistribution of the flow. In order to separate this effect, several interaction parameters are combined with a Hartmann number of $M=100$. Figure 11, when compared to Fig. 3 proofs that there is no significant difference between the flow variables at $N=200$, $N=1000$ and $N=2000$. The same is true for any higher interaction parameter. Of course, inertia will be important for sufficiently low N. This could be investigated using FS only on computers with larger storage the VP50, due to the lower limit on N required by the restriction on mesh Reynolds number. A more general influence of inertia can be detected in Figs. 2 to 4. In the CFS the variation between velocities in the center and near the sidewalls starts slightly further upstream than in the FS. The reason for this is that currents returning through the side layers are not modelled correct in the CFS if the magnetic field changes. But for slowly varying fields it yields acceptable results.[3]

Conclusions

The core flow solution (CFS) and the full solution (FS) are compared for a rectangular duct with a wall conductance ratio typical for fusion reactor blankets under the influence of a variable transverse magnetic field. The difference of the calculated pressure gradient in the fully developed region is about 27 percent for a Hartmann number of $M=50$, decreasing to about 11.5 percent for $M=300$. The potential distributions derived with the two methods agree very well except for the thin side layers. The core flow velocities show an even better agreement. Their difference is less than 1 percent for $M > 100$. It is reasonable to transfer these results to any straight duct with similar wall conductance ratio under constant or slowly varying fields. Bigger computers will permit the simulation of MHD flows at higher Hartmann numbers. In this case the results of the FS and the CFS are expected to fit even better.

The mesh Reynolds number limits the lower values of the interaction parameter N for the FS. None of the interaction parameters studied here, $N>100$ for $M=50$ and $N>2000$ for $M=300$, affect the results significantly. Further comparisons should be done for lower N and for more complex geometries, such as bends and sudden expansions, where inertia and/or viscosity might have a greater influence.

References

[1]Holroyd, R. J., and Mitchell, J.T.D., "Liquid Lithium as a Coolant for Tokamak Fusion Reactors," *Nuclear Engineering and Design/Fusion,* Vol. 1, 1984, pp. 17-38.

[2] Smith, D.L.,"Blanket Comparison and Selection Study — Final Report," ANL/FPP 84-1, Argonne National Laboratory, Argonne, IL

[3] McCarthy, K., "Analysis of Liquid Metal MHD Flow Using a Core Flow Approximation, with Applications to Calculating the Pressure Drop in Basic Elements of Fusion Reactor Blankets," Ph.D. Dissertation, University of California, Los Angeles, CA, February 1989.

[4] Barleon, L., Lenhart, L., Mack, H.J., Sterl, A., and Tomauske, K., "Investigations on Liquid Metal MHD in Straight Ducts at High Hartmann Numbers and Interaction Parameters," Proceedings of the NURETH-4 Meeting, Karlsruhe, October 10-13, 1989, pp. 857-862.

[5] Picologlou, B.F., Reed, C.B., Hua, T.Q., Barleon, L., Kreuzinger, H., and Walker, J., "Experimental Investigation of MHD Flow Tailoring for First Wall Coolant Channels of Self-Cooled Blankets," Eighth Topical Meeting on the Technology of Fusion Energy, Salt Lake City, Utah, October 9-13, 1988.

[6] Sterl, A., "Numerische Simulation magnetohydrodynamischer Flüssig-Metall-Strömungen in rechteckigen Rohren bei großen Hartmannzahlen," KfK-Bericht Nr. 4504, Kernforschungszentrum Karlsruhe, 1989.

[7] Shercliff, J.A., *A Textbook of Magnetohydrodynamics*, Pergamon Press, Oxford, 1965.

[8] Walker, J.S., "Magnetohydrodynamic Flows in Rectangular Ducts with Thin Conducting Walls. Part I: Constant Area and Variable Area Ducts with strong Magnetic Fields," *Journal de Mechanique*, Vol. 20, 1981, pp. 79-112.

[9] Anderson, D.A., Tanhill, J.C., and Plether, R.H., *Computational Fluid Mechanics and Heat Transfer*, Hemisphere Publishing Corporation, New York, 1984, pp. 343-346.

Hydrodynamics and Heat Transfer of Thin Liquid-Metal Films in a Magnetic Field

I. A. Evtushenko,* E. M. Kirillina,* S. Y. Smolentzev,*
and A. V. Tananaev*
Leningrad Polytechnic Institute, St. Petersburg, Russia

Abstract

This paper reports upon a combined theoretical, numerical and experimental investigation of hydrodynamic and heat transfer phenomena in thin layer liquid metal flows in the presence of a strong perpendicular magnetic field. Whereas in the absence of side walls there is no interaction between the field and the flow, there exists interaction in the presence of side walls, unless the liquid metal layer is very thin. The flow stability is found to increase with the increase in the magnetic field strength and the liquid metal layer thickness. Sufficiently strong co-planar magnetic fields suppress surface disturbances and level the free surface along the field lines. The application of a thin liquid metal film flow for the protection of divertor plates of fusion reactors is analyzed. It is found that effective protection of divertor plates is possible with relatively thick (e.g., 8 mm) fast flowing (e.g., 0.9 m/s) liquid metal layers.

Introduction

Analytical, numerical, and experimental results are presented in an investigation of thin-layer liquid metal (TLLM) flows in a strong coplanar magnetic field. The influence of the magnetic field on flow characteristics is simulated by the friction force in Hartmann layers, which develops on sidewalls orthogonal to the coplanar field component. Conditions have been defined at which the coplanar magnetic field influences the heat transfer, free surface configuration, and hydrodynamic stability.

Statement of the Problem

Let us consider a layer of viscous (v), incompressible (div $\overline{V} = 0$), electroconductive (σ) liquid with constant physical properties in an inclined (θ) duct, restricted in the cross-sectional direction by side walls (with $Z = \pm b/2$) (Fig. 1). The duct bottom is an insulator, and the walls are conductive. The analysis is carried out using an inductionless approximation ($Re_m \ll 1$) for a

---

Copyright © 1992 by the American Institute of Aeronautics and Astronautics, Inc. All rights reserved.
*Professor, Faculty of Hydrotechnics.

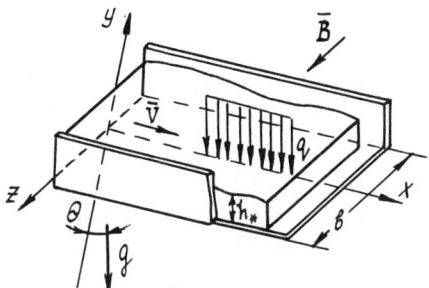

Fig. 1 The problem under consideration

strong coplanar magnetic field ($B_z \gg B_x, B_y$; $M \gg 1$). The influence of the main component (B_z) of the induction vector is simulated by the friction force on the walls restricting the flow region in transverse direction. In the entire flow region, with the exception of thin Hartmann layers, the strong magnetic field tries to eliminate gradients along its direction. For this reason it is possible to average the equations, which describe the flow considered, in the direction of the field supposing the existence of a local Hartmann velocity profile, the error being of the order of $O(M^{-1})$.

The dimensionless equations averaged in the direction Z assume the form:

$$\frac{\partial \overline{V}}{\partial t} + (\overline{V} \cdot \nabla)\overline{V} = -\frac{\text{grad } p}{\text{Fr}} + \frac{1}{\text{Re}} \Delta \overline{V} + \frac{g}{|\overline{g}|} \frac{1}{\text{Fr}}$$

$$+ N_{\|}\overline{j} \times \overline{B} - \frac{2M_{\perp}\beta\overline{V}}{\text{Re}} - \varepsilon N_{\perp} \text{Re} \, \overline{V} \qquad (1)$$

$$(\overline{V} \cdot \nabla)T = \frac{1}{\text{Re Pr}} \Delta T + \frac{\text{Ec}}{\text{Re}} \phi + \text{EcN}j^2 \qquad (2)$$

$$\nabla (\overline{V}) = 0 \qquad (3)$$

$$\nabla (\overline{j}) = 0 \qquad (4)$$

$$(\text{div } \overline{j}) = -\nabla\varphi + \nabla(\text{div } \overline{V}) \times \nabla(\text{div } \overline{B}) \qquad (5)$$

where $\overline{V}$, P, φ, and j, are the velocity, pressure, electric potential, and current density, normalized correspondingly with V_o, $\rho g h_*$, $V_o B_\perp h_*$, and $\sigma V_o B_\perp$. Also, $\text{Re} = V_o h_*/\nu$, $\text{Fr} = V_o^2/gh_*$, $\text{Pr} = \nu/a$, $\text{Ec} = V_o^2/C_p T_o$ are the Reynolds, Froude, Prandtl and Eccert numbers; $M_\perp = B_\perp h_* \sqrt{\sigma/(\rho\nu)}$ is the Hartmann number, based on the coplanar component of the magnetic field: $N_\| = \sigma B_\|^2 h_*/(\rho V_o)$ is the MHD interaction parameter, based on the magnetic induction vector component in the X, Y plane; $\beta = h_*/8$ is the relative film thickness; and ϕ is the dissipation function. For the characteristic dimension the normal film thickness, h_*, is taken.

Equation (2) is obtained by integration of the complete equation of energy across the flow width assuming that there is no heat transfer to the side walls.

Free surface boundary conditions include the absence of tangential and normal stresses for velocity, the absence of the normal current component, and the condition of prescribed heat flux as well as kinematic conditions. On the rigid wall (with y=0) adhesion, electroinsulation, ideal heat removal, or thermal insulating conditions are observed.

Equations (1-5) are taken as the basis of the TLLM behavior analysis, both of the initial part of flow and at the constant depth flow section with developed velocity profile.

Preliminary TLLM Analysis in the Presence of Side Walls

The analysis of Eqs. (1) and (2) makes general conclusions regarding the nature of the effect of a strong coplanar magnetic field on the flow q possible.

In the absence of sidewalls (in the case of strictly two-dimensional flow) in a coplanar magnetic field, the electromagnetic force is vortexless. A uniform magnetic field, applied in the direction of the axis, does not create a volumetric electromagnetic force for this particular flow:

$$\text{rot } \bar{j} = (\bar{B} \cdot \nabla)\bar{V} \equiv 0$$

and the induced potential φ completely balances the volumetric electromotive force $\bar{V} \times \bar{B}$. In the absence of sidewalls, currents do not change in the direction of the field [i.e., $(\bar{B}\nabla)\bar{j} = 0$]; hence, rot $(\bar{j} \times \bar{B}) = 0$, which results in no interaction between the coplanar field and the flow.

The situation radically changes if the flow is restricted by sidewalls. In such a case, rot $(\bar{j} \times \bar{B}) \neq 0$, due to the local Hartmann velocity profile, which witnesses the interaction of the flow and the field in the presence of the sidewalls. In this paper the interaction of the coplanar magnetic field is simulated by the friction force in Hartmann layer [the last term in Eq. (1)].

It should be noted that with a sufficiently thin layer of LM the effect of MHD interaction with the magnetic field vanishes. To provide interaction, a restriction for this parameter β has been introduced:

$$M_\perp^{-1} \ll \beta \ll 1$$

Hydrodynamic Stability in the Region of Developed Flow

The velocity profile of a developed flow in a channel with sidewalls assumes the following form:

$$u(y) = \frac{\text{Re sin}\theta}{\text{FrM}^2}(1-\text{ch}My) + M(\frac{\text{Re}}{\text{FrM}^2}\sin\theta h^*-1)\text{sh } My$$

where

$$M = \{2\beta M_\perp + M_\parallel^2 + \varepsilon N_\perp Re\}^{1/2}$$

and where ε is the relative electroconductivity of the walls. The normal film thickness $h^* = h_*/h_o$ (where h_o is the initial depth) can be obtained from the following equation:

$$h^* = (th Mh^* + M^3 Fr/Re \sin\theta)M^{-1}$$

The asymptotic value of h^* at large values of M ($M \geq 10$) is

$$h^* = \frac{M^2 Fr}{Re \sin\theta} + \frac{1}{M}$$

The analysis of the hydrodynamic stability of the developed film flow is carried out using the linear theory. Equations (1-5) are reduced to the analog of the Orre-Sommerfeld equation, which in this particular case takes into account, besides the terms describing MHD interaction within the layer, the influence of Hartmann layers on the sidewalls as well. Investigation was carried out on the behavior of the two-dimensional disturbances in the X, Y plane as the most important due to the tendency of the magnetic field to suppress disturbances, which have a vorticity vector orthogonal to the magnetic induction vector.

The equation of the neutral stability curve has the following form[1]:

$$Re_{cr} = f_1(M) \cdot M_\parallel^2 ctg\xi + f_2(M) \frac{ctg\theta + \alpha^2 We / \sin\theta - M_\parallel^2 ctg\xi f_3(M)}{1 + M_\parallel^2 f_3(M)}$$

where $f_1(M)$, $f_2(M)$, and $f_3(M)$ are transcendental functions: $M_\parallel = Bh_* \sin\xi \sqrt{\sigma/(\rho\nu)}$ is the Hartmann number, based on the component of the magnetic induction vector in the plane X, Y; $We = \frac{\kappa}{\rho g h_*^2}$ is Weber number; and ξ is the angle between the magnetic field and the bottom in the plane X, Y.

The results of the analysis prove that the flow stability increases with the increase of the coplanar magnetic field (curves 4 and 5 in Fig. 2). The increase of the transverse component of the field brings about the increase of the Reynolds number Re_{cr}, corresponding to the stability losses (curve 3 is higher than curves 1 and 2; see Fig. 2).

The investigation reveals the main criterion for the flow characteristics to be complex, including the effect of the coplanar and normal field components as well as the relative film thickness β. With a strong coplanar field, the main influence on the stability is rendered by Hartmann boundary layers, which develop at the walls orthogonal to the field.

The physical mechanism of stabilization can be explained by the effect of entrainment and retardation within the Hartmann layer of vortices external to the layer. With the increase of the TLLM thickness and the increase of the magnetic field, the stabilizing effect becomes stronger.

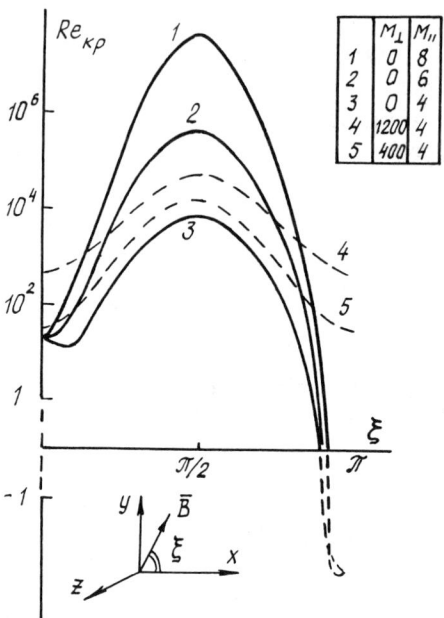

Fig. 2 Critical Re number as a function of ξ angle

The investigation results make it possible to come to a conclusion regarding the feasibility of stabilization of MHD flow of a TLLM with sidewalls by a strong coplanar magnetic field.

Nondeveloped Film Flow

The analysis of the coplanar field effect on the film flow at the initial hydrodynamic zone was carried out on the basis of the numerical solution of Eqs. (1) and (3-5) in the boundary-layer approximation.

To calculate the velocity field, an implicit, absolutely stable method of second-order accuracy is used.[2]

The peculiarity of the dynamic problem of film flow in the initial region necessitates simultaneous calculation of the velocity profile and the free surface curve. To facilitate the calculation, a change of variables is made:

$$\xi = x/h_o, \quad \eta = y/h(x)$$

With the help of this change, the region of integration is transformed to rectangular. The free surface curve is calculated with the help of the kinematic condition.

The coplanar magnetic field leads to retardation of the film, which brings about the decrease of the length of the region of hydrodynamic stability (Fig. 3). Numerical calculations show the influence of the initial velocity profile on the

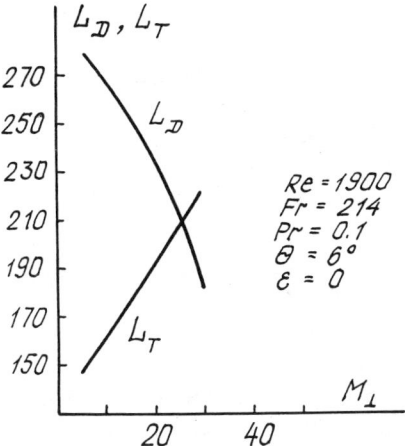

Fig. 3 Hartmann number influence on hydrodynamic and thermal stabilization length.

velocity field and the shape of the free surface, which is felt along the whole length of hydrodynamic stability (Fig. 4a).

With sufficiently large M, the character of the film flow changes: A transition from the condition of decay ($dh/dx < 0$) to the condition of hydrostatic lift ($dh/dx > 0$) is observed. Lines of flow condition divisions in the coordinates Re-Fr are of the form of straight lines starting from the origin with the slope depending on M, ε, and θ.

Experimental Investigation

Experimental tests of the assumptions made in the theoretical analysis were carried out on the alloy In-Ga-Sn. For different flow conditions, measurements were made of the film thickness and of the potential distribution across the width and the depth of the flow. In the flow core the transverse component of the current was found to be constant and to change sharply in the near-wall region. This proves the hypothesis about the local Hartmann velocity profile.

Measurement and visual observations showed that the sufficiently strong coplanar magnetic field suppresses surface disturbances and levels the free surface along the field lines.

Depending on the conditional parameters, decay flow developed (Fig. 4a) or a flow with hydrostatic lift developed (Fig. 4b). In several experiments a sharp thickness increase, a hydraulic jump, was observed (Fig. 4c).

Comparison of the free surface curves obtained theoretically and experimentally showed good agreement (Fig. 4a).

Heat Transfer

In terms of the previously given model of the film flow, an analysis has been carried out of the hydrodynamics and heat exchange in the liquid metal protection of diverter plates of a fusion reactor. Calculations were carried out at constant physical properties of the heat-transfer medium, with the thermoelectromotive force being neglected to make it possible to treat separately dynamic and thermal problems. Taking into consideration the high heat flux in diverters, in Eq. (3) terms characterizing joule and viscous dissipation are omitted.

The coplanar magnetic field affects heat transfer in a TFLM by two mechanisms: 1) the film thickness increase and 2) variations of the velocity profile. The first phenomenon leads to the increase of the heat stabilization zone, and the second leads to the decrease of the heat stabilization zone due to intensification of heat transfer in the transverse direction. The increase of the length of thermal stabilization with the increase of the magnetic field (Fig. 3) is connected to the predominant influence of the first mechanism.

Computations of film cooling were done using $\theta = 45$ degrees ($\varepsilon=0$, B=6T), and heat-flux characteristic of the ITER reactor (Fig. 5, curve 1). Two extreme cases of heat transfer in LM cooling were considered: 1) a film without additional cooling ($\partial T/\partial y|_{y=0} = 0$) and 2) a film with ideal heat transfer from the bottom ($T|_{y=0}$ = const).

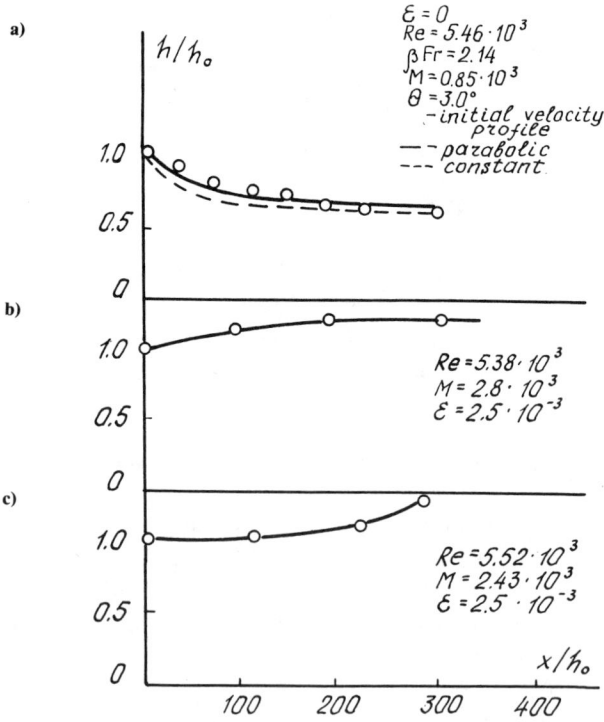

Fig. 4 Change of depth with the longitudinal coordinate

As a criterion of the possibility of the application of film cooling, the condition of small plasma contamination by evaporated metal atoms was used. Similar to Ref. 3, the condition was represented as

$$\int j_s \, dx \geq \int j_v \, dx$$

where j_s, j_v are atom fluxes from the film due to spraying by ions and evaporation. The spraying flux depends on the ion flux falling on the free surface of the film and the coefficient of metal spraying $j_s = j_p \cdot \alpha_s$. In these computations with Ga, $\alpha_s = 0.06$. The flux of evaporated atoms depends on the temperature of the free surface of the film, $j_v = P_o(T_h)/\sqrt{2\pi m k T_h}$ where $P_o(T_h)$ is the metal vapor pressure at the surface temperature.

Thin, slow films of LM are considered the most promising type of protective film cooling, because such flow requires less expenditures on pumping and provides low pressure in feed sections. However, calculations showed that the use of thin, slow films, even with ideal heat removal, leads to high plasma contamination by evaporated metal atoms due to high surface temperature (Fig. 5). In addition, such flows are strongly affected by the ion

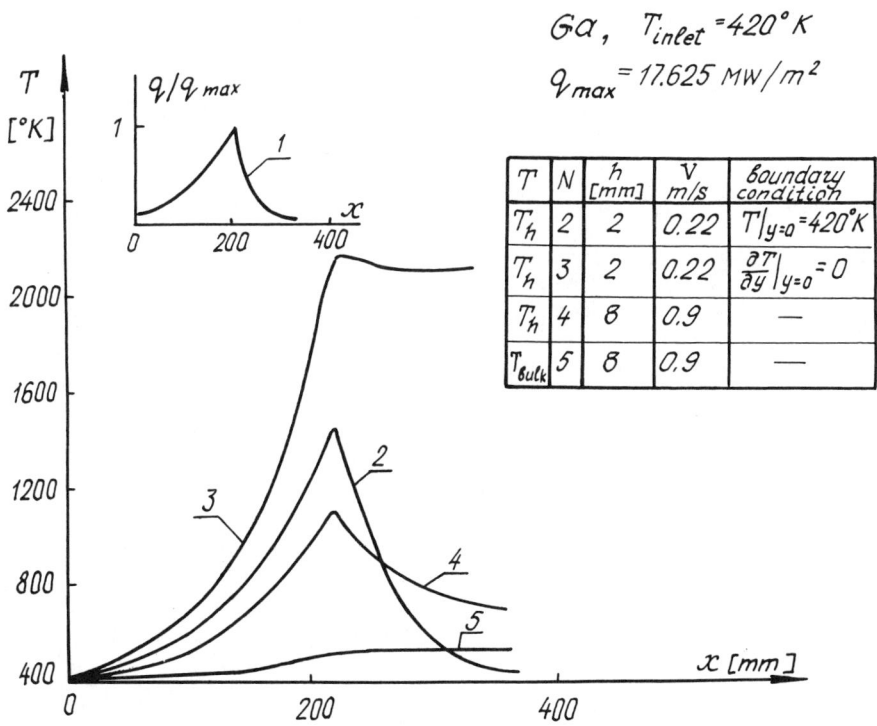

Fig. 5 Temperature distribution

pulses falling on the free surface. Thus, with light LM application, film rupture is possible (in these calculations, only the normal pulse component was taken into consideration). As a result, thin, slow films cannot be used to protect diverter plates.

Effective protection of diverter plates is possible with the help of films of greater thickness and with a greater flow rate of heat-transfer medium. With the preceding considerations it is reasonable to use films of thickness $h_f = 8$ mm and $V = 0.9$ m/s (Fig. 5). It should be noted that, for such films, additional cooling does not bring about an appreciable decrease of evaporation and the improvement of thermophysical characteristics. Only the upper film layers are heated, because of the high flow velocity and large depth (Fig. 5).

References

[1] Kirillina, Y.M., "Stability of Liquid Metal Film in a Strong Magnetic Field of Arbitrary Direction," *Magnetic Hydrodynamics*, 1989, Vol. 1, pp. 133-135.

[2] Paskonov, V.M., Poleznev, V.J., and Chudov, L.A., "Digital Simulation of Heat and Mass Transfer Processes," Moscow, Nauka, 1984, p. 285.

[3] Muravyov, Y.V., "Contact Devices for Diverting and Limiting Systems of "Tokamak"-type Reactors," 1. Devices with Liquid-Metal Work Surface, Voprosy Atomnoy Nauki i Techniki, *Seria Thermonuclear Synthesis*, 1980, Vol. 2, No. 6, pp. 42.

Effects of a Vertical Magnetic Field on Rayleigh-Bénard Convection in Mercury

Joel Stavans*
Weizmann Institute of Science, Rehovot, Israel

Abstract

An experimental study of the effects of uniform vertical magnetic fields on the convective motion of an electrically conducting fluid is presented. The experiment has a Rayleigh-Bénard geometry, and mercury is used as the convective fluid. The convection patterns are visualized by mapping out the temperature field at the upper surface of the mercury with a layer of liquid crystals. The effects of the field on the onset of convection and wavelength selection of straight roll patterns, as well as their stability, are studied.

Introduction

The study of flows of electrically conducting fluids in magnetic fields has been motivated by geophysical and astrophysical questions. A notable example is the way the convective layers and magnetic fields in the sun interact to produce sunspots. In this experiment the influence of a magnetic field on a convective layer with Rayleigh-Bénard geometry is studied, with mercury used as the convective fluid. Rayleigh-Bénard convection is a well-understood problem. Analytical and numerical calculations can be performed, and theoretical models for it agree well with experiment. Moreover, such an experiment can be performed with good experimental accuracy.

The orientation of the magnetic field with respect to the horizontal convective layer is a very important parameter. This is already apparent at the level of a linear stability analysis of the equations of motion.[1]

Whereas a horizontal field only imposes its orientation on convective rolls at onset, a vertical field affects the onset itself as well as the wavelength of the pattern. As the magnitude of the vertical field is increased, the onset is pushed up in Rayleigh number, and the wavelength of the pattern decreases. These

---

Copyright © 1991 by the American Institute of Aeronautics and Astronautics, Inc. All rights reserved.
* Senior Scientist, Department of Electronics.

assertions have been verified experimentally in the case of rigid-bottom stress-free-top boundary conditions, where the convection pattern is cellular.[2,3] No study above the onset was performed at that point. The effects of horizontal fields on instabilities of convection rolls have been investigated theoretically[4] and experimentally.[5,6]

Recently there has been renewed interest in convection patterns from the point of view of nonlinear phenomena. In particular, there is an interest in studying the different states of the convection pattern as a function of the Rayleigh number.

I have studied the effects of vertical magnetic fields on a convection layer of mercury at and above onset. The following differences exist between my experiment and those previously reported[2,3]: 1) My experiment was performed with rigid bottom and top boundary conditions. Visualization of the convective patterns was possible by mapping the temperature field on the upper surface of the convection layer with a layer of cholesteric liquid crystals. 2) I studied the ways in which patterns of convective rolls were destabilized when the Rayleigh number was increased above the onset.

To my knowledge, the only theoretical results to date on this problem are those of Busse and Clever.[7] However, those authors only examined the case of very small Chandrasekhar number (the nondimensional parameter through which the magnitude of the magnetic field enters the problem). With my experiment I am able to probe regimes with a much larger Chandrasekhar number.

Theoretical Aspects

In a Rayleigh-Bénard experiment, a layer of fluid is bounded from top and bottom by two horizontal plates. The temperature of each of the plates is uniform, and the temperture of the lower plate is higher than that of the upper plate. When the temperature gradient is below a certain threshold ΔT_c, the transport of heat through the layer is by diffusion. Above this threshold, convection sets in and the heat transport is enhanced. In addition, in my experiment I impose a vertical uniform magnetic field upon the convection layer. The theoretical description of the problem starts with the Navier-Stokes momentum equation, the Fourier heat equation, and an evolution equation for the magnetic field obtained from

$$1/\rho(\partial_t u + (u \cdot \nabla)u) = -\nabla p + Ra\, T\, k + P_m Q\, (\nabla \times H) \times H + \nabla^2 u \quad (1)$$

$$\partial_t T + (u \cdot \nabla)T = \nabla^2 T \quad (2)$$

$$\partial_t H = \nabla \times (u \times H) + P_m \nabla^2 H \quad (3)$$

These equations are supplemented with the requirements $\nabla \cdot u = 0$ and $\nabla \cdot H = 0$. The pressure and temperature fields (p and T respectively) appearing in these equations are not the real physical fields but are modified variables. The vector k is a unit vector in the vertical direction. In obtaining the equation of evolution of the magnetic field from Maxwell's equations, the displacement current term has been neglected. Its inclusion adds corrections of order $(u/c)^2$. There are four nondimensional numbers in these equations: 1) the Rayleigh number Ra defined by $\alpha g d^3 \Delta T / \nu \kappa$, where α is the coefficient of volume expansion, g the acceleration of gravity, and ν the kinematic viscosity; 2) the Prandtl number Pr, defined by $Pr = \nu/\kappa$; 3) the Chandrasekhar number Q, defined by $Q = H^2 d^2 / \rho \eta \mu \nu$, where ρ is the density μ the vacuum permeability, and η the magnetic diffusivity (the magnetic diffusivity is defined by $\eta = 1/\sigma\mu$, where σ is the electrical conductivity); and 4) the magnetic Prandtl number P_m, defined by η/κ. Notice that when $\eta = 0$, the fluid is a perfect conductor, $P_m = 0$, and no diffusive term appears in the magnetic field evolution equation. In this case, the simplified equation is exactly the same as the equation governing the vorticity in an inviscid fluid. Hence, as the vorticity in the inviscid case, the magnetic field lines are frozen into the fluid. On the other hand, when $P_m \neq 0$, the magnetic field evolves by diffusion as well.

A comparison of the two terms in the right side of the magnetic field evolution equation yields the magnetic Reynolds number Re_m, defined by $Re_m = U/P_m$, where U is a nondimensional velocity scale. Returning to dimensional quantities, it is easy to verify that $Re_m = Vd/\eta$, where V has dimensions of velocity. When Re_m is large, the magnetic field lines are considerably distorted by the velocity field. This situation is prevalent in most geophysical and astrophysical situations. On the other hand, when Re_m is small as in experiments, the field dynamics is dominated by diffusion and the magnetic field in the fluid is slightly different from the applied one.

Chandrasekhar[1] has analyzed these equations at the onset of convection within a linear approximation. He showed that both the onset itself and the wavelength of the pattern are functions of the magnetic field. The onset is pushed up due to the extra mechanism of energy dissipation, namely, the ohmic losses. These losses are produced by the electrical currents induced in the fluid as the fluid moves in the magnetic field. At the same time, the wavelength of the pattern decreases.

Busse and Clever[7] performed a Galerkin analysis of the full nonlinear problem in the case of mercury and sodium. These fluids are characterized by a very small value of the Prandtl number. In particular, for mercury, $Pr = 0.025$. For small Prandtl numbers the dominant nonlinearity in the equations of motion is the $(u \cdot \nabla)u$ term in the Navier-Stokes equation. This term is responsible for the existence of large-scale flows via the generation of small spatial harmonics of the roll periodicity.[9,10] Cross[11] has shown that these flows lead to the skewed varicose[12] and oscillatory instabilities.[13] The skewed varicose instability is a stationary deformation of rolls in which their thickness varies along their axes. The marginal line of this instability limits the stability domain of straight rolls from the large wave number side. The oscillatory instability is a time-dependent periodic mode, in which transverse waves travel along the rolls axes. The typical time scale of the instability is $\tau = d^2/\kappa$, and it appears via a Hopf bifurcation. How are these two instabilities affected by the field? Busse and Clever[7] performed their calculation for Q=20, which is a relatively small value of the Chandrasekhar number. Their results show that for this value of Q, the onset of the oscillatory instability is pushed up by a factor of nearly two relative to the onset of convection. At the same time, the marginal line of the skewed varicose instability is bent toward low wave vectors for a sufficiently high Rayleigh number. Given the fact that most experiments up to date (including ours) can reach a much higher value of Q, a more complete theory is lacking.

Experimental Systems

Figure 1 shows a top and side view of our experimental setup. The bottom plate of the convection cell was made of copper. A heating coil was attached to it from the bottom. The temperature of the plate was measured by a platinum thermistor implanted in the copper with STYCAST. Another thermistor was implanted in the copper, level with the surface in contact with the mercury, so that a local thermal signal could be extracted from the system. The surface of the copper was protected from the mercury by a thin layer of paint. The top plate of the convection cell was made of 1- mm - thick sapphire. The liquid crystals were in thermal contact with the sapphire and mercury. The sapphire's upper surface was cooled by a temperature-controlled turbulent flow of temperature-controlled water (with a long-term stability of 0.02°C). In order to cool homogeneously the sapphire plate, the turbulent water flow was created by 12 equally spaced jets ejected radially into a cooling chamber with stainless steel walls as shown in the top part of Fig. 1. The water was taken out of the cooling chamber from 12 evenly spaced holes as well. The cooling chamber was bounded from the top by a plexiglass window through which the patterns were visualized. The lateral walls of the cell were made of plexiglass. In order to avoid the radial thermal gradient in the edge of the cell, a stainless steel 304 frame was introduced as shown in Fig. 1. This gradient is due to the low thermal conductivity of plexiglass relative to that of mercury. The frame has a thermal conductivity very near that of mercury.

Since mercury is a liquid metal, the temperature difference between hot and cold regions across its surface is small. In order to map the temperature field in the upper surface of the convecting layer, we used commercially available cholesteric liquid crystal sheets active between 14 and 15°C. All of the color spectrum appeared within these two temperatures. The temperature resolution we attained with these sheets was 0.1°C. The sheets used were 7.5×10^{-6} m thick.

In our experiment we used two circular cells differing only in their thickness. We define the aspect ratio G as the ratio between the radius and d. The radius of both cells was 40 mm, and the thickness of the cells was d =6 mm in one case and 8 mm in the other. The stainless steel frame had an outer radius of 40 mm and an inner radius of 35 mm. Taking the stainless steel frame into account, the aspect ratios of our circular cells were G=4.38 and 5.83. The cell as well as the cooling chamber were enclosed in a vacuum can to provide thermal

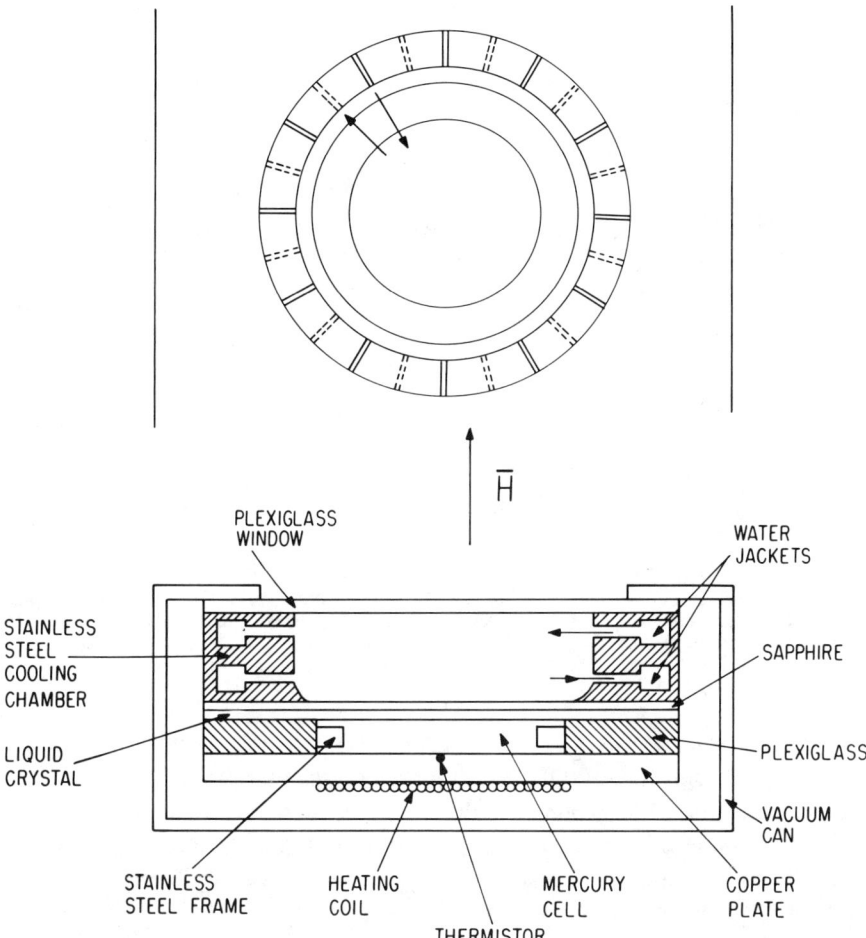

Fig. 1 Top and side views of the experimental setup. In the top part of the figure arrows show the influx and outflux holes through which water flows in the cooling chamber. The innermost circle corresponds to the convective part of the cell (excluding the region lying on top of the stainless steel frame).

isolation. In the experiment only the temperature of the sapphire was controlled, and a constant current was supplied to the heater in the bottom copper plate. Therefore, the power passing through the mercury was effectively controlled. The stability of the setup just described was 20 mK over 2 h. The whole system was held between the poles of a magnet that could rotate in a vertical plane.

Results

In Fig. 2 we use full diamonds to show the Rayleigh number at the onset of convection $Ra_c(Q)$ as a function of the Chandrasekhar number Q. The onset was determined by measuring the thermal conductivity across the mercury layer. The

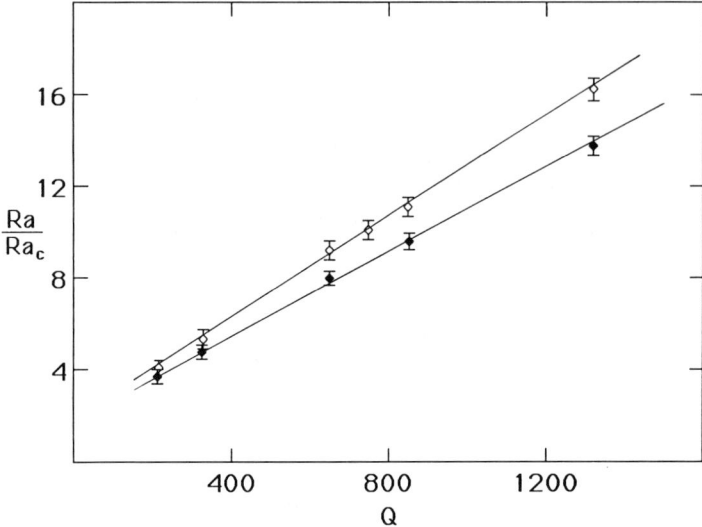

Fig. 2 Parameter space (Ra, Q) obtained from our visualization experiment. Filled diamonds represent the onset of convection $Ra(Q)$, whereas empty diamonds represent the onset of time-dependent pattern instability through the nucleation of defects and their diffusion. Solid lines serve as a guide to the eye.

conductivity is independent of $Ra(Q)$ for $Ra(Q) < Ra_c(Q)$ (heat is transfered by diffusion), whereas for $Ra(Q) > Ra_c(Q)$ it increases linearly with $Ra(Q)$ due to the enhancement of heat transport due to convection. The value of $Ra(Q)$ at which the slope of the conductivity changes determines $Ra_c(Q)$. The line through the experimental points is just a guide to the eye. Within experimental error, $Ra_c(Q)$ increases linearly with Q in agreement with the predictions of linear stability analysis.[1]

Because of the finite resolution of the liquid crystals (0.1°C) and the relatively good thermal conductivity of mercury, we were unable to observe the convection patterns at or just above onset. In order to determine the wave vector a near the onset, we increased $Ra(Q)$ until a pattern was resolved. Typically this occurred when $\varepsilon=0.1$, where $\varepsilon=[Ra(Q)-Ra_c(Q)]/Ra_c(Q)$. In Fig. 3 we use full circles to show the results of these measurements in the range of Q from 0 to 1320. The wave vector is given in units of 1/d. Chandrasekhar's results for rigid-top and bottom boundary conditions are included for comparison (empty circles). The measured values of the wave vector a are smaller than the theoretical ones. There are two possible reasons for this discrepancy: 1) Our measurements were performed for finite ε, whereas the theory is concerned with wavelength selection at the onset of convection. 2) We use a poor thermal conductor for the upper boundary (in comparison to the conductivity of the lower copper plate). As Riahi[15] pointed out, this leads to a decrease in the wave vector proportional to some power of the thermal conductivities ratio between fluid and

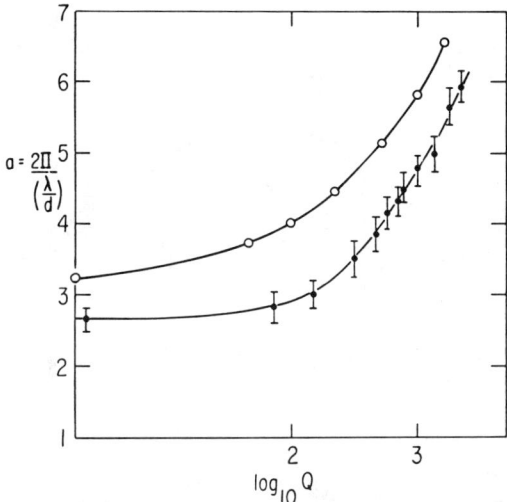

Fig. 3 Dependence of the wave vector of straight roll patterns as a function of the Chandrasekhar number Q. Circles correspond to the theoretical predictions based on high thermal conductivity boundary conditions. Dots represent our results. Solid lines serve as a guide to the eye.

boundary. Despite this discrepancy, the evolution of the wavelength with the Chandrasekhar number Q follows the theoretical one for large values of Q: Whereas the theory predicts the wave vector to grow as $a \sim Q^\gamma$ with $\gamma=1/6$, we obtain $\gamma_{exp} = 0.18 \pm 0.02$ from a least-squares fit to the high-Q portion of our data.

When the Rayleigh number is increased above $Ra_c(Q)$, eventually some value of $Ra_i(Q)$ is reached above which patterns of straight stationary rolls become unstable. This is shown in Fig. 2 by empty diamonds. The line through the experimental points is just a guide to the eye. The salient feature in Fig. 2 is the enhancement of the stability domain of straight stationary roll patterns: As Q increases, $Ra_i(Q) - Ra_c(Q)$ increases. We now describe the nature of this instability. In Fig. 4, we show a time sequence of the system taken at fixed Rayleigh and Chandrasekhar numbers. In the photographs, light regions correspond to downflowing cold currents, whereas dark regions correspond to upflowing hot currents. Therefore, the boundaries between light and dark regions give the positions of individual rolls. Figure 4 was obtained at the onset of time dependence for Q=1319 (ε=0.19). The evolution in this figure starts by the pinching of a downflowing cold current, which creates a dislocation in the convection pattern. Subsequently, the dislocation propagates at 45 deg to the roll's direction through climb and glide motions by pinching and reconnection, eventually crossing all of the pattern. The final pattern in Fig. 4 eventually destabilizes again, without further change in experimental parameters, and the time evolution of the system proceeds through the same mechanism, ending up in rolls that have different directions. This continuing formation and destabilization of straight roll patterns of different orientations is erratic.

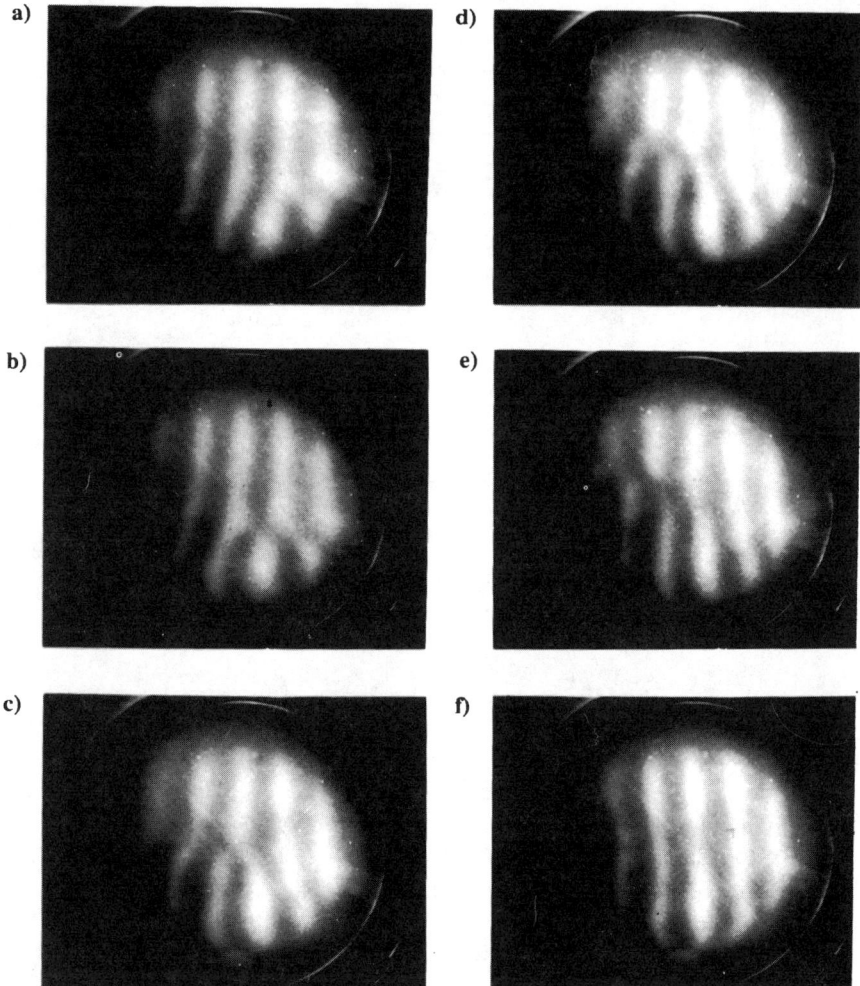

Fig. 4 Climb and glide motion of a dislocation above the onset of time dependence. The dislocation moves at 45 deg to the direction of the roll. These convection states correspond to $\varepsilon = 0.19$ and $Q=1319$.

In conclusion, pinching seems to be the main mechanism through which straight roll patterns are destabilized in large-aspect-ratio cells. The pinching occurs in the bulk of the convective structure, creating dislocation pairs that diffuse in the convective structure by glide and climb motion. Once defects are present, the patterns evolve in time in an erratic fashion.

Increasing $Ra_i(Q)$ above $Ra_i(Q)$, we have found the oscillatory instability, by detecting it with the thermistor in the lower plate. However, the instability is modulated by the lower-frequency signals caused by defects diffusing in the convective structure. The pinching and the ensuing diffusion of defects are much stronger degrees of freedom than the relatively small deformations in the rolls characteristic of the oscillatory instability. In order to study cleanly the effects of the magnetic field on the oscillatory instability, one has to freeze the creation of defects by working with a rectangular cell of smaller aspect ratio. Stavans and Libchaber[18] have done this, and their preliminary studies show a much weaker dependence on Q of the onset of the instability relative to the onset of convection than the theory of Busse and Clever predicts.

To summarize, I have studied the effects of a vertical homogeneous magnetic field on the convection of an electrically conducting fluid. Overall, the field has a stabilizing effect. This is manifested in the increase of both the onset of convection as well as in the stability domain of straight rolls with the field. The enhanced stability of straight rolls is due to the damping by the magnetic field of large-scale horizontal flows, which happen to be very strong in the small-Prandtl-number limit. The onset of convection increases within experimental error linearly with Chandrasekhar number, in agreement with a linear stability analysis.

We have studied the problem of wavelength selection in the presence of the magnetic field. For large values of the Chandrasekhar number, the theory predicts a power law behavior $a \sim Q^{\gamma}$ with $\gamma=1/6$, and my experimental results agree with this: $\gamma_{exp}=0.18 \pm 0.02$. However, the values for the wave vector obtained in the experiment are consistently smaller than those predicted by the theory. The reasons for this are the thermal boundary conditions used in the experiment and the fact that my accuracy allows me to observe rolls a finite amount away from onset. In contrast with this, the theory deals with the problem of wavelength selection at the marginal line. I conjecture that for sufficiently high values of Q, the roll solution must break down. When Q is very large, the rolls are very thin. In this case, horizontal diffusion of heat from upflowing fluid to downgoing fluid within the bulk can be a very important effect.

The main destabilization mechanism of straight roll patterns in my experiment (large aspect ratio) seems to be associated with the skewed varicose instability. When the Rayleigh number is increased above the critical one, patterns eventually destabilize by pinching of rolls in the bulk and evolve erratically through the motion of defects in the structure.

Acknowledgments

This work was supported by the National Science Foundation under Grant DMR-83-16204 and by the Materials Research Laboratory at the University of Chicago under National Science Foundation Grant DMR-82-16892.

This work was performed while the author was a Ph.D. student at the University of Chicago under the supervision of Prof. Albert Libchaber. Professor Lichaber's help and constant support are greatly acknowledged.

References

[1]Chandrasekhar, S., *Hydrodynamic and Hydromagnetic Stability,* Oxford Univ. Press, New York, 1961, Chap 4, pp. 146-195.

[2]Nakagawa, Y., "Experiments on the Inhibition of Thermal Convection by a Magnetic Field," *Proceedings of the Royal Society of London,* Series A, Vol. 240, April 1957, pp. 108-113.

[3] Lehnert, B., and Little, N.C., "Experiments on the Effect of Inhomogeneity and Obliquity of a Magnetic Field in Inhibiting Convection," *Tellus*, Vol. 9, Feb. 1957, pp. 97-103.

[4] Busse, F. H., and Clever, R.M., "Stability of Convection Rolls in the Presence of a Horizontal Magnetic Eld," *Journal de Mécanique Théorique et Appliquée*, Vol. 2, No. 4, 1983, pp. 495-500.

[5] Fauve, S., Laroche, C., and Libchaber, A., "Effect of Horizontal Magnetic Field on Convective Instabilities in Mercury," *Journal Physics Letters Paris*, Vol. 42, No. 21, Nov. 1981, pp. 455-457.

[6] Fauve, S., Laroche, C., Libchaber, A., and Perrin, B, "Chaotic Phases and Magnetic Order in a Convective Fluid," *Physical Review Letters*, Vol. 52, No. 20, May 1984, pp. 1774-1777.

[7] Busse, F. H., and Clever R.M., "Stability of Convection Rolls in the Presence of a Vertical Magnetic FIeld," *Physics of Fluids*, Vol. 25, June 1982, pp. 931-935.

[8] Busse, F. H., "Non-linear Properties of Thermal Convection," *Reports on Progress in Physics*, Vol. 41, No. 12, 1978, pp. 1929-1967.

[9] Zippelius, A., and Siggia, E. D., "Stability of Finite-Amplitude Convection," *Physics of Fluids*, Vol. 26, Oct. 1983, pp. 2905-2915.

[10] Croquette, V., Le Gal, P., Pocheau, A., and Guglielmetti, R., "Large-Scale Flow Characterization in a Rayleigh-Bénard Convective Pattern," *Europhysics Letters*, Vol. 1, No. 8, April 1986, pp. 393-399.

[11] Cross, M. C., "Phase Dynamics of Convective Rolls," *Physical Review A*, Vol. 27, Jan. 1983, pp. 490-498.

[12] Busse, F. H., and Clever, R. M., "Instabilities of Convection Rolls in a Fluid of Moderate Prandtl Number," *Journal of Fluid Mechanics*, Vol. 91, Part 2, 1979, pp. 319-335.

[13] Busse, F. H., "The Oscillatory Instability of Convection Rolls in a Low Prandtl Number Fluid," *Journal of Fluid Mechanics*, Vol. 52, Part 1, 1972, pp. 97-112.

[14] Stavans, J., "Experimental Study of Quasiperiodicity in a Hydrodynamical System," *Physical Review A*, Vol. 35, May 1987, pp. 4314-4328.

[15] Riahi, N., "On Convection with Nearly Insulating Boundaries in a Low Prandtl Number Fluid," *Zeitschrift fuer Angewandte Mathematik und Physik*, Vol. 31, March 1980, pp. 261-266.

[16] Pocheau, A., Croquette, V., and Le Gal, P., "Turbulence in a Cylindrical Container of Argon Near Threshold of Convection," *Physical Review Letters*, Vol. 55, No. 10, Sept. 1985, pp. 1004-1007.

[17] Oswald, P., Stavans, J., and Libchaber, A., private communication.

[18] Stavans, J., and Libchaber, A., private communication.

MHD Flow Around a Cylinder in an Aligned Magnetic Field

J. Josserand,* Ph. Marty,† and A. Alemany‡
Institut de Mécanique de Grenoble, Grenoble, France
and
G. Gerbeth§
Central Institute for Nuclear Research, Rossendorf, Germany

Abstract

RECENT results on the study of a liquid metal flow around an insulating cylinder with constant aligned magnetic field are presented. From the experimental point of view, a special type of differential pressure transducer using strain gauges is described. The results obtained with mercury as liquid metal are presented for an interaction parameter N ranging from 0 to approximately 8. The stabilizing effect of the magnetic field on the boundary layer separation is shown. Pressure distribution around the cylinder as well as the overall pressure drag coefficient C_D are given for different values of N. The last section presents analytical calculations of the flow distribution of an inviscid fluid when $N \ll 1$. The theoretical results are in good agreement with these experimental results.

Introduction

Previous works[1] on MHD flow around an insulating cylinder in magnetic field aligned with the flow have pointed out the stabilizing effect of the magnetic field on the boundary layer separation behind

---
Copyright © 1991 by the American Institute of Aeronautics and Astronautics, Inc. All rights reserved.
* Ph.D. Student, Fluid Mechanics Department, P.O. Box 53 X, 38041.
† Senior Researcher, Fluid Mechanics Department.
‡ Head, Fluid Mechanics Department.
§ Senior Researcher, Sodium Engineering Department, P.O. Box 19, O-8051.

the obstacle for small values of the interaction parameter N. From the theoretical point of view, an Oseen-type approximation has been used. Despite the weakness of such an assumption near the obstacle, very interesting features of the flow have been shown; the velocity in the upstream region has been found to be less than that of the potential flow, whereas at the rear of the cylinder, the flow is faster than the potential flow. Although these calculations do not introduce any separated flow in the downstream region, the boundary layer separation could be expected to occur further downstream. Measurements of velocity and magnetic field perturbations have confirmed these tendencies, and the stabilizing effect of the magnetic forces on the boundary layer has been observed through the disappearance (under certain conditions) of the Von Karman street in the wake. The experimental difficulties encountered in the downstream region have prevented the authors from drawing a completely clear picture of the phenomenon occurring. Consequently, the present study will discuss recent experimental investigations of the pressure distribution around a cylinder with magnetic field aligned with the flow in order to obtain new information on the flow structure. Furthermore, a perturbation method that has been developed to obtain the solution of the problem for small values of the interaction parameter will be presented in the last section.

Experimental Apparatus

One part of the apparatus has been previously described in Ref.1. It is composed of a mercury tank, 20cm in diameter and 2.5m high, located inside a solenoïdal coil. The cylinder is attached to a moving carriage whose velocity can be adjusted between 0 and 1 m/s. A stainless steel cylinder, 34mm in diameter and 112mm long, is fitted to the moving carriage (Fig. 1). The cylinder is insulated by a thin epoxy resin coating and consists of two hollow pieces, one stationary and the other able to rotate, in each of which a 1.5 mm pressure tap is drilled (Fig. 2). The position of one of the pressure taps could thus be varied from zero to 180 deg. with respect to the other, itself aligned with the upstream stagnation point. The hollow parts of the two pieces are separated by a 0.02 mm thick stainless steel diaphragm on which are glued a full bridge of Vishay strain gauges. The signal provided by this differential pressure transducer is finally fed into an electric amplifier and then to an oscilloscope or to an integrating voltmeter.

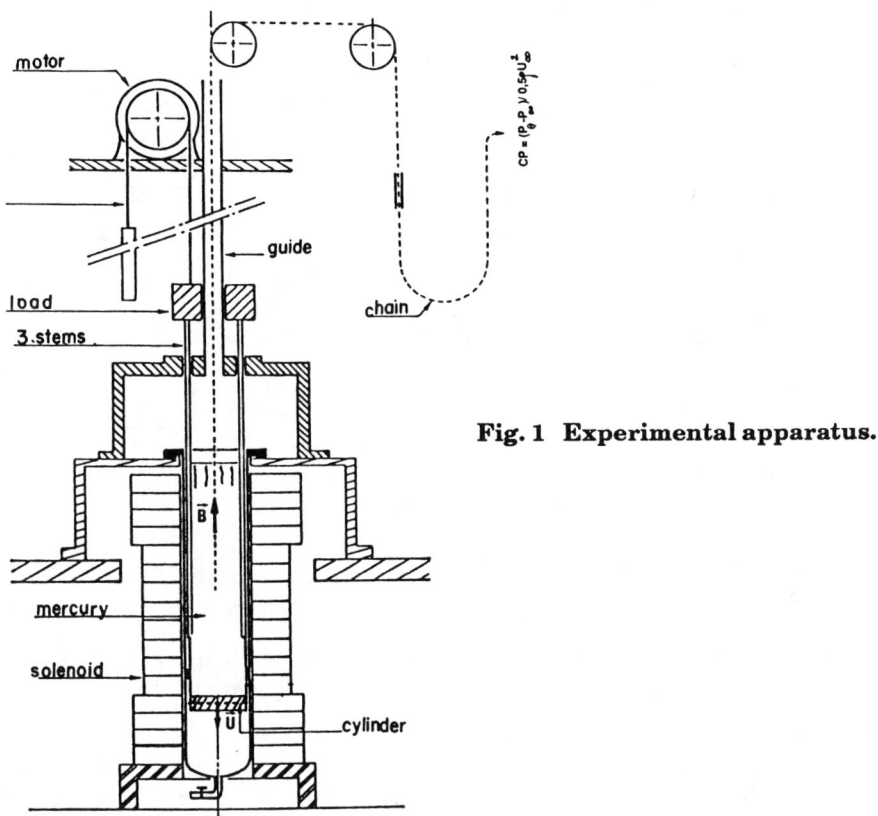

Fig. 1 Experimental apparatus.

Experimental Results

In a first stage the reliability of the experimental apparatus was checked by carefully analyzing the pressure distribution measured with B = 0 (see the N = 0 curve on Fig. 3). Integrating these results leads to a pressure drag coefficient $C_D = 0,8$ (C_D is herein defined as the overall drag coefficient : $D/1/2.\rho.V_\infty^2.d.L$) that is reasonable agreement with experimental results available in the classical hydrodynamic literature[2].

This device is also capable of showing the pressure oscillations due to the formation of Von Karman eddies. The Strouhal number $S = f.d/V_\infty$, where f, d=2a, and V_∞ are the frequency of pressure oscillations, the cylinder diameter, and the undisturbed velocity, respectively, has been found equal to 0.22 for a Reynolds number Re = V.d/ν ranging from 2.10^4 to 5.10^4 (Fig. 4) and, consequently, is in

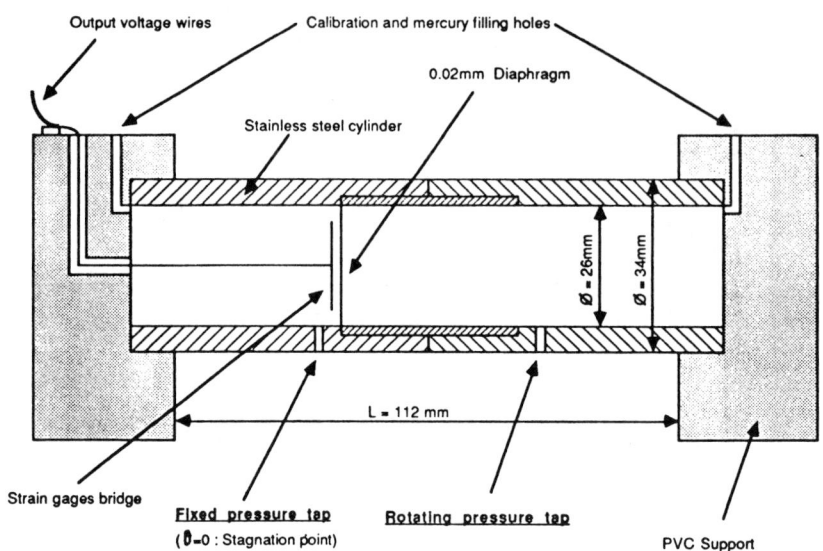

Fig. 2 Sketch of the cylinder with the measurement device.

good agreement with the well-known results of Roshko[3]. Figure 3 shows the angular pressure distribution for several values of the interaction parameter $N = \sigma a B^2 / \rho V_\infty$ ranging from 0 to 8.2.

At the front of the cylinder, the pressure continuously increases with N, indicating the presence of a stagnant fluid region developing near the stagnation point at $\theta = 0$ for large N. This result agrees with what was found at the front of a sphere or a semi-infinite Rankine body in an aligned magnetic field by Maxworthy[4,5]. This behaviour consequently has the effect of increasing the contribution to drag from the front part of the cylinder as N becomes larger. At the rear of the cylinder the pressure is almost constant downstream of the separation angle θ_{sep}, located near 76 deg. for N = 0 and increasing to approximatively 105 deg. for N = 8 (Fig. 5). Contrary to what happens in the upstream part, the base pressure initially increases with N due to the reattachment of the boundary layer. This can be clearly seen on Fig. 6 where pressure histories at $\theta = 90$ deg. with and without magnetic field have been plotted. One can see the suppression of the Von Karman eddies by the magnetic forces, which confirms the reattachment phenomenon. As N is increased for values greater than about one, though, the rearward pressure continuously decreases because of the losses in total pressure suffered by streamlines crossing the magnetic field lines and undergoing Joule effect.

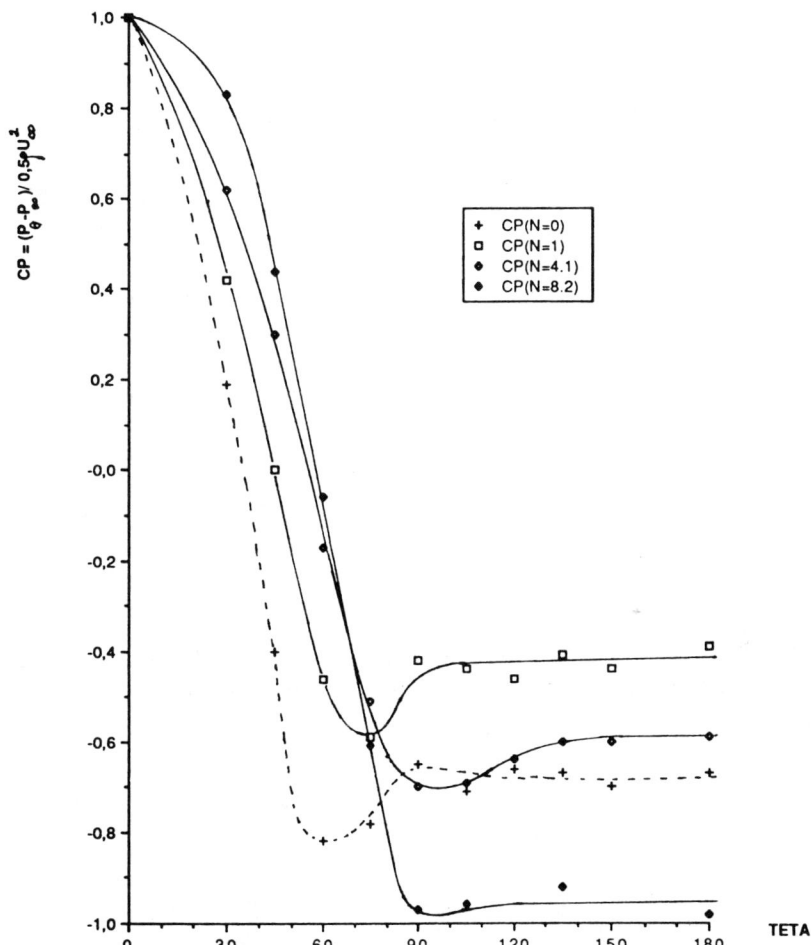

Fig. 3 Angular distribution of the pressure coefficient C_P at the cylinder surface for several values of the interaction parameter (N=0, 1, 4.1, 8.2).

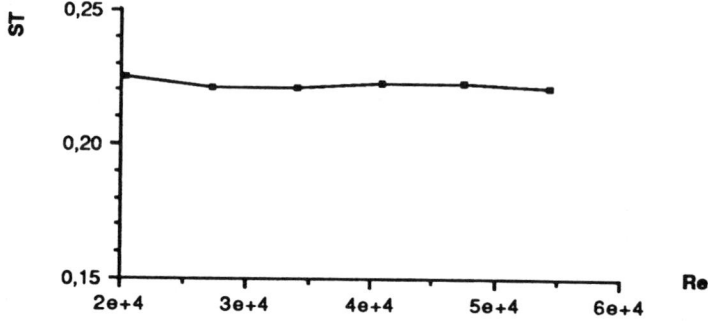

Fig. 4 Strouhal number evolution vs the Reynolds number (B=0).

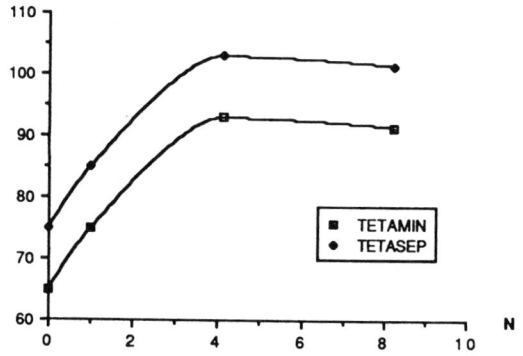

Fig. 5 Minimum pressure angle (θ_{min}) and separation angle (θ_{sep}) evolution vs the interaction parameter N (the separation angle is defined as $\theta_{sep}=\theta_{min}+ 10$ deg4).

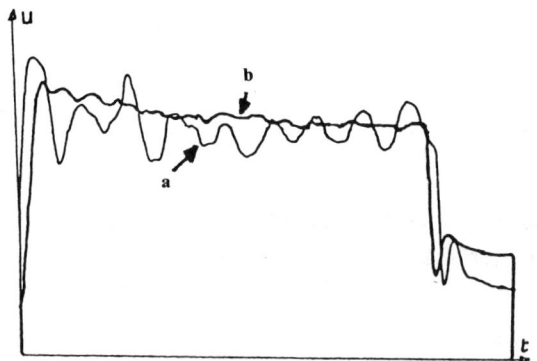

Fig. 6 Typical pressure signal at $\theta= 90$deg: 1) curve a (B=0) shows the oscillations of the Von Karman street; 2) curve b (B=0.325 T) shows the suppression of the VK street due to the boundary layer reattachment.

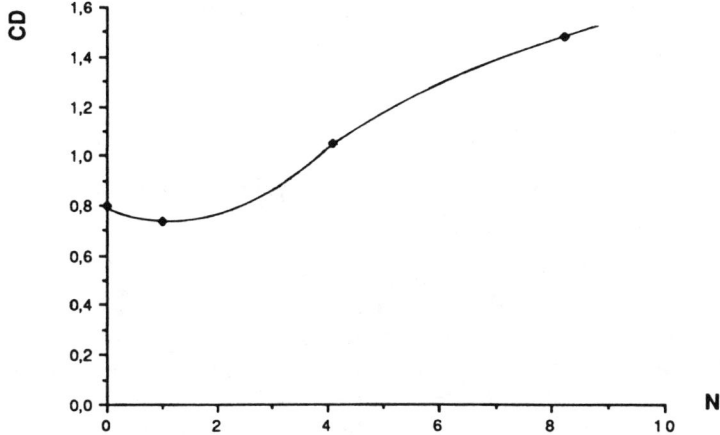

Fig. 7   Pressure drag coefficient evolution vs the interaction parameter.

Figure 7 shows the evolution with N of the pressure drag coefficient C_D previously defined and related to the pressure coefficient by :

$$C_D = \int_0^\pi C_P \cdot \cos\theta \cdot d\theta$$

Between N = 0 and approximately 1.5, C_D undergoes a slight decrease of 7 %, which confirms the previous assumption about the boundary layer reattachment. This is also similar to what was found on a sphere by Motz[6] who noticed that for N ≈ 0.1, C_D was 10 % lower than the potential flow prediction.

For higher values of N, the drag coefficient increases under the influence of the suction effect of the rearward separated flow. The $N^{1/2}$ dependence of C_D has not been clearly shown due to the fact that the maximum value achievable on our facility was N = 8. In this respect we know from previous work on disk[7] or spheres[4] that the high strong field regime only begins to be obtained for N > 10.

Theoretical Aspects

In order to calculate the pressure distribution on the cylinder surface, a two-dimensional flow of a conducting fluid past a cylinder is considered with an external magnetic field B_0 aligned to the uniform flow velocity V_∞ at infinity (Fig. 8). The performed experiments are characterized by the following order of magnitude of nondimensional parameters:

1) Reynolds number $Re = 10^4$ to 10^5.
2) magnetic Reynolds number $Rm = \mu_0 \sigma U_\infty a < 0.01$.
3) interaction parameter $N = 0,..., 8$.

In these expressions μ_0, σ, ρ and ν denote the magnetic permeability, electrical conductivity, density, and kinematic viscosity, respectively, of the liquid metal. The typical length is the cylinder radius a. The values of the parameters suggest the following simplifications of the theoretical descriptions:

1) The external magnetic field is considered as undisturbed (inductionless approximation).
2) The fluid viscosity is neglected. In this way only the outer, inviscid flow is determined which in turn should be an input value of a boundary layer calculation. With this approximation it is clear from the beginning that the obtained surface pressure distribution will be valid at most at the front of the cylinder before the point of separation. On the other hand, it should be noted that the existing papers on MHD flow separation[8] and references in Ref.8 took into account the electromagnetic force in the boundary layer equations only and neglected the influence of the magnetic field on the inviscid external flow. This is obviously not correct.

With these assumptions the Navier-Stokes equation reduces to

$$\overrightarrow{\text{grad}} \, P = \vec{V} \wedge \vec{\omega} - N \, \vec{V}.\vec{e}_y \qquad (1)$$

with $P = p + V^2/2$ as dynamic pressure and $\omega = \text{curl } V = \omega.e_z$ as vorticity. All quantities are made nondimensional using a, V_∞, ρV_∞^2 and $\sigma V_\infty B_\infty$ as units for the length, velocity, pressure and electric current, respectively. Multiplying Eq. (1) with v yields the generalized Bernoulli law for the inviscid MHD flow in an aligned magnetic field

$$\vec{V}.\overrightarrow{\text{grad}} \, P = - Nj^2 = - Nv_y^2 \qquad (2)$$

This means that the dynamic pressure is a constant along streamlines on which $j = - v_y = 0$ and decreases by the amount of the integrated Joule losses along all other streamlines. In particular, the dynamic pressure on the cylinder surface must decrease monotonously from the front to the rear stagnation point. The surface pressure is correlated to the velocity on the cylinder surface

by
$$\frac{\partial P}{\partial \varphi} = - N \cdot V_\varphi \cdot \cos^2 \varphi \text{ at } r = 1 \tag{3}$$

Moreover, the dynamic pressure is constant along $\varphi = \pi$ **due to the flow symmetry.**

Equation (1) was solved by Lahjomri et al.[1] using an Oseen-type approximation, which gives a correct description of the flow-field far from the body. However, this approach becomes incorrect in the vicinity of the cylinder. Here we follow the approach of Tamada[9] who solved Eq. (1) for $N \ll 1$ by use of a series expansion

$$\vec{V} = \vec{V}_0 + N \vec{V}_1 \tag{4a}$$
$$p = p_0 + N \cdot p_1 \tag{4b}$$

where $V_0 = \text{grad} \Phi$; $\Phi = (r + 1/r) \cdot \cos\varphi$ denotes the undisturbed inviscid flow with $\omega_0 = \text{curl } V_0 = 0$. The first correction to the dynamic pressure at the cylinder surface is determined by $P_1(\varphi=\pi) = 0$ and Eq. (3) to

$$P_1(\varphi) = p_1 + V_{0\varphi} \cdot V_{1\varphi} = -\frac{2}{3}(1 + \cos 3\varphi) \tag{5}$$

In order to obtain the static pressure, the first correction $V_{1\varphi}$ to the surface velocity has to be determined. This is accomplished here via the calculation of the first-order contribution ω_1 to the vorticity. Substituting Eq. (4) into the curl of Eq. (1) leads to

$$(\vec{V}_0 \cdot \vec{\text{grad}}) \omega_1 = -\frac{\partial V_{0y}}{\partial x} = \frac{-2}{r^3} \sin 3\varphi \tag{6}$$

This equation was solved by Tamada[9] using the method of characteristics and $\omega_1(\varphi=\pi) = 0$. The result is

$$\omega_1 = \varphi + \sin 2\varphi - \pi + k(3 + 4k^2)I_2 - 4k \cdot I_1 \tag{7}$$

where

$$I_1 = \int_\varphi^\pi \sqrt{k^2 + \sin^2\alpha} \, d\alpha \; ; \; I_2 = \int_\varphi^\pi \left(\sqrt{k^2 + \sin^2\alpha}\right)^{-1} d\alpha$$

with $k = 1/2(r - 1/r) \cdot \sin\varphi$ ($0 \leq \varphi \leq \pi$). On the cylinder surface one obtains

$$\omega_1(\varphi) = \varphi + \sin 2\varphi - \pi \tag{8}$$

which means that ω_1 is discontinuous across the positive half of the x-axis (ω_1 is an odd function with respect to the x-axis). This implies the existence of a vortical trail behind the cylinder, which was discussed in detail by Tamada[9]. It should be noted that ω_1 is always negative in the near vicinity of the cylinder for $0 \leq \varphi \leq \pi$. This implies that the tangential velocity on the cylinder surface is everywhere decreased compared to the undisturbed inviscid flow. At great distances from the cylinder and outside the trail the leading term

$$\omega_1 (r \gg 1) = \frac{\sin 2\varphi}{r^2} \qquad (9)$$

follows from Eq.(7) in complete agreement with the Oseen-type solution of Lahjomri[1].

The velocity field is obtained by introduction of a stream function $V_x = -\partial \psi/\partial y$, $V_y = \partial \psi/\partial x$ and solution of the Poisson equation $\Delta \psi = \omega_1$. This was not carried out by Tamada since the expression of Eq. (7) is fairly complicated and it seems to be impossible to find an analytical solution for ψ_1. Because of this we use a Fourier expansion

$$\omega_1 = \sum_{n=1}^{\infty} \gamma_n (r) \cdot \sin n\varphi \qquad (10)$$

to solve Eq. (6). The stream function

$$\psi_1 = \sum_{n=1}^{\infty} \alpha_n (r) \cdot \sin n\varphi \qquad (11)$$

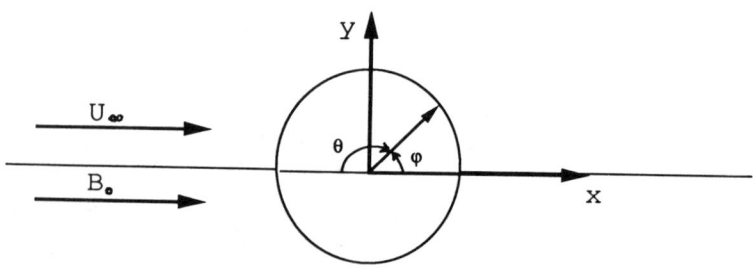

Fig. 8 Definition of the coordinates and angles.

is then determined by the Poisson equation and the tangential velocity on the surface is eventually obtained in the form

$$V_{1\varphi} = \sum_{n=1}^{\infty} S_n \sin n\varphi \qquad (12)$$

where

$$S_n = \alpha'_n (r=1) = -\int_1^{\infty} \gamma_n(r) \cdot r^{1-n} \cdot dr \qquad (13)$$

Details of the calculations are given in the Appendix. It is worth noting that the Fourier series [Eq.(10)] reproduces the vorticity at the surface [Eq.(8)] as well as the leading term [Eq.(9)] at great distances correctly for $0 \leq \varphi \leq \pi$. The pressure loss between front and rear stagnation points due to the generalized Bernoulli law [Eq.(2)] is the source of a force on the cylinder acting in the flow direction. The first contribution to this drag for $N \ll 1$ is obtained by

$$F_1 = -\int_0^{2\pi} p_1 \cdot \cos\varphi \cdot d\varphi = \pi/2 \qquad (14)$$

where the force is in units of $\rho V_\infty^2 a^2$. This is in complete agreement with the result of Zachariah and Singh[10].

The obtained results for the tangential velocity $V_\varphi = V_{0\varphi} + NV_{1\varphi}$ and the surface pressure $p = p_0 + Np_1$ by use of the first six coefficients $S_1, ..., S_6$ are shown in Figs. 9 and 10 for different values of the interaction parameter. Figure 9 shows the previously mentioned decrease of the undisturbed flow everywhere on the cylinder surface. Whether the negative part of V_φ for $N = 0,9$ on the downstream side of the cylinder is realistic or an unjustified extrapolation of the first-order theory for $N \ll 1$ cannot be decided here and needs further investigation. The enhancement of the pressure on the upstream side (see fig. 10) is mainly due to the decrease of the tangential velocity. On the other hand, the pressure decrease at the rear stagnation point is solely caused by the dynamic pressure losses due to the generalized Bernoulli law.

Discussion

Though the theory is only for an inviscid flow with $N \ll 1$ comparison to measurements is interesting, in particular,

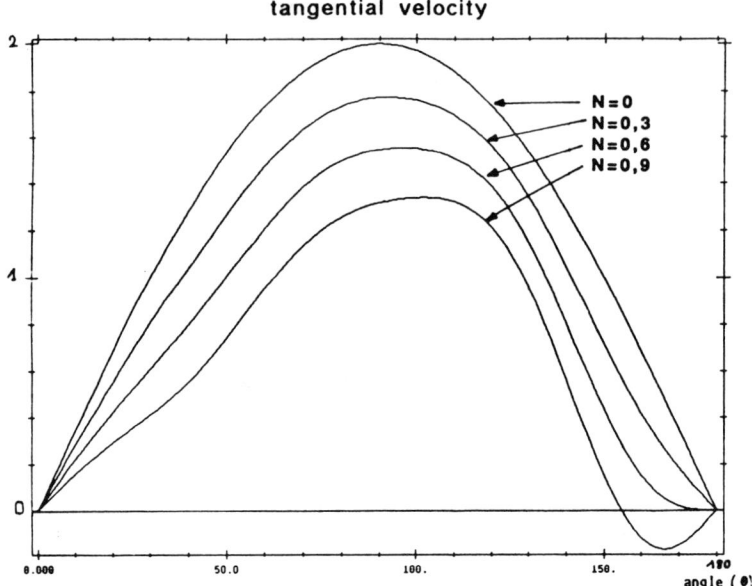

Fig. 9 Tangential velocity vs angle θ for different values of N.

concerning the tendency for increasing N. On the upstream side of the cylinder before separation, there is qualitative agreement between the measured and the calculated pressure for increasing N. The measured pressure distribution around the separation point can be interpreted as a result of two competing effects:

1) Increase of the pressure due to the decrease of the tangential velocity (see Fig. 10).
2) The separation point is shifted downstream with increasing N.

For small values of N the first effect dominates, resulting in the measured increase of the rear stagnation pressure (N = 1). If the interaction parameter is further increased, the second effect becomes more and more important. The minimum of the pressure distribution is shifted downstream and the rear stagnation pressure decreases. The same behavior of the rear stagnation pressure was found by Maxworthy[4] in the corresponding MHD flow around a sphere.

Remark to the Case of a Transverse Magnetic Field

The extension of the calculation to the case of a transverse magnetic field $\mathbf{B} = B_0 \cdot \mathbf{e}_y$ is straightforward. In fact, the corresponding equation for the vorticity is

$$\left(\vec{V}_0 \cdot \overrightarrow{\text{grad}}\right) \omega_1 = \frac{\partial V_{0x}}{\partial y} = + \frac{2}{r^3} \cdot \sin 3\varphi \qquad (6')$$

The difference from Eq. (6) consists in the sign of the right member and, therefore, the first correction to the tangenctial velocity $V_{1\varphi}$ is changed by the sign only. The dynamic pressure on the surface is obtained in the form

$$P_1 = \frac{1}{3} \cdot \left(2 - 3 \cdot \cos \varphi + 2 \cos 3\varphi\right) \qquad (5')$$

This leads to the same result $F_1 = \pi/2$ for the first contribution to the force on the cylinder, as in the aligned field case. It should be emphasized that the front stagnation pressure increases now with N. The results for the surface pressure are shown in Fig. 11. Corresponding measurements by Tsinober[11] are reproduced in Fig. 12. As in the case of an aligned magnetic field a qualitative agreement in the region before separation is found.

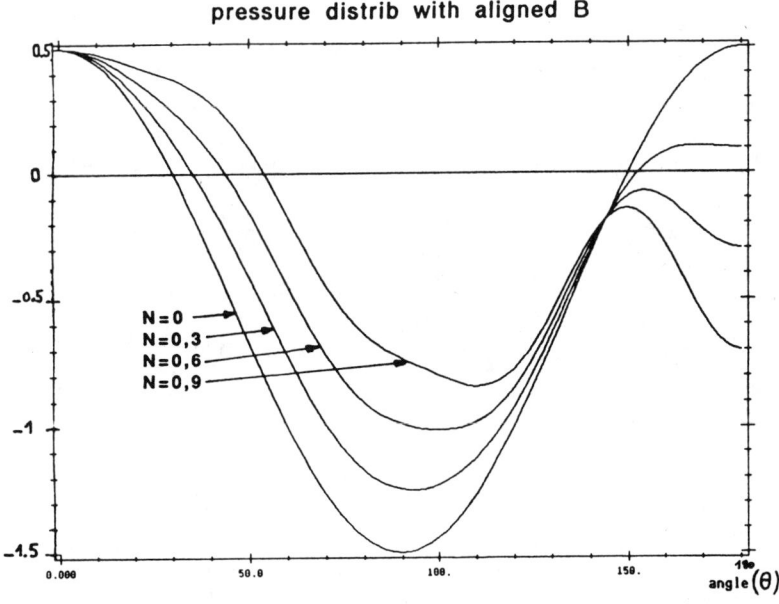

Fig. 10 Angular evolution of the pressure for different N when U_∞ is parallel to B_0.

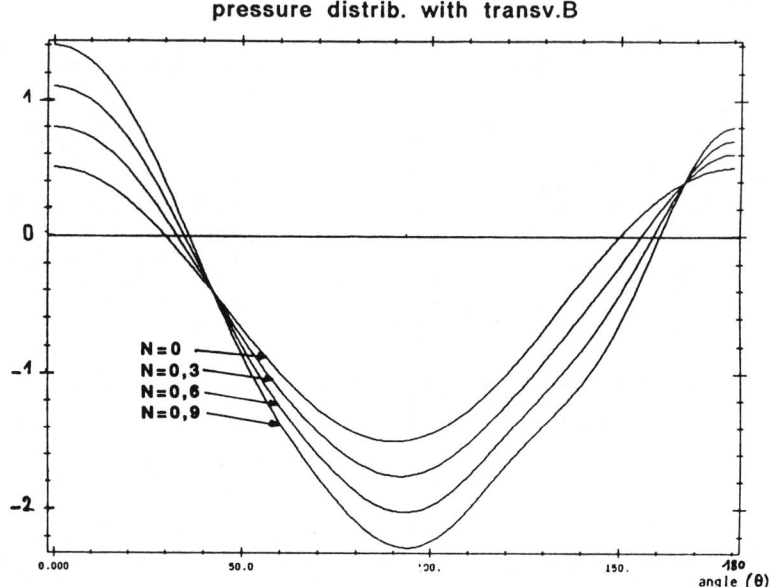

Fig. 11 Angular evolution of the pressure for different N when U_∞ is parallel to B_0.

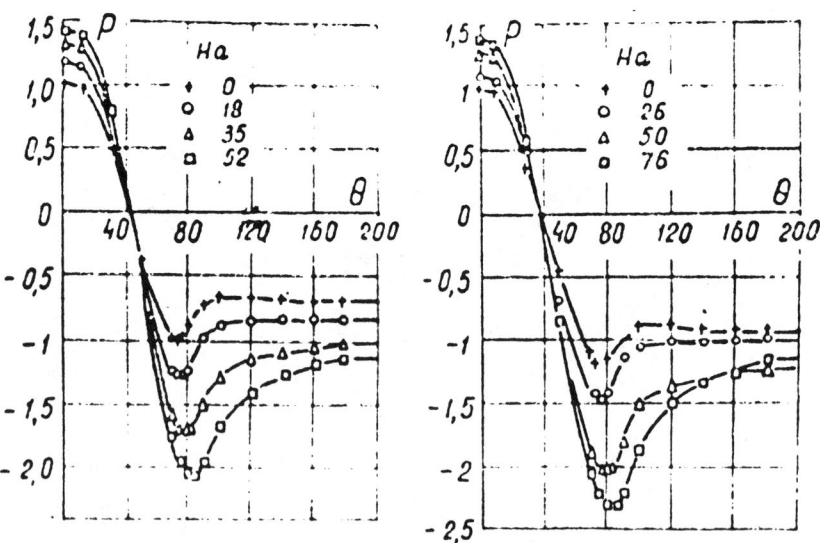

Fig. 12 Experimental results from Tsinober[11] for the case of a transverse magnetic field.

Appendix

Substituting Eq.(10) into Eq.(6) yields $\gamma_2(r) = 0$ and the recurrence relation

$$\gamma_{n+1}(r) = \frac{r^{n+1}}{(r^2-1)^{n+1}} \int_1^r \frac{(x^2-1)^n}{x^{n-1}} \cdot \left[-\frac{4}{3}\delta_{n3} + \frac{n-1}{x}\left(1 + \frac{1}{x^2}\right)\gamma_{n-1} - \left(1 - \frac{1}{x^2}\right)\gamma'_{n-1} \right] dx \qquad (A.1)$$

for $n = 2, 3, 4,...$ (δ_{mn} denotes the Kronecker symbol). This means that all coefficients with even indices are determined uniquely, whereas γ_1 must be chosen before $\gamma_3, \gamma_5...$ can be determined. This situation is similar to the investigation of Ludford and Murray[12] on the flow past a magnetized sphere. Here we determine γ_1 numerically by :

$$\gamma_1(r) = \frac{1}{\pi} \int_0^\pi \omega_1(r, \varphi) \cdot \sin\varphi \cdot d\varphi \qquad (A.2)$$

where the expression in Eq. (7) is used in the integral. The coefficient $\alpha_n(r)$ of the stream function in Eq.(11) are given by

$$\alpha_n(r) = \frac{-1}{2\pi} \int_1^r \left[\gamma_n(x) \cdot \left(\frac{x^{n+1}}{r^n} - \frac{r^n}{x^{n-1}} \right) + A_n\left(r^n - \frac{1}{r^n}\right) \right] \cdot dx \qquad (A.3)$$

with

$$A_n = -\frac{1}{2n} \int_1^\infty \frac{\gamma_n(x)}{x^{n-1}} \cdot dx \qquad (A.4)$$

Eqs. (A.3) and (A.4) lead immediately to the result given in Eq.(13) of the surface velocity coefficients. The values of the first six coefficients S_n are :

$$S_2 = 0, S_4 = 1/6, S_6 = 2.\pi^2 - 59/3,$$

$$S_1 = 0.9360, S_3 = 0.2797, S_5 = 0.1027$$

References

[1] Lahjomri, J., Caperan, Ph., and Alemany, A., "On Local Measurements of the Up and Downstream Magnetic Wake of a Cylinder at Low Magnetic Reynolds Number", Proceedings of the IUTAM Conf. in Riga, USSR, Kluwer, Dordrecht, May 1988, p.381.

[2] Flachsbart, O., Rep. Aerodyn. Versuchsanst - Goett. ser IV, pp 134-138, 1932.

[3] Schlichting, H., "Boundary Layer Theory", Mc Graw-Hill, New York.

[4] Maxworthy, T., "Experimental Studies in Magneto-fluid Dynamics : Pressure Distribution Measurements Around a Sphere", *Journal of Fluid Mechanics*, Vol. 31, Pt 4, 1968, pp. 801-814.

[5] Maxworthy, T., "Experimental Studies in Magneto-fluid Dynamics : Flow over a Sphere with a Cylindrical Afterbody", *Journal of Fluid Mechanics*, Vol. 35, Pt. 2, 1969, pp 411-416.

[6] Motz, R.O., "Magnetohydrodynamics of an Oscillating Dielectric Sphere", Plasma Research Laboratory, Columbia Univ. Rept n° 21, 1965.

[7] Yonas, G., "Measurements of Drag in a Conducting Fluid with an Aligned Field and Large Interaction Parameter", *Journal of Fluid Mechanics*, Vol. 30, Pt. 4, 1967, pp. 813-821.

[8] Buckmaster, J., "Separation and Magnetohydrodynamics", *Journal of Fluid Mechanics*, Vol. 38, N° 3, 1969, pp. 481-498.

[9] Tamada, K., "Flow of a Slightly Conducting Fluid Past a Circular Cylinder with Strong Aligned Magnetic Field", *Physics of Fluids*, Vol. 5, N° 7, 1962, pp. 817-823.

[10] Zachariah, J., and Singh, K.R., "A Two-dimensional MHD Flow of an Inviscid Fluid Past an Elliptic Cylinder", *Journal of the Physical Society* (Japan), Vol. 52, N° 8, 1983, pp. 2751-2760.

[11] Tsinober, A.B., "MHD Flow Around Obstacles", (in Russian), Zinatne, Riga, USSR, 1970, p 248.

[12] Ludford, G.S.S., and Murray, J.D., "On the Flow of a Conducting Fluid Past a Magnetized Sphere", *Journal of Fluid Mechanics*, Vol. 7, 1960, pp. 516-528.

Electromagnetically Driven Flow Around a Cylinder

A. Thess* and G. Gerbeth*
Central Institute for Nuclear Research, Rossendorf, Germany
and
Ph. Marty†
Institut de Mécanique de Grenoble, Grenoble, France

Abstract

The unidirectional flow of an electrically conducting fluid around a cylinder of arbitrary electrical conductivity, which is driven by the interaction of a homogeneous electric current with a homogeneous magnetic field and the resulting force on the cylinder are calculated numerically without any approximation in a large range of parameters. Asymptotic solutions are derived for the case of very strong and very weak magnetic fields respectively. A comparison with experimental results on insulating and highly conducting cylinders leads to a partial agreement although inertial forces are not taken into account in the model. Finally, confinement effects are considered leading to a better agreement between theory and experiment.

Introduction

The flow of viscous fluids around solid bodies of different shape has been the subject of intensive studies from the very beginning of hydrodynamics. In ordinary hydrodynamics as well as in magnetohydrodynamics (MHD) many efforts have been made to understand the "classical" problem where the body is placed in a homogeneous flow. However, a distinctive feature of MHD in current carrying fluids is the existence of electrically or electromagnetically driven flows, which are less studies than their classical counterpart.

Copyright © 1991 by the American Institute of Aeronautics and Astronautics, Inc. All rights reserved.
* Dr. rer. nat., Department of Sodium Engineering, P.B. 19, O-8051.
† Ingénieur de Recherche, CNRS, Fluid Mechanics Department, MHD Division, B.P. 53X.

In the present paper we investigate the flow around a cylinder driven by crossed electric and magnetic fields as an example, which has the advantage that only viscous and electromagnetic forces are present while intertial forces do not participate in the dynamics due to the one-dimensional flow geometry. There are only a few papers devoted to the flow past bodies in crossed electric and magnetic fields. Leonov & Kolin[1] considered the electromagnetically driven flow around an insulating sphere and derived an approximate solution for small Hartmann and small Reynolds numbers, respectively (decoupling of electrodynamic and hydrodynamic equations, neglection of inertial terms). The problem of flow around a cylinder was first studied by Ricou[2] for the particular case of an insulating cylinder. Marty & Alemany[3,4] measured the force on spheres and cylinders and calculated the force on a cylinder for asymptotic values of the Hartmann number. Gerbeth, Thess and Marty[5] performed a comprehensive study of the flow structure for arbitrary values of the conductivity and magnetic fields when the fluid occupies the whole space outside the cylinder (infinite problem). In the present work we summarize the results of [5] and consider the influence of external walls on the structure of the flow.

Phenomenology

Consider a solid cylinder (radius a, electrical conductivity σ_c) which is surrounded by an unbounded incmpressible fluid (viscosity η, conductivity σ_f). This system, as sketched in Fig. 1, is exposed to a magnetic field $B_0^* = B_0 e_x$ and an electric current $J_0^* = J_0 e_y$ which are supposed to be uniform at infinity. The behaviour of the fluid can be understood by a gedanken experiment in which the electrical conductivity of the cylinder can be changed continously. At the first instant let $\sigma_c = \sigma_f$. The current passes through the cylinder without any perturbation and the interaction with the magnetic field gives rise to a uniform Lorentz force $f_0^* = -J_0 B_0 e_z$ parallel to the axis of the cylinder. The fluid is at rest and a pressure distribution $p_0^* = -J_0 B_0 z^*$ balances the Lorentz force. If σ_c is "switched on" and assumes a value $\sigma_c > \sigma_f$ the electric current redistributes immediately in such a way that its density increases within the cylinder and decreases outside. This distribution $J_0^* + J_1^*(r^*)$, governed by classical electrodynamics, leads to an inhomogeneous Lorentz force $f_1^*(r^*)$ with nonzero curl which cannot be balanced by a pressure gradient and, consequently, brings the fluid into motion. Disregarding instabilities and turbulence we assume the velocity field to be parallel to the axis of the cylinder. This assumption is supported by the observation that the rotational part of $f_1^*(r^*)$ has only a z-component. The stationary flow is a balance of between the energy injected by the Lorentz force and the energy dissipated by viscous forces. The final distribution of the electric current $J_0^* + J_2^*(r^*)$ in the stationary flow differs from J_1^* due to the introduction of additional currents by the interaction of the flow with the magnetic field. The flow leads to a frictional force acting on the cylinder. The Lorentz force inside the cylinder gives rise to an electromagnetic force unless the cylinder is an insulator. The same phycsical picture holds for arbitrarily shaped bodies. However, even in the seemingly simple case of a sphere, a quantitative analysis is much more difficult due to the mathematical complexity of the problem.

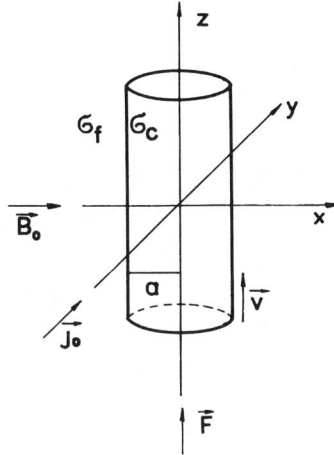

Fig. 1 Geometry of the model.

Basic Equations and Method of Solution

The motion of the fluid is governed by the standard MHD equations (see e.g. Roberts[6] which take a simple form for our geometry. In what follows we use asterisk to distinguish physical quantities from dimensionless ones. We assume the field variables v* and **B*** to be of the following form

$$\mathbf{v}^*(x^*,y^*) = J_o B_o a^2 \eta^{-1} v(x,y) \mathbf{e}_z \quad \text{(for r>1)} \tag{1a}$$

$$\mathbf{v}^*(x^*,y^*) = 0 \quad \text{(for r<1)} \tag{1b}$$

$$\mathbf{B}^*(x^*,y^*) = B_o \mathbf{e}_x + \mu_0 J_0 a M [b_f(x,y) - x/M] \mathbf{e}_z \quad \text{(for r>1)} \tag{2a}$$

$$\mathbf{B}^*(x^*,y^*) = B_o \mathbf{e}_x + \mu_0 J_0 a M [b_c(x,y)] \mathbf{e}_z \quad \text{(for r>1)} \tag{2b}$$

Here $M = B_0 a(\sigma_f/\eta)^{1/2}$ is the Hartmann number, x and y are related to the physical coordinates by x*=ax, y*=ay, and polar coordinates x=r cosφ , y=r sinφ are used. The electric current is obtained from by taking the curl of the magnetic field. Inserting (1) and (2) into the basic MHD equations, the following set of linear partial differential equations is obtained which describes the velocity field v and the magnetic field perturbations b_f and b_c in the fluid and in the cylinder, respectively

$$\Delta v + M \frac{\partial}{\partial x} b_f = 0 \quad \text{(for r >1)} \tag{3}$$

$$\Delta b_f + M \frac{\partial}{\partial x} v = 0 \quad \text{(for } r > 1\text{)} \tag{4}$$

$$\Delta b_c = 0 \quad \text{(for } r < 1\text{)} \tag{5}$$

At the cylinder surface r=1 the no slip condition, the continuity of the magnetic field and of the tangential component of the electric field lead to the following boundary conditions

$$v = 0$$

$$b_f = b_c + \frac{1}{M} \cos \varphi \tag{6}$$

$$\frac{\partial}{\partial r} b_f = \frac{1}{\lambda} \frac{\partial}{\partial r} b_c + \frac{1}{M} \cos \varphi$$

where we have introduced the conductivity ratio parameter $\lambda = \sigma_f/\sigma_c$. Furthermore we require

$$\lim_{r \to \infty} b_f = 0$$

$$\lim_{r \to \infty} v = 0 \tag{7}$$

and b_c to be finite at r=0. From v, b_f and b_c the force on a section of length L of the cylinder can be calculated. Using the expressions for the stress tensor and integrating over the surface of the body the frictional and electromagnetical forces are given by

$$\mathbf{F}_f^* = J_o B_o \pi a^2 L \, f_f(\lambda, M) \, \mathbf{e}_z \quad \text{with} \quad f_f = \frac{1}{\pi} \int_0^{2\pi} d\varphi \left\{ \frac{\partial}{\partial r} v(r, \varphi) \right\}_{(r=1)} \tag{8}$$

$$\mathbf{F}_{em}^* = J_o B_o \pi a^2 L \, f_{em}(\lambda, M) \, \mathbf{e}_z \quad \text{with} \quad f_{em} = \frac{M}{\pi} \int_0^{2\pi} d\varphi \left\{ b_c(r, \varphi) \cos \varphi \right\}_{(r=1)} \tag{9}$$

When the cylinder has a finite length L, an additional expulsive force

$$\mathbf{F}_p^* = J_o B_o \pi a^2 L \, f_p \, \mathbf{e}_z \quad \text{with} \quad f_p = +1 \tag{10}$$

results from the pressure gradient.

The solution of the governing Eqs.(3-7) depends solely on the two parameters λ and M. The Reynolds number does not enter the system since inertial terms in the Navier-Stokes equation vanish identically for the the choice (1). Depending on the value of M three different techniques are employed to find solutions of the equations. First we note that the general solution to the system (3-5), (7) compatible with the symmetry of the boundary conditions (6), can be written in terms of modified Bessel functions K_l as follows

$$v = \frac{1}{2}(\psi_1 - \psi_2) \tag{11}$$

$$b_f = \frac{1}{2}(\psi_1 + \psi_2) \tag{12}$$

$$\psi_{1/2} = (\pm)\, e^{(-/+\, m r \cos\varphi)} \sum_{l=0}^{\infty} (\pm 1)^l A_l\, K_l(mr) \cos(l\varphi) \tag{13}$$

$$b_c = \sum_{l=0}^{\infty} B_l\, r^l \cos(l\varphi) \tag{14}$$

where $M=2m$. When M belongs to the interval $0.01 < M < 30$ we calculate the unknown coefficients A_l and B_l numerically by solving a system of algebraic equations which is obtained upon applying the boundary conditions (6) to (13) and (14). In the limit of high Hartmann numbers this method ceases to be efficient and we turn to an approximation by replacing the modified Bessel functions by their asymptotic formula

$$K_l(x) \approx \sqrt{\frac{\pi}{2x}}\, e^{-x} \tag{15}$$

valid for $x \gg 1$. The range of validity of this assumption was discussed in [5]. In addition, for $M \ll 1$ we use a perturbation expansion in powers of M to construct an approximate solution.

Results for the Unbounded Problem

The results of the analytical and numerical calculations are summarized in Figures 2 - 5 in which the four limiting cases are represented when M and λ are very small and very large, respectively. Figures 2 - 5 are plotted on the basis of approximate solutions of the governing equations (3) - (7) which are derived in [5]. In the case $M \ll 1$ we have obtained

$$v(r,\varphi) = \frac{1}{4}\left(\frac{1-\lambda}{1+\lambda}\right)\left(\frac{1}{r^2} - 1\right)\cos(2\varphi) \tag{16}$$

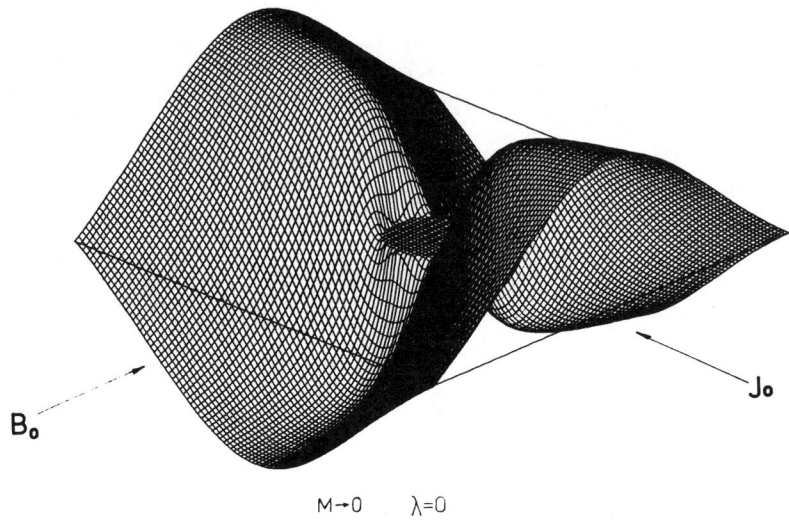

$M \to 0 \quad \lambda = 0$

a) velocity field according to approximation (16)

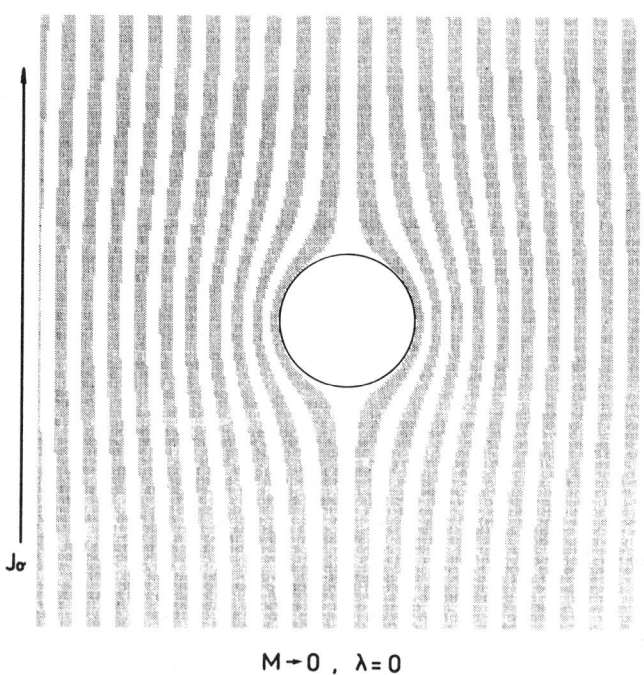

$M \to 0, \quad \lambda = 0$

b) electric current according to approximation (17)

Fig. 2 Electromagnetically induced flow around insulating cylinder in an unbounded fluid for weak magnetic field ($M \ll 1$).

EM FLOW AROUND A CYLINDER

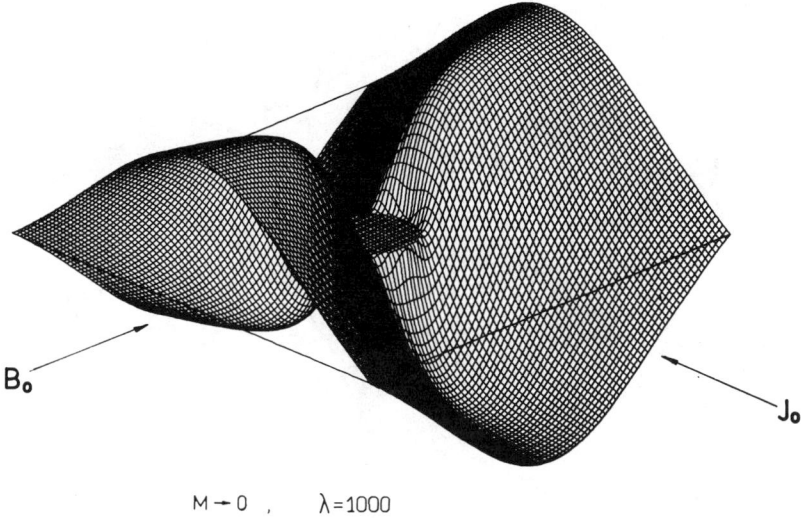

$M \to 0$, $\lambda = 1000$

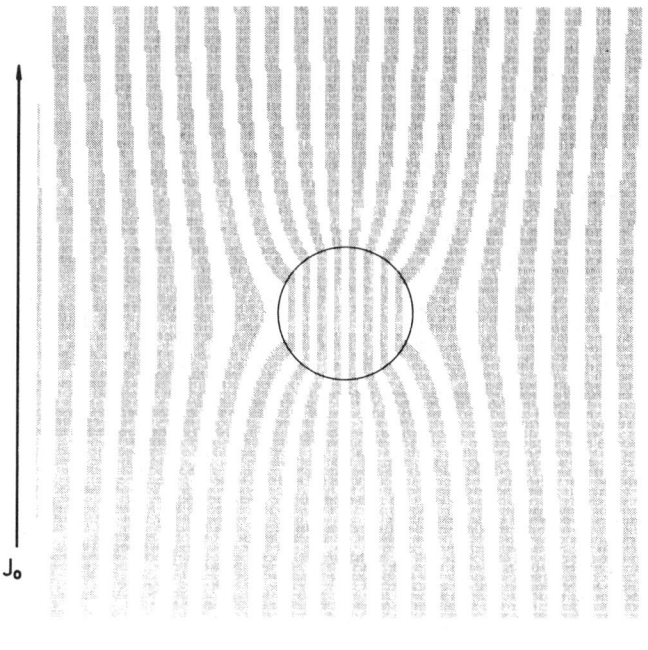

$M \to 0$, $\lambda = 1000$

Fig. 3 Same as Fig. 2 but for highly conducting cylinder

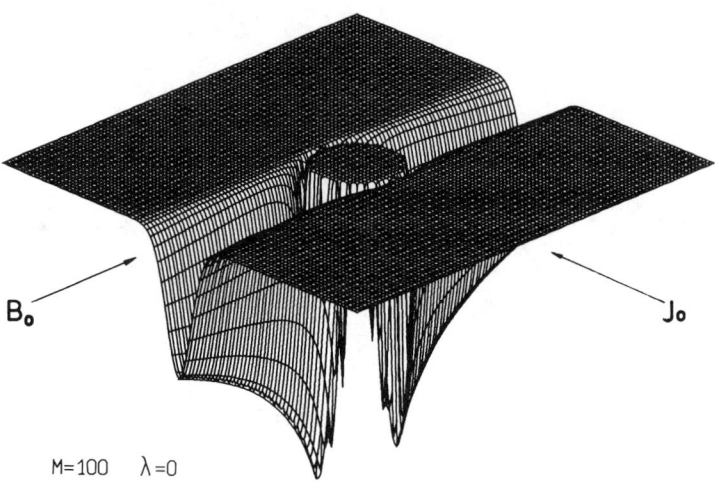

$M=100 \quad \lambda=0$

a) velocity field according to approximation (18)

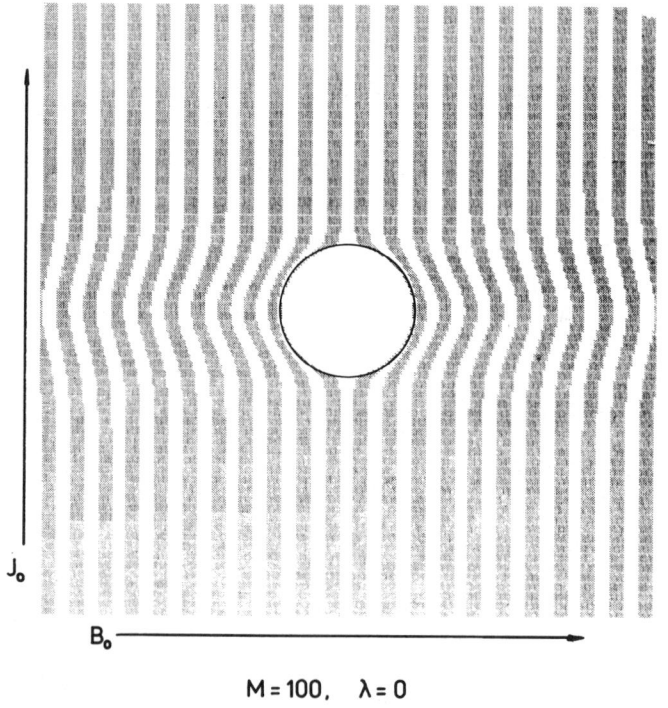

$M = 100, \quad \lambda = 0$

b) electric current according to approximation (19)

Fig. 4 Electromagnetically induced flow around insulating cylinder in an unbounded fluid for strong magnetic field ($M \gg 1$)

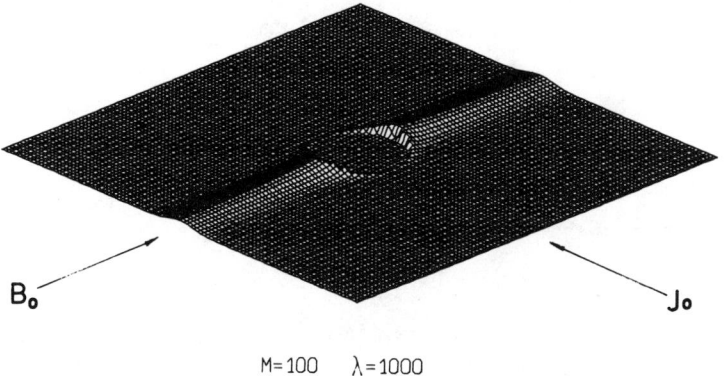

M=100   λ=1000

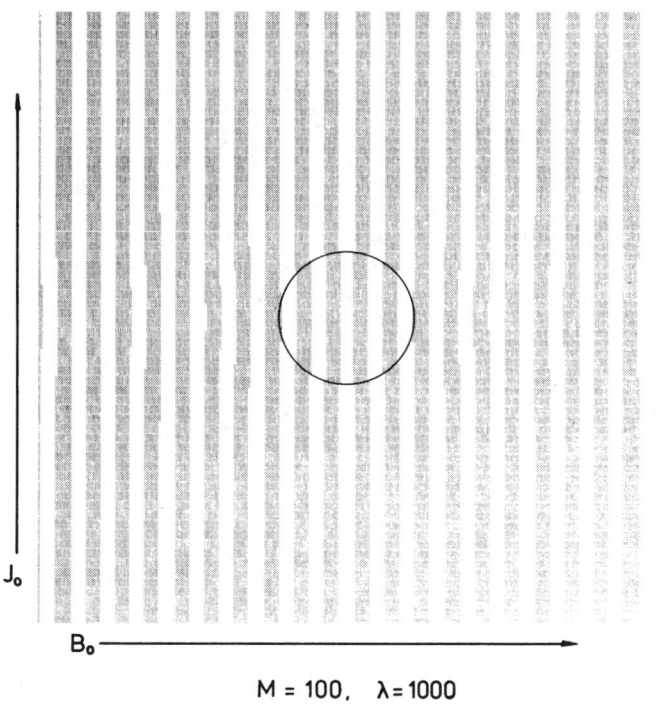

M = 100, λ = 1000

Fig. 5 Same as Fig. 4 but for highly conducting cylinder; the velocity scale is identical to Fig. 4 but different from Figs. 2 and 3.

$$b_f(r,\varphi) = \frac{1}{M}\left(\frac{1-\lambda}{1+\lambda}\right)\frac{1}{r}\cos(\varphi) \qquad (17a)$$

$$b_c(r,\varphi) = -\frac{1}{M}\left(\frac{2\lambda}{1+\lambda}\right) r \cos(\varphi) \qquad (17b)$$

In the limit $M \gg 1$ the following approximations hold

$$v(r,\varphi) = \left(\frac{2(\lambda-1)}{\lambda M + \lambda M^2 + 2M}\right)\frac{1}{\sqrt{r}}\cos(\varphi)\, e^{-m(r-1)}\sinh\{m(r-1)\cos(\varphi)\} \qquad (18)$$

$$b_f(r,\varphi) = \left(\frac{2(\lambda-1)}{\lambda M + \lambda M^2 + 2M}\right)\frac{1}{\sqrt{r}}\cos(\varphi)\, e^{-m(r-1)}\cosh\{m(r-1)\cos(\varphi)\} \qquad (19a)$$

$$b_c(r,\varphi) = \left(\frac{\lambda(M+3)}{\lambda M + \lambda M^2 + 2M}\right) r \cos(\varphi) \qquad (19b)$$

The value of λ has a significant influence on the character of the flow. One observes thath the cases $\lambda < 1$ (weakly conducting or insulating cylinder, see fig. 2 and 4) and $\lambda > 1$ (highly conducting cylinder, see Figs. 3 and 5) differ by the sign. This is a consequence of the fact that the flow is driven by the inhomogeneous part of the Lorentz force

$$\vec{J}_2^* \times \vec{B}_0^* = -J_{2y}^* B_0 \vec{e}_z \qquad (20)$$

being proportional to $-J_{2y}$. Consider the region parallel to the magnetic field (y=0). An insulating cylinder cannot be penetrated by the electric current and consequently $J_{2y} > 0$ in this region. On the other hand a higher conductivity of the body requires $J_{2y} < 0$ there.

For small M, to zero order in the asymptotic expansion, the electromagnetic problem and the hydrodynamic one decouple and the current distribution is governed by classical electrodynamics. This is demonstrated in Figs. 2b and 3b where the isolines of $(b_f - x/M)$ and b_c are plotted which are identical with the direction of the electric current. In fact equation (17) describes the current density around a solid cylinder in a quiescent medium. To first order b_f and b_c remain unchanged, they rather give rise to the velocity field (16) which is shown in Figs. 2a and 3a. As M is increased, the coupling between the flow field and the magnetic field becomes stronger. The velocity scales as $v \sim M^{-1}$ (for $\lambda M \ll 1$)

and $v \sim M^{-2}$ (for $\lambda M \gg 1$) instead of being of the order of unity as in the case $M \ll 1$, which is clearly seen comparing Figs. 4a and 5a. For large M the flow is arranged in four regions. The flow is concentrated upon a *parabolic wake* bounded by $|y| = |x/M|^{1/2}$ and separated from the quiescent fluid by *transition layers* of thickness $\delta_t \sim M^{-1/2}$. The *Hartmann layers* near r=1 with thickness $\delta_h \sim |M \cos(\varphi)|^{-1}$ are characterized by high velocity gradients. Within the *tangential layers* in the vicinity of $|\varphi| = \pi/2$ which are not reflected by approximation (18) in Figs. 4a and 5a but detected in the exact numerical solutions reported in [5], the velocity has the opposite sign compared to the wakes. Moreover, the flow inside the wake can be divided into two regions, namely the so called deep core (where v is approximately constant) and the outer core (where v decreases as $r^{-1/2}$). The former extends from the boundary layer up to $x \sim M^{-1/3}$. For more details as well as a comparison with the general approach to the core variables of Hunt and Ludford we refer to [7].

From the electrodynamical point of view the main effect of the back reaction of the flow on the electric current distribution is the homogenisation of the electric current, demonstrated in Figs. 4b and 5b. While for the exceptional case of an insulating cylinder the homogenisation is inhibited, for any nonzero conductivity the electric current induced by the interaction of the velocity field with the homogeneous magnetic field tends to homogenise the initial current.

A particular property of the velocity field is that the flow rate

$$\phi = \int_{r=1}^{\infty} \int_{\varphi=0}^{2\pi} dr\, d\varphi \{r\, v(r,\varphi)\} \qquad (21)$$

diverges. This is caused by the slow velocity decrease inside the wake and represents an intrinsic feature of the unbounded model. In the experimental facility of Marty & Alemany [3,4] forces were measured on cylinders with length

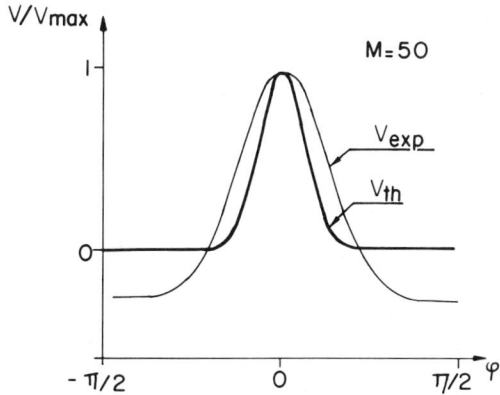

Fig. 6 Velocity profile v-exp measured at r=1.8 and comparison with v-th predicted by the approximation (18) for M=50.

of 6 cm and a radius of 1 cm in a nearly cubic box of size 10 cm containing mercury. Having in mind the diverging flow rate of the present calculations it seems to be difficult to compare our results with measurements. However, a partial agreement for the velocity profiles seen in Figure 6 especially in the region of the parabolic wake. It will be shown in the following section how finite or even a zero flow rate may be obtained taking into account external walls.

The results for the forces f_f and f_{em} are shown in Fig. 7. The effect of homogenization of the electric current is reflected in the right part of the curves for f_{em} which increases from -2 (current density within the cylinder is twice the current density outside) to -1 (homogeneous electric current). The behaviour of f_f for high M becomes evident by inspection of the velocity gradients in the vicinity of the cylinder surface. For decreasing M it vanishes since the radial derivative of v becomes proportional to $\cos(2\varphi)$. On the other hand, for $M \gg 1$ the velocity gradients are mainly positive for $\lambda > 1$ and negative for $\lambda < 1$ respectively. A comparison of the forces with the total force measured in [3] by adding the pressure force (10) reveals that the force on a copper cylinder ($\lambda = 55$) is indeed smaller than the force on an insulating cylinder, which confirmes the effect of homogenisation. However, the absolute values of the forces given in Table 1 are underestimated by the present theory.

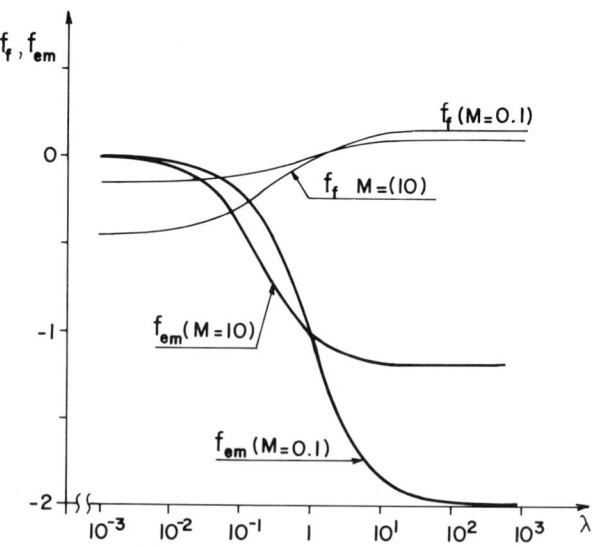

Fig. 7 Frictional force f-f (Eq. 8) and electromagnetic force f-em (Eq. 9) per unit length of a cylinder versus the conductivity ratio λ for M=0.1 and M=10.

Table 1 Comparison of total forces measured in Ref.3 with the theoretical prediction $f = f_f + f_{em} + f_p$

System	Hartmann number	Force (experiment)	Force (theory)
PVC cylinder ($\lambda = 0$)	8	0.57	0.55
	26	0.71	0.52
copper cylinder ($\lambda = 55$)	8	- 0.71	- 0.15
	26	- 0.40	- 0.05

Influence of External Walls

The idealized unbounded model leads to a diverging flow rate due to the slow velocity decrease inside the wakes. Now let us consider how external walls yield a finite flow rate. The simplest way to incorporate confinement effects without dramatic changes in the numerical and analytical complexity of the problem is to suppose the fluid to be bounded by a circular wall at r=R (R>1). In contrast to the infinite problem where the pressure distribution $p_o^* = -J_oB_oz^*$ was uniquely determined by the condition v=0 at infinity, the confined problem admits solutions for any values of the pressure gradient. The ambiguity in the choice of the parameter C which we introduce by $p_o^* = -CJ_oB_oz^*$ can be used to achieve that the flow rate defined in equation (21) be zero. Confining our attention to the case of an insulating cylinder ($b_c=0$) which allows an analytical solution, the flow is described by Eqs. (3) and (4) but with the modified boundary conditions

$$v(1,\varphi) = 0,$$
$$b_f(1,\varphi) = C/M \cos \varphi \quad (22)$$

$$v(R,\varphi) = 0$$
$$b_f(R,\varphi) = R(C-1)/M \cos \varphi. \quad (23)$$

Including the Bessel functions $I_1(mr)$, which increase with increasing distance, into the expansion (13) an exact solution is obtained in the following form

$$\psi_{1/2} = (\pm) e^{(-/+ \, m \, r \cos \varphi)} \sum_{l=0}^{\infty} \left\{ (\pm 1)^l \left[A_l K_l(mr) + \widetilde{A}_l I_l(mr) \right] \cos(l\varphi) \right\}$$

with the coefficients

$$A_l = \left(\frac{2 - \delta_{l0}}{M}\right) \frac{C \, I_l'(m) \, I_l(mR) - (C-1) \, I_l'(mR) \, I_l(m)}{I_l(mR) \, K_l(m) - I_l(m) \, K_l(mR)} \quad (24)$$

$$\tilde{A}_1 = \left(\frac{2-\delta_{10}}{M}\right) \frac{C\, I_1'(m)\, K_1(mR) - (C-1)\, I_1'(mR)\, K_1(m)}{I_1(m)\, K_1(mR) - I_1(mR)\, K_1(m)}$$

$$I_1'(m) = \frac{dI_1(m)}{dm}$$

Unfortunately, a direct numerical analysis is limited to $M < 5$ since some terms of the series grow exponentially with increasing M. A comparison of the velocity magnitude between the infinite problem in Fig. 8a and the confined problem in Fig. 8b for $R=5$ and $M=1$ shows that the walls have no significant influence on this quantity as long as $M=O(1)$. On the other hand it is likely that for $M \gg 1$ the Hartmann layers which develop at the outer wall will lead to a significant velocity decrease compared to the unbounded model. The other main results of the unbounded model (homogenization of the electric current, scaling of the velocities) are not influenced by the presence of external walls as long as $C=1$. However if we calculate C from the condition of zero flow rate $\phi=0$ the flow field as shown in Fig. 8c, is reorganized. Although the value $C=1.0397$ which is necessary for $\phi=0$ differs only slightly from 1, the change in the flow field is fairly important. Indeed, the magnitude of the velocity in the region $x=0$ becomes comparable to that in the wake $y=0$. Although numerical calculations cannot be extended to the experimental value $M=50$ and the above described high M approximation does not work in the confined problem, we may outline some implications of the results shown in the fig. 8 for $M=1$ on the experimental situation. On the one hand the zero flow rate leads to a decrease of the absolute value of the frictional force and therefore to an increased total force in agreement with experiments. On the other hand the nonzero velocity in the tangential layers allows to overcome the discepancy between experiment and theory in Fig. 6.

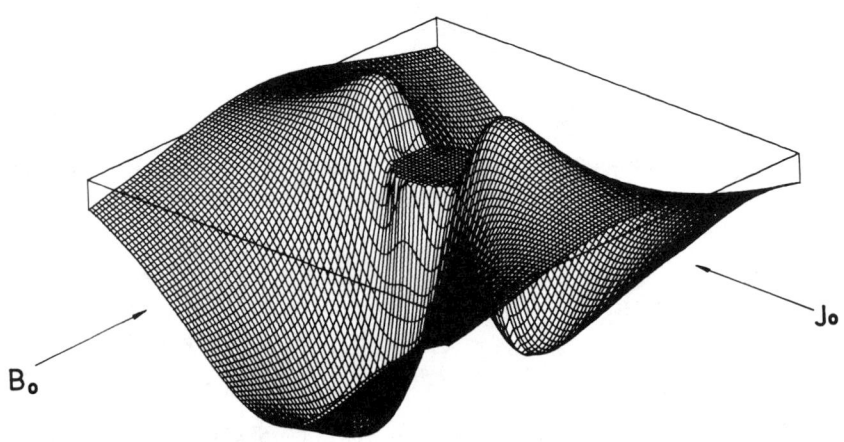

a) unbounded case, infinite flow rate, C=1

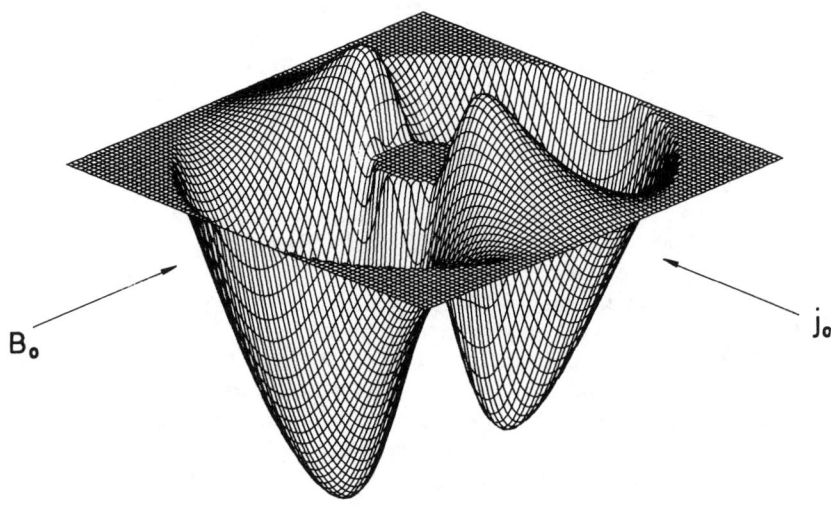

b) bounded case, R=5, finite flow rate, C=1

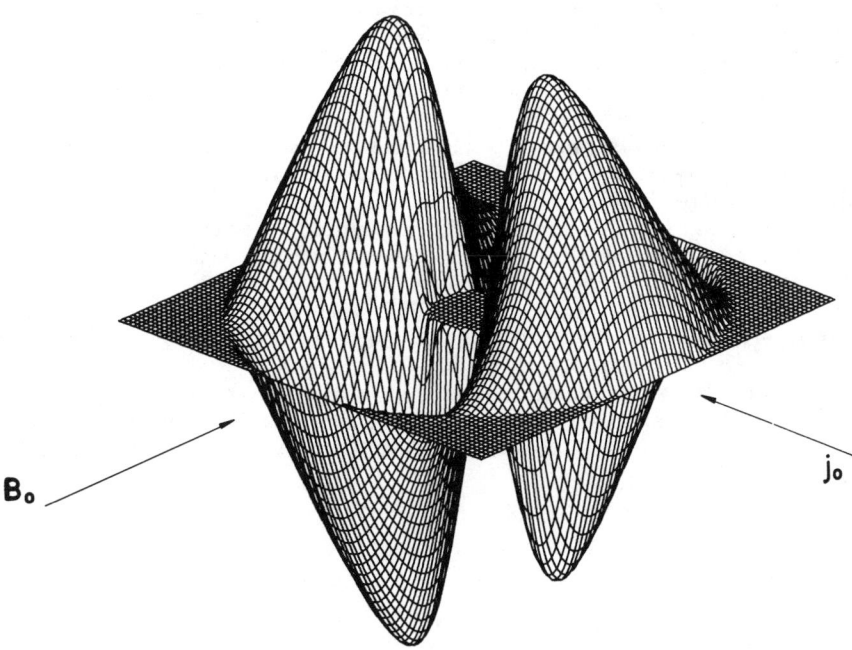

c) bounded case, R=5, zero flow rate, C=1.0397

Fig. 8 Influence of external walls on the velocity field of electromagnetically driven flow around an insulating cylinder for $M=1$.

The velocity scales are the same in b and c but different from a.

Conclusions

In this paper the interplay between electromagnetic and viscous forces in a flow driven by crossed electric and magnetic fields is analyzed. The most important features of the flow in the unbounded model for high Hartmann numbers are

1-) the localization of the flow within a wake parallel to the magnetic field,
2-) the homogenization of the electric current,
3-) the infinite flow rate through each plane z=constant , and
4-) the smallness of the velocity in the tangential layers compared to the magnitude in the wake.

Taking into account external walls we found that the equations describing the confined problem for an insulating inner cylinder

1-) admit solutions with finite or zero flow rate, and
2-) lead to an increased velocity in the tangential layers.

Although we were not able to extend our numerical calculations to the experimental value M=50, these two assertions are believed to hold in this case too. Research along this line is in progress and will be reported elsewhere.

References

[1] Leonov , D., and Kolin , A., "Theory of Electromagnetophoresis", *Journal of Chemical Physics,* Vol. 22, 1954, pp. 683-668.

[2] Ricou, R., "Perturbations desparamètres électriques et mécaniques au sein d'un métal liquide en présence d'un champ de force électromagnétique",These de 3eme cycle, Université d'Aix-Marseille, 1975.

[3] Marty, Ph., and Alemany , A., "Theoretical and Experimental Aspects of Electromagnetic Separation", IUTAM Symposium on Metallurgical Applications of MHD, Cambridge, 1982, pp. 245-259.

[4] Marty , Ph.,and Alemany, A., "Ecoulement du a des champs magnetique et electrique croisés autour d'un cylindre de conductivité quelconque", *Journal de Mécanique,* Vol. 2,n°2, 1985, pp. 227-243.

[5] Gerbeth ,G., Thess, A.,and Marty, Ph., "Theoretical study of the MHD flow Around a Cylinder in crossed electric and magnetic fields", *European Journal of Mechanics,* B/Fluids, Vol. 9,n° 3, pp. 239-257.

[6] Roberts, P., "An introduction to magnetohydrodynamics", Longmans, 1967

[7] Hunt, J.C.R.,and Ludford, G.S.S., "Three-dimensional MHD flow with strong transverse magnetic fields", *Journal of Fluid Mechanics,* Vol. 33,Part 4,1968, pp. 693-714.

New Results for MHD Drag Coefficients

G. Gerbeth*
Central Institute for Nuclear Research, Rossendorf, Germany

Abstract

Theoretical and experimental results of MHD drag coefficients are summarized. Special attention is paid to Stokes flow, where a typical error has been found in the literature. This situation is clarified here, and correct results are presented. Numerical calculations are performed for the MHD Stokes flow around a cylinder in a transverse magnetic field, yielding qualitative agreement for the drag with both measurement as well as a rough asymptotic analysis. The MHD drag coefficient of the cylinder in a transverse magnetic field increases proportionally to $M \cdot \ell n M$ if $M \gg 1$ (where M is the Hartmann number). Finally, the deflection of a rising bubble in a liquid metal is determined if the direction of the magnetic field is inclined relative to the vertical line.

Introduction

Compared to the 1960s, the problem of MHD flow past obstacles has been of less interest recently. However, many questions in this field are still unsolved. This is valid for both the drag coefficients as well as the structure of the MHD fluid flow around the obstacle. Assuming a small magnetic Reynolds number (inductionless approximation), the problem is described by two independent parameters: the Reynolds number and the Hartmann number. Different ranges of their absolute values as well as different ratios between them have to be distinguished. Special attention is paid here to the Stokes approximation in a strong magnetic field (high Hartmann number). In this limit, some contradicting results for the drag forces exist in the literature. This situation is clarified in the present study.

Although there have been a number of drag coefficient calculations, there have been few that were tested in the laboratory. Only one successful comparison of theory with experiment was stated, by Hunt and and Moreau[1] in 1976. A critical analysis of this fact, combined with a reconsideration of previous measurements for a cylinder, is given in this study.

Copyright © 1992 by the American Institute of Aeronautics and Astronautics, Inc. All rights reserved.
*Professor, Department of Sodium Technology.

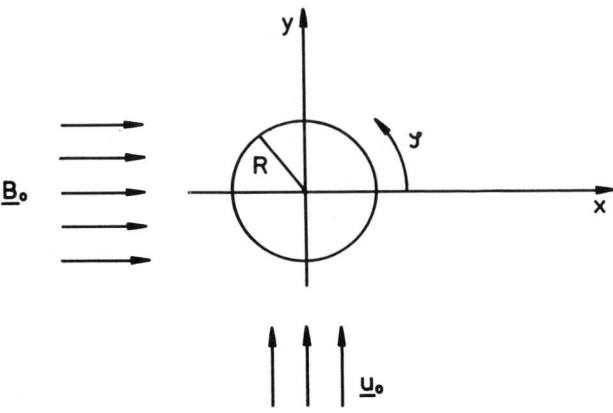

Fig. 1 Sketch of geometry

Basic Relations

The steady flow of an incompressible, viscous, electrically conducting fluid past a solid sphere or cylinder is considered with an external magnetic field $\underline{B}_o = B_o \underline{e}_x$. Concerning the uniform velocity u_o at infinity, we distinguish between the two cases of an aligned field $\underline{u}_o = u_o \underline{e}_x$ and a transverse field $\underline{u}_o = u_o \underline{e}_y$ (see Fig. 1). The body is taken as non-conducting unless it is stated explicitly. The radius R of the body is used in the following discussion as the typical length scale. The fluid is characterized by the magnetic permeability μ, electrical conductivity σ, density ρ, and dynamic viscosity η, respectively. All of these quantities are supposed to be constant throughout the fluid.

The flow field is determined by the modified Navier-Stokes equation

$$\text{Re}(\underline{V} \cdot \text{grad})\underline{V} = -\text{grad } p + \Delta \underline{V} + M^2(\underline{E} + \underline{U} \times \underline{e}_x) \times \underline{e}_x \qquad (1)$$

which is made non-dimensional using R, u_o, $\eta u_o/R$, and $u_o B_o$ as units of the length, velocity, pressure, and electric field, respectively. The characteristic parameters in Eq. (1) are the Reynolds number $\text{Re} = \rho R u_o/\eta$ and the Hartmann number $M = RB_o \sqrt{\sigma/\eta}$. The magnetic Reynolds number $Rm = \mu \sigma u_o R$ is assumed to be small ($Rm << 1$), and, therefore, the magnetic field is taken as unperturbed.

If the flow is in an aligned field, the electric field vanishes due to the symmetry. Flow around a cylinder in a transverse magnetic field leads to the constant value $\underline{E} = \underline{e}_z$ due to the two-dimensional geometry. The only three-dimensional problem is established by the flow around a sphere in a transverse magnetic field. In this case E is expressed in terms of the electric potential, which in turn must be determined by Ohm's law and the continuity of the electric current density.[2]

A simplification of the non-linear Eq. (1) is possible at least in the following two approximations:

1) Stokes (Oseen) approximation. It corresponds to relatively slow motions, where the left-hand side of Eq. (1) can be neglected (Stokes approximation) or can be replaced by its value at large distances Re $(u_o \text{grad})\underline{V}$ (Oseen approximation). It is generally accepted that the magnetic field extends the range of validity of the Stokes approximation (which is restricted to Re $\leq$ 1 in the non-magnetic case). This is supported by the fact that a magnetic field tends to suppress boundary layer separation from the body surface. The problem was discussed in detail by Tsinober,[3] and he concluded that the Stokes approximation is valid up to Reynolds numbers with Re/M $\leq$ const, where the constant is on the order of 1.

2) Inviscid approximation. It corresponds to high Reynolds number flows, where the viscous forces are significant in the boundary layers on the body surface only and are neglected in the outer, inviscid flow region. The pressure has to be scaled by ρu_o^2 in this case. The flow is characterized by the interaction parameter $N = M^2/Re$ alone. It is clear that this approximation cannot be applied to the wake behind the body. On the other hand, the inviscid flow gives a drag force acting on the body in the flow direction because of the influence of the magnetic field on the surface pressure distribution.

The force F on the body is obtained by integrating pressure and viscous stress over the body surface. The drag coefficient is defined by

$$C = \frac{F}{\frac{1}{2}\rho u_o^2 S} \qquad (2)$$

where S represents the area of the body in flow direction.

Literature Review of Theoretical Results

The results summarized in Table 1 are available in the literature on MHD drag coefficients in Stokes approximation. The contradictory expressions for M >> 1 are discussed in the next section. Results in the Oseen approximation were obtained solely for M << 1 and an aligned magnetic field:

sphere[13] $$C_s = C_{so}\left(1 + \frac{3}{16}\frac{Re^2 + 2M^2}{k_1 + k_2}\right) \qquad (3)$$

cylinder[14] $$C_c = C_{co}\left(1 - 2\gamma - \frac{2k_3}{k_1 + k_2}\right)^{-1} \qquad (4)$$

Table 1
MHD drag coefficients in Stokes approximation relative to $C_{so} = 12/Re$ and $C_{co} = 8\pi/Re$. $\gamma = 0.57721...$Euler number

		$M \ll 1$	$M \gg 1$
aligned magnetic field	sphere	$C_{so}(1+\frac{3}{8}M)$ Ref. 4	$C_{so}\begin{cases}\frac{4}{9}M & \text{Ref.8}\\ \frac{1}{3}M & \text{Ref.9}\end{cases}$
	cylinder	$C_{so}(1-2g-2\ell n+\frac{M}{4}M)^{-1}$ Ref.5	$C_{co}\begin{cases}\frac{5}{64}M & \text{Ref.10}\\ \frac{1}{2\pi}M & \text{Ref.11}\end{cases}$
transverse magnetic field	sphere	$C_{so}(1+\frac{9}{16}M)$ Ref. 6	
	cylinder	$C_{co}(-1-2\gamma-2\ell n+\frac{M}{4}M)^{-1}$ Ref. 7	$C_{co}\frac{7}{128}M)^2$ Ref. 12

where

$$4k_1 = \sqrt{Re^2 + 4M^2} - Re, \quad 4k_2 = \sqrt{Re^2 + 4M^2} + Re,$$

$$k_3 = k_1 \ln\frac{k_1}{2} + k_2 \ln\frac{k_2}{2}, \quad \gamma = 0.57721...$$

In the inviscid approximation the drag coefficient of a sphere was obtained by Reitz and Foldy[15] for $N \ll 1$ by calculating the Joule dissipation with the undisturbed inviscid flow. Their results are

$$C_s = \frac{4}{5}N \tag{5}$$

for an aligned field and

$$C_s = \frac{6}{5}N\frac{\sigma + 3\sigma'}{2\sigma + \sigma'} \qquad (6)$$

for a transverse magnetic field (where σ is the electrical conductivity of the sphere). The same method leads in the case of a cylinder to

$$C_c = \frac{\pi}{2}N \qquad (7)$$

for both directions of the magnetic field. A very strong magnetic field $N \gg 1$ was considered by Ludford[16,17] for the transverse geometry. He obtained linear equations by compressing with $\sqrt{N}$ the coordinate in the field direction.

The velocity perpendicular to the field lines must then be magnified at the same rate by reason of continuity. The results for the drag are[16]

$$C_s = 2.76\sqrt{N} \qquad (8)$$

and[17]

$$C_c = 3.66\sqrt{N} \qquad (9)$$

The more realistic problem of a MHD flow around obstacles in a constant area duct with a transverse magnetic field was considered by Hunt and Ludford[18]. In the limits $N \gg 1$, $M \gg 1$, $Re \gg 1$, and $Rm \ll 1$, they analyzed the flow in certain separate regions: the core, the boundary layers, and the shear layers. The solutions of the various regions are matched to each other and are consistent with the original assumptions. They found that the zero-order core flow is not determined uniquely but is controlled by the Hartmann layers at the body as well as at the duct walls. That means that the applicability of the formulas of Eqs. (8) and (9) (obtained for an unconfined flow) to laboratory experiments is reduced significantly. However, this fact represents no restriction on the results in Stokes approximation since in this case viscosity determines the flow structure throughout the fluid.

The influence of confining walls on the results in Stokes approximation remains an open question. The shear layers circumscribing the body and being parallel to the magnetic field can be analyzed analytically[18] in the range $M \gg Re^2$. On this basis Hung[19] calculated the drag of a sphere in a transverse magnetic field with the result

$$C_s = \frac{8M}{Re} \qquad (10)$$

Before comparing some of these theoretical predictions with measurements, the results of Stokes approximation for $M \gg 1$ should be considered in detail in order to overcome the contradicting formulas in Table 1.

High Hartmann Number Stokes Approximation

All formulas given in this section correspond to the two-dimensional flow around a cylinder (Fig. 1). The extension to the flow around a sphere in an aligned magnetic field is straightforward, whereas the three-dimensional problem of a sphere in a transverse field needs further investigation. Eq. (1) is reduced by neglecting the inertial term to

$$\partial_x p = \Delta u_x, \quad \partial_y p = (\Delta - M^2) u_y \qquad (11)$$

where $\underline{u} = \underline{v} - u_o$ is the velocity perturbation.

This system is solved by introduction of a stream function

$$u_y = \partial_x \phi_+, \quad u_x = -\partial_y \phi_+, \quad p = -M \partial_y \phi_- \qquad (12)$$

with

$$(\Delta + M \partial_x)\phi_1 = 0, \quad (\Delta - M \partial_x)\phi_2 = 0 \qquad (13)$$

where $\phi_\pm = \phi_2 \pm \phi_1$. The two functions $\phi_{1,2}$ vanish at infinity and have to provide for the no slip condition at the cylinder surface. The drag force is obtained by integrating pressure and shear stress over the cylinder surface, which leads to

$$F_\perp = M \int_0^{2\pi} d\varphi (\partial_\tau \phi_-)_{\tau=1} \qquad (14a)$$

in the transverse case and

$$F_\parallel = M \int_0^{2\pi} d\varphi (\partial_\varphi \phi_-)_{\tau=1} = M[\phi_-(2\pi) - \phi_-(0)] \qquad (14b)$$

in the aligned field case. Concerning the many-valuedness of the function ϕ_- in an aligned field, we refer to Yosinobu.[14]

The solution of Eq. (13) for the transverse geometry is given by

$$\phi_1 = e^{-mx} \cdot \sum_{n=0}^{\infty} A_n K_n(m\tau) \cos(n\varphi) \qquad (15)$$

and $\phi_2(\varphi) = -\phi_1(\pi - \varphi)$ where $m = M/2$ and K_n denotes the modified Bessel function. The unknown coefficients have to be determined by the no slip condition on the cylinder surface.

The apparently straightforward method to determine ϕ_1 from Eq. (15) in the case $M \gg 1$ is to use the asymptotic formula of the modified Bessel function

$$K_n(x) = \sqrt{\frac{\pi}{2x}} \cdot e^{-x} \cdot \left(1 + O\left(\frac{1}{x}\right)\right) \qquad (16)$$

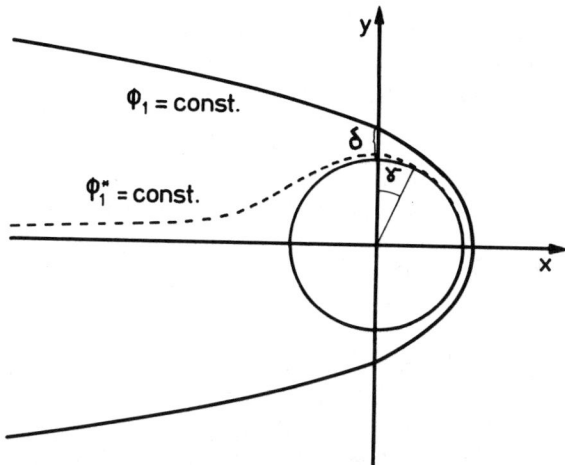

Fig. 2 Streamline ϕ_1 = const. for M >> 1.

if x >> 1. The results in Table 1 marked by an asterisk are obtained in this way. However, this striaghtforward approach is not a correct one. Substituting Eq. (16) into Eq. (15) offers the possibility to contract the set of coefficients A_n together with the trigonometric function to one unknown function of the angle. This function is then determined by the boundary condition at the cylinder surface, which leads by all means to an exponential dependence on m >> 1. From the mathematical point of view, this is in contradiction to the neglect of the higher order terms in Eq. (16). A more physical argument is obtained by consideration of the stream lines ϕ_1 = const.

The described approach leads to the leading term

$$\phi_1^* = \exp[-m(\tau - 1)(1 + \cos\varphi)] \qquad (17)$$

which is shown in Fig. 2. The correct shape of ϕ_1 = const b can be obtained by interpreting Eq. (13) in terms of a heat conduction equation, which leads directly to the curve drawn in Fig. 2.[20] Figure 2 shows that the previously described approximation yields an unphysical radial dependence of ϕ_1 on one side of the cylinder. It should be emphasized that the described asymptotic approach can give correct leading terms in such problems, where the velocity is directly obtained by the stream function and not by its derivative.[6]

The streamline picture for M >> 1 can be used to approximate the leading term of the radial derivative of ϕ_1 by

$$(\partial_x \phi_1)_{\tau=1} = \begin{cases} -M\phi_1 & \text{if} \quad x > 0 \\ 0 & \text{if} \quad x < 0 \end{cases} \qquad (18)$$

Indeed, this crude approximation is a very potential one. Use of Eq. (18) and the corresponding expression of ϕ_2 in combination with the no slip condition determines immediately the values of ϕ_{12} on the body surface and, consequently, the drag force. In this way the drag in an aligned magnetic field can be calculated in the Oseen approximation. The results are

$$C_s = \frac{1}{6} C_{so} \sqrt{Re^2 + 4M^2} \tag{19a}$$

for the sphere and

$$C_c = \frac{1}{4\pi} C_{co} \sqrt{Re^2 + 4M^2} \tag{19b}$$

for the cylinder. These formulas reduce in the limit Re << M to the previous results in Stokes approximation given in Table 1.

Analysis of the flow around a cylinder in a transverse magnetic field by use of Eq. (18) leads to a surface pressure p = tanφ for x < 0, which diverges by x = 0. This is due to the fact that the approximation of Eq. (18) ceases to be valid in the vicinity of x = 0. This region is characterized by a size of the order $\delta \sim M^{-2/3}$ and an extension up to $\gamma \sim M^{-1/3}$.[20] Exclusion of this part from integral Eq. (14) yields

$$C_c = \frac{1}{3\pi} C_{co} \cdot M \cdot \ell nM \tag{20}$$

for the leading term of the drag coefficient of a cylinder in a transverse magnetic field. This expression represents a completely new behavior of the increasing drag with an increasing magnetic field. Of course, it is based on a rough estimation and needs confirmation by a more precise analysis. Therefore, the same problem was solved numerically by transforming it to an algebraic system as described in Ref. 23. The numerical procedure works up to M = 30. The result is shown in Fig. 3 together with the asymptotic formulas of Eq. (20) and Table 1.

Though it is no final answer to the validity of Eq. (20), it clearly excludes the asymptotic formula proportional to M^2. The surface pressure is responsible for the fact that the drag increases stronger than linearly with the Hartmann number. The part of the drag arising from the surface shear stress is proportional to M. It remains the three-dimensional problem of MHD flow around a sphere in a transverse magnetic field. It would be highly beneficial to solve it in Stokes approximation for M >> 1 in order to compare the result of Eq. (10) with that of Hunt[19]. The solution was derived in general by Gotoh[2] but analyzed in detail for M << 1 only. Unfortunately, the simple approach of Eq. (18) fails here because it does not allow one to determine the stream functions on the sphere surface uniquely. Thus, this problem remains for further research.

Comparison with Measurements

In this section we focus on strong magnetic fields. The results for small Hartmann number shave never been tested in the laboratory since the corresponding conditions are difficult to attain experimentally.

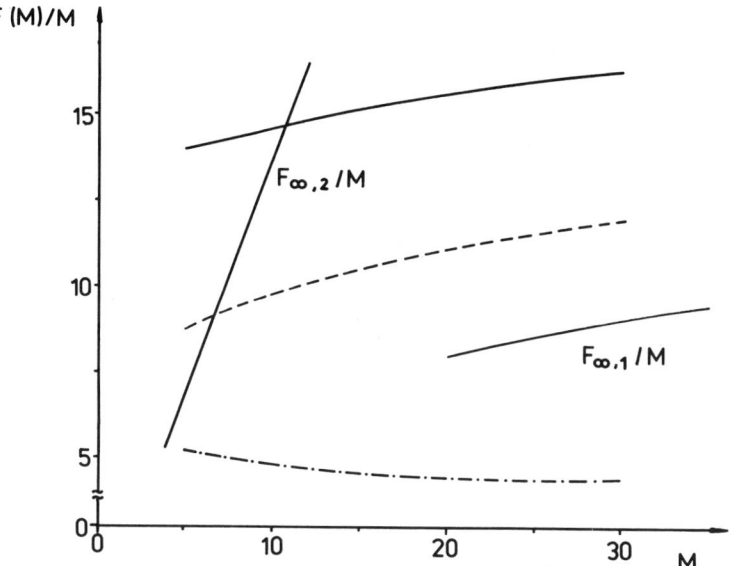

Fig. 3 Drag force on a cylinder in a transverse magnetic field, $F_{\infty,1}$ corresponds to Eq. (20) and $F_{\infty,2}$ to the formula in Table 1.
_____ complete drag;   --------------- pressure-induced drag;
_____ . _____ shear stress-induced drag.

Drag measurements of a cylinder in a transverse magnetic field were performed by Kalis et al.[21] for relatively low Reynolds numbers in combination with high Hartmann numbers. The results of this experiment were approximated by $C/C_o = 1 + 2.2 \sqrt{N_d}$ where C_o denotes the drag without magnetic field and the index d refers to the diameter as typical length. However, this approximation was too rough. A reconsideration of the drag values for the three smallest Re values is shown in Figs. 4a and 4b. It shows that the force increases stronger than linearly but slower than M^2 for increasing M. In order to calculate the drag force from the measured C/C_o ratio, the value of C_o was taken from the standard drag curve of an infinite cylinder since C_o was not given in Ref. 21. Of course, this value is larger than that of the finite cylinder used in the experiment, and, therefore, the measuring points in Figs. 4a and 4b should be shifted to smaller values, improving the agreement between theory and experiment. Though the available measurements as well as the theoretical results up to M = 30 represent no final proof of the asymptotic behavior of Eq. (20), Figs. 4a and 4b are a promising indicator of it.

In this way the presented results are a new candidate for a successful comparison between theory and experiment concerning MHD drag coefficients. It should be noted that the results for MHD drag of a sphere in a transverse magnetic field (which are considered to represent the only successful agreement between theory and experiment up to now) contain a slight uncertainty. The

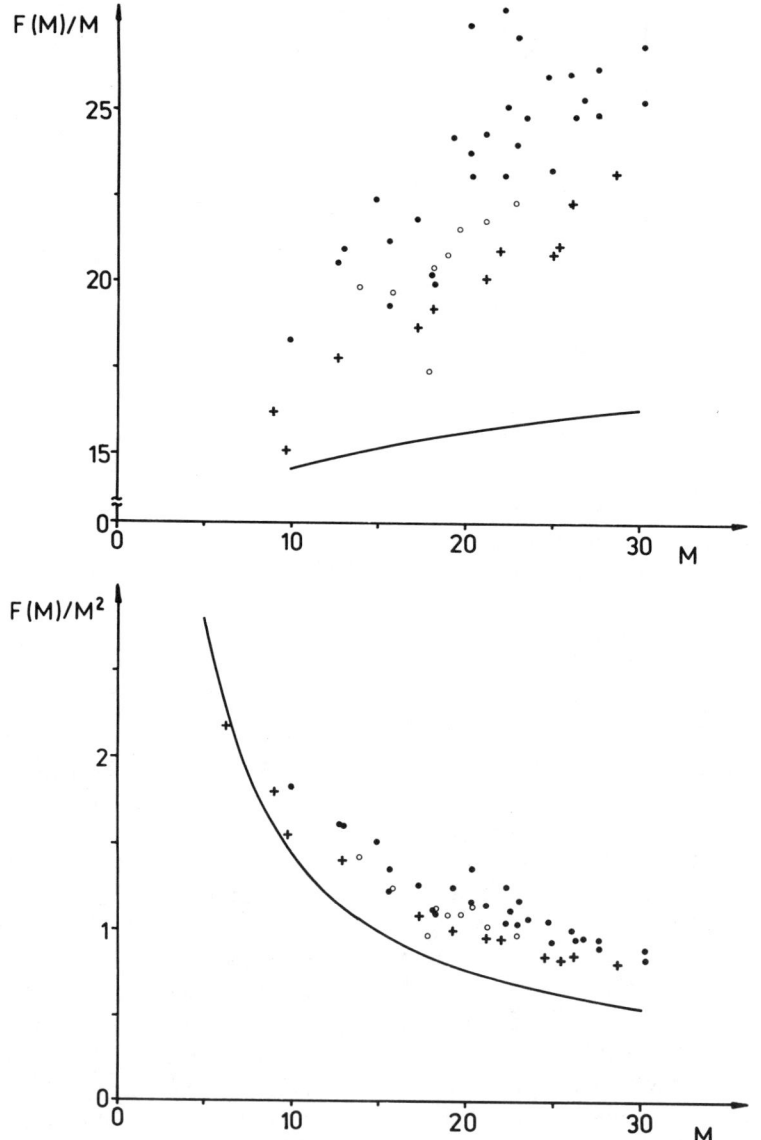

Fig. 4 Cylinder MHD drag coefficients. Comparison between theory and experiment (from Ref. 21). Re = 5.9(o), 6.9(•), 20.3(+).

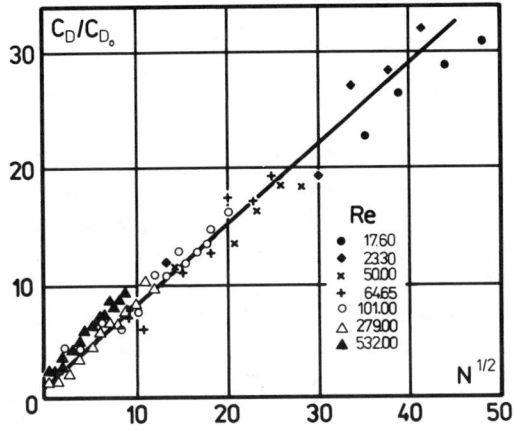
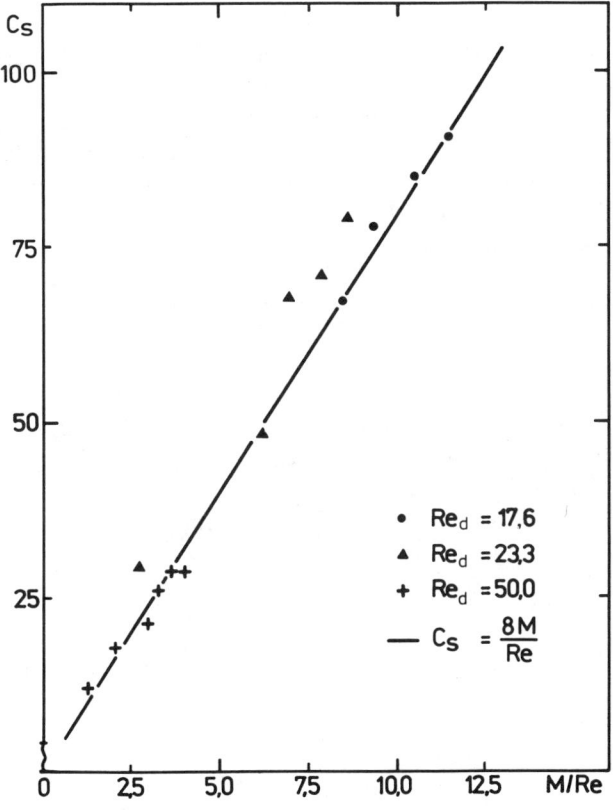

Fig. 5 MHD drag coefficient of a sphere. Experimental values of Kali[21]: a) C/C_o vs. $\sqrt{N_d}$ and b) C vs. M/Re.

measured values are shown in Figs. 5a and 5b in two different ways. Kalis et al.[21] interpreted the measurements by $C/C_o = 1 + 0.7\sqrt{N_d}$ (Fig. 5a), whereas Hunt[19] stated a good agreement with his asymptotic analysis, yielding $C = 8M/Re$ (Fig. 5b). However, the only result is: both interpretations are possible within the experimental scatter. The reason of this ambiguity is that both formulas differ, in principle, only by a factor of $C_o \cdot \sqrt{Re_d}$, which is approximately constant in the range of Reynolds numbers used. To clarify this situation, measurements are necessary for either stronger magnetic fields or smaller Reynolds numbers.

Regarding the results for an aligned magnetic field, there are no measurements with strong magnetic field and relatively low Reynolds number. The values measured by Yonas[22] for a sphere in the range $10 < N_d < 80$, $10^4 < Re < 25 \cdot 10^4$ and $0.1 < Rm < 2.5$ are well approximated by $C = 0.33\sqrt{N_d}$ but cannot be correlated to the results given in Table 1.

Deflection of a Rising Bubble

In this last section an application is given of the presented drag coefficients. In view of the rise of a bubble (which should be considered as a solid sphere due to surfactants) in a liquid metal, the question arises of which way an external magnetic field is able to change the direction of bubble motion. Obviously, corresponding results are valid for a falling solid sphere. Defining the angle between the magnetic field and the direction of buoyancy by β (see Fig. 6), the direction of bubble rise is given by

$$\tan\gamma = \frac{(F_\perp - F_\parallel)\sin\beta\cos\beta}{F_\perp\cos^2\beta + F_\parallel\sin^2\beta} \qquad (21)$$

where $F_\perp$ and $F_\parallel$ denote the drag force of the sphere in a transverse and an aligned field, respectively. Equation (21) emphasizes that a deflection appears only for an inclined magnetic field direction.

Using the result of Eq. (20) and that of Table 1 for $F_\perp$ and $F_\parallel$, respectively, the deflection angle does not depend on the Hartmann number but only on β. A maximum deflection is obtained if

$$\cos^2\beta_m = \frac{F_\parallel}{F_\perp + F_\parallel} \qquad (22)$$

and takes the value

$$\tan\gamma_m = \frac{1}{2}\frac{F_\perp - F_\parallel}{\sqrt{F_\perp \cdot F_\parallel}} \qquad (23)$$

That means a maximum deflection of a rising bubble of about $\gamma_m = 19.58$ is obtained if a strong magnetic field is applied under an angle of $\beta_m = 55$ deg relative to the vertical direction.

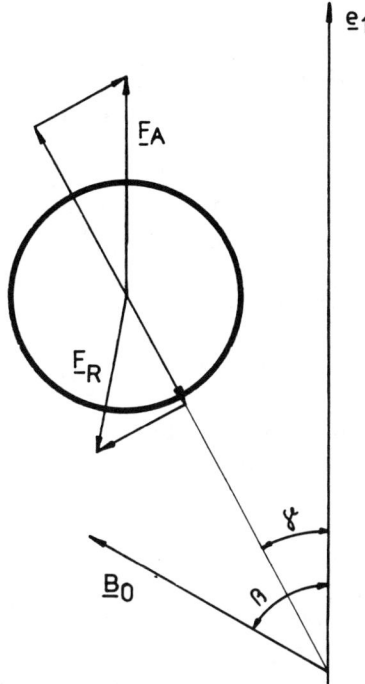

Fig. 6 Deflection of a rising bubble.

Conclusion

A number of new theoretical results [Eqs. (7), (19a) (19b) and (20)] are derived for MHD drag coefficients in an unconfined fluid. For the first time, numerical results of the Stokes drag in a strong magnetic field were obtained, which confirm the somewhat surprising asymptotic dependence of Eq. (20). Regarding the tendency with increasing magnetic field, the exact theoretical results for the cylinder drag in a transverse field are in good qualitative agreement with measurements available in the literature, which were up to now incorrectly interpreted as being proportional to the Hartmann number.

A variety of questions remain unsolved, in particular, the influence of external walls and the range of validity of the Stokes approximation. A serious gap consists in the lack of solution of the high Hartmann number Stokes flow around a sphere in a transverse magnetic field. This will be the subject of a future study since a comparison of the results with the analysis of Hunt and Ludford[18] should substantially improve our knowledge of MHD flow around obstacles in a strong magnetic field.

Acknowledgment

The author is grateful to Dr. A. Gailitis (Institute of Physics, Riga-Salaspils, USSR) for abundant fruitful discussion on this subject.

References

[1] Hunt, J. C. R. and Moreau, R., "Liquid-metal Magnetohydrodynamics with Strong Magnetic Fields: a Report on Euromech 70," *Journal of Fluid Mechanics*, Vol. 78, No. 2, 1976, pp. 261-288.

[2] Gotoh, K., "Stokes Flow of an Electrically Conducting Fluid in a Uniform Magnetic Field," *Journal of Physics Society*, Japan, Vol. 15, 1960, pp. 696-705.

[3] Tsinober, A., "MHD Flow Around Obstacles," *Zinatne*, Riga, USSR, 1970.

[4] Chester, W., "The Effect of a Magnetic Field on Stokes Flow in a Conducting Fluid," *Journal of Fluid Mechanics*, Vol. 3, No. 3, 1957, p. 304.

[5] Damburg, R.J., "Flow of a Viscous, Electrically Conducting Fluid Around an Infinite Cylinder in an Aligned Magnetic Field," *Isvestiya Akademii Nauk & Latvuskoy SSR*, No. 5, 1959, pp. 81-84.

[6] Gerbeth, G., Thess, A., and Marty, Ph., "Theoretical Study of the Flow Around a Cylinder in Crossed Electric and Magnetic Fields," *European Journal of Mechanics* (to be published).

[7] Yosinobu, H. and Kakutani, T., "Two-dimensional Stokes Flow of an Electrically Conducting Fluid in a Uniform Magnetic Field," *Journal of Physics Society*, Japan, Vol. 14, 1959, pp. 1433-1444.

[8] Gerschuni, G. and Shuchovizkij, E., "Flow Around a Sphere of an Electrically Conducting Fluid in a Strong Magnetic Field," *Journal of Technical Physics*, Vol. 30, 1960, pp. 925-926.

[9] Chester, W. and Moore, D. W., "The Effect of Magnetic Field on the Flow of a Conducting Fluid Past a Circular Disk," *Journal of Fluid Mechanics*, Vol. 10, No. 3, 1961, pp. 466-472.

[10] Kotov, J. and Waliev, C., "Flow of an Electrically Conducting Fluid Around an Infinite Cylinder in a Strong Magnetic Field," *Isvestiya Akademii Nauk & Usbeksoy SSR*, No. 5, 1962, pp. 88-89.

[11] Childress, S., "The Effect of a Strong Magnetic Field on Two-Dimensional Flows of a Conducting Fluid," *Journal of Fluid Mechanics*, Vol. 15, No. 3, 1963, pp. 429-441.

[12] Tsinober, A. and Stern, A., "On the Stoke Flow Around a Circular Cylinder in a Transverse Magnetic Field," *Magnitnaja Gidrodinamika*, No. 4, 1967, pp. 146-147.

[13] Gotoh, K., "Magnetohydrodynamic Flow Past a Sphere," *Journal of Physics Society*, Japan, Vol. 15, 1960, pp. 189-196.

[14] Yosinobu, H., "A Linearized Theory of Magnetohydrodynamic Flow Past a Fixed Body in a Parallel Magnetic Field," *Journal of Physics Society*, Japan, Vol. 15, 1960, pp. 175-188.

[15] Reitz, J. R. and Foldy, L. L., "The Force on a Sphere Moving Through a Conducting Fluid in the Presence of a Magnetic Field," *Journal of Fluid Mechanics*, Vol. 11, No. 1, 1961, pp. 133-142.

[16] Ludford, G. S. S., "The Effect of a Very Strong Magnetic Cross-Field on Steady Motion Through a Slightly Conducting Fluid: Three-Dimensional Case," *Archive for Rational Mechanics and Analysis*, Vol. 8, 1961, p. 242-253.

[17]Ludford, G. S. S., "The Effect of a Very Strong Magnetic Cross-Field on Steady Motion Through Slightly Conducting Fluid," *Journal of Fluid Mechanics*, Vol. 10, 1961, pp. 141-155.

[18]Hunt, J. C. R. and Ludford, G. S. S., "Three-dimensional MHD Duct Flows with Strong Transverse Magnetic Fields, Part 1: Obstacles in a Constant Area Channel," *Journal of Fluid Mechanics*, Vol. 33, No. 4, 1968, pp. 693-714.

[19]Hunt, J. C. R., "Bluff Body Drag in Strong Transverse Magnetic Field," *Magnitnaja Gidrodinamika*, No. 1, 1970, pp. 35-38.

[20]Gailitis, A. and Gerbeth, G., "On Stokes Flow Around a Circular Cylinder in a Strong Magnetic Field," *Magnitnaja Gidrodinamika*, No. 2, 1987, pp. 137-139.

[21]Kalis, Ch. E., Slusarjov N. M., Tsinober, A. and Stern, A., "Body Drag for High Stuart Numbers," *Magnitnaja Gidrodinamika*, No. 4, 1966, pp. 152-153.

[22]Yonas, G., "Measurement of Drag in a Conducting Fluid with an Aligned Field and Large Interaction Parameter," *Journal of Fluid Mechanics*, Vol. 30, No. 4, 1967, pp. 813-821.

[23]Gerbeth, G., Numerical Determination of the Stokes Drag of a Circular Cylinder in a Transverse Magnetic Field," *Magnitnaja Gidrodinamika* (to be published).

Heat Transfer in an MHD Flow Inside a Channel with Walls of Finite Thickness

Sergio Cuevas*
*Instituto de Investigaciones Eléctricas,
Cuernavaca, Mor. 62000 Mexico*
and
Eduardo Ramos†
IIM-UNAM, Temixco, Mor. 62580 Mexico

Abstract

The energy balance equation for an MHD flow inside a two-dimensional channel has been solved analytically considering walls of finite thickness. A Hartmann-like velocity profile and the electric density current obtained taking into account the electric conductivity of the walls are used in the convective, viscous, and Joule heating terms. The thermal conductivity of the walls is incorporated through the boundary conditions. The combined influence of both the physical and geometrical properties of the system and boundary conditions on the temperatures are described. It is found that minimum temperatures in the flow can be obtained for an optimum value of the conductance ratio, which is the product of the electric conductivity ratio (wall/fluid) and the thickness ratio (wall/channel).

Introduction

Heat transfer in MHD channel flows where internal heat sources are present is important in the context of the design of various MHD devices. Idealized mathematical models describing temperature fields in channels have appeared in the literature. Nigam and Singh[1] proposed a method to solve the energy balance equation for an MHD channel flow with known

---
Copyright © 1992 by S. Cuevas and E. Ramos. Published by the American Institute of Aeronautics and Astronautics, Inc., with permission.
*Researcher, Dpto. Fuentes No Convencionales de Energía, A.P. 475.
†Researcher, Solar Energy Laboratory, A.P. 34.

wall temperatures; Jain and Srinivasan [2] solved a similar problem considering electrically conducting walls. Singh and Lal [3] worked on a three-dimensional MHD flow in a rectangular duct using a numerical method. Cuevas and Ramos[4] tackled a similar two-dimensional problem but considering boundary conditions of the third kind in order to analyze the effect of heat transfer coefficient from the channel to the surrounding ambient. In that case, the heat transfer from the duct to the surroundings is taken into account via a "Newton's cooling law." This boundary condition is more closely related to the reality. Javeri [5,6] considered this condition for the heat transfer in the thermal entrance region of a flat MHD channel, but all heat dissipation effects and axial heat conduction are neglected.

In the present study, we consider the heat transfer problem of a laminar flow inside an MHD duct with thermally and electrically conducting walls of finite thickness. Likewise, the heat dissipation due to Joule and viscous effects as well as axial heat conduction in the fluid are taken into account. The external ambient temperature is held constant up to a certain location, where it changes to another constant value. These boundary conditions would constitute an idealized model of the flow in a duct with an external heat exchanger whose edge is placed at the location where the ambient temperature changes. The temperature distribution in the flow is described in terms of the thermal and electrical conductivities of the wall, their geometry, and the boundary conditions.

Formulation and Solution of the Problem

Consider a two-dimensional channel infinite in the x' direction and whose inner boundaries are separated a distance $2a$ as shown in Fig. 1. The width of the channel walls is L, and their thermal and electrical conductivity are κ_w and σ_w, respectively. Inside the channel, a viscous and incompressible electrically conducting fluid flows under the combined action of a constant pressure gradient $\partial p / \partial x'$ and an externally imposed magnetic field H_o in the positive y' direction; it is assumed that the hydrodynamic profile is fully developed. The ambient temperature outside the channel is T_{a1} for $x' \leq 0$ and T_{a2} for $x' \geq 0$. The velocity $V(y')$ and the induced magnetic field $H_x(y')$ are governed by the following momentum and magnetic field balance equations:

$$\eta \frac{d^2 V}{dy'^2} + \mu_o H_o \frac{dH_x}{dy'} = \frac{\partial p}{\partial x'} = const. \qquad (1)$$

$$H_o \frac{dV}{dy'} + \frac{1}{\mu_o \sigma_f} \frac{d^2 H_x}{dy'^2} = 0 \qquad (2)$$

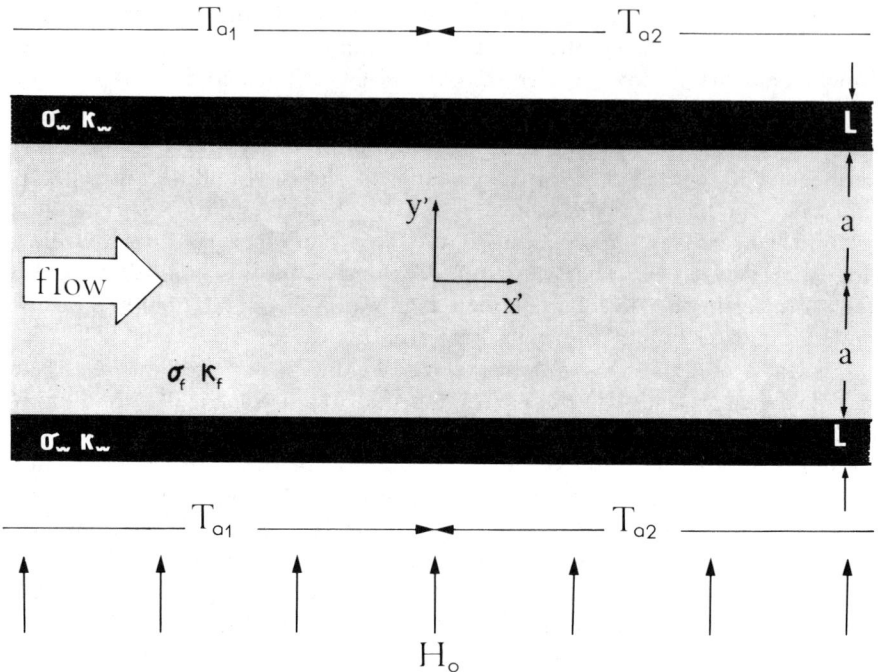

Fig. 1 The geometry considered. Ambient temperature T_{a1} for $x \leq 0$ and T_{a2} for $x \geq 0$. H_0 is the external magnetic field in the y' direction.

with boundary conditions

$$V = 0, \quad \sigma_w L \frac{dH_x}{dy'} \pm \sigma_f H_x = 0 \quad \text{at} \quad y' = \pm a \qquad (3)$$

where η and σ_f are the viscosity and the electrical conductivity of the fluid and μ_o is the vacuum magnetic permittivity. The solution of the system was found by Chang and Yen[7]. Assuming that both walls have the same conductivity, the solution is:

$$V(y') = \frac{\Phi V_m}{A}\left[\cosh M - \cosh\left(\frac{My'}{a}\right)\right] \qquad (4)$$

$$H_x(y') = -\sqrt{\sigma_f \eta}\frac{V_m}{A}\left[\left(\frac{y'}{a}\right)\sinh M - \Phi \sinh\left(\frac{My'}{a}\right)\right] \qquad (5)$$

V_m is the mean velocity of the fluid and $A = \cosh M - M^{-1}\sinh M$. The Hartmann number is defined as $M = a\mu_o H_o \sqrt{\sigma_f/\eta}$. The parameter Φ is given by

$$\Phi = \frac{\varphi + 1}{\varphi M \coth M + 1} \tag{6}$$

where $\varphi = (\sigma_w/\sigma_f)(L/a)$ is the conductance ratio. It should be noted that if $L \to 0$ or $\sigma_w \to 0$, the fully developed Hartmann profile for an open external circuit is recovered. This is one of the cases studied in Ref. 4.

The temperature field $T(x', y')$ can be obtained from the energy balance equation

$$\rho c_p V \frac{\partial T}{\partial x'} = \kappa_f \left(\frac{\partial^2 T}{\partial x'^2} + \frac{\partial^2 T}{\partial y'^2} \right) + \eta \left(\frac{dV}{dy'} \right)^2 + \frac{J_z^2}{\sigma_f} \tag{7}$$

where ρ, c_p, and κ_f are the density, specific heat at constant pressure, and thermal conductivity of the fluid, respectively, and V is given by Eq. (4). J_z is the electric current density and is obtained from

$$J_z = \frac{dH_x}{dy'} = -\sqrt{\sigma_f \eta} \frac{MV_m}{aA} \left[\frac{\sinh M}{M} - \Phi \cosh \left(\frac{My'}{a} \right) \right] \tag{8}$$

The thermal boundary conditions required for the problem can be deduced from Fig. 2. It is assumed that the ambient temperature is T_{aj} where the subindex $j = 1, 2$ denotes the two regions of the problem ($x' < 0$ and $x' > 0$); the wall temperature on the inner and outer faces of the wall are T_w and $T_{w'}$, respectively. Under equilibrium conditions, the heat flow Q transferred from the fluid to the wall should be equal to the heat transferred from the wall to the ambient, assuming that the temperature gradients in the x' direction are much smaller than those in the y' direction. Under these conditions, the following equations must hold:

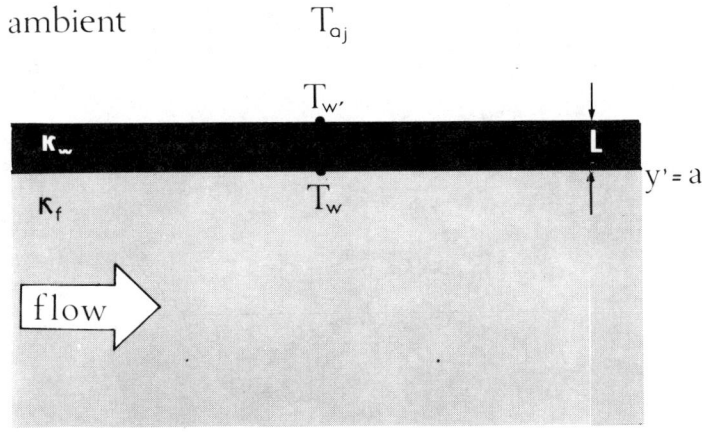

Fig. 2 The thermal boundary conditions.

$$Q = h(T_{w'} - T_{aj}) \quad j = 1, 2 \tag{9}$$

$$Q = -\frac{\kappa_w (T_{w'} - T_w)}{L} \tag{10}$$

$$Q = -\kappa_f \left(\frac{\partial T_j}{\partial y'}\right)_{y'=a} = -\kappa_w \left(\frac{\partial T_w}{\partial y'}\right)_{y'=a} \quad j = 1, 2 \tag{11}$$

h is the heat transfer coefficient from the wall to ambient. From Eqs (9) to (11), Q, T_w, and $T_{w'}$ can be eliminated to give the boundary conditions for the temperature T_j of the fluid in terms of the ambient temperature and the physical and geometrical properties of the wall:

$$\pm \kappa_f \frac{\partial T_j}{\partial y'} + h_{eff}(T_j - T_{aj}) = 0 \quad |x'| < 0, \ y' = \pm a, \ j = 1, 2 \tag{12}$$

h_{eff} is the effective heat transfer coefficient, given by

$$h_{eff} = \frac{\kappa_w}{L + (\kappa_w/h)} \tag{13}$$

which reduces to h when $L \to 0$. Boundary conditions [Eq. (12)] indicate that the heat entering or leaving the system is a linear function of the local wall temperature.

Written in terms of nondimensional variables

$$\Theta_j = \frac{T_j - T_{a2}}{T_{a1} - T_{a2}}, \quad x' = \frac{3}{2} x a Pe, \quad y' = a\left(\frac{2y}{\pi} - 1\right) \tag{14}$$

Eq. (7) is

$$\alpha^2 \frac{\partial^2 \Theta_j}{\partial x^2} + \frac{\partial^2 \Theta_j}{\partial y^2} = \frac{8\Phi}{3A\pi^2}\left[\cosh M - \cosh M\left(\frac{2y}{\pi} - 1\right)\right]\frac{\partial \Theta_j}{\partial x}$$
$$-C\left(\frac{\Phi M}{A}\right)^2\left[\cosh 2M\left(\frac{2y}{\pi} - 1\right)\right.$$
$$\left.-\frac{2\sinh M}{\Phi}\cosh M\left(\frac{2y}{\pi} - 1\right)\right.$$
$$\left.+\left(\frac{\sinh M}{\Phi M}\right)^2\right] \tag{15}$$

where C, α, and the Péclet number Pe are given by

$$C = \frac{4\eta V_m^2}{\pi^2 \kappa_f (T_{a\,1} - T_{a\,2})}, \qquad \alpha = \frac{4}{3\pi Pe}, \qquad Pe = \frac{V_m a \rho c_p}{\kappa_f} \qquad (16)$$

The nondimensional boundary conditions are:

$$\frac{\partial \Theta_1}{\partial y} \pm \frac{2Bi}{\pi}(\Theta_1 - 1) = 0 \qquad x < 0, \; y = \pi, 0 \qquad (17)$$

and

$$\frac{\partial \Theta_2}{\partial y} \pm \frac{2Bi}{\pi}\Theta_2 = 0 \qquad x > 0, \; y = \pi, 0 \qquad (18)$$

where the Biot number Bi, defined as

$$Bi = \frac{a h_{eff}}{\kappa_f} \qquad (19)$$

is the nondimensional expression of the effective heat transfer coefficient. It should be noticed that boundary conditions are discontinuous at $x = 0$, but the fluid temperature and the heat flux are considered continuous everywhere and at $x = 0$ in particular. Then we must have

$$\Theta_1 = \Theta_2 \qquad \frac{\partial \Theta_1}{\partial x} = \frac{\partial \Theta_2}{\partial x} \qquad x = 0, \; 0 \leq y \leq \pi \qquad (20)$$

The solution strategy follows closely that of Ref. 4, and therefore only the main steps will be presented. Full details are given in Refs. 4 and 8. The solution of the governing Eq. (15) with the external boundary conditions (17) and (18) is obtained as follows: a) the two x regions are considered separately; b) in each region the solution is assumed to be the sum of two functions, reflecting the behavior far away from and near to the discontinuity, respectively; and c) the solutions in the two x regions are then coupled using the conditions in Eq. (20). Following the steps just described, the temperature in each region is assumed to be of the form:

$$\Theta_j = \Omega_j(x, y) + \Psi_j(y) \qquad j = 1, 2 \qquad (21)$$

where $\lim_{x \to \pm\infty} \Omega = 0$ and $\Psi_j(y)$ is the temperature profile far away from the origin. Solutions for $|x| \to \infty$ can be found by substituting Eq. (21) in Eq. (15), with $\Omega(x, y) = 0$, to obtain

$$\frac{d^2 \Psi_j}{dy^2} = -C\left(\frac{\Phi M}{A}\right)^2 \left[\cosh 2M\left(\frac{2y}{\pi} - 1\right)\right.$$
$$\left. - \frac{2\sinh M}{\Phi M}\cosh M\left(\frac{2y}{\pi} - 1\right) + \left(\frac{\sinh M}{\Phi M}\right)^2\right] \qquad (22)$$

The boundary conditions are

$$\frac{d\Psi_1}{dy} \pm \frac{2Bi}{\pi}(\Psi_1 - 1) = 0 \quad y = \pi, 0 \tag{23}$$

for $x \to -\infty$ and

$$\frac{d\Psi_2}{dy} \pm \frac{2Bi}{\pi}\Psi_2 = 0 \quad y = \pi, 0 \tag{24}$$

for $x \to \infty$. The solutions to Eq. (22) with boundary conditions (23) and (24) are:

$$\Psi_1(y) = \frac{C}{A^2}F(y) + 1 \quad \text{for } x < 0 \tag{25}$$

$$\Psi_2(y) = \frac{C}{A^2}F(y) \quad \text{for } x > 0 \tag{26}$$

respectively. The function $F(y)$ is given by

$$\begin{aligned}F(y) =\ & -\Phi\frac{\pi^2}{16}\left[\cosh 2M\left(\frac{2y}{\pi}-1\right) - \frac{8\sinh M}{\Phi M}\cosh M\left(\frac{2y}{\pi}-1\right)\right.\\ & + \frac{8\sinh^2 M}{\Phi^2\pi^2}y(y-\pi) - \cosh 2M + \frac{8\sinh M}{\Phi M}\cosh M\Bigg]\\ & -\frac{M}{Bi}\left[2\sinh 2M - \frac{8\sinh M}{\Phi M}\sinh M + \left(\frac{2\sinh M}{\Phi M}\right)^2\right]\end{aligned} \tag{27}$$

Upon substituting Eq. (21) in the governing Eq. (15) and considering Eq. (22), we get the following equation that must be satisfied by Ω_j:

$$\alpha^2\frac{\partial^2\Omega_j}{\partial x^2} + \frac{\partial^2\Omega_j}{\partial y^2} = \frac{8\Phi}{3A\pi^2}\left[\cosh M - \cosh M\left(\frac{2y}{\pi}-1\right)\right]\frac{\partial\Omega_j}{\partial x} \tag{28}$$

The boundary conditions for Eq. (28) are:

$$\frac{\partial\Omega_j}{\partial y} \pm \frac{2Bi}{\pi}\Omega_j = 0 \quad |x| > 0, \ y = \pi, 0, \ j = 1, 2 \tag{29}$$

$$\Omega_1(-\infty, y) = \Omega_2(\infty, y) = 0 \quad 0 < y < \pi \tag{30}$$

$$1 + \Omega_1(0, y) = \Omega_2(0, y) \quad 0 < y < \pi \tag{31}$$

and

$$\frac{\partial \Omega_1}{\partial x} = \frac{\partial \Omega_2}{\partial x} \quad x = 0, \quad 0 < y < \pi \tag{32}$$

The solution of Eq. (28) subject to conditions in Eqs. (29) to (32) involves a rather lengthy procedure that is not reproduced here. Details of the solution procedure can be found in Ref. 4. The solution has the form

$$\Omega_j = \frac{2}{\pi} \sum_{k=1}^{\infty} \sum_{q=1}^{\infty} a_{kj}^{(q)} \exp(\xi_j^{(q)} x)(\sin \lambda_k y + p_k \cos \lambda_k y) \tag{33}$$

where the p_k and λ_k are implicitly determined by substituting Eq. (33) in Eqs. (29) to obtain a transcendental equation whose roots give the values required. Substituting Eq. (33) in Eq. (28) and performing the multiplication by $(\sin \lambda_k y + p_k \cos \lambda_k y)$ and integration with respect to y from 0 to π, a homogeneous system of linear equations for the a_{kj}'s is obtained. The roots of the characteristic associated equation are $\xi_j^{(q)}$. Finally, in order to totally determine the coefficients $a_{kj}^{(q)}$, conditions (31) and (32) are used. The temperature field in the whole flow region is found by substituting Eqs. (25) to (27) and (33) in Eq. (21). The full solution contains an infinite number of terms, but due to the extremely cumbersome algebraic procedure involved in the calculation of large order terms, in this study we work only with the first two terms of the series. With this strategy of solution, the temperature at $x = 0$ can be found using the lefthand ($j = 1$) or righthand ($j = 2$) solutions and taking the limit $x \to 0$. This gives $\Theta(x = 0^-, y)$ and $\Theta(x = 0^+, y)$, which must coincide when an infinite number of terms are taken in the series. All results were calculated for $C = 0.01$ and assuming that $T_{a1} > T_{a2}$. Before discussing the properties of the solution in detail, it must be remarked that making the $Bi \to \infty$, the solution given by Jain and Srinivasan[2] is recovered while the solution presented by Cuevas and Ramos[4] can be obtained by letting $L \to 0$ or $\sigma_w \to 0$.

Results

In order to visualize the influence of the thermally and electrically conducting walls of finite thickness on the temperature of the fluid, we shall describe first the solution in the region $x = \pm \infty$ where the temperature profile is fully developed and a simpler analytical expression is available.

Figure 3 shows the nondimensional temperature Θ for $x \to \infty$ and $y = \pi/2$, which corresponds to the channel center, as a function of the conductance ratio φ for different Biot numbers. In this figure the Hartmann number is equal to 40. In the case $Bi = \infty$, the temperature features a monotonic increasing trend, whereas each of the cases calculated with finite Biot numbers present a minimum. Larger temperatures are found in

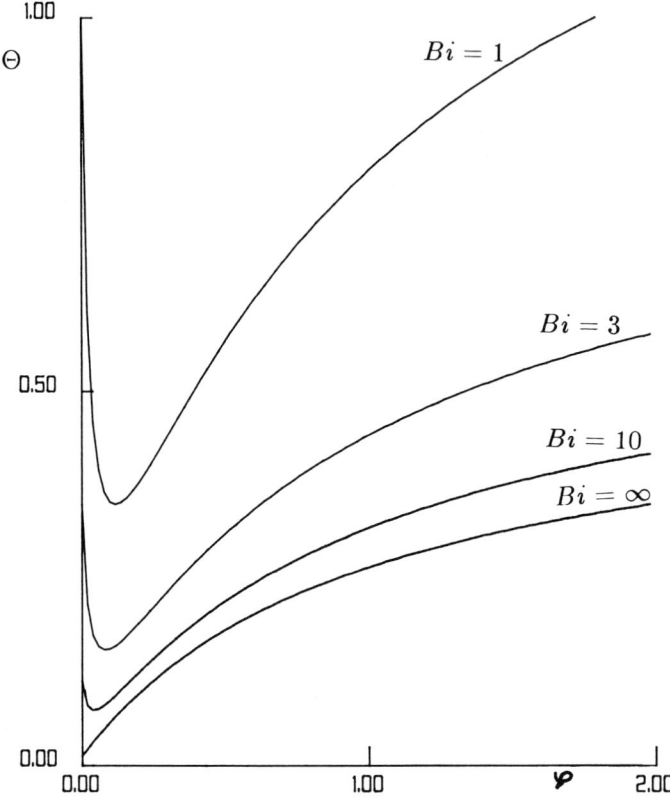

Fig. 3 Temperatures at the channel center as a function of the conductance ratio for $M = 40$ and different Biot numbers.

the channel center for increasing φ using $Bi = \infty$, since Joule heat dissipation is larger for larger φ as can be inferred from Eq. (8). A physical explanation for this behavior can be given examining the electric currents under different values of φ. In all cases the integral of the electric current density must be zero. Insulating walls ($\varphi = 0$) demand that positive and negative currents inside the fluid balance each other while in cases with electrically conducting walls ($\varphi > 0$) negative return currents can flow through the walls themselves, permitting larger positive currents (and consequently larger heat dissipation) in the channel center. The temperature becomes higher as the Biot number decreases, because decreasing the Biot number diminishes the capacity of the system to transfer heat to the environment. In the case $Bi \to 0$, the thermal insulation is perfect and the temperature inside the channel increases without limit [see Eqs. 25 to 27]. All curves in Fig. 3 present similar behavior for large φ, which can be interpreted in like terms as those given for $Bi = \infty$. The minimum in the temperature

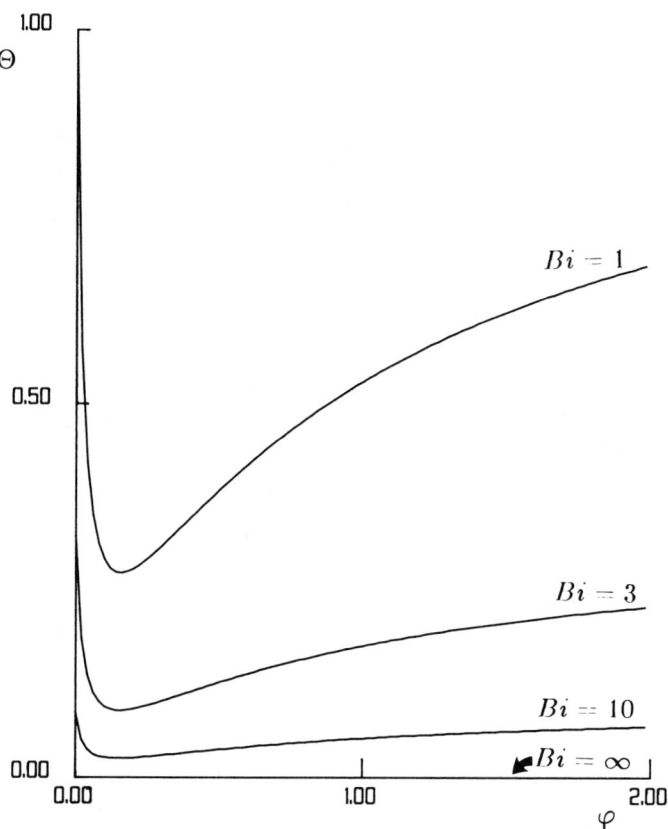

Fig. 4 Temperatures at the channel wall as a function of the conductance ratio for $M = 40$ and different Biot numbers.

curves for finite Biot numbers is located at a larger conductance ratio (φ) for smaller Biot numbers and results from the fact that the heat transfered to the exterior is proportional to the external temperature of the wall. It can be demonstrated that this minimum corresponds to maximum heat transfer to the ambient. This last quantity can be calculated from Eqs. (23) to (27). The conductance ratio φ is the product of the conductivity ratio σ_w/σ_f and the thickness ratio L/a, and, once the fluid and the wall material are fixed, the conditions required to achieve maximum heat extraction (or minimum temperature) are only geometrical. Figure 4 shows the temperature at the channel wall ($y = 0, \pi$) as a function of φ for the same Biot numbers as in Fig. 3. The behavior of the curves is similar to that already described except for the case $Bi = \infty$, where the temperature is always zero. This result follows from the fact that $Bi = \infty$ corresponds to boundary conditions of the first kind, where the wall temperature is fixed.

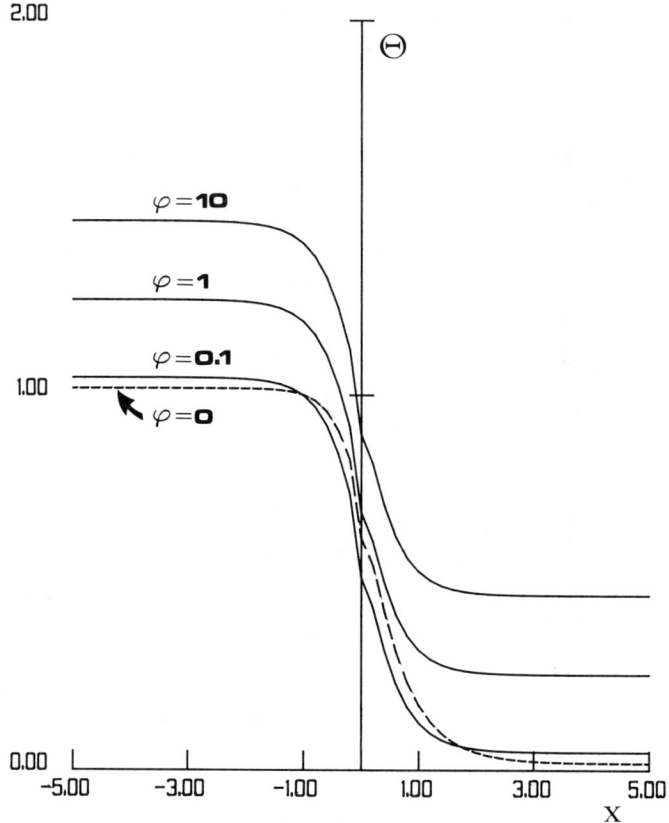

Fig. 5 Temperatures at the channel center as a function of the axial coordinate x for different conductance ratios. $M = 40$, $Pe = 1$, $Bi = \infty$.

This is the particular limit studied in Ref. 2. It must be recalled that a modified version of Biot number is used in this study and $Bi = \infty$ requires $h \to \infty$ and $L \to 0$ simultaneously.

Temperature profiles along the axial coordinate x and different conductance ratios are presented in Fig. 5 and Fig. 6 using $M = 40$ and $Pe = 1$. The temperatures at $y = \pi/2$ and $Bi = \infty$ given in Fig. 5 show the smooth transition from upstream to downstream temperatures. $\varphi = 0$ profile has the lowest temperature value for large $|x|$. In accordance with Fig. 3, for large $|x|$, temperature increases as φ increases, but near $x = 0$, lower temperatures are attained by small φ different from zero. Figure 6 shows the temperatures at the channel center calculated with $Bi = 3$. A similar behavior is observed as for Fig. 5 but the temperature for $\varphi = 0.1$ is always smaller than that for $\varphi = 0$; this feature is in agreement with the results shown in Fig. 3.

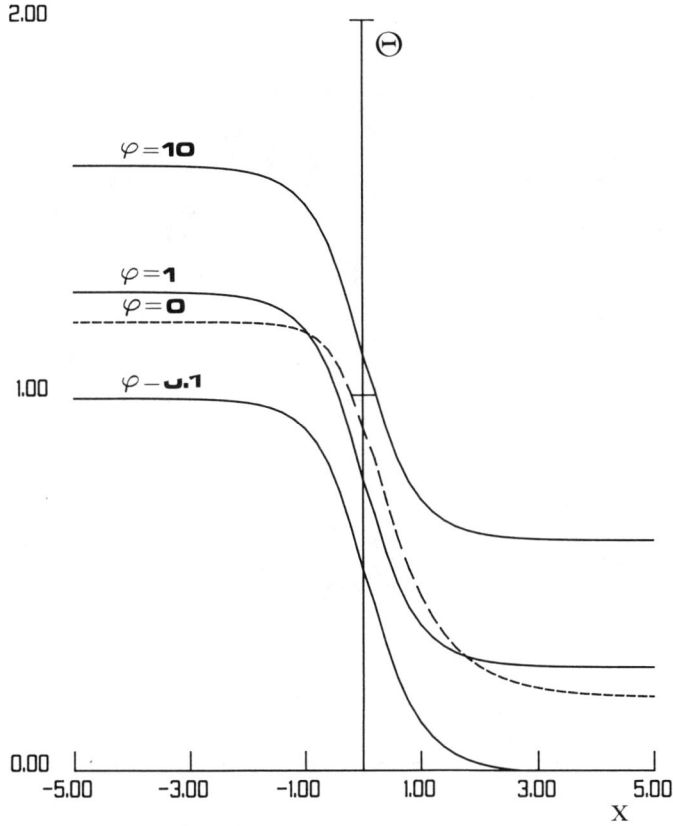

Fig. 6 Temperatures at the channel center as a function of the axial coordinate x for different conductance ratios. $M = 40$, $Pe = 1$, $Bi = 3$.

The temperature as a function of the transverse coordinate y for $M = 40$, $Bi = 10$, $Pe = 10$, and $x = \pm 0.1, \pm 0.5, \pm 1$, and ± 10 is shown in Fig. 7. For $x = 0$ the solution is taken as the average of $\Theta(x = 0^-, y)$ and $\Theta(x = 0^+, y)$ solutions. The two solutions differ due to the limited number of terms in the series. The larger temperature is always found at the channel center. Similar results were found using different values of the conductance ratio and are not shown here. The influence of the Hartmann number on the temperature distribution is similar to that reported in Ref. 4. The larger the Hartmann number, the larger the dissipation and the corresponding temperature increase.

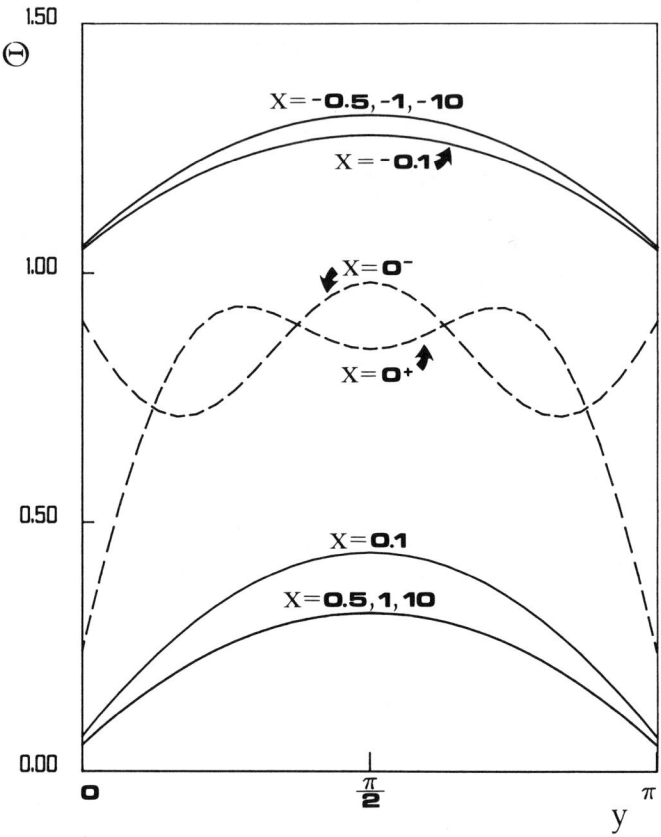

Fig. 7 Temperatures as a function of the transverse coordinate y for different axial positions. $\varphi = 1$, $M = 40$, $Pe = 10$, $Bi = 10$.

Discussion and Conclusions

The heat transfer in an MHD flow through a channel with walls of finite thickness has been studied. A simplified model that takes into account the transverse thermal conductivity but that neglects axial heat conduction through the walls was used. The electrical conductivity of the walls is incorporated through the velocity profile and Joule dissipation term in the fluid; no electrical heat dissipation is considered in the walls.

Results indicate that the temperature distributions are mainly determined by the combined influence of two effects: the electrical and viscous heating and the heat extraction through the walls. Physical and geometrical properties of the walls modify the heat dissipation via the conductance ratio $\varphi = (\sigma_w/\sigma_f)(L/a)$, since this parameter influences both the current density distribution and the velocity profiles. The heat transfered to the ambient is controlled by the Biot number, which in turn is a function of the thermal conductivity and the thickness of the wall. The fully developed temperature inside the channel presents a minimum as a function of the conductance ratio, indicating that for a given set of materials, there is an optimum geometry where the fluid temperature is the lowest. This property is of practical importance for the design of MHD devices. It must be emphasized that this result is independent of the neglect of the thermal axial conduction of the wall. Temperature profiles near the origin of the axial coordinate present a transition from upstream to downstream values, which in general display more similar behavior with respect to the external parameters than do those in fully developed regions. It is expected that taking into account more terms in the series would yield more accurate results; this was not done in the present study due mainly to the extremely laborious algebraic effort.

Acknowledgment

The first author wishes to thank the National Council of Science and Technology (CONACYT) of Mexico, which provided partial support for this work.

References

[1] Nigam, S.D., and Singh, S.N., "Heat Transfer by Laminar Flow between Parallel Plates under Action of Transverse Magnetic Field," *Quart. J. Mech. Appl. Math.* Vol. 13, 1960, pp. 85-97.

[2] Jain, M.K., and Srinivasan, J., "Hydromagnetic Heat Transfer in the Thermal Entrance Region of a Channel with Electrically Conducting Walls," *AIAA Journal*, Vol. 2, 1964, pp. 1886-1892.

[3] Singh, B., and Lal, J., "Heat Transfer for MHD Flow through a Rectangular Pipe with Discontinuity in Wall Temperatures," *International Journal of Heat and Mass Transfer*, Vol. 25, 1982, pp. 1523-1529.

[4] Cuevas, S., and Ramos, E., "Heat Transfer in an MHD Flow with Boundary Conditions of the Third Kind," *Applied Scientific Research*, Vol. 48, 1991, pp. 11-33.

[5] Javeri, V., " Magnetohydrodynamic Channel Flow Heat Transfer for Temperature Boundary Condition of the Third Kind," *International Journal of Heat and Mass Transfer*, Vol. 20, 1977, pp. 543-547.

[6] Javeri, V., "Combined Influence of Hall Effect, Ion Slip and Temperature Boundary Condition of Third Kind on MHD Channel Flow Heat Transfer," *Applied Scientific Resesearch*, Vol. 33, 1977, pp. 11-22.

[7] Chang, C., and Yen, J.T., "Magnetohydrodynamic Channel Flow as Influenced by Wall Conductance," *Zeitschrift für Angewandte Mathematik und Physik*, Vol. 13, 1962, pp. 266-272.

[8] Cuevas, S., "Heat Transfer in an MHD Flow with Boundary Conditions of the Third Kind," (in Spanish), MSc Thesis, Facultad de Ciencias, Universidad Nacional Autónoma de México, México, 1988.

Instability of a Liquid-Metal Surface in a Low-Frequency Alternating Magnetic Field

Jean-Marie Galpin*
*Madylam-Institut National Polytechnique de Grenoble,
St. Martin d'Héres, France*
Alfred Sneyd†
University of Waikato, Hamilton, New Zealand
and
Yves Fautrelle†
*Madylam-Institut National Polytechnique de Grenoble,
St. Martin d'Héres, France*

Abstract

We examine a 190-mm-diameter cylindrical mercury pool located in a low-frequency alternating magnetic field created by a single-phase solenoïdal coil. In the frequency range considered here (1-10 Hz), the oscillating part of the Lorentz forces becomes predominant and generates free surface motions. The experiments show that the free surface deformations consist of various standing wave patterns. We have observed four wave regimes according to the frequency and the strength of the imposed magnetic field. The system exhibits symmetry breakings with subharmonic transitions in some cases. A theoretical analysis of the stability of a free surface in an ac magnetic field shows that instability may occur as in the parametric resonance problem. Under some conditions azimuthal waves become unstable and instability occurs with a subharmonic transition. The predicted most unstable modes correspond to those observed in the experiments.

Copyright © 1991 by the American Institute of Aeronautics and Astronautics, Inc. All rights reserved.
* Doctor.
† Professor.

Introduction

Alternating magnetic fields are commonly used in metallurgy to drive liquid-metal motions. The electric currents induced by an external ac magnetic field interact with it to create electromagnetic body forces. These forces are responsible both for a vigorous motion of the bath and free surface deformations. Single-phase ac fields are commonly used in various processes such as induction furnaces and induction ladles. Electromagnetic stirring has some important advantages over other techniques such as gas bubbling. However, its efficiency is questionable for a large ladle. Industrial supply frequencies are no longer well suited because of the skin effect; thus, lower frequencies must be used.

The effects of a single-phase alternating magnetic field on a liquid metal have been extensively studied.[1-10] However, little attention has been paid to the low-frequency range, namely, $f \leqslant 10 Hz$. In the low-frequency limit, which is characterized by

$$R_\omega \leqslant 1, \qquad R_\omega = \mu\sigma\omega a^2, \qquad \omega = 2\pi f$$

where R_ω, μ, σ, f, and a denote, respectively, the shield parameter, magnetic permeability, electrical conductivity, and frequency of the applied magnetic field and the pool radius, the pulsating part $\widetilde{F}$ of the Lorentz forces becomes predominant with respect to the mean part.[11] The effect of $\widetilde{F}$ is to generate fluctuating velocities in the liquid bulk but also to interact with the pool free surface to excite a wide variety of surface wave patterns. The latter phenomenon, which will be developed throughout this paper, has an analogy with the parametric resonance encountered in related problems, namely, the Faraday problem,[12,13] or the excitation of liquid free surfaces by alternating electric fields.[14] However, the present case seems to be more complex in the sense that many wave patterns may arise according to the values of the two main parameters of the problem: the coil intensity I (or the typical magnetic field strength B_0) and the frequency of the coil current.

This presents some experimental observations and a preliminary theory using a stability model that highlights some of the observed transitions. The experimental results and the stability model are detailed in separate papers,[15,16] and we will focus on a comparison between the experiments and the theory.

Experiments

Apparatus

The experimental ladle consists of a stainless steel cylindrical tank filled with mercury and surrounded by a 360-turn induction coil (see Fig. 1). The radius and height of the mercury pool are a = 95 mm and H = 124 mm. The tank is cooled by maintaining a thin film of cold water over the outside wall. The inductor is supplied with single-phase electrical currents with various frequencies in the range of 2-10 Hz. Liquid-metal velocities are measured by means of a constant-temperature hot-film anemometer. The procedure and discussion of the validity of the results are detailed in a

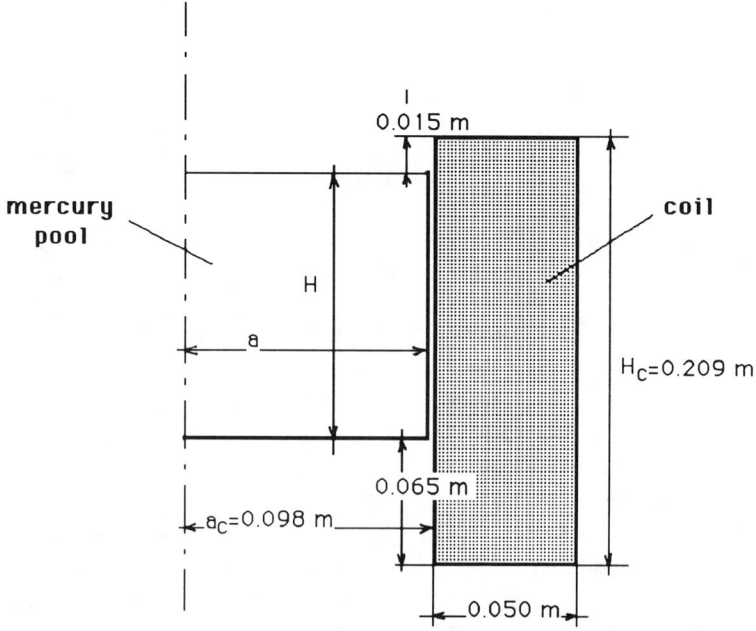

Fig. 1   Scheme of the geometry.

previous paper.[11] A contact probe is used to measure the time deformations for the free surface motion.

Free Surface Patterns

Various complex wave patterns occur according to the values of both the coil current and the frequency. In the (I,f) plane we have qualitatively observed four regions corresponding to four types of standing wave patterns. As shown in the transition diagram in Fig. 2, the four regimes have the following characteristics:

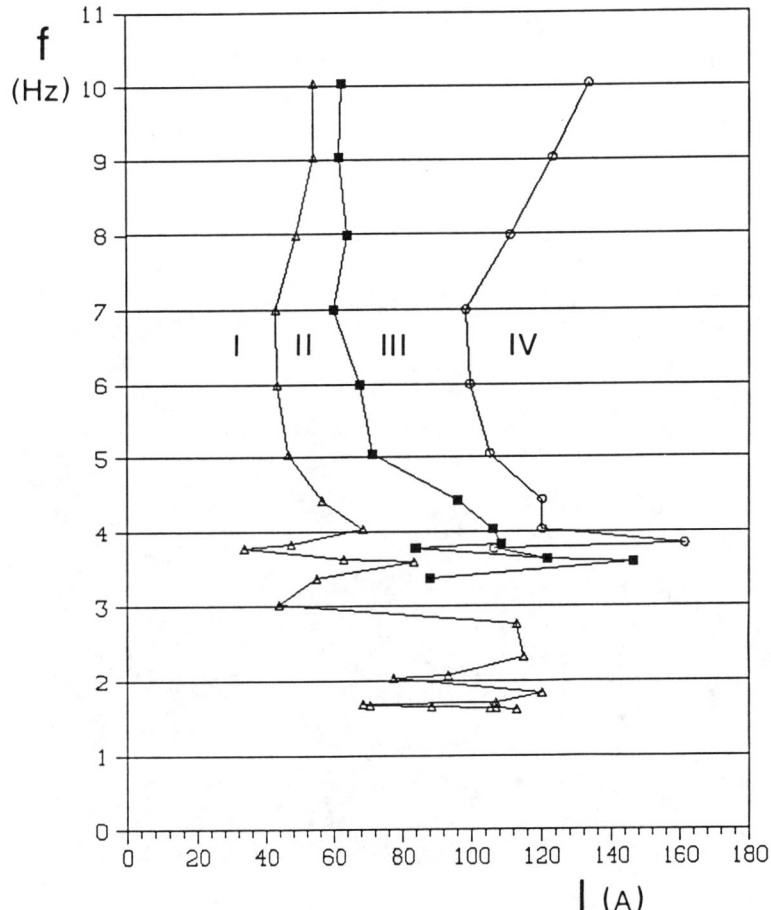

Fig. 2 Stability diagram in the *(I,f)* plane showing the various flow regimes.

1) Type I regime: The free surface motion consists of axisymmetric standing waves whose dominant frequency is 2f.

2) Type II regime: A first symmetry breaking occurs with the emergence of azimuthal waves superimposed to the concentric pattern; the dominant frequency of both concentric and azimuthal waves is still 2f.

3) Type III regime: A second subharmonic transition occurs and low-wave-number azimuthal waves are set up.

4) Type IV regime: The free surface deformation becomes strong and chaotic.

Let (n,m) denote each free mode of the free surface in the Fourier-Bessel decomposition of the free surface deformation (see, for example, the next section). The frequency is shown to select the values of (n,m). The various regimes are illustrated in Figs. 3-6.

Fig. 3 View of the free surface near the transition between the type I and type II regime (f = 4.2 Hz, I = 45 A).

Fig. 4 View of the free surface in the type II regime (f = 4.2 Hz, I = 75 A).

Fig. 5 View of the free surface in the type III regime (f = 4.2 Hz, I = 110 A).

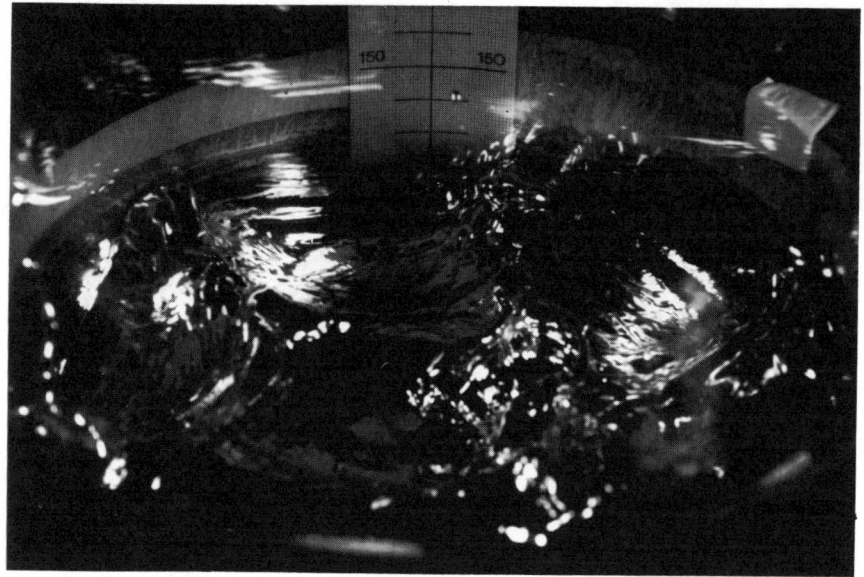

Fig. 6 View of the free surface in the type IV regime (f = 4.2 Hz, I = 180 A).

Evolution of the Amplitude with Respect to the Coil Intensity

Measurements of the wave amplitude have been made at the center and the side of the tank to separate the concentric modes from the azimuthal modes. For a fixed frequency the evolutions are quite complex and depend on the frequency as well as on the nature of the wave regime.

Globally, the amplitude is a growing function of the coil intensity (see Fig. 7). The growth laws exhibit breaks that correspond to the various regime transitions. It is noticeable that the transition between the type II and type III regimes occurs with a sharp increase of the amplitude of the azimuthal waves. That subcritical behavior of the amplitude of the azimuthal systems is consistent with the observed hysteresis phenomena (see, for example, Fig. 2).

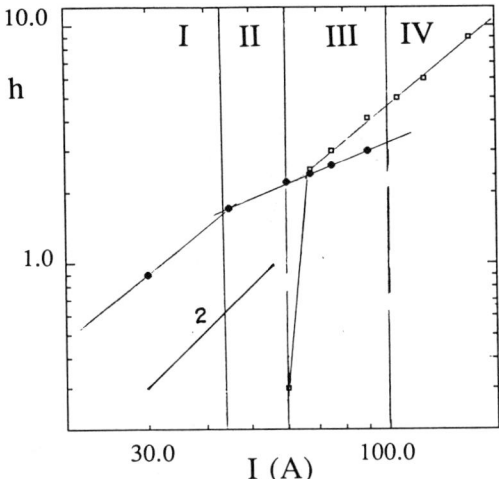

Fig. 7 Evolution of the surface wave amplitude with respect to the coil current at the center (filled circles) and the side of the pool (open squares) for f = 7.02 Hz.

Evolution of the Amplitude with Respect to the Coil Frequency

Analogous amplitude measurements have been performed for a fixed coil current and varying frequencies. The amplitudes are rapidly decreasing functions of the frequency as shown in Fig. 8. Beyond f ≅ 10 Hz the wave amplitudes become negligible. The wavelengths of both azimuthal and concentric waves are decreasing functions of the frequency as well.

The decay laws are not universal in the sense that they depend on the coil current. In the coil current range considered here, the amplitudes behave like $f^{-\alpha}$, with α ranging from 1.8 to 3. according to the coil current.

Parametric Instability Model

Derivation of Equations

We consider a cylindrical tank of liquid metal $0 \leq r \leq a$, $-h \leq z \leq 0$ placed in a uniform alternating magnetic field

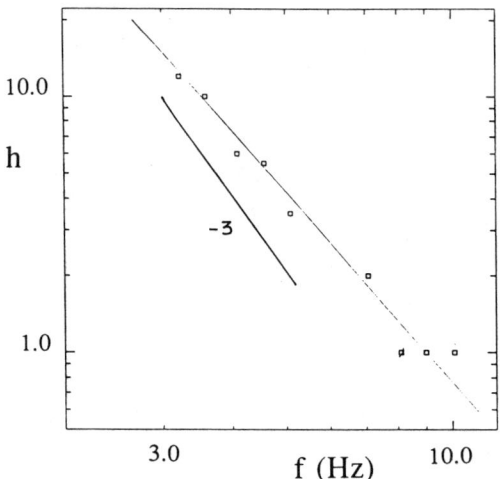

Fig. 8 Evolution of the amplitude of the surface waves at the pool centre with respect to the frequency I = 90 A.

$B_0 \sin\omega t\ \hat{z}$, with the positive z axis being vertically upward (see Fig. 9). The upper (unperturbed) liquid-metal surface is free, and the walls and floor of the container are rigid. We assume that the field frequency is low, i.e., that the magnetic diffusion time is much shorter than the oscillation period, so that $R_\omega \ll 1$. This approximation lets us calculate $\underset{\sim}{J}$ using a quasistatic approximation. Faraday's law gives

$$\underset{\sim}{E} = -1/2\ \omega B_0 r \cos\omega t\ \hat{\theta}$$

and Ohm's law gives

$$\underset{\sim}{J} \times \underset{\sim}{B} = -F_0 \left(\frac{r}{a}\right) \sin(2\omega t)\ \hat{r}$$

$$F_0 = 1/4\ \sigma\omega B_0^2\ a$$

We have also assumed a small magnetic Reynolds number, so that $\underset{\sim}{B}$ is unaffected by fluid motion.

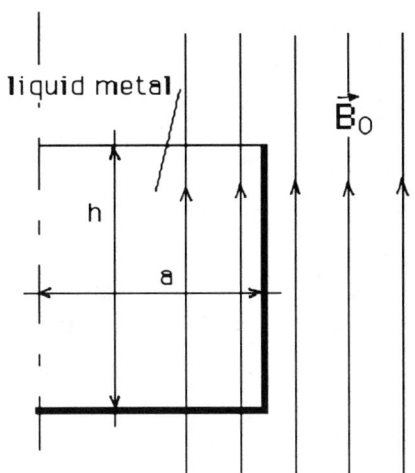

Fig. 9 Geometry of the model.

Axisymmetric State

We assume an axisymmetric free surface perturbation of the following form:

$$z = \eta = \sum_{n=1}^{\infty} A_n(t) J_0(l_n r)$$

and a linearized inviscid equation of fluid motion:

$$\rho \frac{\partial \underset{\sim}{v}}{\partial t} = -\nabla p + \underset{\sim}{J} \times \underset{\sim}{B}$$

To ensure that $\underset{\sim}{v} \cdot \underset{\sim}{n} = 0$ on $r = a$, $l_n a$ must be the nth zero of $J'_0(l_n a)$; then assuming some slight dissipation to guarantee decay of transients, we obtain the quasisteady solution:

$$\eta = \frac{2Fo}{\rho a} \sin(2\omega t) \sum_{n=1}^{\infty} \frac{J_0(l_n r) \tanh(l_n h)}{l_n J_0(l_n a)(4\omega^2 - \Omega_n^2)} \quad (1)$$

where $\Omega_n = [\tanh(l_n h) l_n g]^{1/2}$, $n = 1, 2, \ldots$ are the natural frequencies.

The dominant mode is the mode in which Ω_n is closest to ω; thus the number of surface standing waves increases with ω (see Fig. 10). In the absence of dissipation or nonlinear effects, η becomes very large when ω is close to a natural frequency.

Non-Axisymmetric Perturbations

We now introduce a nonaxisymmetric perturbation that is split in normal modes (n,m) in the following form:

$$\eta_m = ae^{im\theta} \sum_{n=1}^{\infty} \alpha_n(t) \frac{J_m(\lambda_n r)}{J_m(\lambda_n a)} \qquad (2)$$

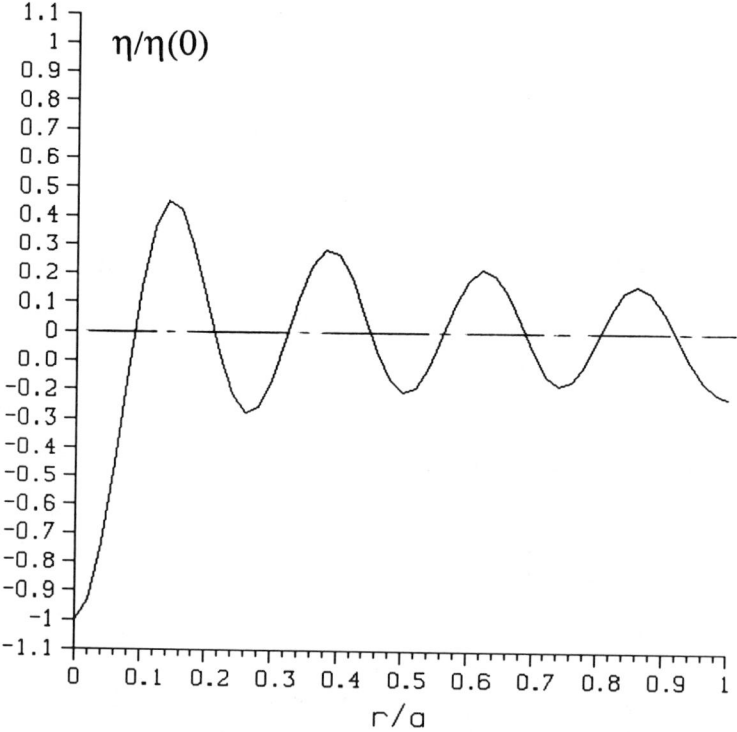

Fig. 10 Free surface profile for f=4.2 Hz according to the linear theory [see Eq (1)].

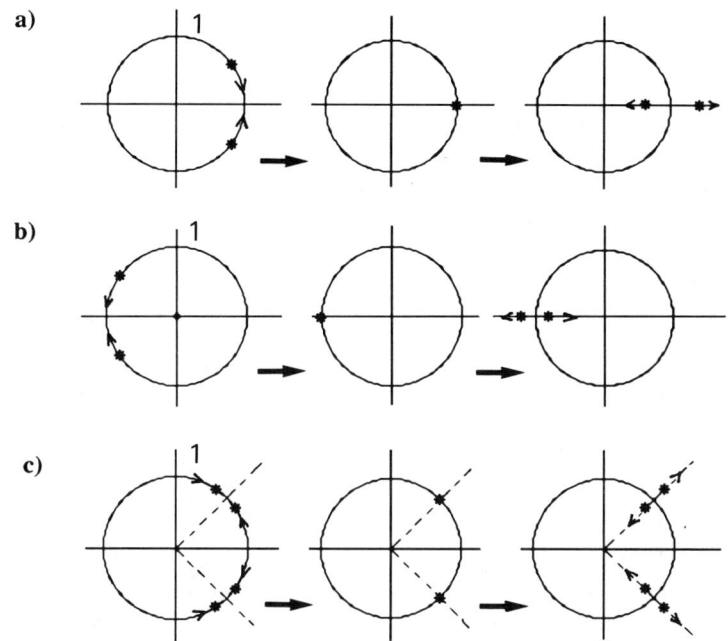

Figure 11 Trajectory of the eigenvalues on the circle of radius r=1 near the onset of instability for the three types of transitions.

where $\lambda_n a$ is the nth zero of $J'_m(\lambda_n a)$. In the axisymmetric state the electric current is unperturbed since it flows parallel to the free surface contours, but for nonaxisymmetric η, $\underset{\sim}{J}$ must be modified to satisfy $\underset{\sim}{J} \cdot \underset{\sim}{n}$ on the free surface, and this current perturbation may cause the disturbance to grow.

We obtain the following system of equations for the $\alpha_n(t)$:

$$\ddot{\alpha}_n + \Omega_n^2 [\alpha_n + L_0 \sin(2\omega t) \Sigma_{nk} \alpha_k] = 0 \qquad (3)$$

where the repeated suffix is summed from 1 to ∞, $L_0 = F_0/\rho g$, and the constant coefficients Σ_{nk} involve sums of integrals of Bessel functions. Equation (3) is a system of coupled Mathieu-type equations, which divides the (f, L_0) plane into regions of stability and instability. Symmetry breaking, due to the growth of perturbation (1), will occur if (f,L_0) lies in an unstable region.

Geometry of Unstable Regions

The stability of Eq. (3) is determined by calculating the Floquet matrix G which advances the solution by one period π/ω. If G has an eigenvalue λ with $|\lambda| > 1$, then the system is unstable. Although Eq. (3) may not necessarily be a Hamiltonian system, the matrix G can be shown to have similar properties to a symplectic matrix: The eigenvalues occur in complex conjugate pairs; G is reciprocal (i.e., if λ is an eigenvalue, so is λ^{-1}); and the product of the eigenvalues is unity. Thus, the eigenvalues of G occur in quadruples of the form

$$r\, e^{i\theta},\quad r\, e^{-i\theta},\quad 1/r\, e^{i\theta},\quad 1/r\, e^{-i\theta}$$

In a stable region all eigenvalues λ lie on the unit circle, and transition to instability occurs when two eigenvalues coalesce, one moving inside and one outside the unit circle. We can divide the transitions into three types, which are illustrated in Fig. 11.

The frequencies most likely to excite nonaxisymmetric disturbances are those near the points of intersection of the unstable region with the $L_0 = 0$ axis. Potential transitions on this axis (i.e., coalescence of eigenvalues) can be listed as follows:

(i) $\omega = \Omega_i/2k$

(ii) $\omega = \Omega_i/(2k-1)$

(iii) $\omega = (\Omega_i \pm \Omega_j)/2k$

where k is a positive integer, and i, j can range from 1 to ∞. A small L_0 analysis of the system shows that transition to instability for arbitrarily small L_0 occurs at points (ii) and (iii) when k=1 (and with a + sign only). Computations indicate that (i) with k=1 and (iii) with k=2 are also unstable transitions. However, the k=1 transitions are "strong" in the sense that in their vicinity $|\lambda|=1+O(L_0)$, whereas the others are "weak" with $|\lambda|=1+O(L_0^2)$. Figure 12 shows a computed stability boundary for m = 5 (a transition frequently observed in practice) with a/h= 1.

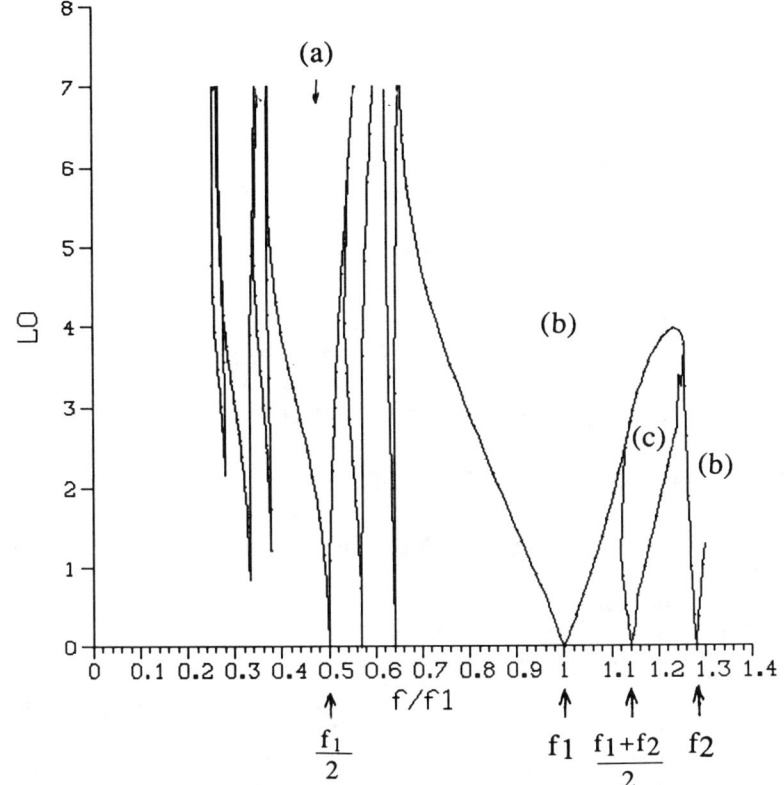

Figure 12 Stability boundary in the (f, L_0) plane for a fixed wave number m = 5.

Spectrum of Solutions

Floquet theory shows that if λ is an eigenvalue of G, there exist solutions of the form

$$\alpha_i = (\lambda)^{t/T} p_i(t)$$

where $T = \pi/\omega$ and p_i is a periodic function of period T. Thus, if $\arg(\lambda) = \theta$, we would expect to observe frequencies

$$(2k \pm \theta/\pi)\omega \quad k = 1, 2, \ldots.$$

For type (i) transitions only the harmonics $2k\omega$ appear; for the strong type (ii) transition subharmonics (2k-

1)ω will also appear, and for type (iii) $\theta=(\Omega_i+\Omega_j)\pi/\omega$ and frequencies

$$2k\omega \pm (\Omega_i + \Omega_j)$$

will occur that are not rational multiples of the fundamental 2ω.

Discussion

Although the observed phenomenon is very complex, some of its features may be understood as a result of the theoretical analysis of the preceding section.

Analysis of the Type I Regime

The quasisteady solution derived in the preceding section consists of a pattern of superimposed concentric waves. Although that solution has been obtained for weak magnetic field values, it possesses all of the characteristics observed in the type I regime:

1) The basic oscillation frequency is 2f.
2) The dominant mode wavelength l_n decreases with the frequency as follows [see Eq. (1)] :

$$l_n = O(\omega^2/g)$$

3) The evolution of the amplitude η in Eq. (1) with respect to the frequency is such that

$$\eta = O(\omega^{-3})$$

which is consistent with some observations (see, for example, Fig. 4).

Analysis of the Type II Regime

We recall that the oscillation frequency of the type II waves is 2f. Moreover, the eigenfrequencies of the observed modes are approximately 2f. Such a transition may be identified with the type (i) instability with k=1. Note that according to the theory that instability is weak. That result is consistent with the fact that the measured wave amplitudes are generally small.

Analysis of the Type III Regime

It is natural to identify the instability of parametric type obtained in the theoretical model with the experimental wave pattern in the type III regime.

The unstable type (ii) solutions of the theoretical model of the preceding section are somewhat analogous to the experimental type III regime. According to the model, the strongest transition is subharmonic, as are the transitions between the type II and type III regimes. It is not possible to make a fully quantitative comparison between the model and the experiments. Indeed, the theory of the preceding section does not take into account viscosity or surface tension. Accordingly, as in the Mathieu equation, instability may arise as soon as the magnetic field is nonzero. That situation is quite different in the experiments where type III regimes occur for quite high magnetic field values. That discrepancy is not well understood. The expected role of viscosity is to increase the magnetic field threshold values, although in the present case the Reynolds number based on the first wavelength and the frequency, i.e.,

$$Re = \frac{\omega l_1^2}{\nu}$$

is large. The role of surface tension on the stability threshold is not obvious. In the classical parametric instability of the oscillating tank (i.e., the Faraday problem) surface tension only modifies the values of the eigenfrequencies of the surface.[13] However, it must be emphasized that the experimental transition curves are somewhat qualitative and may not confidently represent all of the details of the phenomenon.

A semiquantitative comparison has been achieved, nevertheless. For some experimental frequencies given in Table 1, we have compared both the magnetic field threshold values and the most unstable mode.

Table 1 shows that, if the magnetic field threshold values are not well predicted, the observed unstable modes are almost identical to the theoretical most unstable modes that correspond to strong type (ii) subharmonic transition. This seems convincing evidence that the type III regime is an electromagnetically driven parametric

Table 1 Comparison between the experiments and the theoretical model: magnetic field thresholds and most unstable modes

Applied frequency (Hz)	Theory		Experiments	
	magnetic field (T)	mode	magnetic field (T)	mode
2.16	0.17	(1,1)	?	(1,1)
3.19	0.40	(1,3)	?	(1,3)
4.20	0.30	(1,1)	0.21	(1,5)
5.11	0.242	(1,8)	0.16	(1,8)
6.24	0.081	(1,13)	0.13	(1,12)
10.2	0.11	(1,24)	0.17	(1,23)

resonance. It may be observed that the unstable modes correspond to modes whose eigenfrequency is close to the applied magnetic field frequency.

References

[1] Barbier, J.N., Fautrelle, Y.R., Evans, J.W., and Cremer, P., "Simulation numérique des fours chauffés par induction," Journal de Mécanique Théorique et Appliquée, Vol. 1, No 3, 1982, pp. 533-556.

[2] Davidson, P.A., Hunt, J.C.R. and Moros, A., "Turbulent Recirculating Flows in Liquid Metal MHD," Liquid Metal Flow : Magnetohydrodynamics and Applications, edited by H. Branover, M. Mond and Y. Unger eds., Progress in Astronautics and Aeronautics, Vol. 111, AIAA, Washington, DC, 1988, pp. 400-420.

[3] Fautrelle, Y.R., "Analytical and Numerical Aspects of the Electromagnetic Stirring Induced by Alternating Magnetic Fields," Journal of Fluid Mechanics, Vol. 102, 1981, pp. 405-430.

[4] Mikelson, Y.Y., Yakovitch, A.T., and Pavlov, S.I., "Numerical Investigation of Averaged MHD Flow in Cylindrical Regions with

the Adoption of Working Hypotheses for Turbulent Stresses," Magnitnaya Gidrodinamika, Vol. 14, No 1, 1978, pp. 51-58.

[5] Moffatt, H.K., "High Frequency Excitation of Liquid Metal Systems," Proceedings of the IUTAM Symposium on Metallurgical Applications of MHD, The Metals Society London, 1984, pp. 180-189.

[6] Moore D.J. and Hunt, J.C.R., 1984, "Flow, Turbulence and Unsteadiness in Coreless Induction Furnaces," Proceedings of the IUTAM Symposium on Metallurgical Applications of MHD, The Metals Society, London, 1984, pp. 93-107.

[7] Sneyd, A., "Generation of Fluid Motion in a Circular Cylinder by an Unsteady Applied Magnetic Field," Journal of Fluid Mechanics, Vol. 49, 1971, pp. 817-827.

[8] Sneyd, A., "Fluid Flow Induced by a Rapidly Alternating or Rotating Magnetic Field," Journal of Fluid Mechanics, Vol. 92, 1979, pp. 35-51.

[9] Tarapore, E., and Evans, J.W., "Fluid Velocities in Induction Melting Furnaces, Part I: Theory and Laboratory Experiments," Metallurgical Transactions B, Vol. 7, 1976, pp. 343-351.

[10] Trakas, C., Tabeling, P. and Chabrerie, J.P., "Etude expérimentale du brassage turbulent dans le four à induction," Journal de Mécanique Théorique et Appliquée, Vol. 3, No 3, 1984, pp. 345-370.

[11] Taberlet, E. and Fautrelle, Y., "Turbulent Stirring in an Experimental Induction Furnace," Journal of Fluid Mechanics, Vol. 92, 1985, pp. 35-51.

[12] Faraday, M., "On the Forms and States Assumed by Fluids in Contact with Vibrating Elastic Surfaces," Philosophical Transactions of the Royal Society of London, Vol. 121, 1831, pp. 319-340.

[13] Benjamin, T.B. and Ursell, F., "The Stability of a Plane Free Surface of a Liquid in Vertical Periodic Motion," Proceedings of the Royal Society of London, Series A, Vol. 225, 1954, pp. 505-515.

[14] Briskman V.A. and Shaidurov, G.F., "Parametric Instability of a Fluid Surface in an Alternating Electric Field," Soviet Physics Doklady, Vol. 13, No 6, 1968, pp. 540-542.

[15] Galpin, J.M., Fautrelle, Y., "Liquid Metal Flows Induced by Low Frequency Alternating Magnetic Fields," Journal of Fluid Mechanics ,1992, to appear.

[16] Galpin, J.M., Fautrelle, Y., Sneyd, A.D., "Symmetry Breaking in Magnetic Stirring," Journal of Fluid Mechanics ,1992, to appear.

[17] Lamb H., Hydrodynamics, 6th ed., Cambridge Univ. Press, London, 1932.

Chapter 4. Two-Phase Flows

Natural Convection over a Vertical Heated Flat Plate with Gas Injection and in the Presence of a Magnetic Field

Paul S. Lykoudis* and Akira T. Tokuhiro†
Purdue University, West Lafayette, Indiana 47907

Abstract

This paper describes a natural convection experiment in mercury using a vertical plate held at constant heat flux with gas injection and a transverse magnetic field. The experiment was conducted with heat fluxes up to 16 kW/m² ($10^5 \leq Bo_x^* \leq 10^9$), gas injection rates up to 9 cm³/sec and magnetic field intensities up to 0.5 T. Measurements, in the laminar regime, indicate that gas injection enhances the heat transfer coefficient two to three times. In the turbulent regime, where strong stratification was present, the observed enhancement was less pronounced.

In the laminar regime with gas injection and in the presence of a magnetic field of 0.07 T, the heat transfer coefficient decreased six-fold in contrast to its enhancement value observed with injection. At a field intensity of 0.35 T the reduction was fifteen fold. In the turbulent regime, a magnetic field intensity of 0.35 T reduced the heat transfer coefficient by two-fold, whereas a further increase of the magnetic field to 0.50 T reduced the heat transfer three-fold.

The problem of bulk temperature stratification in natural convection of liquid metals was additionally investigated in some detail. A new heat transfer relationship incorporating the stratification parameter S, has been derived and successfully validated.

Copyright © 1992 by the American Institute of Aeronautics and Astronautics, Inc. All rights reserved.
* Professor, School of Nuclear Engineering.
† Doctoral Student, School of Nuclear Engineering.

Nomenclature

B	= magnetic field intensity (T); see Table 1
Bo_x	= local Boussinesq number $[=(g\beta\Delta Tx^3)\alpha^2]$, $(Bo = Ra \cdot Pr)$
Bo_x^*	= modified local Boussinesq number $[=(g\beta q''x^4)/\alpha^k]$, $(Bo^* = Ra^* \cdot Pr)$
D_{ch}	= bubble chord length, mm
Gr_x	= Grashof number $[=(g\beta\Delta Tx^3)/\nu^2]$
Gr_x^*	= modified Grashof number $[=(g\beta q'' x^{4/} \nu^2 k)]$
k	= thermal conductivity, W/m-K
L	= characteristic heated length, m
Nu_x	= local Nusselt number $[=(hx)/k]$
Pr	= Prandtl number $[=\nu/\alpha]$
Q	= gas injection rate, cm^3/s; see Table 1
q''	= heat-flux rate, W/m^2
Ra	= Rayleigh number $[=(g\beta\Delta Tx^{3)/}\alpha\nu)]$
S	= stratification parameter $[=\Delta T_s/\Delta T_w)]$ or $[=(k\Delta T_s/L)/q'']$; Eqs. (1) and (7)
sNu_x^*	= stratified local Nusselt number at constant heat flux $[=Nu_x/(1+Nu_x S)^{2/5}]$; Eq. (8a)
sNu_x	= stratified local Nusselt number at constant wall temperature $[=Nu_x/(1+S)^{1/2}]$; Eq. (8b)
$T_{Top}, T_{Bottom}, T_T, T_B$	= top and bottom bulk temperature, °C
T_{wall}, T_W	= wall temperature, °C
$T_{\infty,x}$	= bulk temperature as function of x, °C
ΔT_s	= $T_{Top} - T_{Bottom}$, °C
ΔT_w	= $T_{wall} - T_{\infty,x}$, °C
U, V	= characteristic velocity in x and y direction respectively, m/s
u, v	= velocity in the x and y directions respectively, m/s
V_{bub}	= bubble rise velocity, m/s
W	= heat-flux rate, W/m^2; see Table 1
X, Y, Z	= x, y, z coordinates, mm
α	= thermal diffusivity, m^2/s; Section 6.0
α	= void fraction
β	= coefficient of thermal expansion [1/°C]
δ	= boundary layer thickness, mm
λ	= free parameter; Eq. (3b)

1.0 Introduction

Natural convection heat transfer has been extensively investigated in the past. In the nuclear engineering industry it is of importance in the design and accident analysis of light water reactors, boiling water reactors, and pool-type reactors. It is also of interest in blanket design studies of proposed Tokamak fusion reactors where the primary heat-transfer medium under consideration is liquid lithium.

There have been many reviews of natural convection heat transfer from vertical geometries. General review articles have been done by Ede,[1] Ostrach,[2] Gebhart,[3] Catton,[4] Churchill,[5] Raithby and Holland,[6] Jaluria,[7] and Yang.[8] For liquid metals, Romig,[9] Lykoudis,[10] Viskanta,[11] Kulacki et al.,[12] and Reed[13] have all provided reviews with and without the presence of a magnetic field.

The particular past investigations relevant to the present work, because of the similarity of their geometry and other experimental conditions, are those by Papailiou and Lykoudis,[14] Julian and Akins,[15] Humphreys and Welty,[16] Sheriff and Davies,[17] and Uotani.[18] Papailiou and Lykoudis[14] investigated laminar convection in mercury and verified experimentally the theoretical solution of Lykoudis.[19] They subsequently studied turbulent natural convection, first detailing the temperature spectra[20] and then the influence of the magnetic field upon the spectra.[21] They found that the transition from turbulent to laminar flow occurs at a constant ratio (Ra/M) of the Rayleigh and Hartmann numbers. In addition, they found that the magnetic field promoted preferentially the suppression of buoyancy forces and reduced the turbulence production. Julian and Akins[15] studied laminar natural convection in mercury and showed agreement between integral and similarity methods and experimental measurements up to a modified Grashof number, $Gr_x^* \sim 10^9$. Humphreys and Welty[16] used mercury and found that, although hot-wire anemometry measurements detected temperature fluctuations at $Gr_x^* \sim 4 \times 10^9$, the laminar correlation of Julian and Akins could be extended up to $Gr_x^* \sim 7 \times 10^{10}$. Sheriff and Davies[17] using sodium, noted the onset of turbulent natural convection at $Bo_x^* \sim 10^7$ based on the deviation of their data points from the laminar theory. They also noted the presence of bulk temperature stratification in their experimental apparatus but did not find a significant difference in their experimental correlation with and without stratification. Uotani[18] using lead-bismuth eutectic, extended the laminar correlation to $Bo_x^* \sim 3 \times 10^8$, slightly beyond that of Humphreys and Welty. Moreover, he found that stratification enhanced the local heat-transfer coefficient. Stratified natural convection will be discussed further in Section 6.0.

Upon reviewing these works, it occurred to us that natural convection heat transfer could be enhanced with gas injection perhaps even in the presence of a magnetic field. In the chemical industry bubble column reactors have long been used to control and augment reaction rates. Schürgel and Lücke[22] have reviewed these phenomena in bubble column reactors in biochemical engineering. Tamari and Nishikawa,[23] in a detailed experiment of natural convection over a vertical heated plate, measured the heat transfer coefficient as it was enhanced with gas injection. In this experiment they used up to three injectors in water and ethyl alcohol and derived an empirical heat-transfer correlation that included factors describing the spatial separation of the injectors from the wall and with each other. Subsequently, at the Liquid Metal Thermal Hydraulics Laboratory at Purdue University, Wachowiak[24] reconfirmed that gas injection in water enhanced the natural convection heat-transfer coefficient. Deckwer[25] reviewed heat-transfer work in bubble column reactors and proposed a theoretical basis to a heat-transfer correlation first suggested by Kast.[26]

We have not been able to find any literature describing liquid metal natural convection with gas injection and in the presence of a magnetic field. Therefore, we set out to investigate this experimentally.

In the folowing sections, the experimental apparatus is described in Section 2.0, followed by the experimental procedure. Then the experimental results are presented in Section 3.0. Subsequently, a discussion of the experimental results is presented in Section 4.0. Finally, stratified natural convection is discussed in Section 5.0 and a conclusion is given in Section 6.0.

2.0 Experimental Apparatus

The present experimental apparatus is the same as that used by Papailiou[14,20,21] in his investigation of natural convection in mercury in the presence of a magnetic field. A schematic of the experimental cell is shown in Fig. 1. The coordinate axes are as indicated. The direction along the height of the plate is taken as the X axis. The direction along the width of the plate (into the schematic) is taken as the Z axis. The direction normal to the plate is taken as the Y axis. The origin is taken as the leading edge of the heated section.

The cell is a $20.3 \times 40.0 \times 7.0$ cm^3 stainless steel rectangular enclosure with one side heated and the opposite side water cooled. The lower portion of the cooled side is tapered. The heat is provided by 14 individually controlled strip resistance heaters bolted onto the back of the stainless steel plate. In the present experiment the heaters were operated under constant-heat-flux conditions rather than at constant temperature. Twenty-eight iron-constantan thermocouples were spot welded onto the heated plate to measure the wall temperature. The heaters and thermocouples were insulated on the exterior by Zonalite insulation.

Inside the cell, at the leading edge, a row of 11 evenly spaced gas injection tubes were installed 2 mm away from the heated wall. The injection tubes are of 0.9-mm-i.d. stainless steel hypodermic tubing. The placement of the tubes was done guided by the investigation of Tamari and Nishikawa.[23]

At the top of the cell, there is a rectangular housing from which mercury vapor is removed with the aid of an exhaust fan. On top of this housing there is a traversing mechanism that moves in increments of 0.80 mm (1/32 in.) in the Y direction. A vertical tube attached to it holds either a thermocouple or double-conductivity probe. The probe is inserted into the cell through a small square opening at the top of the enclosure. The probes used in the experiment were constructed by the experimenters. The double-conductivity probe consisted of two acupuncture needles offset 1-4 mm, electrically insulated and affixed to a larger probe body by epoxy cement. By monitoring the difference in the electrical output signal between mercury and nitrogen gas from each needle, the probe can measure the time-averaged void fraction, chord length, and bubble velocity.

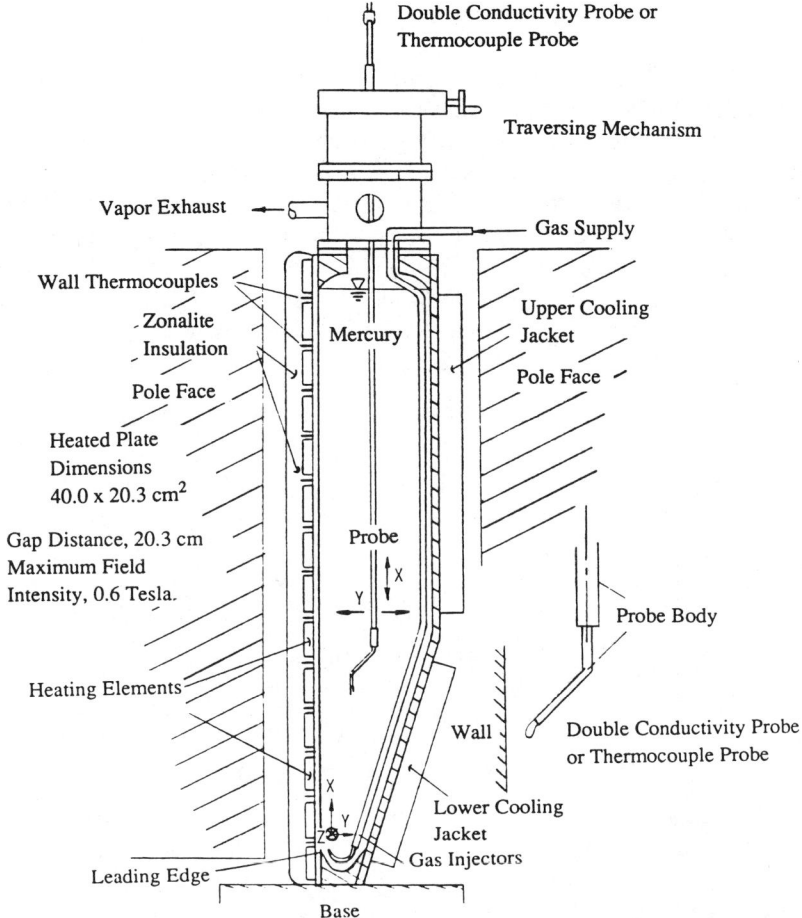

Fig. 1   Schematic of natural convection cell.

The experiment was conducted by setting the wall heat flux at a given level and adjusting the cooling rate so that the bulk temperature could be maintained near room temperature. This was done to minimize heat losses and to keep a low volumetric concentration level of mercury vapor in the lab environment. The bulk temperature was then monitored until steady-state conditions were established. Once the bulk temperature ceased to change roughly 0.5°C over a 40-to 60-min period, steady-state was assumed to be attained. For laminar natural convection this took 6-10 h, whereas for turbulent convection, it took 4-6 h. At this point, heater voltage and current, wall, and bulk temperatures were recorded. The raw wall and bulk temperature data were then curve fitted with a polynomial equation using the SIGMAPLOT[27] software package. The Nusselt and Boussinesq numbers were subsequently calculated. Thermal properties were evaluated at the average bulk temperature.

3.0 Experimental Results

The local heat transfer data of previous experiments from Refs. 15-18 are shown in Fig. 2 along with the laminar correlation. For clarity, only a selected number of data points from past investigations have been plotted in this figure. In some instances the original data were presented in terms of the modified Grashof number. These data were recast for the purpose of this paper in terms of the modified Boussinesq number. Recall that the Boussinesq number is the product of the Rayleigh and Prandtl numbers (Bo = Ra · Pr). The modified Boussinesq number is the product of the Nusselt and Boussinesq numbers (Bo* = Nu · Bo). Note also that the data of Fig. 2 represent natural convection without severe temperature stratification.

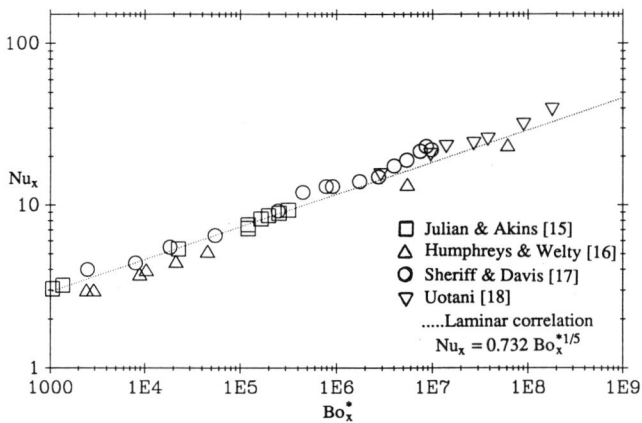

Fig. 2   Local heat-transfer data of previous investigators.

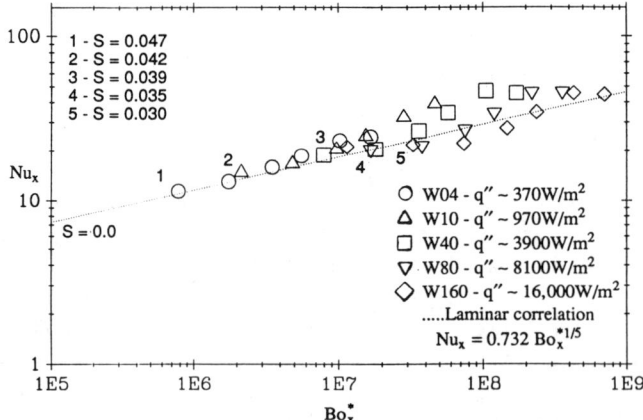

Fig. 3   Local heat-transfer data, single-phase laminar and turbulent regimes.

The present local heat transfer results are shown in Figs. 3 and 4. In contrast to Fig. 2, the modified Boussinesq scale varies from 10^5 to 10^9 because our data points were taken in this range. In Figs. 3 and 4 the different heat fluxes, gas injection rates, and magnetic field intensities have been identified by W, Q, and B, respectively. In the text and figures, for quick recognition by the reader of a certain value of W, Q, and B, we attach to the letter an up-to-three-digit number indicating its value. The numerical value of these labels is shown in Table 1 and also in Figs. 3, 4b, and 4c.

In Fig. 3 we depict our data for both laminar and turbulent single-phase results. Fig. 4a refers to the laminar regime with injection, with and without a magnetic field. Fig. 4b presents the turbulent regime data with three injection rates. In Fig. 4c we show the results for the highest injection rate Q92 with and without the presence of a magnetic field.

Figs. 5 and 6 present the measured experimental chord length, bubble velocity, and void fraction profiles for the laminar and turbulent regime, respectively, with and without the magnetic field. These measurements were taken with a double-conductivity probe at a station 11 cm from the leading edge. This station is located at approximately one-third the length of the heated plate. The curves shown were generated by using the SIGMAPLOT software package.

In Fig. 7 we show a qualitative picture of the heat transfer and void measurement results of Figs. 4-6. Finally, in Figs. 8 and 9 we show the stratified natural convection data of Uotani and our data recast in terms of a modified Nusselt number vs. the modified Boussinesq number. These latter two plots are based on the theoretical work that will be presented in Section 5.0.

4.0 Discussion of Results

4.1 Data in the Absence of Gas Injection and Magnetic Field

Figure 3 shows all of our heat transfer data taken in the laminar and turbulent natural convection regime. Here we only show the data obtained in the middle 20 cm of the test section, thus eliminating data that clearly show entrance and exit effects. However, these data are included and are discussed in Ref. 28. For the heat fluxes W04 and W10 the points at low modified Boussinesq number follow the laminar theory and agree with the data produced by previous investigators, as shown in Fig. 2 up to approximately $Bo_x^* \sim 6 \times 10^6$, where a smooth transition to turbulent convection begins. This trend is similar to that of Sheriff and Davies.[17] The points beyond $Bo_x^* \sim 6 \times 10^6$ approximate the one-third slope, indicating turbulent convection. On the other hand, the data of Humphreys and Welty[16] and Uotani[18] follow the laminar correlation up to about $Bo_x^* \sim 2 \times 10^8$. It should be noted, that Humphreys and Welty report hot-film anemometry data from which they detect transition to turbulence at about the same modified Boussinesq number as we do. However, this transition is not reflected in their heat transfer data. Unfortunately, we do not have similar information from Uotani's work.

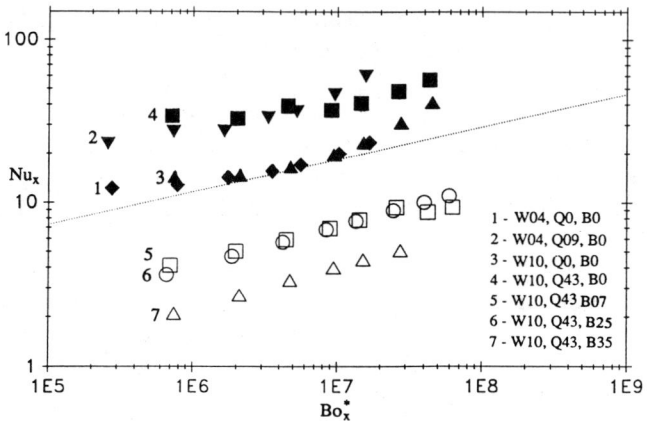

Fig. 4a Local heat-transfer data, laminar regime with injection and magnetic field.

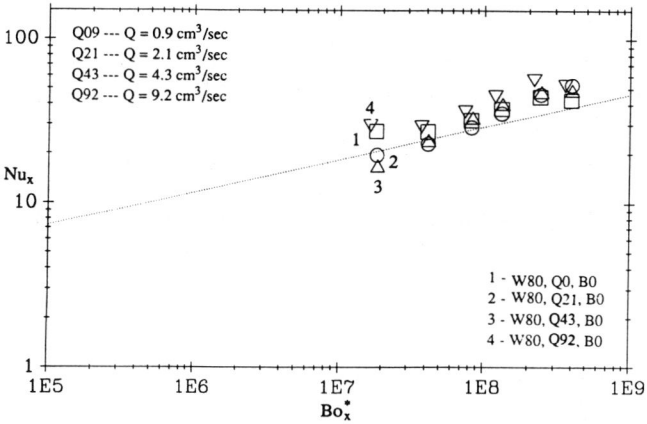

Fig. 4b Local heat-transfer data, turbulent regime with injection only.

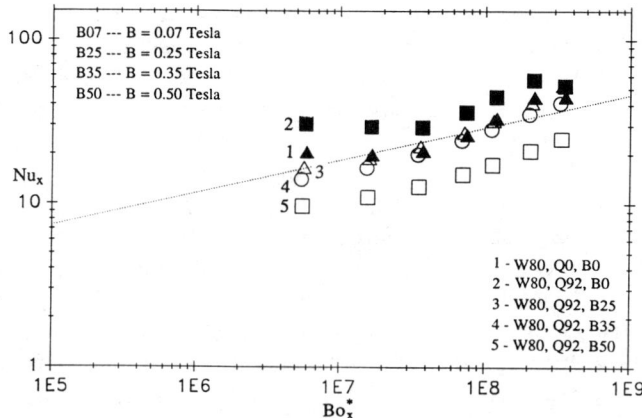

Fig. 4c Local heat-transfer data, turbulent regime with injection and magnetic field.

VERTICAL PLATE HEAT CONVECTION 609

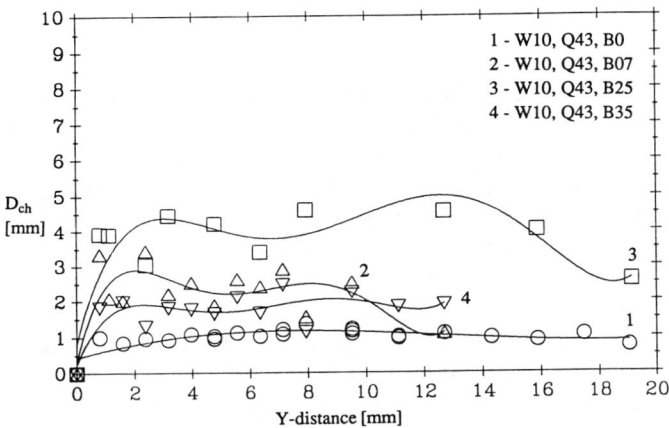

Fig. 5a Bubble chord length vs Y distance at Q43 with magnetic field.

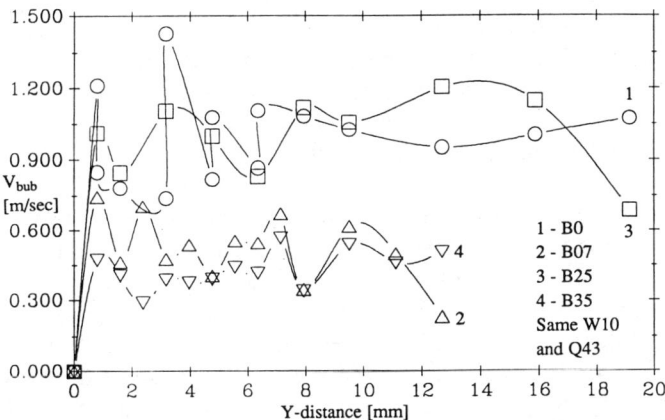

Fig. 5b Bubble rise velocity vs Y distance at Q43 with magnetic field.

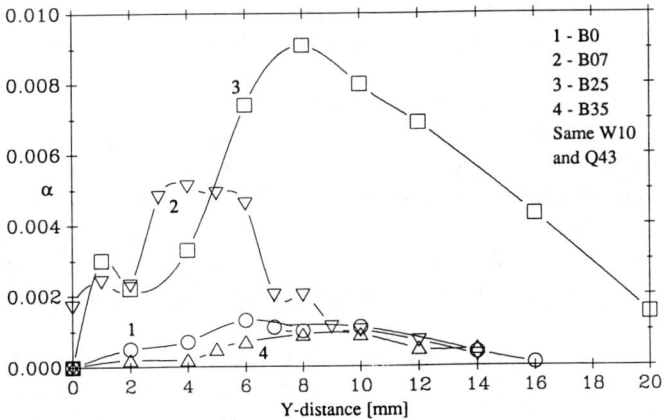

Fig. 5c Void fraction vs Y distance at Q43 with magnetic field.

At this point, we should bring into the discussion the fact that in our experiment we found the transition to turbulence to occur at approximately Ra ~ 2 x 10^7 (Bo_x^* ~ 10^7) rather than Ra ~ 10^9, the conventional accepted value. This lower transition Rayleigh number supports Bejan and Lage,[29] who have suggested that low-Prandtl-number fluids have a lower transition to turbulent convection Rayleigh number. Bejan re-examines the transition criteria in the natural convection problem and argues with an order-of-magnitude analysis that the transition Rayleigh number criteria has a strong Prandtl number dependence for all fluids. In fact, the emerging criterion is that for all fluids, Gr = 10^9 correlates all data.

In our experimental cell at higher heat fluxes, we observed stratification. As one can see from Fig. 3 at heat fluxes above W10, the Nusselt number not only depends on the modified Boussinesq number but also on the heat-flux rate W. This dependence consistently follows the turbulent slope observed at W10 beyond Bo_x^* ~ 6 x 10^6. A possible explanation could be based on the observation that the bulk temperature gradient increased with increasing heat flux. In Fig. 3 we also indicate the values of the stratification parameter S as defined in Section 5.0, Eq. (7). The observed trend in the data in conjunction with stratification parameter S will be discussed in more detail in Section 5.0.

4.2 Data in the Presence of Gas Injection

In the laminar regime, where Bo_x^* ≤ 6 x 10^6, the wall to bulk temperature difference for Q and B zero is about 1°C. For our lowest injection rate of Q09, the measured difference was about 0.5°C or less. Smaller temperature differences were difficult if not impossible to measure considering the inherent fluctuations of the wall temperature. Even so, at the rate of Q09, we measured an enhancement in the heat transfer coefficient of two to three times the single-phase results as shown in Fig. 4a. This was also observed at the heat flux W10 and injection rate Q43 for all but the three uppermost points for which Bo_x^* ≥ 10^7. On the other hand, at points beyond 10^7, the relative enhancement in the heat transfer coefficient begins to decrease. Both trends can be explained as follows. In the laminar regime where the boundary layer is fairly thick (25 mm), the presence of the bubbles within the boundary layer induces turbulence that dramatically enhances the heat transfer. In contrast, in the turbulent regime the boundary layer is thinner (5-10 mm), and fewer bubbles reside within this layer. There are more bubbles exterior to the boundary layer. Hence, the heat transfer coefficient is only moderately enhanced. At the same time, this enhancement is mitigated by the suppression of the bulk temperature gradient due to the mixing provided by the injected gas outside the thermal boundary layer. This results in a less pronounced change in the local Nusselt number.

In Fig. 4b we show the turbulent convection regime for the heat flux W80 at three different injection rates. At the injection rates Q21 and Q43, we see no appreciable heat transfer enhancement. This is consistent with the trend shown for the heat flux W10 and injection Q43, where the upper three points were in

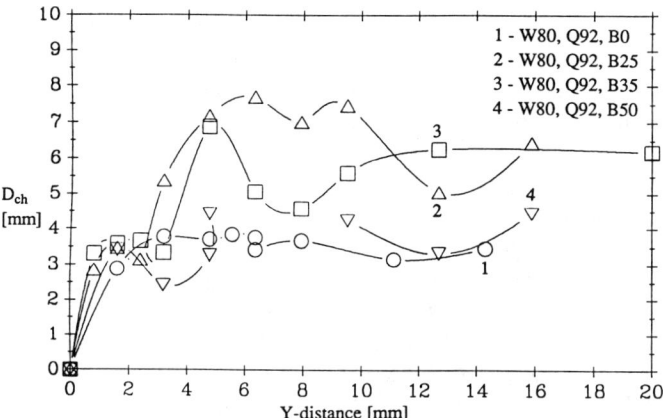

Fig. 6a Bubble chord length vs Y distance at Q92 with magnetic field.

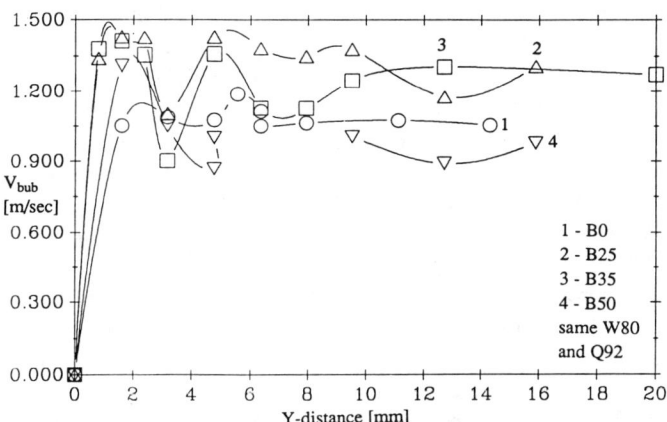

Fig. 6b Bubble rise velocity vs Y distance at Q92 with magnetic field.

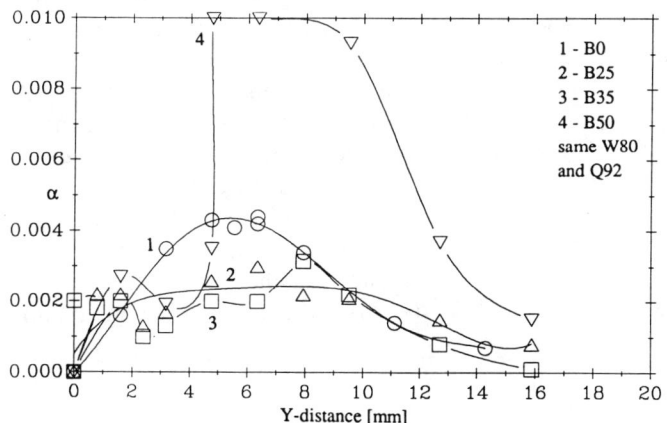

Fig. 6c Void fraction vs Y distance at Q92 with magnetic field.

the turbulent regime. On the other hand, we do observe a small enhancement of about one-third at the highest injection rate of Q92. Here, it appears likely that the small increase in the Nusselt number is due to the high density of bubbles; a few of the bubbles probably penetrate the thin turbulent boundary layer.

4.3 Data with Both Injection and Magnetic Field Present

We first discuss the case of low heat flux (W10) at which the thermal boundary layer is predominantly laminar and, of course thick. Here, as previously stated, bubbles are swimming inside the thermal boundary layer and thus contribute to the higher heat transfer coefficient because of the turbulence generated behind their paths. In fact, when we impose the weakest magnetic fields of B = 0.07 T (B07) and B = 0.25 T (B25), the heat transfer drops six fold in comparison with the nonmagnetic injection case (Q43, B0). This is shown in Fig. 4a. Such a decrease has also been observed by Papailiou[30] (single phase) at the same level of magnetic field intensity; he reports that all of the thermally induced turbulence is damped by the ponderomotive forces. The finding that the bubble-generated turbulence decreases with a magnetic field has been reported through detailed experiments by Gherson and Lykoudis[31] in two different experiments where the turbulence intensity was measured behind nitrogen bubbles injected in a pool of mercury. Figure 4a also shows that, with an increase in the magnetic field to B = 0.35 T (B35), the Nusselt number has dropped 15-fold in comparison with the nonmagnetic case with injection Q43. Here, temperature profiles we have obtained in the boundary layer[28] show the linear change characteristic of the conduction mechanism, thus explaining the reduction in the Nusselt number.

We next consider the case of higher heat fluxes where the flow is predominantly turbulent, a case characterized by a thinner boundary layer inside which few bubbles reside. As one can see in Fig. 4c, at the heat flux of W80, and at the highest injection rate of Q92, the local Nusselt number drops only by approximately one-third for B25, in contrast to the dramatic drop for the low-heat-flux case examined earlier, and remains more or less constant when the magnetic field increases to B = 0.35 T (B35). By increasing the magnetic field intensity to B = 0.50 T (B50) still at the heat-flux level of W80 and injection Q92, turbulence is further suppressed and temperature profiles again indicate that the convection process becomes conduction dominated. This is reflected in a decreased heat transfer coefficient at B50 as shown in Fig. 4c. The steep drop of the Nusselt number profile for small fields, its subsequent flatness for intermediate fields, followed by further decrease as the magnetic field reaches higher values will be related to changes in the bubble size with increasing magnetic intensity. We shall discuss this matter in the following section.

4.4 Void Fraction, Chord Length, and Rise Velocity With and Without a Magnetic Field

The influence of a bubble-generated turbulence by gas injection in both laminar and turbulent natural convection with and without a magnetic field can be further understood by looking at the bubble chord length, rise velocity, and void fraction data as shown in Figs. 5 and 6. Each one of the figures corresponds to a constant gas injection and constant heat-flux rate.

For the purpose of the present discussion, we first define a three-region **boundary-layer** structure based on the observed profiles of D_{ch}, V_{bub}, and α (distinct from the thickness of the thermal boundary layer): The innermost layer (I), adjacent to the heated surface, is single phase. The middle layer (II) is a two-phase layer where most of the gas bubbles are concentrated. The outermost layer (III) is again a single-phase layer.

We first consider Figs. 5a and 5b and 6a and 6b, which depict the change in D_{ch} and V_{bub} in the Y direction for the low- and high-heat-flux rates of W10 and W80, respectively. These figures reveal that the chord length and bubble velocity remain approximately constant along the Y axis at most magnetic field intensities within the scatter of experimental data, but not at all field intensities. For instance, the data for the higher heat flux vary considerably in the Y direction, especially for the intermediate magnetic fields of B25 and B35. In contrast to the chord length and bubble velocity, the void fraction profiles shown in Figs. 5c and 6c exhibit large changes with field intensities in the Y direction. At this point, in order to facilitate the discussion we present graphs in Figs. 7a-7f of D_{ch}, V_{bub}, and α as functions of B by taking the average of the maximum and minimum values along Y from Figs. 5 and 6 for each of these variables. As an example, in Fig. 7a, one can see qualitatively the detailed information provided in Fig. 5a.

Next we discuss the void profiles of Figs. 5c and 6c. Note that in both figures the integration of the void fraction profiles with respect to the Y axis does not yield one, even though they correspond to fixed gas injection rates. This can be explained as follows. In the nonmagnetic case with gas injection, the rising bubbles have a helicoidal motion and hit randomly the conductivity probe.

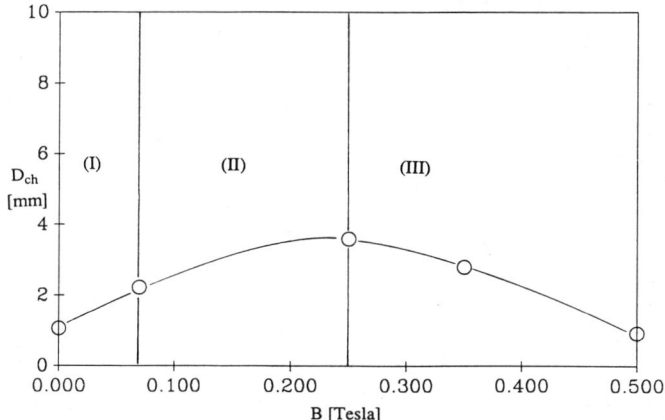

Fig. 7a Qualitative plot of the bubble chord length variation with magnetic field intensity at low heat flux, W10.

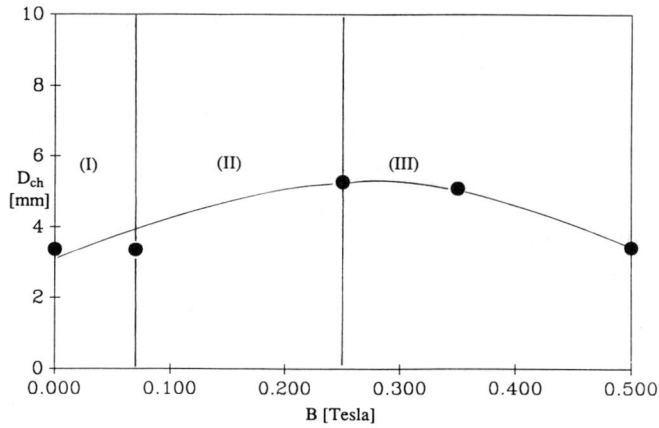

Fig. 7b Qualitative plot of the bubble chord length variation with magnetic field intensity at high heat flux, W80.

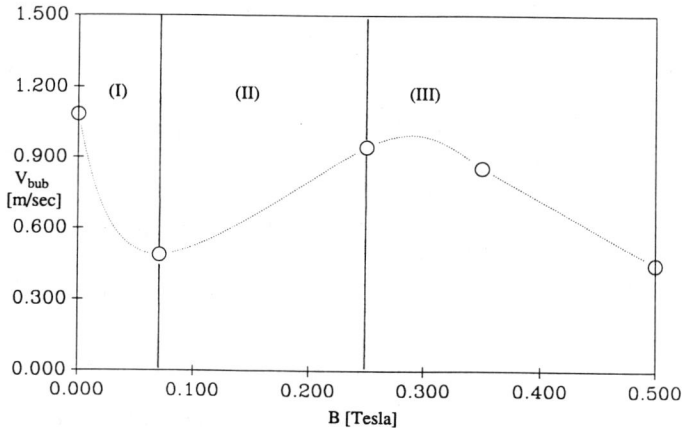

Fig. 7c Qualitative plot of the bubble rise velocity variation with magnetic field intensity at low heat flux, W10.

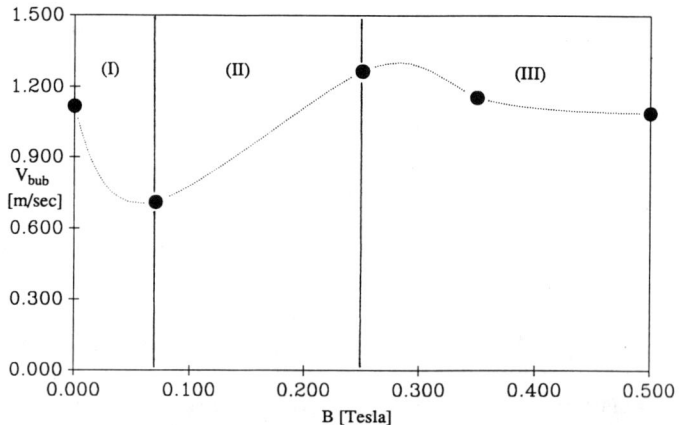

Fig. 7d Qualitative plot of the bubble rise velocity variation with magnetic field intensity at high heat flux, W80.

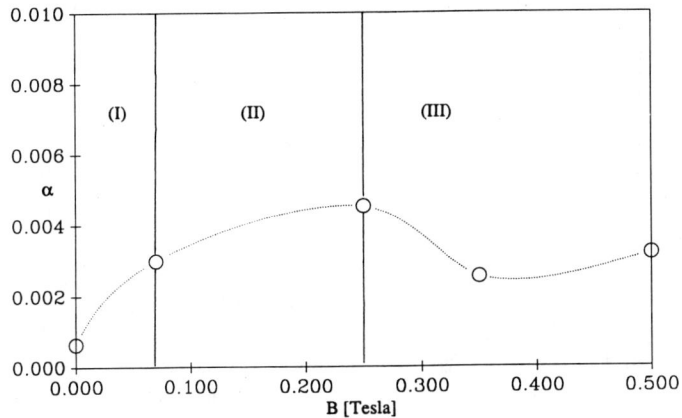

Fig. 7e Qualitative plot of the void fraction variation with magnetic field intensity at low heat flux, W10.

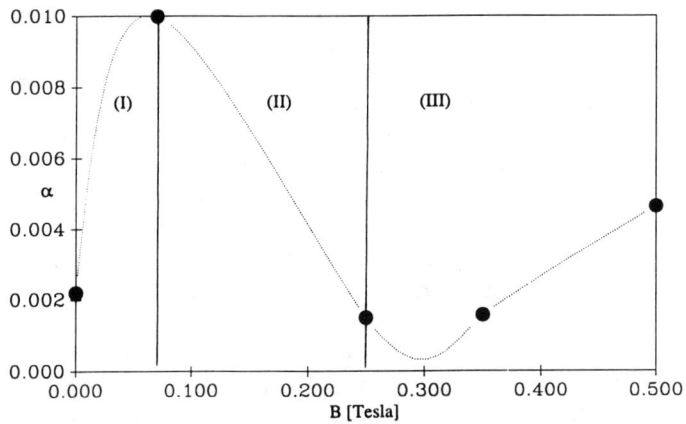

Fig. 7f Qualitative plot of the void fraction variation with magnetic field intensity at high heat flux, W80.

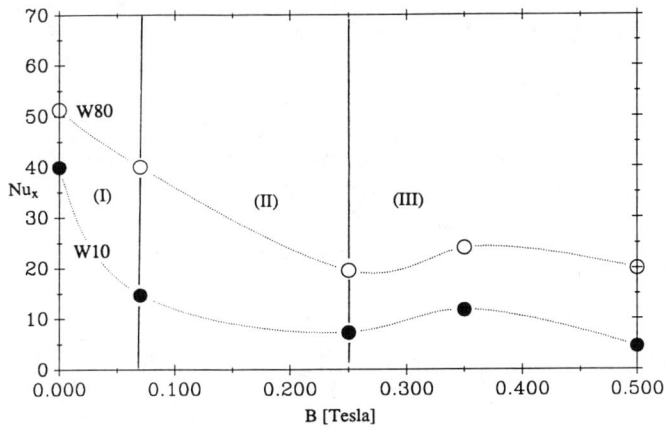

Fig. 7g Bubble chord length, rise velocity, void fraction, and local Nusselt number at W10 (left) and W80 (right) with magnetic field intensity.

However, with the addition of the magnetic field, as Mori et al.[32] have found, the number of bubbles hitting the conductivity probe increases with an increase in field intensity. This occurs because the helical trajectory of a rising bubble changes to a nearly vertical path under the constraint of the ponderomotive force. To be fully able to account for this behavior, the probe would have to be additionally traversed along the Z axis. This would result in both a time- and space-averaged void fraction. This was not done; therefore, Figs. 5c and 6c provide only a qualitative picture of the relative change in $\alpha(Y)$ with the magnetic field. For this reason we shall not attempt to draw general conclusions from the limited information we have on the void fraction.

We continue our discussion of Fig. 7 by defining three regions based on varying trends. Region I marks initial application of the magnetic field with gas injection. In regions I and II the chord length increases with field intensity. Consistent with the findings of Refs. 32 and 31, such an increase is attributed to bubble coalescence facilitated by the change from the helical trajectory to a rectilinear one as discussed previously. In region I the bubble velocity initially decreases. This is due perhaps to the substantial suppression of liquid motion expected even at the low field intensity of $B = 0.07$ T (B07) where the bubbles are still small. On the other hand, as the magnetic field further increases (region II), so does the size of the bubbles and consequently their rise velocity as reflected in Figs. 7c and 7d.

With regard to the heat transfer, in region I, for the low-heat-flux W10, the Nusselt number with increasing field intensity first drops dramatically but less so at the stronger field intensities prevailing in region II. Here the presence of larger bubbles seem to contribute once again to higher turbulence intensities, making the effectiveness of the ponderomotive force less forceful. At the higher heat flux W80, the trend in region I is the same as in the low heat flux W10, but the heat-transfer rate decreases with a smaller slope because, as stated previously, fewer bubbles reside in the mostly turbulent boundary layer.

As we now proceed to region III from region II at a magnetic field of approximately 0.25 T (B25), the chord length and bubble velocity reach a maximum value. In region III, at field intensities larger than 0.25 T the chord length and bubble velocity begin to decrease for both the low and high heat fluxes of W10 and W80. Here the bubbles seem to break up because presumably the surface tension forces cannot sustain them. This was also observed by Gherson and Lykoudis[31] in their experiment at the same range of the magnetic field (~ 0.3 T). Presumably the smaller bubble size induces less turbulence, thus leading to a lower heat-transfer coefficient. That the convection is already conduction dominated is also supported by the linearity of the temperature profiles as reported in Ref. 28.

At this point it is clear that more experiments should be conducted to determine the distribution of void fraction in the Z direction in order to fully clarify the influence of bubble geometry and dynamics to heat transfer.

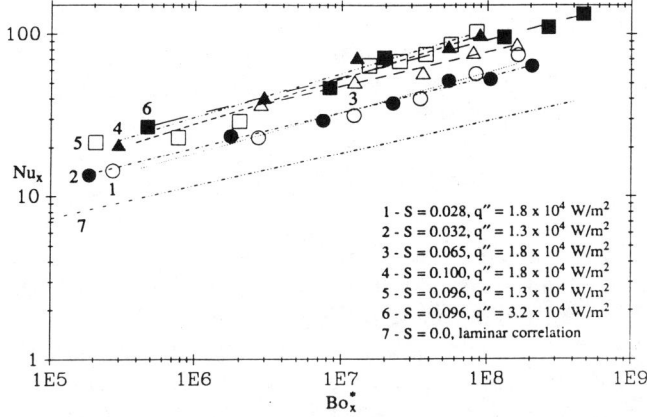

Fig. 8a Uotani's stratified local heat-transfer data.

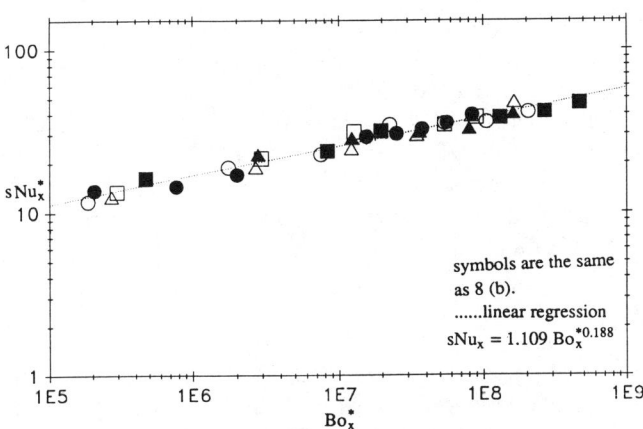

Fig. 8b Uotani's data recast as sNu_x^* vs Bo_x^*.

5.0 Stratified Natural Convection

5.1 Introduction

Thermal stratification occurs when the fluid is spatially confined while it is heated or cooled. Two examples are the stratification of rivers by thermal effluents from industrial facilities and the "temperature inversion" of the Earth's local atmosphere. In both cases there is a vertical temperature gradient and subsequent vertical heat transfer. In large-scale engineering systems, where natural convection heat transfer is the major heat-removal mechanism, we also expect stratification to occur. The pool-type reactor is a prime example. Therefore, we need to understand how bulk temperature stratification influences natural convection heat transfer.

General reviews on stratified flows have been published by Jaluria[33] and Gebhart et al.,[34] where the theoretical aspects of stratification in buoyancy-induced flows have been analyzed. The strength of stratification is measured by the stratification parameter S defined as follows:

$$S = (T_{Top} - T_{Bottom}) / (T_{wall} - T_{\infty,x}) \tag{1}$$

which in the nonstratified case, assumes a value of zero. The influence of stratification on natural convection heat transfer coefficient has been explained as follows. In contrast to the nonstratified case, stratification increases the local ambient temperature. This reduces the thermal buoyancy force, which drives the natural convection because the wall to ambient temperature difference, $\Delta T_w = (T_w - T_\infty)$, is smaller. Consequently, for the constant-heat-flux case the film coefficient increases. Jaluria and Gebhart[35] have numerically demonstrated this behavior for fluids with Pr = 0.733 and 6.7. Chen and Eichhorn[34] also confirmed this experimentally with a cylindrical heater immersed in a stratified water bath. They measured a heat transfer coefficient enhancement up to 1.7 times the nonstratified result.

Uotani[19] has experimentally demonstrated the influence of stratification on the local Nusselt number using lead-bismuth eutectic. With increasing S, as shown in Fig. 8a, he reports an enhancement of as much as four to five times that of the nonstratified case. Likewise, a decrease in S results in a decrease in the heat-transfer coefficient.

5.2 Semiempirical Correlation

Stratified natural convection data, as found at present in the literature, have never been collapsed into one single correlation that absorbs the parameter S. In the following we develop a relationship that attempts to correlate the stratified heat-transfer data through a modified Nusselt number.

Consider natural convection heat transfer from a vertical plate immersed in a stably stratified liquid metal bath. When an order of magnitude analysis of the conservation equations is performed, we have the following equations for mass, momentum, and energy:

From mass conservation:

$$\frac{\partial u}{\partial x} + \frac{\partial v}{\partial y} = 0 \tag{1a}$$

we obtain

$$\frac{U}{L} \sim \frac{V}{\delta} \tag{1b}$$

From momentum conservation,

$$u\frac{\partial u}{\partial x} + v\frac{\partial u}{\partial y} = g\beta(T_w - T_{\infty,x}) \tag{2a}$$

we obtain

$$U\left[\frac{U}{L}\right] + \frac{U\delta}{L}\left[\frac{U}{\delta}\right] \sim g\beta(T_w - T_{\infty,x}) \tag{2b}$$

and from energy conservation,

$$u\frac{\partial T}{\partial x} + v\frac{\partial T}{\partial y} = \alpha\frac{\partial^2 T}{\partial y^2} \tag{3a}$$

we obtain

$$\lambda U\frac{(T_T - T_B)}{L} + \left[\frac{U\delta}{L}\right]\frac{(T_w - T_{\infty,x})}{\delta} \sim \alpha\frac{(T_w - T_{\infty,x})}{\delta^2} \tag{3b}$$

Note that the viscous force term in the momentum equation has been neglected appropriately for liquid metals. The terms of particular interest here are the convective terms of the energy equation. In the nonstratified analysis one considers the characteristic temperature difference as that over the boundary layer, $\Delta T_w = (T_w - T_\infty)$, which is the same along the X and Y axis. This then enables one to solve the energy equation for the characteristic velocity U and boundary-layer thickness δ as the two unknowns of the problem. However, with stratification, we have one additional characteristic temperature difference that describes the vertical temperature difference, ΔT_s. We therefore distinguish between $\Delta T_s = (T_T - T_B)$ along the X axis and ΔT_w across the boundary layer in the Y direction. We add a free parameter, λ, in Eq. (3b) to allow for their relative magnitude. Subsequently, if we solve Eq. (3b) for the characteristic velocity U, we obtain,

$$U \sim \frac{\alpha L}{\delta^2}\left[1 + \lambda\frac{\Delta T_s}{\Delta T_w}\right]^{-1} \tag{4}$$

We next eliminate U in the momentum Eq. (2b) using Eq. (4) and then obtain an expression for the ratio L/δ, which is proportional to the Nusselt number. After some algebra we obtain

$$\left[\frac{L}{\delta}\right] \sim \left[\frac{g\beta\Delta T_w L^3}{\alpha^2}\right]^{1/4}\left[1 + \lambda\frac{\Delta T_s}{\Delta T_w}\right]^{1/2} \tag{5}$$

In terms of the heat flux we have

$$\left[\frac{L}{\delta}\right] \sim \left[\frac{g\beta q'' L^4}{\alpha^2 k}\right]^{1/5}\left[1 + \lambda\frac{L}{\delta}\frac{k\Delta T_s/L}{q''}\right]^{2/5} \tag{6}$$

We next redefine the stratification parameter S for the constant heat flux as follows:

$$S \equiv k\frac{\Delta T_s/L}{q''} \qquad (7)$$

Notice that S now has the physical meaning of a ratio of the vertical heat conduction to the heat input at the wall. Upon inspection of Eqs. (6) and (7), we can define a stratified Nusselt number, sNu, as reflected in the following final equation:

$$sNu_x^* \equiv \frac{Nu_x}{[1 + \lambda Nu_x S]^{2/5}} \sim Bo_x^{*1/5} \qquad (8a)$$

The stratification parameter therefore appears as a modified Nusselt number that reduces to the classical case when there is no stratification. Following the same procedure we can develop sNu_x for all possible cases in terms of the constant wall temperature and heat flux for both liquid metals and ordinary fluids ($Pr \geq 1$). The equivalent results are listed in the following.

For constant wall temperature, $Pr \ll 1$,

$$sNu_x \equiv \frac{Nu_x}{[1 + \lambda S]^{1/2}} \sim Bo_x^{1/4} \qquad (8b)$$

For constant wall temperature, $Pr \geq 1$,

$$sNu_x \equiv \frac{Nu_x}{[1 + \lambda S]^{1/2}} \sim Ra_x^{1/4} \qquad (8c)$$

And for constant heat flux, $Pr \geq 1$,

$$sNu_x^* \equiv \frac{Nu_x}{[1 + \lambda\, SNu_x]^{2/5}} \sim Ra_x^{*1/5} \qquad (8d)$$

The validity of the simple analysis resulting in Eq. (8a) was first tested with the data of Uotani shown in Fig. 8a. One can observe that increasing values of S, which Uotani could impose at will experimentally, are reflected by higher Nusselt numbers. In Fig. 8b, the same data points are shown replotted as sNu_x^* vs Bo_x^*, where the free parameter λ was assigned a value of 1.0. As it can be seen, all of Uotani's data collapse fairly well on the single line of the correlation (8a) of this work.

In similar manner our single-phase data from Fig. 3, where the values of the stratification parameter are shown, are recast in Fig. 9 in terms of the stratified Nusselt number and the modified Boussinesq number. In our case the strength of the stratification could not be controlled as was possible in Uotani's experiment,

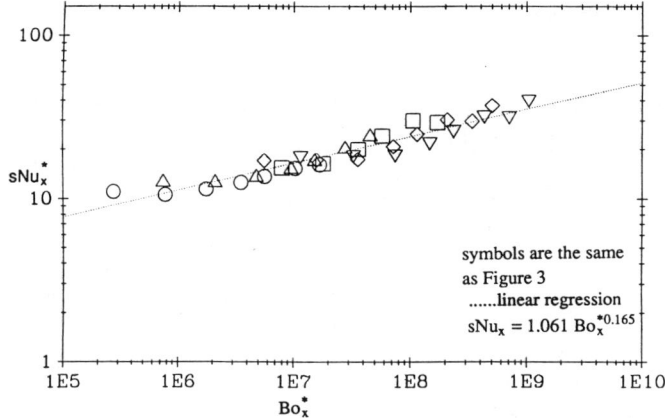

Fig. 9 Local heat-transfer data from Fig. 3 recast as sNu_x^* vs Bo_x^*.

but was determined by the imposed heat flux. Notice in Fig. 3 that for increasing values of the stratification parameter S we also show higher Nusselt numbers.

The search for additional stratified natural convection data to validate further the results of the theory as presented here continues at this time. Tokuhiro's thesis[28] will provide further details and tests of the effectiveness of Eqs. 8a-8d for other fluids such as water.

6.0 Conclusions

We undertook an experimental investigation of natural convection heat transfer in mercury with gas injection and in the presence of a transverse magnetic field. The experiment was conducted with a vertical plate held at constant heat flux up to 16 kW/m^2, gas injection rate up to 9 cm^3/s and magnetic field intensity up to 0.5 T. Local heat transfer and bubble measurements (void fraction, bubble chord length, and rise velocity) were performed in both laminar and turbulent regimes. At the higher heat fluxes, strong stratification was observed.

Our single-phase measurements verified the laminar data of previous investigators for $Bo_x^* \leq 6 \times 10^6$, beyond which stratification influenced the turbulent convection results. With the addition of gas injection, the heat-transfer coefficient was enhanced two- to three-fold in the low-heat-flux, laminar regime. On the other hand, in the turbulent regime this enhancement was less pronounced and was only observed at the highest injection rate. Both trends can be explained as follows: In the laminar regime, the presence of bubble-generated turbulence inside the relatively thick laminar boundary layer promotes the heat transfer process, whereas in the turbulent regime most of the bubbles appear outside the thin turbulent boundary layer. As a result, these bubbles only contribute to the suppression of stratification that mitigates the heat transfer enhancement mechanism.

In the case of gas injection and the presence of a magnetic field, a small field intensity in the laminar regime resulted in a six-fold drop in the heat transfer coefficient compared to the enhancement observed with injection alone. Further increase of the magnetic field reduced the heat transfer coefficient 15-fold. In contrast, in the turbulent regime equivalent magnetic field intensities produced a smaller, twofold reduction in the Nusselt number at 0.25 T and a threefold decrease at 0.35 T. Here the reason is that, in the laminar regime the ponderomotive force suppresses both the thermally induced motion and bubble-generated turbulence within the boundary layer. However, in turbulent convection the magnetic field primarily dampens the thermally induced turbulence, since this is the main source of turbulence in the thin boundary layer where few bubbles reside. The heat transfer trends in the presence of a magnetic field have also been discussed in conjunction with the data on the bubble chord length, bubble rise velocity, and partially with the void fraction profiles.

We also investigated theoretically the influence of stratification on natural convection heat transfer. Using an order-of-magnitude analysis, a new heat transfer correlation was introduced incorporating the stratification parameter S. This new relationship was validated with the data from our experiment and those of Uotani. In both cases the data recast in a modified Nusselt number vs the modified Boussinesq number were collapsed into one single line.

References

[1]Ede, A. J., "Advances in Free Convection," *Advances in Heat Transfer*, Academic Press, New York 1967, Vol. 4, pp. 1-64.

[2]Ostrach, S., "Natural Convection in Enclosures," *Advances in Heat Transfer*, Academic, New York, 1972, Vol. 8, pp. 161-226.

[3]Gebhart, B., "Natural Convection Flows and Stability," *Advances in Heat Transfer*, Academic, New York, 1973, Vol. 9, pp. 273-346.

[4]Catton, I., "Natural Convection in Enclosures," 6th International Heat Transfer Conference, National Research Council of Canada, Vol. 6, 1978, pp. 13-43.

[5]Churchill, S. W., "Free Convection Around Immersed Bodies," *Single-Phase Convective Heat Transfer*, Hemisphere, New York, 1983, pp. 2.5.7-1-2.5.7-30.

[6]Raithby, G. D., and Holland, K. G. T., "Natural Convection," *Heat Transfer Fundamentals*, 2nd ed., McGraw-Hill, New York, 1985, pp. 6.1-6.94.

[7]Jaluria, Y., "Basics of Natural Convection," *Handbook of Single-Phase Convective Heat Transfer*, edited by S. Kakać, R. K. S. Shah, and W. Aung, Wiley, New York, 1987, pp. 12-1-12-31.

[8]Yang, K. T., "Natural Convection in Enclosures," *Handbook of Single-Phase Convective Heat Transfer*, edited by S. Kakać, R. K. S. Shah, and W. Aung, Wiley, New York, 1987, pp. 13-1-13-51.

[9]Romig, M. F., "The Influence of Electric and Magnetic Fields on Heat Transfer to the Electrically Conducting Fluids," *Advances in Heat Transfer*, Academic, New York, 1964, Vol. 1, pp. 268-352.

[10]Lykoudis, P. S., "Natural Convection of Electrically Conducting Fluids in the Presence of Magnetic Fields," Natural Convection: Fundamentals and Applications, edited by S. Kakać, W. Aung, and R. Viskanta, Hemisphere, New York, 1985, pp. 1100-1117.

[11]Viskanta, R, "Electric and Magnetic Fields," *Handbook of Heat Transfer Fundamentals*, 2nd ed., edited by W. M. Rohsenow, J. P. Hartnett, and E. N. Gianić, McGraw Hill, New York, 1985, pp. 10.1-10.45.

[12]Kulacki, F. A., et al., "Convective Heat Transfer with Electric and Magnetic Fields," *Handbook of Single-Phase Convective Heat Transfer*, edited by S. Kakać, R. K. S. Shah, and W. Aung, editors, Wiley, New York, 1987, pp. 9.1-9.49.

[13]Reed, C. B., "Convective Heat Transfer in Liquid Metals," *Handbook of Single-Phase Convective Heat Transfer*, edited by S. Kakać, R. K. S. Shah, and W. Aung, editors, Wiley, New York, 1987, pp. 8-1-8-30.

[14]Papailiou, D. D., and Lykoudis, P. S., "Magneto-Fluid-Mechanic Laminar Natural Convection... An Experiment," *International Journal of Heat Mass Transfer*, Vol. 11, No. 9, September 1968, pp. 1385-1391.

[15]Julian, D. V., and Akins, R. G., "Experimental Investigation of Natural Convection Heat Transfer to Mercury," *Industrial and Engeering Chemistry - Fundamentals*, Vol. 8, No. 4, November 1969, pp. 641-646.

[16]Humphreys, W. W., and Welty, J. R., "Natural Convection with Mercury in a Uniformly Heated Vertical Channel During Unstable Laminar and Transitional Flow," *American Institute of ChemicalEngineering Journal.*, Vol. 21, No. 2, March 1975, pp. 268-274.

[17]Sheriff, and Davies, N. W., "Sodium Natural Convection Form a Vertical Plate," 6th International Heat Transfer Conference, 1978, (Preprint).

[18]Uotani, M., "Natural Convection Heat Transfer in Thermally Stratified Liquid Metal," *Journal of Nuclear Science and Technology*, Vol. 24, No. 6, June 1987, pp. 442-451.

[19]Lykoudis, P. S., "Natural Convection of an Electrically Conducting Fluid in the Presence of a Magnetic Field," *International Journal of Heat and Mass Transfer*, Vol. 5, January-February 1962, pp. 23-34.

[20]Papailiou, D. D., and Lykoudis, P. S., "Turbulent Free Convection Flow," *International Journal of Heat and Mass Transfer*, Vol. 17, No. 2, February 1974, pp. 161-172.

[21]Papailiou, D. D., and Lykoudis, P.S., "Magneto-Fluid-Mechanic Free Convection Turbulent Flow," *International Journal of Heat and Mass Transfer*, Vol. 17, No. 10, October 1974, pp. 161-172.

[22]K. Schürgel and J. Lücke, "Bubble Column Bioreactors", *Advances in Biochemical Engineering*, edited by T. Ghose, A. Fiechter, and N. Blakebrough, Vol. 7, Springer-Verlag, Berlin, 1977.

[23]Tamari, M. and Nishikawa, K., "The Stirring Effect of Bubbles upon the Heat Transfer to Liquids", *Heat Transfer-Japanese Research*, Vol. 5, No. 2, April-June 1976, pp. 31-43.

[24]Wachowiak, R. M., "The Enhancement of Natural Convection by Rising Bubbles," M.S. Thesis, Purdue Univ., West Lafayette, IN, 1986.

[25]Deckwer, W. D., "On the Mechanism of Heat Transfer in Bubble Column Reactors," *Chemical Engeering Science*, Vol. 35, No. 6, June 1980, pp. 1341-1346.

[26]Kast, K., "Analyse des Wärmeubergangs in Blasensäulen," *International Journal of Heat and Mass Transfer*, Vol. 5, March-April 1962, pp. 329-336.

[27]Sigma-Plot, Version 3.1, Jandel Scientific, Sausalito, CA.

[28]Tokuhiro, A. T., Ph.D. Dissertation, "Natural Convection Heat Transfer Enhancement in Mercury with Gas Injection and in the Presence of a Transverse Magnetic Field," Purdue Univ., West Lafayette, IN, 1991, (to be published).

[29]Bejan, A. and Lage, J. L., "The Prandtl Number Effect on the Transition in Natural Convection Along a Vertical Surface", Private Communication, 1989.

[30]Papailiou, D. D., "Magneto-Fluid-Mechanic Turbulent Free Convection Flow," Ph.D. Dissertation, Pt. I, Purdue Univ., West Lafayette, IN, 1971.

[31]Gherson, P., and Lykoudis, P. S., "Local Measurements in Two-Phase Liquid-Metal Magneto-Fluid-Mechanic Flow," *Journal of Fluid Mech.*, Vol. 147, October 1984, pp. 81-104.

[32]Mori, Y., Hijikata, K. and Kuriyama, I, "Experimental Study of aBubble Motion in Mercury With and Without a magnetic Field," *Journal of Heat Transfer*, Vol. 99, August 1977, pp. 404-410.

[33]Jaluria, Y., *Natural Convection Heat and Mass Transfer*, Pergamon, Oxford, UK, 1985, pp. 173-197.

[34]Gebhart, B., Jaluria, Y., Mahajan, R. L., and Sammakia, B., *Buoyancy Induced Flows and Transport*, Hemisphere, New York, 1988.

[35] Jaluria, Y., and Gebhart, B., "Stratified Natural Convection Heat Transfer," *Journal of Fluid Mechanics*, Vol. 66, 1974, p. 593.

[36] Chen, C. C., and Eichhorn, R., "Natural Convection from a Vertical Surface to a Thermally Stratified Fluid," *Journal of Heat Transfer*, Vol. 98, *Journal of Heat Transfer*, Vol. 98, August 1976, pp. 446-451.

Nucleate Boiling of Mercury in the Presence of a Horizontal Magnetic Field

Paul S. Lykoudis* and Minoru Takahashi[†]
Purdue University, West Lafayette, Indiana 47907

Abstract

We present measurements of bubble frequency, size, and rise velocity with the help of a miniature double conductivity probe of our own design, inside a pool boiler in the presence of a horizontal magnetic field. The working fluid was mercury under the following conditions: the cover gas pressure was kept at atmospheric pressure, the maximum heat flux at the heated wall (6 cm in diameter,) was 180 kW/m^2 and the maximum magnetic field intensity was 0.8 Tesla. The experimental apparatus was essentially the same one used in previous work where we reported mercury, pool boiling curves. The main findings of the work are the following: at constant wall heat flux, the bubble generation frequency, which is of the order of 5 Hz at zero field, increases as much as three times for a field intensity of about 0.5 Tesla falling off to less, at our highest field. There were no significant changes in the measurement of bubble chord length and rise velocity when the magnetic field was present except for the case of the highest heat flux. We offer the following explanation for the increased bubble frequency behavior. The presence of the magnetic field diminishes the natural convection field in the vicinity of the nucleation site, so that when a bubble departs, the mercury that takes its place at the vacated spot, is at higher temperature. This means that the amount of energy needed to be provided by the wall for a new bubble to be generated is less, and so is the waiting time. To the extent that our measurements do not show appreciable change in the bubble growth time when the field is present, the bubble generation frequency increases.

Copyright © 1992 by the American Institute of Aeronautics and Astronautics, Inc. All rights reserved.
* Professor, School of Nuclear Engineering.
† Visiting Scholar, School of Nuclear Engineering; on leave of absence from Research Laboratory for Nuclear Reactors, Tokyo Institute of Technology, Japan.

Introduction

This work has been motivated by conceptual designs of blankets of magnetically confined fusion reactors using lithium as the working fluid. Since the presence of the magnetic field necessary for the plasma confinement induces ponderomotive forces on the lithium, it has been suggested that heat exchangers involving the vapor phase of the liquid metal should be used.

This work is a continuation of the theoretical and experimental investigations reported by Lykoudis,[1,2] Wagner,[3] and Wagner and Lykoudis.[4] The main findings of this work are the following: the bubble growth dynamics are controlled by a new nondimensional number called the magnetic boiling interaction parameter Λ, which is proportional to the electrical conductivity, the square of the magnetic field intensity, the superheat, and a combination of other thermophysical properties. The presence of a magnetic field of about 1 T, lowers the heat-transfer rates by as much as 50%.

The experimentally obtained boiling curves (heat flux vs. superheat) maintain the same slope in the presence of the magnetic field. A modified Forster-Zuber correlation developed in Ref. 1, containing the parameter Λ, shows good agreement with the experimental data reported in Ref. 4.

In order to obtain further physical insight into the nucleate boiling mechanism, information is needed on bubble geometry and dynamic behavior, similar to the one we have for non-liquid metal fluids. Such information does not seem to exist for liquid metals since they are not transparent, and also because of the hostile temperature environment during boiling (at atmospheric pressure mercury boils at 357° C).

This work was possible with the development of a miniature double conductivity probe, designed and constructed in our laboratory. The design is credited to the senior author and was constructed and tested extensively by Prabaddh Riddhagni. We believe that the present experimental results are the first of their kind, and for this reason they raise more questions than they answer. As it will become apparent at the end of the paper, more work is needed to settle these questions.

Experimental Apparatus and Procedure

The experimental apparatus was the same one used in the work reported in Ref. 4. The major difference is that the height of the apparatus was shortened in order to accommodate the length of the conductivity probe, which needed to be short in order to be rigid. As shown in Fig. 1, the boiler tube is cylindrical, 60 mm in diameter, and 460 mm high. It is surrounded by an insulation jacket of about 102 mm i.d., and 650 mm high. The gap between the two cylinders is filled with insulating silicon oxide powder. The surface of the heater was blasted with 400 -μ alumina sand for uniform roughness. A type K thermocouple was located at the center of the heater, inserted from its side.

The double-conductivity probe was made of SS-316 wires that were 0.20 mm in diameter. The distance between the two tips was about 2 mm, long enough so that droplets of mercury could not be trapped, thus short-circuiting the two tips, but short enough so that small bubbles could be detected. The probe was located about 2.5 mm from the heating surface. Its stem was slightly bent, so that, with rotation, it could search for a nucleation site.

The heater surface needed to be aged for a considerable length of time as described in Wagner and Lykoudis.[4] More than 24 h of continuous boiling was needed in order to reproduce the bubble data as obtained by the conductivity probe. In addition, more than 1 h of time was needed between the setting of two different magnetic fields in order for a stable thermal state to be established. Our data acquisition device was a MASCOMP MC-500, 32-bit UNIX computer system.

We wish to emphasize that the purest possible mercury needs to be used for this type of experiment, since impurities will deposit on the tips of the probes, resulting in a poor signal just a few hours after boiling has started. We have our own distillation facility that allows us to use quadruple-distilled mercury for this experiment. The inside surfaces of the boiler needed meticulous cleaning every time fresh mercury was used. The cover gas was nitrogen.

The following quantities were measured: the heater current and voltage, the temperature at the center of the heater, the mercury bulk temperature, and the cover gas temperature near the condenser. The sampling frequency chosen was

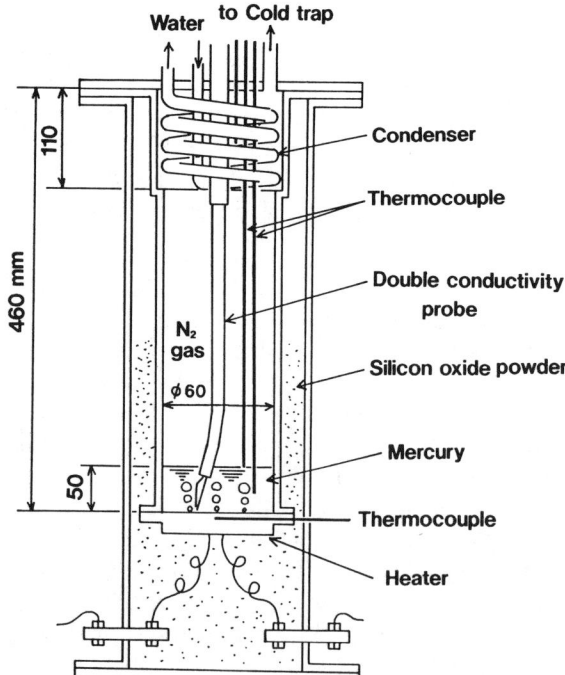

Fig. 1. Test Section.

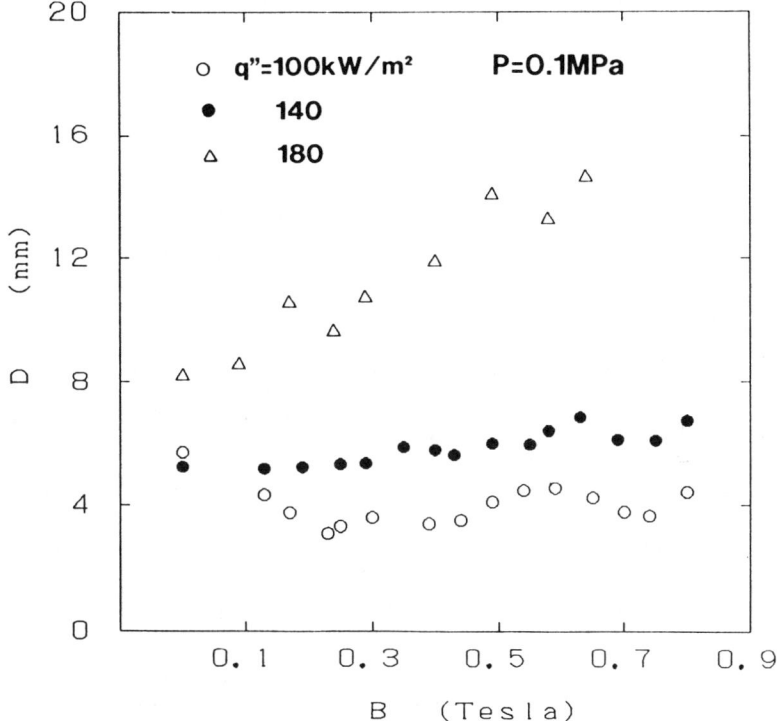

Fig. 2. Bubble Chord Length.

5 kHz. The sampling time increment was 0.2 ms. The sampling time interval between the two probes was set to the computer's minimum value of 1 μs. The data sampling process of 1 s was repeated 100 times for one run, resulting in a total running time of 100 s, deemed to be adequate for statistical purposes.

It is difficult to provide an estimate of the bubble data error. Reproducibility was assumed to exist if data taken at the same location and under the same experimental conditions after more than 24 h of continuous boiling yielded the same mean values.

Experimental Results

Figures 2 and 3 show measured chord lengths D and rise velocities V as functions of the magnetic field for the three heat fluxes. The data for the two lower heat fluxes do not show appreciable dependence on the magnetic field intensity, but this is not so for the case of 180 kW/m^2. The measured growth time t_g, as shown in Fig. 4, does not seem to show strong dependence on the presence of the magnetic field. The opposite is true for the waiting time t_w as shown in Fig. 5. Here, t_w is plotted vs. the magnetic field intensity for different heat fluxes. When the data of Figs. 4 and 5 are used to calculate the frequency of

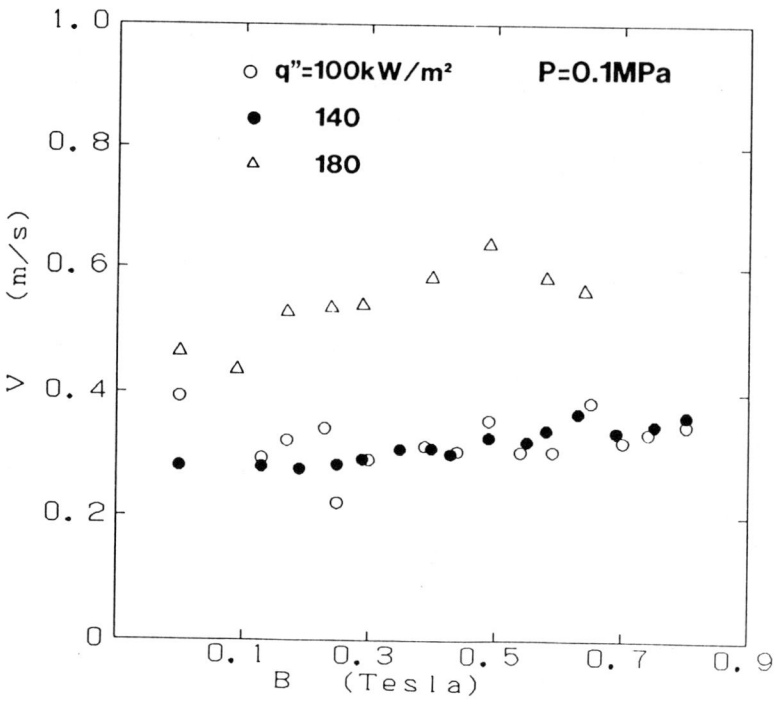

Fig. 3. Bubble Growth Velocity.

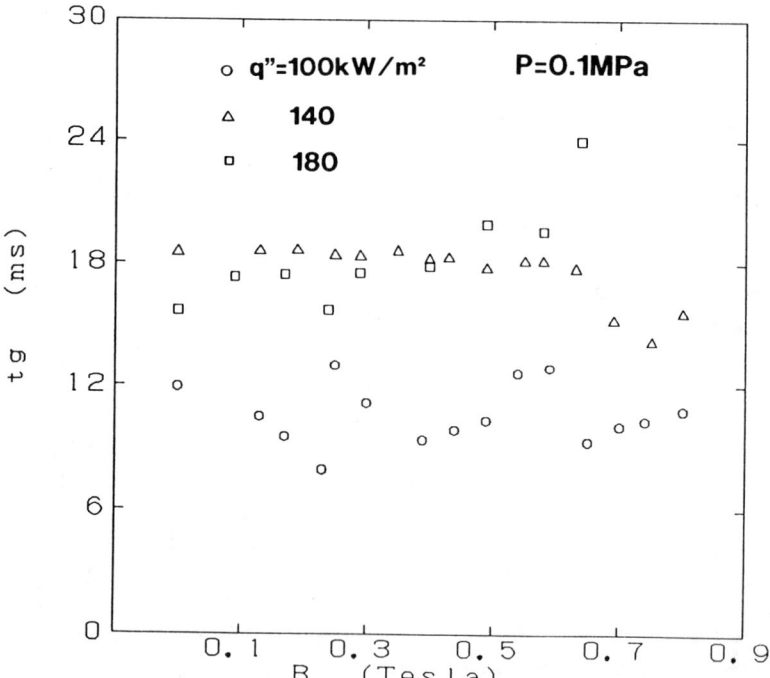

Fig. 4. Bubble Growth Period.

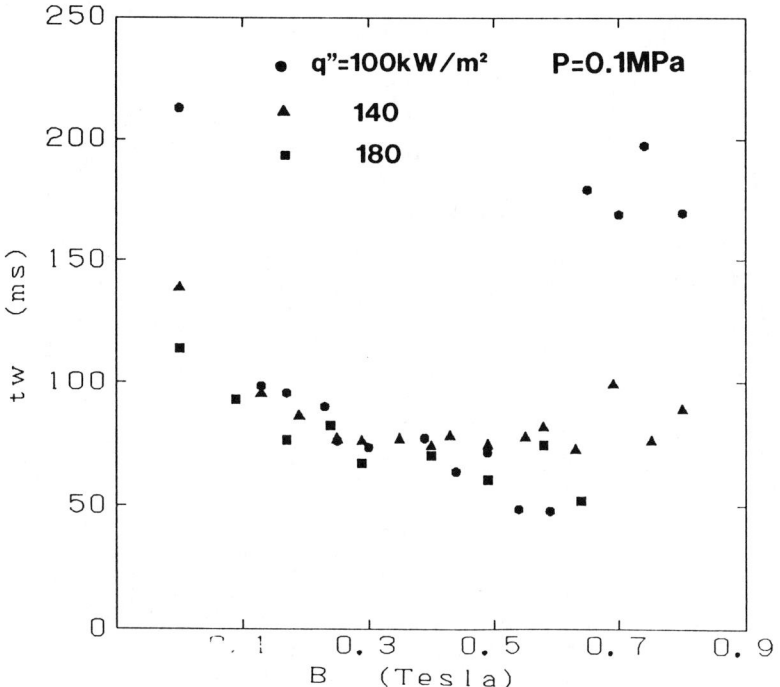

Fig. 5. Waiting Period.

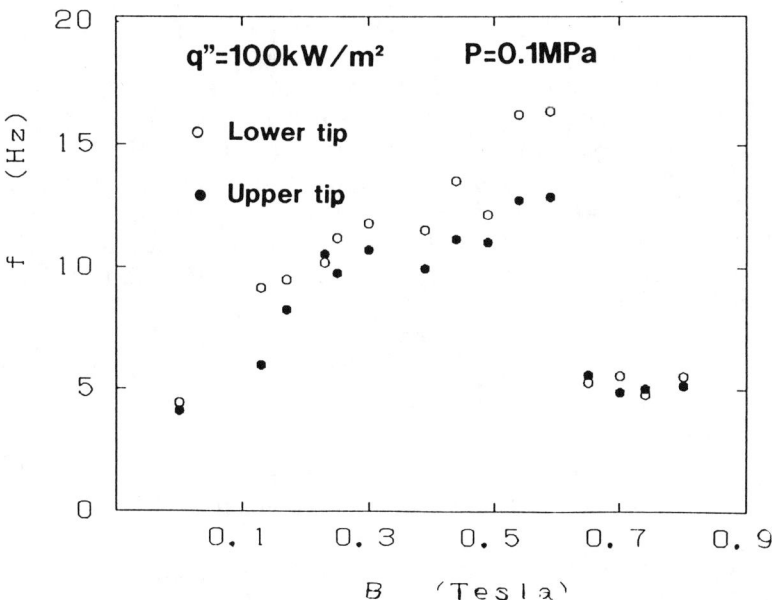

Fig. 6. Frequency of Bubble Departure for $q_w'' = 100$ kw/m^2.

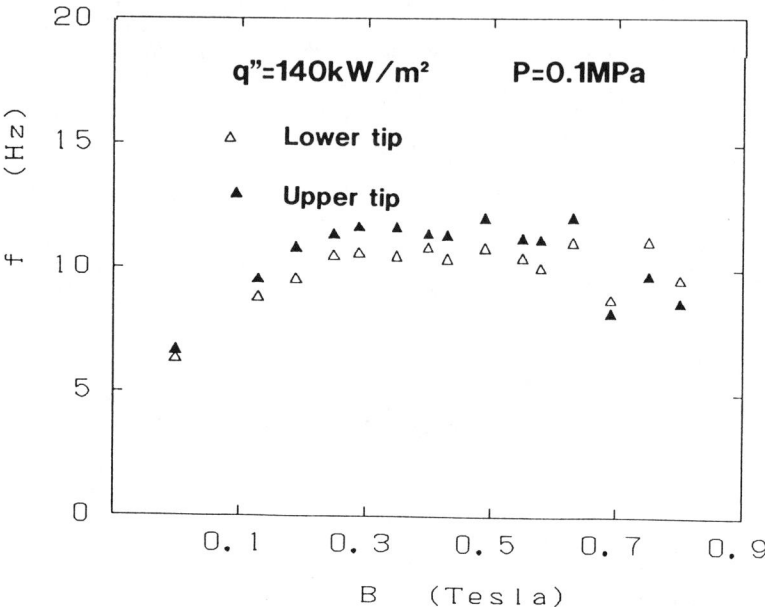

Fig. 7. Frequency of Bubble Departure for $q_w'' = 140$ kw/m².

bubble departure $f = 1/(t_g+t_w)$, one can see a sharp dependence of f with the magnetic field intensity.

Figs. 6-8 show how f varies for the heat fluxes of 100, 140, and 180 kW/m². Each one of these cases has different characteristics. In Fig. 6, at $q_w = 100$ kW/m² the frequency f increases from about 4 Hz at zero field to about 15 Hz or so at a field slightly higher than 0.6 T, with a sudden drop to an f of 5 Hz up to the highest magnetic field we tested (about 0.8 T). The case of $q_w = 140$ kW/m², shows a change in f from about 6 Hz at zero field to a flat value of about 12 Hz between 0.3 and 0.6 T with a clear tendency of dropping off to a value of about 9 Hz at 0.8 T. The behavior is similar to the case of $q_w = 180$ kW/m², where the frequency increases from 8 Hz to a maximum of about 12 at 0.65 T.

Similar, limited data we have obtained at the pressure of 200 mm Hg, but not reported here, exhibit the same trends as those we observed at atmospheric pressure.

Discussion of the Experimental Results

Although we have no explanation for the sudden drop in the bubble frequency f as observed for the case of $q_w = 100$ kW/m² of Fig. 6, we offer a qualitative argument that is in the direction of explaining the increase of f with increasing magnetic field.

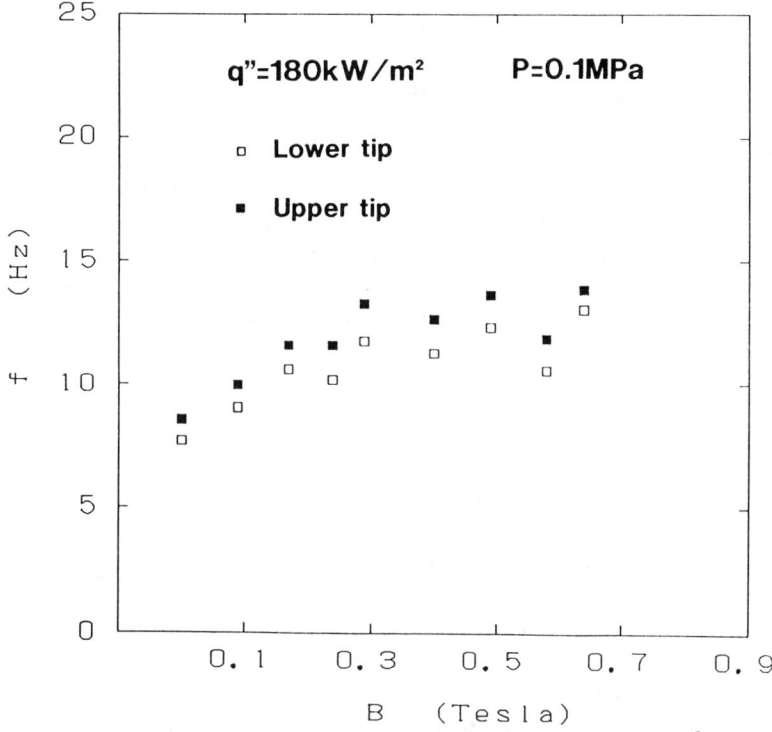

Fig. 8. Frequency of Bubble Departure for $q_w'' = 180$ kw/m^2.

The theory of Lykoudis in Refs. 1 and 2 predicts that the shape of the bubble becomes elongated in the presence of a horizontal magnetic field. Such a change could not be inferred from the data we report here. On the other hand, our data indicate that the ratio of the waiting to the growth time is equal to about 10 for zero field, diminishing to lesser values as the magnetic field increases. This means that the major contribution to the frequency increase with the magnetic field must be sought to a physical mechanism involving the waiting time t_w where the bubble dynamics play a limited role.

We propose the following qualitative sequence of events: a bubble, once generated, departs when its size is such that the buoyancy force overtakes the surface tension and other forces as they keep the bubble bound to the nucleation cavity lips. When this happens, fluid in the vicinity of the cavity rushes in to fill the empty space. The mean temperature of the rushing fluid is very strongly dependent on the vigorousness of the natural convection occurring in the neighborhood of the nucleation site. Natural convection data obtained by Wagner with the same apparatus[3] indicate a strong decrease in the rate of heat transfer due to the presence of the magnetic field, in fact cutting it by more than half for a field of 1 T. It follows that the colder fluid that rushes in to take the departed bubble's space will be at a higher mean temperature when the magnetic field is present. The net result is that a smaller amount of thermal energy needs to be transferred at the wall in order to bring the nucleation site to the thermal regime required for the inception of a new bubble. It follows that the waiting time t_w will be shorter. Since $f = 1/(t_g + t_w)$, with t_g remaining essentially invariable, a decreasing t_w produces a higher frequency.

This mechanism can be substantiated quantitatively by setting a time-dependent energy balance involving both the conduction and natural convection mechanisms in the vicinity of the nucleation site in the manner of Shai and Rohsenow,[5] Ali and Judd,[6] and other workers who have researched the waiting time for regular fluids, modifying these theories by introducing the magnetic field dependence through the parameter Λ (in the heat flux vs. superheat correlation) and the Ly number that correlates Wagner's natural convection data in the presence of a magnetic field. This scheme is currently being implemented, and it might be credible if it stands the test of fair comparison with the experimental data.

We have no explanation to offer for the sudden decrease of the bubble departure frequency as it occurs in the case of Fig. 6. More detailed data are needed to identify this sharp drop. Nor can we explain the bubble size and rise velocity increase for the case of 180 kW/m^2. On the other hand, since the size of our double-conductivity probe is small and the bubble chords are fairly large (especially with boiling at lower than atmospheric pressures), in future experiments we should be able to place more probes in the boiler, so that we can observe in more detail the geometry of a single bubble and also monitor several nucleation sites, both close to the heated wall and in the bulk of the boiling fluid. Such detailed data will help us understand the physics of the geometric and dynamic behavior of bubbles in the presence of the magnetic field.

Acknowledgment

The senior author acknowledges the financial support of the National Science Foundation under Grant 8304743.

References

[1] Lykoudis, P. S., "Bubble Growth in the Presence of a Magnetic Field", *International Journal of Heat and Mass Transfer*, Vol. 19, 1976, pp. 1357-1362.

[2] Lykoudis, P. S., "Bubble Growth in a Superheated Liquid Metal in a Uniform Magnetic Field", 4th Beer-Sheva Seminar on MHD Flows and Turbulence, Ben-Gurion University of the Negev, Beer-Sheva, Israel, Feb. 27-March 2, 1984.

[3] Wagner, L. Y., "Single and Two-Phase Liquid Metal Heat Transfer Under the Influence of a Magnetic Field", PhD Dissertation, Purdue University, West Lafayette, IN, 1981.

[4] Wagner, L. Y. and Lykoudis, P. S., "Mercury Pool Boiling Under the Influence of a Horizontal Magnetic Field", *International Journal of Mass Transfer*, Vol. 24, 1981, pp. 635-643.

[5] Shai, I., and Rohsenow, M., "The Mechanism of and Stability Criterion for Nucleate Pool Boiling of Sodium", *Journal of Heat Transfer*, Vol. 91, 1969, pp. 315-329.

[6] Ali, A. and Judd, R. L., "An Analytical and Experimental Investigation of Bubble Waiting Time in Nucleate Boiling", *Journal of Heat Transfer*, Vol. 103, 1981, pp. 673-678.

Direct Contact Heat Transfer in Two-Phase Gas Liquid Flow

C. Pisoni,* C. Schenone,† and L. Tagliafico*
University of Genoa, Genoa, Italy

Abstract

In this work an investigation is carried out on the heat and mass transfer characteristics of two-phase gas-liquid flow in annular or dispersed annular regimes.

A theoretical analysis of the gas-liquid interactions (heat, mass and momentum), developed on the basis of already known fluid dynamic models, is presented together with a set of direct temperature measurements performed on experimental equipment working with air-water mixtures.

Nomenclature

- A cross section of the duct, m^2
- C_d drag coefficient for spherical droplets
- C_p constant pressure specific heat, J/kgK
- D internal diameter of the duct, m
- D_v mass diffusivity coefficient of the vapor into the gas, m^2/s
- D_0 internal diameter of the liquid injector, m
- F force for unit duct length, N/m
- f friction factor
- G_e evaporating mass flow rate for unit duct length, kg/ms
- g x-component of the specific gravity force, m/s^2
- h thermodynamic enthalpy, J/kg
- h_c convective heat transfer coefficient, W/m^2K
- k thermal conductivity of the gas, W/mK
- k_e evaporation coefficient, kg/sm^2
- k_g gas-side mass transfer coefficient, kg/sm^2Pa
- $\dot{m}$ mass flow rate, kg/s

Copyright © 1992 by the American Institute of Aeronautics and Astronautics, Inc. All rights reserved.

*Professor, Engineering Thermodynamics, Energy Engineering Department.
†Researcher, Thermal Division, Energy Engineering Department.

N number of droplets for unit volume, m^{-3}
Nu Nusselt number, $Nu = h_c D/k$
Pr Prandtl number, $Pr = C_p \mu / k$
p pressure, Pa
p_m mean logarithmic pressure difference, Pa, $p_m = (p_s - p_v)/\ln(\frac{p - p_v}{p - p_s})$
p_s saturation pressure of the liquid at the droplet temperature, Pa
p_v vapor pressure in the gaseous mixture, Pa
q heat flux for unit length exchanged between liquid and gas, W/m
q_{ext} heat flux for unit length from the liquid to the surroundings, W/m
R gas constant, J/kgK
Re Reynolds number, $Re = \rho V D / \mu$
S inter-facial surface for unit duct length, m^2/m
Sh Sherwood number, $Sh = k_g R T p_m D / D_v p$
Sc Schmidt number, $Sc = \mu / \rho D_v$
T temperature, K
V velocity, m/s
x axial coordinate in the flow direction, m
α void fraction
δ liquid film thickness, m
μ dynamic viscosity of the gas, kg/ms
ρ density, kg/m^3
χ humidity ratio, kg of vapor/kg of gas

Subscripts

a   of "dry" gas
c   on the axis (center) of the duct
d   referred to the liquid droplets; or dry
g   of the gaseous mixture
i   inter facial; or at the inlet
l   of the liquid
m   of the gaseous mixture
o   at the outlet
s   smooth pipe
v   of the vapor
w   wall; or wet

I. Introduction

A challenge in the design of efficient energy conversion systems is the exploitation of the maximum thermodynamic potential of the system's heat sources and, in particular, the achievement of the heat transfer at the highest temperatures. To match this requirement, increasing interest has been devoted in recent years to direct contact heat transfer, which is based on the intimate contact between the streams of the fluids involved. This technique permits more efficient use of the heat sources because of the higher working temperatures in the thermodynamic cycles and of the much lower temperature gradients needed to get the desired heat flux between the streams. Furthermore, the absence of a solid separation surface noticeably reduces corrosion problems and avoids the deterioration of heat transfer effectiveness due to fouling.[1]

Direct contact processes have been used in the past mainly for operations associated with mass transfer, and the older heat transfer applications were generally coupled with mass transfer problems. This technique is of primary importance in liquid metal magnetohydrodynamic (LMMHD) energy conversion systems, since in this case a simultaneous heat and momentum exchange between gas and liquid metal is required.[2] Resort to LMMHD generators is essentially due, as is well known, to the absence of moving parts and to the achievement of higher working temperatures with respect to traditional steam or gas turbines, thus assuring a better exploitation of the heat sources usually available. Furthermore, a particular configuration of this generator (the "induction" type[2,3,4]) may directly provide alternating currents, avoiding some of the most crucial drawbacks encountered with direct current generators.

Moreover, the employment of single-phase generators currently is strongly advisable to reduce electrical conductivity problems arising in two-phase gas-liquid metal systems[5] despite the pressure drop across the generator, which becomes in this case an extra pressure load in the liquid metal loop, and requires the velocities in the nozzle and in the separator to become much higher.[6] The mixing of the two phases will generally take place in an appropriate unit (the accelerator), where the heat and momentum transfer between them will occur. The correct design of this unit and the performances of the whole LMMHD plant will be affected by heat and momentum transfer efficiency during expansion in the nozzle.

Indeed, as is well known from basic thermodynamic analyses,[7] the gas expansion taking place in the accelerator with the liquid metal should be quasi-isothermal: the dynamics of the process can strongly affect the real characteristics of this expansion, which could follow, in the worst case, an almost adiabatic trend.

In this study an attempt is made to improve the understanding of all transfer processes occurring in the two-phase flow, by means of a one-dimensional two-component analysis and experimental equipment working with air-water mixtures in thermal nonequilibrium. The aim of this work is then to outline the limits and possibilities of the simple model presented through a comparative investigation of analytical and experimental results and to obtain a deeper insight of heat transfer and flow phenomena for further improvement of the simulation procedure.

II. Analytical Model

The formulation of the model is developed with reference to the mist/annular flow regime in steady-state conditions, assuming in each duct section pressure uniformity and mean values of temperature and velocity for each one of the two phases.

The liquid film effects are considered only from the point of view of the friction losses in the gas, but the film thickness is assumed to be negligible, as well as the correspondent mass flow rate, so that no direct influence results on the heat, mass, and momentum transfer processes. The flow is therefore reduced to a gas mass flow rate dragging a swarm of liquid droplets.

The liquid droplets and the gas streams exchange heat, mass, and momentum through heat transfer, evaporation, and inter facial friction, respectively. The gas phase is then a mixture of air and liquid vapor described by the liquid content per unit mass of gas χ (humidity ratio), which changes along the duct because of the droplet evaporation driven by temperature and pressure variations. Let us note that the influence of χ on the global heat transfer process is of great importance in the case of the couple air-water, but its relevance may be less for liquid metals, since in this case the latent heat of evaporation and the variations of χ during the expansion process may be much lower.

The one-dimensional conservation equations (mass, momentum, and energy) are written in the usual form for each of the two phases moving in the cross section of the duct A

$$\frac{d}{dx}(\rho_g V_g \alpha A) - \dot{m}_a \frac{d\chi}{dx} = 0 \qquad (1a)$$

$$\frac{d}{dx}[\rho_l V_l (1-\alpha)A] + \dot{m}_a \frac{d\chi}{dx} = 0 \qquad (1b)$$

$$\rho_g V_g (1-\alpha) A \frac{dV_g}{dx} - V_v \dot{m}_a \frac{d\chi}{dx} + \alpha A \frac{dp}{dx} =$$
$$= - \rho_g \alpha A g - F_d - F_w \qquad (2a)$$

$$\rho_l V_l (1-\alpha) A \frac{dV_l}{dx} + V_v \dot{m}_a \frac{d\chi}{dx} + (1-\alpha) A \frac{dp}{dx} =$$
$$= - \rho_l (1-\alpha) A g + F_d \qquad (2b)$$

$$\rho_g V_g^2 \alpha A \frac{dV_g}{dx} + \dot{m}_a \frac{dh_m}{dx} - h_v \dot{m}_a \frac{d\chi}{dx} - \dot{m}_a \frac{d\chi}{dx} \frac{V_v^2}{2} =$$
$$= q - (F_d + \rho_g \alpha A g) V_g \qquad (3a)$$

$$\rho_l V_l^2 (1-\alpha) A \frac{dV_l}{dx} + \rho_l V_l (1-\alpha) A \frac{dh_l}{dx} + (h_v - h_l) \dot{m}_a \frac{d\chi}{dx} +$$

$$+ \dot{m}_a \frac{d\chi}{dx} \frac{V_v^2 - V_l^2}{2} = -q + F_d V_l - \rho_l (1-\alpha) A g V_l - q_{ext} \quad (3b)$$

where in particular it has been assumed that the friction force F_w between the liquid film and the wall is balanced by the inter facial drag between the film and the gas, whereas the heat transfer between the system and external sources is assumed to directly involve the liquid phase alone; in particular, the term q_{ext} may assume the physical meaning of the heat loss from the two-phase flow to the surroundings. Furthermore, the vapor phase coming from the liquid is assumed to enter the gas phase with velocity V_v and enthalpy h_v: the first term has been assumed equal to the mean liquid and gas velocities, the second equal to the saturation enthalpy of the vapor at the liquid temperature.

The one-dimensional equations, Eqs. (1-3), obviously need several empirical formulations to define the dynamics of heat, mass, and momentum transfer. Here is a short summary of the correlations adopted:
1) Friction force for unit duct length between the gaseous stream and the "liquid wall":[8]

$$F_w = 1/2 \rho_g V_g^2 \pi f_i \quad (4)$$

$$f_i = (1 + 300 \, \delta/D) f_s \quad (5)$$

$$f_s = 0.079/Re^{0.25} \quad (6)$$

where f_i is the inter facial friction factor between the liquid film and the gas and f_s is the classical Blasius relation for smooth pipes,[9] evaluated with reference to the gas phase. Since the correlation involves the knowledge of the liquid film thickness, it has been used fixing the empirical factor $(1+300 \, \delta/D)$ to a mean value of 1.6.
2) Drag force between liquid flow rate in the droplets and the gas:

$$F_d = 0.5 \, C_d (V_g - V_l)^2 \pi D_d^2 A \quad (7)$$

Assuming the droplets to be rigid, smooth, and spherical particles moving without turning in a non turbulent steady flow, the number of droplets for unit volume N can be easily calculated adopting the correlations proposed by Azzopardi et al.[10] to evaluate the droplet diameter D_d. The same value of the diameter has been also used to calculate the drag coefficient C_d, by means of the empirical equations proposed by Claverie.[11] All of these correlations are available in the literature and are not reported here for the sake of simplicity.
3) Heat flux for unit duct length, exchanged between the phases q: to evaluate the convective heat transfer coefficient h_c the Nusselt number

Nu_d referred to the droplet diameter D_d and to the gas phase has been expressed by the Ranz and Marshall correlation:[12]

$$Nu_d = 2.0 + 0.6\, Re_d^{0.5} Pr^{0.33} \qquad (8)$$

The heat flux q is therefore:

$$q = h_c S (T_l - T_g) \qquad (9)$$

where S is the interface surface between gas and droplets, for unit duct length.

To the six basic conservation equations, Eqs. (1-3), two further relations can be added: the gas phase constitutive equation (here the ideal gas assumption has been made) and a correlation to assign the liquid transfer rate for unit duct length G_e, from the liquid droplets to the gaseous mixture. This last has been introduced on the basis of the heat-mass transfer analogy,[13] which assumes the same Colburn numbers for the two transfer processes, so that:

$$\frac{Nu}{Re\, Pr^{0.33}} = \frac{Sh}{Re\, Sc^{0.33}} \qquad (10)$$

The mass transfer rate results to be:

$$G_e = k_e S\, ln\left(\frac{p - p_v}{p - p_s}\right) \qquad (11)$$

where p_v and p_s are the partial pressure of the liquid vapor in the gaseous mixture and the saturation pressure of the liquid at the droplet temperature, respectively, and k_e an evaporation coefficient linked to the gas-side mass transfer coefficient k_g appearing in the Sherwood number. Finally, one can obtain, with reference to the thermophysical properties of the gaseous mixture:

$$k_e = h_c / C_{pg} (Pr/Sc)^{0.67} \qquad (12)$$

This formulation of the physical problem yields a set of eight first-order differential equations, which has been solved numerically with a Runge-Kutta method to obtain the development of the thermal and fluid dynamic characteristics of the two-phase flow along the x direction.

III. Experiments

Experimental Apparatus

The experimental apparatus can generate, in a suitable test section, a two-phase nonisothermal air-water mixture, with known gas and liquid temperatures, in several flow conditions with particular reference to the annular, dispersed, and mist flow regimes. The equipment, sketched in Fig. 1, can be subdivided in three different sections: the fluid supply, the test section, and the final separation unit. The gas supply consists of a volumetric compressor (A) equipped with a 0.2 m³ tank (B). The air from the compressor (stored at a pressure up to 8 bar) is sent to the inlet section of the tube through a flow meter (C), a regulation valve, and humidifier (D). This device, consisting of a cylinder partially filled with water and equipped with rotating porous baffles, assures almost saturation conditions (relative humidity 90%) to the inlet air stream. Dry and wet bulb temperatures of the air near the inlet section of the test duct are measured by thermocouples indicated with T_{gd} and T_{gw} in the same figure; relative humidity can then be evaluated from these measurements together with the pressure given by the manometer P_{g2}. The water supply section is equipped with a flow meter (F), a pump (L), and a heater fed by tap water at 15°C. The maximum liquid flow rate available from the pump is of 12 kg/s with a pressure head of 20 m of water (lower flow rates are obviously available at higher pressures, up to 5 bar). The heater group consists of a thermostatic bath (E) with an electrical heater (M). The system is provided with a heat storage tank (N), in order to minimize temperature fluctuations of the water supplied to the test section. The injection may take place through different axial nozzles with internal diameters varying from 0.25 mm to 0.9 mm.

The test section (H) is a tube 1.5 m long with a constant internal diameter $D = 11$ mm; a suitable thickness of insulating material reduces the heat losses to the surroundings. Along the tube four measurement ports are provided to introduce the thermocouples T_c and T_w on the axis and near the wall, respectively; each thermocouple is connected to a high-speed data acquisition system (I) for the subsequent post processing of the signals. All thermocouples employed are of type K with an external diameter of 0.5 mm. The pressure measurement at the locations 1, 2 and 4 is made with usual water manometers.

The last separation section (G) operates with a series of concentric metallic cylinders, where the gas and the liquid phases are divided again in two distinct streams, whose temperatures can be measured with the elements T_{god}, T_{gow} and T_{lo} reported in the figure.

Operating Procedure

A controlled mass flow rate of air in the range 1-3 g/s was supplied to the test section, in almost saturated conditions, at ambient temperature (T_{gi}=290 K); the absolute pressure at the inlet of the test duct varied in the range 103-109 kPa.

The water was injected into the tube through one of two central nozzles (0.6 mm or 0.25 mm i.d., D_0) with mass flow rates varying in the range 2-8 g/s, and with temperature varying from 300 to 350 K. During each run, in steady-state conditions, the fluctuations of the controlled inlet temperatures were in the range ±0.1 K, whereas those of the mass flow rates were less then 2%; both figures were indeed comparable to the accuracy of the instrumentation employed. In particular, the given temperature precision has been obtained through a calibration procedure developed for each thermocouple with its own detection and conversion chain.

In each run the pressure measurements were made in the three stations (1, 2, and 4) indicated in Fig. 1, whereas the temperature measurements were carried out in all of the available ports along the tube, obtaining both center and wall temperature values. By using movable thermocouples, the circumferential temperature uniformity at the wall has been verified with good accuracy. Moreover, some qualitative information on the liquid film thickness has been obtained from the detection of the local sharp temperature variation, while moving the thermocouple from the cold gas core to the hot liquid film at the wall.

The temperature signals in the test section, collected with the automatic data acquisition system at the maximum rate of 30 Hz for each channel, were post processed to obtain their probability density function and to develop a statistical analysis of the temperature at the wall and in the central core of the duct.

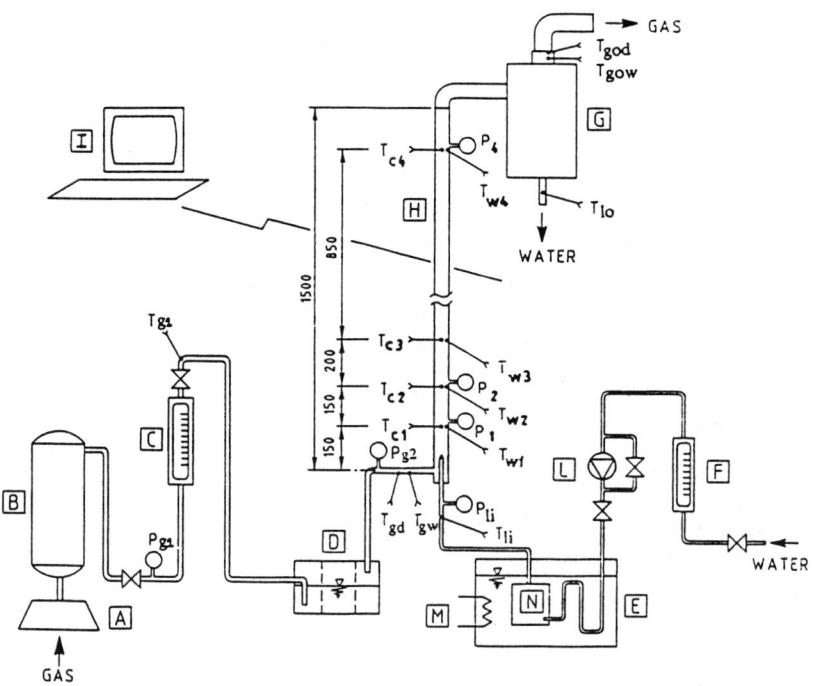

Fig. 1 Schematic diagram of the experimental apparatus: (A) compressor; (B) gas tank; (C) gas flow meter; (D) humidifier; (E) thermostatic bath; (F) liquid flow meter; (G) separator; (H) test section; (I) data acquisition system; (L) pump; (M) electrical heater; (N) temperature fluctuation damper.

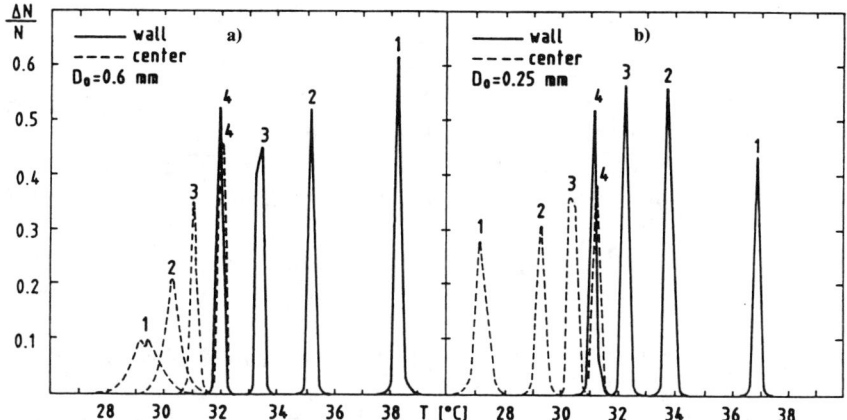

Fig. 2 Temperature signal histograms in the central core (dashed lines) and at the wall (continuous lines) for $\dot{m}_l$=2.09 g/s, $\dot{m}_g$=2.20 g/s: a) T_{li}=48.6°C, T_{gi}=19.8°C, D_0=0.6 mm; b) T_{li}=48.2°C, T_{gi}=19.0°C, D_0=0.25 mm. The characteristic temperature sampling width is 0.15°C.

On the basis of preliminary calibration tests of the whole equipment, the heat losses to the surroundings have been evaluated in different operating conditions, getting a mean global heat loss coefficient of about 7 W/m²K, with reference to the external surface of the duct.

IV. Results and Discussion

The experimental work developed on the test equipment has been mainly devoted to the temperature measurement of the two-phase mixture flowing in the duct in thermal nonequilibrium conditions.

The measuring possibilities of the data acquisition system employed are shown in Figs. 2a and 2b, where the histograms of the probability density of the temperature signals in the central core (dashed lines) and at the wall (continuous lines) are reported. The histograms refer to the four measurement stations sketched in Fig. 1 and are obtained for the same flow conditions with two liquid injectors of different diameter D_0.

A different shape of the histograms results between the "central" and "wall" signals: whereas the thermocouple on the wall is always well in contact with the liquid film, the central one, directly hit by the droplet stream, also may be influenced by the surrounding gas at a different temperature. This effect is even more evident in the case of Fig. 2a at station n. 1, as a consequence of the minor liquid atomization produced by the larger nozzle. The temperature histograms in the other ports, down flow, become sharper and well conditioned even in the central core, because of the better droplet uniformity and the lower temperature gradients. Obviously, the liquid inlet conditions have an influence on measurement quality only in the first stations; indeed, the signals in the upper locations show a very well-defined distribution in a narrow temperature range of only 0.5°C around the mean value. It can then be concluded that a good signal is always given by a wet thermocouple in close thermal contact with the surrounding liquid, in the film or in the droplets.

On this base the temperature distribution inside the test section can be drawn for the liquid film and the droplets, provided the corresponding signals are sufficiently well established. This requirement was fulfilled in most of the tests carried out, and two typical temperature distributions along the duct are reported in Figs 3a and 3b, for two different inlet conditions. The results reported in each diagram were obtained with the two different liquid injectors for the same inlet conditions. The great difference between the liquid temperature in the film (thick upper lines) and in the droplets (thin lower lines) outlines a different heat and mass transfer process between the phases. This effect is enhanced either by an increase of the gas flow rate for the same liquid flow rates (cases a and b) or by the better atomization obtained when the smaller injector is used, as shown in each diagram. Obviously, the inlet dynamics of the two-phase flow has practically no influence on the global energy balance, as can be observed from the asymptotic temperature values which are the same irrespective of the injector geometry; the small differences are only due to small variations in the operating condition at the inlet. In the final section the measured temperatures

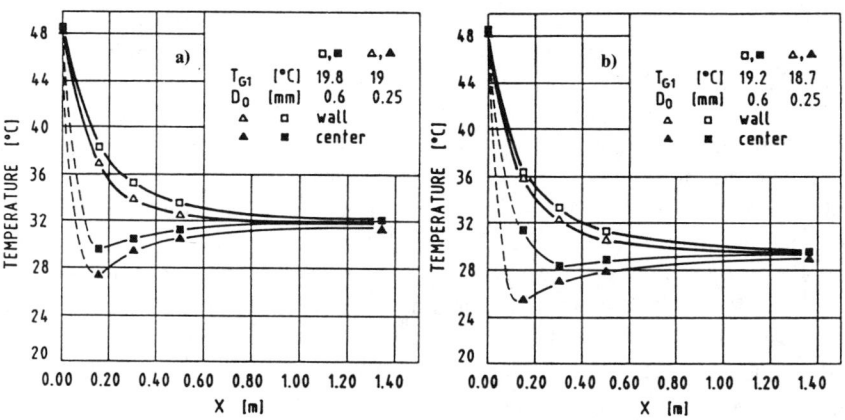

Fig. 3 Temperature variation along the duct for two different inlet conditions (liquid injector diameters: $D_0=0.6$ mm, $D_0=0.4$ mm): a) $\dot{m}_l=2.09$ g/s, $\dot{m}_g=2.20$ g/s; b) $\dot{m}_l=2.09$ g/s, $\dot{m}_g=2.82$ g/s.

in the center and on the wall are always equal, showing that a thermal equilibrium has been reached between the liquid film and the droplets; therefore, a thermal equilibrium with the gas phase must have been reached too, even if an equilibrium condition between the gas and the droplet core could have been reached well before. The resulting measured mean pressure drops along the duct for the two cases (a and b) were of 1200 Pa/m and 1400 Pa/m respectively, for both of the two injectors used.

Because of the lack of direct gas temperature data from the experimental equipment and measurement facilities available, the analytical model previously outlined in Sec. II has been used for the couple air- water to simulate the upward two-phase mist flow in thermal nonequilibrium in the test section.

The results of the numerical simulation for a set of operating conditions already considered in the experimental runs are shown in Figs. 4a-4c, where the velocities and the temperatures of the liquid and the gas, together with pressure data and humidity ratio along the duct, are reported. The two different curves (continuous and dashed) refer to the two injector diameters $D_0=0.6$ mm, $D_0=0.4$ mm which give rise to inlet liquid velocities $V_{li}=7.4$ m/s and $V_{li}=16.6$ m/s respectively. The velocity and temperature curves show that a rather short length is needed to reach dynamic and thermal equilibrium; also in the same length mass transfer is completed, as shown in Fig. 4c. Since the analytical model evaluates the droplet size on the basis of steady-state fluid dynamic conditions, irrespective of the inlet liquid injection system, the differences between the two cases considered are only due to the different liquid initial velocity imposed by the injector size at the same mass flow rate $\dot{m}_l$. When the liquid inlet velocity is low, the high slip ratio between liquid and gas gives rise to a high local pressure drop, which after a short entry length settles around a value of about 1450 Pa/m, in good agreement with the measured mean pressure drop.

In principle, a direct comparison between the analytical results and the experimental data is not feasible, since the model is not consistent with the actual flow conditions, which are influenced by the liquid film at the wall. However, some useful information on the heat and mass transfer processes developing in the test section can be obtained, reporting on the same diagram the temperature distributions given for the liquid and the gas by the model and the temperature measurements carried out along the test section for the same operating conditions. This comparison is shown in Fig. 5 for a liquid mass flow rate $\dot{m}_l=2.09$ g/s and a gas mass flow rate $\dot{m}_g=2.2$ g/s and for inlet temperatures $T_{li}=48°C$ and $T_{gi}=19.8°C$, respectively. As a consequence of the model assumptions, the predicted liquid temperature is always lower than the experimental liquid film temperature, whereas the calculated gas temperature is always higher than

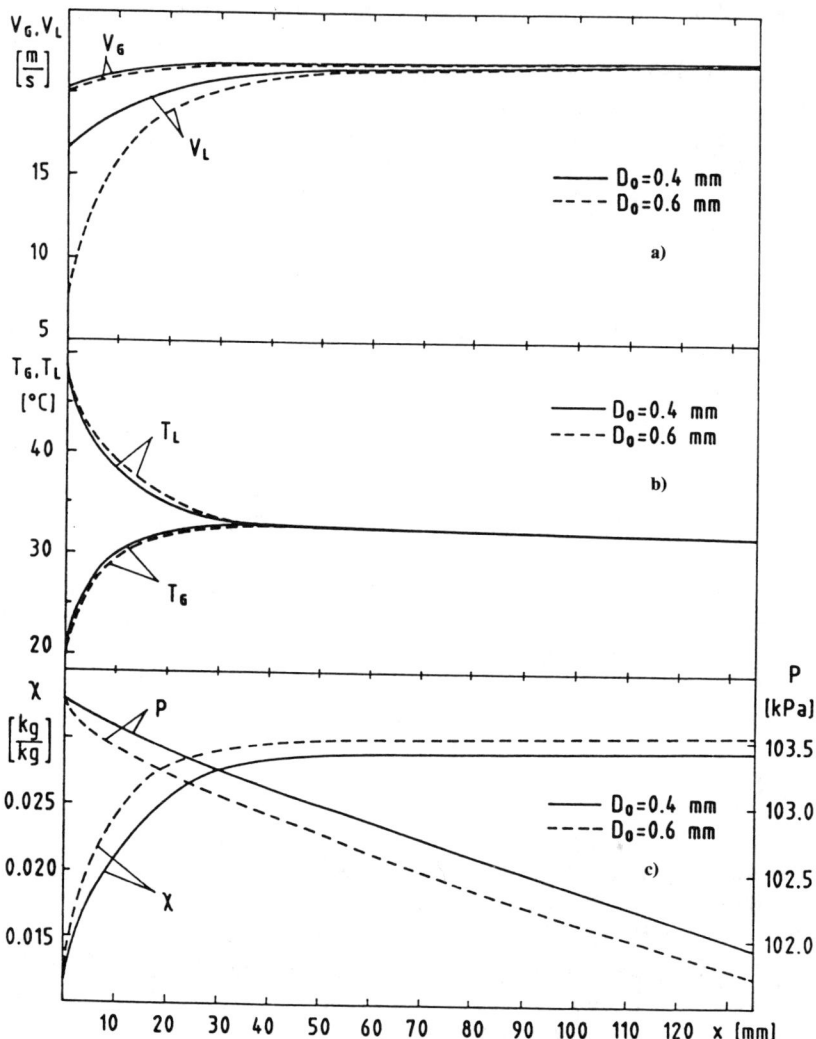

Fig. 4 Numerical solution for upward two phase mist flow in the test section; flow rates $\dot{m}_l$=2.09 g/s, $\dot{m}_g$=2.32 g/s. Distributions of liquid and gas velocity (a), temperature (b), pressure and humidity ratio (c). Continuous lines: D_0=0.6 mm, T_{gi}=19.8°C, T_{li}=48.5°C, p_{g2}=103.8 kPa. Dashed lines: D_0=0.4 mm, T_{gi}=19.1°C, T_{li}=48.2°C, p_{g2}=103.8 kPa.

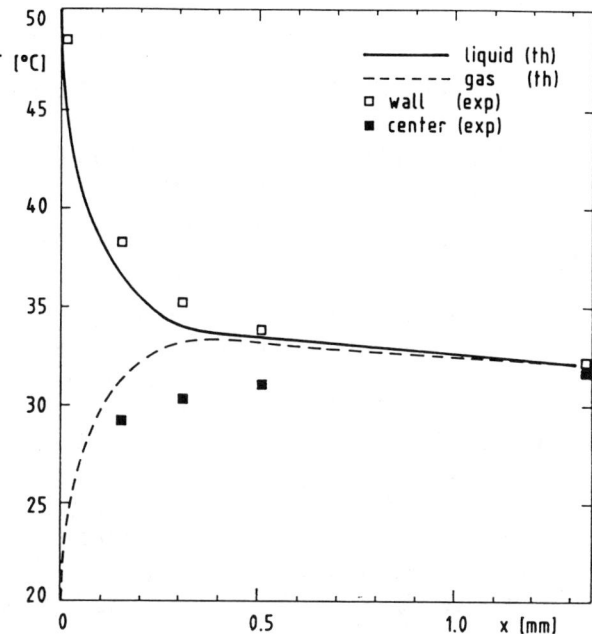

Fig. 5 Calculated and measured temperature distributions along the duct for the following conditions: $\dot{m}_l$=2.09 g/s, $\dot{m}_g$=2.20 g/s, T_{li}=48.6°C, T_{gi}=19.8°C, p_{g2}=103.6 kPa, D_0=0.6 mm.

the measured temperature of the droplets. Because of the direct thermal contact simulated in the model between the gas and the whole liquid phase, the theoretical equilibrium conditions are reached in a rather short entry length. Since it may be expected that heat and mass transfer processes developing in the test section between the liquid droplets and the gas phase are similar to those simulated numerically, it may be concluded that the thermal equilibrium between gas and droplets is also reached quickly in the actual flow. The global heat transfer, however, will need a longer length to be completed, as a consequence of the presence of the warmer liquid film at the wall. Therefore, where the measured film and droplet temperatures are practically equal, a complete thermal equilibrium also will exist, as a matter of fact, between the liquid and the gas phase.

Acknowledgments

The authors wish to thank Dr. Marco Fossa for his valuable help in the realization of the apparatus and in conducting the experimental tests. This work was made possible by the support of Italian Council of Researches in the framework of the national project "Industrial Thermomechanics", sub project "Liquid Metal Magnetohydrodynamic Systems for Energy Conversion".

References

[1] Kreith, F. and Boehm, R.F., *Direct Contact Heat Transfer*, Hemisphere, New York, 1988, pp. 203-242.

[2] Petrick, M. and Branover, H., "Liquid Metal MHD Power Generation - Its Evolution and Status," *Single and Multi Phase Flows in Electromagnetic Fields*, Vol. 100, Progress in Astronautics and Aeronautics, AIAA, New York, 1985, pp. 397-399.

[3] Elliot, D.G., Cerini, D.J., Hays, L.G. and Weinberg, E., "Theoretical and Experimental Investigation of Liquid Metal MHD Power Generation," *Electricity from MHD*, Vol. II, International Atomic Agency, Wien, Austria, 1966, pp. 995-1018.

[4] Alemany, A., Joussellin, F., Werkoff, F. and Marty, Ph., "MHD Induction Generator at Weak Magnetic Reynolds Number," *Enginering Journal of Mechanics: B/Fluids*, Vol. 8, No. 1, 1989, pp. 23-29.

[5] Tanatugu, N., Fujii-E, Y. and Suita, T., "Electrical Conductivity of Liquid Metal Two-phase Mixture in Bubbly and Slug Flow Regime," *Journal of Nuclear Science and Technology*, Vol. 9, No. 12, 1972, pp. 753-755.

[6] Laborde, R., "Convertisseur de Faraday à metal/gaz: contribution à l'etude de l'accelerateur et du separateur," Ph.D. Dissertation, Institut Nationale Polytechnique de Grenoble, Grenoble, France, 1987.

[7] Elliot, D.G., Cerini, D.J. and Weinberg, E., "Liquid-Metal MHD Power Conversion," *Space Power System Engineering*, Vol. 16, Progress in Astronautics and Aeronautics, AIAA, New York, 1966, pp. 1275-1298.

[8] Wallis, G.B., *One Dimensional Two-phase Flow*, McGraw-Hill, New York, 1969.

[9] Douglas, J.P., Gasiorek, J.M. and Swalfield, J.A., *Fluid Mechanics*, Pitman, London, 1979, p. 255.

[10] Azzopardi, B.J., Freeman, G. and King, D.J., "Drop Sizes and Deposition in Annular Two-phase Flow," United Kingdom Atomic Energy Authority, Rept. AERE-R 9634, 1980.

[11] Claverie, J., "Coefficient de trainee d'une sphere," Electricity Dept. of France, Dept. of Machine, Report HO 062 III.1.1., France, 1968.

[12] Ranz, W. and Marshall, W., Jr., "Evaporation from Drops," *Chemical Engineering Progress*, Vol. 48, , 1952, pp. 141-146, 137-180.

[13] Kreith, F., *Principles of Heat Transfer*, 2nd edition, International Textbook, Scranton, Pennsylvania, 1965, pp. 558-560.

Heat and Kinetic Energy Transfer in Two-Phase Flow: Theoretical Aspects

R. Mathes* and A. Alemany†
Institut de Mécanique de Grenoble, Grenoble, France

Abstract

To determine the performance of a LMMHD (Liquid Metal Magnetohydrodynamic) converter a one-dimensional model for annular-dispersed flow has been developed. The distribution of velocity, temperature, void fraction, and entrainment rate is calculated. The results are compared with experimental data of horizontal flow.

Nomenclature

A	=	cross sectional area
c_d	=	drag coefficient
d	=	droplet diameter
d_c, d_t	=	core and tube diameter
$f_{d,g}, f_{f,g}$	=	drag forces
g	=	gravity acceleration
h	=	enthalpies
k_d	=	deposition rate constant
m	=	mass flow
Nu	=	Nusselt number
Pr	=	Prandtl number
q	=	heat transfer term
Sc	=	Schmidt number
r_{vap}	=	latent heat of vaporisation
Sh	=	Sherwood number
u^*	=	shear velocity

Copyright © 1992 by the American Institute of Aeronautics and Astronautics, Inc. All rights reserved.
*Ph.D. Student, B.P. 53X.
†Senior Researcher, B.P. 53X.

We	=	Weber number
X	=	evaporation rate
x	=	coordinate
α	=	void fraction ($A_g/(A_g + A_d)$)
α_c	=	core void fraction
δ	=	film thickness
δ_t	=	thickness of continuous layer
μ	=	dynamic viscosity
ρ	=	density
σ_d, σ_f	=	vaporisation coefficient of droplets and film
σ	=	surface tension
τ	=	shear stress
τ^+	=	particle relaxation time

Subscripts

c	=	core flow
d	=	droplet
eff	=	effective
entr	=	entrainment
f	=	liquid film
g	=	gas flow
gd	=	transfer gas/droplet
gf	=	transfer gas/film
vap	=	vapor

I. Introduction

In the LMMHD converter the mechanical energy is generated from the interaction of a liquid-metal flow with a gas flow, the connection between the phases allowing the transformation of heat (from the liquid metal) into kinetic energy via the expansion of the gas. Thus, optimization of the working conditions of such converter necessitates knowledge of these energy exchanges. This study is devoted to theoretical and experimental analysis of annular dispersed flow with heat and mass transfer under nonequilibrium conditions. The purpose of the study is to determine the conditions that minimize friction losses in order to obtain maximum efficiency.

A. Previous Work

This study pursues the work of R. Laborde,[1] who examined the converter in a theoretical approach and in experiments. The liquid metal/gas flow was simulated by a horizontal water/air flow in the facility as well as in the modelization. Two high-performance ventilators were available, delivering a rapid air flow. The water was injected at the throat of a nozzle and pulverized by the air flow. At the end of the accelerator the two-phase flow showed a high void fraction ($\alpha > 0.99$), and the efficiency of the following separator was relatively low. In the theoretical simulation the gas and the liquid flow were regarded as independent concerning the momentum and heat transfer between the two phases. In addition, the effects of mass transfer (evaporation or condensation and entrainment or deposition of the droplets) were neglected.

B. Objectives of the Study

The objectives of this study are:
1) To develop a model for annular dispersed flow that accounts for heat and mass transfer between the phases, as illustrated in Fig. 1.
2) To compare the prediction of the model with experimental data.
3) To examine the efficiency of the accelerator at lower void fraction in vertical downflow (avoiding the influence of gravity).
4) To examine the influence of nonequilibrium conditions.

II. Mathematical Model

The theoretical aspects are based on a system of conservation equations (mass, momentum and enthalpies) for the gas and droplet flow:

$$\rho_g \alpha \alpha_c A \frac{\partial v_g}{\partial x} + v_g \alpha \alpha_c A \frac{\partial \rho_g}{\partial x} + \rho_g v_g \alpha_c A \frac{\partial \alpha}{\partial x} + \rho_g v_g \alpha \alpha_c \frac{\partial A}{\partial x} = m_g \frac{\partial X_g}{\partial x} \quad (1)$$

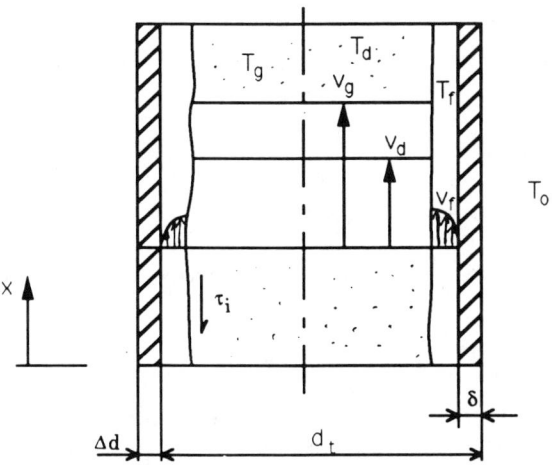

Fig. 1 Schematical illustration of annular dispersed flow in a tube.

$$\rho_d (1-\alpha)\alpha_c A \frac{\partial v_c}{\partial x} + v_d(1-\alpha)\alpha_c A \frac{\partial \rho_d}{\partial x} - \rho_d v_d \alpha_c A \frac{\partial \alpha}{\partial x} + \rho_d v_d(1-\alpha)\alpha_c \frac{\partial A}{\partial x} =$$
$$-m_g \frac{\partial X_d}{\partial x} + m_l \frac{\partial E}{\partial x} \quad (2)$$

$$m_g \frac{\partial v_g}{\partial x} + \alpha \alpha_c A \frac{\partial p}{\partial x} = f_{vap,gd}\frac{\partial X_d}{\partial x} + f_{vap,gf}\frac{\partial X_f}{\partial x} - f_{dg} - f_{fg} - f_g \quad (3)$$

$$m_d \frac{\partial v_d}{\partial x} + (1-\alpha)\alpha_c A \frac{\partial p}{\partial x} = -f_{vap,d}\frac{\partial X_d}{\partial x} + f_{entr}\frac{\partial E}{\partial x} + f_{dg} - f_d \quad (4)$$

$$m_g v_g \frac{\partial v_g}{\partial x} + m_g \frac{\partial h_g}{\partial x} = q_{vap,gd}\frac{\partial X_d}{\partial x} + q_{vap,gf}\frac{\partial X_f}{\partial x} - (f_{dg}+f_{fg}+f_g)v_g - q_{dg} - q_{fg} \quad (5)$$

$$m_d v_d \frac{\partial v_d}{\partial x} + m_d \frac{\partial h_d}{\partial x} = -q_{vap,d}\frac{\partial X_d}{\partial x} + q_{entr,d}\frac{\partial E}{\partial x} + (f_{dg}-f_d)v_d + q_{dg} \quad (6)$$

The heat flux among the liquid, the core-flow, and the tube wall is accounted for with an enthalpy equation for the liquid film in Eq. (7). Here the acceleration terms are neglected:

$$m_f \frac{\partial h_f}{\partial x} = -q_{vap,f}\frac{\partial X_f}{\partial x} - q_{entr,f}\frac{\partial E}{\partial x} + q_{f,g}v_d + q_{dg} \quad (7)$$

Instead of the mass and momentum equation for the liquid film, an analytical model proposed by Dobran[2] is used to calculate the thickness of the liquid film and the velocity in it. In this model the liquid film is divided into a continuous and a wavy layer. The velocity distribution is obtained by the universal velocity profile in the continuous layer and by integration of the momentum equation

$$\tau = \mu_{eff}\frac{\partial v}{\partial y} \quad (8)$$

in the wavy layer with an effective momentum diffusity

$$(\mu_{eff}/\mu_l)_{wl} = 1. + 0.016(\delta^+ - \delta_t^+)^{1.8} \quad . \quad (9)$$

The interfacial friction factor[3] used in the equation for the shear stress

$$\tau_i = \frac{1}{2}f_i \rho_c v_c^2 \quad (10)$$

is expressed in terms of the core Reynolds number Re_c and film thickness δ

$$f_i = 0.078\, Re_c^{-0.25}\left(1 + 24\left(\frac{\rho_l}{\rho_g}\right)^{1/3}\right)\frac{\delta}{D_t} \quad . \quad (11)$$

To close Eqs. (1)-(7), constitutive laws are used for the terms on the right side, and the laws of state are used for the phases. The liquid flow is regarded as incompressible, and for the gas flow we used the ideal gas law.

A. Constitutive Law

1. Interfacial Forces

The drag forces between the droplets and the gas flow are descriped by

$$f_{dg} = \frac{m_d 3 \rho_g c_d |v_g - v_d| (v_g - v_d)}{4 d \, v_d \, \rho_d} \tag{12}$$

with the drag coefficient c_d[4], which includes the influence of the void fraction

$$c_d = 200 \frac{1-\alpha}{\alpha^2 Re_g} + \frac{7}{3\alpha} \tag{13}$$

The interfacial forces between the gas flow and the liquid film follow from (10) i.e. :

$$f_{gf} = \pi \, d_c \, \tau_i \tag{14}$$

2. Heat Transfer

From ref.[5] the drop heat parameter is taken as

$$Nu_{dg} = 2.0 + 0.6 \, Pr^{1/3} \, Re^{1/2} \tag{15}$$

to get the heat flux

$$q_{dg} = m_d 6 \frac{\lambda_g Nu_{dg}}{d} \frac{(T_g - T_d)}{d \, v_g \, \rho_d} \tag{16}$$

To get the heat transfer from the core flow to the liquid film, the core flow was treated like a tube flow with the following Nusselt number[6]:

$$Nu_{gf} = 0.023 \, Pr^{0.8} \, Re^{0.33} \tag{17}$$

The heat flux in the film is expressed by an equivalent Nusselt number[2] taking into account the influence of the continuous and the wavy layer. The heat losses are taken into consideration by a thermal conductivity of the tube wall and a heat transfer coefficient α_0 at the surface of the tube.

3. Evaporation and Condensation

Because of the great change of state in the accelerator, the saturation conditions of the gas flow change drastically. Therefore, the influences of vaporization and condensation on the heat and mass transfer have to be considered. For the vaporization process $X_g < X_{sg}$, it is assumed that the temperature of the droplets and the liquid film is equal to the saturation temperature. The evaporation depends on the difference of vapor concentration on the droplet X_{sg} or liquid film X_{sf} surface and the vapor concentration of the gas X_g. The concentration rate in the gas is calculated as follows:

$$\frac{\partial X_g}{\partial x} = \frac{\partial X_d}{\partial x} + \frac{\partial X_f}{\partial x} \tag{18}$$

The evaporation rate of the droplets can be written as

$$\frac{\partial X_d}{\partial x} = \frac{\sigma_d (X_{sd} - X_g)}{\dot{m}_g} \frac{\partial A_g}{\partial x} \tag{19}$$

with

$$A_d = 6 m_d / (\rho_d v_d d) \tag{20}$$

$$\sigma_d = \frac{\rho_g d_v Sh}{d} \tag{21}$$

and [5]

$$Sh = 2.0 + 0.6 \, Sc^{1/3} \, Re^{1/2} \tag{22}$$

$$Sc = \frac{\mu_g}{\rho_g d_v} \qquad Re = \frac{\rho_g d (v_g - v_d)}{\mu_g} \tag{23}$$

$$\log d_v = \log |0.2 \cdot 10^{-4}| + \log(T_g/273.15) \log(959.0/273.15) \tag{24}$$

For the liquid film a similar equation is applied:

$$\frac{\partial X_f}{\partial x} = \frac{\sigma_f (X_{sf} - X_g)}{\dot{m}_g} \frac{\partial A_f}{\partial x} \tag{25}$$

with

$$\sigma_f = \frac{\lambda_g Nu_{gf}}{d \, c_g} \tag{26}$$

$$A_f = \pi d_c \tag{27}$$

where the Nusselt number of Eq. (16) is used.

In the case of condensation $X_g > X_{sg}$, a simplification was made, assuming that we immediately have a condensation $\Delta X_g = X_g - X_{sg}$, always having saturation conditions.

4. Entrainment and Deposition

At low void fraction, which is expected for a high efficiency of the separator, the thickness of the liquid film at the wall increases and, consequently, so do the pressure losses due to the interfacial friction factor. Therefore, the entrainment of droplets from the liquid film by the core flow (or deposition) is included in the modelization. This process depends on the equilibrium entrainment E_∞ proposed by R.V.A. Oliemanns[7] for an established flow.

The entrainment gradient, for the case where $E < E_\infty$, is given by Kataika and Ishii[8]:

$$\frac{\partial E}{\partial z} = \frac{1}{d_t}\left[2.87 \cdot 10^{-9} \, Re_{ef}^{0.5} \, Re_{elf\infty}^{0.25} \, We \left(1 - \frac{E}{E_\infty}\right)^2 + \right.$$
$$2.64 \cdot 10^{-6} \, Re_f^{-0.075} \, We^{0.925} \left(\frac{\mu_g}{\mu_l}\right)^{0.26} + \qquad (28)$$
$$\left. (1-E)^{0.185} \cdot 0.088 \, Re_f^{-0.26} \left(\frac{\mu_g}{\mu_l}\right)^{0.026} E^{0.74} \right]$$

with

$$Re_f = \rho_l \, j_l \, d_t / \mu_l, \quad Re_{f\infty} = Re_f (1 - E_\infty)$$
$$We = \rho_g \, j_g^2 \, d_t \left[\frac{(\rho_l/\rho_g - 1)^{1/3}}{\sigma}\right]$$
$$j_l = m_l/(\rho_l A) \quad \text{and} \quad j_g = m_g/(\rho_g A)$$

For the case where $E > E_\infty$ the model of McCoy and Hanratty[9] was chosen to calculate the deposition rate

$$\frac{\partial E}{\partial z} = -\frac{4 k_d}{\alpha \, v_g \, d_t} E \qquad (29)$$

with

$$\tau^+ \leq 0.2 \qquad \frac{k_d}{u^*} = 0.0889 \, Sc^{-0.704}$$
$$0.2 \leq \tau^+ \leq 22.9 \qquad \frac{k_d}{u^*} = 3.25 \cdot 10^{-4} \, \tau^{+2} \qquad (30)$$
$$22.9 \leq \tau^+ \leq 1.48 \cdot 10^4 \qquad \frac{k_d}{u^*} = 0.17$$
$$\tau^+ \geq 1.48 \cdot 10^4 \qquad \frac{k_d}{u^*} = 20.7 / \sqrt{\tau^+}$$

The friction velocity is defined by

$$u^* = \sqrt{(\tau_i/\rho_c)} \qquad (31)$$

and the relaxation time by

$$\tau^+ = \frac{d^2 \, u^{*2} \, \rho_g \, \rho_d}{18 \, \mu_g^2} \quad (32)$$

The terms in Eqs. (3) - (7) concerning the mass transfer are written as follows:

Evaporation:

$$f_{vap\,g,d} = m_g (v_d - v_g) \qquad f_{vap\,g,f} = m_g (v_f - v_g)$$

$$f_{vap\,d} = 0$$

$$q_{vap,g} = m_g \left(c_v(T_d - T_g) + \frac{v_g^2 - v_d^2}{2} \right) \qquad q_{vap,f} = m_g \left(c_v(T_f - T_g) + \frac{v_f^2 - v_d^2}{2} \right)$$

$$q_{vap,d} = m_g \, r_{vap} \qquad q_{vap,f} = m_g \, r_{vap}$$

Condensation:

$$f_{vap\,g,d} = 0 \qquad f_{vap\,g,f} = 0$$

$$f_{vap,d} = m_g (v_d - v_g)$$

$$q_{vap,gd} = m_g \left(c_l(T_g - T_d) + \frac{v_g^2 - v_d^2}{2} \right) \qquad q_{vap,gf} = m_g \left(c_f(T_g - T_f) + \frac{v_g^2 - v_f^2}{2} \right)$$

$$q_{vap,d} = m_g \, r_{vap} \qquad q_{vap,f} = m_g \, r_{vap}$$

Entrainment: (33)

$$f_{entr} = m_l (v_f - v_d)$$

$$q_{entr,d} = m_g \left(c_l(T_f - T_d) + \frac{v_f^2 - v_d^2}{2} \right) \qquad q_{entr,f} = 0$$

Deposition:

$$q_{entr,d} = 0 \qquad q_{entr,f} = \dot{m}_l \left(c_l(T_f - T_d) + \frac{v_f^2 - v_d^2}{2} \right)$$

5. Numerical Procedure

A standard fourth order Runge-Kutta scheme is used to integrate Eqs. (1) - (7). As a result, we get the distribution of distribution of pressure, void fraction, entrainment rate, rate of gas saturation, velocities, and temperatures of all of the phases.

III. Experimental Results

A design of the experimental apparatus used by R. Laborde[1] is shown in Fig. 2.

The water was injected upstream and at the throat of the nozzle and dispersed by the high-velocity air flow. The temperature and the static and the absolute pressure of the gas were measured at the entry of the nozzle. In the homogenization part following the accelerator, the static pressure was established at three points. The temperature of the droplets was determined by thermocouples in the core flow and after separation, where the water is collected. The temperature measured by the thermocouples is probably a mean temperature of the droplets and of the hotter gas flow, and they always showed higher values than the temperature of the collected water.

The size and the velocity of the droplets were obtained by the measurements with a dynamic particle analyzer (DPA).

To compare the experimental results with the calculations, the following assumptions were made in the numerical modelization:

1) All of the water is injected upstream of the nozzle.
2) The size of the droplets is determined by the pulverization at the injection. A comparison with the results of M. Pilch and C.A. Erdmann[10] showed that the time during which the droplets stay in the accelerator is much shorter than the breakup time; thus, the assumption of constant droplet diameter is justified.
3) In the apparatus the section changes from a circle in the accelerator to a rectangular form in the homogenization part. In the calculation an equivalent circle diameter was used for the rectangular section.
4) The influence of gravity was neglected.

A typical distribution of pressure, velocity, temperature, and void fraction is presented in Figs. 3 - 6, and a comparison of all of the measured and calculated values is given in Table 1.

The results show the influence of the water flow rate on the temperature and velocity distribution. The greatest part of transformation

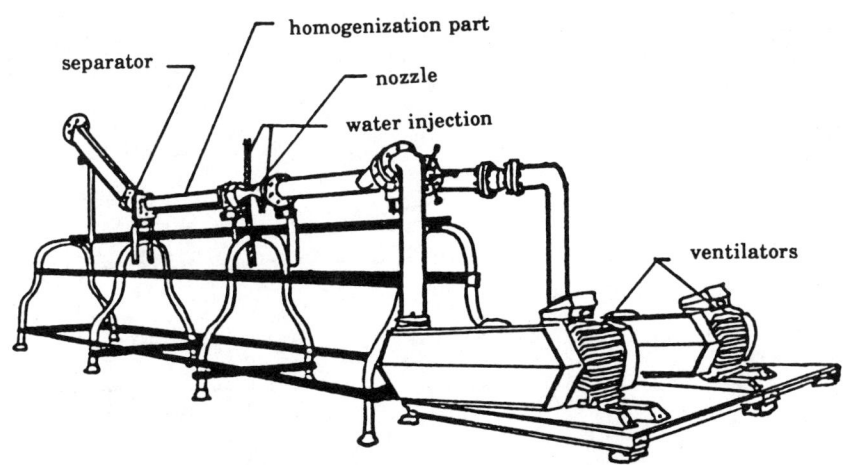

Fig. 2 Scheme of the experimental apparatus.

Table 1 Comparison of experimental and theoretical data

	I exp.	I theor.	II exp.	II theor.	III exp.	III theor.	IV exp.	IV theor.
m_l, kg/s	0.277		0.416		0.555		0.694	
m_g, kg/s	0.334		0.316		0.313		0.304	
P_o, bar	1.414		1.431		1.440		1.442	
P_{acc}, bar	1.227	1.207	1.239	1.200	1.249	1.162	1.248	1.128
$\Delta P_{0,3}$, bar	0.016	0.029	0.016	0.037	0.020	0.047	0.023	0.059
$\Delta P_{0,9}$, bar	0.036	0.072	0.036	0.089	0.043	0.116	0.049	0.146
v_d, m/s	115	133	106	126	94	130	85	134
ΔT_d, K	7.6	5.2	5.6	4.9	5.2	5.4	4.3	3.9
E	>0.94	>0.95	>0.94	>0.96	>0.94	>0.97	>0.94	>0.975
D, µm		386		372		324		320

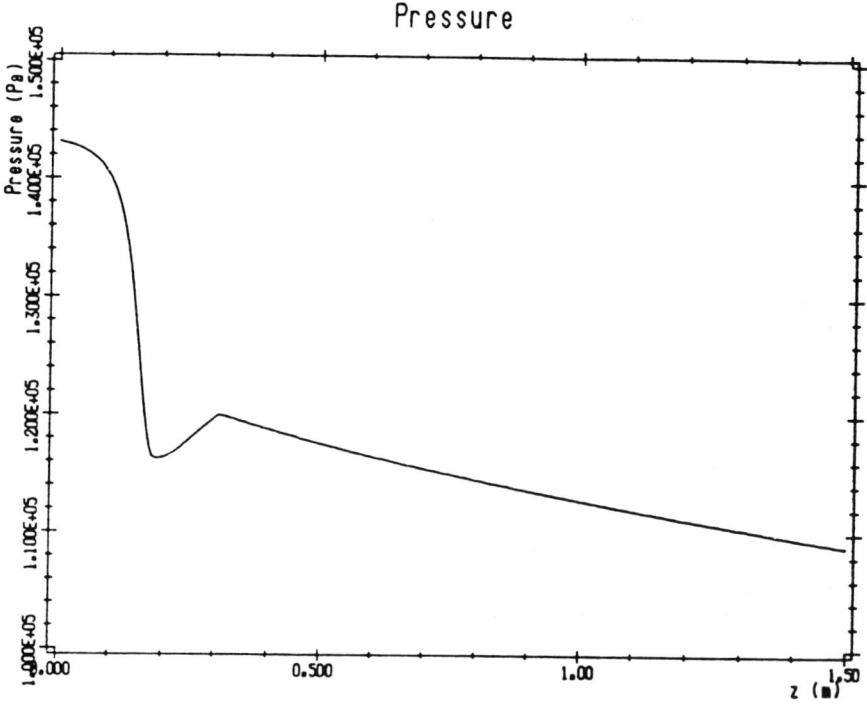

Fig. 3 Pressure distribution.

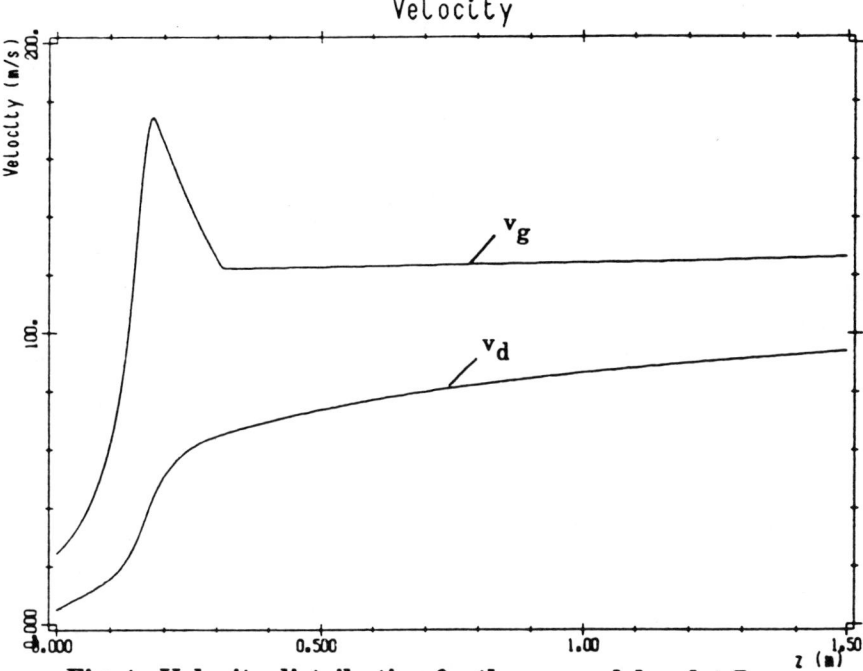

Fig. 4 Velocity distribution for the gas and droplet flow.

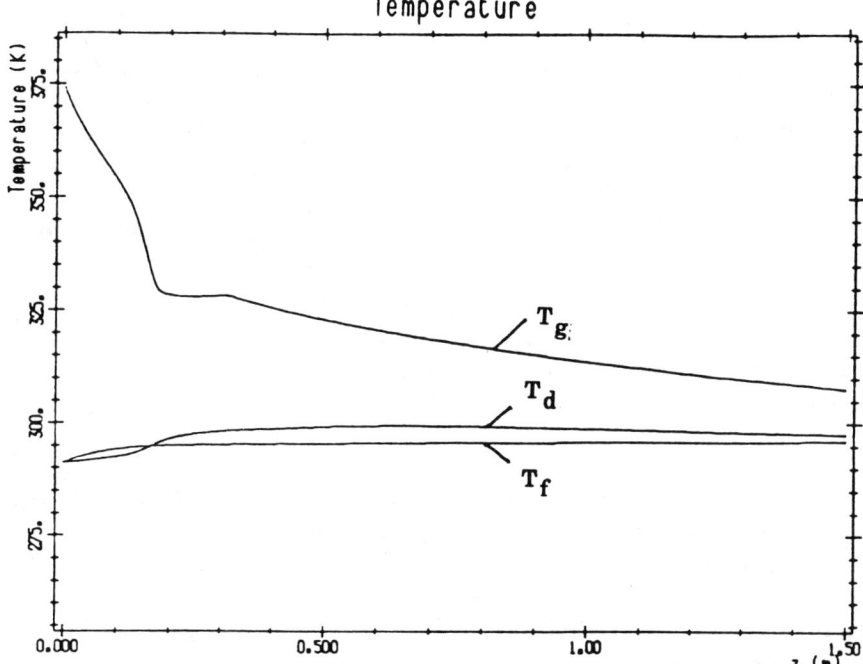

Fig. 5 Temperature distribution for the gas-droplet and film flow.

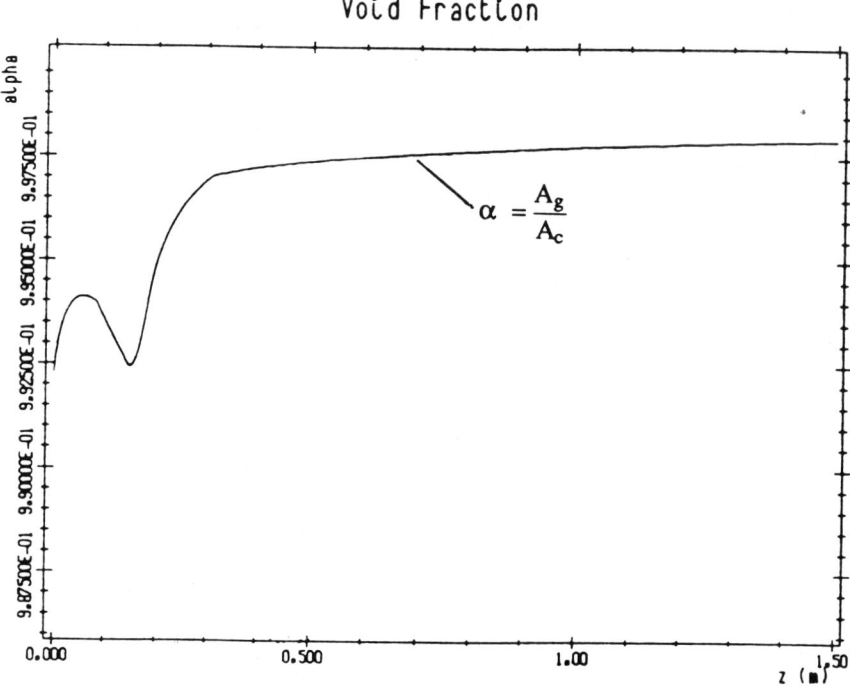

Fig. 6 Distribution of void fraction.

of the thermal and potential energy into kinetic energy (of the gas and especially of the liquid) takes place in the accelerator. The frictional pressure drops are much greater in the accelerator than in the homogenization part due to the high velocity differences. The predictions of the liquid film flow rate are very good and show that the influence of the liquid film on the core flow is small. For much higher liquid flow rates this influence has to be studied.

IV. Conclusion

A model for annular dispersed flow has been used to study the flow conditions in the accelerator of a LMMHD converter. The predictions of the velocity and temperature distribution and the entrainment rate are very good. Further work is required to determine the optimal shape of the accelerator to minimize the interfacial friction losses between the gas and droplet flow. For the next experiment a higher liquid flow rate is envisioned to get a lower void fraction. Therefore, the installation was changed from a horizontal flow to a vertical downflow to avoid the influence of gravity. With the measurements of the DPA a better determination of the distribution of the droplet and film flow rates is expected.

Acknowledgment

This work has been supported by the European Economic Community by a Ph.D. scholarship.

References

[1] Laborde, R., " Faraday Metal/Gas Converter: Contribution to the Study about the Accelerator and the Separator," Masters thesis, Institut National Polytechnique de Grenoble - I.M.G., Grenoble, France, Dec. 22 1987.

[2] Dobran, F., " Hydrodynamic and Heat Transfer Analysis of Two-Phase Annular-Flow with a New Liquid Model of Turbulence," International Journal of Heat Mass Transfer, Vol. 26, NO. 8, 1983, pp. 1159 - 1171.

[3] Bergles, A.E., Collier, J.G., Delhaye, J.M., Hewitt, G.F., and Mayinger, F., "Two-Phase Flow and Heat Transfer in Power and Process Industries," Hemisphere, New York, 1981.

[4] Soo, S.L., "Fluid Dynamics of Multiphase Systems," Blaisdell, Waltham, Massachusetts, 1967

[5] Ranz, W.E., and Marshall, W.R., "Evaporation from Drops," Chemical Engineering Progress, Vol. 48, No. 3, March 1952

[6] Taine, J., and Petit, J.P., "Thermal Transfer, Anisothermic Fluid Mechanic," Bordas, Paris, 1989.

[7] Oliemanns, R.V.A., Pots, B.F., and Trompe, N., "Modelling of Annular Dispersed Two-Phase Flow in Vertical Pipes," International Journal of Multiphase Flow, Vol. 12, No. 5, 1986, PP. 711 - 732.

[8] Kataker, M., and Ishii, "Entrainment and Deposition Rates of Droplets in Annular Two-Phase Flow," Proceedings of the Thermal Engineering Joint Conference, VOL. 1, ASME, New York, 1983, pp. 69 - 80.

[9] McCoy, D.D., and Hanratty, T.J., "Rate of Deposition of Droplets in Annular Two-Phase Flow," International Journal of Multiphase Flow, Vol. 3, No. 4, 1977, pp. 319 - 331.

[10] Pilch, M., and Erdmann, C.A., " Use of Breakup Time Data and Velocity History Data to Predict the Maximum Size of Stable Fragments for Accelerations Induced Breakup of a Liquid Drop," International Journal of Multiphase Flow, Vol. 13, No.6, 1987, pp. 741 - 757.

Nuclear Void Fraction Gaging in Large Two-Phase Organic Liquid-Metal MHD Generators

A. P. Kushelevsky*
Ben-Gurion University of the Negev, Beer-Sheva, Israel

Abstract

Although excellent results are obtained in measuring void fractions by gamma gaging in pipes with diameters of up to 4 in. containing a two-phase organic liquid metal mixture, serious radiation protection problems are encountered when attempting to measure void fractions in pipes of much larger diameters that require very high source activities. The methods available to overcome these problems are discussed. They include: 1) making measurements over extended periods and volumes thus reducing the activity of the source needed at the expense of loss in spatial and temporal resolution; 2) using remotely controlled radiographic projection systems; 3) modifying the measurement geometry by placing the source inside the pipe rather than outside; and 4) using miniature 14 Mev neutron generators to produce a higher penetrating beam which can be switched off when not in use.

Introduction

The void fraction (V_f) is one of the important factors which influence the efficiency of the electrical power conversion of a two-phase organic liquid magnetohydrodynamic(MHD) generator.[1] Measurement by internal probes is problematic. Firstly because the probes measure local conditions that do not necessarily represent the complete flow picture, and secondly because of errors due to residual fluid on the probes and perturbations in the normal flow pattern caused by the probes themselves.[2]

Gamma ray gaging, a noninvasive method that does not suffer from these problems, is therefore preferred.[3] The basic experimental arrangement for gamma ray gaging is simple. It consists of a radioactive source emitting a beam of gamma rays placed on one side of the pipe through which the conducting mixture flows, and a detector facing the source at the opposite side of the pipe.

Copyright © 1991 by the American Institute of Aeronautics and Astronautics, Inc. All rights reserved.
* Chairman, Department of Nuclear Engineering.

For a narrow parallel monoenergetic beam with an incident intensity of I_o, the transmitted intensity I is given by:

$$I = I_o e^{-\mu_g x} \cdot e^{-\mu_L(L-x)} \qquad (1)$$

where μ_L and μ_g are the linear attenuation coefficients of the metal and gas phases, respectively; L the total path length through the two-phase mixture; and x the effective distance traveled in the gas phase alone.

Writing V_f as $\alpha = x/L$ and $I_L = I_o e^{-\mu_L L}$ and $I_g = I_o e^{-\mu_g L}$, it can be shown that:

$$\alpha = \ln(I/I_L) / \ln(I_g/I_L) \qquad (2)$$

Although excellent results are obtained using relatively small gamma sources in small diameter pipes as used in experimental two-phase organic liquid metal MHD generators, very high activity sources are required for corresponding measurements in large diameter pipes of the size which may eventually be used in large commercial two-phase organic liquid metal MHD generators.

Order of magnitude calculations indicate that depending on the nature of the liquid metal, the diameter of the pipe, the source detector geometry, and the required accuracy of the measurement, sources with activities of hundreds of curies or more may be needed. This may be illustrated by calculating the activity of a Co-60 source, a high-energy gamma source widely used in gamma ray radiography, which would be required for the measurement of void fractions in an 8 in. pipe, with lead as the metal phase. Taking the relevant attenuation coefficient to be 0.64 cm^{-1}, we calculate the attenuation of the beam due to the lead to be of the order of 10^{-6} and a further 10^{-4} due to geometric attenuation due to 8 in. interval between source and detector.

Under these conditions for one photon/second to reach the detector, a source emitting 10^{10} photons/second (approximately 1 Ci) is required. Noting that 100 counts are required per measurement to achieve ±10% statistical accuracy activities, of the order of hundreds of curies are evidently required.

The question of how to overcome the problems of using such large sources in an industrial environment is the subject of this paper. Because such large sources are potentially lethal it is doubtful whether radiation protection authorities would authorize work with them even if they were adequately protected.

In this paper we discuss possible solutions to the radiation protection problem under three headings:
1) reduction of source activity by optimizing the measurement procedures;
2) alternative shielding arrangements; and 3) alternative source detector geometry.

Reduction of Source Activity

Extending Measurement Duration

The simplest way of reducing the source activity is to accumulate the required number of counts per measurement at a lower rate with a smaller source rather than a high count rate using a large source. Reducing the activity of the source in this way has its price.[5,6] First, there is a loss of temporal resolution, that is a loss in the ability to detect and measure short-term void fraction fluctuations. Secondly, due to the exponential nature of radiation gaging, averaging the measurement over long periods may cause the measured mean values to deviate considerably from the true mean values by as much as 50%, depending on the size and rate of the fluctuations in the void fraction.

Increasing the Detector Counting Efficiency

Another simple method to reduce the required source activity is by increasing the counting efficiency of the detector by increasing the cross section of the incident gamma beam and correspondingly the size and thickness of the detector in order to intercept the full extent of the beam.[7] This enhances the count rate by at least a factor directly proportional to the area of the detector.

Reducing the required source intensity this way also has its price in lost spatial resolutions meaning that we are no longer able to study the void fraction distribution across the pipe with the same resolution of a narrower beam.

Again, due to the exponential behavior of the radiation gage, this leads to an indeterminate error that depends on the size, rate and spatial distribution of the fluctuations in the void fraction.

Alternative Solutions

Allowing for the difficulties mentioned above, increasing measurement time and enhancing detection efficiency reduces the required activity by one or two orders of magnitude. This on its own however, does not solve our problem since the size of the source although smaller and easier to handle is still formidable and the health hazard still large.

Projection Systems

If the measurements do not need to be made continuously the health hazard problem may be solved by using a source projection system. Projection systems work by guiding the source from its shielded container, which can be placed a considerable distance from the measurement position, to the measurement position by remote control, retracting the source back to its shielded container when the measurement is finished. Commercial systems that handle hundreds of curies are available and may be used provided all persons are removed from the vicinity of the measurement site. The advantages of this arrangement is that, with all the necessary shielding situated away from the pipe, a minimal disturbance of the MHD system is required and, more importantly, commercial projection systems may be purchased that are authorized for immediate use.

Alternative Source Detector Geometry

Changing the geometry by placing the source in the center of the pipe is another way to reduce the activity of the source. This alternative geometry halves the thickness of the attenuating material between the source and detector, compared with the conventional arrangement which places the source and detector at opposite sides of the pipe. This geometry has two more advantages: First, it permits simultaneous void fraction measurements to be carried out at various angles around the pipe using a single source at the center of the pipe with multiple detectors around its circumference. Secondly, the metal phase in the pipe and the construction materials of the rest of the MHD generator act as an effective shield reducing the thickness of additional shielding material required to protect workers in the vicinity of the generator.

Despite these obvious advantages, this solution is not ideal because the source inside the pipe interferes with the normal flow pattern and the measurement is thus no longer non intrusive. It should be noted that provisions must be made to remove the source to a shielded container when the pipe is voided of its lead.

Discussion

As mentioned earlier all of the solutions have their own specific problems. Increasing the measurement duration and using large detectors to increase the detection efficiency reduces the quality of information obtained and introduces errors that are difficult to evaluate. Placing the source within the pipe, altering the conventional source-detector geometry, interferes with the flow pattern, resulting in a measurement that is no longer truly noninvasive . Using a projection system to separate the shielding from the measuring site is a viable solution only if measurements are not required on a continuous basis and only if it is possible to remove all personnel from the site during measurements.

We feel that the optimum solution must carefully draw on all of the possibilities blending them together to take advantage of their individual strengths.

Switching from gamma rays to fast neutrons may provide further flexibility due to the higher penetrability of fast neutrons compared to high energy gammas, as the attenuation coefficient of lead for fast neutrons with energies above 2 MeV is about 20 -25 % of the attenuation coefficient of Co-60 gammas. A miniature high intensity 14 MeV neutron generator as used in oil exploration bore holes would be ideal. Its small size allows it to be placed near the MHD piping, and, in addition, it only emits neutrons when switched on requiring no special shielding, provided all persons can be removed from the vicinity of the measurement site while measurements are carried out.

References

[1] Unger, Y., El Boher, A., Lessin, S. and Branover, H. "Two phase Liquid Metal - Gas Flows in Vertical Pipes" 10th International Conference on MHD Electrical Power Generation, 1986

[2] Hewitt, G.H., Measurement of Two Phase Flow Parameters, New York, Academic Press, 1978

[3] Gardner, R. P. and Ely, R.L. Radioisotopes Measurement Applications in Engineering Reinhold, New York, 1967

[4] D.I. Garber and R.R. Kinsey, Neutron Cross Sections, BNL Vol. 2, Brookhaven National Laboratories, 1976

[5] Harms, A.A. and Forrest, C.F, "Dynamic Effects in Radiation Diagnosis of Fluctuating Voids", Nuclear Science and Engineering, Vol. 46, 1971, pp. 408 -413,

[6] Neal, L.G., Wright, R.W. and Wentz, L.B. "X Ray Measurement of Void Dynamics in Boiling Liquid Metals", Nuclear Applications., Vol. 4, 1968 pp.347-355

[7] Gardner, R.P., Bean, R.H. and Ferrell, J.F. "On The Gamma Ray One Shot Collimator Measurement of Two Phase Flow Void Fractions", Nuclear Applications and Technology. ,Vol. 8, 1969, pp. 88-94,

Liquid-Metal Magnetohydrodynamic Two-Phase Flow Experiment

J.-P. Thibault,* B. Seck,† and A. Cartellier‡
Institut de Mécanique de Grenoble, Grenoble, France

Abstract

An experiment is described in which liquid-metal magnetohydrodynamic (LMMHD) two-phase flow similar to the one of a Faraday two-phase flow generator is instrumented. The choice of a room temperature simulation pair (mercury-nitrogen or mercury-air) is justified. A description of the experimental measurements shows how the data can be used to compare them to the results of existing numerical models that necessitate this validation. Void fraction profiles are obtained from a mobile optical probe that was used for the first time in LMMHD two-phase flow.

Introduction

We develop a joint theoretical and experimental analyses of liquid-metal magnetohydrodynamic (LMMHD) two-phase flow, connected to our program on LMMHD conversion. The object of the analyses is on one hand, to produce an experimental data bank and instrumentation process and, on the other hand, to produce models that are able to predict and to expound the average behavior of the flow. Our first work was theoretical[1], and it mainly consisted of the following: 1) deriving the area average formulation of the thermohydraulic and electromagnetic equations of these flows; 2) expressing the latter in the classical numerically solvable form (i.e., partial differential equations); and 3) extracting from the numerical results a better knowledge of the experimental data

---

Copyright © 1991 by the American Institute of Aeronautics and Astronautics, Inc. All rights reserved.
 * Senior Researcher, MHD Division, B.P. 53X, 38041.
 † Ph.D. Student, MHD Division, B.P. 53X, 38041.
 ‡ Senior Researcher, Two-Phase Flow Division, B.P. 53X, 38041.

needed to obtain the validation of the model. Because the exact area average form of the equations is not presently solvable, the closure of the system of equations necessitates a semi-empirical formulation of the nonreducible terms such as interfacial transfer, wall friction and apparent electrical conductivity. In this first approach we used well-established formulations (i.e., Friedel correlation for the wall friction) and introduced in combination a formulation specific to the MHD case (i.e., apparent electrical conductivity[2]). The validation of this work necessitates a comparison with experimental data. Because the latter are not currently available, we now concentrate our effort on data acquisition. For this purpose, after a first calibration[3] using the Ben Gurion University data collected on ordinary (i.e., non-MHD) liquid-metal two-phase flow[4], we engaged our own experimental capability to collect data on LMMHD two-phase flow. Because the possibilities of instrumentation in high-temperature liquid-metal two-phase flow are limited, we decided to use, in place of the original two-phase flow pair (i.e., lithium-cesium), a simulation pair more convenient at laboratory conditions: liquid-mercury plus air or nitrogen. This choice will be justified in the following. Even if the liquid-metal two-phase flow is at room temperature, the measurements are rather limited. On the other hand, it is not possible to base an extensive analysis of the flow on measurements that are too limited. Thus, based on our parametric studies, we come to select the following as useful area average measurements: pressure, temperature, void fraction, and electrical conductivity. In other words, information is needed on dynamics (pressure), thermodynamics (temperature), and electrics (electrical currents), depending on flow area average topology (void fraction).

Experimental Apparatus

It is now well established that various liquid-metal two-phase flow pairs can be considered for practical applications using, on one hand, a vapor cycle, [i.e., for high temperature lithium (liquid) associated with cesium (vapor), for medium- or low-temperature tin or mercury or lead bismuth associated with water steam] and, on the other hand, a gas cycle [i.e., sodium or sodium-potassium (NaK) associated with nitrogen]. At the laboratory conditions, it is usual to work with a simulation pair, especially if there are some intrinsic technological difficulties connected to high temperature for example. That is the reason we selected a room temperature simulation pair: mercury plus air or nitrogen. This choice is justified on the basis of a selection of the following global and local

Table 1 Physical properties (internatinal units) and dimensionless parameter ratios

	ρ_L	μ_L	σ_L	r_G	γ_G	T_G	t_{sL}
Hg/N$_2$	1.36 10^3	1.5 10^{-3}	10^6	296	1.4	293	0.5
Li/Cs	440	2.4 10^{-4}	2.05 10^6	53	1.7	1300	0.29

Re$_{Li/Hg}$	M$_{Li/Hg}$	$\mathcal{M}_{Cs/N2}$	G$_{Li/Hg}$	Mo$_{Li/Hg}$	Eö$_{Li/Hg}$
0.2	4	1.1	2	5.6	3.4

dimensionless parameters :

1) The liquid-phase Reynolds number:
Re$_L$ = ρ_L V$_L$ e/μ_L (a dynamic parameter)
2) The liquid-phase Hartmann number :
M$_L$ = B e $\sqrt{\sigma_L/\mu_L}$ (an electromagnetic parameter)
3) The gas-phase Mach number :
$\mathcal{M}_G$ = V$_G$/a$_G$ (a thermodynamic parameter)
4) The Morton number :
M$_{0L}$ = G μ_L^4/ρ_L^2 t$_{SL}^3$ (caracterising the liquid gas pair)
5) The gas bubble Eötvos number :
E$_ö$ = G d$_b^2$/t$_{sL}$ (a dimensionless ratio between pressure gradient and surface tension).

The variable ρ is the density; V is the velocity; e is a typical transversal length of the flow (i.e. duct diameter); μ is the dynamic viscosity; B is the magnetic induction; σ is the electrical conductivity; a is the sonic velocity which represents the gas compressibility $\sqrt{\partial P/\partial \rho}$; P is the pressure; G is the pressure gradient (∂P/∂z); d$_b$ is the bubble diameter; t$_s$ is the interfacial tension; the subscript L refers to liquid; and G to gas.

Table 1 shows a comparison of physical properties and typical values of the previous selection of dimensionless parameters for the real pair (lithium-cesium at a temperature of about 1300 K) and the laboratory pair (mercury-nitrogen at room temperature). In the upper half of Table 1 the gas compressibility is considered in term of a perfect gas a$^2{}_G$ = γ_G r$_G$ T$_G$, where γ and r are the constants of the gas at the temperature T. In the bottom half of Table 1, we present the ratio of dimensionless parameter for Li-Cs divided by the one of Hg-N_2, i.e. $Re_{Li/Hg} = \rho_{Li}\mu_{Hg}/\rho_{Hg}\mu_{Li}$. The table demonstrates that, contrary to the physical properties of each component of the two pairs, the typical values of dimensionless parameters are in pretty good agreement in a similarity point of view, which justify our choice.

The schematic of the experimental facility is shown in Fig. 1. The test section (vertical flow) is 0.9 m length, and its constant cross section is 7 cm x 1 cm (see the next paragraph). The magnetic field that is aligned with the smallest length of the flow (i.e., 1 cm), is

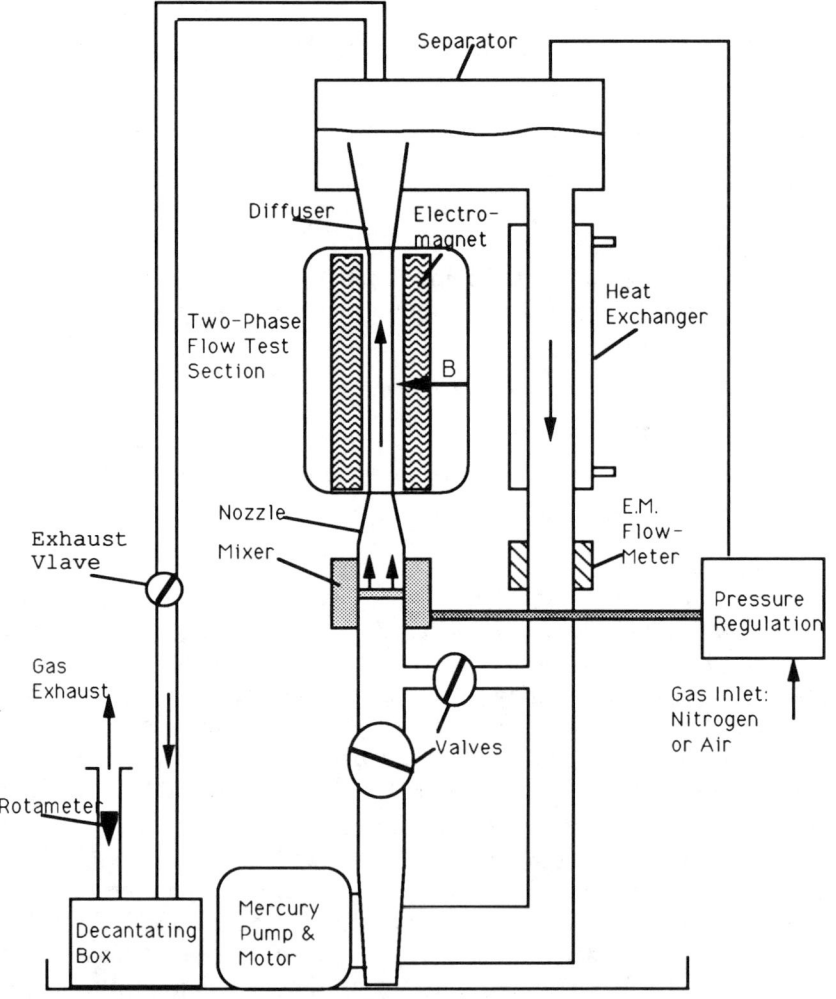

Fig. 1 Schematic of the experimental mercury-air or nitrogen facility.

produced by a large high-quality electromagnet. The magnetic induction is adjustable up to 1.25 T for a 3-cm gap. The mercury flow is possibly 1) generated by a centrifugal pump placed at the bottom of the loop or 2) due to the natural circulation. The latter is produced by the reduction of the apparent density of the vertical two-phase flow in the upcomer in comparison to the density of the vertical pure liquid flow in the downcomer. The control of the flow rate is possibly generated 1) by an air compressor or 2) by the expansion of the nitrogen contained in a tank. The two-phase flow mixer is mainly a group of seven tubes placed in the same cross section and pierced with small holes provided by an external annular chamber that is connected to the injection pressure regulation. Downflow, a nozzle accelerates the flow up to the test section incoming conditions. Between the test section and the gravitational separator, a diffuser allows the transformation of kinetic energy into pressure potential energy; thus, the axial velocity is nearly negligible in the separator. The mercury flow is maintained at constant temperature in a double-wall water-mercury heat exchanger that is placed, in the downcomer, upflow the electromagnetic flowmeter. The gas, after separation, passes through a secondary decantating box that removes any mercury from the gas before exhaust which is downflow the gas flowmeter (rotameter).

Experimental Measurements

Based on the results of our previous analysis, our experimental program attempts to measure, in the test section, the evolution of 1) local quantities (pressure, void fraction, temperature, and load current), and 2) global quantities (liquid and gas flow rate and load

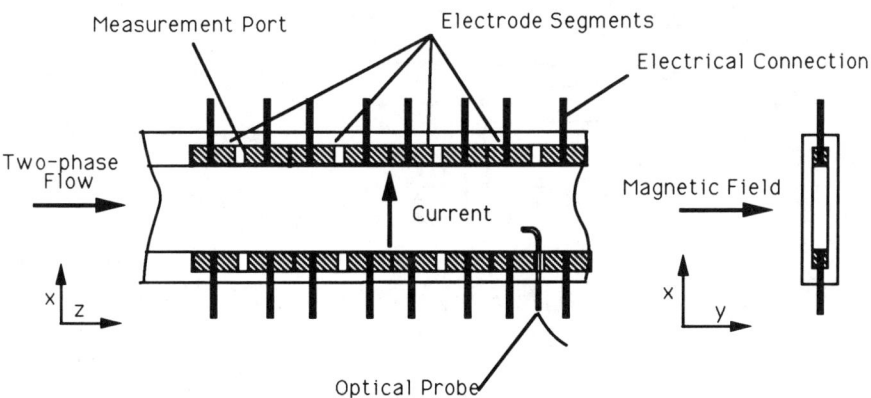

Fig. 2 Partial view and section of the test section and its load.

voltage). In order to do so, two test sections have been constructed with the same internal dimensions (7 cm width w, 1 cm depth d, and 90 cm length l). Both comprise a measurements port every 10 cm on the smaller face of the test section. These ports allow the placement of either a pressure tap connected to a pressure gage or a wall thermocouple or an optical probe. A total of five pressure gages, three thermocouples, and one optical probe (mobile in the transversal direction x) are placed along the test section.

The first test section has no electrode. However, in this case only the electric field induced by the interaction between the two-phase flow and the external magnetic field is present.

The second test section (see Fig. 2) is more similar to a MHD generator, except that electrodes are segmented in order to allow local electrical measurements. It comprises nine identical successive measurements cells. Each of these comprises, for a same 10-cm length (in the flow axis z) two segments of electrode (insulated with regard to other) and two insulated walls. The electrical circuit is presented in Fig. 3, and each of the nine 10-cm-length two-phase flow slice (indexed i) is represented by a voltage generator B w V_i. All the above parameters depend on the average void fraction α_i, for example, $V_i = Q_L/(1 - \alpha_i)$ and $R_i = w/(\sigma_{app}\ e.dx)$ where σ_{app} is an unknown function of α_i. The electrical connections are represented by R_c, including the electrodes and various electrical cables. Finally, the load resistance R_l, which is adjustable, is passed through by the total electrical current I_t. Then for each subcircuit i, including one flow slice,

$$B\ w\ V_i = (R_i + R_C)\ I_i + R_l\ I_t$$

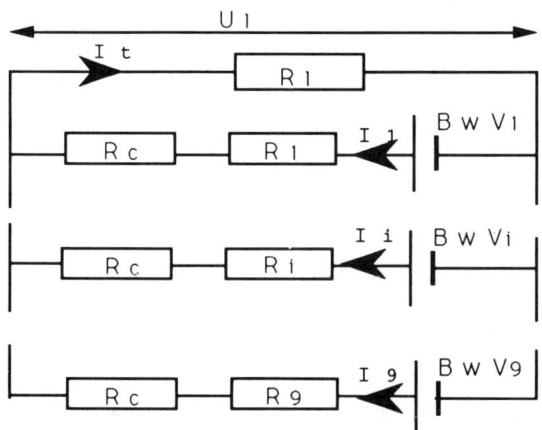

Fig. 3 Equivalent electrical circuit of the test section and its load.

and for the load resistance,
$$U_1 = R_1 I_t \quad \left(\text{as } I_t = \sum_i I_i\right)$$

U_1 is the measured load voltage, which gives with the 9 measured slice currents totally 10 electrical quantities. On the other hand, combining mass conservation and state law for each phase with pressure and temperature evolution, we obtain the evolution of β the volumetric quality:
$$\beta = Q_G/(Q_L + Q_G)$$
where Q are volumetric flow rates.

The void fraction measurement cannot be done with instruments using flow transparency because of mercury. Since the use of a double-conductivity probe is quite difficult[5], we decided to test the optical fiber sensor. The use of this instrument has been well mastered by the two-phase flow team of our laboratory[6]. Except for the fact that we had to reinforce the optical probe, no problem was encontered. For example, Fig. 4 shows a sample of the direct analog output of the optical probe, which presents very steep transitions from gas to liquid level. This very good dynamics of the signal can be explained by the very high reflexion in the mercury and the cancelation of wettability problems between mercury and the optical probe.

The local time fraction is the quantity measured with the optical probe (see Fig. 4). In the case of steady-state flow the probe can be moved in the transversal direction to obtain the one-dimensional space-averaged void fraction R_{G1} by space averaging[7]. The latter can be considered as quasiequal to the area-averaged void fraction α, regarding the very reduced y direction length:
$$\alpha = A_G/A \cong R_{G1}$$
where A is the area of the test section, and A_G is the area occupied by the gas.

Consequently, considering a one-dimensional flow description we introduce the velocity ratio S as: $S = V_G/V_L$ (Ref. 8). We obtain a very simple relation between the area void fraction α, the volumetric quality β, and the velocity ratio S:
$$S = \beta(1-\alpha)/\alpha(1-\beta)$$
Coupling the previous equations with mass conservation, we get:
$$V_L = Q_L/(1-\alpha)A$$
$$V_G = S V_L$$

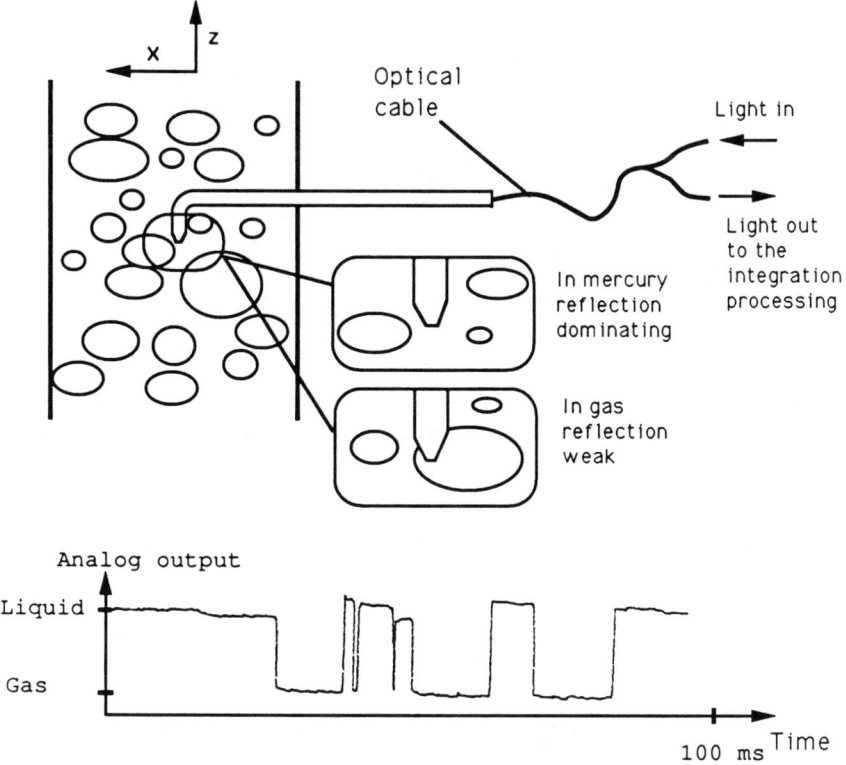

Fig. 4 Optical probe, its principle, and a sample of the analog output.

Thus, it is possible to have a comparison of the two-fluid model computations and experimental results, including mean velocities, because the six main unknowns of the model are the pressure, void fraction, two-phase mean velocities, and two-phase enthalpies.

First Measurements

The actual status of our experimental program consists of the following. All of the measurements collected were obtained on the first test section, without electrodes. This means that we ran the experiment with ordinary and MHD liquid-metal two-phase flow but without an external electric field. Thus, only the induced electric field can exist.

Since a direct physical interpretation of the measured quantities is not possible, we have developed a data treatment program. Some interesting results have already been obtained and are illustrated by the two runs of Table 2, which were obtained running the experiments in similar conditions except the magnetic field. Run 1

Table 2 Experimental measurements (international units): velocity and pressure given at the optical probe position

	Q_G	Q_L	V	P	β	α	S	Re	M	N
Run 1 (B=0)	3.6	2.85	1.45	4.45	0.22	0.15	1.56	$2.21 \cdot 10^5$	0	0
Run 2 (B=0.77)	3.6	2.35	1.25	4.35	0.26	0.16	1.83	$1.91 \cdot 10^5$	345	0.63

is without the magnetic field and run 2 with a 0.77-T magnetic field. Note that these runs were obtained for natural circulation in the loop (the mercury pump was stopped) because, with the maximum magnetic field of 0.77 T achievable with the first test section, a velocity of around 1.4 m/s, as in the present case, is sufficient for obtaining a high enough interaction parameter ($N = M^2/Re$).

Regarding the dimensionless parameter: M/Re that equals 0 for run 1, and 1.8×10^{-3} for run 2, we compared our results (for two-phase flow) with the results obtained by Brouillette and Lykoudis[9] (for liquid-metal flow). Run 2 is placed in the transition region between non-MHD and laminar-MHD (i.e., 0<M/Re<1/225). The comparison between our results with the previous one shows very strong similarity. In fact, between run 1 and run 2, the friction coefficient is multiplied by 1.32. For the same M/Re value , Brouillette and Lykoudis obtained a friction coefficient multiplied by 1.26 (the agreement is satisfied within 6% error). The velocity ratio is also an interesting quantity that can be compared between our numerical simulation and the experimental results. Here the experimental velocity ratios are 1.56 (ordinary flow) and 1.83 (MHD flow) our simulation in the same conditions gives 1.62 (ordinary flow) and 1.9 (MHD flow) (the agreement is satisfied within 4% error). However, we can see from Fig. 5 that the void fraction profiles (in x direction) corresponding to the runs of Table 2 are not one dimensional; in addition the magnetic field does not seem to modify strongly the void fraction profile. Because these runs have been done with moderate velocity (1.4 m/s) and moderate magnetic field (0.77 T), the previous tendency could be modified in the runs that will be done with our second test section. The latter runs will allow a higher magnetic field (1.25 T) to be reached, which corresponds to a 4 m/s velocity for the same typical

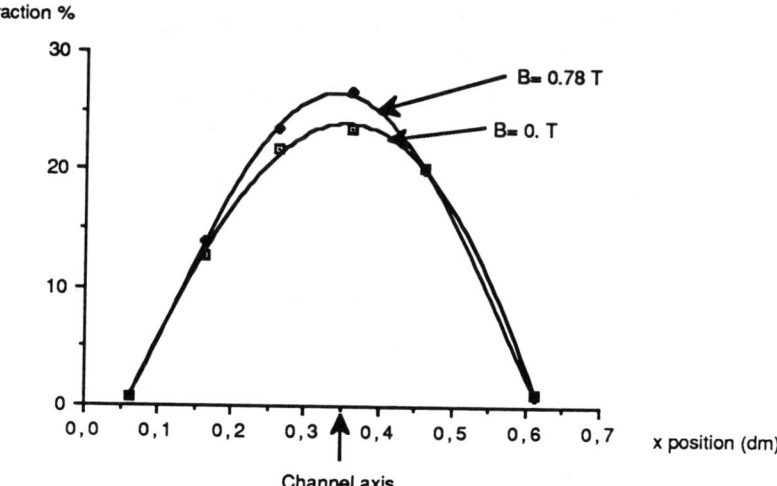

Fig. 5 Void fraction profiles (see Table 2).

value of the interaction parameter as that given earlier. Finally let us mention that the desired working conditions of the second test section are the following: for a load resistance of 200 µΩ, and a velocity of 8 m/s, the total current could be about 3000 A, the load voltage 0.6 V, and the net electric power 1.8 kW.

Conclusions

As a preliminary conclusion, we can mention the pretty good agreement, of this preliminary experimental analysis, to the intuitive physical idea that the two-phase flow has basically the same global MHD behavior as a pure liquid-metal characterized by equal global dimensionless parameters (i.e., Hartmann and Reynolds). On the other hand, a very satisfying agreement between experiment and numerical simulation has been shown. Nevertheless, this work is the necessary preparation to very interesting new experimental aspects of liquid-metal MHD two-phase flow.

References

[1] Thibault, J. P., "Modelling of Two-Phase Magnetohydrodynamic Flows," to be published in International Journal of Multiphase Flow.

[2] Billiotte M., 1987 "Conductivité Electrique d'un Ecoulement Diphasique en Présence d'un Champ Magnétique," Journal of Theoretical and Applied Mechanics, Vol. 6, No. 1, 1987, pp. 145-164.

[3] Thibault, J. P. and Seck, B., 1989, "Modelling of Magnetohydrodynamic Two-Phase Flow in Pipe," Liquid Metal Magnetohydrodynamics, Edited by J. Lielpeteris and R. Moreau, Kluwer Academic Publishers, Dordrecht, 1989, pp.119-125.

[4] Branover, H., "ETGAR 3 and ER4 Experimental Report," Ben-Gurion University, Beer-Sheva, Israel, 1986.

[5] Gherson, P., and Lykoudis, P., "Local Measurements in Two-Phase Liquid-Metal Magneto-Fluid-Mechanic Flow," Journal of Fluid Mechanics, Vol. 147, 1984, pp. 81-104.

[6] Cartellier, A., 1990 "Optical Probes for Local Void Fraction Measurements: Characterisation of Performance," Review of Scientific Instrumentation, Vol. 61, No. 2, 1990, pp 874-886.

[7] Delhaye, J. M., "Fundamental Quantities Describing Two-Phase Pipe Flows," Handbook of Multiphase Systems, Edited by G. Hetsroni, Mc Graw-Hill, New York, 1982, pp. 10.8-10.12.

[8] Hewitt, G. F., "Pressure Drop," Handbook of Multiphase Systems, Edited by G. Hetsroni, Mc Graw-Hill, New York, 1982, pp.1.44-2.46.

[9] Brouillette, E. C., and Lykoudis, P. S., "Magneto-Fluid-Mechanic Channel Flow: I Experiments," Physics of Fluids, Vol. 10, No. 5, 1967, pp. 995-1001.

Investigation of Two-Phase Liquid Gas Mixers for MHD Energy Conversion Systems

D. Farchi,* A. El Boher,† S. Lesin,‡ Y. Unger,§ and H. Branover¶
Ben-Gurion University of the Negev, Beer-Sheva, Israel

Abstract

Analytical and experimental study of two-phase liquid gas mixers is presented. The experiments were conducted in a water-air vertical system, simulating an optimized magnetohydrodynamic conversion (OMACON) system. Two types of mixers were tested, the first was a porous one whereas the second was a jet one. Comparison of the two types shows no difference in the flow pattern, void fraction distribution, and average void fraction in the riser pipe. The energy losses induced by the porous mixers under investigation are higher than those due to the jet mixers. The influence of the mixer type on the general peformance was found to depend on the operation conditions. For the jet mixers a model for the estimation of energy losses was developed based on energy and momentum balance equations. The model uses the following experimental correlations: void fraction at the mixer exit, mixing length, and frictional coefficient for estimation of the liquid pressure losses. The current model and the experimental results are in good fit. Optimization of mixer geometry for minimum energy losses is presented, which gives rise to an optimal construction at a given operational design point.

Introduction

The conversion of heat into electricity by means of liquid metal (LM) magnetohydrodynamics (MHD) is being investigated in depth at the Center for MHD Studies of Ben-Gurion University of the Negev. The concept already adopted for this purpose is the optimized magnetohydrodynamic conversion (OMACON) one, described in detail elsewhere.[1] This concept has been lately adopted also by others.[2]

In such systems, LM and gas streams are mixed together in a mixing element (mixer), flowing up vertically along the riser pipe to the separation tank (separator), there the fluids are separated from one another, the LM flows down along the

---

Copyright © 1992 by the American Institute of Aeronautics and Astronautics, Inc. All rights reserved.
*Project Manager, Center for MHD Studies.
†Program Manager, Center for MHD Studies.
‡Group Leader, Center for MHD Studies.
§Senior Researcher, Center for MHD Studies.
¶Professor and Head, Center for MHD Studies.

downflow pipe (downcomer), passing through the LMMHD generator to the mixer and recirculates. The gas is vented off the separator.

The design and performance of the system is strongly influenced by two-phase flow characteristics. The behavior of two-phase flow along the riser pipe was investigated at a previous stage and the findings were summarized and generalized.[3] A preliminary study on the separation process was also carried out.[4] In this paper the mixture generator — the mixing element — will be addressed. Two main mixer categories were considered for the formation of liquid-gas mixture: one is the porous type mixer where the gas is injected as discrete bubbles into the liquid and creates fine dispersed bubble flow. The other type includes mixers where the gas and the liquid are injected together as continuous jets which finely mix together into a fine dispersed bubble flow pattern.

Related Literature

Mixers of both types were used by different investigators for the formation of liquid-gas mixtures. Herringe and Davis[5] examined mixers of the two types mentioned, where gas injection is directed downstream. The riser diameter is 50.8 mm, and the void fraction profile was measured by a resistivity probe at 8, 36, and 108 diameters downstream the mixer. The authors report that a slight difference in void fraction profile is observed between the different types of mixers used. However, this difference is noticed only close to the entrance, and it almost disappears downstream.

Merchuk and Stein[6] have studied the influence of different mixers on the void fraction evolution along the riser pipe with a diameter of 0.14 m. Their conclusion is that at 30 diameters above the mixer, influence of mixer type on the void fraction is noticed.

Fabris et al.[7] have used an immersed type of air injector in a rectangular channel, which was directed upstream. The void fraction profile was measured in the two planes perpendicular to the flow direction at 7.5-hydraulic channel diameters downstream of the mixer. The profile parallel to the longer side is an M-shaped one, and the second one changes from an M-shaped one at a lower average void fraction value to a parabolic one, and then to a Gaussian one as the void fraction increases. Use of this type of mixer is recommended to achieve a foaming bubbly flow at high void fractions.

El-Boher et al.[3] report on a special experiment performed in a mixture of lead-bismuth alloy and steam in the Etgar-3 facility, where injection direction is first downstream and then upstream. The mixer was an airfoil shaped porous element immersed in the LM flow. The overall performance of the facility was analyzed for both cases, and a conclusion was reached that at low void fraction values upstream steam injection increases liquid metal flow rate, as compared to downstream injection. However, at higher average void fraction values the improvement of facility performance decreases, and then disappears with another increase of void fraction.

Adler[8] tested two types of porous mixers, one with a constant cross section along the mixer and the other with a variable area cross section and no influence on the flow pattern was observed. However, less pressure drop was measured in the second case.

Blumenau et al.[9] were the first to suggest the jet mixer for LMMHD application because of its mechanical advantage. The similarity of jet mixers to jet pumps allows the adoption of already published mathematical models for energy losses prediction and optimization criteria.

In Refs. 10-14 one-dimensional momentum and energy equations for the estimation of the losses and pressure differences in jet pumps are presented, and will be used in our analysis.

Theory

The geometry of an optimal jet mixer should be designed to induce minimal energy losses at the operational design point, and the most important dimensions are the nozzle and throat diameters by which mixer inlet void fraction, α_o, is determined.

Referring to Fig. 1, the jet mixer consists of a gas chamber C, connected at section o-o to the throat pipe, T. The throat diameter is D_T and it ends at the riser entrance at section t-t. A liquid nozzle N penetrates to the chamber base at section i-i and ends at section o-o, where the nozzle diameter is D_N.

The gas enters the chamber at inlet n and exits as an annulus at section o-o. The liquid enters the nozzle at section i-i and leaves the nozzle at section o-o.

Along the throat two zones are defined at the lower part along a path of length L_m, the mixing length, the two streams are not as yet mixed together; and the gas attached to the throat walls surrounds the diverging liquid jet. Downstream of the mixing length the jet breaks up, the liquid wets the walls, and the mixture flows up as a two-phase mixture to the throat exit.

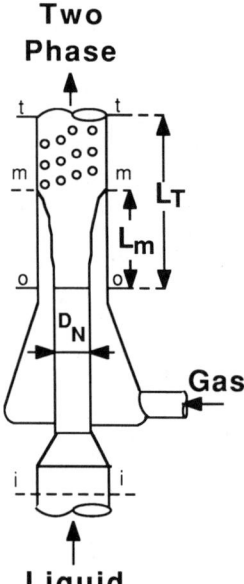

Fig. 1 Schematic of jet mixer installed in Tal facility.

The nozzle energy losses E_N per unit liquid mass is

$$E_N = K_N \, V_{Lo}^2/2 \tag{1}$$

where K_N is the friction coefficient, and is to be found experimentally by

$$K_N = \frac{2(p_i - p_o)}{\rho_L V_{Lo}^2} + \frac{V_{Li}^2 - V_{Lo}^2}{V_{Lo}^2} + \frac{g(z_i - z_o)}{V_{Lo}^2} \tag{2}$$

The energy losses along the throat as a function of α_o are to be analyzed by the energy balance equation. In this equation the unknown parameters are: the pressure drop along the streams flow, and the exit void fraction α_t. The momentum balance equation is used to predict the pressure drop as a function of the mixing length L_m, α_t and the wall friction τ_w. Finally, the energy losses could be presented as a function of the parameter α_o and optimization could be drawn.

The following assumptions are set for the derivation of the one dimensional conservation laws.
1) The flow of gas in the chamber, from entrance n to exit o-o is isothermal and isobaric: $T_n = T_o$; $p_n = p_o$.
2) Because of high heat capacity of the liquid, water in our case, its flow is isothermal.
3) The liquid jet along the mixing length is free of gas bubbles.
4) Along the mixing length the void fraction has a constant value equal to α_o. Along the rest of the throat, two-phase flow exists at a constant void fraction value α_t.
5) Change in the solubility of the gas in the liquid and evaporation of the liquid are neglected.

For the calculation of the throat energy losses one has to consider the momentum and energy conservation laws for the control volume limited by sections t-t, o-o, and the throat pipe walls.

Out of the momentum balance the pressure gradient equation could be presented as the sum of three components,[15] namely, frictional, accelerational, and gravitational pressure gradients:

$$\frac{dp}{dz} = \left(\frac{dp}{dz}\right)_f + \left(\frac{dp}{dz}\right)_{acc} + \left(\frac{dp}{dz}\right)_g \tag{3}$$

where

$$-\left(\frac{dp}{dz}\right)_f = \left(\frac{1}{A}\right) \frac{d}{dz}(F_G + F_L) \tag{4}$$

$$-\left(\frac{dp}{dz}\right)_{acc} = \dot{m}^2 \frac{d}{dz}\left[\frac{x^2}{\alpha \rho_G} + \frac{(1-x)^2}{(1-\alpha)\rho_L}\right] \tag{5}$$

$$-\left(\frac{dp}{dz}\right)_g = g[\alpha \rho_G + (1-\alpha)\rho_L] \tag{6}$$

F_G and F_L are the frictional forces of the gas along the dry zone and of the liquid along the wet zone, respectively.

The integration of Eq. 3 along the throat based on the mentioned assumptions yields

$$p_o - p_t = \int_0^m \frac{\tau_{wG} P}{A} dz + \int_m^t \frac{\tau_{wTP} P}{A} dz + \frac{\dot{m}^2 (1-x)^2}{\rho_L} \left[\frac{1}{1-\alpha_L} - \frac{1}{1-\alpha_o} \right]$$

$$+ \dot{m}^2 x^2 \left[\frac{1}{\rho_{Gt} \alpha_t} - \frac{1}{\rho_{Go} \alpha_o} \right] + g \int_0^t [(1-\alpha)\rho_L + \alpha \rho_G] dz \tag{7}$$

The integration of the last right-hand-side term of Eq. 7 could be done by dividing the throat length to two sections o-m and m-t. Thus,

$$\int_0^t [(1-\alpha)\rho_L + \alpha \rho_G] dz = L_m [(1-\alpha_o)\rho_L + \alpha_o \rho_G] + (L_t - L_m)[(1-\alpha_t)\rho_L + \alpha_t \rho_G] \tag{8}$$

The combined energy balance of both phases is

$$\dot{M}(\delta q - \delta w) = \dot{M} dh + d\left(\frac{\dot{M}_G V_G^2}{2}\right) + \left(\frac{\dot{M}_L V_L^2}{2}\right) + \dot{M} g L_t \tag{9}$$

where δq is the net heat absorbed from the surroundings and δw is the net work done by the fluid on the surroundings, both, per unit mass. In our case δw is equal to zero.

The dissipation of mechanical energy into irreversible heat dE is equal to:

$$dE = Tds - \delta q \tag{10}$$

since the reversible heat δq_{rev} is

$$\delta q_{rev} = Tds \tag{11}$$

By substituting the Maxwell relation

$$dh = Tds + vdp \tag{12}$$

and Eq. 10 into Eq. 9 the energy balance per unit mass in a differential form is[15]:

$$-\frac{dp}{dz}\left[\frac{x}{\rho_G} + \frac{1-x}{\rho_L}\right] = \frac{dE}{dz} + \frac{\dot{m}^2}{2}\frac{d}{dz}\left[\frac{x^3}{\alpha^2 \rho_G^2} + \frac{(1-x)^3}{(1-\alpha)^2 \rho_L^2}\right] + g \tag{13}$$

Note that the term dE/dz in Eq. 13 includes not only the dissipation of mechanical energy due to wall friction but also the irreversibility of the mixing process.

By assuming isothermal expansion of the gas the integration of Eq. 13 yields:

$$\frac{1-x}{\rho_L}(p_o - p_t) + x RT \ln \frac{p_o}{p_t} = E + \frac{\dot{m}^2 x^3}{2}\left[\frac{1}{\alpha_t^2 \rho_{Gt}^2} - \frac{1}{\alpha_o^2 \rho_{Go}^2}\right]$$

$$+ \frac{\dot{m}^2 (1-x)^3}{2\rho_L^2}\left[\frac{1}{(1-\alpha_t)^2} - \frac{1}{(1-\alpha_o)^2}\right] + g L_t \tag{14}$$

By substituting Eq. 7 into Eq. 14 the mixing losses could now be calculated, as well as overall pressure drop if α_t, L_m, and the wall shear stress could be predicted as a function of flow rates, inlet pressure, and α_o. An experimental investigation is required to find out a correlation for the prediction of the three unknowns mentioned: void fraction at the mixer exit, the mixing length, and the wall shear stress.

Experimental Set-up

The water-air experimental facility, Tal, was constructed to study the mixer performance and its influence on the void fraction and phase velocity ratio evolution along the riser pipe, and the integral performance of the facility, as well as the unknown values of void fraction in the mixer outlet, mixing length, and friction coefficient. The facility construction is principally the same as an OMACON type facility, thus a water-air mixture flow serves as a simulation for two-phase processes taking place in LM gas two-phase flows. Figure 2 is a schematic of the facility. The vertical pipe on the left side is the riser and the one on the right side is the downcomer. Both pipes share the separator at the top, and are connected to each other at the bottom. A centrifugal pump is connected in a bypass loop at the bottom to enable forced circulation at a liquid flow rates higher than those achieved at a natural circulation mode of operation.

Two different mixers were installed at the lower part of the riser: A porous mixer which is constructed of a mixer housing, a 79-mm-diam. plexiglass pipe and a sparger immersed in it. The sparger is a stainless-steel air foil shaped element, its upper face is porous and the lower face is solid. The flanges at the ends of the mixing housing enable the installation of the mixer with an upstream or downstream injection position. The second mixer is a jet mixer type. It is constructed of a nozzle through which the water is accelerated. The nozzle is placed at the center of its plexiglass conical housing, through which the air injected surrounds the nozzle. The mixing zone downstream of the nozzle exit is a 45-mm plexiglass pipe, starting at the clearance between the conical housing and the nozzle edge. Different nozzle diameters were constructed and tested: 30, 33, 36 and 40 mm.

The liquid flow rate is controlled by a valve located at the downcomer and in the forced circulation mode a bypass valve is also used. Air is supplied from a compressed air net, passing through a pressure regulator and a high-pressure air capacitor to achieve a stable flow rate at a constant pressure. Pressure taps are connected along the riser pipe and along the downcomer, upstream and downstream the mixer, and along the mixing zone of the jet mixer. The main dimensions of the facility, the measuring instruments, and flow parameters are presented in Table 1.

Comparison Between Porous and Jet Mixers

Void Fraction Distributions

The cross-sectional average longitudinal distribution of the void fraction was calculated from the pressure measurements along the riser pipe. The computer code IDENT[16] was modified and used for this analysis. In this code the pressure distribution is calculated by common numerical iterative methods to fit to the measured values.

The experimental data were reduced using the analysis previously introduced. Values obtained at liquid and gas flow rates of 2.66 and 0.002 kg/s respectively,

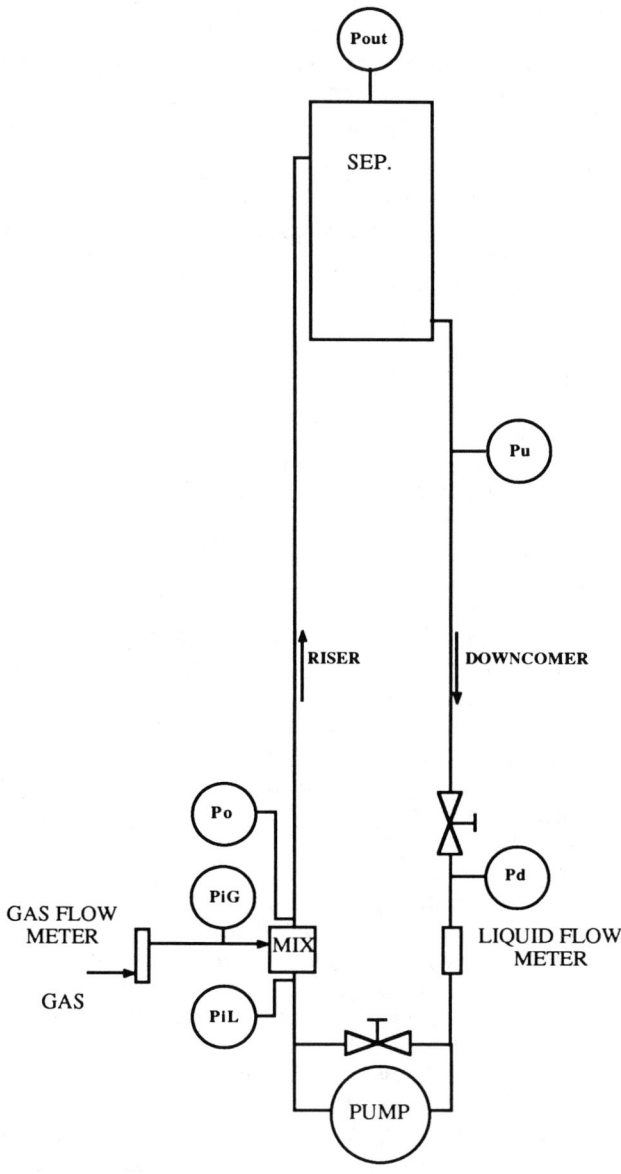

Fig. 2 Schematic of water-air Tal facility.

Table 1 - Parameters of Tal Facility

Dimensions:
    Riser and downcomer diameter, mm   0.045
    Riser height, m   4.225

Instrumentation:

Liquid flow rate	Magnetic flow meter
Gas flow rate	Float type rotameter
Pressure	Gauge and differential electronic pressure transducer
Local void fraction	Single tip resistivity probe
Temperature	Thermoresistance element

Flow parameters:

Superficial liquid velocity, m/s	1-3.3
Gas flow rate, kg/s	$5 \cdot 10^{-4} - 3 \cdot 10^{-3}$
Temperature	Ambient
Separator pressure	Atmospheric

were chosen for the course of typical trends of void fraction, as shown in Figs. 3 and 4. In Fig. 3 the void fraction evolution along the riser is presented. As expected, the values increase vs the height. In Fig. 4 the distribution, as measured by the resistivity probe vs the distance from the pipe center, is presented. A parabolic distribution is noticed. All mixers, used at the specified range of flow rates, are characterized by the same trends.

In Fig. 5 the average riser void fraction values vs liquid flow rate, where gas flow rate is a given parameter, are presented. For all mixers used over the full range of flow rates the values of void fraction are almost identical to each other at the same flow parameters. The differences observed tend to be random and are within the error range.

Mixer Energy Losses

In a previous work,[17] it was shown that the mixer energy losses of the porous mixer tested in the Tal facility were always higher than those induced by the different jet mixers, at the same flow rates.

Integral Performance

The overall performance coefficient presents the ratio between work done by the liquid along the single-phase path and the maximal available work due to an isothermal expansion of the gas from the mixer inlet to the separator air exit. For this analysis the mass flow rate values are required as well as the pressures p_u and p_d at the downcomer upper and lower levels respectively, p_{out} the atmospheric pressure at the separator air exit, and p_{iG} and p_{iL}.

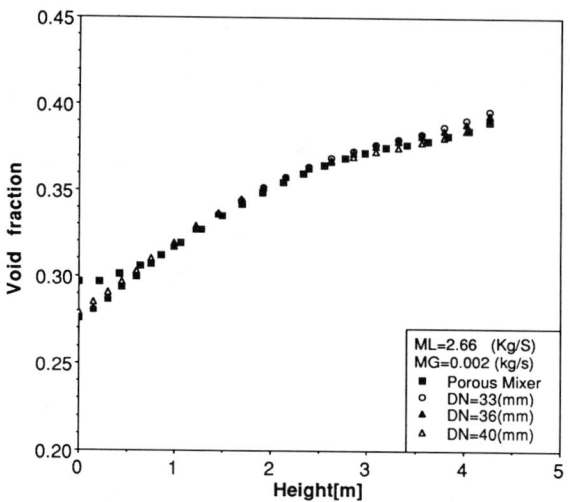

Fig. 3 Void fraction evolution along the riser pipe, with different mixers installed in Tal facility.

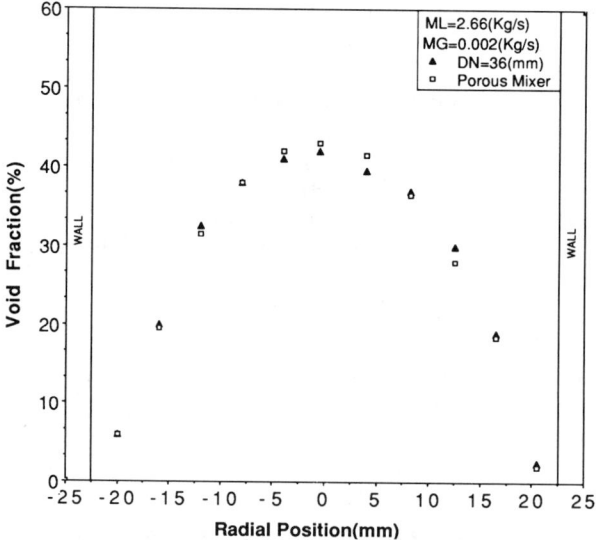

Fig. 4 Local void fraction distribution measured by resistivity probe with different mixers installed in Tal facility.

TWO-PHASE MIXERS

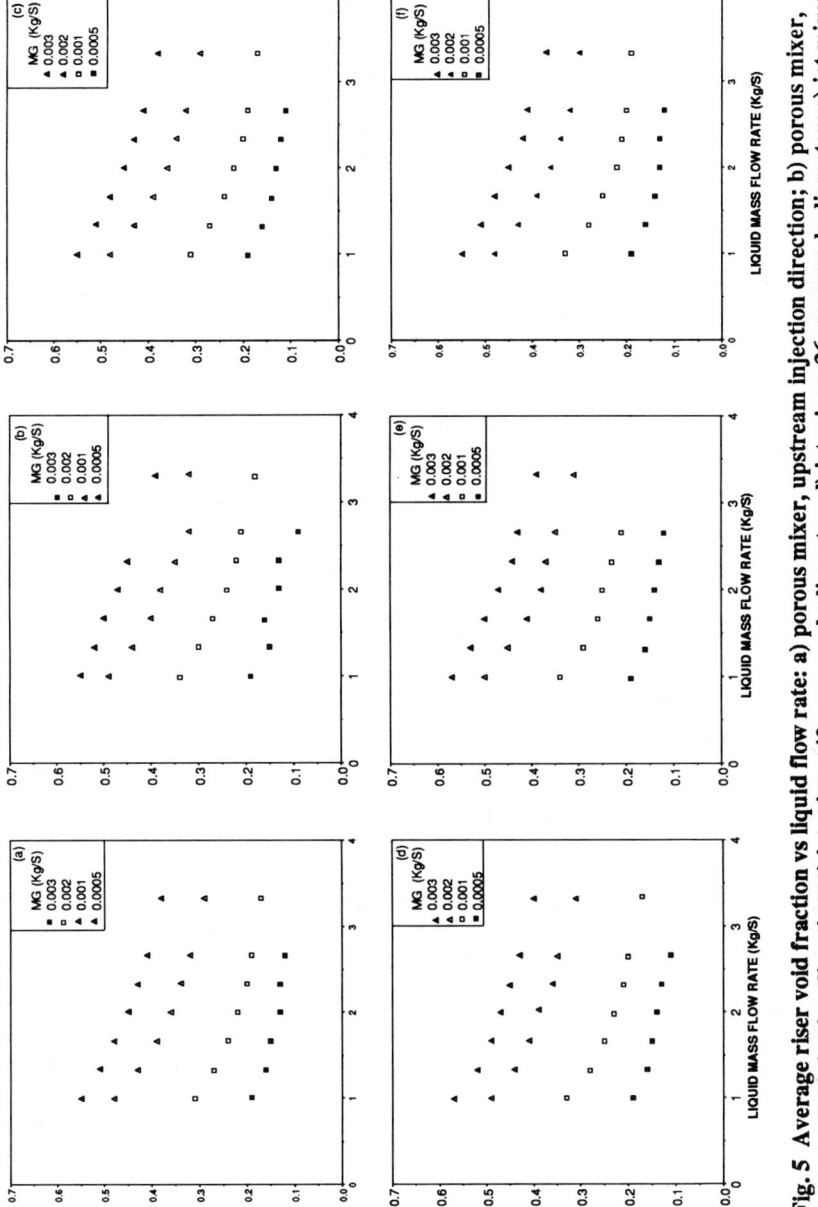

Fig. 5 Average riser void fraction vs liquid flow rate: a) porous mixer, downstream injection direction; b) porous mixer, upstream injection direction; c) jet mixer, 40-mm nozzle diameter; d) jet mixer, 36-mm nozzle diameter; e) jet mixer, 33-mm nozzle diameter; and f) jet mixer, 30-mm nozzle diameter.

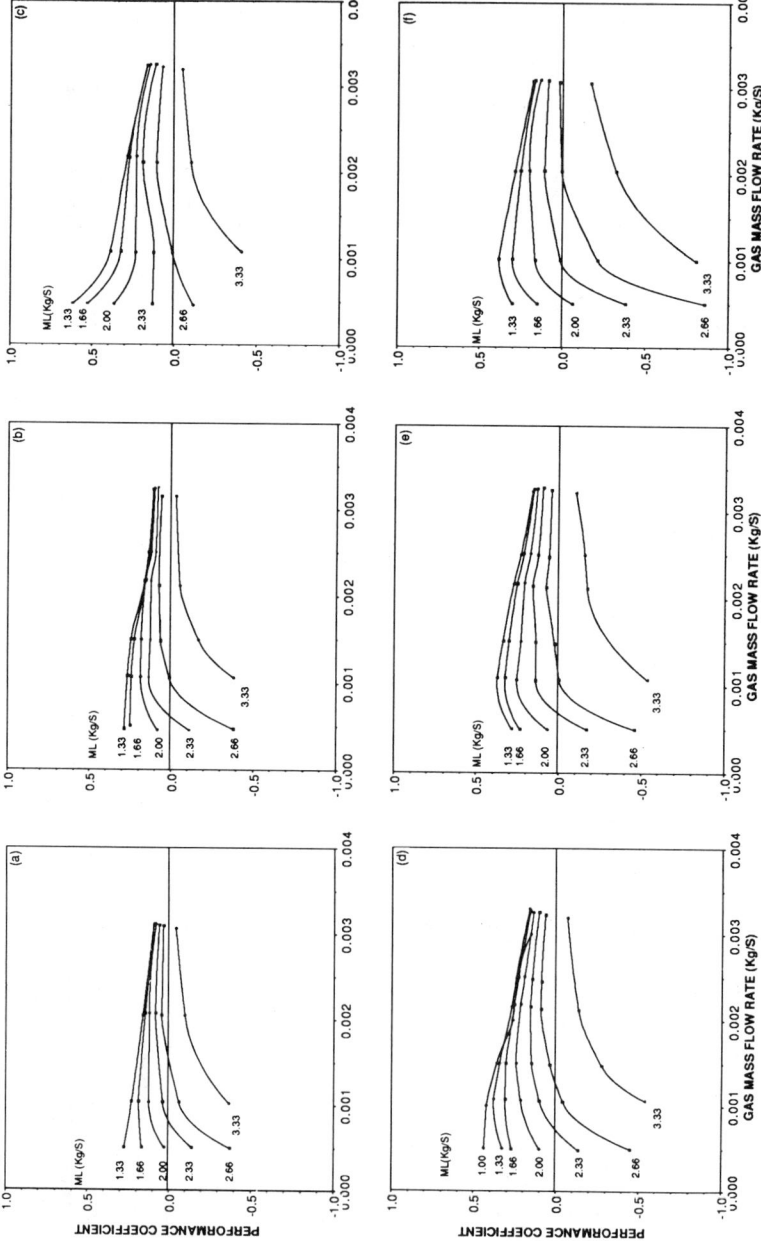

Fig. 6 Overall performance coefficient vs gas flow rate: a) porous mixer, upstream injection direction; b) porous mixer, downstream injection direction; c) jet mixer, 40-mm nozzle diameter; d) jet mixer, 36-mm nozzle diameter; e) jet mixer, 33-mm nozzle diameter; and f) jet mixer, 30-mm nozzle diameter.

The work done by the water along the downcomer is

$$W_{int} = \frac{\dot{m}_L}{\rho_L}[(p_u - p_d) + \rho_L g H_{ud}]$$

The losses include those which are due to the hydraulic resistance of the valve which simulates the pressure drop at an MHD generator in OMACON type systems, and those due to friction.

The work done along the bypass loop is

$$W_p = \frac{\dot{m}_L}{\rho_L}[(p_d - p_{iL}) + \rho_L g H_{id}]$$

The isothermal expansion work W_G is simply calculated by

$$W_G = \dot{m}_G RT \ln \frac{p_{iG}}{p_{out}}$$

Thus,

$$C_p = \frac{W_{int} + W_p}{W_G}$$

C_p has a positive value at lower values of liquid flow rate, and a negative value when the pump operates at high liquid flow rates. In the latter case the work done by the liquid has a negative sign along the pump bypass loop.

Through the performance coefficient introduced, the influence of different components installed in the facility on the overall performance could be examined and in this paper the studied component is the mixer. When gas is injected and circulation is caused by the pump, its head is identical to the overall facility pressure drop. Thus the work along the downcomer itself is less than the pump work, $W_{int} + W_p$ is negative, and C_p goes to negative infinity. When gas flow rate is increased, its contribution to the circulation rate is reflected by an increase in the performance. A maximum is reached, then at larger gas flow rates the performance coefficient tends to asymptotic due to an increase in gas path (from the mixer inlet to the separator exit) pressure drop. At these flow rates the losses are high along the riser due to phase velocity ratio increase. Optimal performance is highest at the maximum C_p value.

The overall performance of the facility is shown in Fig. 6. The performance coefficient C_p is shown vs gas flow rate, where liquid flow rate is a given parameter. For all mixers used one can see that an increase of liquid flow rate reduces the performance coefficient at a given gas flow rate. Along the constant liquid flow rate lines, the tendencies toward negative infinity at low gas flow rate, and toward the asymptotic at high gas flow rate, are noticed.

The influence of the different mixers on the overall performance is observed when one compares the plots of Figs. 6a-6f with one another. Figures 6a and 6b show C_p values obtained when the porous mixer was installed in an upstream and downstream injection direction mode, respectively. Figs. 6c to 6f show the results of the different jet mixers.

Except for the higher $\dot{M}_L$ values, the C_p values are the lowest for porous mixers. The decrease of nozzle diameter of the jet mixers leads to a decrease of C_p values. In the porous mixer, downstream injection is characterized by higher values of C_p than the upstream one. One can see that at low gas flow rates C_p is strongly affected by liquid flow rate, and this effect is dimmed with an increase in gas flow rate.

The previously mentioned trends of C_p are noticed only partially. A maximum is not reached along part of the experimental range due to limits in gas flow rates. As expected, the locus of maxima shifts respective to an increase of the gas flow rate.

It was shown that all mixers used over the whole range of the experiments are characterized by the same average riser void fraction at a given set of gas and liquid flow rates. The isothermal expansion work along the gas path from the mixer outlet to the separator gas exit, defined as C_1, remains the same, and also the work along the path from the separator liquid exit to the mixer exit, defined as C_2. Thus the performance coefficient could be written as

$$C_p = \frac{C_2 - W_{mL}}{C_1 + W_{mG}}$$

where W_{mL} and W_{mG} are the work done by the liquid and the gas on the mixer, respectively.

At given flow conditions, in jet mixers, a decrease of nozzle diameter is proportional to a decrease in C_p and in W_{int}. This yields a decrease in W_{mG} and an increase in W_{mL} and the absolute increase in W_{mG} is larger than the absolute decrease in W_{mL}. The outcome is that the change of nozzle diameter influences W_{mG} more than W_{mL}. This holds true also in cases where the reduction of nozzle diameter does not change the total mixer losses.

In a porous mixer at a given set of flow rates a decrease in C_p is associated with an increase in W_{int}. This could exist if W_{mG} increases and W_{mL} decreases and the absolute increase is larger than absolute decrease, or if both W_{mG} and W_{mL} increase. The latter possibility is physically more feasible, since an upstream injection direction increases the losses of both streams.

In cases where C_p and W_{int} increase, which is observed when the porous mixer is compared to jet mixers at high liquid flow rates, W_{mL} decreases and W_{mG} increases and the absolute increase of W_{mG} is larger than the absolute decrease of W_{mL}. Thus W_{mL} in jet mixers is larger than the respective value for porous mixer.

Developed Correlations

Mixing Length Determination

The pressure distribution along the mixer throat is used to allocate the jet breakup point. Some typical results of throat pressure vs the distance z from its lower pressure tap are presented in Fig. 7 for air flow rate of 1×10^{-3} kg/s. For nozzle diameter of 33 mm the data refers to liquid flow rates of 1.33, 1.66, and 1.97 kg/s and for nozzle diameter of 36 mm the liquid flow rate is 1.97 kg/s. At the

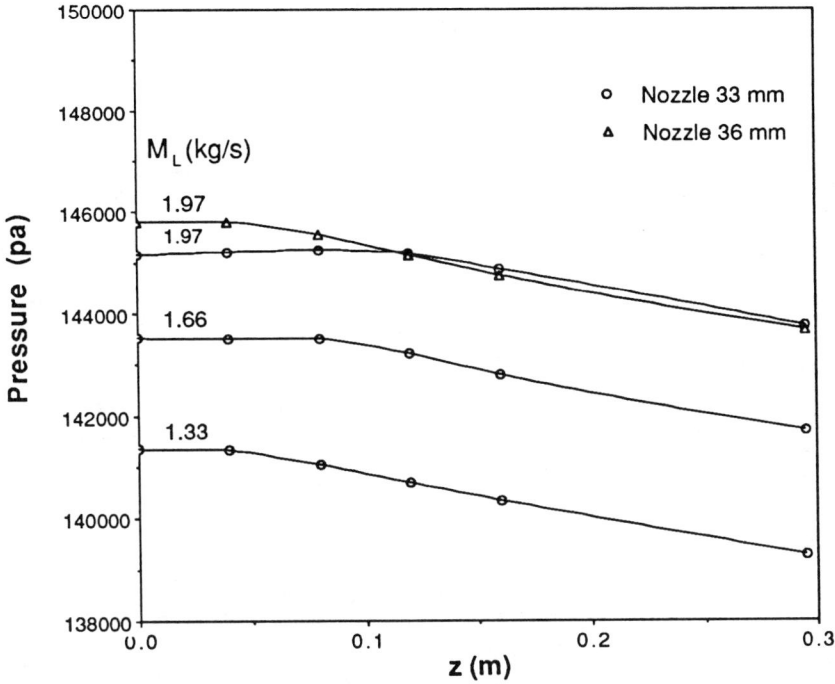

Fig. 7 Throat pressure distribution in jet mixers, $\dot{M}_G = 0.001$ kg/s.

lower part of the throat the pressure is close to constant and a gradient is noticed downstream. The point of deflection is referred as the jet breakup point. Upstream to this point, along the mixing length L_m, the wall is dry and a pressure drop is not noticed. Downstream to this point the wall is wet and the losses are reflected by an appreciated pressure drop. As expected, the increase in liquid flow rate pushes the breakup point downstream. One can also see that at a given set of flow rates ($\dot{M}_G = 1 \times 10^{-3}$ kg/s; $\dot{M}_L = 1.97$ kg/s), the increase of nozzle diameter brings up to a shorter mixing length. It should be noted that at 40 mm nozzle diameter a deflection point was not observed; for this nozzle the gas annulus is so thin that the mixing length is even shorter than the lower pressure tap location.

In Fig. 8 results of pressure distribution in 30-mm nozzle are presented. One can see that at liquid flow rate of 1.33 kg/s an increase of gas flow rate from 10^{-3} to 3×10^{-3} kg/s does not effect the mixing length significantly. This trend is typical for the whole experimental range. For this nozzle a deflection point is noticed only at low liquid flow rate. When $\dot{M}_L \geq 1.64$ kg/s the deflection point could not be observed precisely due to the lack of instruments at the breakup zone. The breakup is expected to appear downstream of the upper pressure tap.

The observations do not fit to jet pump breakup point predictions.[12] Thus, to generalize the results the following nondimensional groups were assumed to govern the nondimensional mixing length L_m/D_N: nozzle Reynolds number

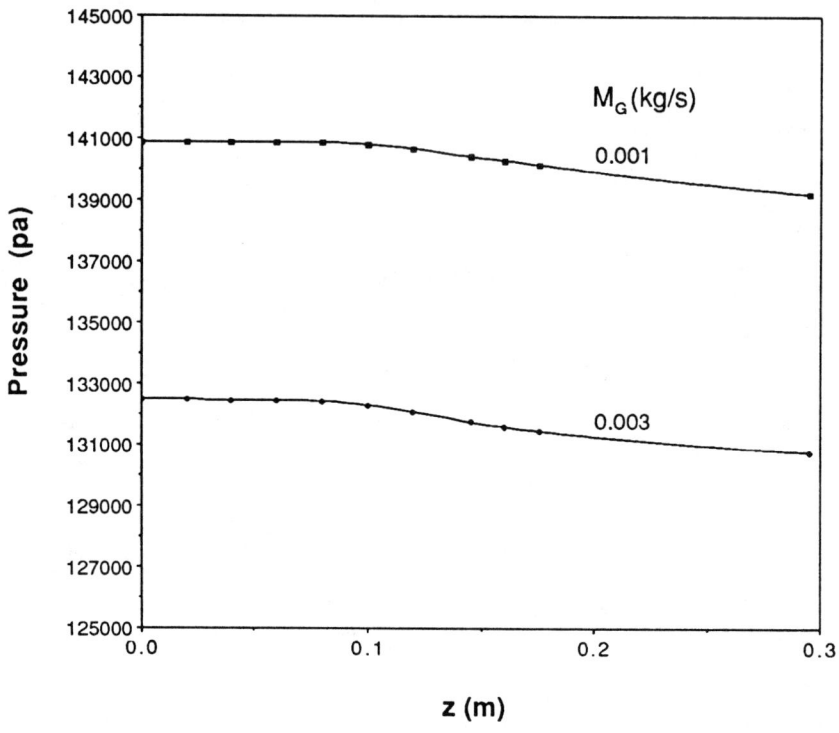

Fig. 8 Throat pressure distribution in jet mixer, Dn = 30 mm, $\dot{M}_L$ = 1.33 kg/s.

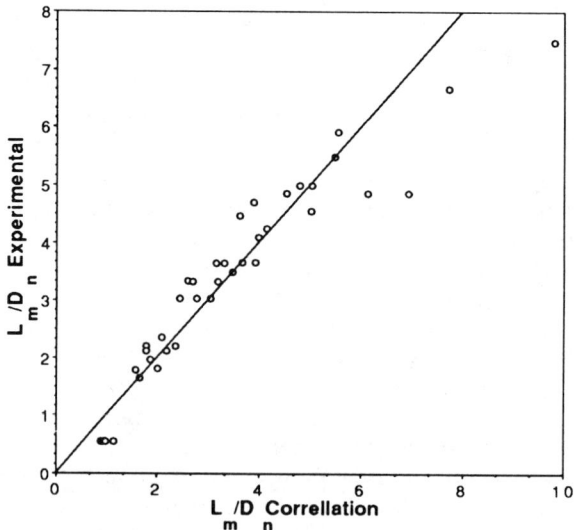

Fig. 9 Nondimensional mixing length correlation, Eq. 15.

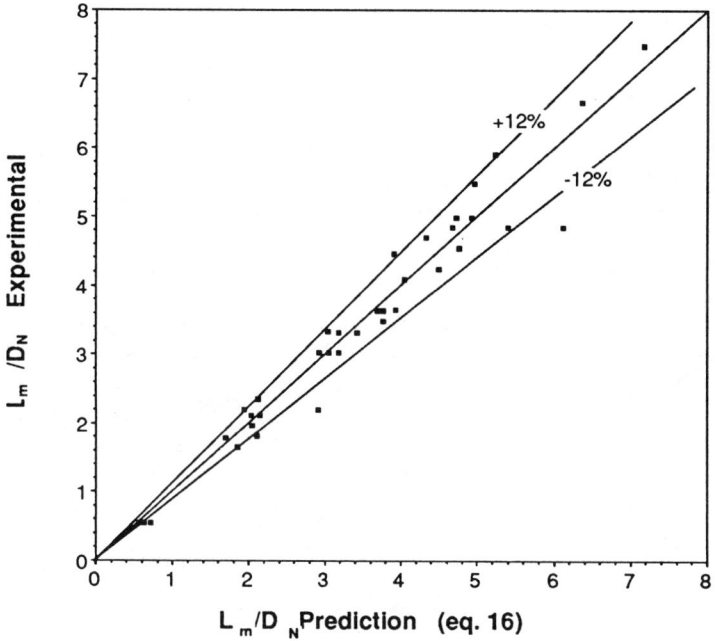

Fig. 10   Modified nondimensional mixing length correlation, Eq. 16.

$Re_N = m_{LN} D_N / \mu_L$, phase velocity ratio S_o at the nozzle exit, diameter ratio, $D_R = D_N/D_T$, and Euler number $Eu = p_o/\rho_L V_N^2$.

The following expression was suggested:

$$\frac{L_m}{D_N} = C_1 Re_N^{C_2} S_o^{C_3} D_R^{C_4} Eu^{C_5} \tag{15}$$

The correlation constants C_1-C_5 were determined using the least square method, which input only those L_m values that could be observed within a reasonable error range and their respective nondimensional parameters.

The comparison between the observed values and those predicted by Eq. 15 are presented in Fig. 9. The experimental values were found to be correlated much better by the form

$$\exp\left(\frac{L_m}{D_N}\right) = 1 + 1028.33 S_o^{-0.33} D_R^{12.52} Eu^{-2.16} \tag{16}$$

and Fig. 10 shows the results. One can see that almost all the data points fall within a band of ±12%. This band will be narrower if one takes into account the error associated with the observed values. The pressure taps are 2 cm apart, and this of course induces some uncertainties. The visual observations of the mixing length were found to be in a relative large deviation from those observed by the pressure distribution, thus the former were not used in this study.

Mixer Exit Void Fraction Determination

The determination of void fraction along the riser pipe in TAL facility is based on pressure distribution measurement and the use of the above mentioned computer code.

To generalize the data, the measured void fraction along the riser pipe was compared to different models. The models selected first for comparison are the homogeneous one, predicting highest values, due to the fact that phase velocity ratio is assumed to be equal to unity; the model of Zivi,[18] predicting lowest values, and that of Harmathy,[19] which considers a constant phase velocity ratio. In Fig. 11 this comparison is presented and it can be seen that the model of Harmathy fits best the experimental data, however, only for lower void fraction values. The well-known correlation of Lockhart and Martinelli[20] and a newly developed one[3] show better agreement to the data (see Fig. 11), as judged by the higher correlational coefficient. Nonetheless all models' results differ significantly from the experimental data. Thus, the following nondimensional groups were selected to govern the void fraction at the mixer exit: throat Reynolds number; $Re_T = \dot{m}_{LT} D_T / \mu_L$, and volumetric flow rate ratio Q_R at mixer exit. A generalizing equation of the form

$$\alpha_{corr} = A_1 Re^{A_2} Q_R^{A_3} \quad (17)$$

was derived, and in Fig. 12 the experimental values are compared to the predictions based on Eq. 17. The parabolic trend of the points on this plot brought up to a final correlation of the form

$$\alpha_{corr} = -3.12 \times 10^{-2} + 1.39 F - 0.907 F^2 \quad (18)$$

where

$$F = 7.12 \times 10^{-3} Re^{0.363} Q_R^{0.693} \quad (19)$$

In Fig. 13 the experimental values are shown vs those predicted by Eq. 18 and the agreement is very good.

Wall Shear Stress In the Mixer Throat

Friction pressure drop of the mixer throat can be defined as

$$-\left(\frac{dP}{dz}\right)_f = \frac{\tau_w P}{A} \quad (20)$$

In addition, in two-phase flow it is convenient to relate to the superficial liquid velocity and a multiplier ϕ_{LO}^2 as follows:

$$\left(\frac{dp}{dz}\right)_{TP} = \phi_{LO}^2 \cdot \left[\frac{dp}{dz}\right]_{LO} \quad (21)$$

where

$$\left[\frac{dp}{dz}\right]_{LO} = \frac{2 f_{LO} \dot{m}}{D \rho_L} \quad (22)$$

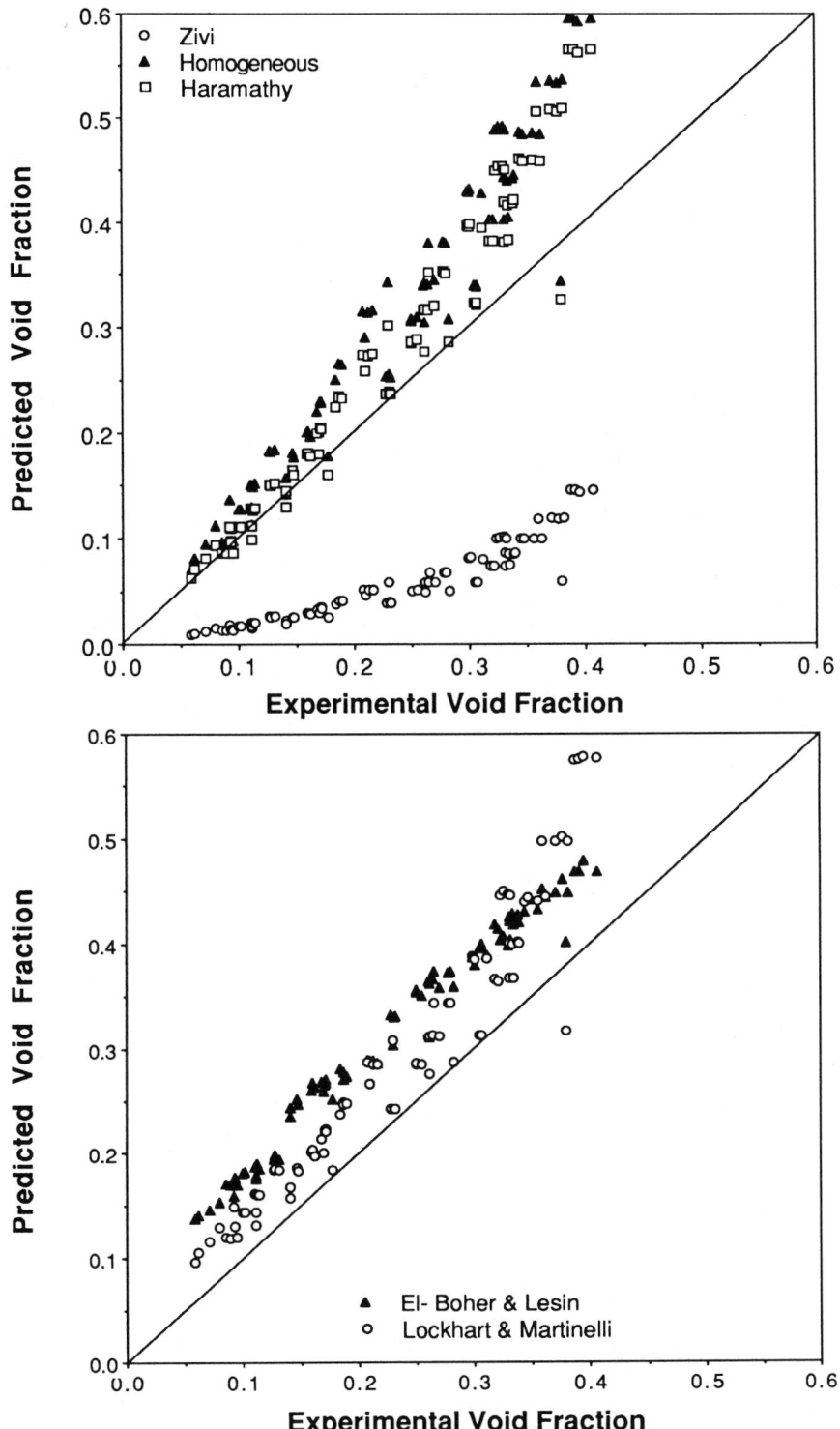

Fig. 11 Comparison of void fraction calculated from the experimental data and value calculated by different models.

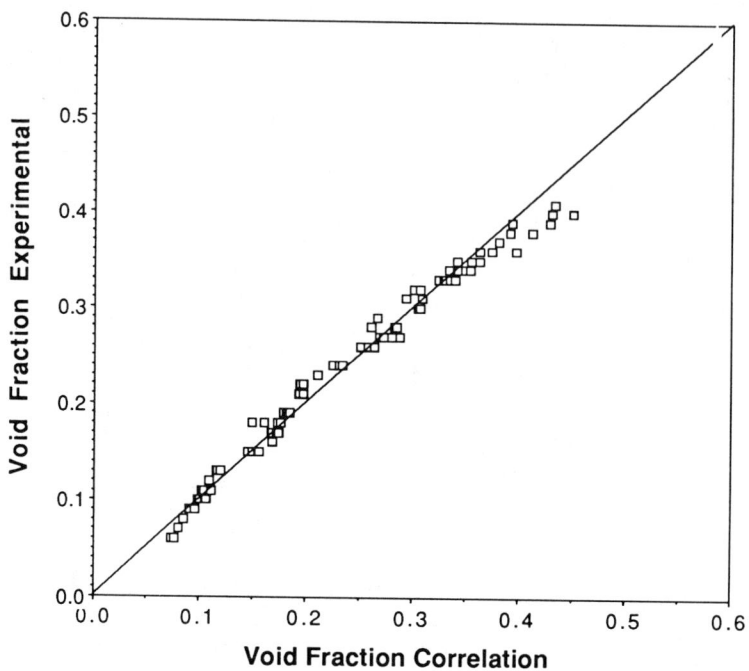

Fig. 12 Comparison of void fraction predicted values (Eq. 17) vs experimental data.

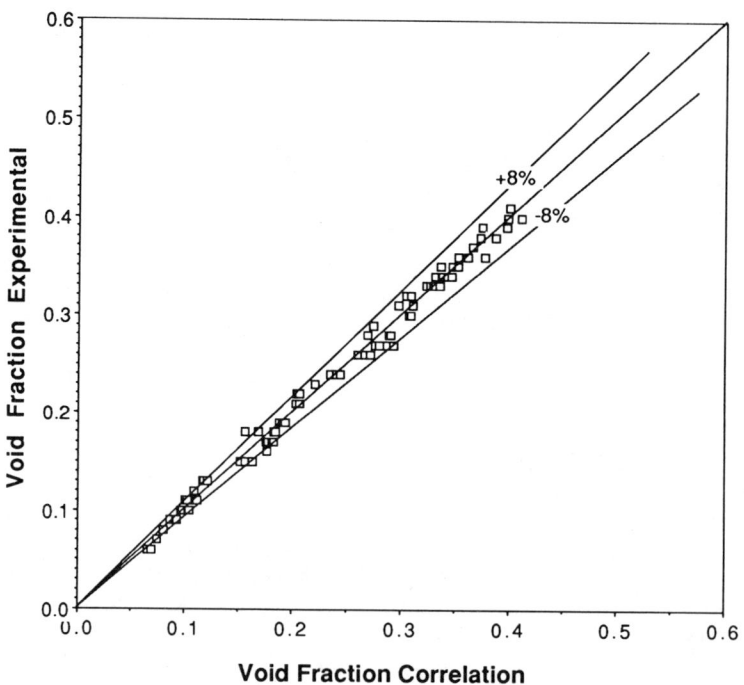

Fig. 13 Comparison of void fraction values, experimental data and modified proposed correlation, Eq. 18.

and f_{LO} is the friction coefficient according to Blasious for smooth turbulent flow region

$$f_{LO} = 0.079 \left(\frac{\dot{m}D_T}{\mu_L}\right)^{-1/4} \tag{23}$$

Different correlations, including the homogeneous flow for ϕ_{LO}^2 were suggested.[20-23] Figure 14 shows a comparison between Friedel[23] multiplier and the others at the experiment flow rates range. One can see that all of the correlations relate to each other by a direct continuous monotonic function, therefore, the experimental data and the correlation of Friedel were compared to each other. It was found that our data is to be generalized independently.

The friction pressure drop can be divided into the following two zones:
1) The zone from the mixing length to mixer outlet where the flow pattern is bubbly, here ϕ_{LO}^2 will be taken as:

$$\phi_{LO}^2 = (1 - \alpha_t)^{-7/4} \tag{24}$$

therefore, $\Delta p_{f(t-m)}$ is the following:

$$\Delta p_{f(t-m)} = (1-\alpha_t)^{-7/4} f_{LO} \frac{L_t - L_m}{D_T} \frac{2\dot{m}^2}{\rho_L} \tag{25}$$

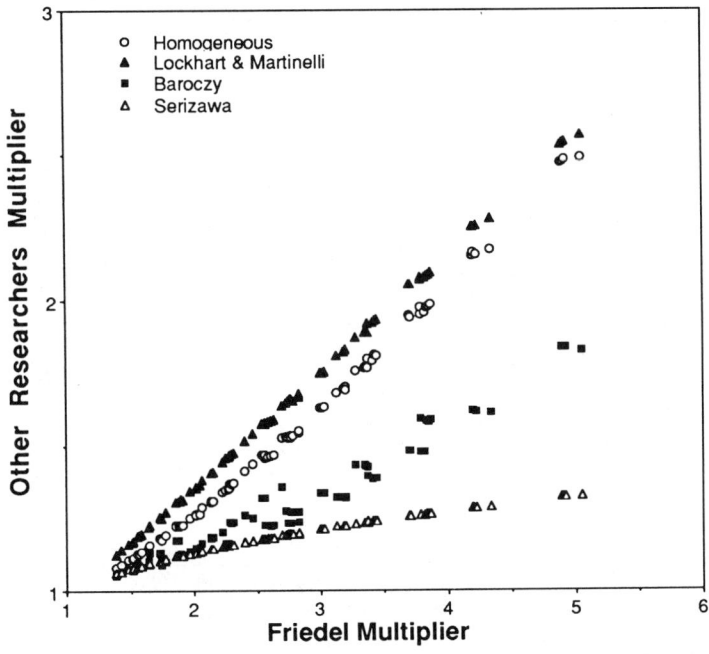

Fig. 14   Different models for frictional multiplier.

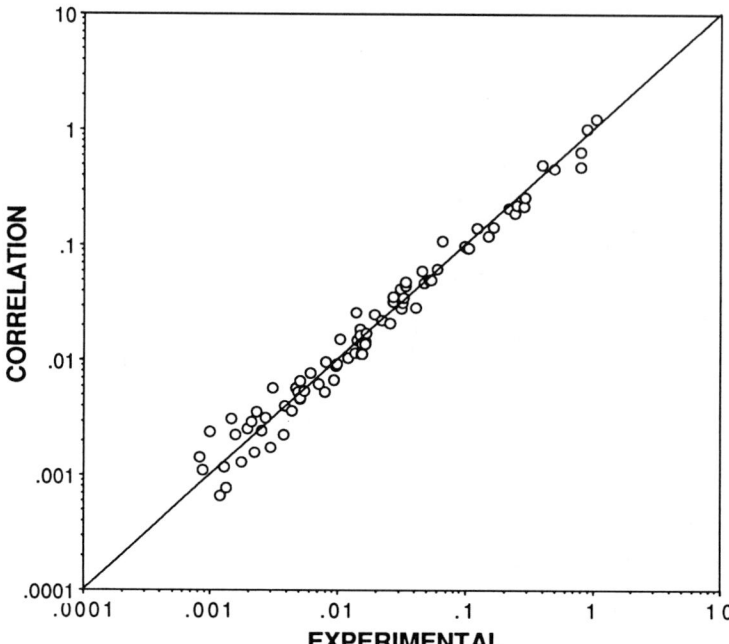

Fig. 15 Comparison between K values calculated from the experimental data and from the model (Eq. 26).

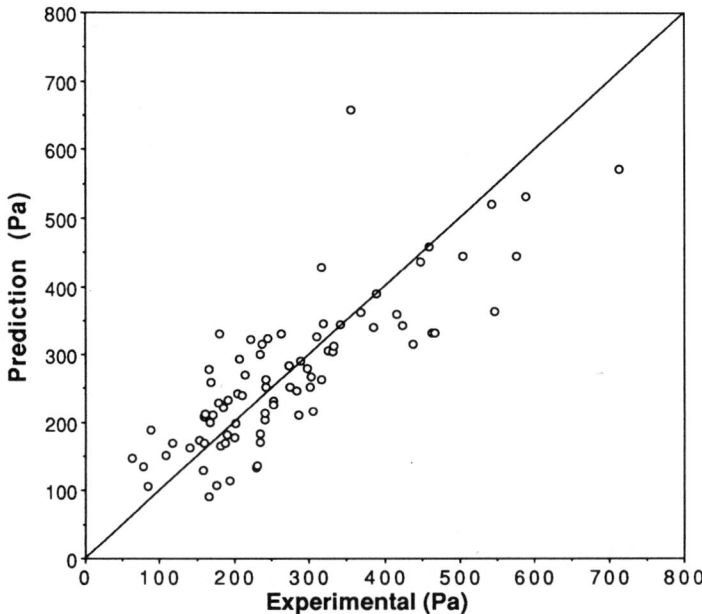

Fig. 16 Comparison of experimental and predicted values of throat pressure drop.

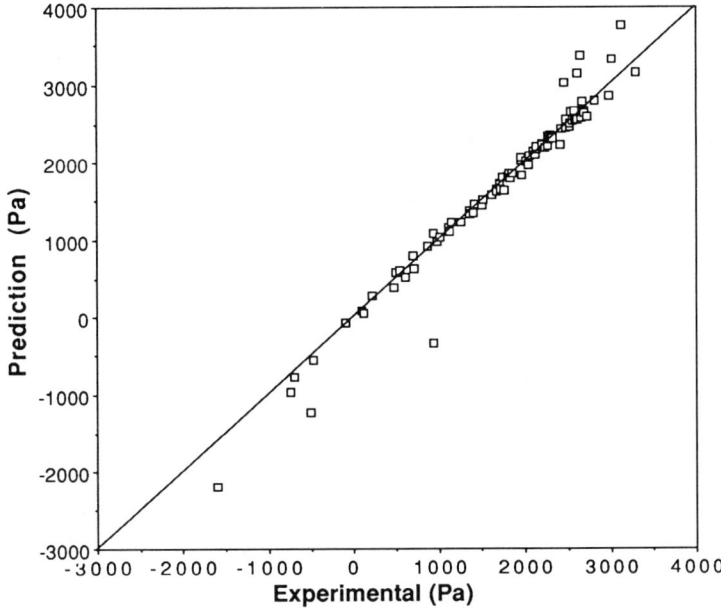

Fig. 17 Comparison of measured and calculated mixer total pressure drop.

2) The zone from the mixer throat inlet to the mixing length, here the friction pressure drop is characterized by the nozzle exit jet velocity V_N

$$\Delta p_{f(0-m)} = K \cdot \frac{L_t}{D_T} \cdot \rho_L \cdot \frac{V_N^2}{2} \qquad (26)$$

where K is a coefficient based on experimental data and can be correlated by the following function:

$$K = 0.042 x^{0.424} D_r^{5.15} \alpha_o^{-2.33} \left[\frac{Re_T}{100000}\right]^{-2} Fr_T^{0.7} \qquad (27)$$

Figure 15 shows a comparison between the generalized value of K and the experimental one, and the fit is good; the correlation coefficient is equal to 0.972.

Figure 16 shows a comparison of the calculated throat friction pressure drop and the experimental values. In Fig. 17 the overall calculated pressure drop vs the measured values are presented. Note that frictional pressure drop is low compared to the other components.

As for the mixing energy losses along the throat, the values calculated from the data and those calculated from Eq. 14, after substituting the developed correlations, fit to each other as presented in Fig. 18.

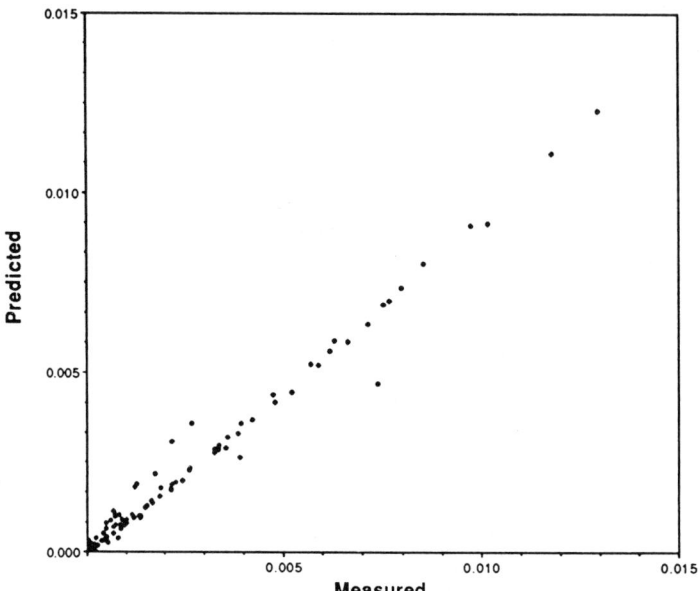

Fig. 18 Comparison of experimental and predicted values of mixing losses.

Based on the discussed analysis a parametric study for the evaluation of energy losses in jet mixers throat was drawn, and the resultant values of the different components of the losses, as well as total values, are presented in Fig. 19. The gas flow rate in these calculations was kept constant: 0.002 kg/s, and an optimal nozzle diameter is noted for liquid flow rates above 1 kg/s. It is evident that the optimal diameter is related to a specific set of flow parameters.

Conclusions

The experimental observations and their analysis permit the drawing of the following conclusions:
1) The different jet mixers and the porous mixer at two modes of operation used in this work does not affect the void fraction distribution in the cross section examined, its longitudinal distribution, nor the average values along the riser.
2) The porous mixers induced higher energy losses compared to the jet mixers under study.
3) The mixer type influences the overall performance, described by the C_p coefficient, of the facility, and the selection of mixer type should consider the operation flow rates.
4) The method used for the analysis of energy losses in jet mixers was found to be appropriate.
5) The correlation derived for the evaluation of mixing length, throat exit void fraction and pressure losses are in good agreement with the experimental data. Thus, the mixing losses could be calculated. Using this analysis enables the design of an optimal jet mixer at a given working zone.

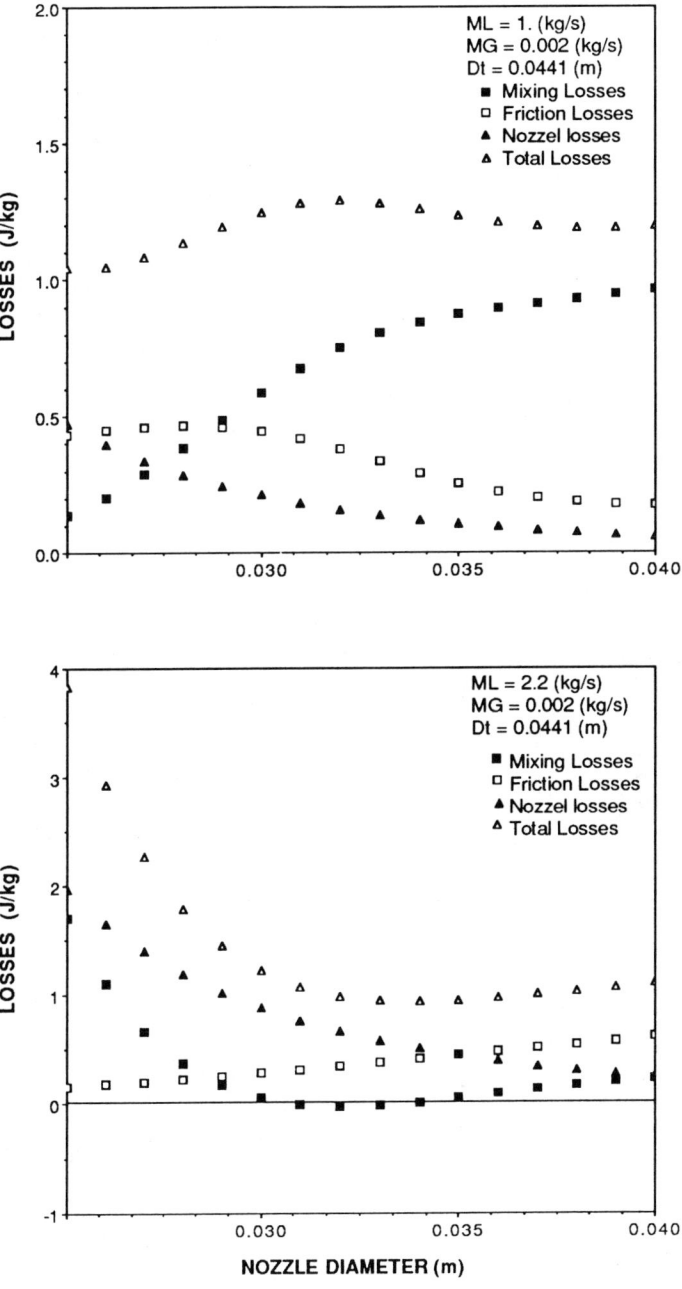

Fig. 19 Parametric study for jet mixer energy losses.

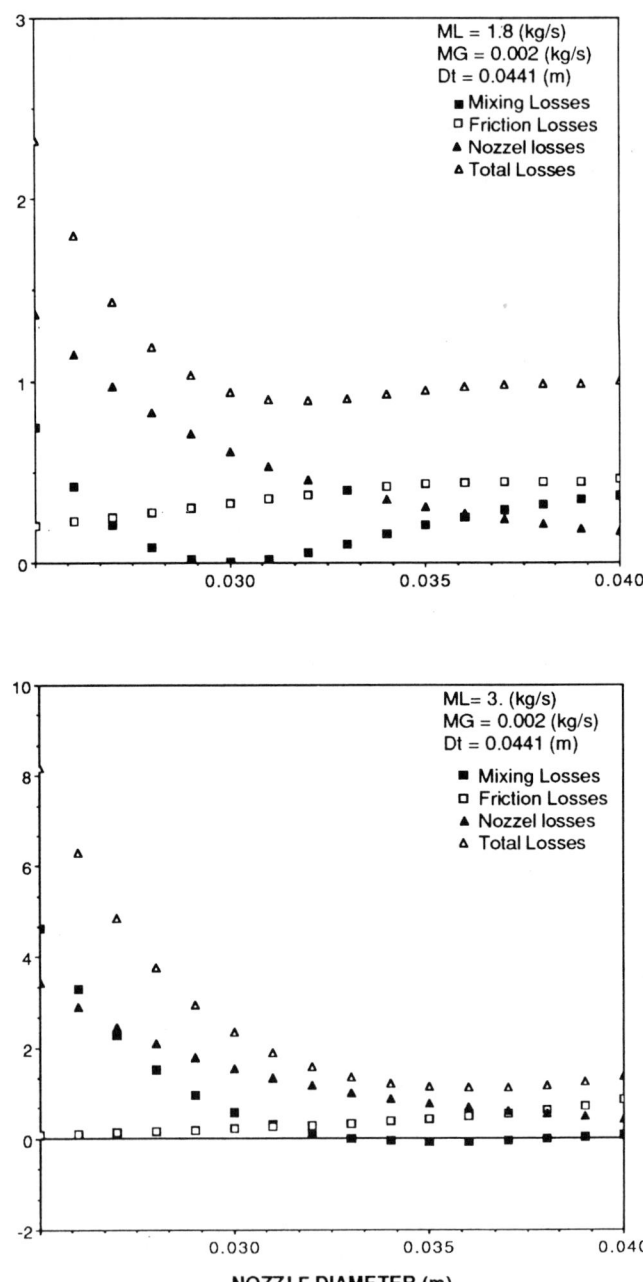

Fig. 19 (continued) Parametric study for jet mixer energy losses.

References

[1]Petrick, M., and Branover, H., "Liquid Metal MHD Energy Conversion — Its Evolution and Status," *Single and Multi-Phase Flow in an Electromagnetic Field — Energy Metallurgical and Solar Applications*, Vol. 100, Progress in Aeronautics and Astronautics, AIAA, New York, 1985, pp. 371-400.

[2]Satyamurthy, P., Thiyagarajan, T.K., Venkatramani, N. and Rohatgi, V.K., "Design of 500 Watt Electrical Steam-Mercury Liquid Metal MHD Experimental System," *Proceedings of the Tenth International Conference on Magnetohydrodynamic Electrical Power Generation*, Vol. 1, Bharat Heavy Electricals Limited, MHD Center, Tiruchirapalli, India, 1989, pp. V.1-7.

[3]El-Boher, A,., Lesin, S., Unger, Y. and Branover, H., "Experimental Studies of Liquid Metal Two-Phase Flows in Vertical Pipes," *Proceedings of the First World Conference on Experimental Heat Transfer, Fluid Mechanics and Thermodynamics*, Dubrovnik, Yugoslavia, 1988, pp. 312-319.

[4]Farchi, D., "Investigation of Mixers and Separators in a Two-Phase Flow of MHD Energy Conversion Systems," M.Sc. Thesis, Department of Mechanical Engineering, Ben Gurion University of the Negev, Beer Sheva, Israel, 1990.

[5]Herringe, R.A. and Davis, M.R., "Structural Development of Gas-Liquid Mixture Flows," *Journal of Fluid Mechanics*, Vol. 73, P.1, 1976, pp. 97-123.

[6]Merchuk, J.C., and Stein, Y., "Local Hold-up and Liquid Velocity in Air Lift Reactors," *American Institute of Chemical Engineers Journal*, Vol. 27, No. 3, 1981, pp. 377-388.

[7]Fabris, G., Dunn, P.F. and Chow, J.C.F., "Two-Phase LMMHD Mixer Development Experiments," *17th Symposium on Engineering Aspects of MHD*, Stanford University, Stanford, CA, 1978.

[8]Adler, P.M., "Formation of an Air-Water Two Phase Flow" *American Institute of Chemical Engineers Journal*, Vol. 23, No. 2, 1977, pp. 185-191.

[9]Blumenau, L., Spero, E. and Branover, H., "Liquid Jet Gas Pump as Mixer Element with Two-Phase Liquid Metal MHD Generator and Compressor Systems," *Liquid-Metal Flows: Magnetohydrodynamics and Applications*, Vol. 111, AIAA, Washington, D.C., 1987, pp. 245-264.

[10]Witte, J.H., "Mixing Shocks in Two-Phase Flow," *Journal of Fluid Mechanics*, Vol. 6, P. 4, 1969, pp. 639-655.

[11]Deych, M.Ye., Tsiklauri, G.V., Kalinin, Yu. F. and Dikiy, N.A., "A Method for Calculating a Two-Phase Jet Injector," *Heat Transfer, Soviet Research*, Vol. 3, No. 6, Nov.-Dec., 1971, pp. 61-68.

[12]Cunningham, R.G. and Dopkin, R.J., "Jet Break-up and Mixing Throat Lengths for the Liquid Jet Gas Pump," *Journal of Fluid Engineering*, 1974, pp. 216-226.

[13]Hongqi, L., "Liquid Gas Phase Flow Theory of Jet Pump," *International Conference of the Physical Modelling of Multi-Phase Flow*, Coventry, England, 1983, pp. 439-451.

[14]Neve, R.S., "The Performance and Modeling of Liquid Jet Gas Pumps," *International Journal of Heat and Fluid Flow*, Vol. 9, No. 2, 1988, pp. 154-164.

[15]Collier, J.G., "Convective Boiling and Condensation," 2nd ed., McGraw-Hill, New York, 1974.

[16]Sukoriansky, S., and Talmage, G., "A Computer Package for Analysis of Liquid Metal MHD Power Conversion Systems," *Proceedings of International Specialist Meeting on Mathematical Modeling of MHD Power Systems*, Eindhoven, The Netherlands, 1986, pp. 91-102.

[17]Branover, H., El-Boher, A., Farchi, D., Lesin, S., and Unger, Y., "Two-Phase Liquid Metal/Gas Flow Characteristics in MHD System," *Proceedings for the 27th Symposium on Engineering Aspects of MHD*, Reno, NV, 1989, pp. 9.3, 1-7.

[18] Zivi, S.M., "Estimation of Steady-State Steam Void-Fraction by Means of the Minimum Entropy Production," *Journal of Heat Transfer*, Transactions of the ASME, 1964, pp. 247-252.

[19] Haramathy, T.Z., "Velocity of Large Drops and Bubbles in Media of Infinite or Restricted Extent," *American Institute of Chemical Engineers Journal*, Vol. 6, No. 2, 1960, pp. 281-288.

[20] Lockhart, R.W. and Martinelli, R.C., "Proposed Correlation Data for Isothermal Two-Phase Two-Component Flow in Pipes," *Chemical Engineering Progress*, Vol. 45, No. 1, 1949, p. 39.

[21] Baroczy, C.J., "A Systematic Correlation for Two-Phase Pressure Drop," *Engineering Progress Symposium Series*, Vol. 62, No. 64, 1966, p. 232.

[22] Serizawa, A., and Michiyoshi, I., "Void Fraction and Pressure Drop in Liquid Metal Two-Phase Flow," *Journal of Nuclear Science and Technology*, Vol. 10, No. 7, 1973, p. 435.

[23] Friedel, L., "Improved Friction Pressure Drop Correlations for Horizontal and Vertical Two-Phase Pipe Flow," European Two-Phase Flow Group Meeting, ISPRA, Paper E2, 1979.

Two-Phase Flow Measurements in Reaction Systems

Yaakov M. Timnat*

Technion—Israel Institute of Technology, Haifa 32000, Israel

Abstract

The development of a phase Doppler anemometry system for measuring simultaneously the velocity and size of particles in two-phase flow is described, and a validation experiment is performed. This method is compared to the pedestal technique, which was previously used, and the two methods are compared critically, discussing under which experimental conditions each one is to be preferred. Measurements of axial and radial temperature distribution and of the length of the recirculation zone in a reactive two-phase flow dump combustor are reported and compared with results of a theoretical calculation. The influence of the fuel injection and the character of the recirculation zone are discussed. It is concluded that the ring-shaped flame observed is basically diffusion controlled and that the combustion processes take place mainly in the volume restricted by the height of the backward-facing step employed in the experimental configuration.

Introduction

This paper describes further developments in the measurement of two-phase turbulent flow, with and without chemical reactions, a topic that I treated in the fourth and fifth seminars.[1,2] In the previous communications a modified laser Doppler anemometer (LDA) technique and further diagnostic measurements in two-phase turbulent flows were described. More recently we have developed phase Doppler anemometry (PDA).[3,4] This technique was introduced by Durst and Zane[5] in 1975, but it was exploited only much later to measure the diameter of spherical droplets in liquid sprays.[6-8]

Copyright © 1991 by the American Institute of Aeronautics and Astronautics, Inc. All rights reserved.

*Professor and Head, Propulsion and Combustion Laboratory, Department of Aerospace Engineering and Space Research Institute.

The PDA technique is explained in detail, as is a validation experiment, which it requires. In this test the diameter of spherical latex particles of known size were measured by PDA, and the results obtained for spheres ranging from 80 to 220 μm gave very good agreement with the manufacturer's specifications. The PDA technique gives more accurate results than the LDA, but it is limited to transparent, spherical liquid drops, whereas the LDA can also be used with solid particles and various geometries. The two experimental systems are compared and the advantages and drawbacks of the two techniques for different situations are discussed; a comparison between predicted and measured gas phase and particulate phase velocity profiles is also carried out.

Measurements of both axial and radial temperature distributions and of the length of the recirculation zone were performed for hollow and full cone configurations and compared with calculated results. It was found that the reactive flowfield is influenced more by the fuel injection than the isothermal one. The droplet paths determining the gas fuel injection cause the appearance of reactive zones, which expand and create temperature and velocity gradients.

It was found that the recirculation zone is very rich in gaseous fuel, even though the droplets did not penetrate through the shear layer, which separates this zone from the main plug flow. This effect is probably due to the rapid rise of the droplets drag caused by the jump of the effective viscosity within the highly turbulent shear layer.

The radial distributions of the velocity, temperature, and gas concentration indicate that the annular flame envelop is located on the shear layer. The results verifying the visualized ring shape of the flame suggest that the combustion is basically diffusion controlled and that the combustion processes take place mainly in the volume restricted by the height of the backward-facing step.

PDA Technique

The measurement principle of the PDA technique is based on the following phenomenon: When a light beam illuminates a transparent sphere with index of refraction greater than that of the surrounding, the light rays can be considered to traverse according to the rules of geometrical optics as shown in Fig. 1. The axial beam crosses through the focal point and forms a cone with divergent angle 2α defined by the focal length and the diameter of the sphere:

$$\alpha \simeq \sin^{-1} \frac{d_p}{2f} \tag{1}$$

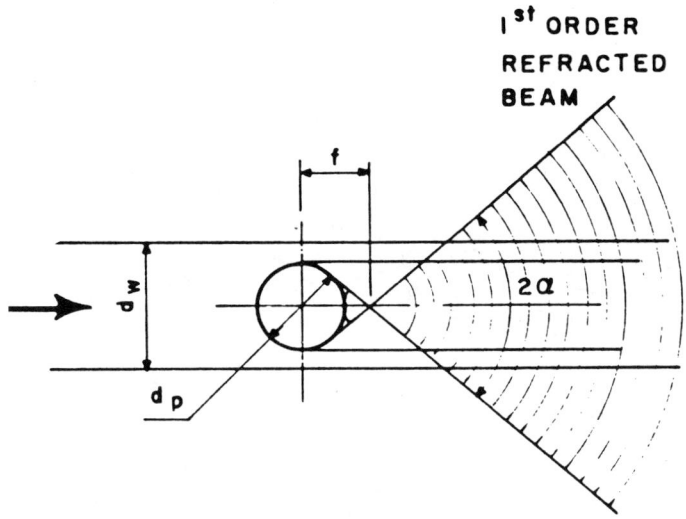

Fig. 1 Transparent sphere illuminated by a laser beam.

where f is the focal length of the sphere:

$$f = \left(\frac{n}{n-1}\right)\frac{d_p}{4} \qquad (2)$$

and n is the index of refraction of the sphere.

It seems that all of the light rays refracted (in their first order) by the sphere pass through the focal point, which can then be considered as a point source of light. When such a sphere is illuminated by two beams from different angles, two point sources of light are created at a fixed location relative to the sphere (see Fig. 2). The distance between the points (Δy) depends on the particle diameter and the angle between the incident beams:

$$\Delta y = 2 \cdot f \cdot \sin\phi \qquad (3)$$

From each of the source points a conical beam of light is emitted with the divergent angle 2α. Whenever the angular condition $\alpha > \phi$ is fulfilled, an overlap region will be formed on the screen (SC). This region will be illuminated by both beams. Within this region the two light beams, originating from source points S_1 and S_2, will form an interference pattern of fringes, linear near the optical axis and diverging as the distance in the y and z direction increases. The spatial distribution of light on the screen (which determines the fringe shape) can be calculated according to

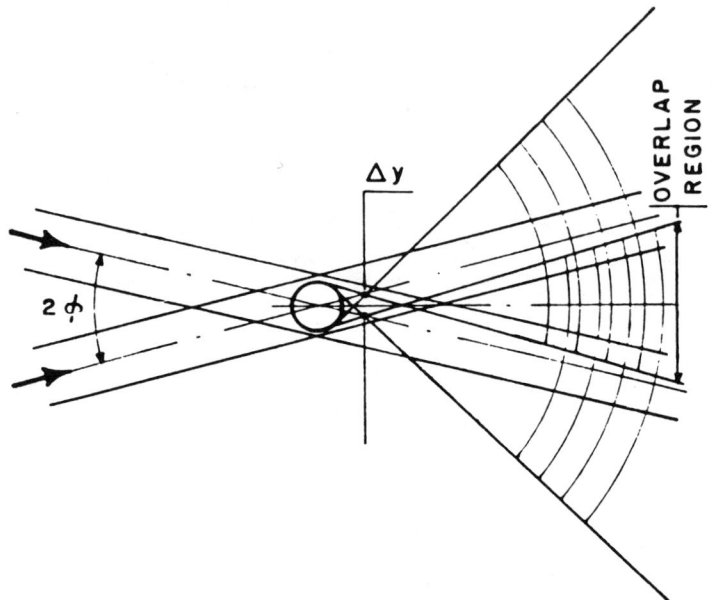

Fig. 2 Transparent sphere illuminated by two laser beams.

a relatively simple geometrical optics calculation. At the center region, where the angle β is symmetrical about the optical axis, the fringe spacing δ can be calculated according to

$$\delta = \frac{\lambda}{2\sin\beta} \quad (4)$$

From Fig. 3 the following relation is determined:

$$\text{tg}\beta = \frac{\Delta y}{2(L - f \cdot \cos\phi)} \quad (5)$$

Combining Eqs. (2-5) and assuming a small β angle, one obtains

$$\delta = \frac{\lambda}{2\sin\phi}\left[\frac{4}{d_p} \cdot \frac{(n-1)}{n} - \cos\phi\right] \quad (6)$$

This demonstrates (for $L/d_p \gg 1$) the inverse relation that exists near the optical axis between the fringe spacing and the sphere diameter. Measuring the fringe spacing can indicate the drop diameter. When a drop traverses through the beam (for $d_p < d_w$), the associated fringe pattern moves along with a certain local velocity (V_f). Two detectors located close to the optical axis but at different angular orientation will record an oscillation of light intensity with a frequency being equal to

$$F = \frac{V_f}{\delta} \tag{7}$$

V_f is obtained through the following relation:

$$V_f = U_\perp \left[\frac{(L-f\cos\phi)\cdot\cos\phi}{f\,\cos} + 1 \right] \tag{8}$$

Thus, obtaining for small spheres,

$$F = \frac{2U_\perp}{\lambda}(\sin\phi + \sin\beta) \tag{9}$$

The signals of the two detectors will show a phase difference θ. The phase difference θ (in angles) will be related to their lateral separation ($\overline{\Delta y}$) according to

$$\theta = 360 \cdot \frac{\overline{\Delta Y}}{\delta} \tag{10}$$

Measuring θ determines δ, which indicates the particle diameter. For small droplets ($L/d_p >> 1$),

$$\theta = d_p \left[\overline{\Delta Y} \frac{180\cdot\sin\phi}{\lambda L} \cdot \frac{n}{n-1} \right] \tag{11}$$

In studies of liquid spray combustion droplet diameters are in the range of $d_p < 300$ μm. If the term in the brackets of Eq. (11) is adjusted, with proper selection of operational parameters, to obtain a value of one million, the phase difference of the signals from the two detectors (in degrees) will equal the diameter of the droplets (in microns). This condition is achieved, for example, in a system containing a HeNe laser with kerosene fuel (n=1.35), where 2ϕ is 8 deg and L=0.5 m, by selecting a value for $\overline{\Delta Y}$ of 6.5 mm.

In order to validate the pattern of the interference phenomena, an experimental setup was assembled. The principal idea of the validation test was to locate a sphere of known diameter at an LDA control volume, measure some parameters of the interference fringes formed on the screen, and compare results with calculated values determined from the given formulas.

In this setup small latex particles were used to simulate liquid droplets. The latex particles (polystyrene divinylbenzene spheres from

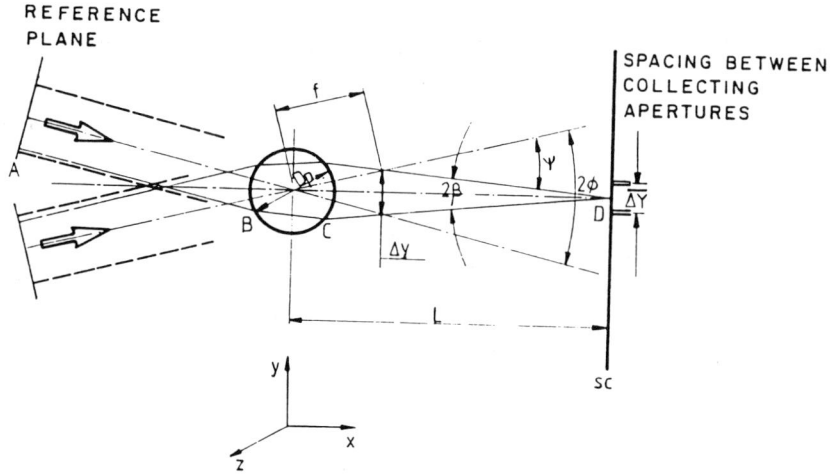

Fig. 3 Geometrical description of the light rays forming the fringes.

Duke Scientific Corporation) are spherical, smooth, and transparent (n=1.59 at λ=540 nm). They can be obtained in samples of uniform diameter in the range of a submicron to hundreds of microns (the typical standard deviation is 5%). The spheres were located on the edge of a fine glass capillary tube with the aid of a vacuum. They were aligned by use of a micrometric x-y-z traversing table in the control volume formed by two incident laser beams originating from LDA modular optics (self-built). Interference fringes were observed on a screen located at some distance L from the measured sphere. For each sphere the distance between the fringes, δ, was measured on the screen, near the optical axis, using a ruler. Some difficulties in the measurements were encountered due to the system's extreme sensitivity to surrounding vibration; however, good results were obtained. Figure 4 demonstrates the relation between the calculated diameter by using the measured δ and Eq. (6), and the diameter of the sphere, as specified by the manufacturer. Results are given for spheres ranging from 80 to 220 µm. The figure clearly demonstrates a very good agreement between the measured diameters of the spheres and the diameters as given by the manufacturer's specifications. By use of currently available techniques it was difficult to locate particles smaller than about 80 µm. One should also mention that at present it is impractical to perform more accurate, independent measurements of the size of the spheres (for example, using a microscope), since they were picked up from a large sample.

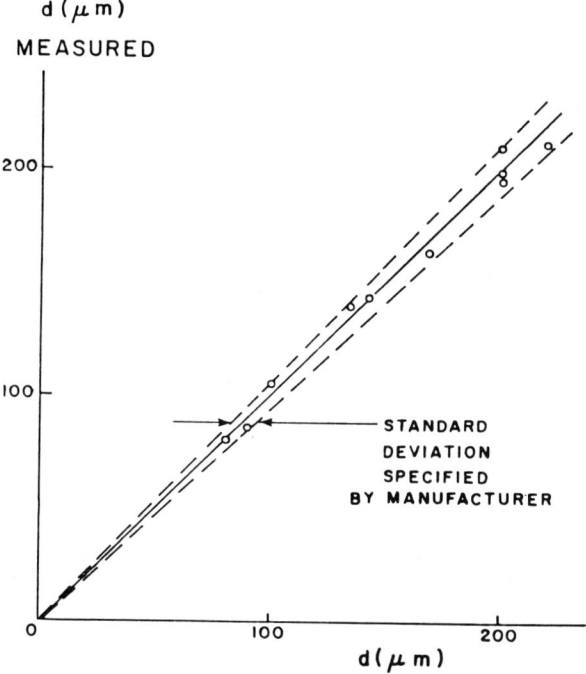

Fig. 4 Measured diameters of the spheres vs the manufacturer's specification.

Comparison of the LDA and PDA Methods

The selection of the optimal measurement technique for a specific application is based on the two-phase flow characteristics. When all the particles fulfill the conditions of sphericity and smoothness, there is a clear advantage in utilizing the PDA technique due to its higher accuracy in determining the particle diameter. This applies to most experimental conditions that include both reacting and nonreacting liquid sprays. The technique can be applied to sprays with a wide size range. It has a lower size limit of about 1-3 µm, due to diffraction phenomena, so that it is not possible to obtain the diameter of smaller droplets. For particles having an arbitrary shape, or in the presence of impurities, PDA is inappropriate; in those cases the pedestal amplitude technique should be used. Its accuracy is lower, depending on the specific experimental conditions. The pedestal technique can determine the size of particles over a large dynamic range; however, it has, a higher size limit that is determined by the diameter of the laser beam waist at the control volume. The lower size limit of the particles can be in the order of 0.1 µm; the limiting factor is the signal-to-noise ratio obtained at the photomultiplier. It should be noted that, whenever the technique is applied,

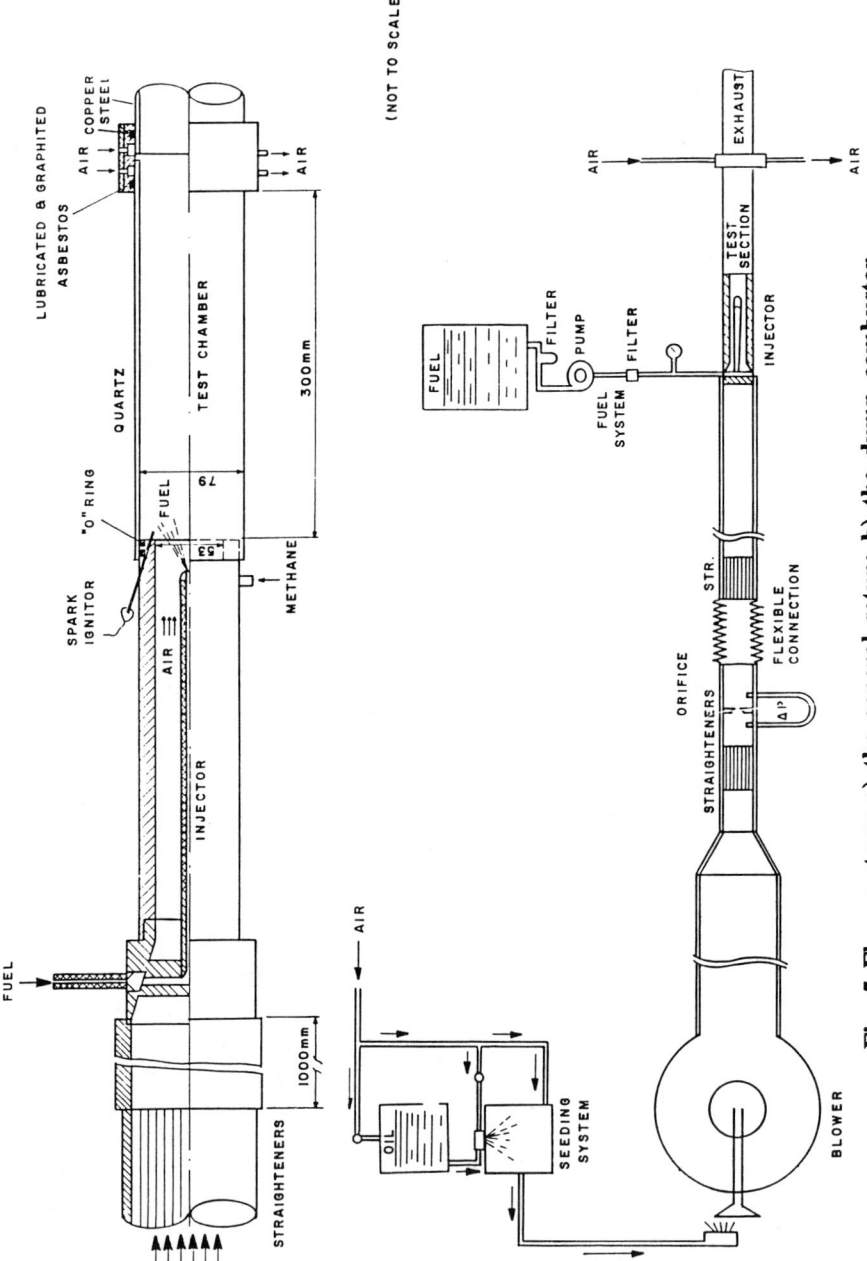

Fig. 5 Flow system: a) the general setup; b) the dump combustor.

calibration curves must be obtained for each experimental condition, and extrapolation of calibration curves cannot be automatically performed. This is due to the fact that particles of various size ranges are affected by different light scattering mechanisms, and the index of refraction also has an important influence on the calibration curve.

Measurements of Temperature and Species Distribution and the Recirculation Zone Length

The measurements were performed in the experimental setup shown in Fig. 5, which presents the general layout and the dump combustor used. The airflow in most experiments was 115 g/s, the fuel (kerosene) flow 2.17 g/s and the diameter ratio of the straightener section to the test section was 1.5. From photographs taken one observes that in the recirculation zone the flame is annular in character and is anchored on the entrance of the dump combustor. Figure 6 shows the static pressure distribution along the wall in cold and reactive flow. Subatmospheric pressure indicates the presence of backflow zones, with the recirculation zone ending when atmospheric pressure is reached. In cold flow the length of this zone equals 11 step heights; this value decreases to 4 in reactive flow. Figure 7 presents the radial temperature distribution of the gases for two values of the airflow (115 and 117 g/s) with the same fuel-to-air ratio. The highest temperature was observed in both cases far from the centerline. The strong gradients indicate the burning zone, and it appears that the chemical reactions take place in the cylindrical volume defined by the expansion step. For the lower flow rate, T_{max}, which reaches 1450 K, is situated 40 mm from the entrance, and farther downstream (after 200 mm) a uniform temperature of 700 K was observed. For the higher flow rate, T_{max} is somewhat higher (1500 K), and the flame extends farther downstream, indicating higher combustion efficiency. This is probably due to a shortening of the recirculation zone caused by the higher entrance Reynolds number and by better mixing due to enhanced turbulence.

This conclusion is supported by measurements showing that oxygen concentration diminishes, whereas the concentration of CO_2 increases (see Fig. 8). The radial distribution of the normalized mean axial velocity, the turbulence intensity, the mean size of the droplets and the normalized variance of the droplets are presented in Fig. 9. It appears that the axial velocity and the turbulence intensity are uniform over the whole cross section, whereas the mean size of the droplets is smaller in the center of the jet. In the combustor middle the spread of the droplet size is appreciable. Figure 10 presents the mean axial velocity of the

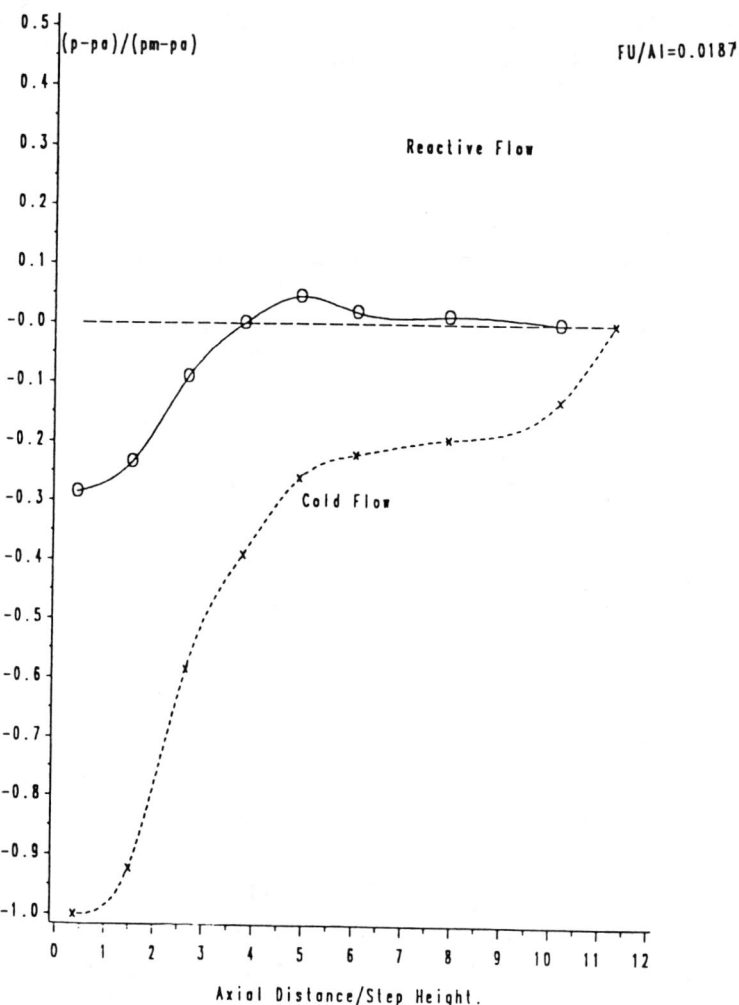

Fig. 6 Static pressure distribution along the combustor wall in cold and reactive flow.

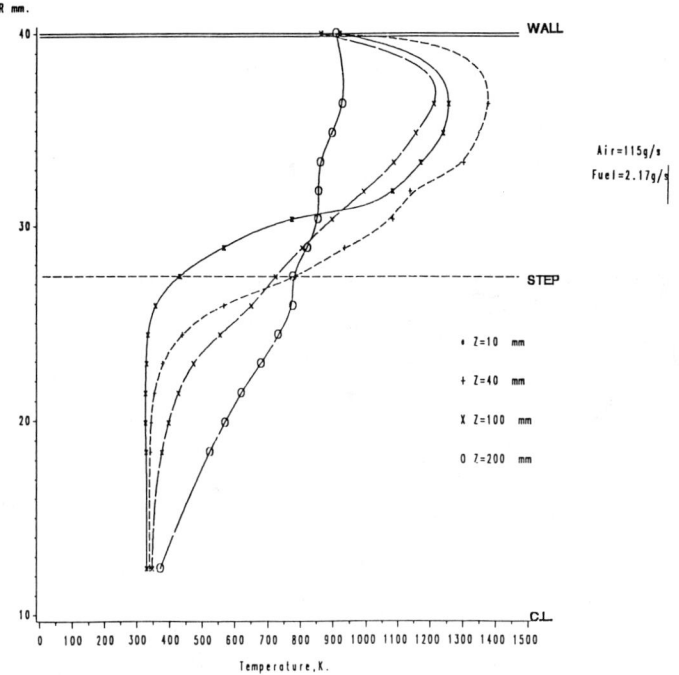

a) Air 115 g/s, fuel 2.17 g/s

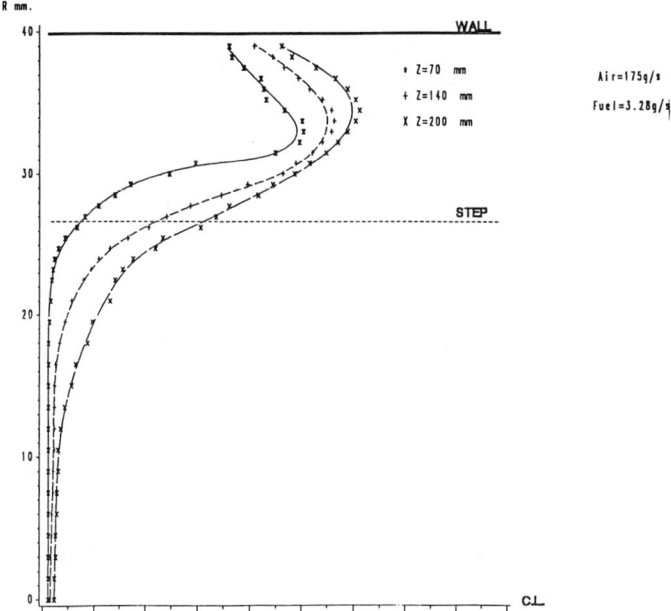

b) Air 1.75 g/s, fuel 3.28 g/s

Fig. 7 Radial distribution of the gas temperature.

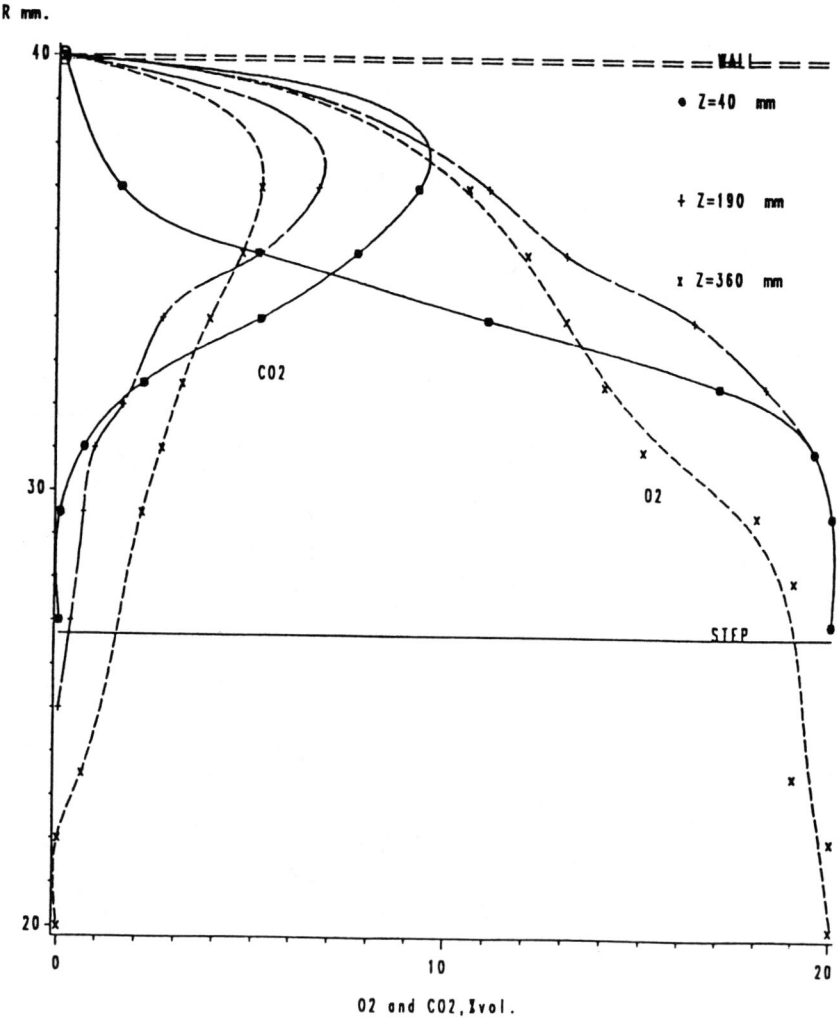

Fig. 8 Radial distribution of carbon dioxide and oxygen 1% volume.

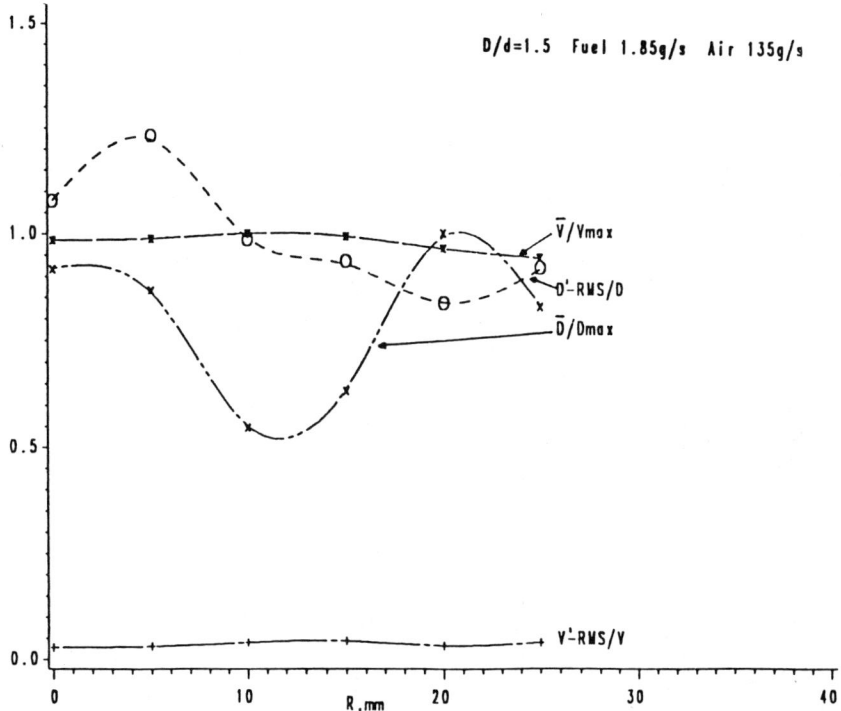

Fig. 9 Radial distribution of the mean axial velocity ($\overline{V}/V_{max}$), turbulence intensity (V'-RMS/V), mean size (D/D_{max}), and normalized variance of the size (D'-RMS/D) for the droplets in reactive flow.

gas for three cases: cold flow, cold flow with fuel injection, and reactive flow, clearly showing the influence of the chemical reactions.

From the results presented it is clear that the flowfield and the droplet trajectories have different character for reacting and non-reacting flows and the injection mode and the injection type strongly influence the spatial distribution and the concentration of the fuel, affecting the stability and the efficiency of the flame.

Numerical Results

A numerical model, in which the gas phase equations based on the conservation of mass, momentum, and thermal energy are Favre averaged, was developed.[10] This is complemented by the k-ε turbulence model. The liquid spray is simulated by a finite number of droplet groups of different sizes, originating at different radial positions and injected at various angles with appropriate initial velocities. Each size group is tracked

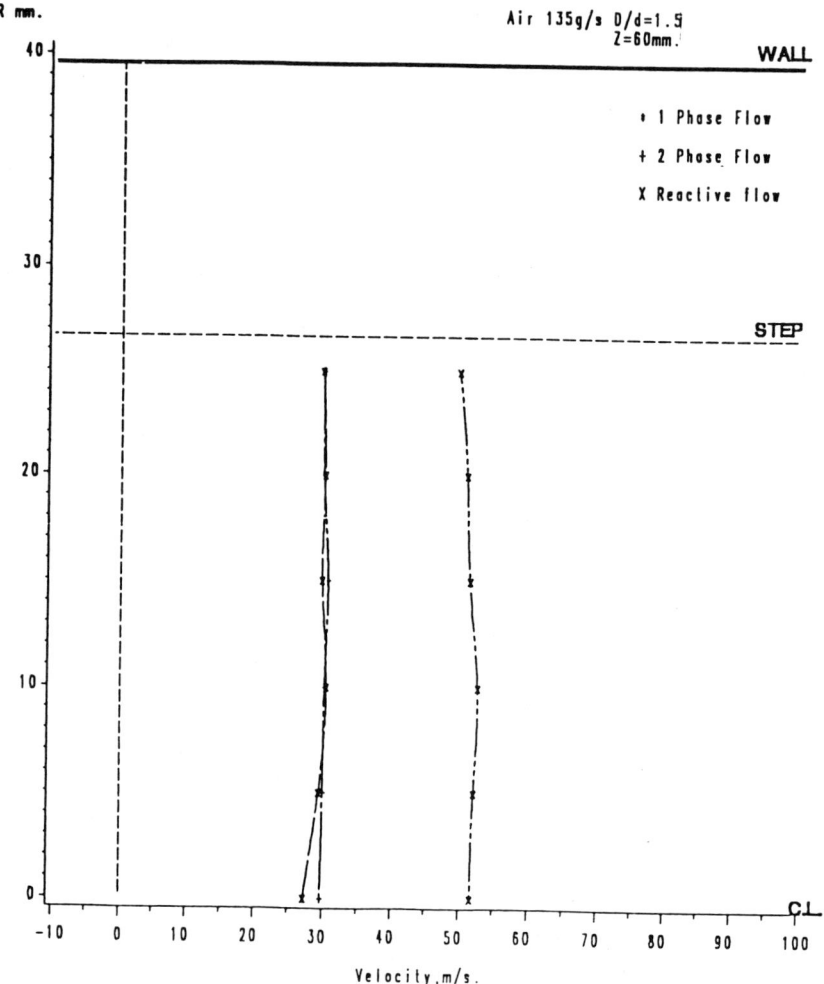

Fig. 10 Mean axial velocity of the gas for cold flow, cold flow with fuel injection and reactive flow.

through the flowfield by solving the Lagrangian equations of motion. The influence of the droplets on the gas field is accounted for by source terms appended to the gas phase equations prior to a calculation of the Eulerian gas field. The two-phase flowfield is computed using the TEACH code,[11] with suitable modifications for the Favre-averaged formulation, the additional liquid phase, and chemical reactions.

Figure 11 compares the velocity vector distribution for the isothermal and the reactive two-phase flowfield, using the experimental injector; one can see that the recirculation zone is shortened very substantially. Figure 12 compares the measured and calculated radial distribution of the

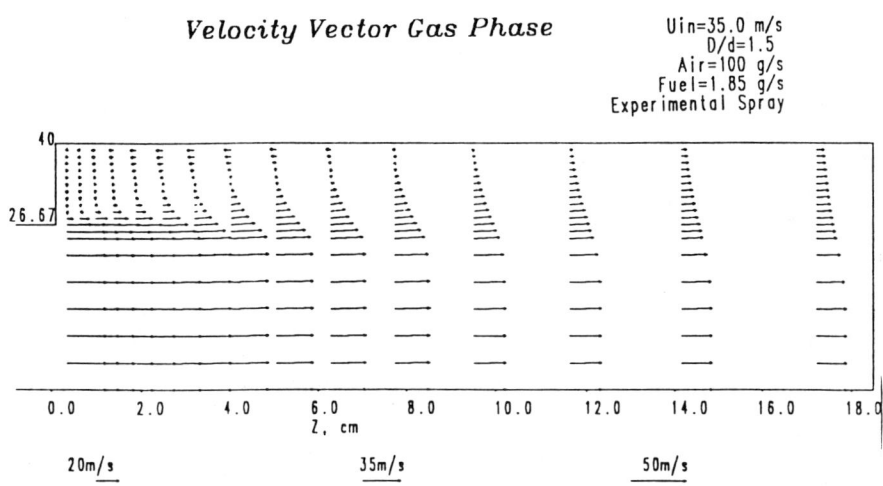

Fig. 11a Isothermal two-phase flowfield with the experimental injector.

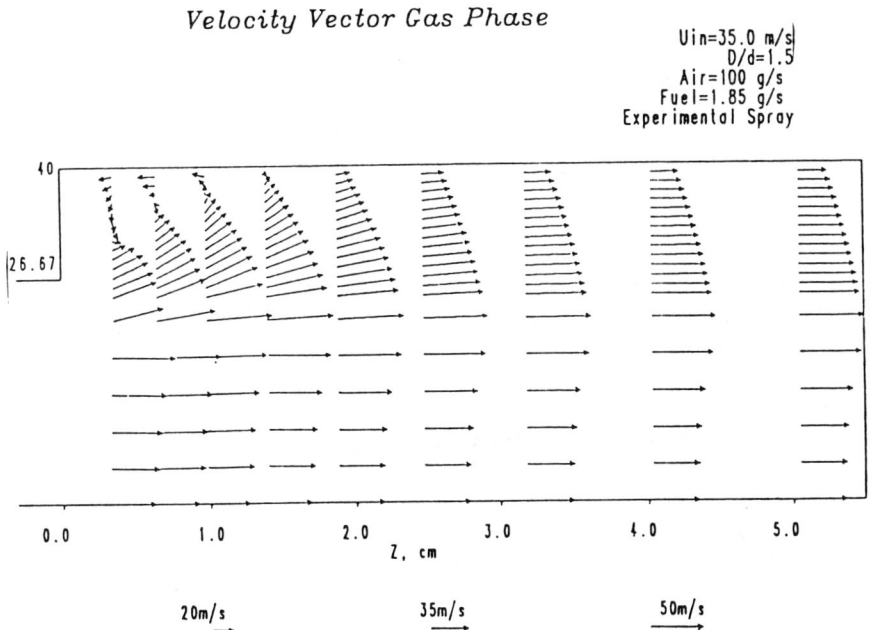

Fig. 11b Reactive flowfield with the experimental injector.

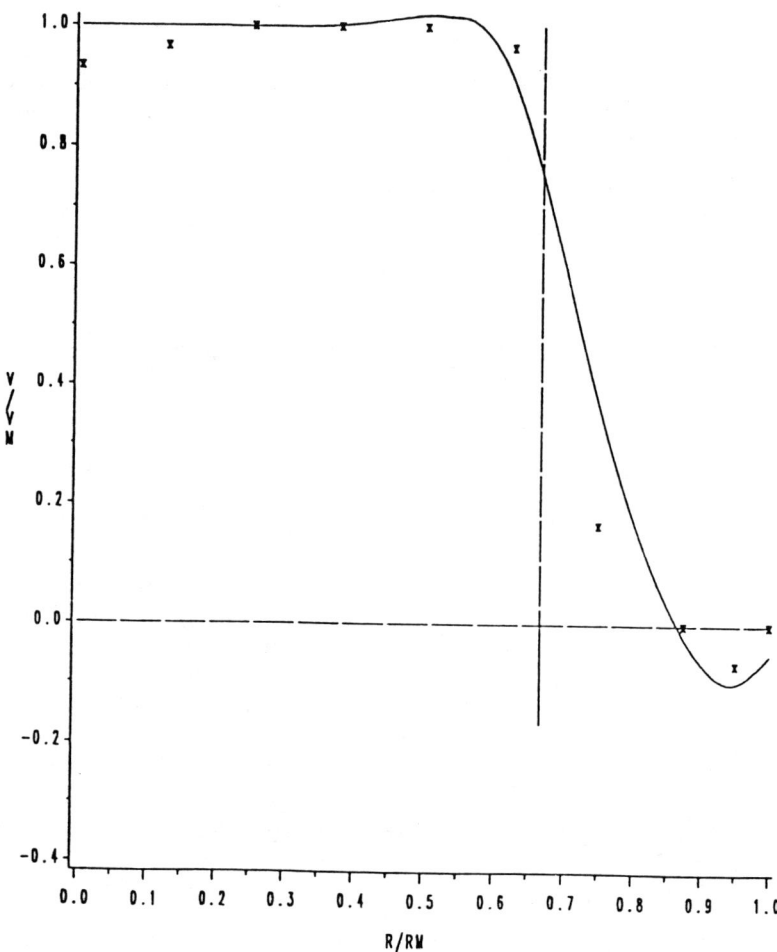

Fig. 12 Measured and calculated velocity profile.

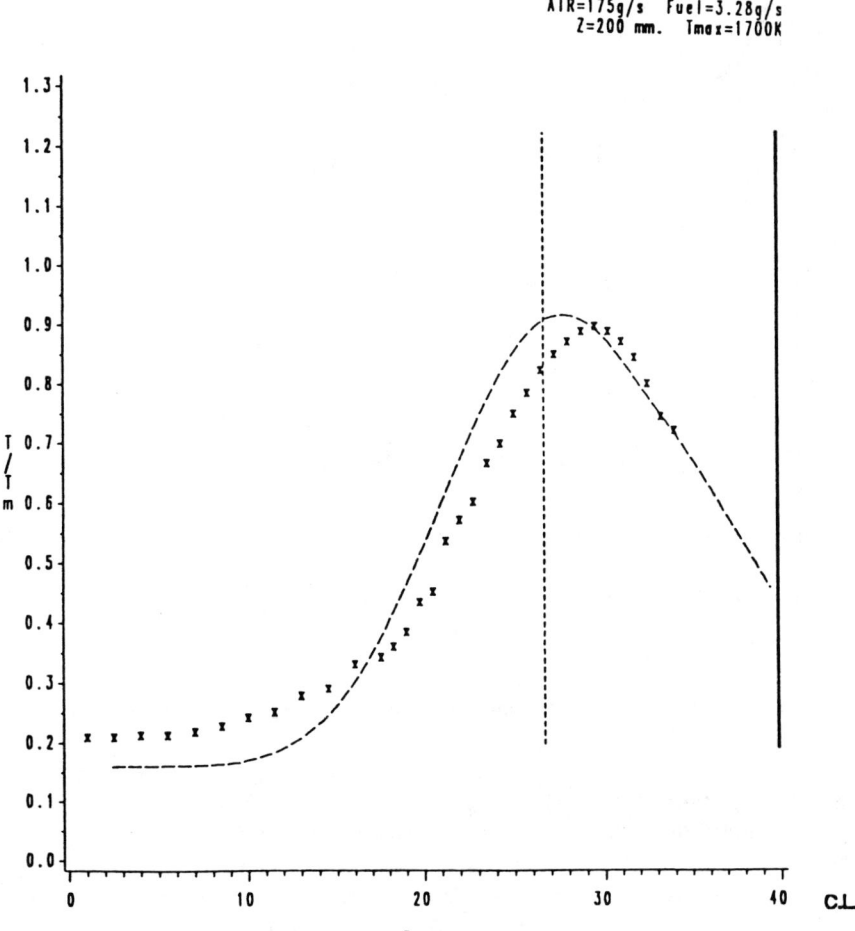

Fig. 13 Measured and calculated temperature profile.

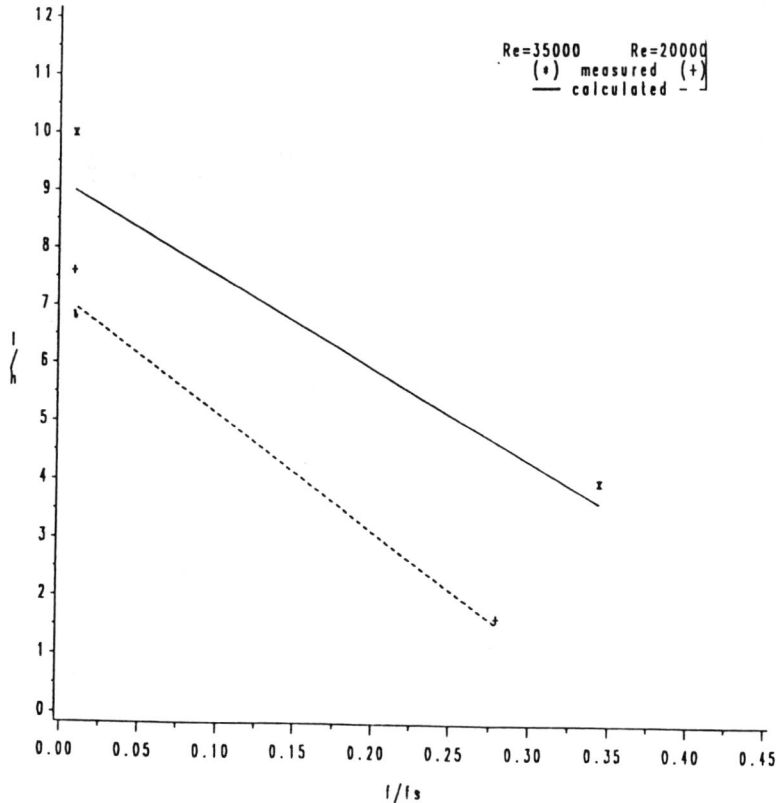

Fig. 14 Length of the recirculation zone vs the fuel-to-air ratio.

axial velocity profile and shows good agreement. This is also the case for the radial distribution of the temperature profile (Fig. 13). The length of the recirculation zone is shown as a function of the fuel-to-air ratio for two Reynolds numbers in Fig. 14 and gives reasonable agreement between experiment and theory for the few points measured.

Conclusions

Combustion causes the flow to accelerate in the hot zones compared with the cold flow situation and changes the turbulent structure. These processes generate strong velocity gradients, which considerably diminish the height of the recirculation zone. The shear layer dividing the recirculation zone from the main stream is conducive to chemical reactions. It was found that this layer is fuel-rich, with low velocities, and contains hot gas products and contributes substantially to an increase in the mixing rate. The mode of full injection has a strong effect on the combustion process, as a consequence of the fuel vapor dispersion, which is governed by the droplet path configuration. Small, fast droplets, which evaporate before reaching the wall, enhance combustion and therefore contribute to diminish the chamber length required for efficient, high-quality combustion.

The volume enclosed by the recirculation zone acts as a generator of hot gases (see Fig. 9). These fuel-rich gases are recirculated and diffused back to the main flow through the shear layer and its detachment and reattachment points (upstream at the step corner and downstream at the end of the recirculation zone). Cold oxygen is convected and diffused from the plug flow toward the shear layer, which acts as an aerodynamic flameholder. Finally, one can conclude that the ring-shaped flame, which was observed in the experiments, is basically diffusion controlled, and the combustion processes take place mainly in the volume defined by the height of the backward-facing step.

References

[1] Levy, Y. and Timnat, Y. M., "Two-Phase Flow Measurements Using a Modified Laser Doppler Anemometry Systems," *Single and Multiphase Flows in an Electromagnetical Field. Energy, Metallurgical, and Solar Applications.* Vol. 100, Progress in Astronautics and Aeronautics Series, AIAA, New York, 1985, pp. 355-367.

[2] Laredo, D., Levy, Y. and Timnat, Y. M., "Two-Phase Flow Diagnostics in Reactive Systems," *Liquid Metal Flows: Magnetohydrodynamics and Applications*, Vol. 110, Progress in Astronautics and Aeronautics Series, AIAA, Washington, DC, 1988, pp. 605-618.

[3] Levy, Y., "The Phase Doppler Anemometry Technique for Measurement of Size and Velocity of Individual Drops in Combustion Systems," *Proceedings of the 3rd International Symposium on Application of Laser Anemometry to Fluid Mechanics*, Lisbon, Paper 22.5, 1986.

[4] Levy, Y., and Timnat, Y. M., "Diagnostics in Reacting Flows," *Dynamics of Reactive Systems. Part I: Flames*, Vol. 113, Progress in Aeronautics and Astronautics Series, AIAA, New York, 1987, pp. 387-402.

[5] Durst, F. and Zane, M., "Laser Doppler Measurements in Two Phase Flows," Univ. of Karlsruhe, Karlsruhe, Germany, Rept. SFB 80/TM/63, July 1975.

[6] Bachalo, W. D. and Houser, M. J., "Development of the Phase/Doppler Spray Analyser for Liquid Drop Size and Velocity Characterization," AIAA Paper 84-1199, June 1984.

[7] Bauckhage, K. and Floegel, H. H., "Simultaneous Measurements of Drop Size and Velocity Distributions," *Proceedings of the 2nd International Symposium on Application of Laser Anemometry to Fluid Mechanics*, Lisbon, July 1984.

[8]Gousbet, G., Grebau, G. and Klein, R., "Simultaneous Optical Measurements of Velocity and Size of Individual Particles," *Proceedings of the 2nd International Symposium on Applications of Laser Doppler Anemometry to Fluid Mechanics*, Lisbon, July 1984.

[9]Durst, F., Melling, A. and Whitelaw, J. H., *Principles and Practice of Laser Doppler Anemometry*, Academic, London, 1981.

[10]Laredo, D., "Reactive Two-Phase Flow with Recirculation," DSc Dissertation, Technion, Haifa, Israel June 1988 (in Hebrew).

[11]Gosman, A. D. and Ideriah, F. G. K., "TEACH-T: A General Computer Program for Two-Dimensional, Turbulent, Recirculating Flows," Imperial College, London, Mechanical Engineering Department Rept., London, June 1976.

Author Index

Alemany, A. 244, 519, 649
Baldwin, M. D. 310
Barak, A. 222
Baranov, N. N. 24
Bityurin, V. A. 373, 398
Blumenau, L. 284
Brancher, J. P. 158
Branover, H. 209, 244, 678
Cartellier, A. 667
Cuevas, S. 566
Davidson, P. A. 50
Del Vecchio, P. 66
Dixit, N. S. 261
Edwards, G. R. 310
Egan, N. 361
El Boher, A. 244, 678
Evtushenko, I. A. 500
Farchi, D. 678
Fautrelle, Y. 3, 580
Felici, T. P. 158
Galpin, J.-M. 580
Gelfgat, Yu. M. 32, 138
Gerbeth, G. . 441, 470, 519, 535, 551
Geri, A. 66
Gorbunov, L. A. 138
Greenspan, E. 222
Hamann, D. 441, 470
Holten, A. P. C. 373
Ievlev, V. M. 24
Josserand, J. 519
Kaplan, Y. 244
Keeton, A. R. 50
Kirillina, E. M. 500
Kirko, I. M. 239
Kirshenbaum, N. W. 335
Kishida, Y. 125
Kolesnichencko, A. 431
Kushelevsky, A. P. 662

Labrie, R. 361
Leboucher, L. 244
Lenhart, L. 482
Lesin, S. 678
Levints, A. M. 422
Lielausis, O. 476
Likhachev, A. P. 373, 398
Lykoudis, P. S. 601, 626
Malakhov, V. 431
Marechal, A. 76
Marty, Ph. 244, 519, 535
Massé, Ph. 244
Mathes, R. 649
McCarthy, K. 482
Meng, J. C. S. 183
Merck, W. F. H. 373
Messerle, H. K. 348
Miyoshino, I. 125
Moreau, R. 457
Pigny, S. 457
Pilaud, A. 244
Pisoni, C. 635
Platacis, E. 476
Platnieks, I. 476
Poinsot, S. 76
Pukis, M. 476
Ramos, E. 566
Rapin, J. 76
Ricou, R. 92
Rohatgi, V. K. 261
Rossignol, R. 76
Satanovsky, V. R. 422
Satyamurthy, P. 261
Schenone, C. 635
Seck, B. 667
Shishko, A. 476
Slepian, R. M. 50
Smolentzev, S. Y. 500

Sneyd, A.	580
Sroka, K.	410
Stavans, J.	509
Stefanov, B.	373
Tagliafico, L.	635
Takahashi, M.	626
Takeda, K.	125
Takeuchi, E.	125
Tananaev, A. V.	500
Thess, A.	535
Thibault, J.-P.	667
Thiyagarajan, T. K.	261
Timnat, Y. M.	705
Tokuhiro, A. T.	601
Uhlmann, G.	470
Unger, Y.	678
van Veldhuizen, E. M.	373
Veca, G. M.	66
Veefkind, A.	373
Venkataraman, R.	310
Venkatramani, N.	261
Vivès, C.	92, 107
Walter, F.	361
Werkoff, F.	76
Zaporowski, B.	343, 410
Zarkova, L.	373
Zatelepin, V. N.	422

PROGRESS IN ASTRONAUTICS AND AERONAUTICS
SERIES VOLUMES

*1. **Solid Propellant Rocket Research** (1960)
Martin Summerfield
Princeton University

*2. **Liquid Rockets and Propellants** (1960)
Loren E. Bollinger
Ohio State University
Martin Goldsmith
The Rand Corp.
Alexis W. Lemmon Jr.
Battelle Memorial Institute

*3. **Energy Conversion for Space Power** (1961)
Nathan W. Snyder
Institute for Defense Analyses

*4. **Space Power Systems** (1961)
Nathan W. Snyder
Institute for Defense Analyses

*5. **Electrostatic Propulsion** (1961)
David B. Langmuir
Space Technology Laboratories, Inc.
Ernst Stuhlinger
NASA George C. Marshall Space Flight Center
J.M. Sellen Jr.
Space Technology Laboratories, Inc.

*6. **Detonation and Two-Phase Flow** (1962)
S.S. Penner
California Institute of Technology
F.A. Williams
Harvard University

———

*Out of print.

*7. **Hypersonic Flow Research** (1962)
Frederick R. Riddell
AVCO Corp.

*8. **Guidance and Control** (1962)
Robert E. Roberson,
Consultant
James S. Farrior
Lockheed Missiles and Space Co.

*9. **Electric Propulsion Development** (1963)
Ernst Stuhlinger
NASA George C. Marshall Space Flight Center

*10. **Technology of Lunar Exploration** (1963)
Clifford I. Cummings
Harold R. Lawrence
Jet Propulsion Laboratory

*11. **Power Systems for Space Flight** (1963)
Morris A. Zipkin
Russell N. Edwards
General Electric Co.

*12. **Ionization in High-Temperature Gases** (1963)
Kurt E. Shuler, Editor
National Bureau of Standards
John B. Fenn,
Associate Editor
Princeton University

*13. **Guidance and Control—II** (1964)
Robert C. Langford
General Precision Inc.
Charles J. Mundo
Institute of Naval Studies

*14. **Celestial Mechanics and Astrodynamics** (1964)
Victor G. Szebehely
Yale University Observatory

*15. **Heterogeneous Combustion** (1964)
Hans G. Wolfhard
Institute for Defense Analyses
Irvin Glassman
Princeton University
Leon Green Jr.
Air Force Systems Command

*16. **Space Power Systems Engineering** (1966)
George C. Szego
Institute for Defense Analyses
J. Edward Taylor
TRW Inc.

*17. **Methods in Astrodynamics and Celestial Mechanics** (1966)
Raynor L. Duncombe
U.S. Naval Observatory
Victor G. Szebehely
Yale University Observatory

*18. **Thermophysics and Temperature Control of Spacecraft and Entry Vehicles** (1966)
Gerhard B. Heller
NASA George C. Marshall Space Flight Center

*19. Communication
Satellite Systems
Technology (1966)
Richard B. Marsten
*Radio Corporation
of America*

*20. Thermophysics of
Spacecraft and Planetary
Bodies: Radiation
Properties of Solids
and the Electromagnetic
Radiation Environment
in Space (1967)
Gerhard B. Heller
*NASA George C. Marshall
Space Flight Center*

*21. Thermal Design
Principles of Spacecraft
and Entry Bodies (1969)
Jerry T. Bevans
TRW Systems

*22. Stratospheric
Circulation (1969)
Willis L. Webb
*Atmospheric Sciences
Laboratory, White Sands,
and University of Texas
at El Paso*

*23. Thermophysics:
Applications to Thermal
Design of Spacecraft
(1970)
Jerry T. Bevans
TRW Systems

24. Heat Transfer
and Spacecraft
Thermal Control (1971)
John W. Lucas
Jet Propulsion Laboratory

25. Communication
Satellites for the 70's:
Technology (1971)
Nathaniel E. Feldman
The Rand Corp.
Charles M. Kelly
The Aerospace Corp.

26. Communication
Satellites for the 70's:
Systems (1971)
Nathaniel E. Feldman
The Rand Corp.
Charles M. Kelly
The Aerospace Corp.

27. Thermospheric
Circulation (1972)
Willis L. Webb
*Atmospheric Sciences
Laboratory, White Sands,
and University of Texas
at El Paso*

28. Thermal
Characteristics
of the Moon (1972)
John W. Lucas
Jet Propulsion Laboratory

*29. Fundamentals
of Spacecraft Thermal
Design (1972)
John W. Lucas
Jet Propulsion Laboratory

30. Solar Activity
Observations and
Predictions (1972)
Patrick S. McIntosh
Murray Dryer
*Environmental Research
Laboratories, National
Oceanic and Atmospheric
Administration*

31. Thermal Control
and Radiation (1973)
Chang-Lin Tien
*University of California
at Berkeley*

32. Communications
Satellite Systems (1974)
P.L. Bargellini
COMSAT Laboratories

33. Communications
Satellite Technology
(1974)
P.L. Bargellini
COMSAT Laboratories

*34. Instrumentation
for Airbreathing
Propulsion (1974)
Allen E. Fuhs
Naval Postgraduate School
Marshall Kingery
*Arnold Engineering
Development Center*

35. Thermophysics and
Spacecraft Thermal
Control (1974)
Robert G. Hering
University of Iowa

36. Thermal Pollution
Analysis (1975)
Joseph A. Schetz
*Virginia Polytechnic
Institute*
ISBN 0-915928-00-0

37. Aeroacoustics: Jet
and Combustion Noise;
Duct Acoustics (1975)
Henry T. Nagamatsu,
Editor
*General Electric Research
and Development Center*
Jack V. O'Keefe,
Associate Editor
The Boeing Co.
Ira R. Schwartz,
Associate Editor
*NASA Ames
Research Center*
ISBN 0-915928-01-9

38. Aeroacoustics: Fan,
STOL, and Boundary
Layer Noise; Sonic
Boom; Aeroacoustics
Instrumentation (1975)
Henry T. Nagamatsu,
Editor
*General Electric Research
and Development Center*
Jack V. O'Keefe,
Associate Editor
The Boeing Co.
Ira R. Schwartz,
Associate Editor
*NASA Ames
Research Center*
ISBN 0-915928-02-7

39. **Heat Transfer with Thermal Control Applications** (1975)
M. Michael Yovanovich
University of Waterloo
ISBN 0-915928-03-5

*40. **Aerodynamics of Base Combustion** (1976)
S.N.B. Murthy, Editor
J.R. Osborn,
Associate Editor
Purdue University
A.W. Barrows
J.R. Ward,
Associate Editors
Ballistics Research Laboratories
ISBN 0-915928-04-3

41. **Communications Satellite Developments: Systems** (1976)
Gilbert E. LaVean
Defense Communications Agency
William G. Schmidt
CML Satellite Corp.
ISBN 0-915928-05-1

42. **Communications Satellite Developments: Technology** (1976)
William G. Schmidt
CML Satellite Corp.
Gilbert E. LaVean
Defense Communications Agency
ISBN 0-915928-06-X

*43. **Aeroacoustics: Jet Noise, Combustion and Core Engine Noise** (1976)
Ira R. Schwartz, Editor
NASA Ames Research Center
Henry T. Nagamatsu,
Associate Editor
General Electric Research and Development Center
Warren C. Strahle,
Associate Editor
Georgia Institute of Technology
ISBN 0-915928-07-8

*44. **Aeroacoustics: Fan Noise and Control; Duct Acoustics; Rotor Noise** (1976)
Ira R. Schwartz, Editor
NASA Ames Research Center
Henry T. Nagamatsu,
Associate Editor
General Electric Research and Development Center
Warren C. Strahle,
Associate Editor
Georgia Institute of Technology
ISBN 0-915928-08-6

*45. **Aeroacoustics: STOL Noise; Airframe and Airfoil Noise** (1976)
Ira R. Schwartz, Editor
NASA Ames Research Center
Henry T. Nagamatsu,
Associate Editor
General Electric Research and Development Center
Warren C. Strahle,
Associate Editor
Georgia Institute of Technology
ISBN 0-915928-09-4

*46. **Aeroacoustics: Acoustic Wave Propagation; Aircraft Noise Prediction; Aeroacoustic Instrumentation** (1976)
Ira R. Schwartz, Editor
NASA Ames Research Center
Henry T. Nagamatsu,
Associate Editor
General Electric Research and Development Center
Warren C. Strahle,
Associate Editor
Georgia Institute of Technology
ISBN 0-915928-10-8

47. **Spacecraft Charging by Magnetospheric Plasmas** (1976)
Alan Rosen
TRW Inc.
ISBN 0-915928-11-6

48. **Scientific Investigations on the Skylab Satellite** (1976)
Marion I. Kent
Ernst Stuhlinger
NASA George C. Marshall Space Flight Center
Shi-Tsan Wu
University of Alabama
ISBN 0-915928-12-4

49. **Radiative Transfer and Thermal Control** (1976)
Allie M. Smith
ARO Inc.
ISBN 0-915928-13-2

50. **Exploration of the Outer Solar System** (1976)
Eugene W. Greenstadt
TRW Inc.
Murray Dryer
National Oceanic and Atmospheric Administration
Devrie S. Intriligator
University of Southern California
ISBN 0-915928-14-0

51. **Rarefied Gas Dynamics, Parts I and II** (two volumes) (1977)
J. Leith Potter
ARO Inc.
ISBN 0-915928-15-9

52. **Materials Sciences in Space with Application to Space Processing** (1977)
Leo Steg
General Electric Co.
ISBN 0-915928-16-7

53. **Experimental Diagnostics in Gas Phase Combustion Systems** (1977)
Ben T. Zinn, Editor
Georgia Institute of Technology
Craig T. Bowman, Associate Editor
Stanford University
Daniel L. Hartley, Associate Editor
Sandia Laboratories
Edward W. Price, Associate Editor
Georgia Institute of Technology
James G. Skifstad, Associate Editor
Purdue University
ISBN 0-015928-18-3

54. **Satellite Communications: Future Systems** (1977)
David Jarett
TRW Inc.
ISBN 0-915928-18-3

55. **Satellite Communications: Advanced Technologies** (1977)
David Jarett
TRW Inc.
ISBN 0-915928-19-1

56. **Thermophysics of Spacecraft and Outer Planet Entry Probes** (1977)
Allie M. Smith
ARO Inc.
ISBN 0-915928-20-5

57. **Space-Based Manufacturing from Nonterrestrial Materials** (1977)
Gerard K. O'Neill, Editor
Brian O'Leary, Assistant Editor
Princeton University
ISBN 0-915928-21-3

58. **Turbulent Combustion** (1978)
Lawrence A. Kennedy
State University of New York at Buffalo
ISBN 0-915928-22-1

59. **Aerodynamic Heating and Thermal Protection Systems** (1978)
Leroy S. Fletcher
University of Virginia
ISBN 0-915928-23-X

60. **Heat Transfer and Thermal Control Systems** (1978)
Leroy S. Fletcher
University of Virginia
ISBN 0-915928-24-8

61. **Radiation Energy Conversion in Space** (1978)
Kenneth W. Billman
NASA Ames Research Center
ISBN 0-915928-26-4

62. **Alternative Hydrocarbon Fuels: Combustion and Chemical Kinetics** (1978)
Craig T. Bowman
Stanford University
Jorgen Birkeland
Department of Energy
ISBN 0-915928-25-6

63. **Experimental Diagnostics in Combustion of Solids** (1978)
Thomas L. Boggs
Naval Weapons Center
Ben T. Zinn
Georgia Institute of Technology
ISBN 0-915928-28-0

64. **Outer Planet Entry Heating and Thermal Protection** (1979)
Raymond Viskanta
Purdue University
ISBN 0-915928-29-9

65. **Thermophysics and Thermal Control** (1979)
Raymond Viskanta
Purdue University
ISBN 0-915928-30-2

66. **Interior Ballistics of Guns** (1979)
Herman Krier
University of Illinois at Urbana-Champaign
Martin Summerfield
New York University
ISBN 0-915928-32-9

*67. **Remote Sensing of Earth from Space: Role of "Smart Sensors"** (1979)
Roger A. Breckenridge
NASA Langley Research Center
ISBN 0-915928-33-7

68. **Injection and Mixing in Turbulent Flow** (1980)
Joseph A. Schetz
Virginia Polytechnic Institute and State University
ISBN 0-915928-35-3

69. **Entry Heating and Thermal Protection** (1980)
Walter B. Olstad
NASA Headquarters
ISBN 0-915928-38-8

70. **Heat Transfer, Thermal Control, and Heat Pipes** (1980)
Walter B. Olstad
NASA Headquarters
ISBN 0-915928-39-6

*71. **Space Systems and Their Interactions with Earth's Space Environment** (1980)
Henry B. Garrett
Charles P. Pike
Hanscom Air Force Base
ISBN 0-915928-41-8

72. **Viscous Flow Drag Reduction** (1980)
Gary R. Hough
Vought Advanced Technology Center
ISBN 0-915928-44-2

73. **Combustion Experiments in a Zero-Gravity Laboratory** (1981)
Thomas H. Cochran
NASA Lewis Research Center
ISBN 0-915928-48-5

74. **Rarefied Gas Dynamics, Parts I and II** (two volumes) (1981)
Sam S. Fisher
University of Virginia
ISBN 0-915928-51-5

75. **Gasdynamics of Detonations and Explosions** (1981)
J.R. Bowen
University of Wisconsin at Madison
N. Manson
Université de Poitiers
A.K. Oppenheim
University of California at Berkeley
R.I. Soloukhin
Institute of Heat and Mass Transfer, BSSR Academy of Sciences
ISBN 0-915928-46-9

76. **Combustion in Reactive Systems** (1981)
J.R. Bowen
University of Wisconsin at Madison
N. Manson
Université de Poitiers
A.K. Oppenheim
University of California at Berkeley
R.I. Soloukhin
Institute of Heat and Mass Transfer, BSSR Academy of Sciences
ISBN 0-915928-47-7

77. **Aerothermodynamics and Planetary Entry** (1981)
A.L. Crosbie
University of Missouri-Rolla
ISBN 0-915928-52-3

78. **Heat Transfer and Thermal Control** (1981)
A.L. Crosbie
University of Missouri-Rolla
ISBN 0-915928-53-1

79. **Electric Propulsion and Its Applications to Space Missions** (1981)
Robert C. Finke
NASA Lewis Research Center
ISBN 0-915928-55-8

80. **Aero-Optical Phenomena** (1982)
Keith G. Gilbert
Leonard J. Otten
Air Force Weapons Laboratory
ISBN 0-915928-60-4

81. **Transonic Aerodynamics** (1982)
David Nixon
Nielsen Engineering & Research, Inc.
ISBN 0-915928-65-5

82. **Thermophysics of Atmospheric Entry** (1982)
T.E. Horton
University of Mississippi
ISBN 0-915928-66-3

83. **Spacecraft Radiative Transfer and Temperature Control** (1982)
T.E. Horton
University of Mississippi
ISBN 0-915928-67-1

84. **Liquid-Metal Flows and Magnetohydrodynamics** (1983)
H. Branover
Ben-Gurion University of the Negev
P.S. Lykoudis
Purdue University
A. Yakhot
Ben-Gurion University of the Negev
ISBN 0-915928-70-1

85. **Entry Vehicle Heating and Thermal Protection Systems: Space Shuttle, Solar Starprobe, Jupiter Galileo Probe** (1983)
Paul E. Bauer
McDonnell Douglas Astronautics Co.
Howard E. Collicott
The Boeing Co.
ISBN 0-915928-74-4

86. **Spacecraft Thermal Control, Design, and Operation** (1983)
Howard E. Collicott
The Boeing Co.
Paul E. Bauer
McDonnell Douglas Astronautics Co.
ISBN 0-915928-75-2

87. **Shock Waves, Explosions, and Detonations** (1983)
J.R. Bowen
University of Washington
N. Manson
Université de Poitiers
A.K. Oppenheim
University of California at Berkeley
R.I. Soloukhin
Institute of Heat and Mass Transfer, BSSR Academy of Sciences
ISBN 0-915928-76-0

88. **Flames, Lasers, and Reactive Systems** (1983)
J.R. Bowen
University of Washington
N. Manson
Université de Poitiers
A.K. Oppenheim
University of California at Berkeley
R.I. Soloukhin
Institute of Heat and Mass Transfer, BSSR Academy of Sciences
ISBN 0-915928-77-9

89. **Orbit-Raising and Maneuvering Propulsion: Research Status and Needs** (1984)
Leonard H. Caveny
Air Force Office of Scientific Research
ISBN 0-915928-82-5

90. **Fundamentals of Solid-Propellant Combustion** (1984)
Kenneth K. Kuo
Pennsylvania State University
Martin Summerfield
Princeton Combustion Research Laboratories, Inc.
ISBN 0-915928-84-1

91. **Spacecraft Contamination: Sources and Prevention** (1984)
J.A. Roux
University of Mississippi
T.D. McCay
NASA Marshall Space Flight Center
ISBN 0-915928-85-X

92. **Combustion Diagnostics by Nonintrusive Methods** (1984)
T.D. McCay
NASA Marshall Space Flight Center
J.A. Roux
University of Mississippi
ISBN 0-915928-86-8

93. **The INTELSAT Global Satellite System** (1984)
Joel Alper
COMSAT Corp.
Joseph Pelton
INTELSAT
ISBN 0-915928-90-6

94. **Dynamics of Shock Waves, Explosions, and Detonations** (1984)
J.R. Bowen
University of Washington
N. Manson
Université de Poitiers
A.K. Oppenheim
University of California at Berkely
R.I. Soloukhin
Institute of Heat and Mass Transfer, BSSR Academy of Sciences
ISBN 0-915928-91-4

95. **Dynamics of Flames and Reactive Systems** (1984)
J.R. Bowen
University of Washington
N. Manson
Université de Poitiers
A.K. Oppenheim
University of California at Bereley
R.I. Soloukhin
Institute of Heat and Mass Transfer, BSSR Academy of Sciences
ISBN 0-915928-92-2

96. **Thermal Design of Aeroassisted Orbital Transfer Vehicles** (1985)
H.F. Nelson
University of Missouri-Rolla
ISBN 0-915928-94-9

97. **Monitoring Earth's Ocean, Land, and Atmosphere from Space — Sensors, Systems, and Applications** (1985)
Abraham Schnapf
Aerospace Systems Engineering
ISBN 0-915928-98-1

98. **Thrust and Drag: Its Prediction and Verification** (1985)
Eugene E. Covert
Massachusetts Institute of Technology
C.R. James
Vought Corp.
William F. Kimzey
Sverdrup Technology AEDC Group
George K. Richey
U.S. Air Force
Eugene C. Rooney
U.S. Navy Department of Defense
ISBN 0-930403-00-2

99. **Space Stations and Space Platforms — Concepts, Design, Infrastructure, and Uses** (1985)
Ivan Bekey
Daniel Herman
NASA Headquarters
ISBN 0-930403-01-0

100. **Single- and Multi-Phase Flows in an Electromagnetic Field: Energy, Metallurgical, and Solar Applications** (1985)
Herman Branover
Ben-Gurion University of the Negev
Paul S. Lykoudis
Purdue University
Michael Mond
Ben-Gurion University of the Negev
ISBN 0-930403-04-5

101. **MHD Energy Conversion: Physiotechnical Problems** (1986)
V.A. Kirillin
A.E. Sheyndlin
Soviet Academy of Sciences
ISBN 0-930403-05-3

102. **Numerical Methods for Engine-Airframe Integration** (1986)
S.N.B. Murthy
Purdue University
Gerald C. Paynter
Boeing Airplane Co.
ISBN 0-930403-09-6

103. **Thermophysical Aspects of Re-Entry Flows** (1986)
James N. Moss
NASA Langley Research Center
Carl D. Scott
NASA Johnson Space Center
ISBN 0-930403-10-X

104. **Tactical Missile Aerodynamics** (1986)
M.J. Hemsch
PRC Kentron, Inc.
J.N. Nielsen
NASA Ames Research Center
ISBN 0-930403-13-4

105. **Dynamics of Reactive Systems Part I: Flames and Configurations; Part II: Modeling and Heterogeneous Combustion** (1986)
J.R. Bowen
University of Washington
J.-C. Leyer
Université de Poitiers
R.I. Soloukhin
Institute of Heat and Mass Transfer, BSSR Academy of Sciences
ISBN 0-930403-14-2

106. **Dynamics of Explosions** (1986)
J.R. Bowen
University of Washington
J.-C. Leyer
Université de Poitiers
R.I. Soloukhin
Institute of Heat and Mass Transfer, BSSR Academy of Sciences
ISBN 0-930403-15-0

107. **Spacecraft Dielectric Material Properties and Spacecraft Charging** (1986)
A.R. Frederickson
U.S. Air Force Rome Air Development Center
D.B. Cotts
SRI International
J.A. Wall
U.S. Air Force Rome Air Development Center
F.L. Bouquet
Jet Propulsion Laboratory, California Institute of Technology
ISBN 0-930403-17-7

108. **Opportunities for Academic Research in a Low-Gravity Environment** (1986)
George A. Hazelrigg
National Science Foundation
Joseph M. Reynolds
Louisiana State University
ISBN 0-930403-18-5

109. **Gun Propulsion Technology** (1988)
Ludwig Stiefel
U.S. Army Armament Research, Development and Engineering Center
ISBN 0-930403-20-7

110. **Commercial Opportunities in Space** (1988)
F. Shahrokhi
K.E. Harwell
University of Tennessee Space Institute
C.C. Chao
National Cheng Kung University
ISBN 0-930403-39-8

111. **Liquid-Metal Flows: Magnetohydrodynamics and Applications** (1988)
Herman Branover,
Michael Mond, and
Yeshajahu Unger
Ben-Gurion University of the Negev
ISBN 0-930403-43-6

112. **Current Trends in Turbulence Research** (1988)
Herman Branover,
Michael Mond, and
Yeshajahu Unger
Ben-Gurion University of the Negev
ISBN 0-930403-44-4

113. **Dynamics of Reactive Systems Part I: Flames; Part II: Heterogeneous Combustion and Applications** (1988)
A.L. Kuhl
R & D Associates
J.R. Bowen
University of Washington
J.-C. Leyer
Université de Poitiers
A. Borisov
USSR Academy of Sciences
ISBN 0-930403-46-0

114. **Dynamics of Explosions** (1988)
A.L. Kuhl
R & D Associates
J.R. Bowen
University of Washington
J.-C. Leyer
Université de Poitiers
A. Borisov
USSR Academy of Sciences
ISBN 0-930403-47-9

115. **Machine Intelligence and Autonomy for Aerospace** (1988)
E. Heer
Heer Associates, Inc.
H. Lum
NASA Ames Research Center
ISBN 0-930403-48-7

116. **Rarefied Gas Dynamics: Space-Related Studies** (1989)
E.P. Muntz
University of Southern California
D.P. Weaver
U.S. Air Force Astronautics Laboratory (AFSC)
D.H. Campbell
University of Dayton Research Institute
ISBN 0-930403-53-3

117. **Rarefied Gas Dynamics: Physical Phenomena** (1989)
E.P. Muntz
University of Southern California
D.P. Weaver
U.S. Air Force Astronautics Laboratory (AFSC)
D. Campbell
University of Dayton Research Institute
ISBN 0-930403-54-1

118. **Rarefied Gas Dynamics: Theoretical and Computational Techniques** (1989)
E.P. Muntz
University of Southern California
D.P. Weaver
U.S. Air Force Astronautics Laboratory (AFSC)
D.H. Campbell
University of Dayton Research Institute
ISBN 0-930403-55-X

119. **Test and Evaluation of the Tactical Missile** (1989)
Emil J. Eichblatt Jr.
Pacific Missile Test Center
ISBN 0-930403-56-8

120. **Unsteady Transonic Aerodynamics** (1989)
David Nixon
Nielsen Engineering & Research, Inc.
ISBN 0-930403-52-5

121. **Orbital Debris from Upper-Stage Breakup** (1989)
Joseph P. Loftus Jr.
NASA Johnson Space Center
ISBN 0-930403-58-4

122. **Thermal-Hydraulics for Space Power, Propulsion and Thermal Management System Design** (1989)
William J. Krotiuk
General Electric Co.
ISBN 0-930403-64-9

123. **Viscous Drag Reduction in Boundary Layers** (1990)
Dennis M. Bushnell
Jerry N. Hefner
NASA Langley Research Center
ISBN 0-930403-66-5

124. **Tactical and Strategic Missile Guidance** (1990)
Paul Zarchan
Charles Stark Draper Laboratory, Inc.
ISBN 0-930403-68-1

125. **Applied Computational Aerodynamics** (1990)
P.A. Henne
Douglas Aircraft Company
ISBN 0-930403-69-X

126. **Space Commercialization: Launch Vehicles and Programs** (1990)
F. Shahrokhi
University of Tennessee Space Institute
J.S. Greenberg
Princeton Synergetics Inc.
T. Al-Saud
Ministry of Defense and Aviation Kingdom of Saudi Arabia
ISBN 0-930403-75-4

127. **Space Commercialization: Platforms and Processing** (1990)
F. Shahrokhi
University of Tennessee Space Institute
G. Hazelrigg
National Science Foundation
R. Bayuzick
Vanderbilt University
ISBN 0-930403-76-2

128. **Space Commercialization: Satellite Technology** (1990)
F. Shahrokhi
University of Tennessee Space Institute
N. Jasentuliyana
United Nations
N. Tarabzouni
King Abulaziz City for Science and Technology
ISBN 0-930403-77-0

129. **Mechanics and Control of Large Flexible Structures** (1990)
John L. Junkins
Texas A&M University
ISBN 0-930403-73-8

130. **Low-Gravity Fluid Dynamics and Transport Phenomena** (1990)
Jean N. Koster
Robert L. Sani
University of Colorado at Boulder
ISBN 0-930403-74-6

131. **Dynamics of Deflagrations and Reactive Systems: Flames** (1991)
A. L. Kuhl
Lawrence Livermore National Laboratory
J.-C. Leyer
Université de Poitiers
A. A. Borisov
USSR Academy of Sciences
W. A. Sirignano
University of California
ISBN 0-930403-95-9

132. Dynamics of Deflagrations and Reactive Systems: Heterogeneous Combustion (1991)
A. L. Kuhl
Lawrence Livermore National Laboratory
J.-C. Leyer
Université de Poitiers
A. A. Borisov
USSR Academy of Sciences
W. A. Sirignano
University of California
ISBN 0-930403-96-7

133. Dynamics of Detonations and Explosions: Detonations (1991)
A. L. Kuhl
Lawrence Livermore National Laboratory
J.-C. Leyer
Université de Poitiers
A. A. Borisov
USSR Academy of Sciences
W. A. Sirignano
University of California
ISBN 0-930403-97-5

134. Dynamics of Detonations and Explosions: Explosion Phenomena (1991)
A. L. Kuhl
Lawrence Livermore National Laboratory
J.-C. Leyer
Université de Poitiers
A. A. Borisov
USSR Academy of Sciences
W. A. Sirignano
University of California
ISBN 0-930403-98-3

135. Numerical Approaches to Combustion Modeling (1991)
Elaine S. Oran
Jay P. Boris
Naval Research Laboratory
ISBN 1-56347-004-7

136. Aerospace Software Engineering (1991)
Christine Anderson
U.S. Air Force Wright Laboratory
Merlin Dorfman
Lockheed Missiles & Space Company, Inc.
ISBN 1-56346-005-5

137. High-Speed Flight Propulsion Systems (1991)
S. N. B. Murthy
Purdue University
E. T. Curran
Wright Laboratory
ISBN 1-56347-011-X

138. Propagation of Intensive Laser Radiation in Clouds (1992)
O. A. Volkovitsky
Yu. S. Sedunov
L. P. Semenov
Institute of Experimental Meteorology
ISBN 1-56347-020-9

139. Gun Muzzle Blast and Flash (1992)
Günter Klingenberg
Fraunhofer-Institut für Kurzzeitdynamik, Ernst-Mach-Institut (EMI)
Joseph M. Heimerl
U.S. Army Ballistic Research Laboratory (BRL)
ISBN 1-56347-012-8

140. Thermal Structures and Materials for High-Speed Flight (1992)
Earl A. Thornton
University of Virginia
ISBN 1-56347-017-9

141. Tactical Missile Aerodynamics: General Topics (1992)
Michael J. Hemsch
Lockheed Engineering & Sciences Company
ISBN 1-56347-015-2

142. Tactical Missile Aerodynamics: Prediction Methodology (1992)
Michael R. Mendenhall
Nielsen Engineering & Research, Inc.
ISBN 1-56347-016-0

143. Nonsteady Burning and Combustion Stability of Solid Propellants (1992)
Luigi De Luca
Politecnico di Milano
Edward W. Price
Georgia Institute of Technology
Martin Summerfield
Princeton Combustion Research Laboratories, Inc.
ISBN 1-56347-014-4

144. Space Economics (1992)
Joel S. Greenberg
Princeton Synergetics, Inc.
Henry R. Hertzfeld
HRH Associates
ISBN 1-56347-042-X

145. Mars: Past, Present, and Future (1992)
E. Brian Pritchard
NASA Langley Research Center
ISBN 1-56347-043-8

146. Computational Nonlinear Mechanics in Aerospace Engineering (1992)
Satya N. Atluri
Georgia Institute of Technology
ISBN 1-56347-044-6

147. Modern Engineering for Design of Liquid-Propellant Rocket Engines (1992)
Dieter K. Huzel
David H. Huang
ISBN 1-56347-013-6

148. Metallurgical Technologies, Energy Conversion, and Magnetohydrodynamic Flows (1993)
Herman Branover
Yeshajahu Unger
Ben-Gurion University of the Negev
ISBN 1-56347-019-5

149. Advances in Turbulence Studies (1993)
Herman Branover
Yeshajahu Unger
Ben-Gurion University of the Negev
ISBN 1-56347-018-7

(Other Volumes are planned.)